Cell Culture and Somatic Cell Genetics of Plants

VOLUME 1

Laboratory Procedures and Their Applications

Edited by

INDRA K. VASIL
Department of Botany
University of Florida
Gainesville, Florida

1984

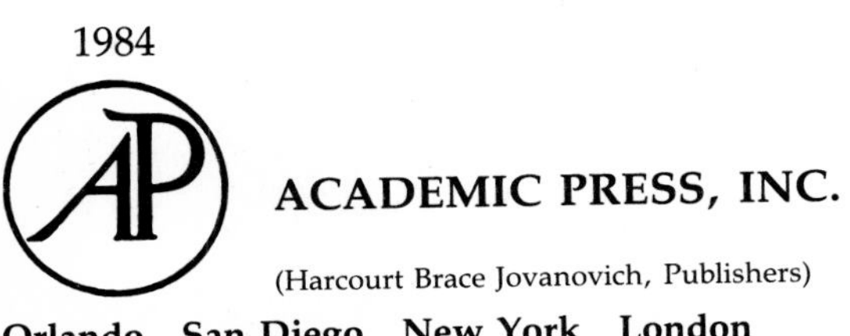

ACADEMIC PRESS, INC.

(Harcourt Brace Jovanovich, Publishers)

Orlando San Diego New York London
Toronto Montreal Sydney Tokyo

ACADEMIC PRESS, INC.
Orlando, Florida 32887

United Kingdom Edition published by
ACADEMIC PRESS, INC. (LONDON) LTD.
24/28 Oval Road, London NW1 7DX

Library of Congress Cataloging in Publication Data
Main entry under title:

Cell culture and somatic cell genetics of plants.

Includes index.
Contents: v. 1. Laboratory procedures and their
applications.
1. Plant cell culture--Collected works. 2. Plant
cytogenetics--Collected works. I. Vasil, I. K.
QK725.C37 1984 581'.07'24 83-21538
ISBN 0-12-715001-3 (alk. paper)

PRINTED IN THE UNITED STATES OF AMERICA

84 85 86 87 9 8 7 6 5 4 3 2 1

To Vimla

Contents

Contributors

Numbers in parentheses indicate the pages on which the authors' contributions begin.

Jenny Aitken-Christie (82), Forest Research Institute, Rotorua, New Zealand

Philip V. Ammirato (139), Department of Biological Sciences, Barnard College, Columbia University, New York, New York 10027, and DNA Plant Technology Corporation, Cinnaminson, New Jersey 08077

D. Aviv (199), Department of Plant Genetics, The Weizmann Institute of Science, Rehovot 76100, Israel

C. A. Beasley (232), Cooperative Agricultural Extension, University of California, El Centro, California 92243

Sant S. Bhojwani (258), Department of Botany, University of Delhi, Delhi 110007, India

Horst Binding (43, 340), Botanical Institute and Botanical Garden, Biology Center, Christian Albrechts University, D-2300 Kiel 1, Federal Republic of Germany

Martin Bopp (96), Botanical Institute, University of Heidelberg, D-6900 Heidelberg, Federal Republic of Germany

Claudia Botti (684), Department of Botany, University of Florida, Gainesville, Florida 32611

Paul J. Bottino (13), Department of Botany, University of Maryland, College Park, Maryland 20742

P. Brodelius (535), Institute of Biotechnology, Swiss Federal Institute of Technology (ETH) Honggerberg, CH-8093 Zürich, Switzerland

Olivia C. Broome (111), Fruit Laboratory, Agricultural Research Service, United States Department of Agriculture, Beltsville, Maryland 20705

Daniel C. W. Brown (1), Ottawa Research Station, Agriculture Canada, Ottawa, Ontario, Canada K1A 0C6

G. B. Collins (241), Department of Agronomy, University of Kentucky, Lexington, Kentucky 40546

F. Constabel (27, 414), Plant Biotechnology Institute, National Research Council, Saskatoon, Saskatchewan, Canada S7N 0W9

Sukhraj S. Dhillon (744), Department of Botany, North Carolina State University, Raleigh, North Carolina 27650

Denes Dudits (391), Institute of Genetics, Biological Research Center, Hungarian Academy of Sciences, Szeged, Hungary H-6701

D. I. Dunstan (123), Plant Tissue Culture Department, Kelowna Nurseries, Ltd., Kelowna, British Columbia, Canada V1Y 7N5

Elizabeth D. Earle (598), Department of Plant Breeding and Biometry, Cornell University, Ithaca, New York 14853

G. E. Edwards (471), Department of Botany and Institute of Biological Chemistry, Washington State University, Pullman, Washington 99164

B. Foroughi-Wehr (293, 311), Institute for Resistance Genetics, Federal Biological Research Center for Agriculture and Forestry, D-8059 Grünbach, Federal Republic of Germany

Larry C. Fowke (728, 785), Department of Biology, University of Saskatchewan, Saskatoon, Saskatchewan, Canada S7N 0W0

M. W. Fowler (167), Wolfson Institute of Biotechnology, University of Sheffield, Sheffield S10 2TN, England

Robert T. Fraley (483), Corporate Research Laboratories, Monsanto Company, St. Louis, Missouri 63167

Tatsuhito Fujimura (159), Mitsuitoatsu Chemicals, Inc., Chigasaki, Kanagawa, Japan

David W. Galbraith (433, 765), School of Biological Sciences, University of Nebraska at Lincoln, Lincoln, Nebraska 68588

E. Galun (199), Department of Plant Genetics, The Weizmann Institute of Science, Rehovot 76100, Israel

Oluf L. Gamborg (18), Genentech, Inc., South San Francisco, California 94080

Burle G. Gengenbach (276), Department of Agronomy and Plant Genetics, University of Minnesota, St. Paul, Minnesota 55108

Y. Y. Gleba (405), Department of Cytophysiology and Plant Cell Engineering, Institute of Botany, Academy of Sciences of the Ukrainian Soviet Socialist Republic, Kiev 252601, Union of Soviet Socialist Republics

Alan R. Gould (698, 753), Plant Genetics Department, Central Research, Pfizer, Inc., Groton, Connecticut 06340

J. W. Grosser (241), Department of Agronomy, University of Kentucky, Lexington, Kentucky 40546

Gyula Hadlaczky (461), Institute of Genetics, Biological Research Center, Hungarian Academy of Sciences, Szeged, Hungary H-6701

Christian T. Harms* (213), Department of Crop Science, Swiss Federal Institute of Technology (ETH) Zürich, Eschikon Experimental Station, CH-8307 Lindau-Eschikon, Switzerland

Kenneth A. Hibberd (571), Molecular Genetics, Inc., Minnetonka, Minnesota 55343

Franz Hoffmann (356, 405), Department of Developmental and Cell Biology, University of California, Irvine, California 92717

Robert B. Horsch (192), Corporate Research Laboratories, Monsanto Company, St. Louis, Missouri 63167

Kerry T. Hubick (659), Plant Physiology Research Group, Department of Biology, University of Calgary, Calgary, Alberta, Canada T2N 1N4

Wolfgang Hüsemann (182), Department of Plant Biochemistry, University of Münster, D-4400 Münster, Federal Republic of Germany

Toshiaki Kameya (423), Institute for Agricultural Research, Tohoku University, Sendai 980, Japan

K. K. Kartha (106, 577, 621), Plant Biotechnology Institute, National Research Council, Saskatoon, Saskatchewan, Canada S7N 0W9

W. A. Keller (302), Genetic Engineering Section, Ottawa Research Station, Ontario Region, Agriculture Canada, Ottawa, Ontario, Canada K1A 0C6

Patrick J. King (130, 547), Friedrich Miescher Institute, CH-4002 Basel, Switzerland

Bernd Knoop (96), Botanical Institute, University of Heidelberg, D-6900 Heidelberg, Federal Republic of Germany

J. Kobza (471), Department of Botany, Washington State University, Pullman, Washington 99164

Hans Willy Kohlenbach (204), Botanical Institute, University of Frankfurt on Main, D-6000 Frankfurt on Main 1, Federal Republic of Germany

Atsushi Komamine (159), Department of Botany, University of Tokyo, Hongo, Tokyo 113, Japan

F. A. Krens (522), Department of Plant Molecular Biology, State University of Leiden, 2333 Leiden, The Netherlands

Gabriela Krumbiegel-Schroeren (43, 340), Botanical Institute and Botanical Garden, Biology Center, Christian Albrechts University, D-2300 Kiel 1, Federal Republic of Germany

John T. Kunisaki (61), Department of Horticulture, University of Hawaii at Manoa, Honolulu, Hawaii 96822

W. G. W. Kurz (644), Plant Biotechnology Institute, National Research Council, Saskatoon, Saskatchewan, Canada S7N 0W9

*Present address: CIBA-Geigy Biotechnology Institute, P.O. Box 12257, Research Triangle Park, North Carolina 27709

Horst Lörz (448), Max Planck Institute for Plant Breeding Research, D-5000 Cologne, Federal Republic of Germany

Joseph H. Lui (637), Biotechnology Department, The Goodyear Tire and Rubber Company, Akron, Ohio 44316

Pal Maliga* (552), Institute of Plant Physiology, Biological Research Center, Hungarian Academy of Sciences, Szeged, Hungary H-6701

László Márton (514), Institute of Plant Physiology, Biological Research Center, Hungarian Academy of Sciences, Szeged, Hungary H-6701

Laszlo Menczel† (428), Institute of Plant Physiology, Biological Research Center, Hungarian Academy of Sciences, Szeged, Hungary H-6701

Jerome P. Miksche (744), Department of Botany, North Carolina State University, Raleigh, North Carolina 27650

Toshiyuki Nagata (328), Department of Biology, Nagoya University, Chikusa-ku, Nagoya 464, Japan

Kamlesh R. Patel (49), Department of Biology, University of Calgary, Calgary, Alberta, Canada T2N 1N4

R. L. Phillips (175, 712), Department of Agronomy and Plant Genetics, University of Minnesota, St. Paul, Minnesota 55108

T. S. Rangan (68, 221, 227), Phytogen, Pasadena, California 91105

David M. Reid (659), Plant Physiology Research Group, Department of Biology, University of Calgary, Calgary, Alberta, Canada T2N 1N4

Minocher Reporter (586), Charles F. Kettering Research Laboratory, Yellow Springs, Ohio 45387

Yoneo Sagawa (61), Harold L. Lyon Arboretum and Department of Horticulture, University of Hawaii at Manoa, Honolulu, Hawaii 96822

Hellmut R. Schenck (356), Institut für Pflanzenbau and Pflanzenzüchtung, Georg August University, D-3400 Göttingen, Federal Republic of Germany

O. Schieder (350), Max Planck Institute for Plant Breeding Research, D-5000 Cologne, Federal Republic of Germany

R. A. Schilperoort (522), Department of Plant Molecular Biology, State University of Leiden, 2333 Leiden, The Netherlands

George Setterfield (785), Department of Biology, Carleton University, Ottawa, Ontario, Canada K1S 5B6

Elias A. Shahin (370, 381), ARCO Plant Cell Research Institute, Dublin, California 94568

V. A. Sidorov (405), Department of Cytophysiology and Plant Cell Engineering, Institute of Botany, Academy of Sciences of the Ukrainian

*Present address: Advanced Genetic Sciences, Inc., P.O. Box 1373, Manhattan, Kansas 66502

†Present address: Department of Botany, University of Florida, Gainesville, Florida 32611

Soviet Socialist Republic, Kiev 252601, Union of Soviet Socialist Republics
Daina Simmonds (785), Department of Biology, Carleton University, Ottawa, Ontario, Canada K1S 5B6
N. Sunderland (283), John Innes Institute, Norwich NR4 7UH, England
Itaru Takebe (328, 492), Department of Biology, Nagoya University, Chikusa-ku, Nagoya 464, Japan
Trevor A. Thorpe (1, 49, 82), Department of Biology, University of Calgary, Calgary, Alberta, Canada T2N 1N4
Brent Tisserat (74), Fruit and Vegetable Chemistry Laboratory, Agricultural Research Service, United States Department of Agriculture, Pasadena, California 91106
K. E. Turner (123), Plant Tissue Culture Department, Kelowna Nurseries, Ltd., Kelowna, British Columbia, Canada V1Y 7N5
Pieter van der Valk (785), Institute of Human Genetics, Free University of Amsterdam, 1007 MC Amsterdam, The Netherlands
Indra K. Vasil (36, 152, 398, 684, 738), Department of Botany, University of Florida, Gainesville, Florida 32611
Vimla Vasil (36, 152, 398, 738), Department of Botany, University of Florida, Gainesville, Florida 32611
Anita Wallin (503), Department of Plant Physiology, Swedish University of Agricultural Sciences, S-750 07 Uppsala 7, Sweden
Jonathan D. Walton (598), Department of Plant Breeding and Biometry, Cornell University, Ithaca, New York 14853
A. S. Wang (175, 712), Department of Agronomy and Plant Genetics, University of Minnesota, St. Paul, Minnesota 55108
G. Wenzel (293, 311), Institute for Resistance Genetics, Federal Biological Research Center for Agriculture and Forestry, D-8059 Grünbach, Federal Republic of Germany
L. R. Wetter (651), Plant Biotechnology Institute, National Research Council, Saskatoon, Saskatchewan, Canada S7N 0W9
J. M. Widholm (563), Department of Agronomy, University of Illinois, Urbana, Illinois 61801
L. Willmitzer (454), Max Planck Institute for Plant Breeding Research, D-5000 Cologne, Federal Republic of Germany
Lyndsey A. Withers (608), Department of Agriculture and Horticulture, School of Agriculture, University of Nottingham, Sutton Bonington, Loughborough LE12 5RD, England
Yasuyuki Yamada (629), Research Center for Cell and Tissue Culture, Kyoto University, Kyoto 606, Japan
Edward C. Yeung (689, 778), Department of Biology, University of Calgary, Calgary, Alberta, Canada T2N 1N4

Maciej Zenkteler (269), Department of General Botany, Institute of Biology, Adam Mickiewicz University, 61-713 Poznan, Poland
Richard H. Zimmerman (111), Fruit Laboratory, Agricultural Research Service, United States Department of Agriculture, Beltsville, Maryland 20705

General Preface

Recent advances in the techniques and applications of plant cell culture and plant molecular biology have created unprecedented opportunities for the genetic manipulation of plants. The potential impact of these novel and powerful biotechnologies on the genetic improvement of crop plants has generated considerable interest, enthusiasm, and optimism in the scientific community and is in part responsible for the rapidly expanding biotechnology industry.

The anticipated role of biotechnology in agriculture is based not on the actual production of any genetically superior plants, but on elegant demonstrations in model experimental systems that new hybrids, mutants, and genetically engineered plants can be obtained by these methods, and the presumption that the same procedures can be adapted successfully for important crop plants. However, serious problems exist in the transfer of this technology to crop species.

Most of the current strategies for the application of biotechnology to crop improvement envisage the regeneration of whole plants from single, genetically altered cells. In many instances this requires that specific agriculturally important genes be identified and characterized, that they be cloned, that their regulatory and functional controls be understood, and that plants be regenerated from single cells in which such gene material has been introduced and integrated in a stable manner.

Knowledge of the structure, function, and regulation of plant genes is scarce, and basic research in this area is still limited. On the other hand, a considerable body of knowledge has accumulated in the last fifty years on the isolation and culture of plant cells and tissues. For example, it is possible to regenerate plants from tissue cultures of many plant species, including several important agricultural crops. These procedures are now widely used in large-scale rapid clonal propagation of plants. Plant cell culture techniques also allow the isolation of mutant cell lines and plants, the generation of somatic hybrids by protoplast fusion, and the regeneration of genetically engineered plants from single transformed cells.

Many national and international meetings have been the forums for discussion of the application of plant biotechnology to agriculture. Neither the basic techniques nor the biological principles of plant cell culture are generally included in these discussions or their published proceedings. Following the very enthusiastic reception accorded the two volumes entitled "Perspectives in Plant Cell and Tissue Culture" that were published as supplements to the *International Review of Cytology* in 1980, I was approached by Academic Press to consider the feasibility of publishing a treatise on plant cell culture. Because of the rapidly expanding interest in the subject both in academia and in industry, I was convinced that such a treatise was needed and would be useful. No comprehensive work of this nature is available or has been attempted previously.

The organization of the treatise is based on extensive discussions with colleagues, the advice of a distinguished editorial advisory board, and suggestions provided by anonymous reviewers to Academic Press. However, the responsibility for the final choice of subject matter included in the different volumes, and of inviting authors for various chapters, is mine. The basic premise on which this treatise is based is that knowledge of the principles of plant cell culture is critical to their potential use in biotechnology. Accordingly, descriptions and discussion of all aspects of modern plant cell culture techniques and research are included in the treatise. The first volume describes every major laboratory procedure used in plant cell culture and somatic cell genetics research, including many variations of a single procedure adapted for important crop plants. Two subsequent volumes in preparation are devoted to the nutrition and growth of plant cell cultures and to the important subject of generating and recovering variability from cell cultures. An entirely new approach is used in the treatment of this subject by including not only spontaneous variability arising during culture, but also variability created by protoplast fusion, genetic transformation, etc. Future volumes are envisioned to cover most other relevant and current areas of research in plant cell culture and its uses in biotechnology.

In addition to the very comprehensive treatment of the subject, the uniqueness of these volumes lies in the fact that all the chapters are prepared by distinguished scientists who have played a major role in the development and/or uses of specific laboratory procedures and in key fundamental as well as applied studies of plant cell and tissue culture. This allows a deep insight, as well as a broad perspective, based on personal experience. The volumes are designed as key reference works to provide extensive as well as intensive information on all aspects of plant cell and tissue culture not only to those newly entering the field but also to experienced researchers.

Indra K. Vasil

Preface to Volume 1

Plant cell and tissue culture techniques are now widely used in a variety of basic research programs as well as in the biotechnology industry. Unfortunately, no comprehensive manual describing the numerous methods and their uses is available. It is, therefore, only fitting that the first volume in this treatise is devoted entirely to laboratory procedures and their applications. All major methods used in modern plant cell culture research are described in considerable detail by individuals who have played key roles in either the development or the application of the techniques they describe. Each chapter provides a brief history of the development of the technique, details of the method actually used, its applications, and key references. Modifications of important procedures such as clonal propagation, anther culture, or protoplast culture for specific crops or plant groups are provided in multiple chapters. It is hoped that the volume will be useful not only to new entrants to the field but also to experienced researchers.

The formidable task of correspondence with authors of the 85 chapters in this volume was made considerably easier by their splendid cooperation in preparation of the manuscripts according to the guidelines provided and strict adherence to time schedules. For this I am grateful. I also thank members of the editorial advisory board for their assistance in organization of the volume.

Indra K. Vasil

CHAPTER **1**

Organization of a Plant Tissue Culture Laboratory

Daniel C. W. Brown

Ottawa Research Station
Agriculture Canada
Ottawa, Ontario, Canada

Trevor A. Thorpe

Department of Biology
University of Calgary
Calgary, Alberta, Canada

I. INTRODUCTION

The term "plant tissue culture" is commonly used to describe the *in vitro* and aseptic cultivation of any plant part on a nutrient medium. However, it has been argued for some time (White, 1941; Bailey, 1943; Street, 1973a) that a more restricted use of the term be adopted to avoid some of the confusion and lack of uniformity that persist in the literature. The technique of plant tissue culture has a long history of controversy, innovation,

CELL CULTURE AND SOMATIC CELL
GENETICS OF PLANTS, VOL. 1

ISBN 0-12-715001-3

and evolution. Yet, ironically, the underlying concept of the technique is simple and remains essentially unchanged since the time when it was first outlined by Haberlandt in 1902 (see Krikorian and Berquam, 1969). As well, many of the procedures used today are firmly based on the original techniques that were detailed mostly by White (1943, 1963) and Gautheret (1959).

Practically all of the plant tissue culture technology, old and new, is based on three fundamental objectives: First, the plant part or explant must be isolated from the rest of the plant body. This effectively disrupts the cellular, tissue, and/or organ interactions that may occur in the intact plant. Second, the explant must be maintained in a controlled and preferably defined milieu. Both the chemical composition of the medium and the physical conditions of the environment should effectively control the expression of any genotypic or phenotypic potential in the cultured plant part. Third, asepsis must be maintained. Most culture media also support the growth of algal, bacterial, and fungal contaminants, which may result in either the overgrowth of the explant or the production of metabolites which can be toxic or influential in the explant's metabolic, growth, or developmental responses. It is the need for asepsis which largely dictates the organization and techniques employed in a laboratory dedicated to the use of plant tissue culture as a research tool. Whether the techniques are being used for simple propagation, as a method to study genetic, metabolic, or developmental changes in a model system, or for the creation of new plant varieties via genetic engineering, there are a number of basic facilities and a minimum level of organization which should be available to either the individual worker or a large team. These basic facilities usually include the following:

1. A wash-up area
2. A media preparation, sterilization, and storage area
3. A transfer area for aseptic manipulations
4. Incubators or culture rooms where light, temperature, and humidity are controlled
5. An observation/experimentation area

This latter area will, of course, be dependent on the nature of the research undertaken and should be organized according to its intended use (i.e., genetics, electron microscopy, biochemistry, or molecular biology).

II. WASHING FACILITIES

An area with large sinks (some lead-lined to resist acids and alkalis), draining boards/racks, and ready access to demineralized, distilled, and double-distilled water is necessary. Sufficient space should be available to set up drying racks and ovens, washing machines, and baths for pipette washing, as well as baths for detergent and acid treatment of glassware.

The conventional method for cleaning laboratory glassware involves a chromic acid–sulfuric acid soak followed by vigorous tap water rinses, distilled water rinses, and a final rinse with double-distilled water (Street, 1973a). However, the particularly corrosive nature of chromic acid and the need for cumbersome protective clothing and goggles has caused chromic acid to be largely replaced by detergents. The routine procedure in our laboratory is to soak glassware in a 2% solution of detergent cleaner (Decon 75, BDH Chemical, Toronto, Ontario, Canada; Liqui-nox, Alconox, Inc., New York, New York) for at least 16 hr, wash with 70°C+ tap water in detergent (Sparkleen, Fisher Scientific), and rinse with 70°C+ tap water, followed by a demineralized water and distilled water rinse. If a mechanical dishwasher is used, glassware is washed with a nonsudsing laboratory detergent (Alcojet, Alconox; Contrad DRI, Canlab) at 70°C for 5–10 min, followed by a 5-min 70°C tap water rinse, a 3-min deionized water rinse, and a final hand rinse with distilled water. The cleaned glassware is inspected and dried at 150°C in a convection drying oven (Despatch, Canlab) for at least 1 hr, capped with aluminum foil, and stored in a closed cupboard to limit dust collection. Highly contaminated glassware and pipettes are routinely first treated in a chromic acid bath (Chromerge, Manostat plus concentrated sulfuric acid) for at least 16 hr, but other procedures can be effective (White, 1963; Coriell, 1973). These methods can include:

1. Boiling in a white soap solution, rinsing thoroughly in water and then in 95% alcohol, followed by drying, wrapping, and dry sterilization
2. Ultrasonic cleaning
3. Washing in a sodium pyrophosphate solution
4. Boiling in metaphosphate (Alconox), rinsing, then boiling in dilute hydrochloric acid and rerinsing in water and then in 95% alcohol
5. Boiling in 80% sulfuric acid containing nitric acid, then washing with water
6. Soaking in sulfuric acid:nitric acid:distilled water (80:15:5) for 48 hr, followed by a 30-min treatment in Calgolac (Calgon Corp., Pittsburgh, Pennsylvania) at 95°C

A. Glassware

Almost any glass container that can be easily cleaned and sterilized can be used for tissue culture, but it is preferable that laboratory glassware be of good-quality borosilicate glass (Pyrex, Corning; Kimax, Kimble), as it is resistant to heat, breakage, and scratching, and can be cleaned with strong solvents and reprocessed repeatedly. Street (1973a) recommends that all glassware should be "broken in" by being filled with distilled water and autoclaved at least twice, with cleaning between autoclavings using a chromic acid–sulfuric acid mixture. Soda lime glass containers such as baby food jars have been widely used and are less expensive, but de Fossard (1976) suggests that they be discarded after 1 year's use or be treated with dimethyl-dichloro-silane (Sigma, St. Louis, Missouri).

Beyond the normal range of glassware and instruments found in a well-stocked laboratory, a plentiful supply of graduated measuring cylinders (10–2000 ml), volumetric flasks (1–2000 ml), burettes (250–1000 ml), large flasks (1–4 liters), beakers (1–4 liters), and pipettes (0.1–25.0 ml) is recommended. Wide-necked Erlenmeyer flasks (50–250 ml), test tubes (25 × 150–250 mm), and glass Petri dishes (60 × 15 and 100 × 20 mm) are particularly useful as tissue culture containers.

B. Plastic Labware

The recent ready availability and wide range of presterilized disposable polystyrene culture containers (Corning, Corning; Costar, Costar; Falcon, Becton Dickinson and Co.; Lux and Lab-tek, Miles Laboratories; Nalgene, Nalge Co.) have greatly reduced the amount of routine washing; however, the increased costs of switching to disposable plastic labware are sometimes prohibitive. Reusable plastic labware comes in many different types, and their chemical and heat resistance vary tremendously. In general, suppliers (Nalge Co.) recommend that plastics be washed with a mild, nonabrasive detergent, followed by a rinse with tap water and then distilled water. Mechanical dishwashers can be used except with polyethylene, acrylic, and polystyrene, which have temperature limitations, and polycarbonate, which is weakened by repeated washings in dishwashers. For more rigorous cleaning, a variety of treatments are available (Nalgene Labware Technical Service, Nalge Co.), but they must be matched to the particular type of plastic under consideration. Methods may include:

1. Rinsing with organic solvents such as acetone, alcohol, chloroform, or methylene chloride. Only alcohols can be used on polycarbonate, polysulfone, polystyrene, or polyvinyl chloride, and organic solvents cannot be used on acrylic.
2. Boiling in dilute sodium bicarbonate. However, polycarbonate, polyethylene, acrylic, and polystyrene cannot be treated in this way.
3. Soaking for less than 4 hr in a chromic acid–sulfuric acid bath.
4. Soaking in 1 *N* HCl for less than 8 hr, followed by soaking in 1 *N* HNO_3 and rinsing in distilled water.

III. MEDIA PREPARATION AREA

The area to be set aside for media preparation should have ample storage and bench space for chemicals, glassware, culture vessels, closures, and all other items needed to prepare media to the point of sterilization. A refrigerator and freezer are essential for safe storage of chemicals, media stocks, etc. Other necessary equipment includes sources of compressed air, vacuum, distilled and double-distilled water, hot plate/stirrers, a pH meter, a top-loading balance (0.001–400.0 g), a semimicro balance (0.00001– 160.0 g), and Bunsen burners with a gas source. An oven and autoclave or domestic pressure cooker is also necessary for sterilizing media, culture vessels, and instruments. A microwave oven is strongly recommended. As Street (1973a) has noted, "More false trails have been laid by mistakes in media preparation than by any other fault of technique." It is particularly important that analytical-grade chemicals be used where possible, and that good weighing habits and an exact step-by-step routine be established for each type of medium. The value of a complete checklist, even for the simplest media, cannot be underestimated.

We routinely prepare 2 liters of 10× concentrated medium, minus the carbohydrates and growth regulators, at a time. It is stored frozen in sterilized 200-ml Whirl-pak (Nasco) bags sealed in plastic (Frig-O-Seal, Plastiques Modernes) containers for future use. At the time of media preparation, the stock solution is thawed in a microwave oven and added, in the proper sequence, to an appropriate volume of distilled water, along with the remainder of the media ingredients. The pH is adjusted as the last step before it is made up to volume, agar added (if needed) and melted (in a microwave oven), dispensed, and sterilized.

A. Water

Tap water is unsuitable for plant tissue culture media and laboratory use because it may contain the following (Pumper, 1973):

1. Cations (ammonium, calcium, iron, magnesium, potassium, sodium, etc.)
2. Anions (bicarbonate, carbonate, chloride, fluoride, nitrate, phosphate, sulfate, etc.)
3. Particulate matter (metalic oxides, oils, organic matter, silica, silt, pipe corrosion products)
4. Microorganisms (algae, bacteria, fungi, viruses)
5. Gases (ammonia, carbon dioxide, chlorine, hydrogen sulfide, oxygen)

Water purification methods should produce at least type II reagent-grade water (i.e., free of pyrogens, gases and organic matter, electrical resistivity >1.0 $M\Omega \cdot cm$, electrical conductivity <1.0 μmho/cm), specified by the American Society for Testing and Materials (ASTM) (Meltzer, 1979). These standards are similar to the College of American Pathologists (CAP) and the National Committee for Clinical Laboratory Standards (NCCLS) guidelines for type II reagent-grade water (Basalik, 1978).

Water purification methods to achieve this standard, as they relate to a tissue culture laboratory, may include one or usually more of the following treatments:

1. Adsorption filtration, which uses activated carbon to remove organic contaminants and free chlorine. Dissolved impurities are not removed by this method.
2. Deionization or demineralization, which removes dissolved ionized impurities by passing water through synthetic resins that exchange H^+ (cations) and OH^- (anions) for the ionized impurities. Since this method is used for removal of dissolved ionized impurities, no sterilization or organic removal is accomplished.
3. Distillation, which removes a broad range of impurities, including most ionic, particulate, and nonvolatile chemical matter, as well as organic matter and microorganisms. Distillation is based on a phase change from liquid to vapor; water is first evaporated, leaving the impurities behind, and then condensed in a water-cooled heat exchanger. Volatile impurities and gases, especially CO_2, are difficult to remove by this method.
4. Membrane filtration, which removes particulate matter and most bacteria. Prefilters may be in the 1–15-μm pore size range, whereas final submicron-size filters are in the 0.20–0.45-μm pore size range. Any dissolved impurities remain unaffected by this method.
5. Reverse osmosis, which removes 99% of the bacteria, organic and particulate matter, and about 90% of the dissolved ionized impurities. The process uses a semipermeable membrane through which a portion of the water is forced under pressure, whereas the balance containing the concentrated impurities is rejected. Little of the gaseous impurity is removed by this method, and often significant levels of ionized matter must be removed by an additional treatment.

The most common and preferred method of water purification for tissue culture use is a deionization treatment followed by one or two glass distillations. The deionization treatment typically removes most of the ionic impurities, but particulate matter, especially organic compounds, pyrogens, and nonionized leachate from the ion-exchange resin, as well as microorganisms and gases, are not efficiently removed by demineralization. Proper distillation will effectively remove large organic molecules, microorganisms, pyrogens, and other ionic and nonionic contaminants, but will not guarantee high purity (Gibbs, 1972a,b; Bonga, 1982), as distillation is a highly complicated science. We have found that by frequently draining the still boiler, operating the condenser hot enough to allow some steam, venting, and discarding the water produced in the first 20–30 min of operation, distillation of deionized water is a reliable method of purifying water to ASTM standards.

The more recently introduced reverse osmosis purifying equipment (Milli-RO, Millipore; RO pure, Barnstead; Bion, Pierce), especially when it is combined with cartridge ion exchange, adsorption, and membrane filtering equipment which claims to produce very pure (type I reagent grade) water, has, in some cases, replaced the traditional glass distillation of water. Some suppliers, however, do not recommend this type of system for analytical work or procedures requiring water low in organics (i.e. tissue culture media) because (1) ASTM standards specify type II water prepared by distillation for these uses, (2) ion-exchange resins may leach organic impurities into the water, and (3) a significant amount of particulate matter, including bacteria and viruses, as well as pyrogens and some dissolved gases, may pass through the system. Regardless of the purification system employed, it is very difficult to eliminate all contaminants from water, especially since stored water can leach various chemicals from glass and plastic tubing and containers. Stored water may also accumulate gases, especially CO_2, atmospheric air pollutants, and microorganisms (Gibbs, 1972a,b; Bangham and Hill, 1972; Pumper, 1973; Bonga, 1982; Favero *et al.*, 1971).

For all of the above reasons, it is advisable that the water purification system be well maintained and monitored closely, and that the water storage time be as short as possible.

B. Sterilization

The need for asepsis requires that all culture vessels, instruments, and media used in handling tissues, as well as the explant itself, be sterilized. Simple precautions such as not sharing areas where tissue culture work is being carried out with microbiologists and pathologists, and maintaining a high level of cleanliness will reduce the risk of widespread as well as occasional contamination. There are a variety of wet and dry heat treatments, radiation, filtration, and gas and chemical treatments available for the direct sterilization of materials (Klein and Klein, 1970), but gas treatments are rarely used in the laboratory and radiation sterilization is usually restricted to ultraviolet light treatments of working surfaces and sterile rooms.

The most popular method of sterilizing both equipment and media is by autoclaving at 121°C with a pressure of 15 psi for 15 min. Modern autoclaves (Barnstead/Castle, Sybron Corp; Amsco, American Sterilizer Comp.) are capable of providing saturated steam treatments ranging from 70° to 132°C, which is a pressure of up to 25 psi (Klein and Klein, 1970). Sterilizers should be versatile, as liquid media require a slow exhaust cycle to avoid boiling over during the time of pressure reduction, and larger volumes of liquid require relatively long autoclaving times (e.g., 2000 ml requires 40 min at 121°C; Biondi and Thorpe, 1981). Glassware and instruments, wrapped in aluminum foil or paper or unwrapped, can be processed more quickly with a rapid exhaust cycle. The temptation to use longer autoclaving times with media should be strictly avoided, as extended times may induce chemical changes in the media and decompose carbohydrates, vitamins, and growth regulators (Bonga, 1982). It is common practice, when autoclaving a large volume of medium, to divide it into several small portions, thereby increasing the surface/volume ratio and the heat exchange properties as well as reducing the exposure time during heat treatment. Vessels, of course, should be fitted with closures to prevent contamination of medium after removal from the autoclave. Plugs of nonabsorbent cotton and aluminum foil, as well as metal and polypropylene plastic closures, can be used for this purpose. Threaded caps, when used, should be placed with the threads totally disengaged during autoclaving to avoid sticking of the plastic sealing rings.

Filter sterilization is the alternative method of media sterilization and the preferred method of handling heat-labile compounds. A range of bacteria-proof membrane filters (Millipore, Millipore; Gelman, Gelman Scientific; Nalgene, Nalge Co., Sartorius, Sartorius Filters, Inc.) and associated equipment is available for volumes in the range of 1–200 ml. Most filters are of the cellulose acetate and/or cellulose nitrate, 0.2-μm type and are available in presterilized, plastic disposable units or as individual autoclavable units for use in a variety of reusable plastic and glass units. Modern plastic filter units are popular but do have disadvantages. Membranes may contain small amounts of detergent and impurities, and extractables have been shown to produce toxic and inhibitory effects, especially in animal cell cultures (Wakeland *et al.*, 1982); in addition, not all plasticware is autoclavable. Polypropylene, polymethylpentene, Teflon, Tefzel, polycarbonate, polyallomer, and polysulfone can be autoclaved, whereas polystyrene, polyvinyl chloride, styrene acrylonitrile, acrylic, and polyethylene cannot (Nalgene Technical Service, Nalge Co.).

Dry heat sterilization, i.e., in an oven, can be used for glassware and instruments that are not damaged by elevated temperatures. We routinely heat sterilize glassware, double-capped or wrapped with aluminum foil, with a 3-hr, 150°C oven treatment. However, as Biondi and Thorpe (1981) have indicated, a systematic study on temperatures and exposure times which are lethal to microorganisms is lacking, and consequently so is a general consensus on the temperature and exposure time required for efficient dry heat sterilization. Unfortunately, the rule in this case is, whatever works is best. Glassware should be thoroughly cleaned to avoid the heat-protective influence of organic matter such as grease, and the time needed for the oven and its contents to reach the sterilizing temperature should be considered in determining proper sterilization times.

Radiation treatments (ultraviolet light) and gas treatments (ethylene oxide) are not commonly used in research laboratories, but various chemical treatments, especially for sterilizing explants, are widely used. The treatments may include (Street, 1973a):

1. 1% solution of sodium hypochlorite (commercial bleach)
2. 7% saturated solution of calcium hypochlorite
3. 1% solution of bromine water
4. 70% alcohol
5. 0.2% mercuric chloride solution
6. 10% hydrogen peroxide solution
7. 1% silver nitrate solution

The type, concentration, and exposure time of the particular sterilant to be used are dependent on the tissue under consideration and are determined empirically. In many cases, a few drops of a wetting agent or liquid detergent (Teepol, BDH; Tween-80, Baker Chemicals) is incorporated into the sterilizing solution to enhance the penetration and effectiveness of the sterilizing agent. In all cases in which sterilizing agents have been in contact with plant material, several changes of sterile distilled water should be used to ensure proper removal of the sterilizing agent. Although it is not recommended that sterile parts of instruments or plant material be handled directly, it is essential that the worker's hands be relatively aseptic during manipulations. An effective procedure to attain an acceptable level of asepsis is to wash with an antibacterial soap followed by a 70% alcohol rinse.

IV. TRANSFER AREA

Tissue culture techniques can be successfully used on an open laboratory bench, especially under very clean and dry atmospheric conditions, but it is advisable that a sterile transfer room and/or a laminar flow work station be available. The most desirable arrangement, for both maintenance of asepsis and flexibility, is a small, dust-free room fitted with overhead ultraviolet light and a positive-pressure ventilation unit with a bacteria-proof high-efficiency particulate air (HEPA) filter. All surfaces should be covered in such a manner that dust and microorganisms do not accumulate and the surfaces can be thoroughly cleaned and disinfected regularly. An air-conditioning unit is advisable, and sources of electricity, gas, compressed air, and vacuum are necessary. One or more laminar flow work stations can be housed in such a transfer area. This type of "clean" or "white" room is especially useful if large numbers of cultures are being manipulated or large pieces of culture equipment are handled.

The laminar flow work station comes in two basic types from many different manufacturers (Canadian Cabinets, Ottawa, Ontario, Canada; Contamination Control, Kulpsville, Pennsylvania; Envirogineering, Toronto, Ontario, Canada; Environmental Air Control, Hagerstown, Maryland; Pure Aire, Van Nuys, California, make cabinets which we have found to be satisfactory), with 0.3-μm HEPA filters of 99.97–99.99% efficiency. Generally, air is forced into the cabinet through a dust filter and then a HEPA filter, and is then directed either downward (vertical flow unit) or outward (horizontal flow unit) over the working surface at a uniform rate. The constant flow of bacteria-free filtered air prevents nonfiltered air and particles from settling on the working area. The simplest transfer cabinet is an enclosed plastic box or shield (available from Fisher Scientific, Ottawa, Ontario; Germfree Labs, Miami, Florida; Scientific Products, Evanston, Illinois) with an ultraviolet light and no airflow. This type of simple tissue culture hood is often sufficient if only limited transfers are performed. These simple transfer cabinets and/or laminar flows are often set up in an infrequently used area of a main laboratory and can be used in this way with a high level of success.

V. CULTURE FACILITIES

Plant tissue cultures, whether on liquid or solid media, should be incubated under conditions of well-controlled temperature, illumination, photoperiod, humidity, and air circulation. All of these environmental factors can affect the growth and differentiation response of tissue cultures directly during culture or indirectly by affecting the subsequent response of donor plants grown under particular environmental conditions. Anther and protoplast cultures are especially sensitive to environmental and media culture conditions (Fujiwara, 1982; Thorpe, 1978; Street, 1974). Incubators, large plant growth chambers, and walk-in environmental rooms can meet these requirements. Incubators of the type supplied by Controlled

Environments (Winnipeg, Manitoba, Canada; Pembina, North Dakota; London, England) have the range of control and flexibility desirable for plant tissue culture use. Typically, they can have the following characteristics:

1. Temperature range, 2–40°C
2. Temperature control, ±0.5°C
3. Safety high- and low-temperature limits
4. Continuous temperature recorder
5. Twenty-four-hour temperature and light programming
6. Adjustable fluorescent lighting up to 10,000 lx
7. Relative humidity range, 20–98%
8. Relative humidity control, ±3%
9. Uniform forced-air distribution
10. Capacity up to 0.7 m^3 of 0.5 m^2 shelf space

If one is willing to sacrifice some flexibility and precision, walk-in environmental control rooms (of the Controlled Environments C-series type) are probably the most cost-effective way of providing adequate culture conditions. These facilities can be constructed almost anywhere using "foamed-in-place" urethane modular panels in a wide variety of shapes and sizes. For static cultures, these rooms can be fitted with open wire shelves to allow for uniform airflow and lighting. These types of culture rooms can also be used to house a wide variety of liquid culture apparatus including gyrotory, tier and reciprocating shakers, and roller tube and tumble tube apparatus (of the type produced by New Brunswick Scientific, Edison, New Jersey), as well as nipple flask culture apparatus or a variety of batch and continuous bioreactor–chemostat units (see Street, 1973b, for detailed descriptions). For detailed information on the effects of controlled environments on plant growth, refer to Krizek (1982) and Tibbitts and Kozlowski (1979).

REFERENCES

Bailey, I. W. (1943). Some misleading terminologies in the literature of "plant tissue culture." *Science* **93,** 539.

Bangham, A. D., and Hill, M. W. (1972). Distillation and storage of water. *Nature (London)* **237,** 408.

Basalik, T. J. (1978). "Reagent Grade Water for Laboratories," Barnstead Publ. RW001-02795MR, pp. 1–4. Barnstead Co., Boston, Massachusetts.

Biondi, S., and Thorpe, T. A. (1981). Requirements for a tissue culture facility. *In* "Plant Tissue Culture: Methods and Applications in Agriculture" (T. A. Thorpe, ed.), pp. 1–20. Academic Press, New York.

Bonga, J. M. (1982). Tissue culture techniques. *In* "Tissue Culture in Forestry" (J. M. Bonga and D. J. Durzan, eds.), pp. 4–35. Nijhoff, The Hague.

Coriell, L. L. (1973). Glassware preparation, sterilizations, and use of laminar flow systems. *In* "Tissue Culture: Methods and Applications" (P. F. Kruse, Jr. and M. K. Patterson, eds.), pp. 671–673. Academic Press, New York.

de Fossard, R. A. (1976). "Tissue Culture for Plant Propagators." University of New England Printery, Armidale, N.S.W., Australia.

Favero, M. S., Carson, L. A., Bond, W. W., and Patersen, N. J. (1971). *Pseudomonas aeruginosa:* Growth in distilled water from hospitals. *Science* **173,** 836–838.

Fujiwara, A., ed. (1982). "Plant Tissue Culture 1982." Maruzen, Tokyo.

Gautheret, R. J. (1959). "La culture des tissues végétaux." Masson, Paris.

Gibbs, E. L. (1972a). "Glass Laboratory Water Stills: Design and Performance," Part A. Ultrascience Inc., Evanston, Illinois.

Gibbs, E. L. (1972b). Glass laboratory water stills: Design and performance. Part B. *In Vitro* **8,** 37–47.

Haberlandt, G. (1902). Culturvesuche mit isolierten Pflanzenzellen. *Sitzungsber. Kais. Akad. Wiss. Wien, Mat.-Naturwiss. Kl., Abt. 1* **111,** 69–92.

Klein, R. M., and Klein, D. T. (1970). "Research Methods in Plant Science." Natural History Press, New York.

Krikorian, A. D., and Berquam, D. L. (1969). Plant cell and tissue cultures: The role of Haberlandt. *Bot. Rev.* **35,** 59–88.

Krizek, D. T. (1982). Guidelines for measuring and reporting environmental conditions in controlled-environment studies. *Physiol. Plant.* **56,** 231–235.

Meltzer, R. L. (1979). "Annual Book of ASTM Standards," Part 31. Water. Am. Soc. Test. Mater., Philadelphia, Pennsylvania.

Pumper, R. W. (1973). Purification and standardization of water for tissue culture. *In* "Tissue Culture: Methods and Applications" (P. F. Kruse, Jr. and M. K. Patterson, eds.), pp. 674–677. Academic Press, New York.

Street, H. E. (1973a). Laboratory organization. *In* "Plant Tissue and Cell Culture" (H. E. Street, ed.), pp. 11–30. Univ. of California Press, Berkeley.

Street, H. E. (1973b). Cell (suspension) culture techniques. *In* "Plant Tissue and Cell Culture" (H. E. Street, ed.), pp. 59–99. Univ. of California Press, Berkeley.

Street, H. E., ed. (1974). "Tissue Culture and Plant Science 1974." Academic Press, New York.

Thorpe, T. A., ed. (1978). "Frontiers of Plant Tissue Culture 1978." Univ. of Calgary Press, Calgary, Canada.

Tibbitts, T. W., and Kozlowski, T. T., eds. (1979). "Controlled Environment Guidelines for Plant Research." Academic Press, New York.

Wakeland, J. R., Crie, J. S., and Wildentha, K. (1982). Toxicity to organ cultured hearts of media prepared with disposable filter units. *In Vitro* **18,** 715–718.

White, P. R. (1941). Plant tissue cultures. *Biol. Rev. Cambridge Philos. Soc.* **16,** 34–48.

White, P. R. (1943). "A Handbook of Plant Tissue Culture." Science Press Printing, Lancaster, Pennsylvania.

White, P. R. (1963). "The Cultivation of Animal and Plant Cells." Ronald Press, New York.

CHAPTER **2**

Educational Services for Plant Tissue Culture

Paul J. Bottino

Department of Botany
University of Maryland
College Park, Maryland

I. INTRODUCTION

In 1978 at the Fourth International Congress of Plant Tissue and Cell Culture in Calgary, a roundtable discussion on teaching plant tissue culture was held. A brief report of this roundtable appeared in the proceedings of the congress. In 1982 at the annual Tissue Culture Association meetings, another roundtable discussion was held on teaching both plant and animal tissue culture. No written report of this discussion is yet available. Therefore, it seemed appropriate to survey the formal course offerings available in plant tissue culture. The present chapter is an attempt to compile and describe these courses.

II. SURVEY METHOD

In November 1982 a series of letters was sent out to determine if the recipients were teaching a course in plant tissue culture or if they knew of such courses being offered. In addition, details on the content of such

CELL CULTURE AND SOMATIC CELL
GENETICS OF PLANTS, VOL. 1

ISBN 0-12-715001-3

courses were requested. In December a second round of letters was sent, and in January 1983 a third. Follow-up letters were sent to those from whom no response had been received. In all, approximately 50 scientists responded to the survey. Although this method may have missed some courses, it has been successful in identifying most. The author wishes to apologize for any who were missed.

III. RESULTS

The results of the survey are summarized in Table I. The courses have been subdivided into academic courses, that is, those which are taught for a semester or quarter for credit at an institution, and short courses, which are taught for 1 or 2 weeks and are more concentrated, but generally grant no academic credit. Included is the name of the institution, the title of the course, and the name of the person responsible for teaching or coordinating the course. In the 1978 roundtable in Calgary, 10 universities in the United States were identified as offering courses. This number has grown considerably since then.

There are many similarities among the courses. First, most are offered in departments directly or indirectly related to agriculture: Botany, Horticulture, Plant Pathology, Plant Science, or Biology. This suggests a recognized need for the use of tissue culture as a research tool in relation to agriculture. Faculty members whose research involved tissue culture would take on the responsibility of developing a formal graduate course and teaching it to a small group, usually once a year or on an alternate-year basis, in their own research laboratory. Most of the courses have approximately the same format: 1–2 hr of lecture per week and at least one 3-hr laboratory period. Most are taught at the graduate level; however, some are undergraduate courses. Virtually all have a limited enrollment of usually 8–20 students.

All of these courses have the same objectives: to emphasize the laboratory techniques of plant cell and tissue culture, and to provide detailed theoretical background. As such, the basic techniques are taught, demonstrated, and practiced, and application of the techniques is emphasized. Most students see several direct applications to their own field by the time they have completed the course. In order to help in this area, time is provided for a special research project of the student's design and choosing. This gives the student experience in working through the first approaches to new material and allows the student to experience success and failure during initial experimentation.

TABLE I

List of Institutions Teaching Courses in Plant Tissue Culture

Institution	Title of course	Instructor/coordinator
A. Academic courses		
SUNY Brockport	Plant Tissue Culture	C. Barr—Biol. Sci.
Univ. Calgary	Experimental Plant Morphogenesis	T. Thorpe—Biol. Sci.
Comm. Coll. Finger Lakes, N.Y.	Micropropagation	P. Pietropaolo—Sci/Tech.
Univ. Conn.	Advanced Plant Propagation	S. Waxman—Plant Sci.
Univ. Conn.	Biotechnical Plant Culture	D. Wetherell/H. Koontz—Biol. Sci.
Univ. Florida	Methods and Applications of Plant Cell and Tissue Culture	I. Vasil—Botany
Univ. Guelph	Plant Cell and Tissue Culture	K. Kasha—Crop Sci.
Univ. Houston	Tissue Culture	S. Venketeswaran—Biol.
Kansas State	Plant Tissue Culture and Regeneration	L. B. Johnson—Plant Path.
Univ. Maryland	Methods in Plant Tissue Culture	P. J. Bottino—Botany
Michigan State	Tissue Culture for Plant Breeding	K. Sink—Horticulture
Univ. Nottingham	Plant Genetic Manipulation	J. B. Power—Botany
SUNY Plattsburgh-Miner Institute	*In Vitro* Cell Biology	W. Graziadei—Biol. Sci.
Queens Univ., Ontario	Plant Cell and Tissue Culture	D. Webb—Biol.
Univ. Rhode I.	Plant Cell, Tissue Organ Culture	B. Krul—Plant/Soil Sci.
Univ. Saskatchewan	Biotechnology	W. Kurz/F. Constabel—Appl. Micro/Food Sci.
So. Dakota State	Plant Morphogenesis	C. H. Chen/G. A. Myers—Botany
Univ. Tennessee	Plant Cell and Tissue Culture	K. Hughes—Botany
Texas A&M	Plant Tissue Culture: Principles and Applications	R. Smith—Plant Sci.
Va. Polytechnic Institute	Plant Tissue Culture	R. Veilleux—Horticulture
B. Short courses		
Catholic Univ.	Plant Cell and Tissue Culture, 9–13 July 1984	R. Nardone—Biol.
Univ. Tennessee	Plant Cell and Tissue Culture, 13–24 August 1984	D. Dougall—Botany
European Mol. Biol. Org.	Somatic Cell Genetics in Plants	O. Schieder—Max Planck Institute, Cologne

Although some of the courses are somewhat specialized, for example, in genetic manipulation approaches, those which deal with basic tissue culture techniques are remarkably similar. This similarity must reflect agreement on the part of those teaching these courses as to what should constitute basic instruction of the technique.

The topics covered by most of the courses usually include:

1. History of the technique
2. Laboratory organization and sterile technique
3. Nutrition of cells in culture and media preparation
4. Establishment of callus culture
5. Establishment and maintenance of suspension cultures
6. Growth measurements of cells in culture
7. Anther culture
8. Morphogenesis from cell cultures
9. Protoplast isolation, culture, and fusion
10. Organ culture—roots, embryos, meristems, etc.

In addition to these major topics, which are common to most of the courses, important additional techniques are taught in some. These include cryopreservation, mutagenesis, secondary products, genetic manipulation, *in vitro* pollination and fertilization, and cytology. In the courses given at the University of Houston and the University of Guelph, some animal cell culture is also included.

A very important point is that in most courses major emphasis is placed on applications of the techniques. This is crucial in teaching this technique. The academic courses have a major advantage in that they last long enough for students to see the results of their efforts and to repeat any laboratory assignment which did not work or was lost due to contamination. This is a distinct disadvantage of the short courses. On the other hand, the short courses make the formal course experience available to virtually everyone, especially those who can invest only a short amount of time in a concentrated course, such as postdoctoral fellows and established scientists in industry or academic institutions.

These courses, no matter where they are offered, are very expensive to teach. They require more facilities than those provided by the standard undergraduate teaching laboratory. Also, not all institutions provide an adequate budget for the course. This means that the course is usually taught in the instructor's research laboratory or at least uses its facilities. In many cases, the instructor's research budget is used to provide supplies. The expense of teaching these courses accounts for the high cost of the short courses. The cost would be even higher if major supply and equipment manufacturers did not donate materials.

IV. BOOKS

The availability of text and laboratory materials for use in tissue culture courses, which was once a problem, has been solved by the appearance of some good publications which are particularly useful to many aspects of teaching tissue culture. Some of them are listed below.

Bottino, P. J. (1981). "Methods in Plant Tissue Culture." Kemtec Educational Corp., Kensington, Maryland.
Dodds, J. H., and Roberts, L. W. (1982). "Experiments in Plant Tissue Culture." Cambridge Univ. Press, London and New York.
Ingram, D. S., and Helgeson, J. P. (1980). "Tissue Culture Methods for Plant Pathologists." Blackwell, Oxford.
Reinert, J., and Yeoman, M. M. (1982). "Plant Cell and Tissue Culture: A Laboratory Manual". Springer-Verlag, Berlin and New York.
Thorpe, T. A., ed. (1981). "Plant Tissue Culture: Methods and Applications in Agriculture." Academic Press, New York.
Tomes, D. T., Ellis, B. E., Harney, P. M., Kasha, K. J., and Peterson, R. L. (1982). "Applications of Plant Cell and Tissue Culture to Agriculture and Industry." Plant Cell Culture Centre, Univ. of Guelph, Guelph, Ontario.
Wetherell, D. F. (1982). "Introduction to *in Vitro* Propagation." Avery Publishing Group, Wayne, New Jersey.
Wetter, L. R., and Constabel, F., eds. (1982). "Plant Tissue Culture Methods," 2nd rev. ed. National Research Council of Canada, Ottawa.

CHAPTER **3**

Plant Cell Cultures: Nutrition and Media

Oluf L. Gamborg

Genentech, Inc.
South San Francisco, California

The development of nutrient media which meet the requirements of plant cells grown aseptically in culture spans several decades, but the greatest progress has been made within the last 20 years. The basic nutritional requirements of cultured plant cells are very similar to those normally utilized by whole plants. However, the nutrient media which are used successfully for cells, tissues, and organs were devised to meet particular requirements (Murashige and Skoog, 1962; Gamborg *et al.*, 1968; Schenk and Hildebrandt, 1972; Phillips and Collins, 1980).

Basic nutritional compositions were adapted with respect to kinds and concentrations for particular plant species and culture systems. Examples of systems include growth of cells on nutrient agar (callus) or in liquid suspension, cells or tissues grown to achieve plant regeneration (morphogenesis), meristem-to-plant regeneration, embryo culture, anther and pollen culture to produce haploids, plant regeneration for micropropagation, and protoplast isolation and culture for growth to complete plants. Any of the above types of culture have unique requirements of one or more nutrient components to achieve the desired results (Gamborg and Wetter, 1975; Thorpe, 1981; Evans *et al.*, 1981; Hughes *et al.*, 1978).

CELL CULTURE AND SOMATIC CELL
GENETICS OF PLANTS, VOL. 1

ISBN 0-12-715001-3

The present chapter discusses the principal components and outlines the compositions and preparations of specific media which have been adopted and proven to be successful for several purposes (see also Street and Shillito 1977; Conger, 1978).

When reference is made to a particular medium, the intent is to identify only the salt composition unless otherwise specified. Any number and concentration of amino acids, vitamins, growth regulator, or organic complex supplements can be added in a nearly infinite variety of combinations to a given salt composition.

I. MEDIA COMPOSITION

A number of basic media are listed in Table I. The Murashige and Skoog (MS) (1962) or Linsmaier and Skoog (LS) (1965) salt composition is used very widely, particularly if the desired objective is plant regeneration. In the development of these and other media, it was demonstrated that not only the presence of the necessary nutrient but also the actual and relative concentrations of, for example, nitrogen, phosphorus, potassium, and perhaps calcium and magnesium are of crucial significance.

The B5 medium, or its various derivatives, have been valuable for cell and protoplast culture (Gamborg *et al.*, 1968, 1981; Kao, 1977). B5 was originally designed for suspension and callus cultures but has also been used effectively in methods of plant regeneration. The major distinction between MS and B5 is the much lower amounts of nitrate and especially ammonium in B5 (Table II). This feature can be significant.

The medium designated N6 was developed for cereal anther culture and is used with success in other types of cereal tissue culture (Chu, 1978). Although N6 is becoming recognized as a suitable medium for cereal tissue culture, other salt compositions have been employed with equally good results (Green and Phillips, 1975). Any success with a medium is in all probability due to the fact that the ratios as well as the concentrations most nearly match the optimum requirement for the cells or tissues for growth and/or differentiation.

The E1 medium (Table I) is a modification of the SL composition developed by Phillips and Collins (1980) for red clover. The medium with the designated additions has been shown to be exceptionally suitable for soybean and related species of *Glycine* (Phillips and Collins, 1981; Gamborg *et al.*, 1983). The E1 composition supports rapid growth of cells for embryogenesis and for the culture of protoplasts.

In the special class of anther culture, the medium devised by Nitsch and Nitsch (1969) is frequently used.

A. Nutrient Media Composition

The nutrient media generally consist of inorganic nutrients, carbon source(s), vitamins, growth regulator(s), and possibly organic supplements.

TABLE I

Inorganic Salt Composition of Plant Tissue Culture Media

Ingredients	Amounts (mg/liter)[a]			
	MS	B5	N6	E1
Macronutrients				
$MgSO_4 \cdot 7H_2O$	370	250	185	400
KH_2PO_4	170	—	400	250
$NaH_2PO_4 \cdot H_2O$	—	150	—	—
KNO_3	1900	2500	2830	2100
NH_4NO_3	1650	—	—	600
$CaCl_2 \cdot 2H_2O$	440	150	166	450
$(NH_4)_2 \cdot SO_4$	—	134	463	—
Micronutrients				
H_3BO_3	6.2	3	1.6	3
$MnSO_4 \cdot H_2O$	15.6	10	3.3	10
$ZnSO_4 \cdot 7H_2O$	8.6	2	1.5	2
$NaMoO_4 \cdot 2H_2O$	0.25	0.25	—	0.25
$CuSO_4 \cdot 5H_2O$	0.025	0.025	—	0.025
$CoCl_2 \cdot 6H_2O$	0.025	0.025	—	0.025
KI	0.83	0.75	0.8	0.8
$FeSO_4 \cdot 7H_2O$	27.8		27.8	
Na_2EDTA	37.3		37.3	
EDTA Na ferric salt		43		43
Sucrose (g)	30	20	50	25
Vitamins				
Thiamine·HCl	0.5	10	1	10
Pyridoxine·HCl	0.5	1	0.5	1
Nicotinic acid	0.05	1	0.5	1
Myo-Inositol	100	100	—	250
pH	5.8	5.5	5.8	5.5

[a] Abbreviations: MS = Murashige and Skoog (1962), B5 = Gamborg *et al.* (1968), N6 = Chu (1978), and E1 = Gamborg *et al.* (1983).

TABLE II

Inorganic Salts: Concentrations of Ions

Salt	Medium			
	MS	B5	N6	E1
Major salts (m*M*)				
Ca	3	1.0	1.1	3.0
Cl	6	2.2	—	—
K	20.1	25	31	0.20
NH_4	20.6	2	6.6	7.5
NO_3	39.4	25	28	27.5
PO_4	1.25	1.1	2.94	1.8
SO_4	1.8	—	0.2	1.6
Mg	1.5	1.0	0.75	1.6
Micronutrients (μ*M*)				
B	100	50	26	50
Co	0.11	0.1	0.11	0.1
Cu	0.1	0.1	—	0.1
Fe	1	1	1	1
I	5	4.5	4.8	4.5
Mn	92.5	60	19.5	60
Mo	1	1	—	1
Zn	30	7	5.2	7

1. Inorganic Nutrients

A variety of salts supply the needed macro- and micronutrients (see Table II). For most purposes, the medium should contain at least 25 m*M* nitrate and potassium. Ammonium can be used in a concentration ranging from 2 to 20 m*M*. The response to ammonium varies from inhibitory to essential depending upon the tissue and the purpose of the culture. The use of ammonium salts of citrate, malate, or succinate is essential if ammonium is to be used as the sole nitrogen source (Gamborg and Shyluk, 1970). When nitrate and ammonium are used in combination, the ammonium is utilized rapidly and before the nitrate.

A concentration of 1–3 m*M* calcium, sulfate, and magnesium is usually adequate. The required micronutrients include iodine, boron, manganese, zinc, molybdenum, copper, cobalt, and iron.

2. Carbon and Energy Source

Without exception, cultured cells utilize sucrose and glucose about equally well. Fructose is less efficient. The sucrose in the medium is rapidly converted to glucose and fructose. Glucose is then utilized first, followed by fructose. Other sugars can be utilized, but poorly. Special cell strains have been isolated which can utilize other metabolites as sole carbon sources (Schaeffer, 1981).

B. Vitamins

Plant cells in culture have a requirement for thiamine. There are also numerous reports on beneficial effects achieved by addition of nicotinic acid, pyridoxine, pantothenate, biotin, and folate. Protoplast media often contain a mixture of most essential vitamins (Kao, 1977; Gamborg *et al.*, 1981).

C. Amino Acids

Although cultured cells normally are capable of synthesizing all of the required amino acids, the addition of L-glutamine (2–8 m*M*) or mixtures of amino acids is frequently beneficial. This is particularly important for establishing cell cultures and the culture of protoplasts.

Enzymatic hydrolysates of proteins such as casein or casamino acids (0.25–1.0 g/liter) are commonly used with success.

D. Growth Regulators

There are four broad classes of growth hormones which are of known importance in tissue culture. They are the auxins, cytokinins, gibberellins, and abscisic acid.

A common feature of auxins is the property of inducing cell division. The compounds include 2,4-dichlorophenoxyacetic acid (2,4-D), indoleacetic acid (IAA), and naphthaleneacetic acid (NAA). Other effective compounds are 2,4-5-trichlorophenoxyacetic acid (2,4,5-T) and picloram (4-amino-3,5,6-trichloropicolinic acid) (Collins *et al.*, 1978). The auxins also stimulate root initiation.

Cytokinins are adenine derivatives which have an important role in shoot induction. The most frequently used cytokinins are kinetin, benzyladenine, BA, zeatin, and isopentenyladenine (IPA). The gibberellins are normally used in plant regeneration after the formation of primordia has occurred.

Ethylene is also an important growth hormone. The compound is produced by cultured cells, but its role in cell and organs in culture is not known.

II. MEDIA PREPARATION

The making of media can become time-consuming. The following procedure has proven to be efficient for a research laboratory operation.

A. Materials

1. Water. The water should be glass distilled or demineralized to high purity.
2. Inorganic nutrients and organic compounds. The compounds should be of the highest grade. Growth regulators may require purification by charcoal decolorization and recrystallization from water–ethanol mixtures. Amino acids should be the L isomer.
3. Protein hydrolysates. The enzymatic digests are preferable because all of the amino acids and glutamine are preserved. Casamino acid preparations are an alternative.
4. Coconut milk. The liquid is collected from several nuts, heated to 80°C with stirring, filtered, and stored frozen.
5. Agar. Bacto from Difco is generally used. New types of agar for special use such as protoplast culture include Prep TM (FMC Corp., Rockland, Maine) (Adams and Townsend, 1983).

B. Procedures

Stock solutions are prepared as shown below. The chemicals are dissolved in distilled or demineralized water. When the salts are dissolved and other ingredients added, the pH is adjusted by using dilute HCl or NaOH.

For convenience, concentrates may be prepared, stored frozen, and used as required.

A 10× concentrate consisting of inorganic salts, vitamins, and sucrose of the media listed in Table I can be prepared.

After the ingredients are dissolved and made to volume, the solution is distributed in Whirl-Pak bags (100 ml into 6-oz bags or 400 ml into 18-oz bags). The bags are stored frozen. One 100-ml bag is sufficient for 1 liter of final medium.

Before autoclaving, the medium is distributed in containers, or the containers and medium may be autoclaved separately. The medium is autoclaved at 120°C for 15–20 min, and the flasks are removed for cooling as soon as possible. Agar media are usually autoclaved in lots of 500 ml and subsequently poured into sterile containers.

Media should be stored at about 10°C.

C. Stock Solutions

1. Stock solutions for B5 medium (for preparation of B5 medium, see below). Note: B5 medium refers to the basic medium with no growth hormone or organic supplements.
   a. Micronutrients (store in freezer)

Nutrient	Amount (mg/100 ml)
$MnSO_4 \cdot H_2O$	1,000
H_3BO_3	300
$ZnSO_4 \cdot 7H_2O$	200
$Na_2MoO_4 \cdot 2H_2O$	25
$CuSO_4 \cdot 5H_2O$	2.5
$CoCl_2 \cdot 6H_2O$	2.5

   b. Vitamins (store in freezer)

Vitamin	Amount (mg/100 ml)
Nicotinic acid	100
Thiamine·HCl	1,000
Pyridoxine·HCl	100
Myo-Inositol	10,000

   c. Calcium chloride
      $CaCl_2 \cdot 2H_2O$ 15 g/100 ml
   d. Potassium iodide (store in amber bottle in refrigerator)
      KI 75 mg/100 ml
   e. 2,4-D (2.2 m*M*)
      Dissolve 50 mg 2,4-D in 2–5 ml ethanol, heat slightly, and gradually dilute to 100 ml with water. Store in the refrigerator.
   f. NAA (2.8 m*M*)
      Prepare the same as for 2,4-D above.
   g. Kinetin, BA
      Dissolve 21.5 mg kinetin (22.5 mg BA) in a small volume of 0.5 *N* HCl by heating slightly and gradually diluting to 100 ml with distilled water. Store in refrigerator. Similar procedures can be used for other cytokinins.

2. Preparation of B5 medium.

Ingredient	Amount (per liter)
$NaH_2PO_4 \cdot H_2O$	150 mg
KNO_3	2500 mg
$(NH_4)_2SO_4$	134 mg
$MgSO_4 \cdot 7H_2O$	250 mg
Ferric EDTA	43 mg
Sucrose	20 g
$CaCl_2 \cdot 2H_2O$, stock solution	1.0 ml
Micronutrients, stock solution	1.0 ml
Potassium iodide, stock solution	1.0 ml
Vitamins, stock solution	1.0 ml

3. MS medium (for composition, see Table I).
   a. MS–micronutrient stock solution (keep frozen).

Ingredient	Amount (mg/100 ml)
H_3BO_3	620
$MnSO_4 \cdot 4H_2O$	2230
$ZnSO_4 \cdot 7H_2O$	860
$Na_2MoO_4 \cdot 2H_2O$	25
$CuSO_4 \cdot 5H_2O$	2.5
$CoCl_2 \cdot 6H_2O$	2.5

   b. All other stock solutions are prepared the same as for B5.
4. Preparation of MS medium.

Ingredient	Amount (per liter)
NH_4NO_3	1650 mg
KNO_3	1900 mg
$MgSO_4 \cdot 7H_2O$	370 mg
KH_2PO_4	170 mg
Ferric EDTA	43 mg
Sucrose	30 g
$CaCl_2 \cdot 2H_2O$ (stock solution)	2.9 ml
MS–micronutrients (stock solution)	1.0 ml
KI (B5 stock solution)	1.0 ml
Vitamins (B5 stock solution	1.0 ml

Adjust pH to 5.8 with 0.2 *N* KOH or 0.2 *N* HCl

REFERENCES

Adams, T. L., and Townsend, J. A. (1983). A new procedure for increasing efficiency of protoplast plating and clone selection. *Plant Cell Rep.* **2,** 165–168.

Chu, C.-C. (1978). The N6 medium and its applications to anther culture of cereal crops. *In* "Proceedings of Symposium on Plant Tissue Culture," pp. 43–50. Science Press, Peking.

Collins, G. B., Vian, W. E., and Phillips, G. C. (1978). Use of 4-amino-3,5,6-trichloropicolinic acid as an auxin source in plant tissue cultures. *Crop Sci.* **18,** 286–288.

Conger, B. V. (1978). Problems and potentials of cloning agronomic crops via in vitro techniques. *In* "Propagation of Higher Plants through Tissue Culture" (K. W. Hughes, R. R. Henke, and M. J. Constantin, eds.), pp. 62–72. Dept. of Energy, Tech. Inf. Cent., Springfield, Virginia.

Evans, D. A., Sharp, W. R., and Flick, C. E. (1981). Plant regeneration from cell cultures. *Hortic. Rev.* **3,** 214–314.

Gamborg, O. L., and Shyluk, J. P. (1970). The culture of plant cells using ammonium salts as the sole nitrogen source. *Plant Physiol.* **45,** 598–600.

Gamborg, O. L., and Wetter, L. R., eds. (1975). "Plant Tissue Culture Methods." National Research Council of Canada, Saskatoon.

Gamborg, O. L., Miller, R. A., and Ojima, K. (1968). Nutrient requirements of suspension cultures of soybean root cells. *Exp. Cell Res.* **50,** 151–158.

Gamborg, O. L., Shyluk, J. P., and Shahin, E. A. (1981). Isolation, fusion and culture of plant protoplasts. *In* "Plant Tissue Culture: Methods and Applications in Agriculture" (T. A. Thorpe, ed.), pp. 115–153. Academic Press, New York.

Gamborg, O. L., Davis, B. D., and Stahlhut, R. W. (1983). Somatic embryogenesis in cell cultures of *Glycine* species. *Plant Cell Rep.* **2,** 209–212.

Green, C. E., and Phillips, R. L. (1975). Plant regeneration from tissue cultures of maize. *Crop Sci.* **15,** 417–421.

Hughes, K. W., Henke, R. R., and Constantin, M. J., eds. (1978). "Propagation of Higher Plants through Tissue Culture." Dept. of Energy, Tech. Inf. Cent., Springfield, Virginia.

Kao, K. N. (1977). Chromosomal behavior in somatic hybrids of soybean—*Nicotiana glauca. Mol. Gen. Genet.* **150,** 225–230.

Linsmaier, E. M., and Skoog, F. (1965). Organic growth factor requirements of tobacco tissue cultures. *Physiol. Plant.* **18,** 100–127.

Murashige, T., and Skoog, F. (1962). A revised medium for rapid growth and bioassays with tobacco tissue cultures. *Physiol. Plant.* **15,** 473–479.

Nitsch, J. P., and Nitsch, C. (1969). Haploid plants from pollen grains. *Science* **163,** 85–87.

Phillips, G. C., and Collins, G. B. (1980). Somatic embryogenesis from cell suspension cultures of red clover. *Crop Sci.* **20,** 323–326.

Phillips, G. C., and Collins, G. B. (1981). Induction and development of somatic embryos from cell suspension cultures of soybean. *Plant Cell, Tissue Organ Cult.* **1,** 123–129.

Schaeffer, G. W. (1981). Mutations and cell selections: Increased protein from regenerated rice tissue cultures. *Environ. Exp. Bot.* **21,** 333–345.

Schenk, R. U., and Hildebrandt, A. C. (1972). Medium and techniques for induction and growth of monocotyledonous and dicotyledonous plant cell cultures. *Can. J. Bot.* **50,** 199–204.

Street, H. E., and Shillito, R. D. (1977). Nutrient media for plant organ, tissue and cell culture. *In* "CRC Handbook in Nutrition and Food" (M. Rechcigl, ed.), Vol. IV, pp. 305–359. CRC Press, Boca Raton, Florida.

Thorpe, T. A., ed. (1981). "Plant Tissue Culture: Methods and Applications in Agriculture." Academic Press, New York.

CHAPTER 4

Callus Culture: Induction and Maintenance*

F. Constabel

Plant Biotechnology Institute
National Research Council
Saskatoon, Saskatchewan, Canada

I. INTRODUCTION

In the context of plant cell culture, callus is a largely unorganized, proliferating mass of parenchyma cells. With age such callus may show meristematic islands or strands and individual or groups of tracheids and pigmented cells. Callus is initiated and maintained on nutrient media *in vitro*. It serves the dual purpose of being studied on plant growth and development and exploited for plant products and propagation. The formation of callus with an explant, i.e., an excised and isolated piece of tissue placed on or in nutrient medium, marks the beginning of successful plant cell culture. Its induction requires three equally important considerations: (1) the selection of an explant, (2) the provision of a suitable nutrient medium and culture conditions, and (3) the isolation and maintenance of callus for subsequent experimentation.

*NRCC No. 22966.

CELL CULTURE AND SOMATIC CELL
GENETICS OF PLANTS, VOL. 1

ISBN 0-12-715001-3

II. THE EXPLANT

The objective of a given research program determines the species to be considered for callus culture. Investigations of growth and developmental phenomena with seed plants are well suited for the use of callus cultures because seed plants readily furnish starting material (explants). Also, tissues from seed plants generally respond well to established culture conditions for callus formation.

Parenchyma is a tissue with considerable developmental plasticity resulting from its low level of differentiation (Esau, 1965). The potential to divide may be retained for many years, as evidenced by callus formation from 25-year-old parenchyma of the *Tilia* stem (Barker, 1953). The pith and cortex of stems and roots, tubers, the mesophyll of leaves, the flesh of succulent fruits, and the endosperm of seeds are examples of plant parts consisting largely or entirely of parenchyma. In phloem and xylem, parenchyma occurs as rays and vertical strands. With seed plants, only a few groups would be difficult to subject to callus formation, i.e., xerophytes, because of limited access to suitable parenchyma, and hydrophytes, because of lack of protection during disinfection. For routine work and demonstrations in classes, carrot root phloem, Jerusalem artichoke tuber, and tobacco stem pith are recommended.

III. DISINFECTION

Once a tissue or plant part has been selected for explantation, it has to be excised and isolated, disinfected, and transferred to a nutrient medium under sterile conditions. Given a healthy plant, only the surfaces of parts harboring the tissues targeted for explantation need be disinfected. Sterilizing agents are listed in Table I. Plant parts are immersed in a disinfectant solution in a beaker or Petri dish of suitable size, on a clean bench of a laminar airflow cabinet, or in a room sterilized by ultraviolet (UV) light or water vapor. After treatment the plant parts are washed in sterile, distilled water two or three times in order to remove completely the sterilizing agent.

A most convenient way to obtain sterile plant material for excision and isolation of explants is the disinfection of seeds and rearing of seedlings in an axenic environment, i.e., on sterilized and moistened filter paper, on cotton, or on 0.8% agar in Petri dishes or test tubes.

TABLE I

Disinfection of Plant Parts for Callus Culture

Sterilizing agent	Concentration used (%)	Duration of treatment (min)	Plant parts[a]
Calcium hypochlorite	9–10	10–40	Young stems Petioles Rhizomes Roots, fruits
Sodium hypochlorite[b]	1.2–2.0[c]	10–40	Same as for calcium hypochlorite
Bromine water	1–2	2–10	Delicate tissue
Ethanol	70	2–3(–5)	Leaves, seeds

[a] A disinfection procedure has to be established for each kind of tissue.
[b] A drop of a wetting agent such as Tween-80 may improve the disinfection.
[c] Twenty percent of commercial bleach (Javex).

The application of antibiotics (10–80 mg/liter streptomycin or tetramycin, 200–400 mg/liter ampicillin, or 20 mg/liter nystatin) should be left to material infected with known bacterial or fungal contaminants, e.g., *Agrobacterium tumefaciens* in crown-gall tissue (Davey *et al.*, 1980). General usage of antibiotics may impair callus formation and culture. For specific recommendations regarding the use of antibiotics see Pollock *et al.* (1983).

Elimination of virus and mycoplasma requires the isolation and culture techniques described in Chapter 65, this volume.

IV. EXCISION AND ISOLATION OF EXPLANTS

The excision of tissue to be explanted usually is performed with dissecting scalpels fitted with regular or miniblades (Beaver Eye Blades), razor blades, or a cork borer and assisted by the use of forceps with pointed tips. Care is taken to rinse the instruments in 70% ethanol and to flame them as often as possible. For multiple explants of equal size, one may use calibrated multiple razor blades, as described by Yeoman and Macleod (1977), or graph paper underlying a Petri dish in which the dissection is being performed.

After disinfection, plant parts are carefully trimmed by cutting away all bleached and dead tissue. Then rectangular or cylindrical pieces approximately 5 mm in diameter are excised. Bigger explants increase the danger of contamination, and smaller ones increase the ratio of wounded to intact cells. A high surface/volume ratio is desirable in order to facilitate exchange of gases and uptake of nutrients.

Explants are transferred to agar plates of nutrient medium in Petri dishes, five to six per dish, or individually to nutrient agar in test tubes. Transfer of explants to liquid nutrient media has two drawbacks: (1) the loss of explants due to contamination can be higher, and (2) callus formation is impeded by lack of gas exchange. Although in most cases explants are simply placed on the agar medium, stem segments may be implanted apical side down in order to simulate the natural flow of auxins and carbohydrates.

V. NUTRIENT MEDIA AND CULTURE CONDITIONS

The induction of callus formation in parenchyma excised from stems, tubers, roots, etc. requires an environment which allows at least some cells to become meristematic and resume mitotic cycles. Aside from suitable humidity, temperature, and aeration, the nutrient medium supplemented with growth hormones is the strongest factor. Formulations of such media have been well reviewed (Gamborg *et al.*, 1976; see also Chapter 3, this volume). Those employed by Murashige and Skoog (1962) for tobacco pith explants (MS medium) and by Gamborg *et al.* (1968) for soybean hypocotyl segments (B5 medium) have found the widest acceptance. The two media differ significantly with respect to nitrogen and phosphorus. Hormone levels are flexible in both and have to be adjusted to the explants in question. Both media may be fortified with 5–15% (v/v) of liquid endosperm of commercial, better noncommercial immature coconuts, or 1–2 g/liter casein hydrolysate in order initially to enhance callus formation. Table II lists a few examples of media used for a variety of explants. A literature search could provide media formulations for a wide range of species. If such a search is of no help, it would be good practice to employ both MS and B5 media with 0.1–2.0 mg/liter auxin, alone or in combination with 0.1–2.0 mg/liter cytokinin, and solidified with 0.6–0.8% agar. For the preparation of media, see Chapters 1 and 3, this volume.

Explants may suffer from oxidations due to wounding and turn brown. This situation may be remedied by increasing the level of hormones and adding coconut water to enhance callus growth, or by supplementing the medium with antioxidants, e.g., 50–100 mg/liter ascorbate.

TABLE II

Nutrient Media for Explants of Selected Species

Species	Explant	Medium[a]
Periwinkle (*Catharanthus roseus*)	Shoot tip meristem	B5 with 1 mg/liter NAA
	Petiole segments	B5 with 0.1 mg/liter NAA
	Hypocotyl segments	B5 with 0.1 mg/liter NAA
Coffee (*Coffea arabica*)	Halved endosperm of immature fruits	MS with 1 mg/liter NAA and 1 mg/1 KIN
Carrot (*Daucus carota*)	Phloem disks	B5 with 1 mg/liter 2,4-D
Pea (*Pisum sativum*)	Epicotyl segments	B5 with 1 mg/liter 2,4-D and 2 g/liter casein hydrolysate
Sorghum (*Sorghum bicolor*)	10–14-day-old embryos	MS with 1–10 mg/liter 2,4-D
Juniper (*Juniperus communis*)	Phloem/cambium/xylem segments	MS with 1 mg/liter 2,4-D, 15% coconut water

[a] Abbreviations: B5 = medium after Gamborg *et al.* (1968), MS = medium after Murashige and Skoog (1962), NAA = 1-naphthalene acetic acid, 2,4-D = 2,4-dichlorophenoxyacetic acid, and KIN = kinetin, 6-furfurylaminopurine.

The temperature for callus formation and culture is usually set at 22–28°C. The humidity should be as high as possible without causing condensation in containers. Strips of Parafilm around Petri dishes and cotton plugs on test tubes generally are sufficient to maintain humidity for up to 4 weeks. Callus formation generally occurs in light (10–40 $W \cdot m^{-2}$) as well as in the dark.

VI. CALLUS FORMATION

Within 2–3 weeks, explants should show new growth as pustules, or protuberances, or as a fine mat across the surface depending on the distribution and mitotic activity of the parenchyma residing in the excised tissue (Figs. 1–4). Continued growth may leave the core of the explants fairly undisturbed, or the explants may disintegrate as the callus grows. In light, the callus may turn green. Callus formation may be accompanied by the formation of roots. A twofold increase in the level of auxins in the medium should prevent rhizogenesis.

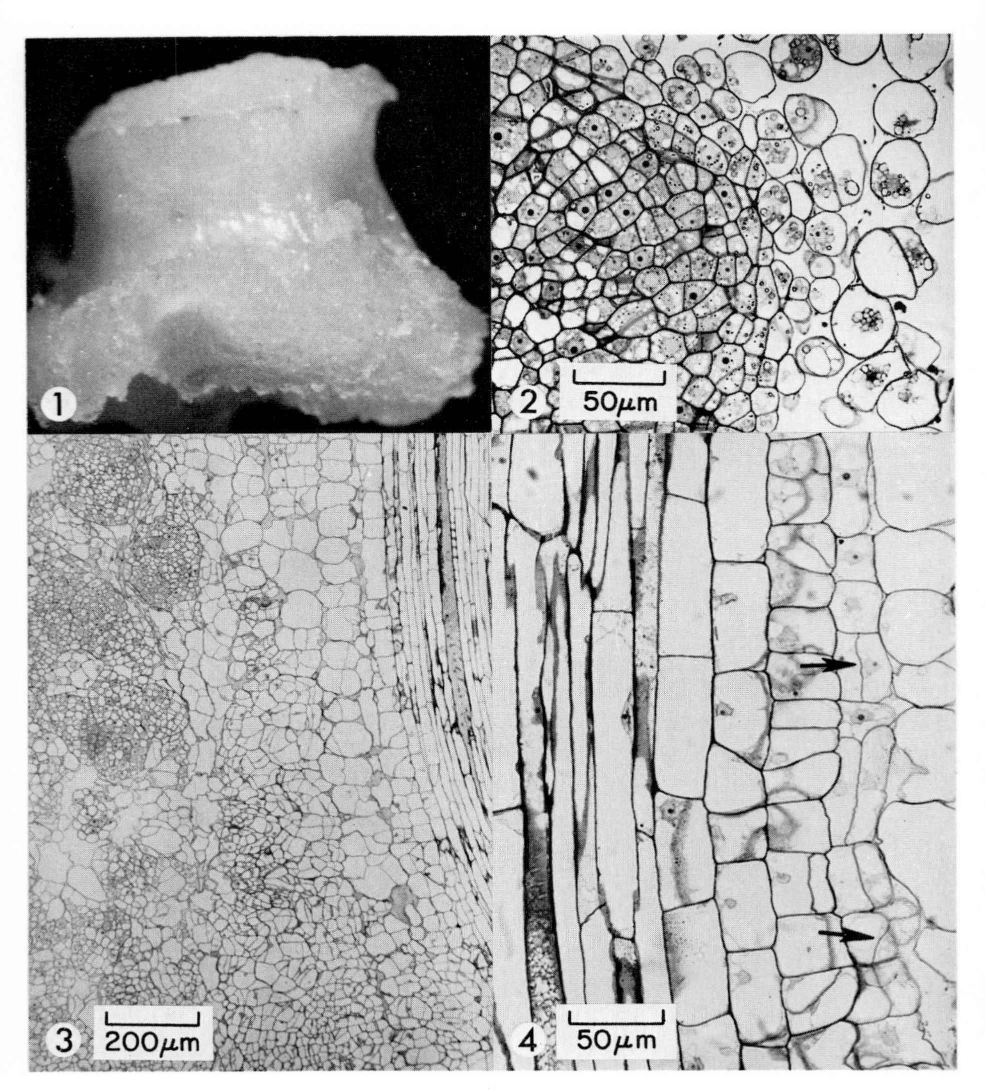

Figs. 1–4. Callus formation with an explanted periwinkle stem segment. **Fig. 1.** Stem segment showing external callus over a pith region at the cut surface and, more vigorously, at the area in contact with agar nutrient medium. **Fig. 2.** Cross section of external callus. **Fig. 3.** Advanced development of callus in pith. **Fig. 4.** Early stage of callus development in the cortex (arrows). (Photographs by L. R. Nesbitt.)

VII. CALLUS SUBCULTURE

It is good practice to allow callus to grow 2–3 cm in diameter before separating it from the explant and/or partitioning it to obtain four to eight inocula for subcultures to be placed on fresh medium. If such initial callus has a rather heterogeneous appearance, one is tempted to prefer as inocula those callus pieces which seem to grow fastest. These usually are pale and relatively soft. It is the ensuing experimentation which dictates whether such a procedure is appropriate. It may well be that more slowly growing callus pieces render material more amenable, for example, to differentiation and regeneration processes.

Cell variation is likely to occur during callus formation (Bayliss, 1980). Serial subcultures of callus pieces showing histological and cytological differences will naturally lead to the establishment of differing callus isolates, strains, and lines. The variation of a desirable trait observed with serial subcultures of callus may be exploited for crop improvement (Heinz and Mee, 1971) or may give rise to material rich in phytoproducts (Ellis, 1982). In a few cases, cell variability has been followed over a number of subcultures. Yamamoto *et al.* (1982) reported consistency of pigment formation (anthocyanin) only after 24 selections (subcultures of 10 days' growth each). On the other hand, Dhoot and Henshaw (1977) demonstrated gradual loss of alkaloids with *Hyoscyamus niger* over 44 subcultures of 21 days' growth each. Procuring rather homogeneous callus from the start, as is required by many experiments related to studies of growth, is dependent on homogeneous explant material such as that found in Jerusalem artichoke tubers, carrot root phloem, or tobacco stem pith (Yeoman and Macleod 1977).

VIII. CALLUS MAINTENANCE

When cultured for several weeks, any callus will show signs of aging, noted as deceleration of growth, necrosis or browning, and finally desiccation. Aging usually is the result of one or several of the following factors: (1) exhaustion of nutrients, (2) inhibition of nutrient diffusion, (3) evaporation accompanied by an increase in the concentration of some constituents of the medium, and (4) accumulation of metabolites, some of which may be toxic. Transfer of healthy, vigorous callus pieces about 5 mm in diameter to 30 ml fresh medium (subculture) in 120-ml jars at intervals of 4–6 weeks will maintain the callus line. Several such lines have been maintained for over 15 years, e.g., a soybean cell line at PRL. Dim light and a temperature of about 24°C generally are suitable for callus storage.

The preservation of callus at low temperature (5–10°C) for long periods of time (6–12 months) without transfer has, in the opinion of this author, too often led to loss and thus cannot be recommended. For a discussion of cryopreservation of cells and small cell aggregates, see Chapters 68 and 69, this volume. The establishment of long-term callus collections has been considered by several laboratories and is, at present, being pursued by the American Type Culture Collection, 12 301 Parklawn Drive, Rockville, Maryland 20852. It appears, however, that laboratory collections have been found unsatisfactory in terms of both supply and demand. Maintenance of a greater than necessary stock of callus cultures by way of monthly subcultures leads to slow change of the material and is labor intensive. Exchange of material with other laboratories or reestablishment of a given culture is preferred. Again, it is the possibility of change of callus over a period of years wich lessens the demand for cultures from collections.

Mailing callus would require the transfer of well-growing callus, about 1–2 cm in diameter, to 50-ml tubes or plastic containers with 5–10 ml liquid rather than agar–medium and the fastest air and courier services. Cross-border regulations covering biological materials must be observed.

IX. PROCEDURES

Specific procedures for the induction and maintenance of callus cultures can be found throughout the scientific literature. Beginners will find laboratory manuals which provide a step-by-step guide to sterilization, isolation, and culture of explants helpful (de Fossard, 1981; Wetter and Constabel, 1982).

REFERENCES

Barker, W. G. (1953). Proliferative capacity of the medullary sheath region in the stem of *Tilia americana*. *Am. J. Bot.* **40,** 773–778.

Bayliss, M. W. (1980). Chromosomal variation in plant tissues in culture. *Int. Rev. Cytol., Suppl.* **11A,** 113–144.

Davey, M. R., Cocking, E. C., Freeman, N., Pearce, N., and Tudor, I. (1980). Transformation of *Petunia* protoplasts by isolated *Agrobacterium tumefaciens* plasmids. *Plant Sci. Lett.* **18,** 307–314.

de Fossard, R. A. (1981). "Tissue Culture for Plant Propagators." University of New England Printery, Armidale, N.S.W., Australia.
Dhoot, G. K., and Henshaw, G. G. (1977). Organization and alkaloid production in tissue cultures of *Hyoscyamus niger*. *Ann. Bot. (London),* [N.S.] **41,** 943–949.
Ellis, B. E. (1982). Selection of chemically-variant plant cell lines for use in industry and agriculture. *In* "Application of Plant Cell and Tissue Culture in Agriculture and Industry" (D. T. Tomes, B. E. Ellis, P. M. Harney, K. J. Kasha, and R. L. Peterson, eds.), pp. 63–80. Univ. of Guelph, Guelph, Ontario.
Esau, K. (1965). "Plant Anatomy." Wiley, New York.
Gamborg, O. L., Miller, R. A., and Ojima, K. (1968). Nutrient requirements of suspension cultures of soybean root cells. *Exp. Cell Res.* **50,** 151–158.
Gamborg, O. L., Murashige, T., Thorpe, T. A., and Vasil, I. K. (1976). Plant tissue culture media. *In Vitro* **12,** 473–478.
Heinz, D. J., and Mee, G. W. P. (1971). Morphologic, cytogenetic, and enzymatic variation in *Saccharum* species hybrid clones, derived from callus tissue. *Am. J. Bot.* **58,** 257–262.
Murashige, T., and Skoog, F. (1962). A revised medium for rapid growth and bioassays with tobacco tissue cultures. *Physiol. Plant.* **15,** 473–497.
Pollock, K., Barfield, D. G., and Shields, R. (1983). The toxicity of antibiotics to plant cell cultures. *Plant Cell Rep.* **2,** 36–39.
Wetter, L. R., and Constabel, F., eds. (1982). "Plant Tissue Culture Methods," 2nd rev. ed. National Research Council of Canada, Ottawa.
Yamamoto, Y., Mizuguchi, R., and Yamada, Y. (1982). Selection of a high and stable pigment-producing strain in cultured *Euphorbia millii* cells. *Theor. Appl. Genet.* **61,** 112–116.
Yeoman, M. M., and Macleod, A. J. (1977). Tissue (callus) cultures - techniques. *In* "Plant Tissue and Cell Culture" (H. E. Street, ed.), pp. 61–102. Univ. of California Press, Berkeley.

CHAPTER **5**

Induction and Maintenance of Embryogenic Callus Cultures of Gramineae

Vimla Vasil
Indra K. Vasil

Department of Botany
University of Florida
Gainesville, Florida

I. INTRODUCTION

Reproducible regeneration of plants is now possible from tissue cultures of all major cereal and grass species (Vasil, 1983). This is due in part to the realization that explants from mature tissues of the Gramineae generally yield either nonregenerable or only root-forming calli, and the consequent widespread use of immature and meristematic tissue explants for the establishment of totipotent cultures. Immature embryos, young inflorescences, and young leaves have proved to be most suitable. Regeneration is said to occur either by *de novo* formation of shoot primordia (Rangan, 1974; Nakano and Maeda, 1979; Shimada and Yamada, 1979; Springer *et al.*, 1979), proliferation of presumptive shoot primordia (King *et al.*, 1978), or

CELL CULTURE AND SOMATIC CELL
GENETICS OF PLANTS, VOL. 1

ISBN 0-12-715001-3

somatic embryogenesis (Vasil, 1982a,b, 1983). Formation of somatic embryos in cereal tissue cultures was first reported in barley (Norstog, 1970), but the embryoids did not give rise to plants. However, somatic embryogenesis recently has been shown to occur in a wide variety of species, leading to the suggestion that it may be a common method of plant regeneration in tissue cultures of the Gramineae (Vasil and Vasil, 1982a). The formation of somatic embryos is preceded by the appearance of a white to pale yellow compact callus that is opaque, often with a nodular and convoluted surface, and rather organized in appearance. Such callus cultures have been described as embryogenic (Vasil and Vasil, 1980, 1981).

II. SELECTION AND STERILIZATION OF EXPLANTS

A. Immature Embryo

Inflorescences bearing young caryopses are obtained from plants 10–15 days after pollination. The immature caryopses are removed and sterilized by a 30-sec rinse in 70% ethanol, followed by 10–20 min in 20% commercial bleach to which a detergent has been added as a wetting agent. In most cases, this procedure is sufficient to remove all surface contaminants. In those species in which sterilization may be a problem, a 30 to 60-sec rinse in mercuric chloride (0.01–0.1%) is recommended. The caryopses are then thoroughly washed in at least three changes of sterile distilled water. For maize, the entire dehusked cob or its segments can be sterilized with ease, and the young embryos can be removed with the unaided eye. However, in most other species in which the caryopses are rather small, an ordinary dissecting microscope is used for the removal of young embryos. As embryo development is greatly influenced by temperature and other environmental conditions, the selection of embryos for culture is made on the basis of both size and developmental stage. The best results are obtained from embryos which range in length from 1 to 2 mm, with the scutellum turning opaque and measuring approximately twice the length of the embryonal axis. The endosperm at this time is nearly cellular.

The manner in which the dissected embryos are placed on the nutrient medium is critical in eliciting the desired response. To obtain proliferation of the scutellum and formation of an embryogenic callus, it is essential that the embryo be so placed that the scutellum is exposed and the embryo axis is in contact with the medium (facedown position). The embryo germinates if placed in the faceup position, in which the embryo axis is exposed.

Whole, mature embryos generally form only soft and root-forming callus. However, Botti and Vasil (1983) obtained embryogenic callus from the shoot meristem, excised along with the two youngest leaf primordia in *Pennisetum americanum,* suggesting that the mature tissues of the explant may have an inhibitory effect on the capacity of the meristematic tissues to form embryogenic callus.

B. Young Inflorescence

Young, unemerged, premeiotic inflorescences in which the primordia of the individual florets are just beginning to be formed have been found to be most suitable. The inflorescences, generally 1–2 cm in length, are sterilized as described in section II,A. Since the inflorescence is sterilized while it is still enclosed in the surrounding leaves, the duration of commercial bleach treatment can be reduced without risking contamination. Following sterilization, the inflorescence is exposed by a vertical incision through the surrounding leaves and then cut into 1- to 2-mm-thick segments. The orientation of the segments on the medium is not critical for initiation of embryogenic callus.

C. Young Leaf

Young, unexpanded leaves can be obtained from seeds germinated under aseptic conditions; in this case, sterilization of leaves is not required. Alternatively, shoots obtained from plants of any age are sterilized, after removal of several of the outermost leaves, as described in Section II,A. Following sterilization, all of the leaves, except the youngest three to five, are removed. The latter are then cut into 1 to 2-mm-thick transverse segments starting from the level of the shoot meristem and going up. Six to eight explants are placed on nutrient medium in a Petri dish.

Irrespective of the type of explant used, its age plays a critical role in determining the response *in vitro.* Very young explants often fail to show any growth, whereas older ones produce either root-forming or nonregenerable callus.

III. NUTRIENT MEDIUM

The most widely used nutrient medium for induction as well as maintenance of embryogenic calli in the Gramineae is that of Murashige and

Skoog (MS) (1962). The synthetic herbicide 2,4-dichlorophenoxyacetic acid (2,4-D) is the only auxin required for induction and maintenance of embryogenic calli. Generally, a concentration of 0.5–2.5 mg/liter has been found to be satisfactory, although many Gramineae can tolerate upto 10 mg/liter. Sucrose requirements vary from 2 to 6%, sometimes higher. Cytokinins (0.1–0.25 mg/liter) are not needed in most species but may be used in conjunction with 2,4-D. Inclusion of coconut milk (5–10%) and/or casein hydrolysate (100–500 mg/liter) has often been found to be helpful, at least during the initiation phase, but is not essential.

For ease of examination, it is preferable to use plastic Petri dishes for culture. Dishes sealed with Parafilm are incubated in light or in dark at about 28°C.

IV. INDUCTION OF EMBRYOGENIC CALLUS

Cell divisions start within 2 days after excision and culture (Vasil and Vasil, 1982b), and visible proliferation can be recognized on the surface of explants after 5–10 days. A compact, opaque, white to pale yellow embryogenic callus is formed within 2 weeks. The callus is nodular and has sectors which are smooth and shiny or rough in appearance. Soon globular to cup-shaped structures, representing somatic embryos, appear on the surface of the callus. Calli obtained from immature embryos, inflorescences, or leaves are similar to and indistinguishable from each other. The rate of growth of the callus varies considerably with the species and the amount of 2,4-D used. For instance, the embryogenic callus of *Panicum maximum* (Lu and Vasil, 1982) and *Pennisetum purpureum* (Wang and Vasil, 1982) grows rapidly and shows early and extensive organization, whereas in *Triticum aestivum* the embryogenic callus appears late (Ozias-Akins and Vasil, 1983).

When immature embryos are cultured, the cells of the embryo axis do divide initially, but it is the peripheral cell layers at the radicular end of the scutellum which proliferate and give rise to embryogenic callus (Vasil and Vasil, 1982b). In some species the callus may be formed over the entire surface of the scutellum.

In cultured inflorescence segments, embryogenic callus is formed primarily by proliferating young floral primordia (Botti and Vasil, 1984). Here too, the quantity, quality, and rapidity of callus formation are particularly dependent upon the age of the inflorescence used, and also on the concentration of 2,4-D in the medium (Rangan and Vasil, 1983; Botti and Vasil, 1984).

Cells of the lower epidermis (abaxial surface), as well as mesophyll cells near the vascular bundles, proliferate in cultured leaf sections to form embryogenic calli (Lu and Vasil, 1981; Haydu and Vasil, 1981; Wernicke and Brettell, 1980; Ho and Vasil, 1983). The developmental age of the leaf in comparison to other leaves, and the position of the explant on the leaf, both determine the response *in vitro*. Leaf sections obtained from the basal (youngest) portions of the youngest one to five leaves are best. The farther the section from the leaf base, the less suitable it is for the formation of embryogenic callus (Ho and Vasil, 1983). The age of the plant does not seem to be important; young leaves obtained even from mature plants have been used successfully (Haydu and Vasil, 1981; Ho and Vasil, 1983).

V. MAINTENANCE OF EMBRYOGENIC CALLUS

The embryogenic callus is almost always associated with at least two other types of callus tissue: (1) a soft, transparent, unorganized callus, which is nonmorphogenic or may form only roots or (2) a soft, mucilaginous, gelatinous callus which is often associated with roots. These calli grow fast and can easily overgrow the embryogenic callus, which is initially slow growing. It is critical, therefore, that the embryogenic callus be separated and subcultured alone after the first few weeks, and care should be taken to remove any nonembryogenic callus at each subculture. The time interval between subcultures varies with the species and the concentration of 2,4-D used. For example, the fast-growing embryogenic calli of *Panicum maximum* (Lu and Vasil, 1981) and *Pennisetum purpureum* (Wang and Vasil, 1982) can be subcultured every 10 days to 2 weeks. In other species this may be done every 4 weeks.

Embryogenic calli often undergo rapid and extensive organization, resulting in the early formation of somatic embryos which germinate precociously. In order to maintain cultures for long periods of time, it is necessary to suppress this tendency. Frequent transfer of calli to media containing levels of 2,4-D used for the initiation of callus is helpful. Use of low concentrations of naphthaleneacetic acid, 2,4-D, and a cytokinin is also recommended. In many species, embryogenic calli have been shown to retain their morphogenetic competence for over 1 year (Vasil *et al.*, 1982).

VI. CONCLUSIONS

Mature, fully differentiated tissues of many species of the Gramineae have been found to be unsuitable for the induction of embryogenic callus. Young, meristematic tissues at a specific but narrow developmental stage—such as immature embryos and young inflorescences or leaves—have proven to be most useful. Identification, selection, and culture of the embryogenic callus at an early stage, and its preferential maintenance, are critical in retaining the long-term morphogenetic potential of such cultures. It is also suspected that in many previous investigations embryogenic tissue cultures of Gramineae were indeed obtained, but these were either inadvertently or deliberately eliminated, and the mode of plant regeneration was either not understood or misinterpreted (Vasil 1982a,b, 1983; Rangan and Vasil, 1983; Ho and Vasil, 1983). Many workers have found that only a few genotypes will give the desired response *in vitro,* but no evidence of this was found in the studies of Vasil *et al.* (1982) involving a number of genotypes from a wide variety of species (Haydu and Vasil, 1981; Lu *et al.*, 1982, 1983; Ozias-Akins and Vasil, 1982, 1983).

REFERENCES

Botti, C., and Vasil, I. K. (1983). Plant regeneration by somatic embryogenesis from parts of cultured mature embryos of *Pennisetum americanum* (L.) K. Schum. *Z. Pflanzenphysiol.* **111,** 319–325.

Botti, C., and Vasil, I. K. (1984). Ontogeny of somatic embryos of *Pennisetum americanum* (L.) K. Schum. II. In cultured immature inflorescences. *Can. J. Bot.* (in press).

Haydu, Z., and Vasil, I. K. (1981). Somatic embryogenesis and plant regeneration from leaf tissues and anthers of *Pennisetum purpureum* Schum. *Theor. Appl. Genet.* **59,** 269–273.

Ho, W., and Vasil, I. K. (1983). Somatic embryogenesis in sugarcane (*Saccharum officinarum* L.). I. The morphology and physiology of callus formation and the ontogeny of somatic embryos. *Protoplasma* **118,** 169–180.

King, P. J., Potrykus, I., and Thomas, R. (1978). In vitro genetics of cereals: Problems and perspectives. *Physiol. Veg.* **16,** 381–399.

Lu, C., and Vasil, I. K. (1981). Somatic embryogenesis and plant regeneration from leaf tissues of *Panicum maximum* Jacq. *Theor. Appl. Genet.* **59,** 275–280.

Lu, C., and Vasil, I. K. (1982). Somatic embryogenesis and plant regeneration from tissue cultures of *Panicum maximum. Am. J. Bot.* **69,** 77–81.

Lu, C., Vasil, I. K., and Ozias-Akins, P. (1982). Somatic embryogenesis in *Zea mays* L. *Theor. Appl. Genet.* **62,** 109–112.

Lu, C., Vasil, V., and Vasil, I. K. (1983). Improved efficiency of somatic embryogenesis and plant regeneration in tissue cultures of maize (*Zea mays* L.). *Theor. Appl. Genet.* **66,** 285–289.

Murashige, T., and Skoog, F. (1962). A revised medium for rapid growth and bioassays with tobacco tissue cultures. *Physiol. Plant.* **15,** 473–497.

Nakano, H., and Maeda, E. (1979). Shoot differentiation in callus of *Oryza sativa* L. *Z. Pflanzenphysiol.* **93,** 449–458.

Norstog, K. (1970). Induction of embryo-like structures by kinetin in cultured barley embryos. *Dev. Biol.* **23,** 665–670.

Ozias-Akins, P., and Vasil, I. K. (1982). Plant regeneration from cultured immature embryos and inflorescences of *Triticum aestivum* L. (wheat): Evidence for somatic embryogenesis. *Protoplasma* **110,** 95–105.

Ozias-Akins, P., and Vasil, I. K. (1983). Improved efficiency and normalization of somatic embryogenesis in *Triticum aestivum* (wheat). *Protoplasma* **117,** 40–44.

Rangan, T. S. (1974). Morphogenic investigations on tissue cultures of *Panicum miliaceum. Z. Pflanzenphysiol.* **72,** 456–459.

Rangan, T. S., and Vasil, I. K. (1983). Somatic embryogenesis and plant regeneration in tissue cultures of *Panicum miliaceum* L. and *Panicum miliare* Lamk. *Z. Pflanzenphysiol.* **109,** 49–53.

Shimada, T., and Yamada, Y. (1979). Wheat plants regenerated from embryo cell cultures. *Jpn. J. Genet.* **54,** 379–385.

Springer, W. D., Green, C. E., and Kohn, K. A. (1979). A histological examination of tissue culture initiation from immature embryos of maize. *Protoplasma* **101,** 269–281.

Vasil, I. K. (1982a). Cell culture and somatic cell genetics of cereals and grasses. *In* "Plant Improvement and Somatic Cell Genetics" (I. K. Vasil, W. R. Scowcroft, and K. J. Frey, eds.), pp. 179–203. Academic Press, New York.

Vasil, I. K. (1982b). Somatic embryogenesis and plant regeneration in cereals and grasses. *In* "Plant Tissue Culture 1982" (A. Fujiwara, ed.), pp. 101–104. Maruzen, Tokyo.

Vasil, I. K. (1983). Regeneration of plants from single cells of cereals and grasses. *In* "Genetic Engineering in Eukaryotes" (P. Lurquin and A. Kleinhofs, eds.), pp. 233–252. Plenum, New York.

Vasil, I. K., Vasil, V., Lu, C., Ozias-Akins, P., Haydu, Z., and Wang, D. (1982). Somatic embryogenesis in cereals and grasses. *In* "Variability in Plants Regenerated from Tissue Culture" (E. Earle and Y. Demarly, eds.), pp. 3–21. Praeger Press, New York.

Vasil, V., and Vasil, I. K. (1980). Isolation and culture of cereal protoplasts. II. Embryogenesis and plantlet formation from protoplasts of *Pennisetum americanum. Theor. Appl. Genet.* **56,** 97–99.

Vasil, V., and Vasil, I. K. (1981). Somatic embryogenesis and plant regeneration from tissue cultures of *Pennisetum americanum* and *P. americanum* x *P. purpureum* hybrid. *Am. J. Bot.* **68,** 864–872.

Vasil, V., and Vasil, I. K. (1982a). Characterization of an embryogenic cell suspension culture derived from inflorescences of *Pennisetum americanum* (pearl millet, Gramineae). *Am. J. Bot.* **69,** 1441–1449.

Vasil, V., and Vasil, I. K. (1982b). The ontogeny of somatic embryos of *Pennisetum americanum* (L.) K. Schum.: In cultured immature embryos. *Bot. Gaz. (Chicago)* **143,** 454–465.

Wang, D., and Vasil, I. K. 1982. Somatic embryogenesis and plant regeneration from inflorescence segments of *Pennisetum purpureum* Schum. (Napier or Elephant grass). *Plant Sci. Lett.* **25,** 147–154.

Wernicke, W., and Brettell, R. (1980). Somatic embryogenesis from *Sorghum bicolor* leaves. *Nature (London)* **287,** 138–139.

CHAPTER **6**

Clonal Propagation: Shoot Cultures

Horst Binding
Gabriela Krumbiegel-Schroeren

Botanical Institute and Botanical Garden
Biology Center
Christian Albrechts University
Kiel, Federal Republic of Germany

I. INTRODUCTION

Shoot cultures have been established in numerous species of higher plants (Murashige, 1978; Vasil and Vasil, 1980); the number is rapidly increasing. The term "shoot culture" describes rootless sprouts growing on agar media. Shoot cultures are being widely used for clonal propagation of ferns, trees, and ornamental flowering plants, for conservation of genetically defined stocks, and as experimental material in biochemical, physiological, and genetic investigations.

CELL CULTURE AND SOMATIC CELL
GENETICS OF PLANTS, VOL. 1

ISBN 0-12-715001-3

II. GENERAL TECHNIQUES

A. Culture Media

MS agar medium (Murashige and Skoog, 1962), with its relatively high concentrations of macro- and micronutrients, is most commonly used for shoot cultures. Evans *et al.* (1981) reported direct shoot formation from various explants on MS medium in about 88% of nongraminaceous species so far tested. Good results for shoot cultures have also been obtained with the media of White (1963), LS (Linsmaier and Skoog, 1965), B5 (Gamborg *et al.*, 1968), NT (Nagata and Takebe, 1971), SH (Schenk and Hildebrandt, 1972), and others.

Organic additives to the basal media are the same as those used for tissue culture media in general (see also Chapter 3, this volume). A minimal supply for spermatophytic shoots is possibly represented by the LS medium with only sucrose, thiamine, and meso-inositol. An abundance is offered by a medium of Kao (1977), which has recently been used for multiple shoot regeneration in *Triticale* (Nakamura and Keller, 1982).

Cytokinins and auxins are added occasionally. Less complex inorganic media, with or without sucrose, have been used for some fern species and for the initiation of shoot cultures from seeds.

B. Culture Vessels

Depending on the growth behavior of the shoots, various kinds of plastic or glass vessels are used. Most species can be grown in Petri dishes. Larger shoots are cultured in larger sterile plastic vessels or in various common glass vessels which are covered by halves of Petri dishes.

C. Control of the Environment

For reliable growth and development, shoots are cultured under controlled conditions concerning the day/night phases, types and intensities of light, and degrees of temperature. Appropriate day length depends on the photoperiodic demands of the species. Usually, cool white fluorescent light is used at 3000–8000 lx; for chlorophyll-deficient mutants, lower intensities (around 1000 lx) are preferable. Temperatures are around 25°C (range, 18–30°C). They are sometimes lowered in the dark phase. Control of humidity depends on the types of closure of the culture vessels; when evaporation is reduced by the use of relatively tight covers or by Parafilm, low humidity is possible and preferred. High humidity in the room retards drying up of unsealed cultures, but more care must then be taken to prevent microbial growth.

III. INITIATION OF SHOOT CULTURES

Shoot cultures are initiated from explants which include meristems (embryos, seeds, or shoot tips and nodes of plants) or which exhibit other types of organization. In the latter cases, morphogenetic pathways include direct formation of shoots or embryoids, as well as indirect organogenesis with an intervening callus phase.

Detailed information on the methods, demands, and processes associated with various species is provided by Evans *et al.* (1981). Selected techniques are described below.

A. Initiation from Seeds

Seedlings are grown on numerous substrates in different types of vessels in the greenhouse or growth chambers. Preference is given to axenic conditions which are indispensable for excised embryos. Usually, seeds are easily decontaminated by procedures described in Chapters 11 and 13, this volume. In some orders, achenes or caryopses are germinated. Success in sterilization of these fruits is usually rather low. The best results are obtained when they are used as long as the pericarps are alive. For sufficient germination, different kinds of treatments are demanded in a number of species. For instance, exposure to low or raised temperatures for several hours, culture in the dark or light, dissection of soaked seeds, excision of embryos, or use of immature seeds are appropriate. Stems of seedlings from germinated seeds are cut at the hypocotyls or epicotyls and transferred to shoot culture conditions.

B. Initiation from Stem Meristems

Stem meristems from apices and axillary buds are nearly as convenient for the initiation of shoot cultures as seeds. Surface sterilization is usually efficient. In cases of systemic infections by pathogens, particularly viruses, only the meristems are excised. Larger explants, including leaf primordia measuring up to 5 mm, can be taken from healthy plants. The sterilized parts are planted in shoot culture media, where they grow up to form shoots. If small isolated meristems are used, often callus production alone is induced on the shoot culture media, or the meristems must be grown on special media. The calluses are then transferred to appropriate media to obtain adventitious shoots or embryos (see Chapters 5, 7–11, this volume).

C. Initiation from Tissues Not Carrying Shoot Meristems

Shoot cultures have been initiated from various types of tissues, including root tips, explants containing cambial or parenchyma cells, microspores, and nucelli, as well as cell and tissue cultures. Preference is given to systems which enable direct organogenesis via embryoids or shoot primordia to get cultures of *at most* controlled genetic uniformities (cf. Section V). This has been established with axes of embryos and seedlings. Choice of appropriate tissue as well as procedures depends on the species and genotypes investigated (Chapter 11, this volume).

IV. SUBCULTURE

Shoot cultures used for the preservation of certain stocks are subcultured at intervals of 3–5 weeks. A more prolonged duration is usually not convenient because impaired growth conditions cause genetic abnormalities and retarded recovery after transfer to fresh media. Subcultures are mainly established by the tips of the primary sprout. The use of this material gives the best guarantee for the conservation of the original genetic constitution. Axillary buds have been found to be nearly as convenient for this purpose. Adventitious shoots, however, have frequently been found to contain cytogenetic variations (Murashige, 1974).

Shoot cultures may be subcultured at shorter periods (e.g., 10 days) to obtain high multiplication rates. For this purpose, the shoots are cut into segments containing a node or the apex. Propagation can be increased in several species by using culture media which induce teratoma-like callus at the basal cutting ends of the shoots. This is achieved, for instance, in most members of the Solanaceae by using culture media with a cytokinin (e.g., B5 or MS with 0.5 mg/liter 6-BA).

V. GENETIC ASPECTS OF SHOOT CULTURE

The aims of shoot cultures with respect to genetic constitution may be the preservation of the genotype of the explant, the establishment of genotypes selected *in vitro*, or the induction of genetic variability.

It has been stated in the preceding chapters that genetic stability is best preserved in organized tissue; hence, the intercalary callus phase must be avoided or at least reduced to a minimum. The probability of the occurrence of aberrant cells increases with the age of explants as well as clones. Explants are, therefore, taken from the youngest possible plants. If morphological variations arise in cultures, this may be due to genetic, epigenetic, or just physiological factors (Chaleff, 1983). Plant regeneration to enable passage through the sexual cycle is needed for clarification.

When genetic clonal variation is suggested, passage through a single cell stage is indicated for the establishment of uniform shoot cultures. This can be managed by protoplast regeneration, by induction of embryoid formation in somatic tissue, or by the use of androgenesis. Clonal separation is also possible, in some cases, by the formation of adventitious shoots. However, this may not be reliable in every case.

VI. PLANT REGENERATION

For maintenance and propagation of shoot cultures, root formation is not desirable, but transfer to the greenhouse assumes *in vitro* rooting of the shoots. Although the medium for initiation and multiplication of shoot cultures can be the same, the addition of root-promoting substances is necessary mostly in the end step of shoot culture (Murashige, 1974). Often, rooting can be achieved on media devoid of phytohormones. In shoots of particular genetic constitutions, roots have been obtained randomly or not at all. In these cases, grafting was a successful means of obtaining the transfer to greenhouse conditions (Schieder, 1980).

VII. CONCLUSIONS AND PERSPECTIVES

Shoot cultures are a means of long-term cultivation of higher plants in an organized condition. In contrast to callus cultures, they retain their regenerative capacities over long culture periods. Hence, they are particularly appropriate as stocks for clonal propagation of crops, ornamentals, and trees by high multiplication rates and sufficient genetic stability, requiring little space per individual. However, there are restrictions in some taxa which have been termed "recalcitrant" with respect to *in vitro* culture. In a number of these species (e.g., the Fabaceae), difficulties in applying tissue culture methods have been partly overcome.

REFERENCES

Chaleff, R. S. (1983). Isolation of agronomically useful mutants from plant cell cultures. *Science* **219,** 676–682.

Evans, D. A., Sharp, W. R., and Flick, C. E. (1981). Growth and behavior of cell cultures: Embryogenesis and organogenesis. *In* "Plant Tissue Culture: Methods and Applications in Agriculture" (T. A. Thorpe, ed.), pp. 45–113. Academic Press, New York.

Gamborg, O. L., Miller, R. A., and Ojima, K. (1968). Nutrient requirements of suspension cultures of soybean root cells. *Exp. Cell Res.* **50,** 151–158.

Kao, K. N. (1977). Chromosomal behavior in somatic hybrids of soybean- *Nicotiana glauca. Mol. Gen. Genet.* **150,** 225–230.

Linsmaier, E. M., and Skoog, F. (1965). Organic growth factor requirements for tobacco tissue cultures. *Physiol. Plant.* **18,** 100–127.

Murashige, T. (1974). Plant propagation through tissue culture. *Annu. Rev. Plant Physiol.* **25,** 135–166.

Murashige, T. (1978). The impact of plant tissue culture on agriculture. *In* "Frontiers of Plant Tissue Culture 1978" (T. A. Thorpe, ed.), pp. 15–26. Univ. of Calgary Press, Calgary Canada.

Murashige, T., and Skoog, F. (1962). A revised medium for rapid growth and bioassays with tobacco tissue culture. *Physiol. Plant.* **15,** 473–497.

Nagata, T., and Takebe, I. (1971). Plating of isolated tobacco mesophyll protoplasts on agar medium. *Planta* **99,** 12–20.

Nakamura, C., and Keller, W. A. (1982). Callus proliferation and plant regeneration from immature embryos of hexaploid triticale. *Z. Pflanzenzuecht.* **88,** 137–160.

Schenk, R. V., and Hildebrandt, A. C. (1972). Medium and techniques for induction and growth of monocotyledonous and dicotyledonous plant cell cultures. *Can. J. Bot.* **50,** 199–204.

Schieder, O. (1980). Somatic hybrids between a herbaceous and two tree *Datura* species. *Z. Pflanzenphysiol.* **98,** 119–127.

Vasil, I. K., and Vasil, V. (1980). Clonal propagation. *Int. Rev. Cytol., Suppl.* **11A,** 145–173.

White, P. R. (1963). "The Cultivation of Animal and Plant Cells," 2nd ed. Ronald Press, New York.

CHAPTER 7

Clonal Propagation: Adventitious Buds

Trevor A. Thorpe
Kamlesh R. Patel

Department of Biology
University of Calgary
Calgary, Alberta, Canada

I. INTRODUCTION

The use of *in vitro* techniques for clonal or asexual mass propagation is the most advanced application of plant tissue culture. Rapid asexual multiplication can be achieved by (1) enhancing axillary bud breaking, (2) production of adventitious buds, and (3) somatic embryogenesis. Adventitious shoots or roots can be induced to form on tissues which normally do not produce these organs. This process is much more common than somatic embryogenesis and has far more potential for mass clonal propagation of plants than multiplication from axillary buds. Adventitious shoots or roots may be produced directly on the explants or on callus derived from primary explants.

In the formation of adventitious buds, there is an interplay between the inoculum, the medium, and the culture conditions. For optimum organogenesis, each of these components has to be critically assessed. In general, the process of plantlet formation via organogenesis requires four distinct stages: (1) induction of shoot buds, (2) development and multiplication of

CELL CULTURE AND SOMATIC CELL
GENETICS OF PLANTS, VOL. 1

ISBN 0-12-715001-3

these buds, (3) rooting of the shoots, and (4) hardening of the plantlets. The optimum requirements for each stage must be experimentally determined.

In this chapter, the general principles used at each stage of plantlet formation will be outlined. This will be followed by the outline of an approach that can be used in working with an unknown tissue.

II. INOCULUM

In manipulation of organ formation, the choice of inoculum is extremely important. The inoculum consists of either explants or callus.

Many factors influence the behavior of the explants in culture (Murashige, 1974). These include (1) the organ that is to serve as the tissue source, (2) the physiological and ontogenetic age of the organ, (3) the season in which the explants are obtained, (4) the size of the explant, and (5) the overall quality of the plant from which the explants are obtained. In some cases, pretreatment of the explant source, e.g., chilling of bulbs or spraying of trees with cytokinins, may also be a requirement for successful organ formation in culture. Some of these variables can be controlled easily, whereas others are more difficult and require experimentation. It is thus important to understand various characteristics of the intended plant material.

Virtually any part of the plant can be used as inoculum, including stem and root segments, leaf and petiole sections, inflorescence portions, seed embryos, and seedling parts such as cotyledons, epicotyls, and hypocotyls. In general, the more juvenile the material, the more easily will organ formation occur *in vitro*. However, the best explant for each species must be determined experimentally. In radiata pine the number of rootable shoots obtained increases in going from excised embryos, to cotyledons separated from cultured embryos, to cotyledons excised from about 1-week-old germinated seed (Aitken *et al.*, 1981; see also Chapter 11, this volume). Similarly, in a comparative study of shoot induction in seedling tissues of white and black spruce, it was found that 25- to 28-day-old epicotyls, consisting of intact cotyledons and a 2-mm stub of hypocotyl for black spruce, and trimmed cotyledons (ca. 1 mm removed from the tip) without hypocotyl tissue for white spruce, were the best explants (Rumary and Thorpe, 1984).

In addition to the bulky explants outlined above, superficial or epidermal explants have served as inoculum in a limited number of species to date (Tran Thanh Van and Trinh, 1978). Here, explants consisting of the epidermis and four to six subepidermal cell layers, when placed in culture, are able to undergo primordium formation directly or give rise to callus.

Sterilization of explants is very important. The usual method involves the following steps: The explant is thoroughly washed in running tap water with or without detergent, followed by a quick dip (30–60 sec) in 70% ethanol and a rinse. The explant is then stirred for 10–20 min in diluted commercial laundry bleach (final concentration 0.5–1.0% NaClO) containing a wetting agent (e.g., Tween-20 or Tween-80) and finally rinsed several times with sterile water under aseptic conditions. H_2O_2 is also an effective sterilant. Other sterilizing agents are discussed in Chapter 1, this volume.

Seeds often require greater treatment (Chapter 11, this volume). In addition, some explants also require more extensive sterilizing, e.g., using a gentle vacuum during the bleach treatment. Jones *et al.* (1977) used a two-step procedure in which the surface-sterilized material is placed on media lacking phytohormones or vitamins for 24 hr, followed by treatment with a fungicide and resterilization before planting on complete culture medium. If seedling parts are to be used as explants, the surface-sterilized seed is often germinated aseptically on an agar medium containing 1–2% sucrose.

Fungal and bacterial contamination usually appears within 5–10 days in culture. Obviously, visually contaminated cultures should be discarded, but the absence of visual contamination does not necessarily mean that the explants are pathogen free. Anderson (1980) has recommended that during reculture, bits of residue tissue be diced and placed in sterile culture containing Bacto nutrient broth for a period of 10 days to determine the pathogen status of the explant. It has also been suggested that difficult to decontaminate material should be grown in media containing antibiotics and fungicides for a short period of time before transfer to a new medium. The usefulness of this approach has not been clearly demonstrated. It is important to recognize that virus-infected plants are likely to give rise to virus-infected tissues and organs.

Callus can be induced from any part of the plant. In general, however, it is easier to obtain callus from juvenile tissue and aboveground plant parts, as mature tissues and belowground parts are often more difficult to decontaminate. Aseptically germinated seedlings can also be used. Once aseptic material is obtained, for most plant species it is relatively easy to maintain on semisolid or in suspension culture through periodic transfer at intervals of 4–6 weeks for callus and 1–2 weeks for suspension cultures. Callus is not a homogeneous tissue and undergoes changes with age in culture (see Section VI).

III. MEDIUM

The major constituents required for successful growth and organogenesis are (1) inorganic macro- and micronutrients, (2) carbon and energy sources, (3) vitamins, (4) reduced nitrogen, and (5) phytohormones. These five classes of compounds are usually sufficient for most plant species. However, natural complexes such as hydrolyzed protein preparations, brewer's by-products, endosperm fluids, fruit pulp and juice, animal by-products, and coconut milk are also often added to the medium. The most frequently used of these addenda is coconut milk (2–15% v/v).

The earliest widely used inorganic salt formulations were those of White (1943) and Heller (1953). However, since the 1960s most researchers have been using Murashige and Skoog's (MS) high-salt formulation (1962) or its derivatives, e.g., B5 (Gamborg *et al.*, 1968), and Schenk and Hildebrandt (SH) (1972) formulations. The major differences of these high-salt media lie in the amount and form of nitrogen, plus the relative amounts of some of the microelements (Gamborg *et al.*, 1976). In many cases, a reduction in the level of salts (usually half) is optimum for rooting of tissue culture-derived shoots.

The energy requirement is usually met with sucrose (2–4% w/v), although this can sometimes be replaced by glucose. Thiamine is the most often added vitamin, followed by nicotinic acid and pyridoxine. The sugar alcohol, inositol, has often aided the growth and differentiation of numerous tissues in culture. Organic nitrogen is most often added during callus initiation but may be beneficial during subculture and organized development. Casein hydrolysate (0.02–0.1%) is a frequent nonspecific organic nitrogen source, whereas the compounds glutamate, asparagine, tyrosine, and adenine are the most frequently used specific reduced nitrogen additives.

Auxins and cytokinins are the two types of phytohormones most often needed in culture. Their concentration and ratio in the medium often control the pattern of differentiation in culture (Skoog and Miller, 1957). A relatively high ratio of cytokinin to auxin favors shoot formation, whereas the reverse favors root formation. Both naturally occurring and synthetic compounds are used. The most frequently used auxins are 2,4-dichlorophenoxyacetic acid (2,4-D), indoleacetic acid (IAA), naphthaleneacetic acid (NAA), and indolebutyric acid (IBA). Kinetin and benzylaminopurine (BAP) are commonly used cytokinins, with isopentenyl adenine and zeatin used less frequently, mainly because of their higher cost. Evans *et al.* (1981) found that for 75% of the species forming shoots, kinetin or benzylaminopurine was used in a concentration of 0.05–46 μM. Auxins such as IAA and NAA were used in concentrations of 0.06–27.0 μM. Graminaceous species tend to have a lower requirement for cytokinins for shoot formation than other species. IBA and NAA are the most commonly used auxins for rooting. In some cases, a mixture of two cytokinins or two auxins has proven superior to a single cytokinin or auxin.

A large number of plant species respond to a suitable auxin–cytokinin balance by forming shoots and roots. In a number of cases, however, this approach leads only to the induction of organogenetic tissue. This tissue will then develop into organs in a medium with an altered phytohormonal balance. In some cases, either exogenous auxin or cytokinin is sufficient to bring about organogenesis (Thorpe, 1980).

Several substituted purines, pyrimidines, and ureas have been used successfully in place of cytokinins to bring about organogenesis. Similarly, various auxinlike compounds can satisfy the auxin requirement. Other phytohormones, including gibberellins and abscisic acid, added to the medium have been shown to play a role in organ formation. No generalizations are possible, as these substances can repress, enhance, or have no effect on different plant species. Finally, other metabolites such as phenolic acids have been shown to enhance organogenesis in some species (Thorpe, 1980).

Explants of many angiosperm tree species turn brown and/or produce large amounts of exudates, which diffuse into the medium and cause necrosis of the tissue. This problem can often be reduced by presoaking the explant in "antioxidants" such as ascorbic and citric acids, cysteine, or glycine (alone or in combination) and by including such compounds (1 μM–10 mM) in the medium, followed by several transfers onto fresh medium, with sequential reduction of the antioxidant content. Activated charcoal (up to 3% w/v) has also been used for this purpose, but it should be noted that this compound will also absorb useful components from the medium, and early transfer to the medium without charcoal is recommended. In addition, the inclusion of activated charcoal (0.1–1.0% w/v) in the medium has been shown to be beneficial during shoot multiplication and development and root formation stages in culture. In some cases, only plant-derived charcoal has been found to be effective (Biondi and Thorpe, 1982; Bonga, 1982a).

Although the size of the explant (provided it is not too small) is rarely a problem in induction of callus or regeneration, its physiological age is extremely important, particularly in woody species. Here, expressions of topophysis such as plagiotropism are not uncommon (Mott, 1981). Approaches which can rejuvenate the explant are useful (Chapter 11, this volume; Bonga 1982b). In all plants, gradients exist along the length of the plant. These gradients, plus the orientation of the explant in culture (polarity effects), must be considered, as these phenomena can influence organogenesis and callus formation (Hughes, 1981a).

IV. CULTURE ENVIRONMENT

There are many aspects of the culture environment that can influence growth and organized development. These include (1) the physical form of the medium, (2) pH, (3) humidity and gas atmosphere, (4) light, and (5) temperature.

The physical form of the medium, i.e., whether it is solidified or liquid, plays an important role in growth and differentiation. Callus maintained on medium solidified with agar (0.6–1.0%) grows slowly, with the new cells formed mainly on the periphery of the existing callus mass. Cell suspension cultures, on the other hand, tend to grow more rapidly, as the single cells or small cell clumps are constantly exposed to the nutrient medium. In general, most success in organogenesis is achieved with explants, callus, or plated cell suspensions on solid medium. However, a liquid phase during plantlet formation, in which the tissue is slowly agitated, may be beneficial (Murashige, 1977). The pH of the medium is usually set at about 5.0 for liquid formulations and at about 5.8 for agar-gel media. Agar substitutes have not yet found widespread use (Bonga, 1982a).

Relative humidity is rarely a problem except in arid climates, where rapid drying of the medium occurs. This can be reduced by the use of tightly closed containers, covering closures such as foam or cotton wool plugs with aluminum foil or another material, and sealing Petri dishes with a household plastic covering (e.g., Handi-Wrap, Saran Wrap) or Parafilm. In climates with high humidity, dehumidifiers in the culture room may be advantageous. Here the major problem is the growth of fungi and other microorganisms. The gas atmosphere above the cultures, which includes ethylene, ethanol, CO_2, and acetaldehyde, can inhibit morphogenesis if gaseous exchange does not occur. Finally, in some urban environments, it may be necessary to filter the air entering the culture room. Such filters, often of charcoal, will also remove dust and spores.

Light is a major factor of the culture environment and has been clearly shown to have an effect on organized development *in vitro*. Light requirements for differentiation involve a combination of several components, including intensity, daily light period, and quality, and are necessary for certain photomorphogenic events (Murashige, 1974). Although maximum callus growth often occurs in darkness, low light intensity (e.g., 90 $nE \cdot m^{-2} \cdot sec^{-1}$) may enhance organogenesis. In some cultures much higher light intensities (e.g., 10–80 $\mu E \cdot m^{-2} \cdot sec^{-1}$) are needed for optimum shoot formation, multiplication, and development. For rooting, lower intensities may be beneficial. During plantlet hardening, even higher light intensities (ca. 300 $\mu E \cdot m^{-2} \cdot sec^{-1}$) may be needed for subsequent survival.

Light sources with electromagnetic spectra closer to those of natural sunlight, e.g. Grolux, are probably best. However, good results have been obtained with normal fluorescent lamps, with or without supplementary incandescent lamps.

Generally, cultures are kept at a constant temperature between 20° and 30°C. However, the optimum temperature for growth and differentiation for a particular species should be determined, as different species have different optima (Hughes, 1981b). Furthermore, thermoperiod and temperature pretreatments (including a chilling temperature) have been shown to affect morphogenesis. Even within the same tissue, different optimum temperatures for shoot formation and rooting have been observed (Cheng and Voqui, 1977; Rumary and Thorpe, 1984).

At present there is a tendency to carry out rooting under nonsterile conditions (Biondi and Thorpe, 1982). Shoots are often dipped in IBA solutions or commercial rooting powders and planted in sterilized soil mixes or supports (e.g., vermiculite, perlite). Watering with half-strength mineral salts and use of fungicides is common. This approach is not only simpler and less costly, but often produces superior roots and facilitates planting out (Chapter 11, this volume).

It is clear that most researchers do not critically evaluate these various factors of the culture environment. Unfortunately, most culture rooms do not have enough flexibility to allow evaluation of most of these factors.

V. PLANTLET FORMATION IN A NEW TISSUE

In working with a new material, it is important to remember that success in plantlet formation results from the interplay of the explant, medium, and culture environment. Thus, the various factors outlined in previous sections must be considered. Initially, it is good to determine if other species of the same genus have been successfully cultured. Useful lists of plants which have been propagated by tissue culture methods can be found in the following references: Murashige (1978); Vasil and Vasil (1980); Bottino (1981) for vegetable crops; Conger (1981) for agronomic crops; Hughes (1981a) for ornamentals; Biondi and Thorpe (1982) and Mott (1981) for trees; and Skirvin (1981) for fruit crops. The experimental approach to be taken will depend on the material available, i.e., seed, greenhouse-grown, or field-grown material. The approach used at Calgary for plantlet formation with juvenile tissue of woody species is outlined below. The aim of this work is to obtain maximum organogenesis with minimum callus formation.

Seeds are surface sterilized and stratified and germinated aseptically in culture tubes or 125-ml Erlenmeyer flasks (Chapter 11, this volume). After an arbitrary time (e.g., 1–4 weeks), which depends mainly on seedling growth, various parts are excised and placed in culture. These parts include cotyledons, epicotyls, hypocotyls, etc. Initial culture is on at least three mineral salts, namely, MS (1962), B5 (Gamborg *et al.*, 1968), and SH (1972). Other salt formulations used will depend on the results of other workers with closely related species. To these salt formulations, White's organics (White, 1943), sucrose (3% w/v), inositol (100 mg/liter), and phytohormones are added. A partial factorial experiment is set up so that cytokinin is tested at three levels (e.g., BAP, 0, 1, and 10 μM) and auxin at two levels (e.g., IAA, 0 and 0.1 μM) in media solidified with agar (ca. 0.8%). The cultures are placed in the light (80 $\mu E \cdot m^{-2} \cdot sec^{-1}$, 16-hr photoperiod) at temperatures of 23° and 28°C ± 1°C for a period of up to 6 weeks using plastic Petri dishes, small vials, or 50-ml Erlenmeyer flasks. The purpose of this initial screening test is to find a mineral salt formulation on which some plant part will stay healthy, grow, and if possible give indications of morphogenesis such as nodular tissue formation, and at the same time produce very little or no callus.

On some occasions one test will give the indications required, but it may be necessary to run further parallel screening tests using two mineral salt formulations and two different explants, and for even a longer period in culture. To date it has always been possible to select a more favorable temperature during the initial test. Additional tests involve the use of a wider phytohormone range (e.g., BAP: 0.5, 1.0, 5.0, 10.0, 30.0 μM; IAA: 0, 0.1, 0.5 μM). This should allow one level of BAP and IAA to be selected, and, hopefully, one salt formulation and one plant part as well. Systematic sequential testing is then carried out to determine (1) the optimum age of the explant; (2) any pretreatments (e.g., cold period, incubation in antioxidant) needed; (3) any manipulations needed on the explant for optimum culture results (e.g., trimming of cotyledons, orientation in culture); (4) level of sucrose; (5) level of and necessity for individual vitamins; (6) effects of amino acids (e.g., glutamate, asparagine, tyrosine) and other organic addenda (e.g., adenine sulfate); (7) effects of mixed phytohormones (e.g., BAP + 2iP); and (8) reinvestigation of the phytohormone level resulting from the above tests.

The second phase of the work involves the development of the nodular tissue into shoots and multiplication with minimum callus formation. This is usually achieved in a medium lacking phytohormones, but several factors have to be tested. These include (1) optimum length of time in phytohormone in the first phase for subsequent shoot development; (2) level of sucrose; (3) effect of charcoal; (4) need for low concentrations of phytohormones at later stages of shoot development; (5) determining if any other mineral salt formulation is better than the one used in the first phase; and (6) frequency of reculture and effect of subdivision on development and multiplication rates. The last factor is important, as the developing shoots often inhibit the formation and development of additional shoots.

The rooting phase is carried out aseptically using auxin. IBA (0.1–100.0 μM) is tested, alone or in combination with other auxins (e.g., NAA or IAA), either in the medium or as a dip prior to planting. Other factors tested are (1) level of sucrose; (2) concentration of mineral salts (e.g., 25%, 50%, 75%, and full strength); (3) length of time in auxin; (4) effect of charcoal; and (5) temperature and thermoperiod requirements. These tests are carried out in Calgary in growth cabinets at lower light intensities (ca. 50 $\mu E \cdot m^{-2} \cdot sec^{-1}$). We can test three different temperature and thermoperiod regimes simultaneously. Finally, rooting is examined under nonsterile conditions using sterilized vermiculite (mainly) in trays.

Using the systematic approach outlined above, it has been possible to achieve plantlet formation in a number of woody plants, both softwood and hardwood, although we do not always examine critically all the factors of the culture environment. The final phase of the work involves hardening of the plantlets so that they will survive transfer to the greenhouse and finally to the field (Sommer and Caldas, 1981; see Chapter 15, this volume).

VI. CONCLUSIONS

By using the principles outlined in this chapter and approaches similar to those outlined in Section V, it has been possible to produce plantlets directly from explants, indirectly from callus formed on the explants, and from subcultured callus in numerous plant species among the angiosperms, gymnosperms, and lower vascular plants. Nevertheless, a few problem areas remain.

One of these is the difficulty encountered with explants from mature tree species. Although some success has been obtained (Chapter 11, this volume), much remains to be done. Finding a means of reversing the developmental phase from maturity to juvenility is the most promising approach. A second problem is obtaining plantlets from legume callus. More success has been achieved with forage legumes than with seed legumes. Another  major problem is the number of aberrant plants produced in culture. Some of these are mutants, and others are epigenetic variants; the frequency of formation varies with the species. In any case, these variants reduce the number of useful plants produced (Chapter 11, this volume), although some of these somaclonal variants may be valuable (Larkin and Scowcroft, 1981).

Callus maintained in culture often undergoes changes with time (Thorpe, 1982). These changes include (1) habituation or loss of the phytohormone requirement, (2) loss of morphogenetic potential, and (3) the emergence of tissue with altered morphological texture, e.g., friable tissue. All of these changes reduce the regenerative capacity of the tissue. The loss in morphogenetic capacity has been restored in a few cases by exposure of the callus to very high levels of cytokinin for short periods, and also by repeated subculture under shoot-forming conditions (Rice *et al.*, 1978). Other problems, which are not as widespread, are the presence of difficult to eradicate internal infestations and the production and accumulation of autointoxicating substances (Murashige, 1978).

Finally, Vasil and Vasil (1980) have pointed out that all species which form organs or somatic embryos in culture are not necessarily suited for large-scale clonal propagation, as for many species the process is too expensive, the rate of multiplication too slow, the mortality rates on transfer to soil too high, and genetic and epigenetic variants (most of which are deleterious) too frequent. Many of these problems would appear to be technical and thus amenable to solution, but they have been around for as long as tissue culture technology itself.

REFERENCES

Aitken, J., Horgan, K. J., and Thorpe, T. A. (1981). Influence of explant selection on the shoot-forming capacity of juvenile tissue of *Pinus radiata*. *Can. J. For. Res.* **11,** 112–117.

Anderson, W. C. (1980). Mass propagation by tissue culture: Principles and techniques. *In* "Proceedings of the Conference on Nursery Production of Fruit Plants through Tissue Culture—Applications and Feasibility," (R. H. Zimmerman, ed.), pp. 1–10. Agric. Res., Sci. Educ. Admin., U.S. Dept. of Agric., Beltsville, Maryland.

Biondi, S., and Thorpe, T. A. (1982). Clonal propagation of forest tree species. *In* "Proceedings of COSTED Symposium on Tissue Culture of Economically Important Plants" (A. N. Rao, ed.), pp. 197–204. COSTED and Asian Network of Biological Sciences, Singapore.

Bonga, J. M. (1982a). Tissue culture techniques. *In* "Tissue Culture in Forestry" (J. M. Bonga and D. J. Durzan, eds.), pp. 4–35. Nijhoff, The Hague.

Bonga, J. M. (1982b). Vegetative propagation in relation to juvenility, maturity and rejuvenation. *In* "Tissue Culture in Forestry" (J. M. Bonga and D. J. Durzan, eds.), pp. 387–412. Nijhoff, The Hague.

Bottino, P. J. (1981). Vegetable crops. *In* "Cloning Agricultural Plants via in Vitro Techniques" (B. V. Conger, ed.), pp. 141–164. CRC Press, Boca Raton, Florida.

Cheng, T.-Y., and Voqui, T. (1977). Regeneration of Douglas-fir plantlets through tissue culture. *Science* **198,** 306–307.

Conger, B. V. (1981). Agronomic crops. *In* "Cloning Agricultural Plants via in Vitro Techniques" (B. V. Conger, ed.), pp. 165–215. CRC Press, Boca Raton, Florida.

Evans, D. A., Sharp, W. R., and Flick, C. E. (1981). Growth and behavior of cell cultures. *In* "Plant Tissue Culture: Methods and Application in Agriculture" (T. A. Thorpe, ed.), pp. 45–113. Academic Press, New York.

Gamborg, O. L., Miller, R. A., and Ojima, K. (1968). Nutrient requirements of suspension cultures of soybean root cells. *Exp. Cell Res.* **50,** 151–158.

Gamborg, O. L., Murashige, T., Thorpe, T. A., and Vasil, I. K. (1976). Plant tissue culture media. *In Vitro* **12,** 473–478.

Heller, R. (1953). Recherches sur la nutrition minérale des tissus végétaux cultivés *in vitro. Ann. Sci. Nat., Bot. Biol. Veg.* [11] **14,** 1–223.

Hughes, K. W. (1981a). Ornamental species. *In* "Cloning Agricultural Plants via in Vitro Techniques" (B. V. Conger, ed.), pp. 5–50. CRC Press, Boca Raton, Florida.

Hughes, K. W. (1981b). In vitro ecology: Exogenous factors affecting growth and morphogenesis in plant culture systems. *In* "Propagation of Higher Plants through Tissue Culture: Emerging Technologies and Strategies" (M. J. Constantin, R. R. Henke, K. W. Hughes, and B. V. Conger, eds.), pp. 281–288. Pergamon, Oxford.

Jones, O. P., Hopgood, M. E., and O'Farrell, D. (1977). Propagation *in vitro* of M. 26 apple rootstocks. *J. Hortic. Sci.* **52,** 235–238.

Larkin, P. J., and Scowcroft, W. C. (1981). Somaclonal variation—a novel source of variability from cell cultures for plant improvement. *Theor. Appl. Genet.* **60,** 197–214.

Mott, R. L. (1981). Trees. *In* "Cloning Agricultural Plants via in Vitro Techniques" (B. V. Conger, ed.), pp. 217–256. CRC Press, Boca Raton, Florida.

Murashige, T. (1974). Plant propagation through tissue culture. *Annu. Rev. Plant Physiol.* **25,** 135–166.

Murashige, T. (1977). Clonal crops through tissue culture. *In* "Plant Tissue Culture and Its Bio-technological Application" (W. Barz, E. Reinhard, and M. H. Zenk, eds.), pp. 392–403. Springer-Verlag, Berlin and New York.

Murashige, T. (1978). The impact of tissue culture on agriculture. *In* "Frontiers of Plant Tissue Culture 1978" (T. A. Thorpe, ed.), pp. 15–26, 518–524. Univ. of Calgary Press, Calgary, Canada.

Murashige, T., and Skoog, F. (1962). A revised medium for rapid growth and bioassays with tobacco tissue cultures. *Physiol. Plant.* **15,** 473–497.

Rice, T. B., Reid, R. K., and Gordon, P. N. (1978). Morphogenesis in field crops. *In* "Propagation of Higher Plants through Tissue Culture—A Bridge Between Research and Application" (K. W. Hughes, R. Henke, and M. Constantin, eds.), pp. 262–278. Tech. Inf. Cent., U.S. Dept of Energy, Springfield, Virginia.

Rumary, C., and Thorpe, T. A. (1984). Plantlet formation in black and white spruce. I. In vitro techniques. *Can. J. For. Res.* **14** (in press).

Schenk, R. U., and Hildebrandt, A. C. (1972). Medium and techniques for induction and growth of monocotyledonous and dicotyledonous plant cell cultures. *Can. J. Bot.* **50,** 199–204.

Skirvin, R. M. (1981). Fruit crops. *In* "Cloning Agricultural Plants via in Vitro Techniques" (B. V. Conger, ed.), pp. 51–140. CRC Press, Boca Raton, Florida.

Skoog, F., and Miller, C. O. (1957). Chemical regulation of growth and organ formation in plant tissue cultures *in vitro. Symp. Soc. Exp. Biol.* **11,** 118–131.

Sommer, H. E., and Caldas, L. S. (1981). *In vitro* methods applied to forest trees. *In* "Plant Tissue Culture: Methods and Applications in Agriculture (T. A. Thorpe, ed.), pp. 349–358. Academic Press, New York.

Thorpe, T. A. (1980). Organogenesis *in vitro:* Structural, physiological and biochemical aspects. *Int. Rev. Cytol., Suppl.* **11A,** 71–111.

Thorpe, T. A. (1982). Callus organization and *de novo* formation of shoots, roots, and embryos *in vitro*. *In* "Application of Plant Cell and Tissue Culture in Agriculture and Industry" (D. T. Tomes, B. E. Ellis, P. M. Harney, K. J. Kasha, and R. L. Peterson, eds.), pp. 115–138. Univ. of Guelph, Guelph, Ontario.

Tran Thanh Van, K., and Trinh, H. (1978). Morphogenesis in thin cell layers: Concept, methodology and results. *In* "Frontiers of Plant Tissue Culture 1978" (T. A. Thorpe, ed.), pp. 37–48. Univ. of Calgary Press, Calgary, Canada.

Vasil, I. K., and Vasil, V. (1980). Clonal propagation. *Int. Rev. Cytol., Suppl.* **11A,** 145–173.

White, P. R. (1943). Nutrient deficiency studies and an improved inorganic nutrient for cultivation of excised tomato roots. *Growth* **7,** 53–65.

CHAPTER **8**

Clonal Propagation: Orchids

Yoneo Sagawa

Harold L. Lyon Arboretum and Department of Horticulture
University of Hawaii at Manoa
Honolulu, Hawaii

John T. Kunisaki

Department of Horticulture
University of Hawaii at Manoa
Honolulu, Hawaii

I. INTRODUCTION

Although tissue culture was initially attempted with orchids for the elimination of viruses, both Morel (1960) and Wimber (1963) quickly realized its greater applicability for clonal propagation. Because orchids are extremely heterozygous and vegetative propagation by division is slow, the commercial application of tissue culture techniques for rapid clonal propagation for some orchids has been extensive (Rao, 1977; Hughes, 1981). However, a few commercially important genera such as *Paphiopedilum* and *Phalaenopsis* remain fairly recalcitrant, with no commercially feasible method for rapid clonal propagation as yet.

CELL CULTURE AND SOMATIC CELL
GENETICS OF PLANTS, VOL. 1

ISBN 0-12-715001-3

Numerous techniques for clonal propagation of orchids are summarized by Arditti (1977). The techniques presented here are recent modifications of our methods published earlier and summarized in Sagawa and Kunisaki (1982).

II. PROCEDURE

A. Sources of Explants

Sources of explants include vegetative shoots with unexpanded leaves, stems with axillary buds, young inflorescences still enclosed in bracts, mature inflorescences with nodes, young expanded leaves, and root tips.

B. Disinfestation and Culture

1. Vegetative shoots with unexpanded leaves
   a. Remove two to three leaves.
   b. Soak in 10% commercial bleach with 2–3 drops of Tween-20 for 10 min; shake occasionally.
   c. Rinse in sterile water.
   d. Remove the remaining leaves.
   e. Soak in 5% commercial bleach with 2–3 drops Tween-20 for 5 min; shake occasionally.
   f. Transfer to sterile water in a transfer hood.
   g. Dissect a 2- to 5-mm cube of shoot tip or axillary buds.
   h. Culture in solid or liquid medium.
2. Stems with axillary buds
   a. Trim expanded leaves.
   b. Cut the stem into segments with two to three nodes.
   c. Soak in 10% commercial bleach with 2–3 drops of Tween-20 for 20 min; shake occasionally.
   d. In a transfer hood, completely remove the leaf base.
   e. Soak in 5% commercial bleach with 2–3 drops of Tween-20 for 10 min; shake occasionally.
   f. Rinse in sterile water.
   g. Dissect a 2- to 3-mm cube with an axillary bud.
   h. Culture it in liquid medium.

3. Young inflorescences
   a. Carefully collect young inflorescences which are still enclosed in bracts.
   b. Soak in 10% commercial bleach with 2–3 drops of Tween-20 for 15 min; shake occasionally.
   c. In a transfer hood, remove the bracts. Soak in 5% commercial bleach with 2–3 drops of Tween-20 for 10 min; shake occasionally.
   d. Rinse in sterile water.
   e. Transfer to a liquid medium.
4. Mature inflorescences
   a. Trim flowers from the tips of the inflorescences.
   b. Wipe with cheese cloth soaked in 70% ethanol.
   c. Cut into 3.0- to 3.5-cm segments each with a node.
   d. Rinse in running water for 2 hr.
   e. Soak in 10% commercial bleach with 2–3 drops of Tween-20 for 20 min; shake occasionally.
   f. In a transfer hood, completely remove the bract.
   g. Soak in 5% commercial bleach with 2–3 drops of Tween-20 for 10–12 min; shake occasionally.
   h. Rinse in sterile water.
   i. Trim the bottom with a sterile blade.
   j. Culture on a solid medium.
5. Leaves
   a. Remove the young, recently expanded leaves.
   b. Wipe with cheese cloth soaked in 70% ethanol.
   c. Rinse in running water for 2 hr.
   d. Soak in 10% commercial bleach with 2–3 drops of Tween-20 for 15 min; shake occasionally.
   e. In a transfer hood, transfer to sterile water and rinse.
   f. Cut 1-cm squares.
   g. Culture on a solid medium. (Another source of young leaves is plantlets in sterile culture. In this case, disinfestation is not required. Therefore, the procedure is initiated at step f after the leaves are removed aseptically from the plantlets. If the leaves are small, the entire leaf may be cultured.)
6. Root
   a. Remove a 5- to 10-mm root tip from actively growing aerial roots.
   b. Soak in 10% commercial bleach with 2–3 drops of Tween-20 for 10 min on an ultrasonic cleaner.
   c. In a transfer hood, rinse in sterile water.
   d. Transfer to a solid medium. (Another source of root tips is plant-

lets in sterile culture. Entire plantlets may be transferred to an appropriate solid medium.)

C. Medium

1. Modified Vacin and Went formulation
   Constituents for Vacin and Went medium (Vacin and Went, 1949) as modified are:

Constituents	Chemical form	Amount per liter
Tricalcium phosphate	$Ca_3(PO_4)_2$	0.20 g
Potassium nitrate	KNO_3	0.525 g
Monopotassium acid phosphate	KH_2PO_4	0.25 g
Magnesium sulfate	$MgSO_4 \cdot 7H_2O$	0.25 g
Ammonium sulfate	$(NH_4)_2SO_4$	0.50 g
Manganese sulfate	$MnSO_4 \cdot H_2O$	0.0068 g
Sucrose		20.00 g
Agar[a]		8.00 g
Water		845 ml
Coconut water[b]		150 ml
Iron chelate (Sequestrene 330 Fe) stock solution[c]		5 ml

[a] If cultures are incubated in greenhouses or at temperatures above 32°C, the agar content may be increased up to 15 g/liter to maintain medium solidity.
[b] Liquid from immature coconuts is filtered through Whatman CC31 Cellulose Powder and the Whatman Glass Microfibre Filter (GF/B) and stored in the freezer. Defrost in a microwave oven.
[c] Iron chelate stock = 1.14 g/100 ml.

Modified Vacin and Went medium can also be prepared by the use of stock solutions as follows:

Solutions	Chemical form	Amount per liter
Stock solution A:		
Potassium nitrate	KNO_3	5.25 g
Monopotassium acid phosphate	KH_2PO_4	2.50 g
Ammonium sulfate	$(NH_4)_2SO_4$	5.00 g
Manganese sulfate	$MnSO_4 \cdot H_2O$	0.068 g
Stock solution B		
Magnesium sulfate	$MgSO_4 \cdot 7H_2O$	2.50 g

a. To 250 ml distilled water add 5 ml iron chelate stock.
b. Add 100 ml stock solution A and 100 ml stock solution B.
c. Dissolve 0.2 g calcium phosphate with 1 *N* HCl. Add.

d. Add 150 ml coconut water.
e. Add 20.0 g sucrose.
f. Bring the volume to 1000 ml.
g. Adjust the pH to 4.8–5.0.
h. Add 8.0–9.0 g agar (depending on the solidity of the medium desired).
i. Boil to dissolve agar on a hotplate, in an autoclave, or in a microwave oven. (Caution: Agar burns easily.)
j. Dispense into containers.
k. Autoclave for 10 min at 15 psi and 121°C.
l. Cool. (If slants are required, containers should be slanted at this time.)

2. Addition of other natural complexes
   a. Green banana
      i. Harvest, wash, and store green bananas in the freezer.
      ii. Just prior to use, defrost at room temperature or in a microwave oven, and peel (fresh green bananas are very difficult to peel).
      iii. For each liter weigh the appropriate amount, macerate in a blender for 30–60 sec at high speed, and add to the medium.
   b. Potato extract
      i. Weigh 100 g fresh potato.
      ii. Dice into 1-cm cubes.
      iii. Boil in 200 ml water for 5 min.
      iv. Add supernatant to the medium.

D. Culture Conditions

1. Medium
   a. For initiation and multiplication, use a liquid medium. Cultures in a liquid medium respond better when agitated on a horizontal gyrotory or vertical wheel-type shaker.
   b. For differentiation, use a solid medium.
2. Light and temperature
   a. Approximately 120 foot-candles of continuous light is adequate.
   b. Temperatures may range from 25° to 28°C.

III. RESULTS AND CONCLUSIONS

Clonal propagation of orchids has been successful with explants upon modification of modified Vacin and Went medium as follows:

Orchid	Initiation[a]	Multiplication	Differentiation
Nonsarcanthine orchids			
Apical shoot/axillary buds			
Cymbidium	+C	+C	+C
Dendrobium	+C	+C	+C
Oncidium	—	—	—
Leaves			
Cymbidium	+C (15–50%)	+C (15–50%)	+C (15–50%)
Dendrobium	+C (15–50%)	+C (15–50%)	+C (15–50%)
Root			
Dendrobium	2,4-D	2,4-D	—
Sarcanthine orchids			
Apical shoot/axillary buds			
Aranda	+C	+C−S	+C−S+B
Aranthera	+C	+C−S	−S
Phalaenopsis	+C	+C−S	+C
Vanda	+C	+C−S	+C
Inflorescence			
Ascofinetia	+C	+C−S	$+C+B^1+P$
Neostylis	+C	+C−S	$+C+B^1+P$
Vascostylis	+C	+C−S	$+C+B^1+P$
Leaves			
Doritaenopsis	+C (15–50%)	+C (15–50%)	+C (15–50%)
Doritis	+C (15–50%)	+C (15–50%)	+C (15–50%)
Phalaenopsis	+C (15–50%)	+C (15–50%)	+C (15–50%)
Vanda	+C (15–50%)	+C (15–50%)	+C (15–50%)
Root			
Phalaenopsis	2,4-D	2,4-D	+C

[a] Abbreviations: +C = 15% coconut water unless specified in parentheses; −S = minus sucrose; +B = 50 g/liter macerated green banana; $+B^1$ = 100 g/liter macerated green banana; +P = 200 ml potato extract/liter; 2,4-D (2,4-dichlorophenoxyacetic acid) = 1 mg/ml.

Sucrose, generally a constituent of all synthetic orchid seed germination and seedling media, inhibited proliferation and chlorophyll formation in protocorm-like bodies of sarcanthine orchids, which are generally monopodial. This may account for the earlier and more general success in clonal propagation of nonsarcanthine orchids.

Apical shoots or axillary buds are by far the best source of explants for successful clonal propagation, followed by inflorescences, leaves, and roots.

REFERENCES

Arditti, J. (1977). Clonal propagation of orchids by means of tissue culture—a manual. *In* "Orchid Biology, Reviews and Perspectives, I" (J. Arditti, ed.), pp. 203–293. Cornell Univ. Press, Ithaca, New York.

Hughes, K. W. (1981). Ornamental species. *In* "Cloning Agricultural Plants via in Vitro Techniques" (B. V. Conger, ed.), pp. 5–50. CRC Press, Boca Raton, Florida.
Morel, G. M. (1960). Producing virus-free *Cymbidium*. *Am. Orchid Soc. Bull.* **29,** 495–497.
Rao, A. N. (1977). Tissue culture in the orchid industry. *In* "Applied and Fundamental Aspects of Plant Cell, Tissue, and Organ Culture" (J. Reinert and Y. P. S. Bajaj, eds.), pp. 44–69. Springer-Verlag, Berlin and New York.
Sagawa, Y., and Kunisaki, J. T. (1982). Clonal propagation of orchids by tissue culture. *In* "Plant Tissue Culture 1982" (A. Fujiwara, ed.), pp. 683–684. Maruzen, Tokyo.
Vacin, E. F., and Went, F. W. (1949). Some pH changes in nutrient solutions. *Bot. Gaz. (Chicago)* **110,** 605–613.
Wimber, D. D. (1963). Clonal propagation of cymbidium through tissue culture of the shoot meristem. *Am. Orchid Soc. Bull.* **32,** 105–107.

CHAPTER **9**

Clonal Propagation: Somatic Embryos of *Citrus*

T. S. Rangan

Phytogen
Pasadena, California

I. INTRODUCTION

In angiosperms, the nucellus is a tissue of limited life and morphogenetic ability. However, in species such as *Citrus, Mangifera, Opuntia,* and *Trillium,* the nucellus is known to form adventive embryos. The adventive embryos are of interest to citriculturists in that they are genetically uniform and reproduce the characters of the maternal parent without inheriting the variations brought about by gametic fusion.

In contrast to plants established from cuttings or other conventional methods of propagation, plants obtained from nucellar embryos are free of most viruses, as it is generally believed that most of the viruses of citrus are not transmitted through seedlings, whether of zygotic or nucellar origin. Although the polyembryonic citruses pose no problem in yielding nucellar embryos and seedlings, many of the commercially important monoembryonic varieties produce no nucellar embryos under normal conditions. It was therefore believed that the monoembryonic citrus could neither be freed from viruses nor rejuvenated through nucellar seedlings. Research in the last 10–15 years, however, has demonstrated the morphogenetic potential of the nucellar cells of both poly- and monoembryonic citruses and their relatives, and the role of tissue and cell culture in the

CELL CULTURE AND SOMATIC CELL
GENETICS OF PLANTS, VOL. 1

ISBN 0-12-715001-3

breeding and improvement of *Citrus* (Spiegel-Roy and Kochba, 1980; Rangan, 1982; Rangaswamy, 1981).

II. PROCEDURE

To initiate nucellar embryogenesis *in vitro*, the nucellus is excised at an appropriate stage of development and cultured on a suitable nutrient medium. To ensure the uniform age of the explant, the flower buds are tagged on the day of anthesis. In some studies, the flowers are emasculated and bagged 2 days before anthesis and 24 hr later are pollinated with pollen of trifoliate orange (*Poncirus trifoliata* Raf.). The reason for controlled pollination is to stimulate ovule development and to provide a readily distinguishable marker, so that any zygotic seedlings could be easily identified from nucellar seedlings by their trifoliate leaves. Following pollination, preliminary samples of the developing ovules are taken at weekly intervals to determine the suitable stage of culture, which is known to vary with different species (Bitters *et al.*, 1970). At the proper stage of development, the zygotic embryo is easily recognizable by its position and the fact that it is the only embryo present. In *Citrus* the nucellus is relatively persistent as a fleshy tissue and is located immediately beneath the integuments. Depending on the cultivar, the developing fruits are harvested 70–120 days postpollination, washed in tap water, surface sterilized in 10–15% calcium hypochlorite or commercial bleach for 10–15 min, and rinsed twice in sterile distilled water. The fruits are cut open under aseptic conditions, and the ovules are removed.

The isolation of the zygotic embryo and the nucellus is carried out under a binocular dissecting microscope equipped with 20–60× magnification. To accomplish this, the ovules are held at the chalazal end with a microforceps. An incision from the chalazal to the micropylar end is made through the integuments with a sharp scalpel. The integuments are peeled away, exposing the nucellus, which is then split open. The endosperm and the zygotic embryo are discarded. At this stage the nucellus does not show any sign of degeneration or adventive embryo formation. The nucellus is severed at the chalazal end, where it is attached to the integuments, and is placed on the nutrient medium.

To obtain unfertilized ovules, the unopened flower buds are emasculated, bagged, and allowed to develop for 8–12 weeks (Button and Bornman, 1971a). Subsequently, the fruits are harvested and the nucellus is isolated and cultured, as mentioned earlier.

The nutrient medium generally consists of Murashige and Skoog's (MS)

(1962) salt mixture. The basal medium is supplemented with auxins, cytokinins, and other growth substances of natural origin such as casein hydrolysate, malt extract, and yeast extract as required and is gelled with 0.8% agar. Cultures are maintained at 25 ± 2°C, 50–60% relative humidity, and under a 16:8-hr light:dark period in diffuse light (ca. 1000–1500 lx).

The nucellar seedlings obtained *in vitro* and growing in sterile, humid conditions on nutrient medium require careful conditioning prior to transfer to soil; the protocol developed by Button and Bornman (1971b) is quite satisfactory. The plantlets in culture are exposed to high light intensity (4400 lx) for a few days at 28°C. The culture closures (cotton plugs or Kaputs) are then removed; the relative humidity is maintained at 85% and the high light intensity is reduced by 50%. The cultures are kept under these conditions for 2 days before being exposed again to high light intensity, and the relative humidity is gradually reduced to 45% over a week. The plantlets are subsequently removed, planted in pots with sandy soil, and placed under low light intensity for a few days in the greenhouse.

III. RESULTS AND CONCLUSIONS

Rangaswamy (1961) cultured the ovules and nucelli of *Citrus microcarpa* (a polyembryonic sp.) to study their morphogenetic response to growth substances and the factors controlling nucellar embryony. In ovule cultures the nucellar embryos showed normal growth and differentiation on a medium containing only sucrose, and the ovules matured into seeds. The micropylar half of the nucellus cultured on a medium supplemented with casein hydrolysate proliferated into a callus mass and differentiated into embryolike structures (termed "pseudobulbils") which eventually developed into plants. Sabharwal (1963) obtained similar results with *C. reticulata* but believed that the pseudobulbils arise not from the nucellus *per se* but from the preexisting nucellar embryos. Rangan *et al.* (1968, 1969) induced the development of nucellar embryos in monoembryonic species of *Citrus,* but unlike those in *C. microcarpa* or *C. reticulata,* the nucellar explants did not produce callus or pseudobulbils; rather, they gave rise directly to organized somatic embryos. Although in earlier studies (Rangaswamy, 1961; Sabharwal, 1963; Rangan *et al.*, 1968, 1969) nucelli were taken from fertilized ovules, it was subsequently observed that nucellar embryos can be initiated in nucelli of unfertilized ovules (Bitters *et al.*, 1970; Button and Bornman, 1971a; Kochba *et al.*, 1972; Mitra and Chaturvedi, 1972).

Kochba and Spiegel-Roy (1973) obtained an embryogenic callus from the nucellus of *C. sinensis* cv. "Shamouti" which, under suitable conditions, rapidly proliferated and formed numerous green embryos. After several passages, the embryogenic callus was found to be habituated with regard to its auxin and cytokinin requirements. It has been observed that addition of auxins indoleacetic acid (IAA), naphthaleneacetic acid (NAA), and 2,4-dichlorophenoxyacetic acid 2,4-D) suppresses embryogenesis, whereas inhibitors of auxin synthesis stimulate this process (Spiegel-Roy and Kochba, 1980). Aging the callus by increasing the interval between subcultures also enhances embryogenesis (Kochba and Button, 1974). The embryogenic callus converts a greater amount of IAA very quickly into IAA–aspartate (considered to be a stable conjugate), whereas a nonembryogenic line forms very little IAA–aspartate. This ability to form the conjugate probably reduces auxin to a level more conducive for embryogenesis (Epstein *et al.*, 1977).

Like auxins, cytokinins and gibberellic acid suppress embryogenesis, whereas abscisic acid stimulates it (Kochba *et al.*, 1978a). The omission of sucrose from the medium for a single passage caused cessation of growth but stimulated embryogenesis during subculture on a sucrose-supplemented medium (Kochba and Button, 1974). Subsequently, Kochba *et al.* (1978b) observed that galactose and galactose-yielding sugars stimulate embryogenesis.

Other complex growth substances of natural origin such as malt extract (Rangan *et al.*, 1968, 1969; Button and Bornman, 1971a; Kochba *et al.*, 1972), casein hydrolysate, and yeast extract (Rangaswamy, 1961) are also known to initiate embryogenesis in nucellar explants.

Nucellar embryogenesis *in vitro* can be traced back to single cells. The embryos develop mainly from single surface cells of existing nucellar embryos, as well as from thick-walled, plasma-rich cells embedded within the callus (Button *et al.*, 1974).

Button and Botha (1975) succeeded in obtaining embryogenesis from enzymatically isolated single cells. Vardi *et al.* (1975, 1982) induced it in protoplast-derived callus of *Citrus*.

Although totipotency in *Citrus* nucellar callus has been established, it is not clear whether all cells are totipotent or whether the totipotent cells are found along with nontotipotent ones. The nucellar callus has many features which make it a suitable system for breeding and mutation studies. The embryos originate from single cells, the capacity for embryogenesis is maintained in subcultures, and plants can be regenerated from colonies developed from protoplasts. Recent attempts to isolate NaCl-tolerant and 2,4-D-resistant cell lines have proved encouraging (Spiegel-Roy and Kochba, 1980; Spiegel-Roy *et al.*, 1983).

Efforts to achieve embryogenesis in nucelli from strictly monoembryonic cultivars, in spite of the suggested presence of substances inhibiting embryogenesis in nucellar cells (Tisserat and Murashige, 1977), have been quite successful (Rangan *et al.*, 1968; Button and Bornman, 1971a; Rangan, 1982). Because of its high regenerative ability, the nucellus is the most promising source for culture in *Citrus,* and a cell to complete plant protocol has been established. Tissue and cell culture studies of *Citrus* have also contributed significantly to our understanding of the factors regulating nucellar embryony.

REFERENCES

Bitters, W. P., Murashige, T., Rangan, T. S., and Nauer, E. (1970). Investigations on established virus-free *Citrus* plants through tissue culture. *Calif. Citrus Nurserymen's Year.* **9,** 27–30.

Button, J., and Bornman, C. H. (1971a). Development of nucellar plants from unpollinated and unfertilized ovules of the Washington Navel orange *in vitro. J. S. Afr. Bot.* **37,** 127–134.

Button, J., and Bornman, C. H. (1971b). Development of nucellar plants from unfertilized ovules of the Washington Navel Orange through tissue culture. *Citrus Grow. Sub-Trop. Fruit J.* No. 451, pp. 11–14.

Button, J., and Botha, C. E. J. (1975). Enzymatic maceration of *Citrus* callus and the regeneration of plants from single cells. *J. Exp. Bot.* **26,** 723–729.

Button, J., Kochba, J., and Bornman, C. H. (1974). Fine structure of and embryoid development from embryogenic callus of "Shamouti" orange (*Citrus sinensis* Osb.). *J. Exp. Bot.* **25,** 446–458.

Epstein, E., Kochba, J., and Neumann, H. (1977). Metabolism of indoleacetic acid by embryogenic and non-embryogenic callus lines of "Shamouti" orange (*Citrus sinensis* Osb.). *Z. Pflanzenphysiol.* **85,** 263–268.

Kochba, J., and Button, J. (1974). The stimulation of embryogenesis and embryoid development in habituated ovular callus from the "Shamouti" orange (*Citrus sinensis*) as affected by tissue age and sucrose concentration. *Z. Pflanzenphysiol.* **73,** 415–421.

Kochba, J., and Spiegel-Roy, P. (1973). Effect of culture media on embryoid formation from ovular callus of "Shamouti" orange (*Citrus sinensis*). *Z. Pflanzenphysiol.* **69,** 156–162.

Kochba, J., Spiegel-Roy, P., and Safran, H. (1972). Adventive plants from ovules and nucelli in *Citrus. Planta* **106,** 237–245.

Kochba, J., Spiegel-Roy, P., Neumann, H., and Saad, S. (1978a). Stimulation of embryogenesis in *Citrus* ovular callus by ABA, Ethephon, CCC and Alar and its suppression by GA_3. *Z. Pflanzenphysiol.* **89,** 427–432.

Kochba, J., Spiegel-Roy, P., Saad, S., and Neumann, H. (1978b). Stimulation of embryogenesis in *Citrus* tissue culture by galactose. *Naturwissenschaften* **65,** 261.

Mitra, G. C., and Chaturvedi, H. C. (1972). Embryoids and complete plants from unpollinated ovaries and from ovules of in vivo-grown emasculated flower buds of *Citrus* spp. *Bull. Torrey Bot. Club* **99,** 184–189.

Murashige, T., and Skoog, F. (1962). A revised medium for rapid growth and bioassays with tobacco tissue cultures. *Physiol. Plant.* **15,** 473–497.

Rangan, T. S. (1982). Ovary, ovule and nucellus culture. *In* "Experimental Embryology of Vascular Plants" (B. M. Johri, ed.), pp. 105–129. Springer-Verlag, Berlin and New York.

Rangan, T. S., Murashige, T., and Bitters, W. P. (1968). In vitro initiation of nucellar embryos in monoembryonic *Citrus*. *HortScience* **3,** 226–227.

Rangan, T. S., Murashige, T., and Bitters, W. P. (1969). In vitro studies on zygotic and nucellar embryogenesis in *Citrus*. *Proc. Int. Citrus Symp., 1st, 1968* **1,** 225–229.

Rangaswamy, N. S. (1961). Experimental studies on female reproductive structures of *Citrus microcarpa* Bunge. *Phytomorphology* **11,** 109–127.

Rangaswamy, N. S. (1981). Nucellus as an experimental system in basic and applied tissue culture research. *In* "Proceedings of the COSTED Symposium on Tissue Culture of Economically Important Plants" (A. N. Rao, ed.), pp. 269–286. COSTED and Asian Network of Biological Sciences, Singapore.

Sabharwal, P. S. (1963). In vitro culture of ovules, nucelli and embryos of *Citrus reticulata* Blanco var. Nagpuri. *In* "Plant Tissue and Organ Culture—A Symposium" (P. Maheshwari and N. S. Rangaswamy, eds.), pp. 265–274. Univ. of Delhi, Delhi, India.

Spiegel-Roy, P., and Kochba, J. (1980). Embryogenesis in *Citrus* tissue cultures. *Adv. Biochem. Eng.* **1,** 27–48.

Spiegel-Roy, P., Kochba, J., and Saad, S. (1983). Selection for tolerance to 2,4-dichlorophenoxyacetic acid in ovular callus of orange (*Citurs sinensis*). *Z. Pflanzenphysiol.* **109,** 41–48.

Tisserat, B., and Murashige, T. (1977). Probable identity of substances in *Citrus* that repress asexual embryogenesis. *In Vitro* **13,** 785–789.

Vardi, A., Spiegel-Roy, P., and Galun, E. (1975). *Citrus* cell culture: Isolation of protoplasts, plating densities, effect of mutagens and regeneration of embryos. *Plant Sci. Lett.* **4,** 231–236.

Vardi, A., Spiegel-Roy, P., and Galun, E. (1982). Plant regeneration from *Citrus* protoplasts: Variability in methodological requirements among cultivars and species. *Theor. Appl. Genet.* **62,** 171–176.

CHAPTER **10**

Clonal Propagation: Palms

Brent Tisserat

Fruit and Vegetable Chemistry Laboratory
Agricultural Research Service
United States Department of Agriculture
Pasadena, California

I. INTRODUCTION

The Arecaceae (Palmae) is a family of woody perennial monocots and consists of about 200 genera and 2500 species (Corner, 1966; McCurrach, 1960; Tomlinson, 1961). Palms are common to the tropical regions of the world and occur to a lesser extent in the subtropical regions (e.g., *Phoenix dactylifera* L., date palm). Palms have had numerous agricultural and economic uses since the dawn of history and constitute the second most important plant family to humans aside from the Gramineae. The importance of the palm as a multiuse plant employed in subsistence agriculture has diminished in the last 200 years. Today, palm crops provide edible and nonedible commodities, often produced through large-scale plantation systems, that are consumed both domestically and internationally.

Improvement in palm crops has traditionally been slow due to their long-lived nature, growth habit, habitat, and absence in many species of an adequate vegetative propagation method. Many species, such as the coconut palm (*Cocos nucifera* L.) and oil palm (*Elaeis guineensis* Jacq.), can be propagated only by seed. Progeny resulting from seed usually exhibit het-

CELL CULTURE AND SOMATIC CELL
GENETICS OF PLANTS, VOL. 1

ISBN 0-12-715001-3

erozygosity. Also, propagation of palms through suckering, branching, or teratological events is limited.

Micropropagation offers an alternative for the rapid clonal production of palms. Since the 1970s, several reports on palm tissue culture have appeared in the literature (Ammar and Benbadis, 1977; Corley *et al.*, 1977; de Guzman *et al.*, 1979; Eeuwens and Blake, 1977; Poulain *et al.*, 1979; Rabéchault *et al.*, 1970; Reuveni, 1979; Reynolds and Murashige, 1979; Tisserat, 1979, 1981). Free-living palms have been obtained from embryogenic callus in oil (Corley *et al.*, 1977) and date palms (Tisserat, 1979, 1981). Date palm plantlets derived from rooted shoot tips and lateral buds have also been transferred to soil (Poulain *et al.*, 1979; Tisserat, 1981). The following information is based on research by the author since 1977 on rapid propagation of date palms.

II. MATERIALS AND METHODS

Methods and procedures used to obtain clonal and zygotic tissue cultures from adult, offshoot, and seedling date palms have been previously described (Tisserat, 1979, 1981). A detailed step-by-step procedure is offered to outline the protocol employed to obtain plantlets from a variety of palm explant sources. Nutrient media employed to establish and maintain palm cultures and plantlets *in vitro* are summarized in Table I.

A. Procedure Used to Obtain Plantlets from Shoot Tip and/or Lateral Bud Callus

1. Dissect offshoots, seedlings, or trees, using a hatchet or serrated knife when appropriate. Remove leaves acropetally, exposing the lateral buds at the axil of each leaf. Shoot tips are removed from the shoot terminal after all mature leaves are peeled away. Store buds and tips in cold antioxidant solution (150 mg/liter citric acid and 100 mg/liter absorbic acid). Keep explants in the refrigerator at 0°C until the surface sterilization procedure is performed.
2. Trim the outermost leaves of the buds and tips to obtain explants that are 0.5 cm^2.
3. Sterilize the explants by wrapping them in cheesecloth to prevent loss during handling procedures in a 25 × 150-mm culture tube. Sterilize them

TABLE I

Palm Tissue Culture Media

Components	Media types (mg/liter)				
	Callus production	Plantlet germination	Shoot tip	Shoot proliferation	Adventitious rooting
Inorganic salts					
Murashige and Skoog (MS) (1962)	+	+	+	+	+
Carbohydrate source					
Sucrose	30,000	30,000	30,000	30,000	30,000
Vitamin sources					
Meso-inositol dihydrate	100	100	100	100	100
Thiamine·HCl	0.4	0.4	0.4	0.4	0.4
Complex addenda					
Phytagar	8,000	8,000	8,000		8,000
Charcoal, activated neutralized	3,000	3,000	3,000		
Phytohormones					
2,4-Dichlorophenoxyacetic acid (2,4-D)	100		10		
N^6-(Δ^2-isopentyl)adenine (2iP)	3			10	
α-Naphthaleneacetic acid (NAA)				0.1	0.1

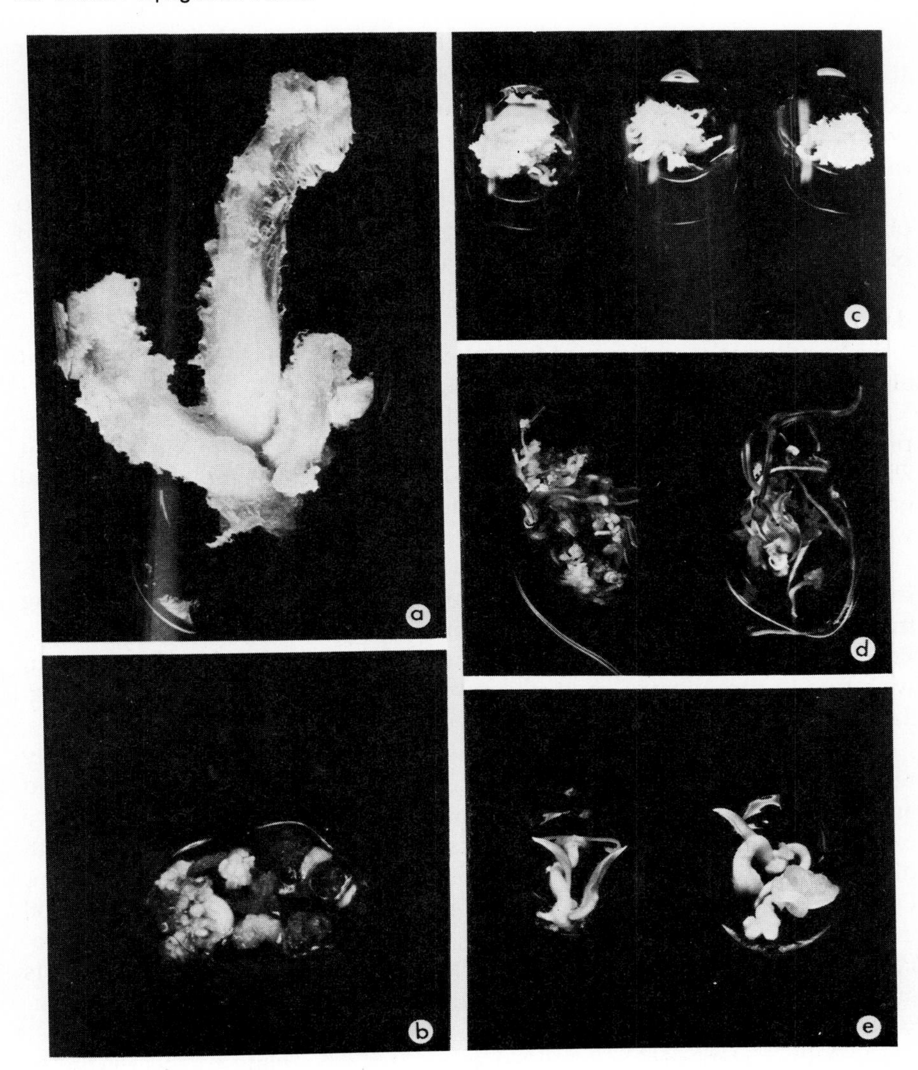

Fig. 1. Asexual embryogenesis in date palms from cultured shoot tips. (a) Shoot tip about 4 weeks old cultured on a medium containing 100 mg/liter 2,4-dichlorophenoxyacetic acid (2,4-D), 3 mg/liter $N^6(\Delta^2$-isopentyl)adenine (2iP), and 0.3% charcoal. (b) Initiation of nodular callus from leaves. (c) Friable callus producing asexual embryos on media devoid of hormones. (d) Production of asexual embryos and plantlets from callus. (e) Recultured isolated plantlets.

in 2.6% sodium hypochlorite solution (containing 1 drop of Tween-20 per 100 ml solution) for 15 min. Dislodge air bubbles from the tissues by periodic agitation of the tube. Pour off the bleach solution and rinse three times with sterile water. Remove the explants and transfer them aseptically to the sterile Petri dish (15 × 150 mm in diameter).

4. Remove additional leaves from the shoot tip and bud explants to obtain a culture that is 1–3 mm^2 in size. A 10-sec dip of this explant into bleach solution may reduce contaminates prior to planting.

5. Plant the explant on the surface of callus production agar medium, as described in Table I.

6. Reculture the explants at 8-week intervals. Callus initiation is evident after two to three culture passages (Fig. 1a,b).

7. When white, friable, nodular callus is prominent, subculture 1-cm^2 pieces to plantlet germination media (Table I) and incubate the cultures at 28°C under a 16-hr photoperiod of 50 foot-candle (fc) intensity. Asexual embryos and green plantlets will usually become apparent within 2–4 weeks in culture (Fig. 1b–e).

8. To enhance adventitious root formation, reculture the seedling with the primary root trimmed to 1–2 cm in length to adventitious rooting medium (Table I). Continue the reculturing procedure every 8 weeks for one to three culture passages until the plantlets reach a length of 10 cm, with two to three leaves, and possesses an adventitious root system (Fig. 2a).

9. Transfer of plantlets to free-living conditions: Plantlets are carefully removed from the agar medium without damage to the root system, and are soaked in distilled water for 15 min to avoid dehydration and remove excess adhering media. The plantlets are then rinsed three times with distilled water, sprayed with 0.5% benolate (du Pont Co., Wilmington, Delaware) fungicide solution, and transferred to a soil medium. The soil medium consists of sterile peat moss and vermiculite in a 1:1 (v/v) ratio. Plantlets are planted in either 3-in.-diameter plastic pots or jiffy peat pots and enclosed within a transparent tent composed of two interlocking clear polystyrene tumblers.

10. Administer weekly applications of 0.5% benolate to the foliage to minimize fungal growth. Water the pots every day with distilled water and once a week with one-fourth strength Hoagland's solution during the first 2 months of development. Incubate the plants initially in an environmentally controlled chamber under 800 fc light intensity, 16-hr photoperiod at 28°C for 2 weeks. Transfer the plants to a shaded greenhouse. Gradually acclimate the plantlets to the greenhouse humidity conditions by punching holes in the plastic cover. After 2 months, the covers may be removed and the plant treated as a normal palm seedling (Fig. 2b).

Fig. 2. Plantlets obtained from callus. (a) Plantlets producing adventitious roots in media containing 0.1 mg/liter α-naphthaleneacetic acid (NAA). Note that the second and fourth cultures from the left are also producing axillary bud outgrowths. (b) Free-living date palm plantlet.

B. Procedure Used to Propagate Palms by Rooting Shoot Tips

1. Repeat the protocol for excision and planting of shoot tips and/or lateral bud explants as described in steps 1–4 in Section II,A.
2. Plant the explants on the surface of shoot tip media as described in Table I.
3. Incubate the cultures under 50 fc intensity of a 16-hr photoperiod at 28°C in an environmental chamber.
4. The cultures will initiate leaves and increase considerably in size within the next 4–6 weeks in culture. Reculture the explant to fresh medium at the end of the 8-week culture passage.
5. Axillary shoot outgrowths from cultured shoots and buds can be obtained by transfer to shoot proliferation media (Table I). Divide the shoot structures to encourage further axillary bud differentiation as desired.
6. Follow steps 8–10 in Section II,A to root buds and tips and obtain free-living palms.

III. RESULTS AND CONCLUSIONS

The procedure presented for the tissue culture of palms should be considered a general guide since these methods were developed almost exclusively for the date palm. The author has applied these techniques to produce embryogenic callus from various explants of a number of palm species (*Phoenix reclinata, Erythea armata, P. canariensis, P. pusilla,* and *P. sylvestris*). The tissue of choice in palms for obtaining embryogenic callus is either the actively growing lateral bud (when available) or the shoot tip (Tisserat, 1981). Older tissues and organs do not respond well in culture, and callus produced from them usually has limited morphogenetic potential (e.g., mature leaf and flower bud callus produces roots only).

Browning of the original palm tissue is common. However, this problem can be minimized either through inclusion of adsorbents such as charcoal or frequent reculturing of small-size explants.

Each piece of embryogenic date palm callus contains thousands of proembryos. Subdivision of this callus to media devoid of hormones will yield hundreds of visible asexual embryos and plantlets within a few weeks (Tisserat, 1981). Successful transfer of plantlets to soil depends on potting plantlets with adequate photosynthetic shoot and adventitious root systems. Production of plantlets from offshoots initiated on cultured shoot tips and buds is a less well-developed technique compared to the asexual embryogenesis process. The parameters involved in maximizing offshoot initiation *in vitro* still remain to be explored. However, plantlets produced by the organogenesis process should be clonal and produced with less risk of genetic aberrance than callus-derived plantlets (Tisserat, 1981).

REFERENCES

Ammar, S., and Benbadis, A. (1977). Multiplication végétative du Palmier-dattier (*Phoenix dactylifera* L.) par la culture de tissus de jeunes plantes issues de semis. *C. R. Hebd. Seances Acad. Sci., Ser. D* **284,** 1789–1792.

Corley, R. H. V., Barrett, J. H., and Jones, L. H. (1977). Vegetative propagation of oil palm via tissue culture. *Int. Dev. Oil Palm, Proc. Malays. Int. Agric. Oil Palm Conf., 1976* pp. 1–7.

Corner, D. J. H. (1966). "The Natural History of Palms." Univ. of California Press, Berkeley.

de Guzman, E. V., del Rosario, A. G., and Ubalde, E. M. (1979). Proliferative growths and organogenesis in coconut embryo and tissue cultures. *Philip. J. Coconut Stud.* **7,** 1–10.

Eeuwens, C. J., and Blake, J. (1977). Culture of coconut and date palm tissue with a view to vegetative propagation. *Acta Hortic.* **78,** 277–286.

McCurrach, J. C. (1960). "Palms of the World." Harper, New York.

Murashige, T., and Skoog, F. (1962). A revised medium for rapid growth and bioassays with tobacco tissue cultures. *Physiol. Plant.* **15,** 473–497.

Poulain, C., Rhiss, A., and Beuchesne, G. (1979). Multiplication végétative en culture *in vitro* du palmier-dattier (*Phoenix dactylifera* L.). *C. R. Acad. Agric.* **11,** 1151–1154.

Rabéchault, H., Ahée, J., and Guénin, G. (1970). Colonies cellulaires et formes embryoïdes obtenues *in vitro* à partir de cultures d'embryons de palmier à huile (*Elaeis guineensis* Jacq. var. *dura* Becc.) *C. R. Hebd. Seances Acad. Sci., Ser. D* **270,** 233–237.

Reuveni, O. (1979). Embryogenesis and plantlets growth of date palm (*Phoenix dactylifera* L.) derived from callus tissues. *Plant Physiol.* (S) **63,** 138.

Reynolds, J. F., and Murashige, T. (1979). Asexual embryogenesis in callus cultures of palms. *In Vitro* **15,** 383–387.

Tisserat, B. (1979). Propagation of date palm (*Phoenix dactylifera* L.) *in vitro*. *J. Exp. Bot.* **30,** 1275–1283.

Tisserat, B. (1981). Date palm tissue culture. *U.S., Agric. Res. Serv., Adv. Agric. Technol.* **ATT-W-17,** 1–50.

Tomlinson, P. B. (1961). "Anatomy of the Monocotyledons," Vol. II. Oxford Univ. Press (Clarendon), London and New York.

CHAPTER **11**

Clonal Propagation: Gymnosperms

Jenny Aitken-Christie

Forest Research Institute
Rotorua, New Zealand

Trevor A. Thorpe

Department of Biology
University of Calgary
Calgary, Alberta, Canada

I. INTRODUCTION

Many excellent recent reviews cover the history of gymnosperm micropropagation, the wide variety of species and explants used, the number of species in which plantlets have been produced, and the different methods used to produce them (Mott, 1978, 1981; Sommer and Brown, 1979; Biondi and Thorpe, 1982; Karnosky, 1981; David, 1982; Thorpe and Biondi, 1983).

CELL CULTURE AND SOMATIC CELL
GENETICS OF PLANTS, VOL. 1

ISBN 0-12-715001-3

Since the first plantlet was produced in 1975 (Sommer *et al.*, 1975), much progress has been made. As an example of gymnosperm micropropagation, a procedure which has been used to produce plants for field trials will be described in this chapter. The fact that field trials have been established provides evidence for practical application of such procedures. Reference should also be made to "Tissue Culture in Forestry" (Bonga and Durzan,

TABLE I

Nutrient Media Used for Radiata Pine Micropropagation (mg/liter)

Nutrient	SH[a]	GD[b]	LP[c]
Major			
KNO_3	2,500	1,000	1,800
$MgSO_4 \cdot 7H_2O$	400	250	360
$NH_4H_2PO_4$	300		
$CaCl_2 \cdot 2H_2O$	200	150	
$(NH_4)_2SO_4$		200	
KCl		300	
$NaH_2PO_4 \cdot 2H_2O$		100	
$Na_2HPO_4 \cdot 12H_2O$		75	
NH_4NO_3			400
KH_2PO_4			270
$Ca(NO_3)_2 \cdot 4H_2O$			1,200
Minor	[d]	[d]	
$FeSO_4 \cdot 7H_2O$	15	30	30[d]
$Na_2EDTA \cdot 2H_2O$	20	40	40[d]
$MnSO_4 \cdot 4H_2O$	20	20	1
H_3BO_3	5	5	6.2
$ZnSO_4 \cdot 7H_2O$	1	1	8.6
KI	1	1	0.08
$CuSO_4 \cdot 5H_2O$	0.2	0.2	0.025
$Na_2MoO_4 \cdot 2H_2O$	0.2	0.2	0.25
$CoCl_2 \cdot 6H_2O$	0.2	0.2	0.025
Vitamins		[d]	
Thiamine·HCl	5	5	0.4
Nicotinic acid	5	5	
Pyridoxine·HCl	0.5	0.5	
Inositol	1,000	1,000	1,000
Sucrose	20,000 or 30,000	20,000	30,000

[a] SH = based on Schenk and Hildebrandt (1972) and modified by Reilly and Washer (1977).
[b] GD = based on Gresshoff and Doy (1972) and modified by Horgan and Aitken (1981).
[c] LP = based on Quoirin and Lepoivre (1977) and modified by J. Aitken-Christie (unpublished).
[d] Modification.

1982), the first book to be published on this topic. This volume is a direct result of the interest in this area over the last 10 years and provides a comprehensive background to the subject.

II. NUTRIENT MEDIA

Fully defined culture media have been used to produce large numbers of shoots in *Pinus radiata* D. Don (radiata pine) (Table I). Nutrient media preparation is facilitated by the use of stock solutions of the various groups of constituents, namely, 1-liter concentrated stock solutions of the major elements (50×), minor elements (100×), vitamins (100×), and iron (100×). These are stored in the refrigerator if in regular use or frozen in aliquots, that will make up 1 liter of nutrient medium, in plastic screw-top containers if they are not being regularly used. The $CaCl_2 \cdot 2H_2O$ in the Gresshoff and Doy (GD) medium (Table I) is made up as a separate stock solution because of precipitation problems. Stock solutions of the phytohormones benzylaminopurine (BAP), indolebutyric acid (IBA), and naphthaleneacetic acid (NAA) are prepared at a concentration of 100 mg/liter, i.e., BAP at 4.44×10^{-4} *M*, IBA at 4.92×10^{-4} *M*, and NAA at 5.37×10^{-4} *M*. Commercial-grade sucrose is used. Inositol, 1 g/liter, and Difco Bacto agar, 8 g/liter, are used for all nutrient media. Medium is autoclaved at 100–140 kPa (15–20 psi) for 15 or 20 min for jars (100 ml medium each) and conical flasks (1 liter) (from which Petri dishes are poured), respectively.

III. PROCEDURE FOR THE MICROPROPAGATION OF JUVENILE RADIATA PINE

The steps involved in the production of micropropagated planting stock, currently used at the Forest Research Institute (FRI), Rotorua, New Zealand, are shown in Fig. 1. The details of this figure are found under the corresponding numbers in the text below. This procedure is based on the methods of Reilly and Washer (1977), Aitken *et al.* (1981), Horgan and Aitken (1981), Horgan (1982), and Aitken-Christie *et al.* (1982), updated with more recent improvements and findings.

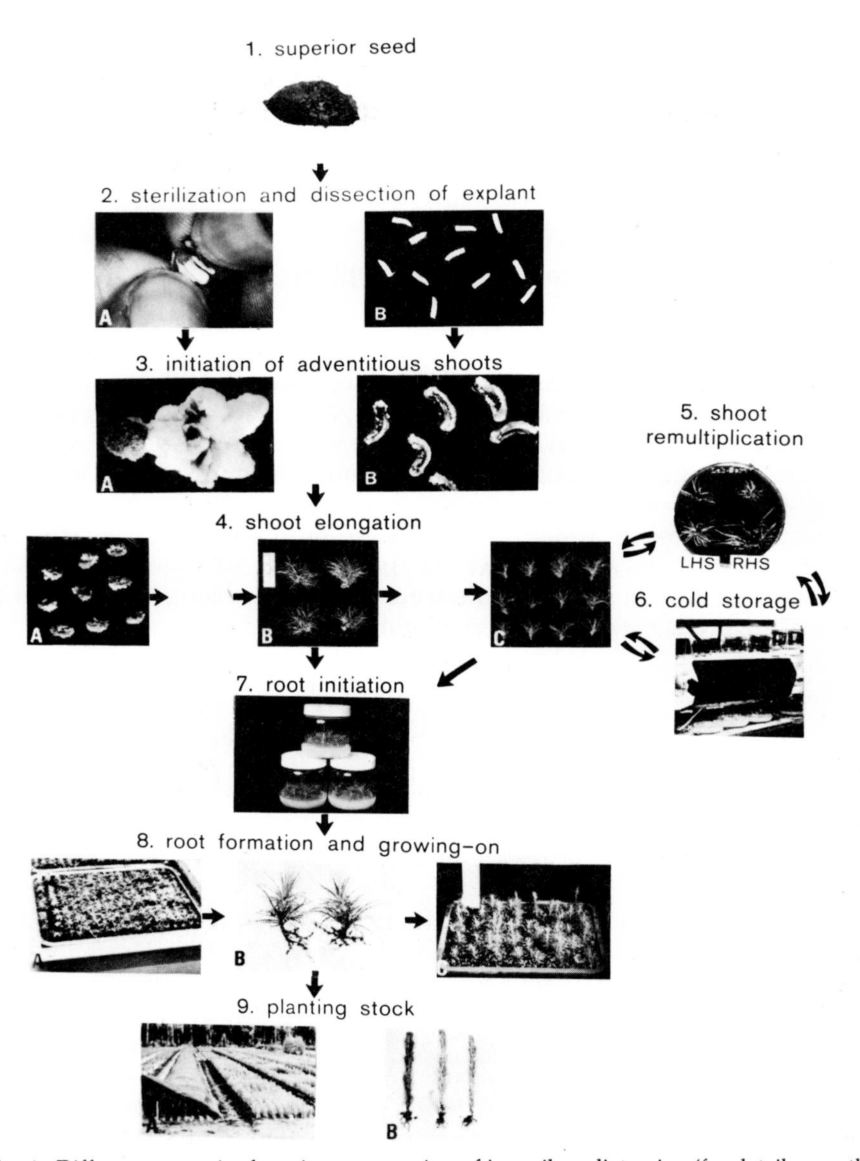

Fig. 1. Different stages in the micropropagation of juvenile radiata pine (for details, see the text).

A. Superior Seed

Methods have been developed for micropropagating control-pollinated seed (Fig. 1, part 1). The expense of micropropagation is worthwhile only if the seed to be used is superior to that which is currently available from the best seed orchards.

B. Sterilization and Dissection of the Explant

Excised embryos and cotyledons from germinated seeds are used as explants. The latter produce more shoots. Seeds are placed in a loosely tied muslin bag and surface sterilized for 15–20 min in a 50/50 solution of commercial bleach (2.5% available chlorine) with a few drops of Tween-80. The seeds are stirred throughout. After being rinsed in running water for 24 hr, the seeds are surface sterilized again in 6% H_2O_2 for 5–10 min, rinsed twice in sterile water, and stratified in an empty sterile jar in the refrigerator for 2 days. The seeds are again surface sterilized in 6% H_2O_2 for 5–10 min and rinsed in sterile water prior to excising the embryo or placing the seeds on a sterile germinating medium.

1. Seeds are split carefully in half with thumbs which have been dipped in 70% ethanol (Fig. 1, part 2A). The embryos are then extracted with cooled flamed tweezers. The embryos are placed in a Petri dish of sterile water until 12–24 have been excised. Usually six are placed in each Petri dish for shoot initiation. Viable embryos are white. Yellow embryos or embryos with damaged cotyledons are discarded. Embryos are laid on top of the agar lengthwise. Contact of the cotyledons with the medium is essential.
2. Seeds are germinated in the dark at 25°C on a prewetted filter paper disk on top of a capillary matting[1] disk in a glass Petri dish. The amount of water present is critical. The capillary matting disks are soaked, drained until no further water drips off, and then placed in the Petri dishes with an additional 7.5 ml of distilled water and the filter paper disk on top. Seeds take 5–14 days to germinate depending on the seed source. Generally, 70–90% germination is obtained within 7–10 days. Cotyledons are excised from seeds which have germinated for 1 day (Fig. 1, part 2B) (i.e., the radicle has emerged for 1 day). This is the best stage of development for maximum shoot formation. Cotyledon length and number vary from seed to seed. The cotyledons from four seeds are placed on top of the agar of each Petri dish, as for embryos.

[1]Capillary matting is a polyester capillary felt manufactured by Alex Harvey Industries, Auckland, New Zealand.

C. Initiation of Adventitious Shoots

A controlled environment culture chamber is necessary for all the sterile shoot formation stages. The "day" temperature varies between 25 and 28°C, with the higher temperatures favored for shoot initiation (Fig. 1, part 3) and the cooler ones for shoot elongation (Fig. 1, part 4) and shoot remultiplication (Fig. 1, part 5). There is a drop in temperature of 5°C with a daily change to darkness. A 16-hr photoperiod is employed for all stages, and the cultures receive an average light intensity of 80 $\mu Em^{-2}s^{-1}$ (approximately 6700 lx) under cool white fluorescent light.

Embryos and cotyledons from germinated seeds are placed on a Quoirin and Lepoivre (LP) medium (see Table I) containing 5 mg/liter (2.22×10^{-5} *M*) BAP in a shallow Petri dish (90 × 15 mm). The cotyledons of the excised embryos can be cut from the embryos after 1 week and replaced on the same medium for greater surface area contact. The cotyledons from germinated seeds sometimes curl away from the medium and should be pressed back down. On average, with the excised embryo method, 10–20 shoots per embryo are formed. When the cotyledons are cut after 1 week, this is increased to 20–50 shoots per embryo. The cotyledons from the germinated seed method produce 180–250 shoots per seed on average. Some clones have produced in excess of 1000 shoots.

After 3 weeks a nodular, yellow meristematic tissue forms on the side of the cotyledons in contact with the medium (Fig. 1, parts 3A and 3B). This tissue, which is a mass of dividing cells with no organized meristems, gives rise to adventitious shoots.

D. Shoot Elongation

Further steps are common to both types of explants. Cotyledons with meristematic tissue are placed on an LP medium (Table I) without BAP in a deep Petri dish (90 × 25 mm) for 4–6 weeks to promote the elongation of adventitious shoots (Fig. 1, part 4A). They are then transferred to fresh medium of the same composition, in 580-ml jars every 4–6 weeks until shoots have elongated to approximately 1.0–1.5 cm (base of the shoot to the apex). During elongation, clumps of shoots are cut into smaller clumps on top of paper towels which have been folded like a book and sterilized in an oven bag. Shoots are also grown in jars for remultiplication and cold storage. Previously, the Schenk and Hildebrandt (SH) medium (Table I) containing 3% sucrose was used for the first shoot elongation transfer in deep Petri dishes. Thereafter, the SH medium containing 2% sucrose in jars was used until final elongation on GD medium.

Shoots frequently elongate at different rates. Generally, the shoots from clones with low shoot numbers (20–50) grow to a rootable size faster than those with high shoot numbers (500+), and a difference of 18 weeks compared with 34 weeks from initiation on SH medium is not uncommon. With the LP medium, fewer shoots (i.e., approximately 25%) per clone than those on the SH medium are initiated, and the average time for shoots growing out to a rootable size is reduced to 12 weeks (J. Aitken-Christie, unpublished). Thus, fewer transfers are required until shoots have elongated fully on the LP medium than on the SH medium. Nutrients are often limiting in clones with large numbers of shoots because of shoot competition. Extra splitting of large clumps of shoots into smaller ones is necessary for continuous growth (Fig. 1, part 4B). When shoots have reached an optimum stem size of 1.0–1.5 cm, they are either remultiplied by topping, cold stored, or rooted (Fig. 1). Shoots are either left in clumps, which is less labor intensive, or separated into single shoots (Fig. 1, part 4C).

There are three different types of adventitious shoots produced during the shoot elongation stage: "waxy" shoots, "wet" shoots, and translucent shoots. Translucent shoots have been called "vitrified," "water-soaked," and "waterlogged" shoots in other species. These are discarded when they appear because they do not survive when removed from sterile conditions. The wet shoots often have needles which are stuck together and lack epicuticular waxes, whereas the waxy shoots look like shoots from a seedling. Both the waxy and wet shoots appear to be normal and can be rooted, although the wet type is more prone to rot during rooting.

E. Shoot Remultiplication

Rootable shoots (1.0–1.5 cm) growing on the LP medium without BAP are remultiplied by a topping method (J. Aitken-Christie, unpublished). The shoot apex and apical tuft of the primary needles ("top") are severed from each shoot with as little damage as possible to the primary needles (Fig. 1, part 5, LHS). New shoots form from the axillary meristems in the basal portions in 3 months (Fig. 1, part 5, RHS); these are either rooted, remultiplied again, or cold stored (Fig. 1). The tops are grown further and either topped again or rooted. At least a fourfold remultiplication every 3 months is possible. In general, there is a direct relationship between shoot size and the number of axillary buds and needles, and thus bigger shoots produce more new shoots than do smaller shoots when they are topped. Also, if more top is cut out during topping, fewer axillary meristems are left in the basal portion and thus fewer new shoots grow out.

F. Cold Storage of Shoots

Rootable shoots (1.0–1.5 cm) growing on the LP medium without BAP are cold stored at 2°C in a refrigerator or cold store (Fig. 1, part 6) under a low light intensity (8 $\mu Em^{-2}s^{-1}$ or 670 lx) and a 16-hr photoperiod. Shoots are removed from the cold store and allowed to come to room temperature before being transferred to fresh medium and placed under normal culture conditions for growing up, rooting, or further remultiplication (Fig. 1). Shoots have been cold-stored for up to 17 months to date (J. Aitken-Christie, unpublished).

G. Root Initiation

Single shoots (1.0–1.5 cm) are placed in an 0.8% Difco Bacto water agar medium in 580-ml jars containing 1.0 mg/liter (4.92×10^{-6} *M*) IBA and 0.5 mg/liter (2.68×10^{-6} *M*) NAA for 5 days (Fig. 1, part 7). The jars are placed in the culture chamber at 20°C (day temperature) to be conditioned for the rooting chamber. The separation of shoots from shoot clumps need not be done under sterile conditions. Wet shoots are more affected by the auxin treatment than waxy shoots. With some clones, the needles at the tip of the shoot and the stem become yellow and twisted after 5 days. This often causes the shoots to rot during root formation. Four days on auxin is sufficient for the wet shoot.

H. Root Formation and Growing On

After 5 days in the water–agar–auxin medium, shoots are stirred at room temperature in a large beaker to wash off the agar. The shoots are then planted in a nonsterile peat–pumice mix in a plastic tray (45 × 30 × 6 cm) at a density of 90–100 per tray. Each tray contains approximately 3 cm of coarse pumice[2] for drainage in the base of the tray topped with 3 cm of prewetted, finely sieved peat and pumice in a ratio of 60:40.

Root formation is best under the same photoperiod and light intensity as that for shoot formation, but at 20–25°C day temperature (5°C lower during darkness) and a high relative humidity (90–95%). A controlled environ-

[2]Pumice is a porous volcanic rock found in New Zealand which under intense heat can be made into perlite.

ment chamber, similar to that used for shoot formation, is used for root formation. Two trays of shoots are placed into a larger plastic tray with a glass lid (Fig. 1, part 8A). Daily misting and aeration are necessary. Eighty percent of the shoots develop roots in 2–8 weeks. Rooting ability (number of roots and speed of rooting) varies with the clone. In some clones shoots form three to six roots in 4 weeks, whereas in others, one or two roots are formed in 8 weeks. New needle growth at the top of the shoot indicates that the shoots are rooted. This is easy to distinguish on the wet shoots because the new needle growth is of the waxy type. When the shoots are rooted they are carefully dug up, the roots are pruned to 1–2 cm from the base of the shoot, and the plantlets are replanted in a peat–pumice mix containing Magamp at a rate of 14 g per tray under high humidity for a further 2–3 weeks. Magamp is a slow-release fertilizer containing magnesium, ammonium, and phosphate. Lateral roots elongate from the pruned roots, giving the plantlet more root surface area (Fig. 1, part 8B). Plantlets are gradually hardened off to a lower humidity under a mist bench in a glasshouse, grown up in the glasshouse, and then transferred outside while still in the rooting trays (Fig. 1, part 8C). Weekly applications of liquid nutrients are given to maintain good growth.

I. Planting Stock

After being root pruned again, plantlets are lined out in the nursery bed in late spring under shade cloth (Fig. 1, part 9A). The shade cloth is removed after several months, and quality planting stock is ready for planting out in the forest the following winter (Fig. 1, part 9B). Micropropagated planting stock is treated the same way as normal bare-rooted radiata pine seedlings in the nursery bed and at outplanting. This includes a conditioning program of undercutting, wrenching, and lateral root pruning to stimulate fibrous root development and control height growth. Plantlets are then able to withstand the stresses of transplantation, and field survival and growth rates are improved.

IV. MICROPROPAGATION OF MATURE GYMNOSPERMS

The micropropagation of mature trees using mature parts of the tree has been achieved only on a small scale, with a few plantlets being produced (David, 1982). In many cases, small numbers of buds, shoots, and somatic embryos have been produced and no plantlets have been formed (von

Arnold and Eriksson, 1979; Bonga, 1977, 1981; Boulay, 1979b). This suggests that the ideal conditions for shoot growth, shoot remultiplication, and rooting from mature explants have not been elucidated, and thus a reliable procedure cannot be given. The biggest success story of a micropropagated gymnosperm, either juvenile or mature, is the redwood tree.

The work of Ball (1950), Ball *et al.* (1978), and Boulay (1977) triggered the research on the practical application of redwood micropropagation. Details of the procedure can be found in Boulay (1979a) and Poissonier *et al.* (1980). Over 30,000 plants from 200 clones have been produced at Association Forêt-Cellulose (AFOCEL) in Nangis, France, for field trials. Redwood is unique in that juvenile shoots sprout from the base of the mature tree. These shoots are easier to micropropagate than are mature shoots from the upper bole of the tree. Most gymnosperm species do not produce such sprouts. From phenotypically selected mature trees, newly developed shoots collected in summer (June) and lignified shoots collected in autumn (October) are taken from the annual basal sprouts and used as explants. Explants taken from 2-year-old basal sprouts are not as reactive. Shoots collected in summer take 1.0–1.5 months to form buds, whereas winter shoots take only 15 days–1 month. Shoot pieces are planted vertically in a modified Murashige and Skoog (MS) (1962) medium containing adenine sulfate and kinetin, but no vitamins. For shoot elongation, half-strength medium containing 2% activated charcoal is used. Shoot multiplication is achieved by replanting in medium with the phytohormone. For rooting, 40- to 50-mm shoots are given a cold treatment for 1–2 weeks, followed by an IBA dip before planting in a rooting mixture of vermiculite and perlite. Regular applications of fungicides are needed.

The best example of micropropagation from mature explants is the work of David *et al.* (1979) and Franclet *et al.* (1980), in which the regeneration of plantlets from 11-year-old *Pinus pinaster* Sol. was achieved. Their procedure is similar to that described for redwood in that shoots are grown on a medium with phytohormones for shoot remultiplication, followed by growth on a medium without phytohormones but with charcoal for shoot elongation. MS medium (1962) without ammonium nitrate or phytohormones is used for the shoot induction phase, followed by cultivation on Campbell and Durzan (CD) medium (1975) containing 5×10^{-6} *M* BAP for remultiplication and the same medium containing 2% activated charcoal (Merck 2186) for elongation (David *et al.*, 1979). Rooting was achieved after an IBA/NAA treatment. The explants used were the fascicle shoots, induced by pruning, of well-fertilized glasshouse-grown rooted cuttings (David, 1982). However, the characteristics of the mother trees and the positions from which the cuttings were taken were not stated.

M. Abo El-Nil (1982) of the Weyerhaeuser Co., Tacoma, Washington, has recently patented a method for the asexual reproduction of mature coniferous trees. Cytokinin is sprayed on the trees and juvenile-looking adventitious shoots are produced. These, when placed in culture, give rise to further juvenile-looking shoots. No indication of whether plantlets resumed their mature form after further growth is given.

Recently at the FRI, over 6000 radiata pine shoots from 22 clones from 4-, 8-, 12-, 20-, and 50-year-old trees have been produced and the use of activated charcoal has also played a major part (K. Horgan, unpublished). Each clone has been remultiplied three times per year. Only the 12-, 20-, and 50-year-old explants were considered to be mature; however, the same procedure was used for all ages. Although some shoots have rooted, rooting has not yet been attempted on a large scale.

V. CONCLUSIONS

Using techniques similar to those used for the micropropagation of juvenile radiata pine, a variety of gymnosperm species have been micropropagated and are now in field trials (Table II). Once field trials have demonstrated that micropropagated trees perform well, this approach can be used to multiply rapidly special families or clones from control-pollinated crosses. In this way, the faster establishment of improved genotypes could mean real economic gains from tree breeding. In the meantime, methods

TABLE II

Some Examples of Micropropagated Planting Stock Planted and to Be Planted in the Forest

Species	Number of trees by 1982	Expected number of trees by 1983
Pinus taeda L. (loblolly pine)	2,100[a]	4,000[a]
Pinus pinaster Sol. (maritime pine)	1,200[b]	Unknown
Pseudotsuga menziesii Mirb. (Douglas fir)	5,800[c]	10,800[d]
Sequoia sempervirens Endl. (redwood)	30,000[e]	Unknown
Pinus radiata D. Don (radiata pine)	2,700[f]	7,700[f]

[a] H. V. Amerson and R. L. Mott, North Carolina State University, Raleigh, North Carolina; also see Mott and Amerson (1981).

[b] A. H. David, Laboratoire de Physiologie Végétale, Université de Bordeaux I, Talence, France. (Note: Approximately 1000 of these were produced at AFOCEL, Nangis, France.)

[c] Sommer and Wetzstein (1984).

[d] M. Abo El-Nil, Weyerhaeuser Co., Tacoma, Washington.

[e] Poissonier *et al.* (1980).

[f] D. R. Smith, Forest Research Institute, Rotorua, New Zealand.

need to be developed to make the system commercially viable (Smith *et al.*, 1981, 1982).

Another possible application of micropropagation which offers large increments in genetic improvement is the use of cold storage of clones. Some shoots of each of a large number of clones from controlled pollinations could be cold stored in a juvenile state while the remaining shoots are rooted to produce planting stock and then field tested. Later (e.g., after 8 years), the best families and clones could be propagated from the cold-stored juvenile shoots by remultiplying and rooting them. This is one of the major rationales at the FRI for doing micropropagation research.

The use of activated charcoal has revolutionized the micropropagation of mature trees. All of the credit is due to the French researchers for their methods, which have been successful when applied to the micropropagation of mature trees of other species (e.g., radiata pine). The micropropagation of trees from mature explants is expected to play an important role in New Zealand in the near future, producing trees for seed orchards as well as providing an excellent system for rejuvenation studies. Many researchers elsewhere hope to use micropropagation for clonal afforestation.

ACKNOWLEDGMENTS

We thank Kathy Horgan and Dale Smith for their helpful comments.

REFERENCES

Abo El-Nil, M. (1982). U.S. Patent 4,353,184, to Weyerhaeuser Company, Tacoma, Washington.

Aitken, J., Horgan, K. J., and Thorpe, T. A. (1981). Influence of explant selection on the shoot-forming capacity of juvenile tissue of *Pinus radiata*. *Can. J. For. Res.* **11,** 112–117.

Aitken-Christie, J., Horgan, K., and Smith, D. R. (1982). Micropropagation of radiata pine. Mikroformering af skovtraeer. *Ugeskr. Jordbrug* **19,** 375–382.

Ball, E. A. (1950). Differentiation in a callus culture of *Sequoia sempervirens*. *Growth* **14,** 295–325.

Ball, E. A., Morris, D. M., and Reydelius, J. A. (1978). Cloning of *Sequoia sempervirens* from mature trees through tissue culture. *In* "In Vitro Multiplication of Woody Species" (P. Boxus, ed.), pp. 181–226. Centre de Recherches Agronomiques de l'Etat Ministère de l'Agriculture, Gembloux.

Biondi, S., and Thorpe, T. A. (1982). Clonal propagation of forest tree species. *In* "Proceedings of the COSTED Symposium on Tissue Culture of Economically Important Plants" (A. N. Rao, ed.), pp. 197–204. COSTED and Asian Network of Biological Sciences, Singapore.

Bonga, J. M. (1977). Organogenesis in *in vitro* cultures of embryonic shoots of *Abies balsamea* (balsam fir). *In Vitro* **13,** 41–48.

Bonga, J. M. (1981). Organogenesis *in vitro* of tissues from mature conifers. *In Vitro* **17,** 511–518.

Bonga, J. M., and Durzan, D. J., eds. (1982). "Tissue Culture in Forestry." Nijhoff, The Hague.

Boulay, M. (1977). Multiplication rapide du *Sequoia sempervirens* en culture *in vitro. Ann. AFOCEL, 1977* pp. 37–66.

Boulay, M. (1979a). Multiplication and rapid cloning of *Sequoia sempervirens* by culture "in vitro." *In* "Micropropagation d'Arbres Forestiers," AFOCEL etudes et recherches, No. 12, 6/79, pp. 49–56. AFOCEL, Nangis.

Boulay, M. (1979b). In vitro propagation of Douglas fir by micropropagation of aseptic germination and culture of dormant buds. *In* "Micropropagation d'Arbres Forestiers," AFOCEL etudes et recherches, No. 12, 6/79, pp. 67–75. AFOCEL, Nangis.

Campbell, R. A., and Durzan, D. J. (1975). Induction of multiple buds and needles in tissue cultures of *Picea glauca. Can. J. Bot.* **53,** 1652–1656.

David, A. (1982). In vitro propagation of gymnosperms. *In* "Tissue Culture in Forestry" (J. M. Bonga and D. J. Durzan, eds.), pp. 72–108. Nijhoff, The Hague.

David, A., David, H., Faye, M., and Isemukali, K. (1979). The cultivation "in vitro" and the micropropagation of the maritime pine (*Pinus pinaster* Sol.). *In* "Micropropagation d'Arbres Forestiers," AFOCEL etudes et recherches, No. 12, 6/79, pp. 33–40. AFOCEL, Nangis.

Franclet, A., David, A., David, H., and Boulay, M. (1980). Première mise en évidence morphologique d'un rajeunissement de meristèmes primaires caulinaires de Pin maritime (*Pinus pinaster* Sol.). *C. R. Hebd. Seances Acad. Sci.* **290,** 927–929.

Gresshoff, P. M., and Doy, C. H. (1972). Development and differentiation of haploid *Lycopersicon esculentum* (tomato). *Planta* **107,** 161–170.

Horgan, K. J. (1982). The tissue culture of forest trees. *Proc. Int. Symp. Natl. Acad. Sci., 9th, 1982* pp. 105–120.

Horgan, K. J., and Aitken, J. (1981). Reliable plantlet formation from embryos and seedling shoot tips of radiata pine. *Physiol. Plant.* **53,** 170–175.

Karnosky, D. F. (1981). Potential for forest tree improvement via tissue culture. *BioScience* **31,** 114–120.

Mott, R. L. (1978). Tissue culture propagation of conifers. *In* "Propagation of Higher Plants through Tissue Culture—A Bridge between Research and Application" (K. W. Hughes, R. Henke, and M. Constantin, eds.), pp. 125–133. Tech. Inf. Cent., U.S. Dept. of Energy, Springfield, Virginia.

Mott, R. L. (1981). Trees. *In* "Cloning Agricultural Plants via *in Vitro* Techniques" (B. V. Conger, ed.), pp. 217–254. CRC Press, Boca Raton, Florida.

Mott, R. L., and Amerson, H. V. (1981). A tissue culture process for the clonal production of loblolly pine plantlets. *N.C., Agric. Res. Serv., Tech. Bull.* **271,** 1–14.

Murashige, T., and Skoog, F. (1962). A revised medium for rapid growth and bioassays with tobacco tissue cultures. *Physiol. Plant.* **15,** 473–497.

Poissonier, M., Franclet, A., Dumant, M. J., and Gautry, J. Y. (1980). Enracinement de tigelles *in vitro* de *Sequoia sempervirens. Ann. AFOCEL, 1980* pp. 231–253.

Quoirin, M., and Lepoivre, P. (1977). Etudes de milieux adaptes aux cultures *in vitro* de *Prunus. Acta Hortic.* **78,** 437–442.

Reilly, K. J., and Washer, J. (1977). Vegetative propagation of radiata pine by tissue culture: Plantlet formation from embryonic tissue. *N. Z. J. For. Sci.* **7,** 199–206.

Schenk, R. U., and Hildebrandt, A. C. (1972). Medium and techniques for induction and growth of monocotyledonous and dicotyledonous plant cell cultures. *Can. J. Bot.* **50,** 199–204.

Smith, D. R., Aitken, J., and Sweet, G. B. (1981). Vegetative amplification: An aid to optimizing the attainment of genetic gains from *Pinus radiata? In* "Proceedings of the Symposium on Flowering Physiology at the XVII IUFRO World Congress" (S. L. Krugman and M. Katsuta, eds.), pp. 117–123. Japan Forest Tree Breeding Association, Tokyo.

Smith, D. R., Horgan, K. J., and Aitken-Christie, J. (1982). Micropropagation of *Pinus radiata* for afforestation. *In* "Plant Tissue Culture 1982" (A. Fujiwara, ed.), pp. 723–724. Maruzen, Tokyo.

Sommer, H. E., and Brown, C. L. (1979). Application of tissue culture to forest tree improvement. *In* "Plant Cell and Tissue Culture: Principles and Applications" (W. R. Sharp, P. O. Larsen, E. F. Paddock, and V. Raghavan, eds.), pp. 451–491. Ohio State Univ. Press, Columbus.

Sommer, H. E., and Wetzstein, H. Y. (1984). Application of tissue culture to forest tree improvement. *Long Ashton Symp.* [*Proc.*] (in press).

Sommer, H. E., Brown, C. L., and Kormanik, P. P. (1975). Differentiation of plantlets in longleaf pine (*Pinus palustris* Mill.) tissue cultured *in vitro. Bot. Gaz. (Chicago)* **136,** 196–200.

Thorpe, T. A., and Biondi, S. (1983). Conifers. *In* "Handbook of Plant Cell Culture" (D. A. Evans, W. R. Sharp, and P. V. Ammirato, and Y. Yamada, eds.), Vol. 2. Macmillan, New York (in press).

von Arnold, S., and Eriksson, T. (1979). Induction of adventitious buds on buds of Norway spruce (*Picea abies*) grown *in vitro. Physiol. Plant.* **45,** 29–34.

CHAPTER **12**

Culture Methods for Bryophytes

Martin Bopp
Bernd Knoop

Botanical Institute
University of Heidelberg
Heidelberg, Federal Republic of Germany

I. CULTURE OF WHOLE PLANTS

One of the advantages of bryophyte cultures for physiological experiments is the possibility of keeping a great number of whole plants in small culture vessels under aseptic conditions. With the application of normal sterile techniques, there is usually no difficulty in maintaining such cultures aseptic for a long time, but the introduction of new wild material for sterile cultures needs some discussion.

A. Starting Material

1. Spores

The most suitable way to start a bryophyte culture is with spores. Inside the closed capsules, the spores usually are not contaminated by fungi or bacteria. If necessary, capsules can be surface sterilized (e.g., with 1% sodium hypochloride for 1–3 min or absolute ethanol for a few minutes).

CELL CULTURE AND SOMATIC CELL
GENETICS OF PLANTS, VOL. 1

ISBN 0-12-715001-3

They are then washed with sterile water and opened by a needle, scalpel, or forceps over a Petri dish with the culture substrate. Some light strokes on the holding forceps will scatter the spores. This should be done in an ultraviolet light-sterilized chamber rather than in a laminar airflow bench because the air stream would blow the spores away.

This method results in a relatively dense lawn of germinating protonemata, which later have to be isolated and transferred to new dishes as single plants if desired. Most species need light for spore germination.

Another way to gain single moss plants from spores is to distribute the capsule contents on a sterile microscope slide and to transfer single spores (under microscopic control) to the substrates by means of fine, freshly drawn glass needles to which the spores adhere. This method, however, requires manual skill and is less effective, because not all spores can germinate and sometimes spores are lost during transfer.

Especially when large numbers of dishes have to be inoculated to produce masses of plant material, one can distribute the contents of several spore capsules in sterile water or nutrient solution and spray them onto the substrates. In this case, the appropriate spore density first has to be adjusted. Usually a detergent must be added to keep the spores with their waxy surface suspended (e.g., polyvinyl–pyrrolidone, MW 25,000, 250 mg/liter).

The germination time of the spores can be quite different depending on the species and on environmental conditions such as light and temperature. A relevant table is given by Sussman (1965, p. 980).

2. Active Tissues

To start a sterile bryophyte culture from living turgescent tissues is more difficult for two reasons:

1. Surface sterilization of such tissues is usually not tolerated by the cells and leads to severe damage. As the vegetative bryophyte material is often only one or a few cells thick, the whole organ or plant will die easily. This applies to leaflets, thalli of Jungermaniales, and gemmae (organs of vegetative propagation).

2. Thicker plants or organs, e.g., the thalli of Marchantiales, stems, or sporophytes, may survive surface sterilization with the cells of inner layers. However, in our experience (with *Marchantia, Lunularia,* and *Fontinalis* shoots and buds), the contamination of wild material with fungal spores and bacteria often extends to the inner parts of the plant. In this case, one has to cut such surface-sterilized plants into very small pieces, culture them, and select uncontaminated pieces from these cultures. The best results will be obtained with the youngest parts of a plant similar to isolated meristems of higher plants (Basile, 1972).

The application of antibiotics may prove advantageous, but only a few relevant tests have been performed up to now. Once introduced, aseptic bryophyte cultures can easily be propagated by explants of vegetative material, because most of the living tissues and even single isolated cells are able to regenerate whole plants. It seems to be a general law that small explants redifferentiate to an earlier stage of ontogeny than larger ones regenerating a new plant.

The high regenerative capacity also makes it easy to build up clones. Such strains are often grown from mutants with special features, such as endogenous phytohormone deficiency or overproduction (Ashton *et al.*, 1979). It should, however, be mentioned that in several laboratories, vegetatively propagated clones of standard strains on agar showed clear deviations after continued passages, compared with plants derived from spores of the same harvest (spores of *Funaria* keep their germinability for 10 years or more; cf. Sussman, 1965). Similarly, a liquid culture of the same species, which originally grew as nearly pure chloronema, showed increasing amounts of spontaneously differentiating caulonema after 1 year. The reasons for these "degenerations" are unknown but have been sporadically described for higher plant tissues too (Street, 1977).

Long-term vegetative propagation thus bears the danger of sneaking alterations. This must be taken into account especially in the case of sterile races and mutants, which cannot be grown from spores.

B. Culture Vessels

Bryophytes as small plants can usually be cultured for the whole life cycle in Petri dishes, test tubes, or Erlenmeyer flasks on either solid or liquid substrates. Attention should be paid to ensure a sufficient gas exchange with the environment because the bryophytes are photosynthetic organisms. Although Parafilm or similar products allow gas exchange to a certain extent, a double layer of this material around a Petri dish can disturb plant development, probably through CO_2 deficiency. The resulting phenomena are regenerative processes similar to those induced by gasing the plants with CO_2-free air (Knoop, 1973).

With unsealed Petri dishes, however, another problem may arise. Especially in climate chambers with rapid air circulation, cultures often exhibit secondary contaminations which appear during a culture period of some weeks along the edge of the dish. This can be avoided by wrapping the dishes with transparent paper in order to prevent their direct exposure to the circulating air. The simultaneously reduced light intensity has to be compensated for by appropriately stronger illumination.

Under certain conditions, ion-free cultures are desired. In this case, one has to avoid glass dishes because, Ca^{2+} ions, for instance, are released from glass in an amount sufficient to support Ca^{2+}-dependent processes.

C. Physical Conditions

If the culture conditions of a bryophyte species are unknown, one should at first copy the average conditions of its natural habitat. However, most of the bryophytes cultured today for experimental use (only a small number of species) have remarkably uniform physical requirements.

1. Temperature

Optimal growing temperatures for bryophytes range from 10°C for a cold water moss such as *Fontinalis* (Glime and Acton, 1979) to more than 25°C for tropical species (Chopra and Rawat, 1973). Most species, especially the few used for experimental work, grow best at temperatures of around 20°C. In contrast to natural conditions, a day–night temperature rhythm should be avoided for technical reasons: All oscillations of more than ± 1°C (this may also occur with slow thermostats in climate chambers) will lead to water condensation under the lid of the culture dish, which not only prevents direct microscopic observation within the closed vessel but also favors secondary contamination by forming a liquid film around the zone of contact between the dish and the lid. Finally, the growth of protonema can be changed drastically by a liquid film formed by the dropping water.

2. Light

Except for some mosses usually growing in unshaded areas such as *Sphagnum* (Rudolph, 1978), most bryophytes have a low light saturation. In these cases, an energy flux rate of 1–3 W/m^2 suffices for optimal growth, whereas higher intensities often are harmful (e.g., with *Funaria, Polytrichum,* and *Marchantia*).

If the culture is to deliver "standardized" plant material of reproducible shape and development, close attention must be paid to provide a uniform light field. The intensity requirements of the individual species can be tested quite simply, e.g., by covering the cultures with varying layers of parchment paper as gray filters.

The moss *Funaria hygrometrica* grows and develops best under a daily light–dark rhythm (20:4 hr). However, as a consequence, its physiological

TABLE I

Culture Media for Some Bryophytes

Species	Ingredient								
	$NaNO_3$	Na_2HPO_4	K_2SO_4	KNO_3	KCl	K_2HPO_4	KH_2PO_4	$Ca(NO_3)_2 \times 4H_2O$	$CaCl_2 \times 2H_2O$
Hepatics									
Lunularia cruciata	—	—	—	—	—	—	0.1	—	0.1
Riccia fluitans	—	—	—	1.2	—	0.8	—	—	0.3
Marchantia polymorpha	—	—	—	—	—	—	0.1	—	0.1
Bazzania albicans	0.375	—	—	—	0.06	—	0.125	—	0.125
Sphaerocarpos donellii	—	—	—	—	—	0.1	—	—	0.1
Mosses									
Funaria hygrometrica	—	—	—	—	0.25	—	0.25	1.0	—
Polytrichum juniperum	—	0.05	0.25	—	—	—	0.20	0.25	—
Physcomitrella patens	—	—	—	—	—	—	0.1	—	0.1
Ceratodon purpureus	—	—	0.25	—	—	—	0.25	0.25	—
Physcomitrium turbinatum	—	—	—	—	—	—	0.2	—	0.1
Pohlia nutans	—	—	—	—	0.06	—	0.175	0.5	—
Aquatic bryophyte									
Riella helicophylla	—	—	—	—	0.025	—	0.025	0.1	—
Liquid culture of terrestrian bryophytes									
Sphaerocarpos donnellii	—	—	—	1.5	—	—	1.37	—	0.15
Pylaisiella selwynii	—	—	—	—	—	—	0.2	—	0.1
Gymnocolea inflata	—	—	—	—	—	—	0.2	—	0.1
Funaria hygrometrica I	—	—	—	0.813	0.25	—	0.25	0.05	—
Funaria hygrometrica II	—	—	0.1	0.77	0.25	—	0.22	0.05	—

[a] Point (·) means no data cited. Fe is sometimes part of trace elements.

performances such as rate of regeneration and bud formation behave rhythmically too.

As bryophytes respond to blue and red light at which the corresponding receptor for red light is phytochrome (Fredericq and De Greef, 1968; Gay, 1980), the light quality also plays a role in such cultures. It has, however, been found sufficient to illuminate the plants with fluorescent tubes of the daylight type, although all of these tubes are known to emit almost no far red as it is present in real daylight. The light direction is unimportant as long as the bryophytes are grown only for biochemical work on tissue extracts. For developmental studies, however, the cultures should be illuminated from the top, because of a pronounced phototropic response that may differ with the stage of development (Bopp, 1959).

3. Humidity

As aseptic cultures are kept in closed vessels, the humidity of the culture room need not be regulated.

Summarizing these three points, a climate chamber for the culture of bryophytes must be adjustable for light intensity, day length, and tempera-

$MgSO_4 \times 7H_2O$	NH_4NO_3	$FeCl_3 \times 6H_2O$[a]	$FeSO_4 \times 7H_2O$	Trace elements	Glucose	Agar (%)	pH	References
0.1	0.2	0.004	—	—	—	1–2	·	Nehira (1964)
0.3	—	Trace	—	—	—	1–2	·	Klingmüller (1959)
0.1	0.2	0.005	—	—	—	1.5	·	Nehira (1964)
0.125	—	Trace	—	—	—	1–2	·	Nehira (1964)
0.1	0.2	0.01	—	—	—	1.3	6.5–7	Kurz (1974)
0.25	—	Trace	—	—	—	2.0	5.5–6	Bopp (1952)
0.25	—	0.001	—	+	—	1.2	·	Kofler (1959)
0.1	0.2	0.001	—	—	—	2.0	6.8	Bopp and Fell (1976)
0.25	—	0.001	—	+	—	0.8	·	Schneider and Szweykowska (1974)
0.2	0.5		—	+	—	1.5	·	Nebel and Naylor (1968)
0.175	—	Trace	—	—	—	1–2	·	Nehira (1964)
0.025	—	—	—	—	—	—	·	Stange (1957)
0.25	—	·	—	+	20	—	5.0	Machlis (1962)
0.2	0.5	·	—	+	—	—	·	Spiess *et al.* (1971)
0.2	0.5	·	—	+	10	—	4.6	Basile and Basile (1980)
0.25	—	·	—	+	10	—	·	Johri (1974)
0.25	—	—	0.01	+	20	—	·	Saxena and Rashid (1981)

ture. Even the temperature requirement can be obtained without great expense. Because of the optimal temperature of about 20°C, an air-conditioning system used for offices and living rooms is sufficient, provided that the temperature does not oscillate by more than ± 1°C.

D. Substrates

Bryophytes are cultured as whole green photosynthetic plants. Therefore, culture media usually require no supplemental organic compounds such as sugars or vitamins. Generally, macrosalts, trace elements, and a species-dependent optimal pH are sufficient. On solid substrates the trace elements (except iron) can even be omitted unless the agar is highly purified. In liquid cultures, however, trace elements are essential, e.g., Hoagland, Hellers, or Murashige and Skoog (MS).

Whether solid or liquid substrates are used depends on the scientific question. The morphogenesis of a terrestrial moss should, of course, always be studied on a solid medium because thigmic stimuli, the fixed situation, and as a consequence a gravitational continuity strongly affect the development (Schmiedel and Schnepf, 1979). This phenomenon is clearly demonstrated, e.g., with *Riccia,* which grows well on substrates with both low and high water content but forms quite different shapes (Klingmüller, 1959). For mass production of plants to obtain extracts of lipids, secondary substances, enzymes, etc., liquid cultures (Naef and Simon, 1978) with or without shaking are sometimes more effective. Liquid cultures of *Funaria* do not bear shaking in our own experiments.

When mosses are grown on a solid substrate, a layer of cellophane on the medium prevents the plants from growing into it and facilitates their transfer from one medium to another without damage (Bopp *et al.*, 1964). As molecules up to MW 15,000 can pass this membrane, it is no barrier to nutrients from the substrate (we use Kalle Einmachcellophan, 20 μm thick, normally sold to seal jam jars).

The addition of sugars as a carbon source is usually not necessary in agar cultures except to accelerate growth and development (this does not always work) or for use in experiments without light. However, callus formation and the induction of apogamous sporophytes of mosses seem to depend on additional sugars (Menon and Lal, 1977).

In contrast to agar cultures, liquid media need the supplement of sugar to maintain a sufficient growth rate and density. This is perhaps due to a limited gas exchange in the culture flasks that results in a CO_2 deficiency and therefore insufficient photosynthesis.

The substrate compositions summarized in Table I have been found to allow satisfactory growth and development of the cited species, but they can also be used for other species.

The liquid culture of terrestrial mosses still requires some discussion. It may be started from spores or from growing tissues. When moss protonemata are used for initiation, only those that divide during culture into subunits, e.g., by fragmentation of the filaments at preformed sites, can be successfully cultivated.

The shorter these fragments are, the better and more homogeneous the culture will grow as fragments elongate only at new and mainly polar formed apical cells. The substrate and culture conditions should be adjusted accordingly.

In liquid cultures of the moss *F. hygrometrica,* for example, the absence or low concentration of calcium causes short fragments, higher concentrations, longer chloronema, and occasionally caulonema, which never show fragmentation (Saxena and Rashid, 1981).

Because all bryophytes have a high regeneration capacity, small parts of stem, thalli, or sporophytes can also grow out to new plants either by activation of resting tissues or by regeneration of single cells in liquid media (Stange, 1965).

II. PRODUCTION AND CULTURE OF PROTOPLASTS

In principle there is no methodological difference between protoplast production in bryophytes and higher plants (chapters 38–45, this volume). The protoplasts are released mechanically (Binding, 1966; Gay, 1976), or the walls are removed by enzymes. The optimal enzyme and buffer composition varies depending on the species (Bopp, 1984). In the moss *F. hygrometrica,* the following method has been successful (Bopp *et al.*, 1980): (1) preincubation and plasmolysis in glycine (5.0% w/v) at pH 8.0 for 10–15 min; (2) premaceration in glycine (5.0% w/v) with PATE (pectine transeliminase, 0.2% w/v, pH 8.0) for 2 hr; (3) maceration in glycine (5.0% w/v) with Cellulase Onozuka R-10 (5% w/v, pH 5.0) for 2–3 hr. All solutions were adjusted to the desired pH with KOH or HCl and sterile filtered through a 0.15-μm pore size filter (5 *M* Sartorius). To stop maceration, the enzymes were diluted with the plasmolyticum. The protoplasts were removed from the remaining parts of the protonema by centrifugation and transferred to Knop-agar, also containing 5.0% glycine. After formation of new cell walls, the substrate is changed to simple Knop-agar without plasmolyticum, on which a rapid and regular development of new protonemata starts (Batra and Abel, 1981).

The bryophyte protoplast has one great advantage over that of higher plants: its high capacity and readiness to regenerate. After the formation of new walls, the cells behave similarly to spores and germinate without extended callus formation to develop new plants. During this step, the substrate requirements are those of normal cultures without exogenous phytohormones or other organic substances (Burgess and Linstead, 1982).

Before the new cell wall is formed, the protoplasts can be fused by the aid of $CaCl_2$ or polyethylene glycol. In this way, somatic hybrids between different mutants could be obtained successfully (Grimsley *et al.*, 1977; Schieder, 1974).

Diploid and tetraploid regenerates appear even in the absence of fusion (Bopp *et al.*, 1980). These certainly derive from already endopolyploid cells of the original plant material (Knoop, 1978).

REFERENCES

Ashton, N. W., Cove, D. J., and Featherstone, D. R. (1979). The isolation and physiological analysis of mutants of the moss, *Physcomitrella patens,* which over-produce gametophores. *Planta* **144,** 437–442.

Basile, D. V. (1972). A method for surface-sterilizing small plant parts. *Bull. Torrey Bot. Club* **99,** 313–316.

Basile, D. V., and Basile, M. R. (1980) Ammonium ion-induced changes in form and hydroxyproline content of wall protein in the liverwort *Gymnocolea inflata. Am. J. Bot.* **67,** 500–507.

Batra, A., and Abel, W. (1981). Development of moss plants from isolated and regenerated protoplasts. *Plant Sci. Lett.* **20,** 183–189.

Binding, H. (1966). Regeneration und Verschmelzung nackter Laubmoosprotoplasten. *Z. Pflanzenphysiol.* **55,** 305–321.

Bopp, M. (1952). Entwicklungsphysiologische Untersuchungen an Laubmoosprotonemen. *Z. Bot.* **40,** 119–152.

Bopp, M. (1959). Versuche zur Analyse von Wachstum und Differenzierung des Laubmoosprotonemas. *Planta* **53,** 178–197.

Bopp, M. (1984). Developmental physiology of bryophytes. *In* "Manual of Bryology" (R. M. Schuster, ed.), pp. 276–324. Hattori Botanical Lab, Nishinan, Japan.

Bopp, M., and Fell, J. (1976). Manifestation der Cytokinin abhängigen Morphogenese bei der Induktion von Moosknospen. *Z. Pflanzenphysiol.* **79,** 81–87.

Bopp, M., Jahn, H., and Klein, B. (1964). Eine einfache Methode, das Substrat während der Entwicklung von Moosprotonemen zu wechseln. *Rev. Bryol. Lichenol.* **33,** 219–223.

Bopp, M., Zimmermann, S., and Knoop, B. (1980). Regeneration of protonema with multiple DNA content from isolated protoplasts of the moss *Funaria hygrometrica. Protoplasma* **104,** 119–127.

Burgess, J., and Linstead, P. J. (1982). Cell-wall differentiation during growth of electrically polarised protoplasts of *Physcomitrella. Planta* **156,** 241–248.

Chopra, R. N., and Rawat, M. S. (1973). In vitro production of secondary gemmae on the protonema of *Bryum klinggraeffii* Schimp. *Bryologist* **76,** 183–185.

Fredericq, H., and De Greef, J. (1968). Photomorphogenic and chlorophyll studies in the bryophyte *Marchantia polymorpha.* I. Effect of red, far-red irradiations in short and long-term experiments. *Physiol. Plant.* **21,** 346–359.

Gay, L. (1976). The development of leafy gametophytes from isolated protoplasts of *Polytrichum juniperinum* Willd. *Z. Pflanzenphysiol.* **79,** 33–39.

Gay, L. (1980). Etude de la régénération du gamétophyte feuillé des Polytrichacées. Thesis, Univ. of Lyon.

Glime, J. M., and Acton, D. W. (1979). Temperature effects on assimilation and respiration in the *Fontinalis duriaei - Periphyton* association. *Bryologist* **82,** 382–392.

Grimsley, N. H., Ashton, N. W., and Cove, D. J. (1977). The production of somatic hybrids by protoplast fusion in the moss, *Physcomitrella patens. Mol. Gen. Genet.* **154,** 97–100.

Johri, M. M. (1974). Differentiation of caulonema cells by auxins in suspension cultures of *Funaria hygrometrica.* "Plant Growth Substances 1973," pp. 925–933. Hirokawa Publ. Co., Tokyo.

Klingmüller, W. (1959). Zur Entwicklungsphysiologie der Ricciaceen. *Flora (Jena)* **147,** 76–122.

Knoop, B. (1973). Untersuchungen zum Regenerationsmechanismus bei *Funaria hygrometrica* Sibth. I. Die Auslösung der Caulonemaregeneration. *Z. Pflanzenphysiol.* **70,** 22–33.

Knoop, B. (1978). Multiple DNA contents in the haploid protonema of the moss *Funaria hygrometrica* Sibth. *Protoplasma* **94,** 307–314.

Kofler, L. (1959). Contribution à l'étude biologique des mousses cultivées in vitro: Germination des spores, croissance et développement du protonéma chez *Funaria hygrometrica. Rev. Bryol. Lichenol.* **28,** 1–202.

Kurz, E. H. (1974). Zur Regulation der Keimung beim Lebermoos *Sphaerocarpos.* Dissertation, Univ. of Heidelberg.

Machlis, L. (1962). The effects of mineral salts, glucose, and light on the growth of the liverwort *Sphaerocarpos donnellii*. *Physiol. Plant.* **15,** 354–362.

Menon, M. K. C., and Lal, M. (1977). Regulation of a subsexual life cycle in a moss. Evidence for the occurrence of a factor for apogamy in *Physcomitrium*. *Ann. Bot. (London)* [N. S.] **41,** 1179–1189.

Naef, J., and Simon, P. (1978). Etude de la croissance de protonémas de mousse (*Funaria hygrometrica*) en miliéu liquide. *Saussurea* **9,** 51–56.

Nebel, B. J., and Naylor, A. W. (1968). Initiation and development of shoot-buds from protonemata in the moss *Physcomitrium turbinatum*. *Am. J. Bot.* **55,** 33–37.

Nehira, K. (1964). Culture media for bryophytes. *Bull. Biol. Soc. Hiroshima Univ.* **31,** 15–21.

Rudolph, H. J. (1978). 15 Jahre Kultur von Sphagnum unter definierten Bedingungen: Eine Übersicht über Resultate, Probleme und Perspektiven. *Bryophytorum Bibl.* **13,** 279–309.

Saxena, P. K., and Rashid, A. (1981). High frequency regeneration of *Funaria hygrometrica* protoplasts isolated from low calcium protonemal suspension. *Plant Sci. Lett.* **23,** 117–122.

Schieder, O. (1974). Fusion zwischen Protoplasten aus Mutanten von *Sphaerocarpos donnellii* Aust. *Biochem. Physiol. Pflanz.* **165,** 433–435.

Schmiedel, G., and Schnepf, E. (1979). Side branch formation and orientation in the caulonema of the moss *Funaria hygrometrica:* Normal development and fine structure. *Protoplasma* **100,** 367–383.

Schneider, J., and Szweykowska, A. (1974). Changes in enzyme activities accompanying cytokinin-induced formation of gametophore buds in *Ceratodon purpureus*. *Z. Pflanzenphysiol.* **72,** 95–106.

Spiess, L. D., Lippincott, B. B., and Lippincott, J. A. (1971). Development and gametophore initiation in the moss *Pylaisiella selwynii* as influenced by *Agrobacterium tumefaciens*. *Am. J. Bot.* **58,** 726–731.

Stange, L. (1957). Untersuchungen über Umstimmungs- und Differenzierungsvorgänge in regenerierenden Zellen des Lebermooses *Riella*. *Z. Bot.* **45,** 197–244.

Stange, L. (1965). Zelldifferenzierung und Embryonalisierung bei dem Lebermoos *Riella*. *Ber. Dtsch. Bot. Ges.* **78,** 411–417.

Street, H. E. (1977). Old problems and new perspectives. *In* "Plant Tissue and Cell Culture" (H. E. Street, ed.), 2nd ed., pp. 501–511. Blackwell, Oxford.

Sussman, A. S. (1965). Physiology of dormancy and germination in the propagules of cryptogamic plants. *In* "Handbuch der Pflanzenphysiologie" (W. Ruhland, ed.), 2nd ed., Vol. XV, pp. 933–1025. Springer-Verlag, Berlin and New York.

CHAPTER **13**

Culture of Shoot Meristems: Pea*

K. K. Kartha

Plant Biotechnology Institute
National Research Council
Saskatoon, Saskatchewan, Canada

I. INTRODUCTION

Shoot apical meristems are located at the extreme growing tip of principal or lateral shoots and contribute to their continued vegetative growth. The apical meristem usually is a dome of tissue, 0.1 mm in diameter and 0.25–0.30 mm in length, consisting of different cell layers grouped into zones such as tunica, corpus, central mother cells, flank, and rib meristems. The constituent cells of apical meristems are less differentiated and are genetically more stable than those of other mature tissues (D'Amato, 1952; Partanen *et al.*, 1955), with the result that the progeny regenerated from *in vitro* cultured meristems exhibit greater genetic stability. Since plant regeneration can readily be induced from meristems cultured *in vitro*, and since the regenerated plants, in some cases, proliferate adventitiously into a large population, the meristem culture technique has found application in the commercial propagation of a wide range of plant species

*NRCC No. 22967.

CELL CULTURE AND SOMATIC CELL
GENETICS OF PLANTS, VOL. 1

ISBN 0-12-715001-3

(Murashige, 1974). Another important application of the technique is in the elimination of systemic viral infections (Chapter 65, this volume) and in the longer-term preservation of germplasm by cryogenic methods (Chapters 68 and 69, this volume).

Subsequent to the detection of internally seed-borne viral infection in large collections of pea (*Pisum sativum*) germplasm, a technique was developed to regenerate plants from *in vitro* cultured meristems of pea (Kartha *et al.*, 1974). This technique was successfully applied to the production of virus-free plants and to the preservation of meristems by cryogenic methods (Kartha *et al.*, 1979). The following account discusses the methods employed to regenerate plants from *in vitro* cultured meristems of pea isolated from germinated seeds.

II. STERILIZATION OF SEEDS

Seeds collected from field-grown plants usually carry externally a number of microorganisms; therefore, disinfection of seeds is the first step in establishing a culture. Seeds are thoroughly rinsed in 70% ethanol for 60 sec and surface sterilized with a 20% (v/v) solution of commercial bleach (1.2% sodium hypochlorite) for 20 min on a gyrotory shaker at 200 rpm, followed by four rinses with autoclaved distilled water. The seeds are then germinated aseptically in humidified glass jars on moistened cotton at 28°C in darkness. Immediately upon germination, the apical meristems measuring approximately 0.4–0.5 mm in length and containing a pair of leaf primordia are isolated from the emerging shoot.

III. CULTURE MEDIA

A. Shoot Regeneration Medium

The B5 medium (Gamborg *et al.*, 1968) has been found to be best suited for culturing pea meristems. The shoot regeneration medium is prepared by supplementing the B5 basal medium with 0.5 μM benzyladenine (BA) and 0.8% Difco Bacto agar. Prior to the addition of agar, the pH of the medium is adjusted to 5.7 with 0.1 N KOH. After the addition of agar, the medium is gently heated with constant stirring until the agar is completely dissolved. The medium (2.5 ml) is then dispensed into 10 × 1.2-cm Pyrex test tubes, plugged with absorbent cotton, and autoclaved at 120°C for 20 min. While dispensing medium into the tubes, one should take care that the medium does not come in contact with the cotton plug.

B. Root Regeneration Medium

The root regeneration medium consists of B5 basal medium, but with the media components reduced to half their strength and supplemented with 1.0 μM naphthaleneacetic acid (NAA) and 0.8% agar. Large screw-capped culture jars containing 30 ml medium are satisfactory for root regeneration purposes.

IV. ISOLATION OF MERISTEMS

The isolation of meristems is to be performed in an aseptic environment and preferably in a laminar airflow cabinet. It can also be carried out under semisterile conditions, if so warranted, but the contamination of cultures might outweigh any benefits of the latter approach.

A. Requirements

1. A stereomicroscope with adequate light sources.
2. A pair of knives, needles, and forceps. The knives should have very sharp, thin cutting edges. These are easily made by mounting razor blade chips on stainless steel holders. Similarly, sewing needles are mounted on stainless steel holders.
3. A stack of 10–15 sheets of autoclaved absorbent paper tissues of convenient size.
4. Two sterile 80- or 100-ml beakers, one containing 30–40 ml 70% ethanol and the other a similar volume of autoclaved distilled water.

B. Procedure

Wipe the microscope, especially the stage and the control knobs, with a clean sterile paper towel moistened with 70% ethanol. This process should be repeated before each operation and as frequently as possible. Extreme care should be taken to ensure that ethanol does not penetrate into the optical components. All the dissection instruments should be sterilized in absolute ethanol immediately prior to the commencement of the experiment. Passing the instruments on the flame is not recommended. Subsequent sterilization, to be repeated as often as possible, should be carried out by immersing the instruments briefly in 70% ethanol contained in the beaker, followed by immersion in a beaker containing autoclaved distilled water. The instruments are then kept dry in folds of sterile tissue paper. The process is continued until the meristem isolation is completed.

The emerging shoot from the germinated seed is aseptically removed with a pair of forceps and held under the microscope, using a convenient magnification. The outer whorls of the leaves are gently removed, one at a time, with the aid of the needle held in the other hand. The process is repeated until the meristematic dome and the first pair of leaf primordia are reached. A V-shaped cut is applied with the knife approximately 0.4–0.5 mm below the tip of the dome so as to include the pair of primordia and some subjacent tissue. The explant is immediately removed and aseptically inoculated onto the culture medium contained in the tubes. The cotton plugs are then wrapped with strips of Parafilm, and the cultures are incubated under appropriate physical environmental conditions. Efficient shoot regeneration is accomplished at a diurnal temperature of 20°C day/15°C night, 16-hr photoperiods at 4000 lx combined intensity from fluorescent and incandescent lights, and 60% relative humidity. The same conditions are optimal for efficient root regeneration as well.

V. RESULTS AND CONCLUSIONS

The size of the meristem explants has a direct bearing on the success of the plant regeneration process; the larger the meristems, the greater the proportion of plants regenerated, and vice versa. In usual practice, a meristem explant of 0.4–0.5 mm responds well to the imposed *in vitro* conditions, resulting in high frequency plant regeneration, provided that at least a pair of leaf primordia and some subjacent tissues are included on the explant. Contrary to the established principles of plant regeneration *in vitro* as influenced by a critical balance of exogenously applied cytokinins and auxins, pea meristems differentiate only into shoots, irrespective of the application of cytokinins alone or in combination with auxins (Kartha *et al.*, 1974). Whole plant regeneration, therefore, has to be accomplished in two sequential steps: (1) induction of shoots from cultured meristems and (2) induction of roots on the differentiated shoots. In order to induce shoot regeneration, only the cytokinin BA at a level of 0.5 μ*M* was necessary. The meristems turn green within 2–3 days, followed by swelling and slight callus formation from the basal cut ends within 20 days. All the meristems (100%) developed into well-formed shoots after about 3 weeks of culture. The same morphogenetic responses were observed after the meristems were cultured on media supplemented with various concentrations and combinations of BA and NAA. However, if the meristems are cultured using 1 μ*M* NAA as the only hormone, about 50% of them develop into complete plantlets. Since the plant regeneration frequency was less than optimum, this approach was not pursued further; instead, the two-step method was followed. A pronounced effect of temperature on the plant regeneration frequency was also observed. At a constant temperature of 26°C, only 80% of the meristems regenerated shoots (Kartha *et al.*, 1974), whereas with a diurnal regime of 20°C day/15°C night, 100% of the meristems formed shoots (Kartha *et al.*, 1979). Prolonged incubation of shoots on the same shoot regeneration medium or on basal medium supplemented with various concentrations of NAA did not lead to root differentiation. On the other hand, by reculturing the regenerated shoots, 2–3 cm long, on the root regeneration medium (0.5 B5 + 1.0 μ*M* NAA), root formation could readily be achieved in about 2 weeks and the plantlets could be successfully transferred to soil and grown to maturity.

The technique described here has been successfully extended to the elimination of internally seed-borne viral infection (pea seed-borne mosaic virus) from over 200 genetic lines of *Pisum sativum* and also to the long-term storage of pea meristems by cryogenic methods.

REFERENCES

D'Amato, F. (1952). Polyploidy in the differentiation and function of tissues and cells in plants. *Caryologia* **4**, 311–358.

Gamborg, O. L., Miller, R. A., and Ojima, K. (1968). Nutrient requirements of suspension cultures of soybean root cells. *Exp. Cell Res.* **50**, 151–158.

Kartha, K. K., Gamborg, O. L., and Constabel, F. (1974). Regeneration of pea (*Pisum sativum* L.) plants from shoot apical meristems. *Z. Pflanzenphysiol.* **72**, 172–176.

Kartha, K. K., Leung, N. L., and Gamborg, O. L. (1979). Freeze-preservation of pea meristems in liquid nitrogen and subsequent plant regeneration. *Plant Sci. Lett.* **15**, 7–15.

Murashige, T. (1974). Plant regeneration through tissue cultures. *Annu. Rev. Plant Physiol.* **25**, 135–166.

Partanen, I., Sussex, I. M., and Steeves, T. A. (1955). Nuclear behavior in relation to abnormal growth in fern prothalli. *Am. J. Bot.* **42**, 245–256.

CHAPTER 14

Culture of Shoot Meristems: Fruit Plants

Olivia C. Broome
Richard H. Zimmerman

Fruit Laboratory
Agricultural Research Service
United States Department of Agriculture
Beltsville, Maryland

I. INTRODUCTION

Meristem tip culture of strawberries was originally developed to produce virus-free plants (Belkengren and Miller, 1962). Later, the technique was used as the starting point for large-scale micropropagation of this crop (Boxus, 1974). The methodology has since been adapted for the micropropagation of other fruit crops. When the goal includes producing pathogen-free plants, one must remember that meristem tip culture by itself does not assure freedom from virus or other disease organisms; pathogen status must be established by appropriate testing of the resulting plants. The methods detailed here are those used satisfactorily for rapid micro-

CELL CULTURE AND SOMATIC CELL
GENETICS OF PLANTS, VOL. 1

ISBN 0-12-715001-3

propagation of mature plant materials of apple (*Malus domestica* Borkh.), thornless blackberry (*Rubus* sp.), highbush blueberry (*Vaccinium* sp.), and strawberry (*Fragaria* × *ananassa* Duch.) in this laboratory. Techniques used elsewhere for these crops, or ones used for other fruit crops, will not be described here. Extensive references to tissue culture of fruit crops have been summarized recently (Lane, 1982; Zimmerman, 1983).

II. SOURCE AND CONDITION OF EXPLANT MATERIAL

In the cases mentioned here, "shoot tip explant" refers to a meristematic tip 1–2 cm long, and "meristem tip explant" indicates a tip 1 mm long or less. Lengths between 1 mm and 1 cm are not used because they offer fewer axillary buds than the longer explant, and the use of lengths greater than 1 mm defeats the purpose of selecting a meristem tip, i.e., the possible exclusion of viruses or other microorganisms that could be carried in the vascular system or tissues older than the meristematic tip.

Selection of explant material varies with the plant used. The apple explants used are either the tips of actively growing shoots or meristem tips from lateral buds. The active shoots are taken from orchard trees during spring and summer or collected throughout the year from greenhouse trees. The buds for meristem tips are usually taken from dormant trees in the orchard. Explants for thornless blackberry and highbush blueberry are active shoot tips from field or greenhouse plants. For strawberry, meristem tips are removed from the runners of greenhouse-grown plants. All greenhouse plants are grown under long days to maintain active growth, water is kept off the leaves and shoots for sanitary reasons, and plants are sprayed, as needed, for disease and pest control. Whenever terminal bud set occurs on the woody plants, new growth for tissue culturing is stimulated by pruning.

III. TISSUE COLLECTION AND PREPARATION FOR CULTURE

A. Shoot Tip Explants

Shoots are collected only when actively growing. Apple and blackberry shoots 7–10 cm long are collected in plastic bags; 2- to 7-cm-long blueberry shoots are collected in distilled water since they wilt readily. During collection and further handling, surgical gloves and instruments cleaned with 75% ethanol are used to decrease the possibility of contamination by human-carried organisms. In the laboratory, leaves are carefully removed at the bases of their petioles, leaving a few tightly furled leaves enclosing the apex. When shoot tips are collected in the field, rinsing them under running tap water for 30–90 min (Jones *et al.*, 1979) facilitates cleanup. Otherwise, shoot tips are washed with detergent (2 drops/100 ml water) for about 1 min, briefly rinsed in distilled water, and disinfested for 10 (blueberry) to 20 (apple and blackberry) min using calcium hypochlorite [$Ca(ClO)_2$] with 0.01% polyoxyethylene (20) sorbitan monolaurate (PSM) as the wetting agent. A saturated solution of $Ca(ClO)_2$ is prepared by mixing 60 g in 1 liter of distilled water, stirring for 15 min, and filtering under vacuum. Before being used, the solution is diluted to half strength. Although apple and blackberry shoots can be agitated during cleanup using a magnetic stirrer, the stirring bar damages blueberry tissues, resulting in eventual death of the explant. Therefore, blueberry shoots are gently agitated by hand. The sterilant is removed by several brief rinses of sterile distilled water. After disinfestation, the shoot tips are trimmed to their final size and transferred to establishment medium. Shoot tip explants are placed horizontally on the medium and slightly embedded to permit maximum tissue–medium contact.

B. Meristem Tip Explants

Two distinctly different techniques are used in this laboratory for extracting meristem tips from apple and strawberry.

1. Apple

Apple meristem tips are dissected from lateral buds of both active and dormant shoots. Dormant shoots are cut, just above each bud, into one-node sections. These stem sections are agitated for 10 min in 95% ethanol plus PSM, 1 drop/100 ml, then for 20 min in a saturated $Ca(ClO)_2$ solution with 0.01% PSM. After being drained and briefly rinsed with sterile water, the buds are ready for dissection. Holding the basal end of the stem with gloved fingers rather than with forceps aids manipulation during dissection, and the instruments are changed frequently so that only a sterile surface touches the bud. Portions of the outer bud scales and thin layers of bark are removed by longitudinal cuts (numbers 1–3 in Fig. 1A and B), and shorter trimming cuts (numbers 4–6 in Fig. 1B) remove additional bark and portions of the outer bud scales. Then a shallow cut across the base of the

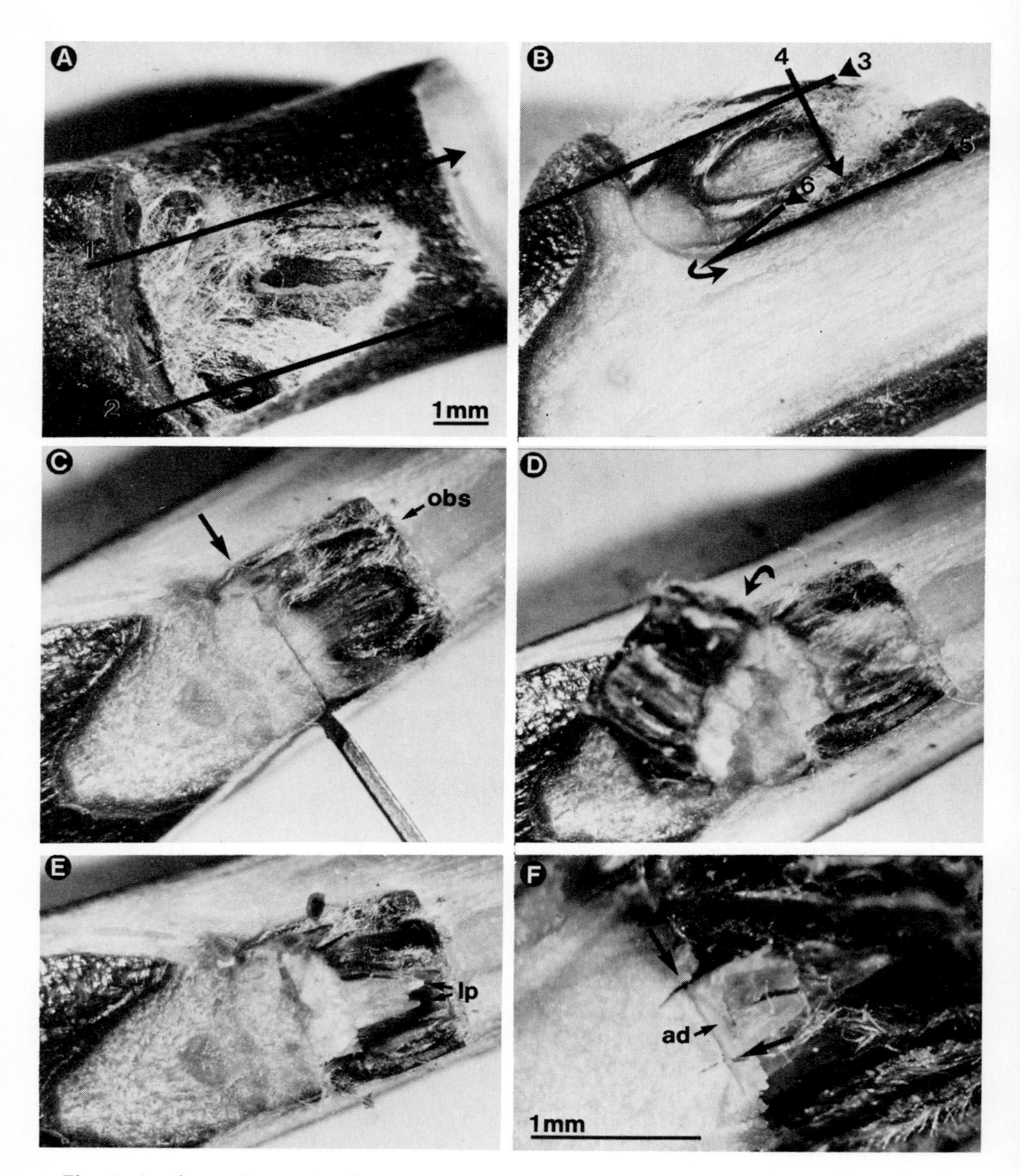

Fig. 1. Apple meristem tip dissection. (A) Intact dormant bud on a stem piece. Arrows indicate the first two cuts of the dissection. (B) Appearance of the bud from the side after cuts 1 and 2. Arrows show the location and sequence of cuts 3–6. (C) Remains of the outer bud scales (obs) after cuts 1–6. A shallow cut is made across the base of the outer bud scales. (D) Removal of scales. (E) Shallow cuts and scale removal are repeated until leaf primordia (lp) are visible. (A–E are depicted at the same scale.) (F) Enlargement shows the meristem tip explant containing an apical dome (ad) plus a few leaf primordia, delineated by cuts.

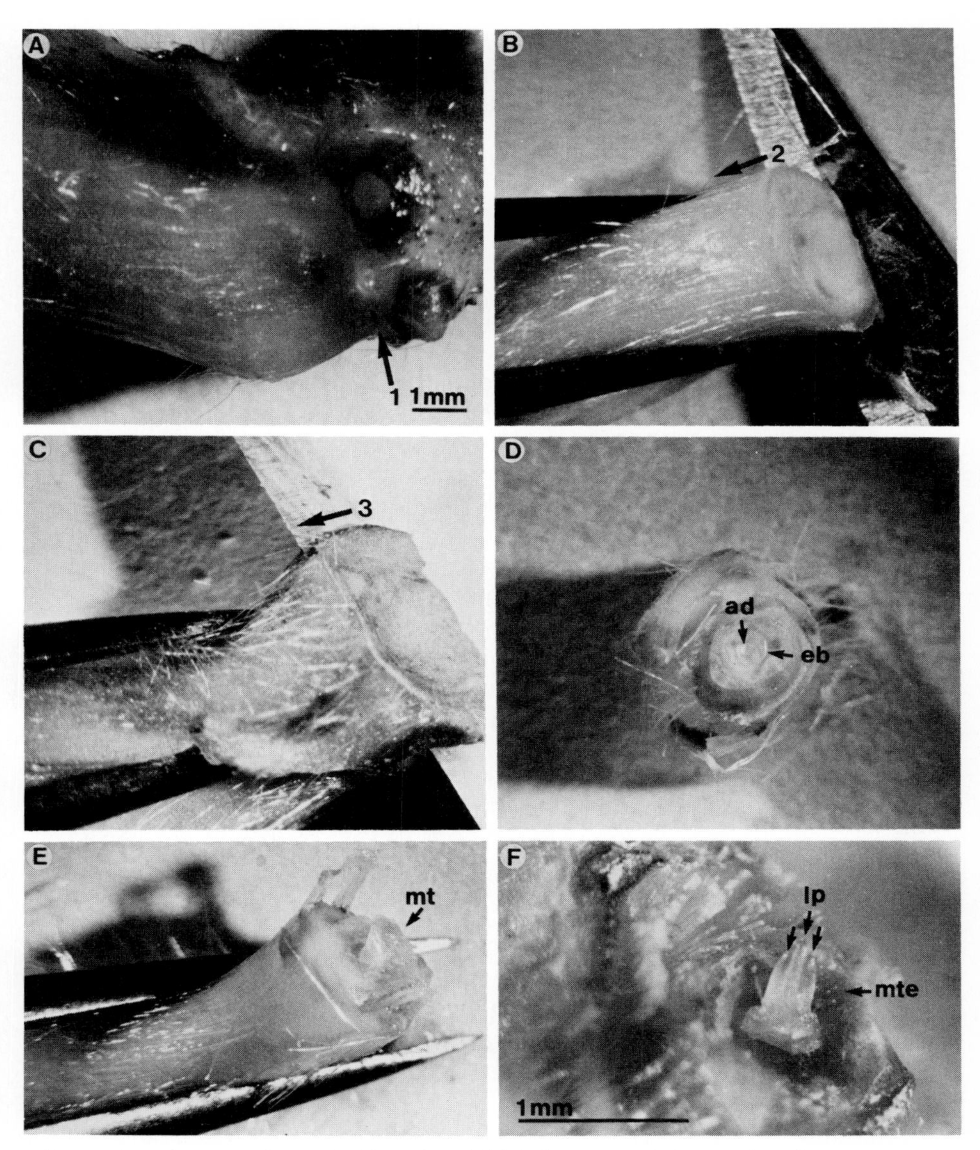

Fig. 2. Strawberry meristem tip dissection. (A) Meristem-containing portion of a young runner plantlet with the leaves removed, apical end to left. The protuberances are young roots. The arrow indicates the site of the first cut, a cross section just above the roots. (B) Cut 2 removes a thin longitudinal slice from the side. (C) A cross-sectional cut is made starting from the side already cut. (D) Cut 2 is repeated once and cut 3 several times, taking very thin slices, until the apical dome (ad) is apparent. Then the circumference of the explant base (eb) is lightly scored. (E) The meristem tip (mt) is lifted out from the surrounding tissue. (A–E are depicted at the same scale.) (F) Enlargement shows the upright meristem tip explant (mte) with the apical dome enclosed in several leaf primordia (lp).

bud scales on the front of the bud (Fig. 1C) and another on the back of the bud will allow one or more pairs of bud scales to be removed (Fig. 1D). This process is repeated to expose the leaf primordia surrounding the apical dome (Fig. 1E). After a few of these primordia are removed, the meristem tip explant (Fig. 1F), composed of the apical dome and two or more leaf primordia, can then be excised and explanted.

2. Strawberry

Strawberry runners are usually collected soon after roots appear on the new runner plant (Fig. 2A). After the outer leaves are removed, the runner tips are held between layers of moist paper towels until cleanup. The runner tip plus a short piece of the stolon is agitated for a few seconds in 0.5% NaOCl plus PSM, 2 drops/100 ml, and drained on paper toweling. During dissection, watch makers forceps and no. 11 scalpel blades are exchanged frequently for sterile ones. Dissection proceeds by trimming off thin slices of tissue on one side on the young crown (Fig. 2B) and by removing cross sections sequentially from the base (Fig. 2C). If the microscope light is placed so that it illuminates the cut surface at about 45° with respect to the tissue surface, the separation of concentric tissue layers is distinctive (Fig. 2D). When the apical dome is only slightly below the surface, it is visible as a pearlescent zone (Fig. 2D). Once the dome is visible, the explant can be extracted by removing very thin cross sections until a cut frees the meristem tip, or by scoring around an area including the apical dome and a few leaf primordia and lifting out the explant (Fig. 2E). The latter method yields a slightly larger explant (Fig. 2F). Then the meristem tip is explanted to the surface of an agar medium; using medium cooled on a slant facilitates release of the tissue from the scalpel blade.

IV. MEDIUM PREPARATION AND CULTURE CONTAINERS

A. Medium Preparation

Because several different medium formulas are used for the different explants, single-salt stock solutions of the macronutrients are convenient. Some stocks are light or temperature sensitive and must be stored in amber glass and/or refrigerated (Huang and Murashige, 1976). In addition, the container must be considered since certain plastics impart phytotoxic plasticizers to solutions stored in them (Robbins and Hervey, 1974). Micronutrients, except iron, are supplied as 100× combined stocks. Iron chelate and all organic stocks are prepared separately. The only solutions which are not 100% aqueous are N^6-γ,γ-dimethylallylamino purine (2iP), 3-indoleacetic acid (IAA), and 3-indolebutyric acid (IBA). Enough cytokinin for a 100-ml solution is first dissolved in 1 ml of 1 *N* HCl, and enough auxin is dissolved in 2 ml 95% ethyl alcohol for a 100-ml solution before each is diluted to volume. Distilled water is used throughout stock solution and medium preparation. Aliquots of stock solutions and the carbon source are added to distilled water. The solution is then brought up to the desired volume with more distilled water. The pH of the medium is adjusted with 1 *N* KOH or 1 *N* HCl prior to adding agar whenever agar is used. All components are autoclaved at 1.1 kg/cm^2 for 15 min.

B. Culture Containers

Of the many possible culture containers, the following have optimum openings through which to work and have proven to be the most useful in this laboratory: for meristem tip and shoot tip establishment, 25 × 75-mm scintillation vials and 34 × 60-mm French squares, both closed with autoclavable plastic screw caps; for shoot tips in liquid medium, 125-ml widemouth Erlenmeyer flasks capped with aluminum foil; for proliferation and mass rooting, medium (125-ml) and large (500-ml) straight-side, widemouth jars capped with glass Petri dish lids or bottoms and held in place with polyvinylchloride film; and for rooting single shoots, 25 × 95-mm shell vials and 25 × 150-mm test tubes closed with polypropylene caps. A particular container is chosen with respect to the size and number of explants inserted and anticipated explant development in that container.

V. CULTURE ESTABLISHMENT, PROLIFERATION, AND ROOTING

A. Establishment

The establishment media for meristem tips contain lower concentrations of minerals and growth regulators than those for shoot tip explants. Apple meristem tips are established on a medium including the Lepoivre macronutrients (Quoirin *et al.*, 1977), Murashige and Skoog (MS) micro-

TABLE I

Media for Establishment and Proliferation of Fruit Plant Cultures

Component	Apple meristem: establishment[a]		Apple and blackberry shoot tip: establishment and proliferation[b]		Strawberry meristem				Blueberry shoot tip: establishment and proliferation			
					Establishment[c]		Proliferation[d]		Zimmerman[e]		WPM[f]	
	mM or μM	mg/liter	mM or μM	mg/liter	mM or μM	mg/liter	mM or μM	mg/liter	mM or μM	mg/liter	mM or μM	mg/liter
NH_4NO_3	5.0 mM	400	20.6 mM	1650	—	—	20.6 mM	1650	2 mM	160	5.0 mM	400
$(NH_4)_2SO_4$	—	—	—	—	0.95 mM	125	—	—	1.5 mM	198	—	—
$Ca(NO_3)_2$ $4H_2O$	5.1 mM	1200	—	—	2.12 mM	500	4.2 mM	1000	3 mM	708	2.4 mM	556
$CaCl_2$ $2H_2O$	—	—	2.99 mM	440	—	—	—	—	—	—	0.7 mM	96
KNO_3	17.8 mM	1800	18.8 mM	1900	—	—	2.5 mM	250	2 mM	202	—	—
K_2SO_4	—	—	—	—	—	—	—	—	—	—	0.6 mM	990
KH_2PO_4	1.98 mM	270	1.25 mM	170	0.92 mM	125	1.8 mM	250	3 mM	408	1.25 mM	170
KCl	—	—	—	—	1.68 mM	125	—	—	—	—	—	—
$MgSO_4$ $7H_2O$	1.46 mM	360	1.50 mM	370	0.51 mM	125	1.0 mM	250	1.5 mM	370	1.5 mM	370
$FeSO_4$ $7H_2O$	0.1 mM	27.8	0.1 mM	27.8	1.3 μM	0.375	0.1 mM	27.8	0.2 mM	55.7	0.1 mM	27.8
Na_2EDTA	0.1 mM	37.2	0.1 mM	37.2	1.0 μM	0.375	0.1 mM	37.2	0.2 mM	74.4	0.1 mM	37.2
$MnSO_4$ H_2O	0.1 mM	16.9	0.1 mM	16.9	0.5 μM	0.08	0.1 mM	16.9	0.1 mM	16.9	0.1 mM	16.9
$ZnSO_4$ $7H_2O$	0.03 mM	8.6	0.03 mM	8.6	—	—	0.03 mM	8.6	0.03 mM	8.6	0.03 mM	8.6
H_3BO_3	0.1 mM	6.2	0.1 mM	6.2	0.5 μM	0.03	0.1 mM	6.2	0.1 mM	6.2	0.1 mM	6.2
KI	5.0 μM	0.83	5.0 μM	0.83	0.36 μM	0.06	5.0 μM	0.83	5.0 μM	0.83	—	—

$Na_2MoO_4 \cdot 2H_2O$	1.0 μM	0.25	1.0 μM	0.25	0.01 μM	0.0025	1.0 μM	0.25	1.0 μM	0.25	1.0 μM	0.25
$CoCl_2 \cdot 6H_2O$	0.1 μM	0.025	0.1 μM	0.025	0.02 μM	0.005	0.1 μM	0.025	0.1 μM	0.025	—	—
$CuSO_4 \cdot 5H_2O$	0.1 μM	0.025	0.1 μM	0.025	—	—	0.1 μM	0.025	0.1 μM	0.025	1.0 μM	0.25
myo-Inositol	0.56 mM	100	0.56 mM	100	—	—	0.56 mM	100	0.56 mM	100	0.56 mM	100
Nicotinic acid	4.06 μM	0.5	—	—	8.1 μM	1.0	4.1 μM	0.5	—	—	4.06 μM	0.5
Pyridoxine HCl	2.43 μM	0.5	—	—	0.97 μM	0.2	2.43 μM	0.5	—	—	2.43 μM	0.5
Thiamine HCl	1.2 μM	0.4	1.2 μM	0.4	0.59 μM	0.2	0.30 μM	0.1	1.2 μM	0.4	2.96 μM	1.0
Adenine sulfate $2H_2O$	—	—	—	—	—	—	—	—	0.2 mM	80	—	—
Glycine	—	—	—	—	—	—	27 μM	2.0	—	—	27 μM	2.0
Ca pantothenate	—	—	—	—	0.04 μM	0.02	—	—	—	—	—	—
Biotin	—	—	—	—	0.08 μM	0.02	—	—	—	—	—	—
BA	0.44 μM	0.1	4.44 μM	1.0	—	—	4.44 μM	1.0	—	—	—	—
2iP	—	—	—	—	—	—	—	—	73.8 μM	15	73.8 μM	15
IAA	—	—	—	—	—	—	—	—	22.8 μM	4	22.8 μM	4
IBA	0.05 μM	0.01	0.49 μM	0.1	—	—	2.46 μM	0.5	—	—	—	—
Gibberellic acid, K salt	—	—	1.3 μM	0.5	—	—	0.26 μM	0.1	—	—	—	—
Glucose	—	—	—	—	167 mM	30,000	222 mM	40,000	—	—	—	—
Sucrose	87.6 mM	30,000	87.6 mM	30,000	—	—	—	—	87.6 mM	30,000	58.4 mM	20,000
Agar (Difco Bacto)	—	7,000	—	7,000	—	8,000	—	—	—	—	—	—
Agar (Phytagar)	—	—	—	—	—	—	—	4,800	—	5,000	—	4,800
pH (without agar)	5.0		5.2		4.3		5.7		4.8		5.2	

[a] Lepoivre (Quoirin *et al.*, 1977) macronutrients; MS (1962) micronutrients; Walkey (1972) vitamins.
[b] MS (1962) macro- and micronutrients.
[c] McGrew medium, modified half-strength Knop's macronutrients (Galzy, 1970); pH not adjusted.
[d] Boxus (1974) macronutrients + NH_4NO_3; MS (1962) micronutrients.
[e] Zimmerman and Broome (1980b).
[f] Lloyd and McCown (1980); substituted Phytagar.

nutrients (1962), and the vitamins of Walkey (1972) (Table I). Apple and blackberry shoot tips are established on a modified MS medium (Table I). Establishment of shoot tips is facilitated by a few days' culture in flasks of liquid medium tilted on their sides and rotating in a vertical plane. During this time, the tissue is given maximum contact with nutrients and contaminants show up readily. Additional time in liquid results in vitrification of apple tissue but favors rapid growth of blackberry shoots. However, geotropic response causes the blackberry shoots to develop as intertwined masses. Although blueberry shoot tips react adversely to any time spent rotating in liquid medium, they will both establish and proliferate on a gelled medium with the same composition for both stages. Also, their growth is similar on either of two different media, both containing much less nitrogen than MS (Table I). Strawberry meristem tips are established on half-strength modified Knop's medium plus micronutrients and vitamins, glucose, agar, and no hormones (McGrew, 1980). Standard growing conditions for all cultures are 16 hr of deluxe warm white fluorescent light at 50–80 $\mu Em^{-2}s^{-1}$ at a constant 25 ± 1°C.

B. Proliferation

Once an explant shows leaf expansion and shoot elongation, it is transferred to a proliferation medium that contains higher concentrations of minerals and growth regulators (Table I). (In the case of strawberry, the developing shoot is divided in two by severing the meristem tip. When a plant develops from the severed meristem tip or from a lateral bud of the remaining part, it must be tested and confirmed to be virus free before the other part is proliferated.) During proliferation, tissue is subdivided and transferred to fresh medium at specific intervals, usually 3–4 weeks. After that time, apple shoots begin to senesce, exhibiting terminal dormancy, leaf yellowing and drop, and a darkened petiole abscission zone. The other fruits can tolerate a 1- to 3-week delay in transfer before the culture deteriorates, or cultures can be held at 4°C until they can be transferred. The tissue transferred can be a shoot, shoot tip, bud, stem section, or stem base. Roughly, sufficient explants are transferred to cover up to one-fourth of the surface area of the medium. Cultures are subdivided numerous times to provide quantities of propagules for rooting.

C. Rooting

Apple, blackberry, and blueberry shoots 1–2 cm long are collected from proliferating cultures for rooting. Attention is given to choosing a shoot that is not senescing (see section V,B) and to cutting the base just below a node. The standard rooting medium for apple is the same as for proliferation (Table I), but with the minerals and sucrose reduced by half and no benzyladenine (BA) or gibberellic acid. The optimal response for the apple cultivars examined has been to a low concentration of auxin, 4.9×10^{-7} *M* to 1.47×10^{-6} *M* IBA (Zimmerman and Broome, 1980a). Many factors affecting *in vitro* rooting of apple have recently been summarized (Zimmerman, 1984). In contrast to apple, rooting blackberry and blueberry cuttings directly under mist after treating the cuttings with IBA on talc is simpler and more effective than rooting *in vitro* (Zimmerman and Broome, 1980b,c). Strawberry crowns and clumps of crowns root readily *in vitro* on the proliferation medium minus cytokinin (Boxus, 1974; Damiano, 1977), although the IBA concentration should be reduced by half for the everbearing type of strawberry.

VI. ACCLIMATIZATION

Once an entire plantlet has been produced *in vitro*, it cannot simply be moved to low-humidity air because of the resulting desiccation. From 1 to 2 weeks in a fog tent or under mist provides an adequate transition for strawberry plants. Apple plants require a more gradual acclimatization. Removing the test tube closure exposes the rooted shoot gently to open air at standard culture conditions. After a few days, the plant is sufficiently conditioned so that it can withstand being transplanted to a porous artificial soil mix and moved to a humid chamber at 65% relative humidity and 21–29°C. After 1–2 weeks in the humid chamber, plants are ready to be transferred to the greenhouse. Blackberry and blueberry plants from *in vitro* cuttings that are rooted directly under mist require no acclimatization period.

REFERENCES

Belkengren, R. O., and Miller, P. W. (1962). Culture of apical meristems of *Fragaria vesca* strawberry plants as a method of excluding latent A virus. *Plant Dis. Rep.* **46,** 119–121.

Boxus, P. (1974). The production of strawberry plants by *in vitro* micro-propagation. *J. Hortic. Sci.* **49,** 209–210.

Damiano, C. (1977). La ripresa vegetativa di piantine di fragola provenienti da colture in vitro. *Frutticoltura* **39**(2), 3–7.

Galzy, R. (1970). Recherches sur la croissance de la vigne saine et court-nouée cultivée "in vitro." Ph.D. Thesis, Univ. of Clermont.

Huang, L., and Murashige, T. (1976). Plant tissue culture media: Major constituents, their preparation and some applications. *Tissue Cult. Assoc. Man.* **1,** 539–548.

Jones, O. P., Pontikis, C. A., and Hopgood, M. E. (1979). Propagation *in vitro* of five apple scion cultivars. *J. Hortic. Sci.* **54,** 155–158.

Lane, W. D. (1982). Tissue culture and *in vitro* propagation of deciduous fruit and nut species. *In* "Application of Plant Cell and Tissue Culture to Agriculture and Industry" (D. T. Tomes, B. E. Ellis, P. M. Harney, K. J. Kasha, and R. L. Peterson, eds.), pp. 163–186. Univ. of Guelph, Guelph, Ontario.

Lloyd, G., and McCown, B. (1980). Commercially feasible micropropagation of mountain laurel, *Kalmia latifolia,* by use of shoot-tip culture. *Proc. Intern. Plant Prop. Soc.* **3,** 367–396.

McGrew, J. R. (1980). Meristem culture for production of virus-free strawberries. *U.S., Dep. Agric., Sci. Educ. Admin.* **ARR-NE-11,** 80–85.

Murashige, T., and Skoog, F. (1962). A revised medium for rapid growth and bioassays with tobacco tissue cultures. *Physiol. Plant.* **15,** 473–497.

Quoirin, M., Lepoivre, P., and Boxus, P. (1977). Un premier bilan de 10 années de recherches sur les cultures de meristèmes et la multiplication "in vitro" de fruitiers ligneux. *In* "Compte Rendu des Recherches, 1976-1977," pp. 93–117. Station des Cultures Fruitières et Maraichères, Gembloux.

Robbins, W. J., and Hervey, A. (1974). Toxicity of water stored in polyethylene bottles. *Bull. Torrey Bot. Club* **101**(5), 287–291.

Walkey, D. G. (1972). Production of apple plantlets from axillary-bud meristems. *Can. J. Plant Sci.* **52,** 1085–1087.

Zimmerman, R. H. (1983). Tissue culture. *In* "Methods in Fruit Breeding" (J. N. Moore and J. Janick, eds.), pp. 124–135. Purdue Univ. Press, Lafayette, Indiana.

Zimmerman, R. H. (1984). Apple tissue culture. *In* "Handbook of Plant Cell Culture" (W. R. Sharp, D. A. Evans, P. V. Ammirato, and Y. Yamada, eds.), Vol. 2, pp. 369–395. Macmillan, New York.

Zimmerman, R. H., and Broome, O. C. (1980a). Micropropagation of thornless blackberry. *U.S., Dep. Agric., Sci. Educ. Admin.* **ARR-NE-11,** 23–27.

Zimmerman, R. H., and Broome, O. C. (1980b). Blueberry micropropagation. *U.S., Dep. Agric., Sci. Educ. Admin.* **ARR-NE-11,** 44–47.

Zimmerman, R. H., and Broome, O. C. (1980c). Apple cultivar micropropagation. *U.S., Dep. Agric., Sci. Educ. Admin.* **ARR-NE-11,** 54–58.

CHAPTER **15**

The Acclimatization of Micropropagated Plants

D. I. Dunstan
K. E. Turner

Plant Tissue Culture Department
Kelowna Nurseries, Ltd.
Kelowna, British Columbia, Canada

I. INTRODUCTION

Acclimatization has been defined as a process, controlled by humans, to adapt an organism to an environmental change (Brainerd and Fuchigami, 1981). The acclimatization of micropropagated fruit trees from the *in vitro* to the *in vivo* environment is concerned with the following factors: (1) rooting, either *in vitro* or *in vivo*, and (2) transfer to nonsterile conditions with humidity control and temperature control.

Acclimatization is necessary because *in vitro* plant material is not adapted for *in vivo* conditions. Several authors have indicated that the waxy cuticle and stomata on leaves of *in vitro*-grown plants are inadequate or inoperative (Grout and Aston, 1977a; Sutter and Langhans, 1979, 1982; Fuchigami *et al.*, 1981; Brainerd and Fuchigami, 1981, 1982; Wetzstein and Sommer, 1982). Such leaves are incapable of preventing or reducing the water loss that can occur in the variable humidity of the *in vivo* environment. Similarly, there is reason to believe that the high sucrose and salt medium that is

CELL CULTURE AND SOMATIC CELL
GENETICS OF PLANTS, VOL. 1

ISBN 0-12-715001-3

often used with *in vitro*-grown cultures limits the photoautotrophic capacity of leafy shoots (Grout and Aston, 1977b; Wetzstein and Sommer, 1982). Published rooting procedures can be broadly categorized into one of two main routes based upon either *in vitro* rooting, e.g., begonia (Bigot, 1981), olive (Rugini and Fontanazza, 1981), pine (Francelet *et al.*, 1980), plum (Howard and Oehl, 1981), potato (Goodwin *et al.*, 1980), raspberry (Snir, 1981), and rose (Hyndman *et al.*, 1982) or *in vivo* rooting, e.g., potato (Goodwin *et al.*, 1980) and rose (Davies, 1980).

In vitro rooting occasionally makes use of growing conditions that differ from the multiplication stages of culture. The changes may include reduced salt concentration (Hyndman *et al.*, 1982) and altered light intensity (Goodwin *et al.*, 1980). Such differences may not only affect the rooting characteristics of shoots but may also prepare them for transplantation by stimulating autotrophism and changes in leaf morphology.

In vivo rooting combines the rooting process with acclimatization to the *in vivo* environment in one step. In either case, shoots will die if the main acclimatization step from the *in vitro* to the *in vivo* environment has not been successful. The following technique is an acclimatization procedure for the transfer of M7 apple rootstock plantlets or shoots from the *in vitro* to the *in vivo* environment.

II. PROCEDURE

A. Shoot Material

Shoots are collected from *in vitro* multiplying shoot cultures that have been grown for 28 days on agar-solidified nutrient medium [Murashige and Skoog (MS) salts and vitamins, 1962, less glycine, containing 3.75 μM 6-benzyl aminopurine, Sigma Chemical]. Cultures are maintained at 23°–27°C (night-day) under a 16-hr photoperiod with an incident 750 lumen (high-output cool white fluorescent lights).

Shoots used in rooting (either by an *in vitro* or an *in vivo* technique) are 1.5–2.0 cm from cut base to shoot tip, 1.0–2.0 mm in stem caliper, and have 9–12 nodes with expanding leaves.

Shoots are treated in one of the two following ways.

1. *In vitro* rooting. Shoots are rooted in an agar-solidified, three-quarter-strength solution of the MS salt and vitamin formula containing 1.25 μM indole-3-butyric acid (IBA) (Sigma Chemical). Thirty-five shoots are placed in 100 ml medium contained in 500-ml jam jars with metal screw caps (the

surface area of the medium is approximately 42 cm^2). Cultures are maintained for 17 days under standard growth room conditions. At day 17, plantlets are removed from their jars and collected in plastic bags until ready for planting. Occasional fogging or misting is used to prevent the plantlets from wilting. Provided the bags are inflated and excess moisture is drained, humid shoots can be kept refrigerated for up to 3 days.

2. *In vivo* rooting. The shoots are collected directly from *in vitro* multiplying cultures under clean, nonsterile conditions and stored in plastic bags. Occasional fogging or misting is done as necessary to prevent wilting. After collection, the shoots are immersed for about 5 sec in a 7.39-μM solution of IBA. After immersion, the solution is thoroughly drained.

B. Planting Details

Plantlets or shoots are transplanted into a soilless 1:2:1 mixture of sand, peat, and perlite (v:v:v) containing osmocote 18:6:12 (nitrogen:phosphorus:potassium) (Sierra Chemical). Water is added until the mix appears moist, but without forming a soil ball when hand squeezed. The mix is used to fill open flats (30.5 × 55.9 × 6.4 cm). One hundred plantlets, or shoots, are placed into each flat. Mist is manually applied as needed to prevent wilting. The flats are kept in a high-humidity environment (relative humidity 90% or above). In winter this environment is created by covering moistened, bottom-heated sand beds with polyethylene ridge tents which are housed in polyethylene-covered hoop houses. Bottom heat is adjusted to give a temperature of 18–20°C in the soilless mix. Ambient temperature around the shoot is kept between 13 and 18°C during acclimatization (35–40 days). The tent is raised during the day to prevent buildup of excess surface soil moisture. At this time, automatic misting is used as needed to prevent wilting. Occasional light watering is necessary to maintain the consistency of the original mix. The tents are lowered at night. After the initial 28–35 days, the tent and bottom heat are no longer needed. An increase in ambient air temperature (e.g., 20–25°C) at this time results in the onset of shoot growth.

During the summer, humidity is created by overhead misting (flora mist foggers) controlled by a time mechanism. The misting frequency and duration must be monitored. Generally, a gradual reduction should occur over the acclimatization period. The ambient air temperature during summer is controlled by a typical greenhouse fan-pad cooling system. Shading of 50% (Saran cloth) is used over the hoop houses to reduce heat buildup. No ridge tent is used in summer. Bottom heat is maintained as in winter. *In vivo*-rooted shoots remain under mist for a minimum of 28 days, and *in*

TABLE I

Rooting and Transplant Survival of Micropropagated M7 Apple Shoots Using Two Methods of Rooting[a]

A. *In vitro*

Replication	Day 1 (number inoculated)	Day 17 (number rooted/unrooted)	Day 42 (number surviving)
1	200	186/14	193
2	200	188/12	196
3	200	192/8	190

B. *In vivo*

Replication	Day 1 (number inoculated)	Day 21 (number surviving)	Day 42 (number surviving)
1	200	183	179
2	200	174	155
3	200	179	167

[a] Acclimatization procedure as noted for winter. Fresh weights of sample shoots are 56.17 ± 6.46 mg.

vitro-rooted material remains for 2–3 weeks. At this time plants are removed from the propagation area to the hardening-off area.

III. RESULTS

Summarized results for a typical winter planting (Table I) show that a high figure for rooting and survival can be achieved using the *in vitro* rooting system. After 17 days in the rooting medium, plantlets are taller and have a larger leaf surface area than the original inoculum (Fig. 1). Most shoots are already well rooted (up to eight roots per rooted shoot, 0.2–1.0

Fig. 1. A typical shoot collected from a multiplying shoot culture of M7 apple rootstock. Bar = 2.0 cm.

Fig. 2. *In vivo:* rooted shoot 62 days after collection. Ambient air temperature was held below 10°C. Bar = 2.0 cm.

Fig. 3. *In vitro:* rooted shoot 62 days after collection. Ambient air temperature was held below 10°C. Bar = 2.0 cm.

Fig. 4. *In vitro:* rooted shoot 90 days after collection. Ambient air temperature was raised to 25°C on day 62. Bar = 2.0 cm.

1
2
4
3

cm long). All material from the rooting jars is planted. Originally unrooted shoots often root and are represented in the total survival figure (Table I).

In comparison, at 21 days, *in vivo* rooting shoots are similar in size to the original inoculum. Random observations at this time indicate that a majority of the surviving shoots have already developed root initials. Survival is lower, however, with the *in vivo* rooting technique (Table I). Surviving plantlets are little changed from the originally excised shoot, except that they appear more hardened (less succulent; Fig. 2). In contrast, *in vitro* rooted plants are more like apple seedlings (Fig. 3).

A gradual rise in ambient air temperature promotes the growth of either shoot type. This is usually more rapid for the *in vitro* rooted shoots (Fig. 4).

IV. CONCLUSIONS

The acclimatization procedure can be successfully used when transplanting M7 apple shoots from the *in vitro* to the *in vivo* environment. Essentially similar survival results have been obtained for M2 and M26 apple rootstocks, and for Myrobolan B plum rootstock (*in vitro*) and Mazzard F12/1 cherry rootstock (*in vitro*). Modifications to these procedures will be required when other plant material is used or when other transplantation facilities or environments are present. For example, the control of air temperature to maintain a gradient (from cool at the shoot tip to warm at the shoot base) is more difficult in summer months. Humidity control and the need for irrigation are also more demanding. There is a fine line between too much humidity and too little. Excessive soil moisture promotes anaerobic conditions, and moisture applications need to be carefully monitored. This is especially important when changing the size and shape of the container. Such changes will affect the drainage properties of the soil mix. Generally, a deep container is better than a shallow one. The need for antifungal sprays (and their rates of application) should be individually assessed.

As soon as the acclimatization process has been accomplished, plantlets should be transplanted and/or allowed to make shoot growth by raising the ambient air temperature as described, with a reduction in bottom heat. The acclimatization process may take longer during summer months for the *in vivo* system if it is not possible to maintain the desired soil–air temperature gradient.

An additional acclimatization step may be required when transferring plants from the winter hoop house environment to full sun. This can often be achieved by placing plants in an area covered with a 50% shade cloth.

In general, *in vivo* rooting can be of greater economic benefit at times when a temperature gradient from shoot tip to base can be maintained; *in vitro* rooting may be of greater benefit when the gradient cannot be maintained.

REFERENCES

Bigot, C. (1981). Multiplication végétative *in vitro* de *Begonia x hiemalis* ("Rieger" et "Schwagenland") 1. Méthodologie. *Agronomie* **1,** 433–440.

Brainerd, K. E., and Fuchigami, L. H. (1981). Acclimatization of aseptically cultured apple plants to low relative humidity. *J. Am. Soc. Hortic. Sci.* **106,** 515–518.

Brainerd, K. E., and Fuchigami, L. H. (1982). Stomatal functioning of *in vitro* and greenhouse apple leaves in darkness, mannitol, ABA, and CO_2. *J. Exp. Bot.* **33,** 388–392.

Davies, D. R. (1980). Rapid propagation of roses *in vitro. Sci. Hortic. (Amsterdam)* **13,** 385–389.

Francelet, A., David, A., David, H., and Boulay, M. (1980). Première mise en évidence morphologique d'un rajeunissement de méristèmes primaires caulinaires de Pin maritime âgé (*Pinus pinaster* Sol.). *C. R. Hebd. Seances Acad. Sci.* **290,** 927–930.

Fuchigami, L. H., Cheng, T. Y., and Soeldner, A. (1981) Abaxial transpiration and water loss in aseptically cultured plum. *J. Am. Soc. Hortic. Sci.* **106,** 519–522.

Goodwin, P. B., Kim, Y. C., and Adisarwanto, T. (1980) Propagation of potato by shoot-tip culture. 2. Rooting of proliferated shoots. *Potato Res.* **23,** 19–24.

Grout, B. W. W., and Aston, M. J. (1977a). Transplanting cauliflower plants regenerated from meristem culture. I. Water loss and water transfer related to changes in leaf wax and to xylem regeneration. *Hortic. Res.* **17,** 1–7.

Grout, B. W. W., and Aston, M. J. (1977b). Transplanting cauliflower plants regenerated from meristem culture. II. Carbon dioxide fixation and the development of photosynthetic activity. *Hortic. Res.* **17,** 65–71.

Howard, B. H., and Oehl, V. H. (1981). Improved establishment of *in vitro* -propagated plum micropropagules following treatment with GA_3 or prior chilling. *J. Hortic. Sci.* **56,** 1–7.

Hyndman, S. E., Hasegawa, P. M., and Bressan, R. A. (1982). Stimulation of root initiation from cultured rose shoots through the use of reduced concentrations of mineral salts. *HortScience* **17,** 82–83.

Murashige, T., and Skoog, F. (1962). A revised medium for rapid growth and bio assays with tobacco tissue cultures. *Physiol. Plant.* **15,** 473–497.

Rugini, E., and Fontanazza, G. (1981). *In vitro* propagation of "Dolce Agogia" Olive. *HortScience* **16,** 492–493.

Snir, I. (1981). Micropropagation of red raspberry. *Sci. Hortic. (Amsterdam)* **14,** 139–143.

Sutter, E., and Langhans, R. W. (1979). Epicuticular wax formation on carnation plantlets regenerated from shoot tip culture. *J. Am. Soc. Hortic. Sci.* **104,** 493–496.

Sutter, E., and Langhans, R. W. (1982). Formation of epicuticular wax and its effect on water loss in cabbage plants regenerated from shoot-tip culture. *Can. J. Bot.* **60,** 2896–2902.

Wetzstein, H. Y., and Sommer, H. E. (1982). Leaf anatomy of tissue-cultured *Liquidambar styraciflua* (Hammelidaceae) during acclimatization. *Am. J. Bot.* **69,** 1579–1586.

CHAPTER 16

Induction and Maintenance of Cell Suspension Cultures

Patrick J. King

Friedrich Miescher Institute
Basel, Switzerland

I. INTRODUCTION

The reasons for attempting to induce plant cell suspension cultures usually include the wish to generate rapidly large amounts of cell material, to sample a homogeneous cell population, or to examine quantitatively the growth and/or metabolism of a homogeneous culture. Thus, for practical purposes, a cell suspension culture may be defined as a rapidly dividing, homogeneous suspension of cells from which samples can be removed accurately using a normal laboratory glass pipette. Unfortunately, there is no guaranteed method for producing such a culture in any given species (King, 1980).

II. INDUCTION

Methods used to "induce" suspension cultures fall into three main categories:

CELL CULTURE AND SOMATIC CELL
GENETICS OF PLANTS, VOL. 1

ISBN 0-12-715001-3

1. Transfer of obviously friable sectors of callus to one of the liquid media commonly reported to favor suspension culture growth, e.g., Heller's (Heller, 1953), B5 (Gamborg *et al.*, 1968), CC (Potrykus *et al.*, 1979), LS (Linsmaier and Skoog, 1965), Blaydes (Blaydes, 1966), or MX (Nash and Davies, 1972), in which the cells disperse within one to two passages to produce the required culture when shaken continuously.

2. Transfer of the only available culture material, regardless of friability, to a shaken liquid medium, followed by repeated transfer using a spoon or wide-bore pipette until the suspension culture (or the parent callus) reaches the required degree of friability. Often, after many passages, the friability of the culture suddenly increases.

3. The same as category 2, but with the application of various treatments to reduce aggregate size and to select for a lower mean aggregate size. Mechanical disruption of aggregates can often be achieved by, e.g., use of culture flasks with baffles (Fig. 1), by forcing the culture through a sieve, or by pumping the culture repeatedly through a syringe with a wide-bore cannula (Fig. 2). Selective subculture can be achieved by, e.g., allowing the cultures to settle briefly and pipetting only suspended material, gradually reducing the bore size of the pipette, or sieving before subculture. Some selection for suspension cultures has been reported by plating the early

Fig. 1. A DeLong flask (Pyrex, 300 ml) with a Morton cap, baffles, and side arm, suitable for cell suspension cultures.

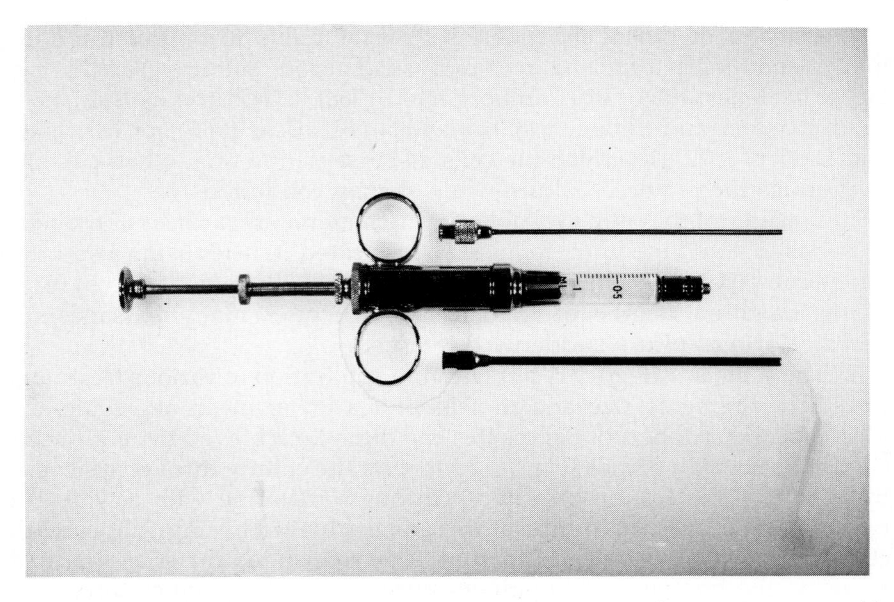

Fig. 2. A spring-loaded syringe with stainless steel cannulae (max. diameter 3 mm), suitable for cell suspension cultures (ARH pipetting unit—see text).

suspension cultures and picking only the most rapidly growing, friable colonies for further culture in liquid medium (Wilson and Street, 1975). A reduction in mean aggregate size can be brought about by inclusion of cellulases and pectinases in the medium (King *et al.*, 1973) (Fig. 3). There are no rules about medium composition, although high auxin concentrations are generally beneficial, especially in the form of 2,4-dichlorophenoxyacetic acid) (2,4-D). Suspension culture media generally contain little or no ammonium ion.

Rapid subculture is usually beneficial to the establishment of suspension cultures, and cultures should not be overdiluted initially. Production of a "cloud" of single, extravagantly shaped cells (usually a sign that all is not well) sloughed off from large aggregates should not be confused with growth of a good suspension culture. In the latter, the cells are often isodiametric and part of opaque aggregates, with an average of 20–100 cells per aggregate (Fig. 3). The supernatant, after settling or filtration, should be plated from time to time to check for dividing cells (Strauss and King, 1980).

It is by no means clear whether any of the recommended treatments actually induce suspension cultures. It seems more likely that the culture regimes described favor the growth of a certain cell type either present in

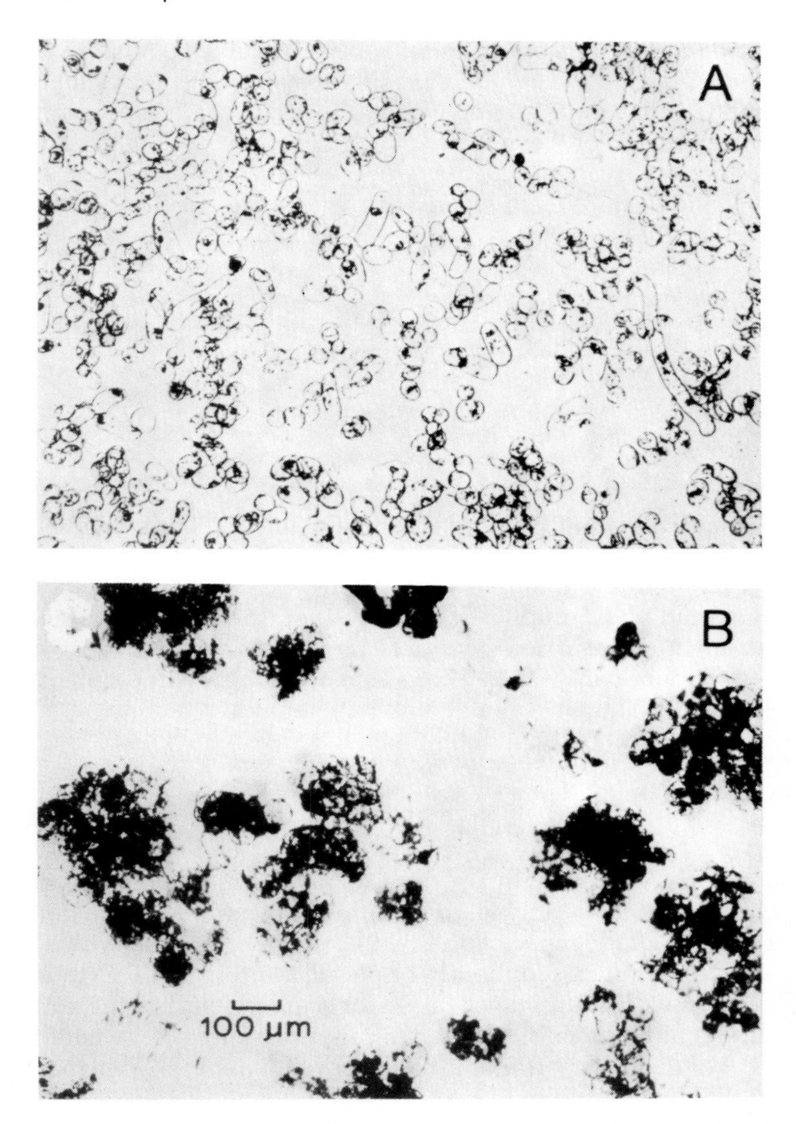

Fig. 3. Enhanced cell separation in suspensions of *Acer pseudoplatanus* cultured in the presence (A) or absence (B) of crude macerozyme (0.05%), crude Cellulase Onozuka P-1500 (0.05%), and sorbitol (8.0%). (From King *et al.*, 1973. Reproduced by permission of the National Research Council of Canada from the *Canadian Journal of Botany*, 1973.)

the original callus or appearing with time in culture. As a general rule, conditions favoring the evolution of fine suspension cultures tend (almost by definition) to reduce the organization within aggregates and the morphogenic potential of the cultures. The changes are less marked when the cultures are embryogenic, perhaps because such cells are also small and rapidly dividing and embryogenic aggregates are small and compact (see Chapter 18, this volume).

III. MAINTENANCE

A. Containers

Erlenmeyer flasks of Pyrex glass and 100-ml or 250-ml volume containing, respectively, 20 ml or 50 ml of culture are best. Wide-necked flasks (to avoid touching the rim during inoculation) with an intact rim (to ensure a complete seal) and an optically clear bottom (to allow culture inspection with an inverted microscope) should be used. The glass can be worked easily to produce baffles (Fig. 1) or a side arm for growth estimates (Fig. 1) (see Section III,G).

B. Closures

Flask closures must maintain sterility, allow gas exchange, and reduce evaporation. Either a double layer of autoclaved aluminium foil drawn down over the flask rim (not chipped!) or a DeLong flask with a straight neck and firmly fitting aluminium or plastic Morton cap (Fig. 1) should be used. Cotton wool plugs may be used for sealing flasks during autoclaving but not for culturing cells. They are a common source of contamination in flasks that are sitting on a shaker for several weeks. When packed too tightly, they may also limit O_2 solution rates (King *et al.*, 1973).

C. Shakers

Most suspension cultures are shaken on rotary shakers with a throw about 8 cm in diameter at 100–150 rpm both to aerate the cultures and to reduce aggregate size. The shakers must run forever, and it is especially important that the effort required to drive the fully loaded platform be well within the capacity of the motor. One should avoid switching the shaker off each time flasks are removed or added. The machine should have an emergency power supply, although cultures at medium density with a low culture volume/flask volume ratio can survive standing for at least 12 hr.

D. Pipettes

Suspension cultures can be transferred using normal laboratory pipettes, but spring-loaded syringes are highly recommended. The advantages are (1) rapid and thus representative culture sampling; (2) rapid discharge, avoiding settling of cells and blockage of the pipette; (3) uniform sample volume by rapid pipetting; (4) short opening period of donor and recipient flasks; and (5) pumping action reducing aggregate size. A good example of a reliable syringe is the ARH pipetting unit (A. R. Howell, Ltd., Kilburn High Road, London, England) (Fig. 2).

E. Subculture Routine

Suspension cultures should be transferred frequently and regularly. Optimal passage length is 1–2 weeks, but the exact time and the dilution required must be determined for each cell line. Dilutions of 1:4 after 1 week of 1:10 after 2 weeks are commonly used. The most common source of contamination during subculture is the rim of the culture flask. Thoroughly flame the rim after removing the flask closure and avoid further rim contact with fingers or pipettes. Use cotton wool, rolled paper or foam rubber stoppers during autoclaving of flasks and discard them just prior to application of the aluminium foil. To avoid serious loss of cultures by contamination, two independently cultured sublines should be maintained. Many rapidly dividing, finely dispersed suspension cultures can now be freeze preserved at −196°C in liquid nitrogen (Withers and King, 1980).

F. Sterility Checks

The following parameters can be used to test cultures routinely for sterility: (1) color; (2) pH—perhaps by including a pH indicator, e.g., chlorophenol red (F. Meins, FMI); (3) smell; (4) clarity of the meniscus (should be water clear); (5) presence of microorganisms as seen with the phase contrast microscope; and (6) growth of microorganisms in nutrient broth to which culture supernatant was added. Tests 1–5 can be performed rapidly at each subculture and when in doubt or at regular intervals, followed by tests 5 and 6.

G. Growth Measurements

1. Cell Number

This is an indispensable growth parameter for suspension cultures. Add 1 volume of culture to 4 volumes of 12% (w/v) aqueous chromium trioxide (filtered before use). Heat at 70°C until the cells are stained and plasmolyzed, as shown in Fig. 4A. Disrupt the partially macerated aggregates by vigorously pumping the culture with a syringe (Fig. 4B). The incubation time and maceration required will vary with culture age. Recently inoculated cultures with a high sugar content in the medium and small, closely aggregated cells require more treatment than older cultures with less medium sugar and expanded, loosely aggregated cells. Cells are best counted using a large-capacity counting slide constructed in the laboratory (Fig. 5). Count a minimum of 10 random fields (mag. 100–200 ×) in sets of 10. Dilute the macerated culture so that there are about 5–10 cells per field. Measure the field volume of the microscope used. Use an old or inexpensive, simple microscope, as chromium trioxide is destructive. The maceration technique works best with uniform cultures. With cultures containing

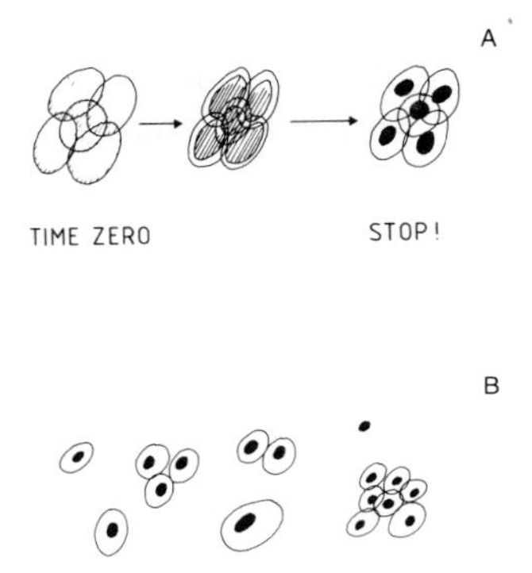

Fig. 4. Plasmolysis and staining of cells incubated in chromium trioxide solution at 70°C. (A) To establish the optimal incubation time, take samples every 5 min and examine the cells for the changes shown in the diagram. On average, 15–20 min incubation is necessary with an 8–12% CrO_3 solution. Shrinkage of the protoplasts makes counting of the cells in the aggregates easier. (B) Repeated pipetting of the preparation with a syringe will separate the cells. It is not necessary to produce single cells; aggregates of about 10 cells can be counted easily.

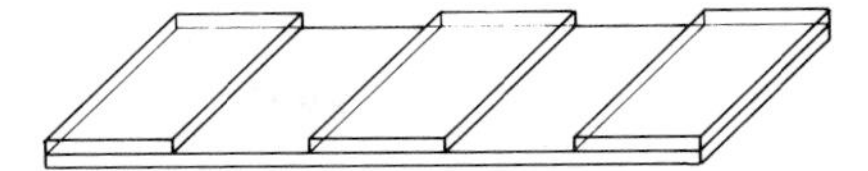

Fig. 5. Cell-counting slide constructed from good-quality microscope slides 0.8–1.0 mm thick. Sections (1.0–1.5 cm wide) of one microscope slide are glued onto a second complete slide with a very thin layer of Epoxy resin to give two to three channels 1.0–1.5 cm wide and 0.9–1.1 mm deep. Surplus adhesive should be removed from the edges of the channels. A coverslip (24 × 60 mm) is pressed firmly across the channels using a smear of glycerol as an adhesive, and the channels are loaded with macerated cell suspension using a Pasteur pipette. The cells should be allowed to settle for 1–2 min before counting. To compensate for the nonuniform distribution of cell clumps on the slide, at least 100 fields should be examined. Such homemade slides are far more suitable than hemocytometer slides for counting plant cells or protoplasts.

a mixture of large single and small aggregated cells, the former are likely to be destroyed by a treatment macerating the latter (King *et al.*, 1974).

2. Dry Weight/Packed Cell Volume

Centrifuge an appropriate volume of culture in a graduated conical 15-ml glass centrifuge tube at 2000 × *g* for 5 min. Note the pellet volume [packed cell volume (PCV)]. Discard the supernatant and wash the pellet onto a preweighed 2.5-cm Whatman GF/C filter paper. Wash the residue with 2 × 10 ml water and dry overnight at 80°C. Cool in a desiccator and reweigh (dry weight). If the PCV is not required, it is in any case worth centrifuging the sample and washing once by resuspension in order to remove polysaccharides in the medium that will otherwise block the filter.

With suspensions that are difficult to sample with a pipette, growth can be measured by allowing cells to settle into side arms in the walls of the culture flasks (Fig. 1). Flasks can be supported by placing the side arms in a test-tube rack and allowing the cells to settle for a fixed time. The height of the cell mass is measured with a ruler.

REFERENCES

Blaydes, D. F. (1966). Interaction of kinetin and various inhibitors on the growth of soybean tissue. *Physiol. Plant.* **19,** 748–753.

Gamborg, O. L., Miller, R. A., and Ohyama, K. (1968). Nutrient requirements of suspension cultures of soybean root cells. *Exp. Cell Res.* **50,** 151–158.

Heller, R. (1953). Recherches sur la nutrition minérale des tissus végétaux cultivés *in vitro. Ann. Sci. Nat., Bot. Biol. Veg.* [11] **14,** 1–22.

King, P. J. (1980). Cell proliferation and growth in suspension cultures. *Int. Rev. Cytol. Suppl.* **11A,** 25–53.

King, P. J., Mansfield, K. J., and Street, H. E. (1973). Control of growth and cell division in plant cell suspension cultures. *Can. J. Bot.* **51,** 1807–1823.

King, P. J., Cox, B. J., Fowler, M. W., and Street, H. E. (1974). Metabolic events in synchronised cell cultures of *Acer pseudoplatanus* L. *Planta* **117,** 109–122.

Linsmaier, E. M., and Skoog, F. (1965). Organic growth factor requirements of tobacco tissue cultures. *Physiol. Plant.* **18,** 100–127.

Nash, D. T., and Davies, M. (1972). Some aspects of growth and metabolism of Paul's Scarlet rose cell suspensions. *J. Exp. Bot.* **23,** 75–91.

Potrykus, I., Harms, C. T., and Lörz, H. (1979). Callus formation from culture protoplasts of corn (*Zea mays* L.). *Theor. Appl. Genet.* **54,** 207–214.

Strauss, A., and King, P. J. (1980). Analysis of plating technique for viability and drug-sensitivity determinations using Rosa cells. *Physiol. Plant.* **51,** 123–129.

Wilson, H. M., and Street, H. E. (1975). The growth, anatomy and morphogenetic potential of callus and cell suspension cultures of *Hevea brasiliensis*. *Ann. Bot. (London)* [N.S.] **39,** 671–682.

Withers, L. A., and King, P. J. (1980). A simple freezing unit and routine cryopreservation method for plant cell cultures. *Cryo-Lett.* **1,** 213–220.

CHAPTER **17**

Induction, Maintenance, and Manipulation of Development in Embryogenic Cell Suspension Cultures

Philip V. Ammirato

Department of Biological Sciences
Barnard College, Columbia University
New York, New York
and
DNA Plant Technology Corporation
Cinnaminson, New Jersey

I. INTRODUCTION

A. Stationary versus Suspension Cultures

Somatic embryos were first observed to arise from cultured carrot (*Daucus carota*) tissues in suspension cultures by Steward *et al.* (1958) and in callus cultures on semisolid medium by Reinert (1958, 1959). Investigations of somatic embryogenesis using carrot cultures, of both the wild and cultivated varieties, have been widespread (cf. Tisserat *et al.*, 1979, pp. 33–35), and carrot has become to a large extent the proverbial model system for somatic embryogenesis (Ammirato, 1983b). Similarly, a range of culture

CELL CULTURE AND SOMATIC CELL
GENETICS OF PLANTS, VOL. 1

ISBN 0-12-715001-3

modes have continued to be successfully employed in studies of somatic embryogenesis from stationary cultures using a variety of supports to suspension cultures whereby the liquid medium is contained in a number of different culture vessels and variously agitated.

Suspension cultures offer several distinct advantages over stationary cultures. When grown in liquid medium, cells and embryos are bathed by the culture medium and evenly exposed to nutrients, hormones, etc. This allows more precise manipulation of media components, handling of cells and embryos, and control of development. In stationary cultures, gradients develop from the medium to the top of the tissue. Uniform tissue response and control of development become more difficult. In addition, in suspension cultures, proembryogenic clusters and embryos usually separate from each other and float freely in the medium. Cells can easily be sieved, centrifuged, or otherwise manipulated. Large numbers of cells can be moved easily from one vessel to another and through the various treatments whereby cells grow into embryos and then into plantlets. Embryogenic suspension cultures, then, offer the prospect of large-scale cloning of plants and, through proper staging and control of development, the imposition of artificial dormancy and the creation of artificial seeds and/or the use of mechanized delivery systems.

B. Basic Requirements

During the early years of research, the basic requirements for the induction and promotion of somatic embryos were established, primarily by the use of suspension cultures.

1. Auxin Supply

In most cases (but not all), the presence of an auxin or auxinlike substance is critical for embryo initiation. The lowering of the auxin concentration or its complete absence fosters maturation (Steward *et al.*, 1958, 1967; Halperin and Wetherell, 1964; Halperin, 1966). In a number of cases, somatic embryos have formed without an external auxin supply, but these were on explanted tissue that was either embryonic (cotyledons of *Ilex aquifolium*, Hu and Sussex, 1971; hypocotyls of *Albizzia*, Gharyal and Maheshwari, 1981) or reproductive (ovules of *Citrus*, Kochba and Spiegel-Roy, 1973). In contrast to vegetative cells of mature plants, embryonic or reproductive tissue shows a propensity for direct embryogenesis without cell proliferation (cf. Sharp *et al.*, 1980). However, for most tissues, an external auxin supply is necessary for the induction and maintenance of embryogenic suspension cultures, and its removal leads to embryo maturation.

2. Nitrogen Supply

A substantial amount of nitrogen, usually in reduced form such as ammonium salts, is necessary for both embryo initiation (Halperin and Wetherell, 1965; Halperin, 1966) and maturation (Ammirato and Steward, 1971). This requirement, in the case of carrot somatic embryos, can be satisfied by high concentrations of inorganic nitrogen in the form of nitrate (Reinert, 1967). However, the benefits of reduced nitrogen, in addition to nitrate, seem well established. The source can be in the form of complex addenda such as coconut milk or water, which also supply nutrients and stimuli (Steward and Shantz, 1959), or casein hydrolysate (Ammirato and Steward, 1971), a mixture of amino acids (Kato and Takeuchi, 1966), or a single amino acid such as L-glutamine or L-alanine (Wetherell and Dougall, 1976), or the presence of ammonium ion (Halperin, 1966; Ammirato and Steward, 1971.) The Murashige–Skoog (MS) (1962) salt formulation, which is commonly employed, contains high levels of nitrogen in the form of ammonium nitrate.

II. PROCEDURES

This section outlines some of the experimental procedures useful for initiating and maintaining embryogenic suspension cultures and for fostering and controlling development. Carrot, as the model system, will be used to illustrate the technique (for embryogenic suspension cultures of cereals and grasses, see Chapter 18, this volume). *Dioscorea,* the yam of agriculture and medicine, provides a brief example of the technique applied to a monocotyledonous plant along with a few problems typical of more recalcitrant species. The reader is directed to a recent review of somatic embryogenesis (Ammirato, 1983b) and references there cited.

A. Carrot Somatic Embryogenesis

1. Initiation of Embryogenic Tissue

For carrot, almost any part of the plant body taken at any stage of development can successfully produce somatic embryos in culture: excised zygotic embryos (Steward *et al.,* 1964), hypocotyl (Fujimura and Komamine, 1979), young roots (Smith and Street, 1974), taproots (Steward *et al.*, 1958), petioles (Halperin, 1966), peduncle (Halperin and Wetherell, 1964), and protoplasts (Kameya and Uchimiya, 1972). This is apparently also the case for the other Umbelliferae and many other taxa. In carrot, petiole or taproot segments are the most common explants.

Sections of petiole or taproot are rinsed in tap water and surface sterilized for 15–20 min in 20% (v/v) sodium hypochlorite (commercial bleach). Following at least six rinses in sterile distilled water to remove all traces of bleach, 0.5–1.0 cm petiole explants or 0.5 cm^3 storage root explants are aseptically removed and placed on MS (1962) basal agar medium containing 4.5 μM 2,4-dichlorophenoxyacetic acid (2,4-D). Alternatively, the medium devised by White (1963), supplemented with 200 mg/liter casein hydrolysate (CH) and containing 10.7 μM naphthaleneacetic acid (NAA), can be used. The cultures can be incubated in either light or darkness and respond favorably at temperatures ranging from 20 to 27°C. The most commonly used temperature is 25°C. At this temperature, there should be sufficient growth after 4 weeks.

The MS salt and vitamin formulation is the one most commonly used. In their survey of somatic embryogenesis in crop plants, Evans *et al.* (1981) noted that 70% of the explants were cultured on MS medium or a modification of it. In this author's experience, the initiation and maintenance of umbellifer cultures on White's (W) medium, supplemented with casein hydrolysate (200 mg/liter) to provide a source of reduced nitrogen, appears to preserve the potentiality of the embryogenic suspensions better than growth on the MS medium. This has also been observed by Mouras and Lutz (1980).

Many auxins will successfully initiate embryogenic cultures in carrot, e.g., NAA (Ammirato and Steward, 1971), 2,4-D (Halperin, 1966), and indoleacetic acid (IAA) (Kato and Takeuchi, 1963). Of all the auxin or auxinlike plant growth regulators, 2,4-D has proven to be extremely beneficial, being used in 57.1% of successful embryogenic cultures (Evans *et al.*, 1981). However, it is this author's experience that cultures maintain their embryogenic potential longer if initiated and maintained with NAA in the medium. For long-term cultures, then, the preferred medium in W + CH + NAA. For excellent cultures to be used on a short-term basis, MS + 2,4-D is recommended.

As in many other examples given in earlier chapters, growth initiation appears to fare better on semisolid medium. This can be contained in a variety of vessels: test tubes, bottles, or Petri plates.

2. Establishment of Embryogenic Suspension Cultures

To initiate the suspension culture, callus tissue is excised from the explant and transferred to liquid medium of the same composition used to start growth (i.e., MS + 4.5 μM 2,4-D or W + CH + 10.7 μM NAA). If the proliferations on the explants are minimal, then the entire explant can be transferred.

There are a number of suspension culture techniques employing different culture vessels and methods of agitation. The most common one uses the standard Erlenmeyer flask placed on a gyrotory or orbital shaker. Typically, about 2.5–3.5 g callus tissue is placed in 50 ml of growth medium in each 250-ml Erlenmeyer flask. These flasks are agitated at 125–160 rpm.

An alternative technique, the auxophyton, developed by F. C. Steward and colleagues, provides a more gentle method of agitation and aeration. Tumble tubes (Steward *et al.*, 1952) hold 10 ml medium, and culture (nipple) flasks (Steward and Shantz, 1956) contain 250 ml. Both are rotated slowly on a klinostat at 1 rpm and, due to the asymmetry in the vessels as they rotate, alternately lift up the tissue to aerate and bathe the cells in medium. Because of the large amount of medium in the culture flasks, the general procedure is to transfer the primary callus tissue to tumble tubes for the first passage and then, when there are abundant cells, to the larger culture flask.

There are more elaborate methods for growing suspension cultures, e.g., spinning cultures, stirred cultures, and various continuous culture systems. The reader is referred to Street (1977) for an excellent discussion.

3. Maintenance of Suspension Cultures

With either culture flasks on an auxophyton or Erlenmeyer flasks on a shaker platform, an embryogenic cell suspension will develop in 2–3 weeks. As with all cultures, tissue must be transferred to fresh medium periodically. The timing is especially critical with suspension cultures in that they rapidly senesce at the end of the growth phase. In addition, there are a number of studies demonstrating that frequent subculturing can effectively minimize the extent of chromosomal changes in cell cultures and maintain embryogenic potentiality (Bayliss, 1977; Sunderland, 1977; Evans and Gamborg, 1982). Therefore, to maintain active growth and embryogenic potential, it is best to transfer the cells during the log phase of growth.

In addition, as with all suspension cultures, a particular density must be achieved when cells are transferred to fresh medium. This "minimum effective density" (Street, 1977) is particularly important if embryogenic potential is to be maintained. A fairly dense inoculum with minimal carryover of old medium is the key to maintaining embryogenic suspension cultures.

To subculture embryogenic suspensions, allow the cells to settle for several minutes and decant or aspirate almost all the medium, leaving enough to resuspend the suspension. Transfer one-fourth to one-sixth of the entire population to fresh medium. For example, the contents of 250 ml medium in a culture flask is allowed to settle, and all but 20 ml is decanted off. A 5-

ml aliquot is then inoculated into each culture flask with fresh medium. Subcultures are performed every 21 days. For a finer suspension, cells may be coarsely sieved (using 200-μm stainless steel mesh sieve or one layer of cheesecloth) at the time of transfer.

4. Promotion of Somatic Embryo Maturation

The transfer of the embryogenic cell suspension to medium lacking auxin fosters the maturation of somatic embryos. For carrot somatic embryos, the MS basal medium is excellent. For low-density cultures or finely sieved cultures, a low level of a cytokinin such as zeatin (0.1 μM) may prove beneficial (Fujimura and Komamine, 1980), especially in combination with abscisic acid (ABA) (Ammirato, 1983a). For the best embryo maturation, stale medium must be removed and the cells resuspended in a small amount of MS medium devoid of any auxin prior to inoculation.

Populations of somatic embryos typically show a wide range of sizes and stages of development. At the time of transfer from the maintenance medium to the medium that will allow development, there is a mixture of proembryonic cell clusters ranging from those with just a few cells to those substantially larger. In addition, in some cultures, such as caraway and, to some extent, carrot, somatic embryogenesis is repetitive and new embryogenic centers will arise from maturing embryos.

The most common method used to obtain some degree of uniformity, at least in terms of the initial population, is to sieve the inoculum. A graded series of stainless steel mesh (Halperin, 1966; Ammirato, 1974) or nylon mesh sieves (Fujimura and Komamine, 1975) has proven adequate. Passing the suspension through glass beads has also been effective (Warren and Fowler, 1978). Sieving followed by centrifugation in 16% Ficoll solution containing 2% sucrose isolated a population of cell clusters of only 3–10 cells each that developed synchronously when moved to growth regulator-free medium (Fujimura and Komamine, 1979).

In addition to repetitive embryogenesis, somatic embryos may develop abnormally, producing multiple cotyledons or callused shoots. Somatic embryos may also germinate precociously, producing long hypocotyls and extended radicles, often at the expense of good cotyledon and shoot development. This is typical in carrot cultures. The addition of ABA at moderate concentrations (usually 0.1–1.0 μM) has been shown to normalize somatic embryo development, to limit severely abnormal proliferations typical of caraway somatic embryos (Ammirato, 1974, 1977), and to prevent precocious germination found in carrot somatic embryos (Ammirato, 1983a). Short, dicotyledonous embryos free of extraneous proliferations dominate the populations.

A variety of suspension culture techniques have been used for embryo maturation, including tumble tubes (Ammirato, 1974), Erlenmeyer flasks

(Halperin and Wetherell, 1964), and test tubes that rotate slowly around their longitudinal axes (Halperin, 1966). All have successfully promoted embryo maturation. However, recent work has shown that the suspension culture techniques used to grow the somatic embryos can affect their morphology (Ammirato, 1983a). Distinctly different populations of somatic embryos developed in Erlenmeyer flasks, tumble tubes, and test tubes. The flasks promoted extensive proliferations and callusing of embryos; the test tubes did not promote good cotyledonary development. Interestingly, it was also found that ABA could normalize maturation in all cases. The most effective concentration, however, varied from vessel to vessel. The reasons for the differences remain to be elucidated. For the moment, however, it appears that the conditions of culture can affect the pattern of somatic embryo maturation.

On this point, little attention has been directed to the effects of light, darkness, or photoperiod on embryo development. Two studies, one with carrot (Ammirato and Steward, 1971) and the other with caraway (Ammirato, 1974), have shown that somatic embryo maturation proceeds more normally in complete darkness. Light was reported to be essential for somatic embryogenesis in tobacco (Haccius and Lakshmanan, 1975) and eggplant (Gleddie *et al.*, 1983).

For carrot somatic embryo maturation, aseptically aspirate and/or decant the stale medium and resuspend the cells onto culture medium devoid of 2,4-D or NAA. Pass the proembryo suspension through a 200-μm sieve and then a 74-μm sieve. Allow the suspension that has passed the sieves to settle and decant off most of the medium to prepare a dense source of cells. Transfer the washed and sieved suspension to 10 ml MS medium in tumble tubes or 50 ml medium in 250-ml Erlenmeyer flasks. Embryos should first appear in about 8 days and reach mature size in 10–15 days.

5. Plantlet Development

Somatic embryos can be plated out on agar medium devoid of any growth regulators for growth into plantlets. Single embryos may profit by the inclusion of low levels (0.1 μ*M*) of zeatin in the medium. After a number of leaves and some roots have formed, the small plantlets can be transferred to Jiffy pots or vermiculite for subsequent development. It is best not to wait until the plants become too large before removing them from cultures to soil.

B. *Dioscorea* Somatic Embryogenesis

The monocotyledonous genus *Dioscorea* contains plants of both agricultural and medicinal importance (Coursey, 1967). A number of species such as *D. alata* and *D. esculenta* provide edible tubers, whereas a number of wild species including *D. floribunda, D. composita,* and *D. deltoidea* provide a source of the steroid-containing diosgenin that is the synthetic precursor for the commercial preparation of corticosteroids and sex hormones.

By adapting the procedures developed for carrot cultures, somatic embryos have been grown from suspension cultures of *D. floribunda* derived from excised zygotic embryos (Ammirato, 1978) and, more recently, from axillary bud tissue of *D. bulbifera,* a species that produces edible aerial tubers or bulbils (Ammirato, 1982). A brief outline of the procedure for *D. floribunda* will be presented. The development and application of cell culture techniques for *Dioscorea* have recently been reviewed (Ammirato, 1984).

As with many monocotyledonous plants, there are parts of the vegetative plant that do not readily proliferate in culture. These include mature stem internodes and mature leaves. Explants for culture are best taken from embryonic or meristematic regions. *Dioscorea floribunda,* in contrast to many of the species cultivated for food, is fertile and produces large, winged seeds and viable embryos. The seeds are exceptionally hard and, following surface sterilization, need to soak for several days in sterile distilled water before embryo excision. The embryo is extremely small, about 1 mm in length, and is located in a small depression at one side of the seed. By cutting away the rim of the seed opposite the embryo, the two halves of the persistent endosperm can be separated and the embryo exposed.

The excised embryo is transferred to semisolid medium containing MS + 18 μM 2,4-D and incubated in complete darkness. A small, slow-growing proliferation will appear in 4–6 weeks. After 12 weeks, the tissue is removed to tumble tubes containing the same medium and rotated in complete darkness. After 8 weeks, there should be a sufficient suspension to permit subculturing and scale-up. At this time only the embryogenic tissue, composed of small, densely cytoplasmic cells, is selectively transferred. Mixtures of cell types are typical of most cultures, and the growth medium, environmental conditions, and transfer techniques may need to be modified to selectively foster the growth of embryogenic cells.

Once a good suspension culture has been established, somatic embryos can be obtained in two ways—by allowing the culture to age or by transferring the tissue to 2,4-D-free medium. On unsupplemented basal medium, there is often root development and little else. The most normal maturation occurs on medium supplemented with 500 mg/liter glutamine and either 0.1 μM zeatin or 0.1 μM ABA. Culture flasks filled with somatic embryos can be grown (Fig. 1A). *Dioscorea floribunda* somatic embryos (Fig. 1B) appear as short conical structures, each with a collar of cotyledonary tissue. The tissue initiates as a complete ring (Fig. 1D) but often develops asymmetrically; when mature, these embryos often have a visible first leaf (Fig.

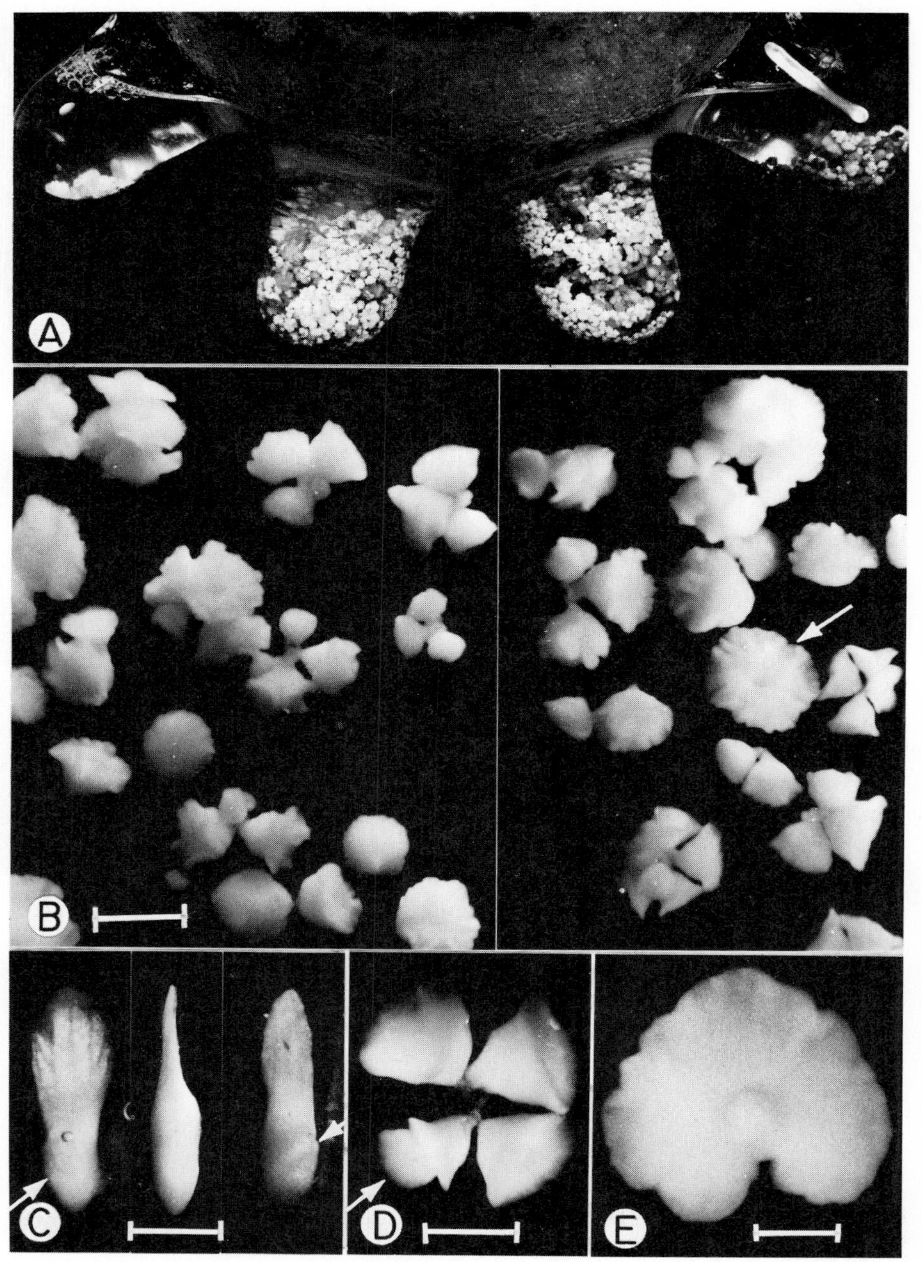

Fig. 1. Somatic embryos of *Dioscorea floribunda*. (A) A culture flask showing a copious crop of dense white tissue, somatic embryos. (B) A portion of the crop shown in (A) with numerous somatic embryos, some occurring singly and others joined at the tip of the radicle. Most have a ring of cotyledonary tissue. Scale = 1 mm. (C) Three views of an excised zygotic embryo. Note the single, fan-shaped cotyledon with a flattened, laminar apical portion and a radially

1E). The typical zygotic embryo (Fig. 1C) contains a single fan-shaped cotyledon with a flattened apical end and a round basal end. This single cotyledon has a sheathing leaf base enclosing the shoot apex (Fig. 1C, arrow). The somatic embryos show the same single cotyledon and sheathing base but, having developing while floating freely, the cotyledon is not flattened.

If somatic embryos are placed in groups of five on unsupplemented MS semisolid medium, plantlet development will proceed. Individual embryos will grow on MS medium supplemented with 500 mg/liter glutamine and 0.1 μM zeatin. The plants are hardy and transfer readily to soil.

III. RESULTS AND CONCLUSIONS

As illustrated by the results with *Dioscorea* cultures, the procedures outlined for carrot can be suitable for other embryogenic cultures. However, as detailed in a recent review of somatic embryogenesis (Ammirato, 1983b), the specific requirements for each species are often different and must be determined empirically (see Chapter 18, this volume). In fact, recent studies have shown that the frequency of response within a species may vary considerably from one genotype to another (Lu *et al.*, 1982) and that different conditions may be required for each genotype (Kao and Michayluk, 1981). Theoretically, at least, it should be possible, if embryogenic callus can be obtained, to initiate and maintain suspension cultures.

Embryogenic suspension cultures present an excellent tool for both theoretical studies and practical applications. For example, selection schemes with embryogenic cultures have given rise to temperature-sensitive carrot variants for studies of differentiation (Breton and Sung, 1982) and stable salt-tolerant *Citrus* embryos and plants (Kochba *et al.*, 1982). The biosynthetic potentiality of cell cultures has long been of interest. Embryogenic suspension cultures and populations of somatic embryos may be another alternative for the production of important chemicals. Somatic embryos of celery produce the same flavor compounds present in the mature plant but absent in celery callus cultures (Al-Abta *et al.*, 1979). Somatic embryos of cacao produce the same lipids, including cocoa butter,

symmetrical basal portion. The shoot apex and first leaf are located asymmetrically beneath the sheath (arrow). The radicle is extremely short. Scale = 0.5 mm. (D) A cluster of four somatic embryos. Three are very immature and have initiated an annular ridge of cotyledonary tissue. The fourth (arrow) appears to have developed two cotyledons. Later study showed the structure to be two fused somatic embryos. Scale = 0.5 mm. (E) A mature somatic embryo showing a single, asymmetric cotyledon and the leaf primordium in the center. Scale = 0.5 mm.

as their zygotic counterparts (Pence *et al.*, 1981). The synthesis of storage proteins in legume somatic embryos may be possible.

That substantial populations of somatic embryos can be raised in a small volume of liquid medium offers the distinct possibility that somatic embryogenesis can be used for large-scale clonal propagation. To achieve this, procedures need to be perfected for both large-scale cultures and the maintenance of genetic integrity. Suspension cultures of somatic embryos, in which embryos separate and float freely in the medium, appear to be especially amenable to mechanical handling. Fluid drilling (Gray, 1981) has been suggested as one way to deliver large quantities of somatic embryos from the culture flask to the field (Evans *et al.*, 1981). This could be achieved even more readily if development could be synchronized and staged. The results seen with ABA and with the effects of different suspension culture techniques suggest that this is possible.

Embryos are natural organs of perennation, many of which typically become dormant. Because of their innate properties, somatic embryos may prove useful for long-term storage such as in germplasm banks. Cold storage, dry storage, or cryogenic preservation may play a role here. If dormancy could be induced in somatic embryos, they could be incorporated into artificial seeds by coating or encapsulation (Murashige, 1980). There are a number of problems to be solved, including the induction, maintenance, and breaking of dormancy, but the prospects are intriguing.

REFERENCES

Al-Abta, S., Galpin, I. J., and Collin, H. A. (1979). Flavour compounds in tissue cultures of celery. *Plant Sci. Lett.* **16,** 129–134.

Ammirato, P. V. (1974). The effects of abscisic acid on the development of somatic embryos from cells of caraway (*Carum carvi* L.). *Bot. Gaz. (Chicago)* **135,** 328–337.

Ammirato, P. V. (1977). Hormonal control of somatic embryo development from cultured cells of caraway: Interactions of abscisic acid, zeatin and gibberellic acid. *Plant Physiol.* **59,** 579–586.

Ammirato, P. V. (1978). Somatic embryogenesis and plantlet development in suspension cultures of the medicinal yam, *Dioscorea floribunda. Am. J. Bot.* **65,** Suppl., 89.

Ammirato, P. V. (1982). Growth and morphogenesis in cultures of the monocot yam, *Dioscorea. In* "Plant Tissue Culture 1982" (A. Fujiwara, ed.), pp. 169–170. Maruzen, Tokyo.

Ammirato, P. V. (1983a). The regulation of somatic embryo development in plant cell cultures: Suspension culture techniques and hormone requirements. *Bio/Technology* **1,** 68–74.

Ammirato, P. V. (1983b). Embryogenesis. *In* "Handbook of Plant Cell Culture" (D. A. Evans, W. R. Sharp, P. V. Ammirato, and Y. Yamada, eds.), Vol. 1, pp. 82–123. Macmillan, New York.

Ammirato, P. V. (1984). Yams. *In* "Handbook of Plant Cell Culture" (P. V. Ammirato, D. A. Evans, W. R. Sharp, and Y. Yamada, eds.), Vol. 3. Macmillan, New York (in press).

Ammirato, P. V., and Steward, F. C. (1971). Some effects of environment on the development of embryos from cultured free cells. *Bot. Gaz. (Chicago)* **132,** 149–158.

Bayliss, M. W. (1977). Factors affecting the frequency of tetraploid cells in a predominantly diploid suspension of *Daucus carota*. *Protoplasma* **92,** 109–115.

Breton, A. M., and Sung, Z. R. (1982). Temperature-sensitive carrot variants impaired in somatic embryogenesis. *Dev. Biol.* **90,** 58–66.

Coursey, D. G. (1967). "Yams." Longmans, Green, New York.

Evans, D. A., and Gamborg, O. L. (1982). Chromosome stability of cell suspension cultures of *Nicotiana* species. *Plant Cell Rep.* **1,** 104–107.

Evans, D. A., Sharp, W. R., and Flick, C. E. (1981). Growth and behavior of cell cultures: Embryogenesis and organogenesis. *In* "Plant Tissue Culture: Methods and Applications in Agriculture" (T. A. Thorpe, ed.), pp. 45–113. Academic Press, New York.

Fujimura, T., and Komamine, A. (1975). Effects of various growth regulators on the embryogenesis in a carrot cell suspension culture. *Plant Sci. Lett.* **5,** 359–364.

Fujimura, T., and Komamine, A. (1979). Synchronization of somatic embryogenesis in a carrot cell suspension culture. *Plant Physiol.* **64,** 162–164.

Fujimura, T., and Komamine, A. (1980). Mode of action of 2,4-D and zeatin on somatic embryogenesis in a carrot cell suspension culture. *Z. Pflanzenphysiol.* **99,** 1–8.

Gharyal, P. K., and Maheshwari, S. C. (1981). In vitro differentiation of somatic embryoids in a leguminous tree, *Albizzia lebbeck* L. (Short comm.). *Naturwissenschaften* **68,** 379–380.

Gleddie, S., Keller, W., and Setterfield, G. (1983). Somatic embryogenesis and plant regeneration from leaf explants and cell suspensions of *Solanum melongena* (eggplant). *Can. J. Bot.* **61,** 656–666.

Gray, D. (1981). Fluid drilling of vegetable seeds. *Hortic. Rev.* **1,** 1–27.

Haccius, B., and Lakshmanan, K. K. (1975). Adventiv-embryonen aus *Nicotiana* Kallus, der bei hohen Lichtintensitäten kultiviert wurde. *Planta* **65,** 102–104.

Halperin, W. (1966). Alterative morphogenetic events in cell suspensions. *Am. J. Bot.* **53,** 443–453.

Halperin, W., and Wetherell, D. R. (1964). Adventive embryony in tissue cultures of the wild carrot, *Daucus carota*. *Am. J. Bot.* **51,** 274–283.

Halperin, W., and Wetherell, D. R. (1965). Ammonium requirement for embryogenesis in vitro. *Nature (London)* **205,** 519–520.

Hu, C. Y., and Sussex, I. M. (1971). In vitro development of embryoids on cotyledons of *Ilex aquifolium*. *Phytomorphology* **21,** 103–107.

Kameya, Y., and Uchimiya, H. (1972). Embryoids derived from isolated protoplasts of carrot. *Planta* **103,** 356–360.

Kao, K. N., and Michayluk, M. R. (1981). Embryoid formation in alfalfa cell suspensions from different plants. *In Vitro* **17,** 645–648.

Kato, H., and Takeuchi, M. (1963). Morphogenesis in vitro starting from single cells of carrot root. *Plant Cell Physiol.* **4,** 243–245.

Kato, H., and Takeuchi, M. (1966). Embryogenesis from the epidermal cells of carrot hypocotyl. *Sci. Pap. Coll. Gen. Educ., Univ. Tokyo* **16,** 245–254.

Kochba, J., and Spiegel-Roy, P. (1973). Effect of culture media on embryoid formation from ovular callus of 'Shamouti' orange (*Citrus sinensis*). *Z. Pflanzenphysiol.* **69,** 156–162.

Kochba, J., Ben-Hayyim, G., Spiegel-Roy, P., Saad, S., and Neumann, H. (1982). Selection of stable salt-tolerant callus cell lines and embryos in *Citrus sinensis* and *C. aurantium*. *Z. Pflanzenphysiol.* **106,** 111–118.

Lu, C., Vasil, I. K., and Ozias-Akins, P. (1982). Somatic embryogenesis in *Zea mays* L. *Theor. Appl. Genet.* **62,** 109–112.

Mouras, A., and Lutz, A. (1980). Induction, repression et conservation des propriétés embryogénètiques des cultures de tissus de Carotte sauvage. *Bull. Soc. Bot. Fr., Actual. Bot.* **127,** 93–98.

Murashige, T. (1980). Plant growth substances in commercial uses of tissue culture. *In* "Plant Growth Substances" (F. Skoog, ed.), pp. 426–434. Springer-Verlag, Berlin and New York.

Murashige, T., and Skoog, F. (1962). A revised medium for rapid growth and bioassays with tobacco tissue cultures. *Physiol. Plant.* **15,** 473–497.

Pence, V. C., Hasegawa, P. M., and Janick, J. (1981). Sucrose-mediated regulation of fatty acid composition in asexual embryos of *Theobroma cacao. Physiol. Plant.* **53,** 378–384.

Reinert, J. (1958). Morphogenese und ihre Kontrolle an Gewebekuluren aux Carotten. *Naturwissenenschaften* **45,** 344–345.

Reinert, J. (1959). Uber die Kontrolle der Morphogenese und die Induktion von Advientiveembryonen an Gewebekuluren aus Karotten. *Planta* **58,** 318–333.

Reinert, J. (1967). Some aspects of embryogenesis in somatic cells of *Daucus carota. Phytomorphology* **17,** 510–516.

Sharp, W. R., Sondahl, M. R., Caldas, L. S., and Maraffa, S. B. (1980). The physiology of in vitro asexual embryogenesis. *Hortic. Rev.* **2,** 268–310.

Smith, S. M., and Street, H. E. (1974). The decline of embryogenic potential as callus and suspension cultures of carrot (*Daucus carota* L.) are serially subcultured. *Ann. Bot. (London)* [N.S.] **38,** 223–241.

Steward, F. C. and Shantz, E. M. (1956). The chemical induction of growth in plant tissue cultures. *In* "The Chemistry and Mode of Action of Plant Growth Substances" (R. L. Wain and F. Wightman, eds.), pp. 165–187. Academic Press, New York.

Steward, F. C. and Shantz, E. M. (1959). The chemical regulation of growth: Some substances and extracts which induce growth and morphogenesis. *Annu. Rev. Plant Physiol.* **10,** 379–404.

Steward, F. C., Caplin, S. M., and Millar, F. K. (1952). Investigations on growth and metabolism of plant cells. I. New techniques for the investigation of metabolism, nutrition and growth in undifferentiated cells. *Ann. Bot. (London)* [N.S.] **16,** 58–77.

Steward, F. C., Mapes, M. O., and Mears, K. (1958). Growth and organized development of cultured cells. II. Organization in cultures grown from freely suspended cells. *Am. J. Bot.* **45,** 705–708.

Steward, F. C., Mapes, M. O., Kent, A. E., and Holsten, R. D. (1964). Growth and development of cultured plant cells. *Science* **143,** 20–27.

Steward, F. C., Kent, A. E., and Mapes, M. O. (1967). Growth and organization in cultured cells: Sequential and synergistic effects of growth regulating substances. *Ann. N. Y. Acad. Sci.* **144,** 326–334.

Street, H. E. (1977). Cell (suspension) cultures—Techniques. *In* "Plant Tissue and Cell Culture" (H. E. Street, ed.), 2nd ed., pp. 61–102. Univ. of California Press, Berkeley.

Sunderland, N. (1977). Nuclear cytology. *In* "Plant Tissue and Cell Culture" (H. E. Street, ed.), 2nd ed., pp. 177–205. Univ. of California Press, Berkeley.

Tisserat, B., Esan, B. B., and Murashige, T. (1979). Somatic embryogenesis in angiosperms. *Hortic. Rev.* **1,** 1–78.

Warren, G. S., and Fowler, M. W. (1978). Cell number and cell doubling times during development of carrot embryoids in suspension culture. *Experientia* **34,** 356.

Wetherell, D. F., and Dougall, D. K. (1976). Sources of reduced nitrogen supporting growth and embryogenesis in cultured wild carrot tissue. *Physiol. Plant.* **37,** 97–103.

White, P. R. (1963). "A Handbook of Plant and Animal Tissue Culture." Jacques Cattell Press, Lancaster, Pennsylvania.

CHAPTER **18**

Isolation and Maintenance of Embryogenic Cell Suspension Cultures of Gramineae

Vimla Vasil
Indra K. Vasil

Department of Botany
University of Florida
Gainesville, Florida

I. INTRODUCTION

Cell suspension cultures capable of regenerating plants by the process of somatic embryogenesis were first reported in carrot 25 years ago (Steward *et al.*, 1958, 1964). Such embryogenic cell suspensions have since been established in many species of angiosperms. However, in the Gramineae, only nonmorphogenic suspensions have been obtained and studied in a few species (Gamborg and Eveleigh, 1968; Oswald *et al.*, 1977; Shannon and Liu, 1977; Vasil and Vasil, 1979; Brar *et al.*, 1979; Polikarpochmina *et al.*, 1979; Potrykus *et al.*, 1979). Even this has proved to be very difficult, and only in rare instances are true cell lines established in suspension. In most

CELL CULTURE AND SOMATIC CELL
GENETICS OF PLANTS, VOL. 1

ISBN 0-12-715001-3

cases, the cultures consist of proliferating root meristems that constantly slough off large, vacuolated, and generally nondividing cells from their surface (King *et al.*, 1978; King, 1980) and are kept in suspension because of rapid agitation. Such cultures cannot be truly described as suspensions and are not useful for cell culture, plant regeneration, or the study of somatic cell genetics. Embryogenic cell suspensions in the Gramineae were reported early in *Bromus inermis* (Gamborg *et al.*, 1970), but these were not stable and formed only albino plantlets.

The recent availability of embryogenic callus cultures of the Gramineae has led to the establishment of totipotent embryogenic cell suspensions of several species in our laboratory (Vasil and Vasil, 1981, 1982; Lu and Vasil, 1981; Vasil *et al.*, 1983; Ho and Vasil, 1983). Such suspensions provide a means of rapid, large-scale clonal propagation but, more importantly, can be used effectively for the isolation of variant (Rangan and Vasil, 1983) or mutant cell lines, and currently are the only source of totipotent protoplasts in the Gramineae (Vasil and Vasil, 1980; Lu *et al.*, 1981; Vasil *et al.*, 1983).

II. ESTABLISHMENT OF EMBRYOGENIC CELL SUSPENSIONS

Embryogenic calli, obtained from cultured immature embryos, and young leaf or inflorescence segments (see Chapter 5, this volume, for details) are sliced or teased apart into small pieces. Any nonembryogenic callus present is discarded. The embryogenic callus is then transferred to an Erlenmeyer flask containing Murashige and Skoog's (MS) (1962) medium supplemented with 2,4-dichlorophenoxyacetic acid (2,4-D) (1.0–2.5 mg/liter), sucrose (3%), and inositol (100 mg/liter), with or without coconut milk (5–10%). The amount of embryogenic callus available determines the size of the culture flask and the volume of liquid medium used. For small amounts of callus (50–100 mg), it is advisable to use a 50-ml flask with 10–15 ml liquid medium. We have generally used 250-ml flasks containing 30–35 ml of the medium and 250–500 mg of freshly obtained embryogenic callus. The cultures are placed on a gyrotory shaker at 150 rpm and incubated in the dark at 27°C. It is useful to start with several such cultures for any given species.

For the first several (7–10) days, no manipulation of the culture is necessary. However, the nutrient medium must be slowly drained off and replaced with fresh medium at 2- to 3-day intervals if it rapidly turns dark due to the release of phenolic substances, and contains much debris caused by the slicing and teasing of the original callus. After about 2 weeks, the flask is removed from the shaker, the cells are allowed to settle for 1 min, and 10–15 ml of the supernatant medium is removed by a spring-loaded automatic pipette or by slow draining of the flask and replaced by an equal volume of fresh medium. This procedure is repeated every 3–5 days until the original pieces of callus start to break up in smaller pieces but continue to grow. The draining of supernatant medium removes most of the large, vacuolated, and nondividing cells which are sloughed off into the medium along with other cellular debris.

Under the best conditions, the suspension becomes thick and mucilaginous in consistency in 3–4 weeks, and it becomes increasingly difficult to remove the medium by draining. It contains a large population of enlarged and elongated, thick-walled cells which are devoid of starch. In addition, it contains a few groups of small, round, richly cytoplasmic, and starch-containing cells. These are described as embryogenic cells. For the next few subcultures, the contents of the culture vessels are divided into two halves after addition of 30–35 ml of fresh medium. The rate of growth of the suspension increases markedly as the embryogenic cell groups start to divide faster. This allows subculture at 4- to 6-day intervals by dividing the contents of each flask into two halves and transferring each half to an Erlenmeyer flask containing 30–35 ml fresh medium. The volume of the inoculum added to fresh medium at the time of subculture is reduced gradually until a stable embryogenic cell suspension culture has been established. Such a culture is composed of some elongated thick-walled, nondividing cells and a large population of small, richly cytoplasmic, starch-filled embryogenic cells which are always present in small groups.

III. MAINTENANCE OF EMBRYOGENIC CELL SUSPENSIONS

The maintenance of suspensions is determined by the rate of cell division characteristic of each species, the concentration of 2,4-D in the medium, and the specific purpose for which the suspensions are to be used. For recovery of embryoids upon plating, the suspension can be maintained by subculture at a 1:2 dilution (suspension inoculum:fresh medium) at 4- to 7-day intervals. However, if the suspension is to be used for the isolation of protoplasts, the dilution ratio as well as the duration of each subculture is decreased over a period of several weeks to obtain cultures that are composed almost entirely of groups of embryogenic cells (Lu *et al.*, 1981; Vasil and Vasil, 1982; Ho and Vasil, 1983; Vasil *et al.*, 1983). Such suspensions grow rapidly, are finely dispersed, and do not contain any organized meristems or meristemoids. These predominantly embryogenic suspensions are maintained by subculture at a 1:5–1:7 dilution every 4–5 days. Concentrations of 2,4-D of 1.0–2.5 mg/liter do not seem to affect the nature of the suspension or its rate of growth. Higher levels of 2,4-D are inhibitory to growth, whereas at lower concentrations the cells enlarge rapidly, become vacuolated, and show reduced amounts of starch. Unless carefully controlled, reduced levels of 2,4-D and the presence of large, vacuolated cells that are devoid of starch can rapidly lead to the loss of the embryogenic nature of the suspension. A range of 1.0–2.0 mg/liter 2,4-D is generally suitable for routine maintenance.

IV. IMPORTANT PARAMETERS

A number of factors have been found to be important for the establishment and maintenance of embryogenic cell suspension cultures.

A. Nature of the Embryogenic Callus Used

This is probably the most critical factor in determining the outcome of attempts to establish embryogenic suspensions. There is a great deal of variation in the nature as well as the amount of embryogenic callus formed based on the quality of the explants, particularly if the latter are obtained from field-grown plants. The problem becomes even more acute because plants grown in the greenhouse have not proven to be ideal sources of explants. In general, we have found the shiny, compact, opaque, and white embryogenic callus to give the best results. The presence of too much soft and nonmorphogenic callus is not desirable. The amount of callus available for the initiation of the suspension determines how rapidly it may be obtained. Larger amounts are helpful.

B. Concentration of 2,4-D

A range of 1.0–2.5 mg/liter has proved to be most satisfactory for the initiation as well as the maintenance of suspensions in most of the species used by us. The amount of 2,4-D is particularly critical during the period of initiation since at low concentrations many roots are formed which may lead to suspensions consisting of proliferating root meristems. Higher concentrations support neither rapid proliferation nor the breakup of the embryogenic calli.

C. Subculture

Removal of supernatant medium containing phenolic substances, enlarged and vacuolated nonembryogenic cells, and other cellular debris is critical during the early stages of the establishment of embryogenic suspensions. Aging of the suspensions during the early stages, and the formation of a viscous, mucilaginous substance in the medium, are helpful. Dilution ratios used during subculture are very important. Use of different nutrient formulations or the addition of various growth substances, etc. did not prove useful.

D. Examination of Cultures

Frequent microscopic examination of cultures during the early stages, and then at the time of each subculture to determine the nature of the embryogenic cells, is critical. An experienced investigator may change dilution ratios and the duration of each subculture based on the condition of the available suspension.

V. CONCLUSIONS

The procedures described above, although highly empirical, have nevertheless allowed the successful establishment of embryogenic cell suspension cultures in *Pennisetum americanum* (Vasil and Vasil, 1981, 1982), *P. purpureum* (Vasil *et al.*, 1983), *Panicum maximum* (Lu and Vasil, 1981), and *Saccharum officinarum* (Ho and Vasil, 1983). The establishment of an embryogenic suspension is not an isolated and uncontrolled event, as attested to by the fact that such suspensions bave been repeatedly established in our laboratory over a period of several years, and more recently by others (Jones and Dale, 1982) using procedures described here. Depending on the planned use of the suspension, it may be maintained in a condition in which somatic embryos up to the globular or the early scutellar stage are formed in the suspension (Vasil and Vasil, 1981; Lu and Vasil, 1981; Ho and Vasil, 1983), or it may consist almost entirely of embryogenic cells for the formation of somatic embryos upon plating (Vasil and Vasil, 1982; Ho and Vasil, 1983) and the isolation of totipotent protoplasts (Vasil and Vasil, 1980; Lu *et al.*, 1981; Vasil *et al.*, 1983).

REFERENCES

Brar, D. S., Rambold, S., Gamborg, O. L., and Constabel, F. (1979). Tissue culture of corn and *Sorghum. Z. Pflanzenphysiol.* **95,** 377–388.

Gamborg, O. L., and Eveleigh, D. E. (1968). Culture methods and detection of glucanases in suspension cultures of wheat and barley. *Can. J. Biochem.* **46,** 417–421.

Gamborg, O. L., Constabel, F., and Miller, R. A. (1970). Embryogenesis and production of albino plants from cell cultures of *Bromus inermis. Planta* **95,** 355–358.

Ho, W., and Vasil, I. K. (1983). Somatic embryogenesis in sugarcane (*Saccharum officinarum* L.). II. The growth of and plant regeneration from embryogenic cell suspension cultures. *Ann. Bot. (London)* [N.S.] **51,** 719–726.

Jones, M. G. K., and Dale, P. J. (1982). Reproducible regeneration of callus from suspension culture protoplasts of the grass *Lolium multiflorum. Z. Pflanzenphysiol.* **105,** 267–274.

King, P. J. (1980). Cell proliferation and growth in suspension culture. *Int. Rev. Cytol., Suppl.* **11A,** 25–53.

King, P. J., Potrykus, I., and Thomas, E. (1978). In vitro genetics of cereals: Problems and perspectives. *Physiol. Veg.* **16,** 381–399.

Lu, C., and Vasil, I. K. (1981). Somatic embryogenesis and plant regeneration from freely suspended cells and cell groups of *Panicum maximum* in vitro. *Ann. Bot. (London)* [N.S.] **47,** 543–548.

Lu, C., Vasil, V., and Vasil, I. K. (1981). Isolation and culture of protoplasts of *Panicum maximum* Jacq. (Guinea grass): Somatic embryogenesis and plantlet formation. *Z. Pflanzenphysiol.* **104,** 311–318.

Murashige, T., and Skoog, F. (1962). A revised medium for rapid growth and bioassays with tobacco tissue culture. *Physiol. Plant.* **15,** 473–497.

Oswald, T. H., Nicholson, R. L., and Bauman, L. F. (1977). Cell suspension and callus culture from somatic tissue of maize. *Physiol. Plant.* **41,** 45–50.

Polikarpochmina, R. T., Gamburg, K. Z., and Khavkin, E. E. (1979). Cell suspension culture of maize (*Zea mays* L.). *Z. Pflanzenphysiol.* **95,** 57–67.

Potrykus, I., Harms, C. T., and Lorz, H. (1979). Callus formation from cell culture protoplasts of corn (*Zea mays* L.). *Theor. Appl. Genet.* **54,** 209–214.

Rangan, T. S., and Vasil, I. K. (1983). Sodium chloride tolerant embryogenic cell lines of *Pennisetum americanum* (L.) K. Schum. *Ann. Bot. (London)* [N.S.] **52,** 59–64.

Shannon, J. C., and Liu, J. W. (1977). A simplified medium for the growth of maize (*Zea mays*) endosperm tissue in suspension culture. *Physiol. Plant.* **40,** 285–291.

Steward, F. C., Mapes, M. O., and Mears, K. (1958). Growth and organized development of cultured cells. II. Organization in cultures from freely suspended cells. *Am. J. Bot.* **45,** 705–708.

Steward, F. C., Mapes, M. O., Kent, A. E., and Holsten, R. D. (1964). Growth and development of cultured plant cells. *Science* **143,** 20–27.

Vasil, V., and Vasil, I. K. (1979). Isolation and culture of cereal protoplasts. I. Callus formation from pearl millet (*Pennisetum americanum*) protoplasts. *Z. Pflanzenphysiol.* **92,** 379–383.

Vasil, V., and Vasil, I. K. (1980). Isolation and culture of cereal protoplasts. II. Embryogenesis and plantlet formation from protoplasts of *Pennisetum americanum*. *Theor. Appl. Genet.* **56,** 97–99.

Vasil, V., and Vasil, I. K. (1981). Somatic embryogenesis and plant regeneration from suspension cultures of pearl millet (*Pennisetum americanum*). *Ann. Bot. (London)* [N.S.] **47,** 669–678.

Vasil, V., and Vasil, I. K. (1982). Characterization of an embryogenic cell suspension culture derived from inflorescences of *Pennisetum americanum* (pearl millet, Gramineae). *Am. J. Bot.* **69,** 1441–1449.

Vasil, V., Wang, D., and Vasil, I. K. (1983). Plant regeneration from protoplasts of *Pennisetum purpureum* Schum. (Napier grass). *Z. Pflanzenphysiol.* **111,** 233–239.

CHAPTER **19**

Fractionation of Cultured Cells

Tatsuhito Fujimura

Mitsuitoatsu Chemicals, Inc.
Chigasaki, Kanagawa, Japan

Atsushi Komamine

Department of Botany
University of Tokyo
Hongo, Tokyo, Japan

I. INTRODUCTION

The population of cells in a suspension culture is usually heterogeneous in terms of size and specific gravity. However, the cell population used in biochemical experiments is required to be as homogeneous as possible, because biochemical parameters measured in a heterogeneous cell population show only average values of different types of cells at different levels of biochemical activity. Physical fractionation of cells is one of the most efficient methods of obtaining a homogeneous cell population. Two methods have generally been used for fractionation of cells in suspension cultures: (1) fractionation by difference in size, i.e., by sieving, and (2) fractionation by difference in specific gravity, i.e., by discontinuous density

ISBN 0-12-715001-3

gradient centrifugation. The two methods are often used successively. Recently, fused protoplasts labeled by fluorescent dyes have been the subject of attempts to fractionate them manually under a microscope (Patnaik *et al.*, 1982) or by using a cell sorter (Redenbaugh *et al.*, 1982; Galbraith and Harkins, 1982). As this sophisticated method (see Chapter 50, this volume) cannot be used for fractionation of cells with cell walls, it will be not discussed here.

The method described here is suitable for the fractionation of cells at stationary phase cultured in 150 ml of medium. The packed cell volume is about 5 ml/150 ml culture medium.

II. FRACTIONATION OF CULTURED CELLS BY SIEVING

Fractionation of cells of different sizes can be performed by passing them through beds of glass beads (Warren and Fowler, 1977) or by passing them through a nylon screen (Fujimura and Komamine, 1979) or a stainless steel screen (Kamada and Harada, 1979). The method using a glass bead bed is not suitable for experiments on a large scale, and the results of the fractionation are not always reproducible. We describe here the method using nylon screen sieves. This method can be applied to fractionation of suspension culture up to 10 liters with changes of scale.

A. Materials and Equipment

The following sterile items are required:

1. Sieves. Nylon sieves for this procedure are prepared as follows: Nylon cloth of a definite size of mesh is cut into a circle (50 mm in diame-

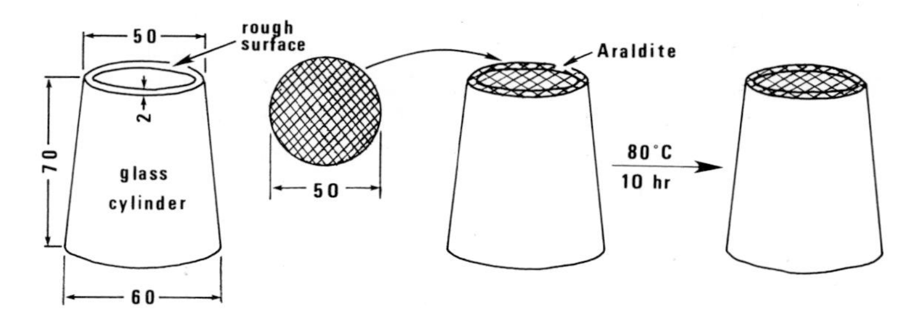

Fig. 1. Preparation of the nylon sieve. The unit of length is the millimeter.

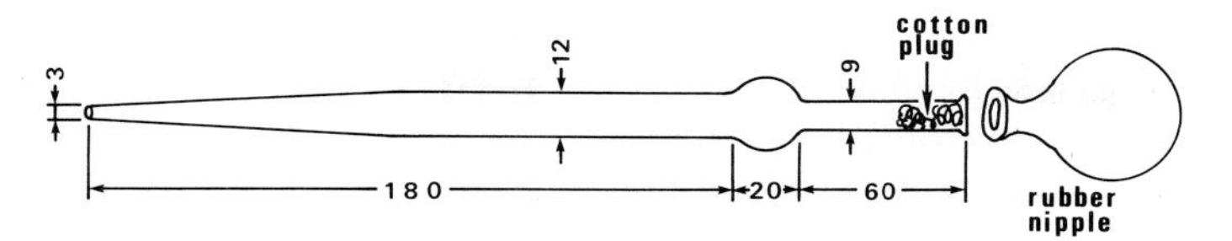

Fig. 2. "Komagome" pipette. The unit of length is the millimeter.

ter). It is attached to the smaller end of a tapering glass cylinder (Fig. 1) with an epoxy resin cement, Araldite (standard, Ciba-Geigy, Basel, Switzerland). The Araldite is hardened at 80°C for 10 hr. The end of the cylinder is ground with carborundum to make a rough surface before the nylon screen is attached. Sieves made by this procedure can be autoclaved more than 10 times. The size of the sieve mesh must be selected according to the purpose of the experiment (e.g., 26-, 31-, 47-, 81-, 161-, and 270-μm openings).

2. Three conical beakers (Pyrex, 500 ml).
3. Five "Komagome" pipettes (5 or 10 ml). The end of the pipette should be plugged with cotton wool (Fig. 2).
4. Rubber nipples (5 or 10 ml) for the "Komagome" pipettes, sterilized by dipping in 70% ethanol.
5. Three Erlenmeyer flasks (200 ml) containing 150 ml culture medium.

B. Procedure

1. Cultured cells in 150 ml medium are subjected to coarse sieving: A cell suspension is poured onto a sieve of coarse mesh (e.g., 47-, 81-μm or larger

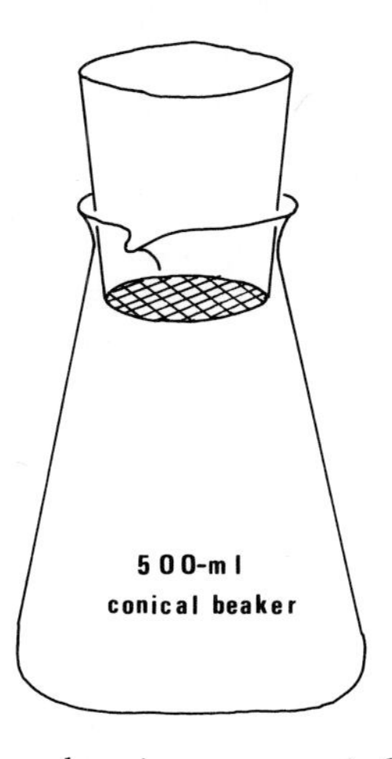

Fig. 3. The nylon sieve on a conical beaker.

openings), which is put on a 500-ml conical beaker (Fig. 3). The size of the sieve mesh should be selected according to the purpose of the experiment.

2. When the sieve becomes choked with cells, the cell suspension on the sieve is agitated for smooth filtration by repeated sucking up and discharging of the cell suspension using a "Komagome" pipette with a rubber nipple.

3. After the culture medium containing small cells has totally passed through the sieve, 150 ml of fresh medium is poured onto the sieve to pass the remaining small cells through the sieve.

4. This procedure should be repeated once more.

5. The cell suspension in the conical beaker which has passed through the coarse sieve is poured onto a sieve of fine mesh (e.g., 47-, 31-, 26-μm or smaller openings). The suspension is agitated with a "Komagome" pipette, as described in step 2 for easy sieving.

6. The cells trapped on the fine mesh sieve are washed with fresh medium (150 ml) to remove smaller cells or cell debris as described in step 3.

7. The washed cells are transferred to an appropriate medium, and the uniform-size cells are cultured in the medium or are successively subjected to further fractionation.

III. FRACTIONATION OF CELLS BY DISCONTINUOUS DENSITY GRADIENT CENTRIFUGATION

If cells in a heterogeneous population have different inclusions such as starch granules and vacuoles, resulting in a different specific gravity in each cell, they can be fractionated by the differences in specific gravity. Discontinuous density gradient centrifugation in a solution of appropriate osmoticum is a suitable method for fractionation of cells of different specific gravity.

Ficoll (type 400, Pharmacia, Uppsala, Sweden) is the most useful solute for the preparation of solutions of various densities. The reason is that this polymer of sucrose is stable when autoclaved and shows low osmolarity and low viscosity even at high concentrations (10–20%).

It is recommended that the cells be fractionated by sieving before fractionation by density gradient centrifugation.

A. Materials and Equipment

The following sterile items are required:

1. A series of Ficoll solutions (10, 12, 14, 16, and 18%) containing 2%

sucrose as an osmoticum. These solutions of various concentrations of Ficoll (2 ml each) are autoclaved spearately. Ficoll is sometimes toxic to cells, especially protoplasts. It is desirable, therefore, to dialyze Ficoll in a cellophane tube against distilled water at 4°C for more than 10 hr before use. Forty percent of the Ficoll solution is dialyzed, and the concentration of the dialyzed Ficoll is adjusted to 20% Ficoll with distilled water following the specific gravity. The specific gravity of 20% Ficoll is 1.066. The dialyzed Ficoll solution can be stored at −20°C in a deep freezer until use.

2. Eight glass centrifuge tubes (10 ml) with screw caps or aluminum foil caps.

3. Two hundred milliliters of culture medium.

4. Five "Komagome" pipettes (5 or 10 ml) and rubber nipples as described in section II,A.

One nonsterile item, a low-speed centrifuge (1000–4000 rpm), is required.

B. Procedure

1. A discontinuous density gradient of Ficoll solutions (12–18%) containing 2% sucrose is made in a centrifuge tube (10 ml) under aseptic condition by successive addition in order of Ficoll solutions to the tube (Fig 4). The interfaces of the solutions are agitated slightly with a glass rod to make the interfaces unclear. The cells can be sedimented easily through the interfaces during centrifugation by this procedure.

2. The cells (less than 500 μl packed cell volume) are suspended in 10% Ficoll solution containing 2% sucrose.

3. The suspension is layered carefully onto the top of the density gradient. An interface is made between the 10% and 12% unclear Ficoll solutions, as described in step 1. The centrifuge tube is sealed with a screw cap or with aluminium foil.

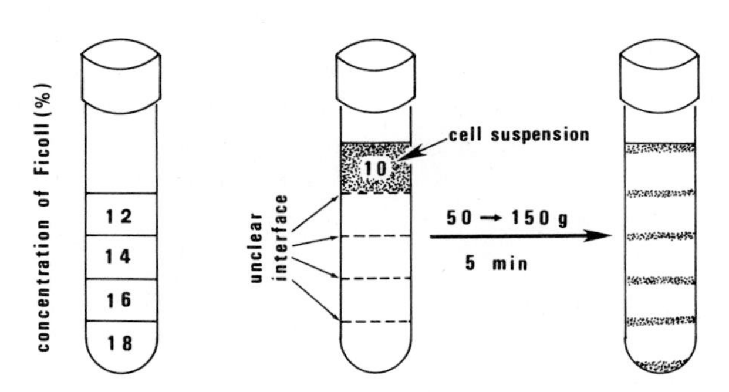

Fig. 4. Cell fractionation by discontinuous density gradient centrifugation.

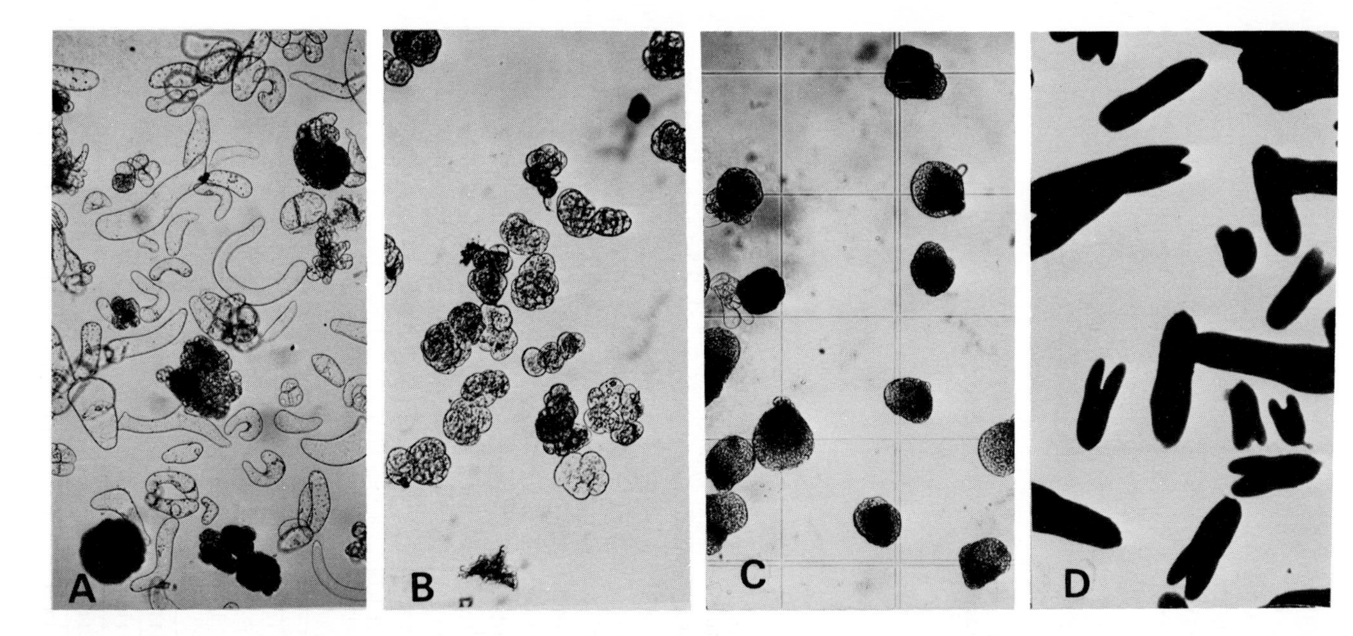

Fig. 5. Photographs of carrot cultured cells. (A) Cell population of stock suspension culture 7 days after subculture. (B) Cell clusters obtained after fraction by sieving with nylon screens of 31-μm and 47-μm pores, 10 to 18% Ficoll density gradient centrifugation at 150 *g* for 5 min and repeated (five times) centrifugation at 50 *g* for 5 sec. (C) Culture on seventh day after transfer to the embryo-inducing medium showing many globular stage embryos. (D) Culture on thirteenth day after transfer to the embryo-inducing medium showing many heart and torpedo stage embryos. (From Fujimura and Komamine, 1979.)

4. The tube is centrifuged at 50 *g* for the first 1 min, and then centrifugation is accelerated to 150 *g* for the subsequent 4 min.

5. The cells floated in the solution of 10% Ficoll and those sedimented on the interfaces and at the bottom of the tube are transferred separately to centrifuge tubes.

6. The cells in each fraction are suspended in 10 ml of the culture medium and centrifuged at 50 *g* for 30 sec. This procedure should be repeated more than three times to remove the Ficoll solution completely from the cells.

7. The fractionated cells are more homogeneous populations and can be used for further experiments, such as the induction of embryogenesis (Fujimura and Komamine, 1979).

IV. RESULTS AND CONCLUSIONS

A carrot suspension culture at the stationary phase (10 days after transfer) was subjected to fractionation by sieving using nylon screens of 31 and 47 μm. The collected cells 31–47 μm in size were subsequently subjected to fractionation by density gradient centrifugation in Ficoll solutions (10–18%). Figure 5A shows the initial heterogeneous cell population, and Fig. 5B indicates the cell clusters which were sedimented in 18% Ficoll solution by density gradient centrifugation. The fraction obtained was a homogeneous population of cell clusters composed of fewer than 10 cells, which differentiated to embryos synchronously in the auxin-free medium as shown in Figs. 5C and 5D, indicating that the attempt to obtain a cell population at a physiologically homogeneous level using fractionation by sieving and subsequently by density gradient centrifugation was successful (Fujimura and Komamine, 1979).

These results indicate that the procedures for cell fractionation described here are effective in fractionating cells of different physiological and biochemical properties in a cell population.

REFERENCES

Fujimura, T., and Komamine, A. (1979). Synchronization of somatic embryogenesis in a carrot cell suspension culture. *Plant Physiol.* **64,** 162–164.

Galbraith, D., and Harkins, K. (1982). Cell sorting as a means for isolating somatic hybrids. *In* "Plant Tissue Culture 1982" (A. Fujiwara, ed.), pp. 617–618. Maruzen, Tokyo.

Kamada, H., and Harada, H. (1979). Studies on organogenesis in carrot tissue cultures. I. Effects of growth regulators on somatic embryogenesis and root formation. *Z. Pflanzenphysiol.* **91,** 255–266.

Patnaik, G., Cocking, E. C., Hamill, J., and Pental, D. (1982). A simple procedure for the manual isolation and identification of plant heterokaryons. *Plant Sci. Lett.* **24,** 105–110.

Redenbaugh, K., Ruzin, S., Bartholomew, J., and Bassham, J. A. (1982). Characterization and separation of plant protoplasts via flow cytometry and cell sorting. *Z. Pflanzenphysiol.* **107,** 65–80.

Warren, G. S., and Fowler, M. W. (1977). A physical method for the separation of various stages in the embryogenesis of carrot cell culture. *Plant Sci. Lett.* **9,** 71–76.

CHAPTER **20**

Large-Scale Cultures of Cells in Suspension

M. W. Fowler

Wolfson Institute of Biotechnology
University of Sheffield
Sheffield, England

I. INTRODUCTION

With the development of cell lines which accumulate high levels of desirable products (Fowler, 1982a), attention has begun to turn to the question of large-scale growth of plant cells and the goal of industrial application. Plant cells have now been cultured in a wide range of bioreactors over a range of about 2–20,000 liters. The purpose of this chapter is to discuss some of the practical details which need to be considered when beginning the large-scale culture of plant cells, to indicate the range of systems that have been used to date, and to provide some practical insights into the operation of a large-scale growth system by looking in some detail at a model system.

II. THE NATURE OF PLANT CELLS IN CULTURE

Although there have been a number of successful attempts to grow plant cells in various forms of microbial bioreactors, there are properties of plant

CELL CULTURE AND SOMATIC CELL
GENETICS OF PLANTS, VOL. 1

ISBN 0-12-715001-3

cells which impose conditions of operation different from those encountered with microbial systems.

Plant cells are much larger than microbial cells, being on average some 30- to 100-fold greater in diameter at about 20–150 μm. This difference in size poses its own problems in terms of mixing of biomass and nutrient and gas transfer from the gas to the liquid phase. An additional problem with plant cells is that they rarely grow as free cell suspensions. More generally, they occur as a heterogeneous collection of clumps of cells, ranging in cell number per clump from 2 to 200 and up to 2 mm in diameter, depending upon such factors as origin of the cell line, nutrient formulation, and culture age. These clumps may arise in at least two ways. The first is through lack of cell separation following cell division and is of general occurrence. The second mode of cell clump formation frequently seen occurs when, in late log growth in batch culture, cells begin to excrete polysaccharide and protein and also in some way develop a sticky surface. Cells then begin to come together, forming larger and larger aggregates. At high biomass and viscosity levels, this can lead to major problems of mixing and biomass circulation. The other morphological feature of plant cells which needs to be carefully considered is the cellulose cell wall, which provides the cell with a fairly rigid exoskeleton. This wall has a high tensile strength but relatively low shear resistance (Mandels, 1972). In consequence, the high shear forces used in conventional microbial turbine-stirred reactors to aid high gas transfer may have deleterious effects on many plant cell cultures (Smart and Fowler, 1981). This is a point to which I will return later.

Plant cells require a complex nutrient medium, the formulation of which often provides an excellent growth regime for fungi, although not so much for bacteria. Fungal growth rates are much higher than those for plant cells, and so the microorganism quickly takes over the culture. In consequence, close attention to aseptic technique is required both in the preparation of the culture system and in the design of the culture rig. The slow growth rate of plant cells is also an important point when considering the length of equipment operation. Even batch culture runs may take a minimum of 2–3 weeks to complete; semicontinuous and continuous culture experiments may take 2–3 months. Equipment around the bioreactor, pumps, probes, valves, monitoring devices, and so on, therefore needs to be of the highest quality and reliability.

Plant cell culture media have a comparatively high viscosity, principally because of the carbohydrate load. The viscosity tends to rise exponentially with biomass increase and is also affected by the propensity of many cell cultures to excrete polysaccharide. The whole system has non-Newtonian characteristics in terms of liquid–solid interphase mixing and rheology about which comparatively little is known. This is an area greatly in need of study in practical terms of mass plant cell culture.

Another aspect of the nutritional requirements of growing plant cells concerns the gas regime. All plant cells are aerobic and therefore require a continuous oxygen supply. However, in contrast to microbial systems, the requirement is not for a very high gas transfer rate from the gas to the liquid phase but rather for a controlled input, often at comparatively low dissolved oxygen values. There are indications that plant cell cultures are highly sensitive to variations in oxygen tension; too high can be as deleterious as too low. In consequence, careful monitoring of the oyxgen supply to the vessel, with analysis of the effluent stream, if possible, is an important practical consideration when setting up large-scale systems. CO_2 also poses problems. The majority of plant cell cultures are run in the pH range 5–7. At a high gassing rate there is a tendency at these pH levels for CO_2 to be "stripped off," limiting culture growth. To some extent, this problem can be relieved by applying a "CO_2 bleed" to the input gas stream. Unfortunately, little is known about this effect, and continuous online analysis of CO_2 in both liquid and gas phases is either problematical or exceedingly difficult. There are other gases which may be important, such as ethylene; however, comparatively little is known of them.

The mention of CO_2 brings us to photosynthetic cell cultures. Comparatively few plant cell cultures have been maintained photosynthetically (Yamada *et al.*, 1978; see Chapter 22, this volume), and where they have been, efficiencies are not high. Furthermore, the majority of cultures have been maintained on a small scale, e.g., in 250- to 500-ml Erlenmeyer flasks. There are one or two examples of growth in larger systems, but not much above 10 liters. A particular problem here is not the supply of CO_2 and the general nature of the gas regime but the penetration of incident light into the culture. This is especially problematical at high biomass levels, where effective light penetration at a uniform level is difficult to achieve and where the level of intensity required may result in overheating problems. This is not an area to be entered by workers new to large-scale culture without very careful consideration.

Two problems which affect all cell culture systems to some degree are foaming and surface adhesion. Foaming is a particular problem with microbial cultures but does not appear to affect plant cell cultures in quite the same way. Antifoam agents such as propylene glycol- and silicone-based systems have been used to great effect in microbial cultures, but do not appear to have been very successful in plant cell cultures mainly because of deleterious effects on cell metabolism. With plant cell cultures the nature of the foam is rather different; the constituent bubbles are large compared with the microbial situation and are typically coated in protein and excreted polysaccharide. This makes the bubble very sticky and cells soon become almost encapsulated within the foam, thereby being removed from the circulating nutrient medium. This, of course, leads to a highly heterogeneous culture. Eventually, if not controlled by chemical or mechanical means, the accretion of foam and cells begins to affect mixing and the general stability of the culture. Well before this happens, other problems may be encountered. Typically these affect orifices into and out of the bioreactor which are not filled with liquid. Here bubbles of foam pass into the orifice (the condenser outlet for the gas effluent stream is usually the most problematical), and cells are carried over from the culture and deposited in the outlet lines. As more and more cells are deposited, the walls of the outlet become occluded and eventually blocked if not cleared. Careful equipment design is required to overcome such problems.

Surface adhesion is also to some degree a problem. In particular, cells adhere to the vessel walls just above the surface of the culture, and also to the surface of probes and baffles in the vessel. The layer of cells above the culture can usually be removed mechanically. The cell material around probes is more problematical and may lead to probe deterioration. Little is known about the nature of surface adhesion and wall growth. Coating the sides of the vessel and probes with silicone has been attempted with variable effect. Japanese workers have also claimed some success through modification of the ionic composition of the nutrient media.

III. LARGE-SCALE CULTURE SYSTEMS

A wide variety of plant cells have now been grown in an equally wide range of bioreactor configurations and sizes. This subject has recently been reviewed in detail by Martin (1980) and Fowler (1982a), and readers are referred to these articles for the detailed references to individual systems. A summary of the different systems is provided in Table I.

A variety of rig configurations have been tested, ranging from conventional stirred tank reactors to bubble column and airlift systems, the latter reflecting the desire to achieve good mixing and homogeneous growth conditions without high shear rates which may be deleterious to cell growth. Vessels have been constructed principally from glass and/or stainless steel and have been both proprietary and homemade in origin. Many of the papers referred to in Table I do give detailed descriptions of the vessel systems. The majority of large-scale cultures have been run as batch systems; however, there is an increasing number of reports on semi-continuous and continuous cultures (Wilson, 1981; Fowler, 1982b). These latter approaches pose practical rather than conceptual problems, particularly in terms of run time, often 2–3 months, with consequential effects on probes and ancillary equipment around the vessel, the maintenance of

TABLE I

Some Examples of Large-Scale Systems for the Growth of Plant Cells

Capacity (liters)	Configuration	Material of construction	Cell lines	Reference
57	Inverted Erlenmeyer flask-stirring bar	Glass	*Ipomoea* sp.	Veliky and Martin (1970)
5	Round-bottom vessel; stirring bar	Glass	*Acer pseudoplatanus*	Wilson *et al.* (1971)
10	Roller bottle system	Glass	*Acer pseudoplatanus*	Lamport (1964)
10	Conventional microbial vessel	Glass	*Morinda citrifolia*	Wagner and Vogelmann (1977)
10	Airlift reactor with draught tube	Glass	*Morinda citrifolia*	Wagner and Vogelmann (1977)
30	Conventional tank	Stainless steel	*Lolium* and rose	Tulecke and Nickell (1959)
65	Bubble column reactor	Glass/stainless steel	Tobacco	Kato *et al.* (1977)
100	Airlift loop	Glass	*Catharanthus roseus*	Fowler (1982a)
1,500	Bubble column reactor	Stainless steel	Tobacco	Kato *et al.* (1977)
20,000	Conventional stirred tank	Stainless steel	Tobacco	Kato *et al.* (1977)

sterility, and the removal of excess nutrient and biomass on a regular basis. Specific ways of overcoming this latter problem have been described by a number of workers (for references, see Fowler, 1982a).

Because of space limitations, it is not possible to describe individual systems. However, for those wishing to enter the field, it is of value to go through the stages of setting up a model system. This will now be done.

IV. GENERAL POINTS IN SETTING UP A MODEL SYSTEM FOR LARGE-SCALE PLANT CELL CULTURE

Our laboratory has adopted the maxim that simplicity is a virtue. This applies particularly to vessel design and construction. Given the problems

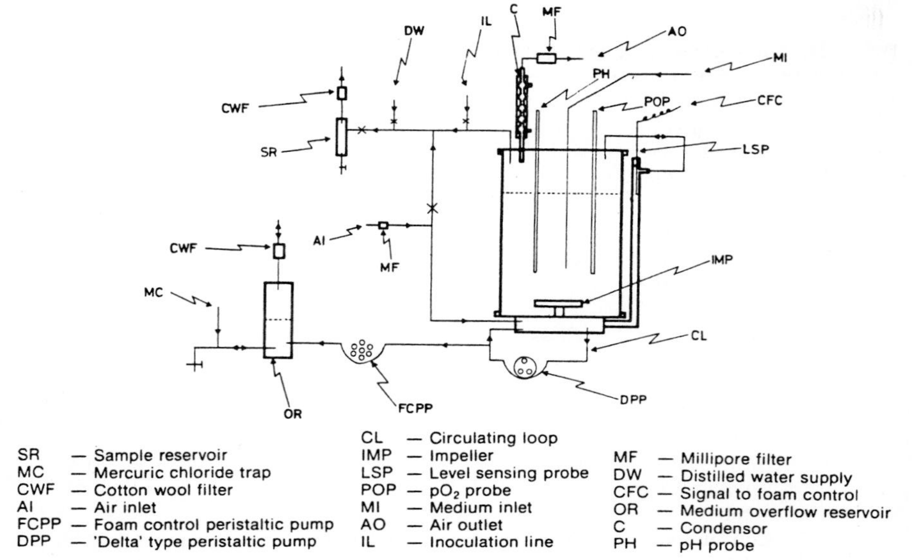

Fig. 1. Outline diagram of an LKB Ultraferm bioreactor modified for use with plant cell cultures. This vessel may be used in both batch and continuous modes. Further details of its construction and operation are given in Fowler (1976).

of maintaining asepsis with plant cell cultures, the fewer the inlets and outlets into the vessel through which probes pass the better, thereby reducing the number of passages through which microbial attack may be effected. For a number of years, we have used a bioreactor design based on the LKB Ultraferm system. A detailed description of the use of this vessel is contained in Fowler (1976). We have used this vessel in both a turbine-stirred and a draught tube configuration. The same general considerations apply to both.

The vessel is very simple (Fig. 1), consisting of a single glass cylinder with base and head plates. In a turbine-stirred mode, the base plate supports a turbine system coupled magnetically to an external electric motor. This has the advantage of not requiring a drive shaft to penetrate the base plate with attendant glands and seals, all of which may provide entry for microbial spores. The drawback of a magnetically coupled system is that it develops comparatively low torque in the coupling mode. Slippage may therefore occur at high viscosity and biomass levels.

The vessel contains two probes for pH and pO_2. These are proprietary probes and may be obtained from a range of suppliers. Also, they are not too difficult to make in the laboratory, and details are given in Aiba *et al.* (1973). Air is provided through a Millipore filtration system (1-μm exclusion) and passes into the vessel and nutrient through a ring sparger system. Spent air is vented through a condenser placed on the head plate.

To prepare the vessel for culture, the head and base plate seals are checked for fit and the various probes, output lines, and inlet lines are connected to the vessel. It should be noted that as the vessel is to be sterilized by autoclaving or steam passage, appropriate outlets to the vessel need to be left open and provided with backup filters to prevent poststerilization entry of microbes through the development of a slight vacuum in the vessel.

For vessels up to about 10 liters, it is usually practical to autoclave at 120°C, 15 psi, for 15 min, bearing in mind the possible need to autoclave parts of the nutrient media separately (Chapter 3, this volume). In the case of larger systems, it is usual to steam-sterilize the vessel with line-piped steam and to sterilize the media separately. Steam sterilization of the vessel requires a high-capacity steam generation system which provides high-quality steam. In general, large vessels are steamed for two or three separate periods of about 2 hr each to remove fungal spores. Practice does, however, vary considerably. Media sterilization can be achieved by either autoclaving, which is not really suitable for a large bulk, or by membrane filtration, followed by addition to the vessel. It is crucial that close attention be paid to maintaining sterility during medium input into the vessel.

Once the medium has been added together with ancillary components, the vessel is brought to the appropriate temperature and the various probes are tested and calibrated. This, again, is a crucial part of the operation. Only when the operator is fully satisfied that the system is sterile, with fully operational probes and services, should cell inoculation begin. As a general rule, an inoculum ratio of 1:10 has been found suitable for plant cell cultures. If the main reactor is 1000 liters, careful consideration must be given to the time taken to prepare the inoculum through the "seed" stages of 1, 10, and 100–1000 liters. At each stage, the same quality of activity is vital if the series is to be achieved. Such a series could take 1 month or so with present cell generation times. Once the culture has been inoculated, close attention must be paid to the following: pH, pO_2, gas flow (in and out), culture volume, cell accretion, and foaming. If a continuous culture is being operated, additional consideration needs to be given to the rate of medium supply, pump performance, and consistency of removal of the nutrient and biomass. The constant maintenance of sterile technique cannot be overstressed.

V. FINAL COMMENTS

I have not attempted, in the space available, to detail any particular system for mass cell growth. As more and more data are presented in the literature, it is becoming increasingly apparent that probably no one system is generally applicable to all plant cell cultures. This is perhaps not surprising given the range of metabolic activities and properties of plant cells. Instead, I have tried to describe the problems involved in growing plant cells on a large scale and to provide an entry into the increasingly diverse literature in this area. Regardless of the approach to large-scale plant cell culture, there are certain underlying principles governing all of them. I hope that I have been able to elucidate these principles in this chapter and to prepare better the research worker who wishes to enter an increasingly fascinating area of plant cell culture.

REFERENCES

Aiba, S., Humphrey, A. E., and Millis, N. F. (1973). "Biochemical Engineering," 2nd ed. Academic Press, New York.

Fowler, M. W. (1976). Continuous (chemostat) culture of plant cells using the LKB Ultroferm Fermentation System. *LKB Appl. Note* **252,** 1–5.

Fowler, M. W. (1982a). The large scale cultivation of plant cells. *Prog. Ind. Microbiol.* **16,** 207–230.

Fowler, M. W. (1982b). Substrate utilisation by plant cell cultures. *J. Chem. Tech. Biotechnol.* **32,** 338–346.

Kato, A., Shiozawa, Y., Yamada, A., Nishida, K., and Noguchi, M. (1977). A jar fermentor culture of *Nicotiana tabacum* L. cell suspensions. *Agric. Biol. Chem.* **36,** 899–904.

Lamport, D. T. A. (1964). Cell suspension culture of higher plants: Isolation and growth energetics. *Exp. Cell. Res.* **33,** 195–206.

Mandels, M. (1972). Culture of plant cells. *Adv. Biochem. Eng.* **2,** 201–215.

Martin, S. M. (1980). Mass culture system for plant cells. *In* "Plant Tissue Culture as a Source of Biochemicals" (E. J. Staba, ed.), pp. 149–166. CRC Press, Boca Raton, Florida.

Smart, N. J., and Fowler, M. W. (1981). Effect of aeration on large-scale cultures of plant cells. *Biotechnol. Lett.* **3,** 171–176.

Tulecke, W., and Nickell, L. G. (1959). Production of large amounts of plant tissue by submerged culture. *Science* **130,** 863–864.

Veliky, I., and Martin, S. M. (1970). Fermenter for plant cell suspension cultures. *Can. J. Microbiol.* **16,** 223–226.

Wagner, F., and Vogelmann, H. (1977). Cultivation of plant tissue cultures in bioreactors and formation of secondary metabolites. *In* "Plant Tissue Culture and Its Biotechnological Applications" (W. Barz, E. Reinhard, and M. H. Zenk, eds.), pp. 245–252. Springer-Verlag, Berlin and New York.

Wilson, G. (1981). Continuous culture of plant cells using the chemostat principle. *Adv. Biochem. Eng.* **16,** 1–26.

Wilson, G., King, P. J., and Street, H. E. (1971). Studies on the growth on culture of plant cells: A versatile system for the large scale batch or continuous culture of plant cell suspensions. *J. Exp. Bot.* **22,** 177–207.

Yamada, Y., Sato, F., and Hagimori, M. (1978). Photoautotrophism in green cultured cells. *In* "Frontiers of Plant Tissue Culture 1978" (T. A. Thorpe, ed.), pp. 453–462. Univ. of Calgary Press, Calgary, Canada.

CHAPTER **21**

Synchronization of Suspension Culture Cells

A. S. Wang
R. L. Phillips

Department of Agronomy and Plant Genetics
University of Minnesota
St. Paul, Minnesota

I. INTRODUCTION

Cultured plant cells vary greatly in size and shape, nuclear volume and DNA content, and cell cycle time (Wang *et al.*, 1982). This variation complicates studies of biochemical, genetic, and physiological aspects of cell division, metabolism, and other cellular properties. Several attempts have been made to achieve a high degree of synchronized cell division by manipulating growth conditions or adding DNA synthesis inhibitors to asynchronous suspension cultures. Progress has been slow. Low percentages of actively dividing cells (Wang *et al.*, 1982) and the tendency of cells to form aggregates in plant suspension cultures are major obstacles.

The cell cycle can be divided into interphase and mitosis (M). Interphase is divided into a DNA synthesis phase (S) preceded by a presynthetic

CELL CULTURE AND SOMATIC CELL
GENETICS OF PLANTS, VOL. 1

ISBN 0-12-715001-3

phase (G-1) and followed by a postsynthetic phase (G-2). The mitotic phase generally represents a small proportion of the total cell cycle. Mitosis is bordered by G-1 and G-2. Complete synchronization would require that all cells proceed simultaneously through specific stages of the cell cycle. This goal is seldom achieved in plant cell systems. Mitosis is one of the most visible means for measuring the degree of synchronization. The mitotic index is calculated as the percentage of cells in the mitotic phase of the cell cycle. The S period is much longer than M and can be determined by [^{3}H]thymidine autoradiography. King (1980) reviewed results on synchronization of plant cell suspension cultures. This chapter will outline the methodology of selected procedures that have particular merit. Only a general procedure is given in each case. Specific details are generally omitted because they vary with the species, media, and purpose of the experiment; such details may be found in the references cited.

Physical and chemical methods are both used for cell synchronization. Regarding physical methods, synchronization is based on physical cellular properties such as size of individual cells or aggregates, or on environmental growth conditions such as light and temperature. With chemical methods, inhibitors are often used to prevent cells from completing a cell cycle. Cells accumulate at a particular point in the cell cycle depending on the inhibitor.

II. PHYSICAL METHODS

A. Selection by Volume

Cultured plant cells are of irregular shape and volume and tend to form aggregates. These variables make selection by cell volume nearly impossible, but it is possible to select on the basis of aggregate size. Fujimura and Komamine (1979) synchronized somatic embryogenesis in carrot (*Daucus carota*) suspension cultures to the extent that 90% of the cell aggregates were in early embryogenic stages after the following procedure.

1. Subculture suspensions for 7 days in basal medium supplemented with 0.5 μM 2,4-dichlorophenoxyacetic acid (2,4-D).
2. Pass the cells through a 47-μm nylon screen first, and then filter through a 31-μm screen.
3. Collect cells and cell aggregates that are retained on the 31-μm screen and add an equal volume of liquid medium to the collected cells.
4. Gently layer 1 ml of cell suspension in a 10-ml centrifuge tube contain-

ing a 10–18% (w/v) Ficoll discontinuous density gradient (8 ml) containing 2% sucrose.

5. Centrifuge at 180 *g* for 5 min.
6. Collect 1.5-ml fractions and resuspend the heaviest fraction in 8 ml basal medium.
7. Centrifuge at 50 *g* for 5 sec; repeat this step four more times.
8. Transfer the cells in the lower portion to an embryo-inducing medium. Embryo formation can be detected in 4 or 5 days.

B. Cold Treatment

Temperature shocks can often be employed to increase synchronization. Okamura *et al.* (1973) combined a cold treatment and nutrient starvation by the following procedure to synchronize carrot suspension culture cells.

1. Transfer 10 ml cultured cells to 100 ml fresh medium.
2. Maintain cultures on a shaker (155 strokes/min) at 27°C until cell number plateaus; keep on the shaker for another 40 hr.
3. Place the culture in a cold room (4°C) and allow it to stand for 3 days.
4. Add 10-fold warm (27°C) fresh medium to the culture and allow growth for 24 hr at 27°C.
5. Repeat the cold treatment for another 3 days.
6. Incubate the culture at 27°C; increased cell division frequency and number of cells should be detected in 2 days.

III. CHEMICAL METHODS

A. Starvation

Deprivation of an essential growth compound from suspension cultures eventually leads to a stationary growth phase. Resupplying the missing compound or subculturing to a fresh complete medium will allow growth to resume and may result in synchronization of cell growth. The following procedure is simple; it is especially effective for sycamore (*Acer pseudoplatanus*) suspension cultures (King *et al.*, 1974). Gould *et al.* (1981) reported that sycamore cells are arrested in G-1 and G-2 by deprivation of phosphorus and carbohydrate. Nitrogen-starved cells accumulated in G-1 only. The general procedure is as follows:

1. Grow suspension cells in complete medium (King *et al.*, 1974) to the stationary phase.
2. Wait 1–2 weeks.
3. Transfer the culture to 10-fold fresh complete medium. The cultured cells may grow synchronously for two to five cell cycles.

Cell division in carrot (Nishi *et al.*, 1977) and sycamore (Everett *et al.*, 1981) suspension cultures has been synchronized by deprivation and readdition of auxin and cytokinin. Everett *et al.* (1981) reported that auxin starvation caused a periodic increase in the mitotic index (MI) and cell number. The general procedure is as follows:

1. Harvest cultured cells in stationary growth.
2. Wash three times with auxin- or cytokinin-free medium.
3. Subculture in auxin- or cytokinin-free medium until the MI declines to 0%.
4. Transfer to fresh complete medium containing 4.5 μM auxin or 1.2 μM cytokinin. A 5- to 10-fold increase in the MI should be observed in 3 days.

Jouanneau (1971) indicated that the best results are obtained in tobacco (*Nicotiana tabacum*) suspension cultures when the starvation method is coupled with volume selection. The general procedure is as follows:

1. Grow suspension cultures in complete medium to the stationary phase.
2. Pass cells through a 350-μm nylon screen.
3. Transfer cell aggregates larger than 350 μm to basal medium.
4. Adjust cell concentration to 25–30 $\times$ 10^3 cells/ml. Most (90%) cell aggregates should consist of approximately 25 cells.
5. Add 0.37 μM 2,4-D after 25 hr and 0.25 μM kinetin after 30–35 hr of growth.

A peak MI (8%) should be obtained in 60–70 hr after transfer.

B. Block and Release

Synchronization is achieved by temporarily blocking the progression of events in the cell cycle and accumulating cells in a specific stage. Upon release of the block, cells will synchronously enter the next stage. Generally, DNA synthesis inhibitors, such as 5-fluorodeoxyuridine (FudR), excess thymidine (TdR), and hydroxyurea (HU), are media supplements that accumulate cells at the G-1/S interface. FudR has been used to synchronize *Haplopappus gracilis* (Eriksson, 1966), soybean (*Glycine max*) (Chu and Lark,

1976), tobacco, and tomato (*Lycopersicon esculentum*) (Malmberg and Griesbach, 1980) suspension cultures. The general procedure is as follows:

1. Add FudR (2 μg/ml, final concentration) and uridine (1 μg/ml) to suspension cultures 1 day after subculture. The cell concentration should be between 5×10^3 and 3×10^6 cells per milliliter of medium.
2. Incubate for 12–24 hr.
3. Remove FudR and uridine by washing three times with fresh medium.
4. Subculture in fresh medium containing thymidine (2 μg/ml) for 12–24 hr.
5. Add colchicine (0.005%) to the culture.

This procedure may increase the MI 5- to 10-fold in one cell cycle.

Excess TdR was used by Eriksson (1966) to synchronize *H. gracilis* suspension cultures. One-day-old cultures were treated with 6 m*M* TdR for 24 hr, washed with fresh medium to remove the TdR, and maintained in fresh medium. A peak in the MI may be observed in 16 hr. A second block may be applied after the first one. In this case, the entire procedure is repeated.

Hydroxyurea has been used to increase synchronization of *H. gracilis* (Eriksson, 1966), wheat (*Triticum monococcum*) and parsley (*Petroselinum hortense*) (Szabados *et al.*, 1981), and corn (*Zea mays*) (Mi *et al.*, 1982) suspension cultures. The general procedure is as follows:

1. Add 3–5 m*M* HU to exponentially growing cultures for 24–36 hr (equivalent to one to one and a half cell cycles).
2. Wash culture three times with fresh medium.
3. Resuspend cells in fresh medium supplemented with 0.01–0.05%

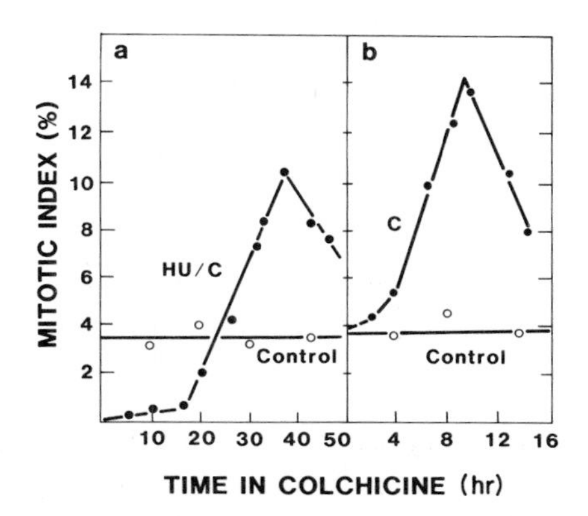

Fig. 1. Effect of hydroxyurea (HU) plus colchicine (C) and colchicine on the MI of Black Mexican sweet corn suspension cultures. (a) A 4-day-old culture was treated with 5 m*M* HU for 36 hr and subcultured to fresh medium supplemented with 0.02% (w/v) C. (b) C (0.02%, w/v) was added to exponentially growing cultures not previously treated with HU. Samples were taken at the intervals indicated. Each value is based on analysis of 1000 cells.

(w/v) colchicine. The final growth medium may be made with 20–50% conditioned medium.

Synchronized cells should be in mitosis after 15–36 hr. Using this technique, an MI of 30% in wheat suspension cultures and 80% in parsley suspension cultures was obtained (Szabados *et al.*, 1981). A five- to sixfold increase in MI is observed in corn suspension cultures compared to untreated control cultures (Fig. 1a).

Most DNA synthesis inhibitors should be filter sterilized; thymidine can be autoclaved.

IV. MITOTIC ARREST

Among the various spindle inhibitors (Phillips, 1981), colchicine is one of the most effective in arresting cells at metaphase. Suspension cultures in exponential growth are supplemented with 0.02% (w/v) colchicine. For corn, the MI should increase in 4 hr and reach a peak at 10–12 hr (Fig. 1b). The method is simple to use, but prolonged colchicine treatment should be avoided, as it leads to an increased frequency of abnormal mitoses in which metaphase chromosomes tend to become sticky. Chromosome stickiness has been observed in corn suspension cultures after an 8-hr colchicine treatment. To avoid chromosome stickiness, a shorter period (4–6 hr) of colchicine treatment is recommended. Colchicine should be filter-sterilized.

V. CONCLUSIONS

Difficulties in achieving a high degree of synchronization in plant cell suspension cultures are mainly due to the low frequency of dividing cells and the formation of cell aggregates. Frequent subculturing of exponentially growing cultures may reduce these problems. Procedures effective for synchronization with one species may not be useful for another. Combinations of these methods are often used. Nutrient starvation combined with colchicine or HU combined with colchicine, for example, are satisfactory for synchronization of corn suspension cultures. Most of the procedures listed here can be applied to callus and root tips with slight modifications.

REFERENCES

Chu, Y., and Lark, K. G. (1976). Cell-cycle parameters of soybean (*Glycine max* L.) cells growing in suspension culture: Suitability of the system for genetic studies. *Planta* **132,** 259–268.

Eriksson, T. (1966). Partial synchronization of cell division in suspension cultures of *Haplopappus gracilis*. *Physiol. Plant.* **19,** 900–910.

Everett, N. P., Wang, T. L., Gould, A. R., and Street, H. E. (1981). Studies on the control of the cell cycle in cultured plant cells. II. Effect of 2,4-dichlorophenoxyacetic acid (2,4-D). *Protoplasma* **106,** 15–22.

Fujimura, T., and Komamine, A. (1979). Synchronization of somatic embryogenesis in a carrot cell suspension culture. *Plant Physiol.* **64,** 162–164.

Gould, A. R., Everett, N. P., Wang, T. L., and Street, H. E. (1981). Studies on the control of the cell cycle in cultured plant cells I. Effects of nutrient limitation and nutrient starvation. *Protoplasma* **106,** 1–13.

Jouanneau, J. P. (1971). Contrôle par les cytokinines de la synchronisation des mitoses dans les cellules de tabac. *Exp. Cell Res.* **67,** 329–337.

King, P. J. (1980). Cell proliferation and growth in suspension cultures. *Int. Rev. Cytol., Suppl.* **11A,** 25–53.

King, P. J., Cox, B. J., Fowler, M. W., and Street, H. E. (1974). Metabolic events in synchronised cell cultures of *Acer pseudoplatanus* L. *Planta* **117,** 109–122.

Malmberg, R. L., and Griesbach, R. J. (1980). The isolation of mitotic and meiotic chromosomes from protoplasts. *Plant Sci. Lett.* **17,** 141–147.

Mi, C. C., Wang, A. S., and Phillips, R. L. (1982). Partial synchronization of maize cells in liquid suspension culture. *Maize Genet. Coop. News Lett.* **56,** 142–144.

Nishi, A., Kato, K., Takahashi, M., and Yoshida, R. (1977). Partial synchronization of carrot cell culture by auxin deprivation. *Physiol. Plant.* **39,** 9–12.

Okamura, S., Miyasaka, K., and Nishi, A. (1973). Synchronization of carrot cell culture by starvation and cold treatment. *Exp. Cell Res.* **78,** 467–470.

Phillips, R. L. (1981). Plant cytogenetics. *In* "Staining Procedures" (G. Clark, ed.), 4th ed., pp. 341–360. Williams & Wilkins, Baltimore, Maryland.

Szabados, L., Hadlaczky, G., and Dudits, D. (1981). Uptake of isolated plant chromosomes by plant protoplasts. *Planta* **151,** 141–145.

Wang, A. S., Phillips, R. L., and Mi, C. C. (1982). Cell cycle parameters and doubling time of Black Mexican Sweet corn suspension cultures. *Maize Genet. Coop. News Lett.* **56,** 144–146.

CHAPTER **22**

Photoautotrophic Cell Cultures

Wolfgang Hüsemann

Department of Plant Biochemistry
University of Münster
Münster, Federal Republic of Germany

I. INTRODUCTION

Photoautotrophic plant cell cultures provide a very useful tool for studying chloroplast-associated metabolic pathways and the relationship of cell growth and chloroplast development. Continuous photoautotrophic chemostat cultures will also aid in understanding how the mode of nutrition determines the photosynthetic assimilation of carbon dioxide and the productivity of the green cell.

Sustained photosynthetic growth of cultured plant cells was first re-

CELL CULTURE AND SOMATIC CELL
GENETICS OF PLANTS, VOL. 1

ISBN 0-12-715001-3

ported by Bergmann (1967). Since then, photoautotrophic growth has been obtained in many species. The cells are propagated as callus (Corduan, 1970; Berlyn and Zelitch, 1975), as cell suspensions (Chandler *et al.*, 1972; Hüsemann and Barz, 1977; Yamada and Sato, 1978; Katoh *et al.*, 1979; Hüsemann, 1981), as batch fermenter cultures (Bender *et al.*, 1980; Yamada *et al.*, 1981; Hüsemann, 1982), and as continuous cultures of the chemostat (Dalton, 1980; Hüsemann, 1983) or turbidostat type (Peel, 1982).

Some of the requirements of culture conditions and the methods for selecting and propagating photoautotrophic cultures of *Chenopodium rubrum* are presented here.

II. CULTURE CONDITIONS AND SELECTION PROCEDURES

The establishment of photoautotrophic cell cultures represents a process of serial selection by the replacement of sucrose as the normal carbon source in the nutrient medium with elevated levels of carbon dioxide and light.

A. Illumination

The cells are illuminated with continuous white light (19 W/m^2; 8000 lx) provided by fluorescent tubes or HQL de Luxe lamps (Osram, Federal Republic of Germany).

B. Carbon Dioxide

Two different methods are used for establishing high CO_2 concentrations of 1–2% (v/v) in the gaseous atmosphere above the cells in the culture flasks.

1. The cell cultures are gased with a continuous stream of a gas mixture of known composition enriched with CO_2 to 1–2% (v/v).
2. High CO_2 partial pressures can be established in a closed culture flask by a bicarbonate–carbonate mixture, according to Warburg *et al.* (1962). Some $KHCO_3/K_2CO_3$ ratios for establishing different CO_2 partial pressures are given in Table I.

TABLE I

The Establishment of High CO_2 Partial Pressures by Bicarbonate–Carbonate Mixtures

Stock solutions: $KHCO_3$–K_2CO_3 (moles/liter)	Buffer mixtures: ml $KHCO_3$/ml K_2CO_3	CO_2 partial pressure[a]	
		mm Brodie solution	% (v/v)
3.0	50/50	51	0.5
3.0	62/38	104	1.0
3.0	73/27	203	2.0
2.0	79/21	202	2.0

[a] The CO_2 partial pressures above the buffer solutions are calculated for 25°C after the formula $(KHCO_3)^2/(K_2CO_3 \times CO_2) = K$. The value for $K_{25°C}$ has been determined by interpolating between $K_{20°C} = 3.35 \times 10^{-2}$ and $K_{38°C} = 1.78 \times 10^{-2}$ $\left(\frac{\text{mol/liter}}{\text{mm Brodie}}\right)$ according to Warburg *et al.* (1962).

C. Sugar

Sugar may be omitted at once or in several steps from the culture medium. Both strategies have proved to be satisfactory in selecting photoautotrophic cell cultures. But the slow adjustment of the cells to reduced growth rates at low sugar concentrations in the presence of elevated levels of carbon dioxide is obviously more beneficial and even perhaps necessary for some plant cell cultures to survive.

D. Growth Regulators

In general, auxin concentrations in the culture medium should be kept below 10^{-6} *M*. In some cases, the replacement of 2,4-dichlorophenoxyacetic acid (2,4-D) by naphthaleneacetic acid (NAA) will favor chlorophyll formation.

E. Selection Procedures

The change from the photoheterotrophic (i.e., light-grown chlorophyllous cells requiring an exogenous sugar as the main carbon source) to the photoautotrophic mode of nutrition, with carbon dioxide as the sole carbon source, initially results in a drastic reduction of culture growth and the chlorophyll content of the cells. Only the cells with the desired photosynthetic capacity will survive this selection pressure. Therefore, rapidly growing (more than a 300% increase in fresh weight within 10 days) and highly chlorophyllous (more than 70 μg chlorophyll per gram of fresh weight) cell cultures are necessary prerequisites for the selection of cell lines which are capable of photoautotrophic growth.

Fig. 1. A two-tier culture flask for photoautotrophic growth of plant cell suspensions at elevated CO_2 concentrations. The lower compartment contains 50 ml of a $KHCO_3$–K_2CO_3 buffer solution as a CO_2 reservoir. The liberated CO_2 passes through a glass tube into the upper compartment containing a 30-ml cell suspension. (From Hüsemann and Barz, 1977.)

For example, green cell suspensions from *C. rubrum* are propagated photoheterotrophically under continuous illumination (19 W/m^2) in 40 ml of Murashige and Skoog's (MS) (1962) nutrient medium supplemented with 2% sucrose and 10^{-7} *M* 2,4-D. The cells are transferred every 10 days into fresh culture medium. Highly chlorophyllous photoheterotrophic cell suspensions, resting in the early phase of stationary growth, are the starting material for the selection of photoautotrophic cell lines. Two-gram cells are suspended into 30 ml sugar-free nutrient medium of a two-tier culture flask (Fig. 1). The CO_2 concentration in the gaseous atmosphere of the culture vessel is adjusted to approximately 2% (v/v) using the $KHCO_3$–K_2CO_3 buffer system. The cell suspensions are grown under continuous illumination (19 W/m^2) on a gyrotory shaker (120 rpm) at 25 ± 1°C.

Careful visual monitoring during subculture is necessary to select fast-growing, dark green photoautotrophic cell suspensions. Small green cell groups and single cells are separated by filtering through nylon gauze (pore size: 100 μm) and subsequently collected on a 10-μm close-meshed nylon gauze. Only the greenest cells are transferred into fresh culture medium. In general, the establishment of photoautotrophic cell cultures is a long process, taking, for example, about 6 months for cell suspensions from *C. rubrum, Peganum harmala,* and *Morinda lucida.*

III. A TWO-TIER CULTURE FLASK FOR PHOTOAUTOTROPHIC CULTURE GROWTH

The two-tier culture flask has been developed for the photoautotrophic growth of cell suspensions of *C. rubrum* (Hüsemann and Barz, 1977). This culture method has also been used successfully for the propagation of photoautotrophic cell suspensions from *P. harmala* and *M. lucida.*

The two-tier culture flask is constructed of two 200-ml small-necked Erlenmeyer flasks (Fig. 1). The lower compartment contains 50 ml of a 2 *M* or 3 *M* $KHCO_3$–K_2CO_3 buffer mixture for establishing enriched CO_2 concentrations in the gaseous atmosphere of the culture vessel. The liberated CO_2 passes through the central glass tube [2.0 cm inner diameter (i.d.), 3.5 cm in length] into the upper compartment of the flask, establishing the desired CO_2 partial pressure for photosynthetic CO_2 assimilation. The upper compartment functions as the culture vessel for propagating the cells in 30 ml sugar-free nutrient medium.

The two-tier culture flask (openings closed by aluminum foil), the culture medium, and the bicarbonate–carbonate buffer mixtures are separately sterilized by autoclaving at 120°C for 20 min. The buffer mixture must be filled into a screw-capped flask made of Duran glass (Schott, Federal Republic of Germany), that will stand the pressure of the boiling buffer solution during autoclaving and thus prevent the loss of CO_2.

For routine subculturing of photoautotrophic cell suspensions of *C. rubrum*, 30 ml sugar-free MS medium and approximately 1 g cells are introduced into the upper compartment of the two-tier culture flask to give an initial cell culture density of about 400,000 cells/ml. For transferring the cells, a stainless steel spoon or pipette may be used. Fifty milliliters of the 2 *M* or 3 *M* $KHCO_3$–K_2CO_3 mixture is introduced into the lower compartment of the culture flask by its side arm (1.5 cm i.d., 3.0 cm in length). The openings of the culture flask are tightly sealed with sterile aluminum foil and finally with Parafilm to reduce gas exchange with the atmosphere. The cultures are mounted on a gyrotory shaker, (120 rpm), illuminated with continuous white light at 19 W/m^2 at 25 ± 1°C.

IV. BATCH FERMENTER AND CONTINUOUS CULTURE OF PHOTOAUTOTROPHIC CELL SUSPENSIONS

A. Construction of the Airlift Culture Unit

Air bubble turbulence is well suited for mixing and aerating cells in suspension if culture volumes of 1–3 liters are desired (Kurz, 1973; Wilson, 1976).

Here, the construction and operation of an airlift culture system that can be made by the local glass blower will be described (Fig. 2). It has been developed for the photoautotrophic growth of *C. rubrum* cell suspensions (Hüsemann, 1982, 1983).

The culture vessel is made of a 2-liter small-mouthed Erlenmeyer flask used in the inverted position. The mouth of the flask is sealed off. The upper side now carries six ports made of ground glass joints (NS ground glass, Schott) with sockets inserted.

The central port (no. 4; NS 24/29) is fitted with the air inlet tube. The lower end of the tube is set 1.5 cm from the bottom of the culture vessel. Five other ports are located off center.

Port no. 3 (NS 14/23) carries a stainless steel tube (4 mm i.d., 18 cm in length) that is closed at its lower end for holding a thermometer to control the temperature of the cell suspension.

Cells and nutrient medium are introduced into the culture vessel through port no. 5 (NS 29/32).

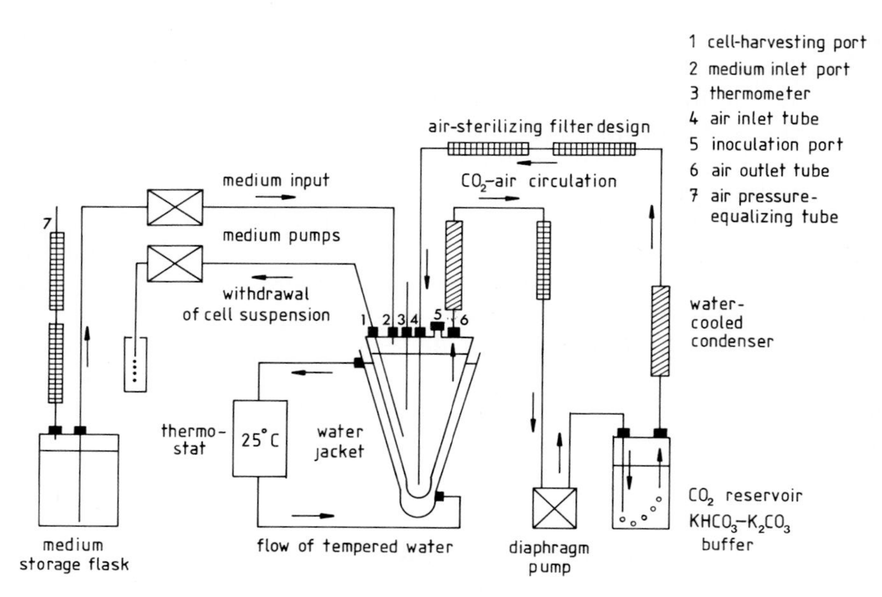

Fig. 2. Sectional view of a continuous-culture system working on the airlift principle for photautotrophic culture growth. (From Hüsemann, 1983.)

Port no. 6 (NS 14/23) functions as the air outlet port and is connected to a water-cooled condenser (25 cm in length) that prevents evaporation losses from the culture medium as well as moistening of the air-sterilizing filter assembly. Sterile filters made of cotton wool enclosed in glass tubes (2.5 cm i.d., 20 cm in length) are set into the silicon rubber tubings of the air inlet and air outlet systems to keep the cell culture aseptic.

For a continuous operation, the culture unit is equipped with two oil-free diaphragm pumps (FE 211, B. Braun, Melsungen, Federal Republic of Germany) for the simultaneous addition of fresh nutrient medium and withdrawal of equal amounts of cell suspension. Both medium pumps are *controlled* by an electric timer (W. Järsch, Nordwalde, Federal Republic of Germany) for synchronous operation at preset time intervals to give the desired dilution rate.

Port no. 2 (NS 14/23) functions as the medium inlet port in the case of continuous operation of the culture system and is connected by viton tubes (2 mm i.d.) to the medium pump and to a 10-liter medium storage flask. In addition, sterile filters fitted to the medium storage flask function as air pressure-equalizing tubes.

The cell harvesting system (port no. 1; NS 14/23) is made of a glass tube (2 mm i.d., 20 cm in length) that is connected by a viton tube (2 mm i.d.) to a diaphragm pump for withdrawal of cell suspension.

B. Carbon Dioxide Supply and Mode of Aeration

1. A CO_2 concentration of approximately 2% (v/v) in the gaseous atmosphere of the culture unit is provided by 1.5 liters of a 3 *M* $KHCO_3$–K_2CO_3 buffer mixture (ratio 73/27). To ensure a sufficient supply of the cells with CO_2, the buffer mixture should be renewed every sixth day.

2. The air inlet and air outlet ports of the culture vessel and the buffer reservoir are connected by silicon rubber tubes to an oil-free diaphragm air pump (WISA 300, W. Sauer, Wuppertal, Federal Republic of Germany) by which the CO_2–air mixture is set into continuous circulation through the culture system. The air input pipe discharges the gas mixture at the bottom of the buffer reservoir. The rising air bubbles, enriched in CO_2, leave the flask, passing first a water-cooled condenser by which evaporation losses from the buffer solution as well as moistening of the sterile filters will be prevented. After the gas mixture has been made aseptic by passing through the air-sterilizing filter assembly, it finally enters the culture vessel by the air inlet tube and emerges from the aeration shaft at the bottom of the culture flask. The air bubbles, together with the conical shape of the culture flask, induce enough turbulence for mixing and aeration of the cell suspension. Before flowing back to the air pump and finally to the buffer reservoir, the CO_2–air mixture must pass a water-cooled condenser and an air-sterilizing filter assembly.

C. Temperature Control

The temperature of the cell suspension is controlled by the flow of tempered water through the water jacket of the culture unit. The water jacket is built by placing the fermenter vessel into a 5-liter Erlenmeyer flask. The temperature can thus be precisely adjusted (± 0.5°C) by using a thermostat, type Frigomix 1495/Thermomix 1441 BKU (B. Braun, Melsungen, Federal Republic of Germany).

D. Aseptic Manipulations

The culture vessel including all silicon rubber tubes, the air-sterilizing filter assemblies, the water-cooled condensers, the pump pistons of the medium pumps including the viton tubes, and the culture medium (introduced into separate flasks) are sterilized by autoclaving for 30 min at 120°C. All open connections are closed by aluminum foil. All manipulations for assembling the culture unit are carried out in a laminar flow hood.

E. Batch Fermenter or Continuous-Culture Growth under Photoautotrophic Conditions

Photoautotrophic cell suspensions from *C. rubrum*, propagated in two-tier culture flasks and resting in the early phase of stationary growth, are introduced into 1.5-liter sugar-free culture medium of the fermenter vessel to give an initial culture density of about 600,000 cells per milliliter of suspension. The CO_2 concentration in the gaseous atmosphere is adjusted to approximately 2% (v/v), and the airflow rate is set at 300 ml/min. The cells are illuminated with continuous white light at 19 W/m^2 provided by three HQL de Luxe lamps. The temperature of the cell suspension is kept at 25 ± 0.5°C. For a continuous operation of the culture unit, the dilution rate can be set to a maximum 0.16/day (10 ml/hr), taking 1.5 liters as the working volume of the culture.

Some problems may arise with regard to floating of cells and their accretion on the glass walls and tubes. They may be resolved by daily wiping of cells back into the culture medium by gently swirling the culture vessel. Culture densities above 1,500,000 cells/ ml increase the risk of blocking the cell-harvesting system.

V. CONCLUSIONS

Using the two-tier culture flask as well as the airlift fermenter system, cell suspensions from *C. rubrum* have successfully been grown photoautotrophically in a simple mineral salt medium, with the maintenance of high photosynthetic capacities. In addition, in continuous photoautotrophic culture growth, steady-state conditions have been established with respect to cell growth, cell composition, and photosynthetic as well as enzymatic activities of the cells.

The methods used for selection and propagation of photoautotrophic cell cultures, as well as the mode of CO_2 supply described here, have proved satisfactory for routine culture work.

ACKNOWLEDGMENTS

The author's studies on photoautotrophic cell cultures have been supported by the Deutsche Forschungsgemeinschaft.

REFERENCES

Bender, L., Kumar, A., and Neumann, K. H. (1980). Photoautotrophe pflanzliche Gewebekulturen in Laborfermentern. *In* "Fermentation" (R. M. Lafferty, ed.), pp. 193–203. Springer-Verlag, Berlin and New York.

Bergmann, L. (1967). Wachstum grüner Suspensionskulturen von *Nicotiana tabacum* var. "Samsun" mit CO_2 als Kohlenstoffquelle. *Planta* **74,** 243–249.

Berlyn, M. B., and Zelitch, I. (1975). Photoautotrophic growth and photosynthesis in tobacco callus cells. *Plant Physiol.* **56,** 752–756.

Chandler, M. T., Tandeau de Marsac, N., and De Kouchkovsky, Y. (1972). Photosynthetic growth of tobacco cells in liquid suspension. *Can. J. Bot.* **50,** 2265–2270.

Corduan, G. (1970). Autotrophe Gewebekulturen von *Ruta graveolens* und deren $^{14}CO_2$-Markierungsprodukte. *Planta* **91,** 291–301.

Dalton, C. C. (1980). Photoautotrophy of spinach cells in continuous culture: Photosynthetic development and sustained photautotrophic growth. *J. Exp. Bot.* **31,** 791–804.

Hüsemann, W. (1981). Growth characteristics of hormone and vitamin independent photoautotrophic cell suspension cultures from *Chenopodium rubrum*. *Protoplasma* **109,** 415–431.

Hüsemann, W. (1982). Photoautotrophic growth of cell suspension cultures from *Chenopodium rubrum* in an airlift fermenter. *Protoplasma* **113,** 214–220.

Hüsemann, W. (1983). Continuous culture growth of photoautotrophic cell suspensions from *Chenopodium rubrum*. *Plant Cell Rep.* **2,** 59–62.

Hüsemann, W., and Barz, W. (1977). Photoautotrophic growth and photosynthesis in cell suspension cultures from *Chenopodium rubrum*. *Physiol. Plant.* **40,** 77–81.

Katoh, K., Ohta, Y., Hirose, Y., and Iwamura, T. (1979). Photoautotrophic growth of *Marchantia polymorpha* L. cells in suspension culture. *Planta* **144,** 509–510.

Kurz, W. G. W. (1973). A chemostat for single cell cultures of higher plants. *In* "Tissue Culture: Methods and Applications" (P. F. Kruse, Jr. and M. K. Patterson, eds.), pp. 359–363. Academic Press, New York.

Murashige, T., and Skoog, F. (1962). A revised medium for rapid growth and bioassay with tobacco cultures. *Physiol. Plant.* **15,** 473–479.

Peel, E. (1982). Photoautotrophic growth of suspension cultures of *Asparagus officinalis* L. cells in turbidostats. *Plant Sci. Lett.* **24,** 147–155.

Warburg, O., Geissler, A. W., and Lorenz, S. (1962). Neue Methode zur Bestimmung der Kohlensäuredrucke über Bicarbonat-Carbonatgemischen. *In* "Weiterentwicklung der zellphysiologischen Methoden" (O. Warburg, ed.), pp. 578–581. Thieme, Stuttgart.

Wilson, G. (1976). A simple and inexpensive design of chemostat enabling steady-state growth of *Acer pseudoplatanus* L. cells under phosphate-limiting conditions. *Ann. Bot. (London)* [N.S.] **40,** 919–932.

Yamada, Y., and Sato, F. (1978). The photoautotrophic culture of chlorophyllous cells. *Plant Cell Physiol.* **19,** 691–699.

Yamada, Y., Imaizumi, K., Sato, F., and Yasuda, T. (1981). Photoautotrophic and photomixotrophic culture of green tobacco cells in a jar-fermenter. *Plant Cell Physiol.* **22,** 917–922.

CHAPTER **23**

Quantitative Plating Technique

Robert B. Horsch

Corporate Research Laboratories
Monsanto Company
St. Louis, Missouri

I. INTRODUCTION

This chapter will focus on several methods for plating plant cells on agar-solidified medium using a defined and dispersed suspension of cells as inoculum. Three quantitative techniques will be described, each designed to fulfill different goals: (1) selection of rare variant colonies that are resistant to a toxic agent in the medium, (2) measurement of growth rate, and (3) low-density plating of cells.

The characteristic growth of plant cells in clusters and the requirement of a minimum cell density for growth can complicate the selection of rare resistance mutants from a huge excess of inhibited wild-type cells. Reconstruction experiments (Horsch and Jones, 1980a) have demonstrated the need to evaluate and control various parameters during selection, including overall density and size of cell clusters at the time of plating.

Measurement of culture growth is very simple for pieces of callus on agar plates but quite variable due to nonuniform exposure to the medium. Suspension culture provides more uniform contact between cells and medium but requires more space, time, and materials for growth measurements. The filter paper growth assay (FPGA) (Horsch *et al.*, 1980) combines

CELL CULTURE AND SOMATIC CELL
GENETICS OF PLANTS, VOL. 1

ISBN 0-12-715001-3

the ease of callus culture with uniformity comparable to that of a suspension culture by using a fine suspension of cells spread over a large surface of a movable substrate laid over agar medium.

In general, single plant cells or small clusters of cells will not grow below a critical density of 10^3 or 10^4 cells per milliliter. This density is too high to maintain the clonal identity of separate colonies except by special treatments such as microdrop culture or use of nurse tissues. For large-scale experiments, however, cells must be dealt with in large groups while still maintaining independence. The double filter paper plating technique (Horsch and Jones, 1980b) provides efficient, density-independent growth of single cells on a movable substrate over a nurse culture. This method depends upon the presence of a nurse culture only in the initial phases of growth, allows easy counting of colonies, and permits rapid, convenient transfer of groups of colonies to other media without loss of individual colony identity or spatial distribution.

II. PROCEDURES

A. High-Density Plating of Sized Suspension Cultures

Plant cells grow in suspension in units ranging from single cells to clusters of more than 1000 cells. To deal with such a broad distribution, the population of units can be sized by using different screens (e.g., Teflon screen, Albert Godde Bedin Sales, Elmsford, New York). Cell clusters smaller than a given screen size can be collected on a Miracloth (Chicopee Mills, New York, New York) filter after passage through the screen. Cell clusters larger than a given size can be obtained by exhaustively washing smaller clusters through the proper size screen. A desired range of cell cluster sizes can be obtained by a combination of the two approaches.

Effective sieving was obtained using the Millipore Sterifil Aseptic System (Millipore, Bedford, Massachusetts). A 12-in. square of Teflon screen was draped into the upper funnel so that it almost touched the Miracloth filter in the bottom, and was securely bent over the rim of the funnel at the top. Clogging was prevented by capping the outlets in the collection flask to prevent drainage of liquid through the filter and then filling the funnel with water or medium to keep the cells suspended. Mechanical agitation facilitated movement of small units through the screen; more than half of the desired units present could be sieved in a single attempt.

Opening a side vent in the collection flask permitted the liquid to drain from the funnel; the sieved cells collected on the Miracloth filter. These cells were then weighed into a sterile beaker with a stir bar, and medium was added to create the desired suspension of sieved cells. With gentle stirring on a magnetic stir plate, the defined, uniform suspension was ready for inoculation onto agar plates. Typically, 1 ml was spread over the surface of 25 ml medium in 100 × 15-mm plastic Petri plates. The concentration of cells in the suspension determined the amount of inoculum. For example, a 10% (w/v) suspension of cells will provide 0.1 g wet weight of cells per plate. This is a reasonable starting point for tailoring the technique to a new situation. For *Haplopappus gracilis,* a 350-μm filtrate contained clusters of up to 120 cells and 0.1 g wet weight contained about 10^6 cells.

Several important parameters to explore in new applications include size of cell clusters, density of plating, concentration of inhibitor, and general culture conditions. It is important that cells be able to grow to form a uniform confluent layer in the absence of an inhibitor. It should be kept in mind that everything that grows on selective medium is not stable or clonal in many cases. Cloning and rigorous retesting are always prudent.

B. Filter Paper Growth Assay

Suspension cultures were sieved through either 350-μm or 500-μm screens, collected, and resuspended in medium as described above at a density of 20% (w/v). One-half-milliliter aliquots were inoculated onto the surface of 7-cm disks of Whatman no. 1 (or other) qualitative filter paper that had been placed on 25 ml of agar medium in 100 × 15-mm plastic Petri plates. The plates were carefully tilted and rotated to distribute the cells without spilling off the filters.

Culture wet weight was measured as follows: The entire assay plate was weighed, and then the filter was removed with forceps and held in the air while the plate alone was weighed. The filter was then replaced. The difference equaled the weight of the filter paper plus that of the adhering cells. By subtracting the weight of the inoculum, it was possible to calculate the tare weight of the filter paper alone (including the weight of medium it had absorbed). At daily intervals each plate was weighed again, with and without the filter paper and cells. The difference in weight was recorded on the lid of each plate. The filter paper tare weight was subtracted from each daily reading to give the net weight of the culture.

All the operations were performed under aseptic conditions in a laminar flow hood, using a top-loading electronic balance with an accuracy of ±10 mg. The plates were stored in plastic bags to prevent excessive drying.

For dry weight determination of FPGA cultures, filters were dried at 60°C, weighed and autoclaved before use. After the desired culture period, filters with cells were removed from agar medium, placed in the lids of the corresponding plates and dried at 60°C to constant weight (overnight). The filters were then weighed again (with the dry cells), and the tare weight of the paper was subtracted. The average dry weight of medium trapped by a filter disk was also subtracted.

Growth phase and density of the inoculum, medium, and culture conditions can affect the growth rate of the cells and the reproducibility of the assay. For example, low-density inoculum can result in large variation between replicates due to sporadic growth, whereas high-density inoculum reduces the sensitivity of the assay.

C. Double Filter Paper Plating System

To facilitate the growth of large numbers of isolated cells, feeder plates were prepared by layering filter paper disks over feeder cells over agar medium in Petri plates. The feeder cells were obtained from suspension cultures of either the same species as the cells to be plated on top or a different species found (by experiment) to be compatible. If necessary, large clumps were removed by sieving the culture as described above. Typically, 0.1–0.2 g of feeder cells were spread over the surface of 25 ml of agar medium in 100 × 15-mm plastic Petri plates.

Circles of Whatman no. 1 (or other) filter paper (guard disks) trimmed (or special ordered) to 8.5 cm in diameter were placed over the feeder cells. A second disk (transfer disk) of 7-cm-diameter paper was placed over the center of each guard disk. The double filter paper plates were ready to use immediately or could be stored for a few days. They were usually prepared just prior to use.

Plating suspensions were obtained by filtering suspension cultures through a Teflon screen with 74-μm openings. Each cell or clump in the filtrate was referred to as a "plating unit." Plating unit titers were determined by counting cells in a hemocytometer. The plating suspension was diluted serially in medium to the titer desired for plating. One-half milliliter of a dilute cell plating suspension typically containing 20–200 plating units was then pipetted onto the inner filter of each feeder plate, and the plates were gently swirled so that cells were distributed over the surfaces of the transfer disks without spilling onto the guard disks.

Plates were incubated under suitable conditions for cultures of the species until visible colonies appeared (about 14–28 days after plating). Absolute plating efficiency was defined as the percentage of plating units that developed into visible colonies.

To be certain of the origin of specific colonies, plating units were selected and transferred to feeder plates in the following way: A dilute plating suspension was spread over the surface of a thin layer of agar medium. Isolated, individual units were detected visually, removed along with the surrounding agar medium, and transferred to a sterile Petri plate. The number of cells in a unit was determined using an inverted microscope, and the unit with its supporting medium was placed on the transfer disk of a feeder plate. Growth of each colony was easily monitored.

Protoplast-derived cells, and perhaps even freshly isolated protoplasts, can be readily grown using this system, although better results were obtained with cells that had re-formed a wall before plating.

The medium and culture conditions must be appropriate for the species and cell origin of the plating units. Specific conditions for *H. gracilis* or *Datura innoxia* are given in the referenced publications. Extrapolation to other species, genotypes, or tissue sources must be optimized by experimentation. The source of feeder cells, the hormones in both the feeder plates and in the culture used as a source of plating units, and environmental conditions such as light and temperature can have a large influence on plating efficiency.

III. RESULTS AND DISCUSSION

Resistance mutants can be selected from high-density cultures spread over agar medium (Horsch and Jones, 1980a). In the case of azauracil resistance, it was possible to isolate quantitatively single resistant cells from a huge excess of wild-type cells in a single step. However, in the case of azaguanine resistance, it was possible to select resistant colonies only by gradual enrichment. The efficacy and optimal conditions of selection must be determined by experiment for each application. Cultures that do not plate well from suspension cannot be used with this technique.

The filter paper growth assay has been used to measure culture growth effects under a wide variety of experimental conditions, including the response of strains to metabolite analogs (Horsch and Jones, 1980a) and antibiotics (Horsch and King, 1983), and the response of auxotrophic mutants to different concentrations of required metabolites (King *et al.*, 1980). Standard errors from triplicate cultures were usually less than 5% of the mean. Three basic kinetic responses to suboptimal conditions have been observed: reduced growth rate, longer lag periods followed by growth at similar rates, and initially normal growth kinetics followed by early cessation of growth.

The double filter paper plating system has been used to obtain large numbers of isolated colonies from finely sieved suspensions (Horsch and Jones, 1980b) and protoplast-derived cells (Fraley *et al.*, 1983). These colonies have been counted to determine the cell titer before and after mutagenesis or selection, isolated to obtain clones, or scored or selected for mutant phenotypes.

Many other useful plating techniques have been published for a wide variety of plant cell cultures; the techniques described in this chapter may not be suitable or optimal for many situations. Sources of further information on plating plant cells include Bergmann (1977), Caboche (1980), Gleba (1978), Jones *et al.* (1960), Muir *et al.* (1954), Raveh *et al.* (1973), Street (1977), Weber and Lark (1979), and Werry and Stoffelsen (1978).

The key to successful application of these techniques to any system includes (1) determination of the suitability of the technique to answer the question(s) being asked, (2) experimentation to test the efficacy of the technique within the limitations of the culture being used, and (3) optimization to find the best size, density, and age of the inoculum and the best medium, hormones, and environmental conditions for incubation.

REFERENCES

Bergmann, L. (1977). Plating of plant cells. *In* "Plant Tissue Culture and Its Bio-Technological Application" (W. Barz, E. Reinhard, and M. H. Zenk, eds.), pp. 213–225. Springer-Verlag, Berlin and New York.

Caboche, M. (1980). Nutritional requirements of protoplast-derived haploid tobacco cells grown at low cell densities in liquid medium. *Planta* **149,** 7–18.

Fraley, R. T., Rogers, S. G., Horsch, R. B., Sanders, P. R., Flick, J. S., Adams, S. P., Bittner, M. L., Brand, L. A., Fink, C. L., Fry, J. S., Galluppi, G. R., Goldberg, S. B., Hoffmann, N. L., and Woo, S. C. (1983). Expression of bacterial genes in plant cells. *Proc. Natl. Acad. Sci. U.S.A.* **80,** 4803–4807.

Gleba, Y. Y. (1978). Microdroplet clture: Tobacco plants from single mesophyll protoplasts. *Naturwissenschaften* **65,** 158–159.

Horsch, R. B., and Jones, G. E. (1980a). The selection of resistance mutants from cultured plant cells. *Mutat. Res.* **72,** 91–100.

Horsch, R. B., and Jones, G. E. (1980b). A double filter paper technique for plating cultured plant cells. *In Vitro* **16,** 103–108.

Horsch, R. B., and King, J. (1983). A covert contaminant of cultured plant cells: Elimination of a *Hyphomicrobium sp.* from cultures of *Datura innoxia* (Mill.). *Plant Cell, Tissue Organ Cult.* **2,** 21–28.

Horsch, R. B., King, J., and Jones, G. E. (1980). Measurement of cultured plant cell growth on filter paper discs. *Can. J. Bot.* **58,** 2402–2406.

Jones, L. E., Hildebrandt, A. C., Riker, A. J., and Wu, J. H. (1960). Growth of somatic tobacco cells in microculture. *Am. J. Bot.* **47,** 468–475.

King, J., Horsch, R. B., and Savage, A. (1980) Partial characterization of two stable autotrophs of *Datura innoxia* (Mill.). *Planta* **149,** 480–484.

Muir, W. H., Hildebrandt, A. C., and Riker, A. J. (1954). Plant tissue cultures produced from single isolated cells. *Science* **119,** 877–878.

Raveh, D., Huberman, D. E., and Galun, E. (1973). *In vitro* culture of tobacco protoplasts: Use of feeder techniques to support division of cells plated at low densities. *In Vitro* **9,** 216–222.

Street, H. E., (1977). "Plant Tissue and Cell Culture." Univ. of California Press, Berkeley.

Weber, G., and Lark, K. G. (1979). An efficient plating system for rapid isolation of mutants from plant cell suspensions. *Theor. Appl. Genet.* **55,** 81–86.

Werry, P. A. T. J., and Stoffelsen, K. M. (1978). Conditions for a high plating efficiency of free cell suspensions of *Haplopappus gracilis* (Nutt.) Gray. *Theor. Appl. Genet.* **51,** 161–167.

CHAPTER **24**

The Feeder Layer Technique

D. Aviv
E. Galun

Department of Plant Genetics
The Weizmann Institute of Science
Rehovot, Israel

I. INTRODUCTION

Plant cells, especially plant protoplasts, are very sensitive to plating densities. Below a minimal cell density, which is often around 10^4/ml, the cells do not divide and usually disintegrate rapidly. This phenomenon interferes significantly with plant cell cultures, especially in those cases in which only a small number of surviving cells are expected, e.g., mutagenized cells and protoplasts of somatic fusion products.

A similar phenomenon of density dependence was observed with cultured mammalian cells before its detection in plant cells. The former was overcome successfully by the use of an underlay of X-irradiated cells (Puck and Marcus, 1955). Ten years ago, a similar technique based on X-irradiated protoplasts was developed in our laboratory (Raveh *et al.*, 1973); since then, it has been successfully employed in a wide range of experiments. It will henceforth be referred to as the "feeder layer technique."

CELL CULTURE AND SOMATIC CELL
GENETICS OF PLANTS, VOL. 1

ISBN 0-12-715001-3

II. THE FEEDER LAYER TECHNIQUE

A. Protoplast Irradiation

The protoplasts can be X-irradiated by either a roentgen apparatus (e.g., Dermavolt, Rich Seifert and Co.) or a gamma ray source (e.g., ^{60}Co, G. B. 150A, Atomic Energy of Canada). The optimal dose is the lowest dose which "completely" inhibits (i.e., by over 99.99%) protoplast division. This dose may differ from one species to another, but the source of tissue from which the protoplasts were derived has a far greater effect on the optimal dose. Thus, the dose which prevents division of *Nicotiana tabacum* mesophyll protoplasts is about 5 krad (Galun and Raveh, 1975; Sidorov *et al.*, 1981) and is similar to the dose required for other Solanaceae mesophyll protoplasts. In contrast, the dose required to prevent mitosis in cell suspension-derived *N. tabacum* protoplasts is 20 krad (D. Aviv, unpublished) and is similar to the dose required for other cell suspension- or callus-derived protoplasts, e.g., *Citrus* protoplasts (Vardi *et al.*, 1975; Vardi and Raveh, 1976).

The irradiated protoplasts should be washed once or twice. This probably removes toxic free radicals. For simplicity, the protoplasts can be irradiated while still in the enzyme solution, just before collection, and thereafter washed as usual.

B. Plating of Irradiated Protoplasts

The irradiated protoplasts can be used in one of three ways: (1) mixed in agar medium together with the unirradiated protoplasts (the latter will henceforth be referred to as the "target protoplasts"); (2) mixed in agar medium and plated in Petri plates as an underlay; upon solidification, the target protoplasts are plated on top as an overlay in a softer agar medium; or (3) plated in liquid medium to which the target protoplasts are added.

In the first report of the use of a feeder layer (Raveh *et al.*, 1973), it was noted that the closer the contact between irradiated and target protoplasts, the better the recovery of the latter. Therefore, procedures 1 and 3 have certain advantage over procedure 2. On the other hand, in procedure 2 the feeder plates can be prepared and stored in a cool place ahead of time (up to at least 1 week before use). Procedure 3 is the procedure of choice if the target protoplasts should preferably be cultured in liquid medium. In that case, the contact is maximal and the irradiated protoplasts can be prepared ahead of time.

The plating density of the irradiated protoplasts is crucial, as shown by

Raveh *et al.* (1973). If their density is too high, they have an inhibitory effect on the target protoplasts; if their density is too low, they do not have a feeding effect. Their optimal concentration is equivalent to the lowest concentration in which the same protoplasts would divide satisfactorily when not irradiated.

C. Source of Protoplasts for the Feeder Layer

Mesophyll-, cell suspension-, and callus-derived protoplasts can all be used as a feeder layer. It should be recalled, however, that their respective optimal radiation doses are very different (see Section II,A). For microscopic observations, it is recommended that dissimilar protoplasts be used, i.e., when the target protoplasts are mesophyll derived, the feeder layer protoplast should be derived from callus or cell suspension and *vice versa*. This will facilitate the microscopic follow-up of target protoplasts, especially during the first few days after plating.

The irradiated protoplasts need not be of the same species as the target protoplasts because an interspecific feeding effect has been repeatedly observed (Aviv and Galun, 1980; A. Vardi and E. Galun, unpublished). Even very distant species can serve as feeder, e.g., tobacco feeder for orange protoplasts (Vardi and Raveh, 1976) or carrot feeder for tobacco protoplasts (Cella and Galunn 1980), provided both species can share the same medium.

Actually, there is an advantage in using a feeder of a different species from the target protoplasts. It enables the identification of X-irradiated escapees, which sometimes do occur.

For target cells and protoplast-derived microcolonies, irradiated cell suspensions can successfully serve as a feeder layer (Cella and Galun, 1980).

D. Procedure

The following procedure for the preparation of a feeder layer composed of *N. tabacum* mesophyll protoplasts shall serve as an example of the preparation of such a layer.

1. Collect protoplasts from an enzyme solution by centrifugation.
2. Resuspend protoplasts in a wash solution, e.g., MSS (Zelcer *et al.*, 1978) to reach a density of about 2×10^5/ml.
3. Irradiate protoplasts (5 krad) by ^{60}Co or X-rays.
4. Wash protoplasts twice and resuspend in culture medium (Nagata and Takebe, 1971).

5. Plate protoplasts in agar medium (the final agar concentration is 0.6%), 10^4 protoplasts per mililiter, 4 ml per 5-cm Petri dish. The plates can be kept for up to a week, at relatively low temperatures, in a plastic bag.
6. Overlay 2–3 ml of a dilute density target protoplast suspension in agar medium (the final agar concentration is 0.4–0.5%).
7. After 2–4 weeks, the upper layer can be easily separated from the feeder layer and transferred to another plate, with or without an agar underlayer.

E. Nonirradiated Feeder Layer

Metabolically active protoplasts which divide very slowly or not at all may serve as feeder. Thus, nitrate reductase-deficient protoplasts in NO_3 medium may serve as efficient feeder (Hein *et al.*, 1983).

Phenotypically different protoplasts may also be used as feeder. For example, individually isolated heterokaryons were transferred into albino protoplasts which could later be reisolated as green colonies in the background of albino colonies (Menczel *et al.*, 1978).

Finally, Weber and Lark (1979) reported a more elaborate feeder technique in which cells are grown on a solid support system which can be transferred from one feeder suspension underlay to another. This technique should be particularly useful for mutant selection experiments in which the feeder cells, affected by the selection conditions, can be replaced frequently. Using a similar procedure, these authors recently reported the isolation of nucleoside requiring auxotrophs (Roth *et al.*, 1982).

III. RESULTS AND CONCLUSIONS

The feeder layer procedure was developed to overcome the difficulty of culturing a small number of plant cells and protoplasts. Another approach to the same problem is the single cell culture (Chapter 46, this volume).

Irradiated feeder layers were very useful in the recovery of small numbers of somatic fusion calli (Zelcer *et al.*, 1978; Aviv and Galun, 1980). They were also used successfully in promoting divisions in otherwise poorly dividing protoplast systems. Thus, protoplast division and colony formation in *Solanum chacoense* and *S. tuberosum* were substantially improved by plating over a *Nicotiana* feeder layer (Huchko and Butenko, 1977). This method is also routinely used in *Citrus* protoplast work in which feeder layers of one species efficiently promote protoplast division of other species (A. Vardi and E. Galun, unpublished). It could be used for mutant isolation, especially when a feeder layer is prepared from a resistant species and the target protoplasts of another species are screened for the same resistance.

REFERENCES

Aviv, D., and Galun, E. (1980). Restoration of fertility in cytoplastmic male sterile (CMS) *Nicotiana sylvestris* by fusion with X-irradiated *N. tabacum* protoplasts. *Theor. Appl. Genet.* **58,** 121–122.

Cella, R., and Galun, E. (1980). Utilization of irradiated carrot cell suspensions as feeder layer for cultured *Nicotiana* cells and protoplasts. *Plant Sci. Lett.* **19,** 243–252.

Galun, E., and Raveh, D. (1975). *In vitro* culture of tobacco protoplasts: Survival of haploid and diploid protoplasts exposed to X-ray radiation at different times after isolation. *Radiat. Bot.* **15,** 79–82.

Hein, T., Przewozny, T., and Schieder, O. (1983). Culture and selection of somatic hybrids using an auxotrophic cell line. *Theor. Appl. Genet.* **64,** 119–122.

Kuchko, A. A., and Butenko, R. G. (1977). The culture of protoplasts isolated from *Solanum tuberosum* and *Solanum chacoense*. *In* "Use of Tissue Cultures in Plant Breeding" (F. J. Novak, ed.), pp. 441–461. Czech. Acad. Sci., Inst. Exp. Bot., Prague.

Menczel, L., Lazar, G., and Maliga, P. (1978). Isolation of somatic hybrids by cloning *Nicotiana* heterokaryons in nurse cultures. *Planta* **143,** 29–32.

Nagata, T., and Takebe, I. (1971). Plating of isolated tobacco mesophyll protoplasts on agar medium. *Planta* **99,** 12–20.

Puck, T. T., and Marcus, P. I. (1955). A rapid method for viable cell titration and clone production with Hela cells in tissue culture: The use of X-irradiated cells to supply conditioning factors. *Proc. Natl. Acad. Sci. U.S.A.* **41,** 432–437.

Raveh, D., Huberman, E., and Galun, E. (1973). *In vitro* culture of tobacco protoplasts: Use of feeder techniques to support division of cells plated at low densities. *In Vitro* **9,** 216–222.

Roth, E. J., Weber, G., and Lark, K. G. (1982). Use of isopropyl - N(3-chlorophenyl)carbamate (CIPC) to produce partial haploid cells from suspension cultures of soybean (*Glycine max*). *Plant Cell Rep.* **1,** 205–208.

Sidorov, V. A., Menczel, L., Nagy, F., and Maliga, P. (1981). Chloroplast transfer in *Nicotiana* based on metabolic complementation between irradiated and iodoacetate treated protoplasts. *Planta* **152,** 341–345.

Vardi, A., and Raveh, D. (1976). Cross feeder experiments between tobacco and orange protoplasts. *Z. Pflanzenphysiol.* **78,** 350–359.

Vardi, A., Spiegel-Roy, P., and Galun, E. (1975). Citrus cell culture: Isolation of protoplasts, plating densities, effect of mutagens and regeneration of embryos. *Plant Sci. Lett.* **4,** 231–236.

Weber, G., and Lark, K. G. (1979) An efficient plating system for rapid isolation of mutants from plant cell suspensions. *Theor. Appl. Genet.* **55,** 81–86.

Zelcer, A., Aviv, D., and Galun, E. (1978). Interspecific transfer of cytoplasmic male sterility by fusion between protoplasts of normal *Nicotiana sylvestris* and X-ray irradiated protoplasts of male sterile *N. tabacum*. *Z. Pflanzenphysiol.* **90,** 397–407.

CHAPTER **25**

Culture of Isolated Mesophyll Cells

Hans Willy Kohlenbach

Botanical Institute
University of Frankfurt on Main
Frankfurt on Main, Federal Republic of Germany

I. INTRODUCTION

About 80 years ago, Haberlandt (1902) predicted the formation of "artificial embryos" from single vegetative cells. He considered every cell as an elementary organism and was convinced of the totipotency even of differentiated cells. By his fundamental experiments on the culture of mesophyll cells of *Lamium purpureum, in vitro* research was initiated. But Haberlandt and his successors failed to achieve more than limited survival and extension growth of isolated cells due to the lack of favorable media at that time. For a certain time, development of *in vitro* research in plants was based on the culture of multicellular explants. Haberlandt's aim was first achieved about 60 years later: Differentiated vegetative (mesophyll) cells of *Macleaya cordata* directly isolated from the organism developed into cell colonies, calluses, and later somatic embryos (Kohlenbach, 1959, 1965, 1966). These early experiments were performed with complex media, but they were later repeated and extended with defined medium (Lang and Kohlenbach, 1978). In 1965, Ball and Joshi reported division growth of

CELL CULTURE AND SOMATIC CELL
GENETICS OF PLANTS, VOL. 1

ISBN 0-12-715001-3

isolated leaf cells of *Arachis hypogaea*. They scraped out cells from the mesophyll with microscalpels, whereas Kohlenbach isolated the cells by manually shaking thin leaf sections in culture medium. In 1969, Rossini described an isolation method using a glass homogenizer which yielded large numbers of cells in a short time. In addition to mechanical methods, Takebe *et al.* (1968) developed a method for enzymatic isolation by macerating the mesophyll with a pectinase.

II. PROCEDURE

A. Cell Isolation

1. Mechanical isolation by:
   a. Disintegration of mesophyll of leaf pieces with hand-operated glass homogenizers, according to Potter-Elvejhem or Eppendorf (Rossini, 1969; Kohlenbach and Schmidt, 1975; Jullien and Rossini, 1977; Lang and Kohlenbach, 1978)
   b. A Potter homogenizer at 300 rpm (Miksch and Beiderbeck, 1976)
   c. A rotating "vortex impinger" (Schwenk, 1981)
   d. Treatment of leaf pieces in a blender (Fukuda and Komamine, 1982)
   e. Grinding leaf tissue in a smooth porcelain mortar with a pestle (Gnanam and Kulandaivelu, 1969)
2. Enzymatic isolation by treating small sections or small leaf pieces with pectinases (Takebe *et al.*, 1968; Otsuki and Takebe, 1969)

Only a few culture experiments were performed with enzymatically isolated mesophyll cells (Usui and Takebe, 1969; Jullien, 1971; Bidney and Shepard, 1980). The enzyme treatment of mesophyll tissue usually is employed to get isolated protoplasts, either in a one-step or a two-step procedure, with a pectinase and a cellulase (Evans and Cocking, 1974). Grinding tissue in a mortar has so far been used only to obtain cells for short-term experiments on metabolism. Currently, the most widely used method for obtaining isolated mesophyll cells for culture experiments is glass homogenization. This mechanical method, like the others listed above, however, is by no means universally applicable. Whether it is successful at all, and to what extent, depends mainly on the plant species. Obviously, a loose, not too compact structure of the mesophyll and resistant, robust cell walls are necessary prerequisites.

B. Plant Material

Several authors have examined the suitability of different plant species to yield isolated mesophyll cells when treated with a manual or electrical glass homogenizer. The two lists which have been published so far (on manual isolation: Jullien, 1975; Jullien and Rossini, 1977; with an electrical homogenizer: Miksch and Beiderbeck, 1976) overlap to some extent. The list presented by Jullien and Rossini contains 208 plant species (monocots and dicots), 48 of which were found to yield cell suspensions with at least 5% of intact cells. Among those which allow a much higher yield (up to 70% intact cells in two species) are *Asparagus officinalis* L., *Aster novae-angliae* L., *Dianthus barbatus* L., *Fragaria vesca* L., *Ipomoea batatas* Lam., *Scabiosa columbaria* L., and *Tilia sylvestris* Desf. According to the list of Miksch and Beiderbeck, of 98 dicots examined, about 30% turned out to be suitable. In some cases, the authors used culture experiments to determine whether the isolated cells could be induced to divide. The species in which divisions were obtained are listed in Section II,D. Among those examined which did not divide under the conditions of the authors are *Anthemis tinctoria* L., *Aquilegia vulgaris* L., *Campanula glomerata* L., *Centaurea jacea* L., *Coronilla varia* L., *Daucus carota* L., *Galium mollugo* L., and *Teucrium chamaedrys* L. The paper by Gnanam and Kulandaivelu (1969) contains a small list of plant species which yield intact mesophyll cells after mild grinding in a mortar. To these lists, further plant species may be added in the future.

The fact that a plant species has yielded a good suspension of intact mesophyll cells does not mean that this result is always reproducible, and the failure of a newly examined species on the basis of only a few experiments does not allow the conclusion that this species cannot deliver single cells. After the author's intensive experiences in many trials, it appears that the success of an isolation procedure depends not only on the species but also on the physiological status of the leaves and their physical consistency. Both depend on the developmental stage and on environmental factors. All of the mechanical isolations reported were done with leaf material from outdoors, greenhouses, and growth chambers, but not with material from axenic shoot cultures. Their leaves will be too tender. This imposes the use of nonsterile material, which has two disadvantages:

1. The leaves must be surface-sterilized. This treatment may damage the tissue. Antibiotics have not yet been tried.

2. Plants which are not cultivated in insect-proof greenhouses may be attacked by insects which introduce microorganisms into the intercellular spaces, making it impossible or at least very difficult to obtain sterile suspensions. Treating the plants with pesticides shortly before the isolation

obviously has a negative influence on the growth responses of the cultured cells.

C. Cell Culture

The initiation of cell culture, culture medium, and culture conditions are described according to the procedure used for the experiments with *Zinnia elegans* (Kohlenbach and Schmidt, 1975), *Macleaya cordata*, *Convolvulus cantabrica*, *Ipomoea purpurea*, and *Quamoclit coccinea* (Lang and Kohlenbach, 1978).

1. Initiation of Cell Culture

Leaves are cleaned with tap water, surface sterilized in a solution of NaOCl, and finally rinsed with sterile distilled water. Under aseptic conditions, the midrib and the main veins are removed and the leaves dissected in small pieces of about 5 × 5 mm. Leaf pieces are put into the tube of a glass Eppendorf homogenizer together with 20 ml culture medium and macerated manually. The pestle should fit loosely into the tube. After the disintegration of the tissue, the suspension is passed through a nylon cloth to separate the cells from larger cell groups and tissue fragments. Then the suspension is centrifuged at 50 g and the pellet resuspended in culture medium. This step may be repeated to remove all or most of the debris. It results in a clean suspension of mesophyll cells, mostly palisade cells which are obviously more resistant than spongy mesophyll cells. After the cell density has been adjusted to 5×10^4/ml, 1.5 ml of suspension is pipetted per 3.5 cm Petri dish. The Petri dishes are sealed with Parafilm.

2. Culture Medium (Modified after Rossini, 1969)

All measurements are in milligrams per liter: KNO_3 (950), NH_4NO_3 (720), $MgSO_4 \cdot 7H_2O$ (185), $CaCl_2 \cdot 2H_2O$ (166), KH_2PO_4 (68), $MnSO_4 \cdot H_2O$ (25), $ZnSO_4 \cdot 7H_2O$ (10), BO_3H_3 (10), $Na_2MoO_4 \cdot 2H_2O$ (0.25), $CuSO_4 \cdot 5H_2O$ (0.025), $FeSO_4 \cdot 7H_2O$ (27.85), Na_2EDTA (37.25), meso-inositol (100), adenine (20.25), glutamine (14.7), nicotinic acid (5), glycine (2), pyridoxine–HCl (0.5), thiamine–HCl (0.5), folic acid (0.5), biotin (0.05), 2,4-dichlorophenoxyacetic acid (1), kinetin (1), and sucrose (10,000). The pH is adjusted to 5.5 with NaOH. The medium is autoclaved at 112°C for 20 min or by filter sterilization.

3. Culture Conditions

The cells may be plated in agar medium as an alternative to liquid culture and incubated at 25°C in a light-dark period of 12:12 of 600 lx provided by daylight fluorescent tubes. The plating cell density is 5×10^4/ml. One gram fresh weight of leaf material yields about 1×10^4 cells. In the case of *Macleaya*, 70% of the cells are normally intact (Fig. 1).

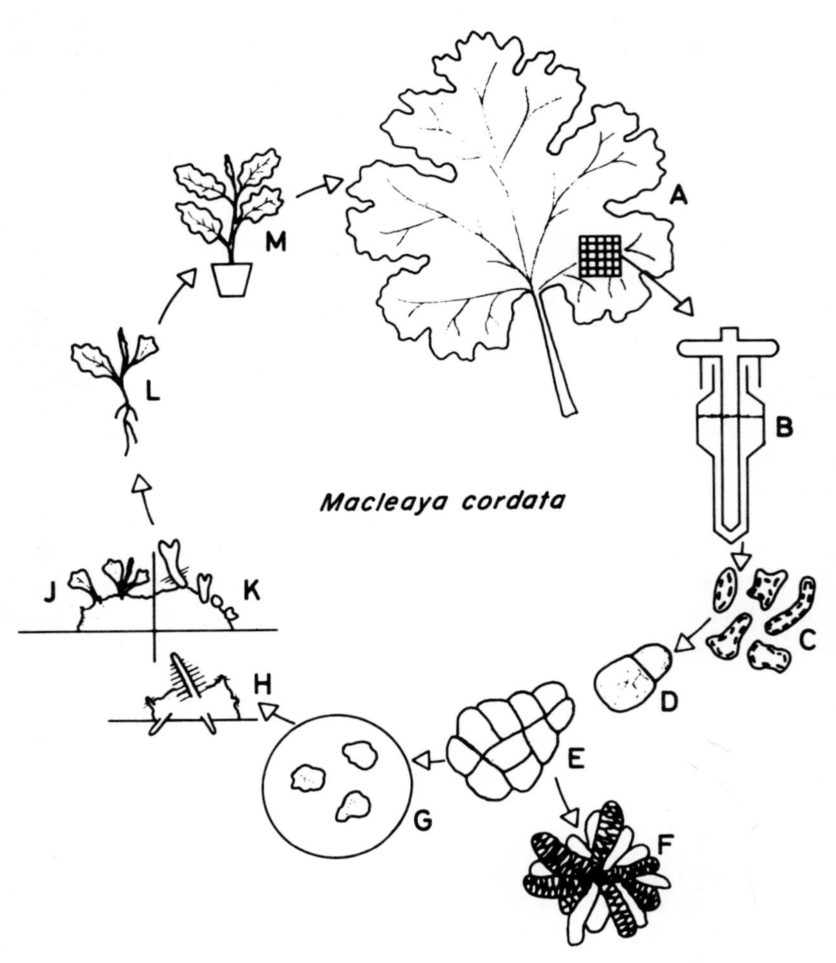

Fig. 1. *Macleaya cordata*, vegetative cycle. (A) Mature leaf, (B) homogenizer, (C) isolated cells, (D) first division, (E) cell cluster, (F) tracheid differentiation, (G) callus growth, (H) root formation, (J) shoot formation, (K) embryogenesis, (L) plant regeneration, and (M) plant in soil. (From Lang and Kohlenbach, 1978.)

D. Growth Responses

In experiments using the procedure described above, isolated mesophyll cells of *M. cordata, C. cantabrica, I. purpurea, Q. coccinea,* and *Z. elegans* were induced to perform cell divisions which gave rise to cell clusters and calluses. In *Q. coccinea* the highest percentage of dividing cells was about 30% of inoculated intact cells. In a culture experiment with microelements, $FeSO_4$, and Na_2EDTA in a dilution of 1:10 in *Macleaya,* a plating efficiency of 78% of inoculated intact cells could be obtained (H. Lang, unpublished).

Below are listed all plant species in which mechanically isolated mesophyll cells could be induced to cell division growth:

Asparagus officinalis L.	Jullien (1974)
Asparagus plumosus	Albinger and Beiderbeck (1982)
Calystegia sepium (L.) R. Br.	Rossini (1969)
Calystegia soldanella	Schmidt (1975)
Convolvulus arvensis L.	Miksch and Beiderbeck (1976)
Convolvulus cantabrica L.	Lang and Kohlenbach (1978)
Ipomoea batatas Lam.	Kohlenbach *et al.* (1971)
Ipomoea purpurea (L.) Roth	Lang and Kohlenbach (1978)
Ipomoea quamoclit L.	Bhatt and Mehta (1974)
Quamoclit coccinea (L.) Moench	Lang and Kohlenbach (1978)
Quamoclit lobata (Llav. ex Lex.) House	H. Lang (unpublished)
Potentilla argenta L.	Miksch and Beiderbeck (1976)
Rosa Hybrida "Baby Masquerade"	Miksch and Beiderbeck (1976)
Arachis hypogaea L.	Ball and Joshi (1965)
Astragalus cicer L.	Miksch and Beiderbeck (1976)
Glycine max (L.) Merr	Schwenk (1981)
Sophora japonica L.	Jsi *et al.* (1975)
Aster novae-angliae L.	Jullien (1975)
Chrysanthemum corymbosum L.	Miksch and Beiderbeck (1976)
Zinnia elegans L.	Kohlenbach and Schmidt (1975)
Macleaya cordata	Kohlenback (1959, 1966); Lang and Kohlenbach (1978)

Very recently, it has been found that the donor plants in the experiments with *Macleaya* were a hybrid between *M. cordata* (Willd.) R. Br. and *M. microcarpa* (Maxim.) Fedde.

In *Macleaya* (Lang and Kohlenbach, 1978) and *Asparagus officinalis* (Jullien, 1974), whole plants were regenerated from mechanically isolated single cells.

III. RESULTS AND CONCLUSIONS

The totipotency of even differentiated single cells could for the first time be demonstrated with isolated mesophyll cells of *Macleaya* (Kohlenbach, 1959, 1965, 1966). From these experiments, a direct line leads to the first regeneration of plants from isolated leaf protoplasts of *Nicotiana* reconverted to cells (Takebe *et al.*, 1971). But besides mesophyll protoplasts, which without doubt offer a broader spectrum of application in somatic cell genetics and which can be widely obtained compared with mechanically isolated cells restricted to a limited number of plant species, mechanically isolated mesophyll cells will still be suitable for the selection and clonal mass propagation of genotypes. Further, they give the opportunity to study fundamental developmental processes: the dedifferentiation of cells, wound reactions (Wernicke and Kohlenbach, 1976), extension growth of single cells, the transition from former quiescent cells (G_0) to dividing ones (Kohlenbach, 1966; Jullien *et al.*, 1980), modes of cytodifferentiation including the direct transformation of one specialized cell type to another one, and the formation of shoot buds, root primordia, somatic embryos, and plants out of unorganized tissue derived from isolated leaf cells (Kohlenbach, 1965; Jullien, 1974; Lang and Kohlenbach, 1978).

In 1975, Kohlenbach and Schmidt reported that mechanically isolated mesophyll cells of *Z. elegans* can be directly transformed into tracheary elements without a preceding cell division. Meanwhile, the *Zinnia* system has become an efficient tool for studying biochemical mechanisms involved in tracheary element formation (Dodds, 1980; Fukuda and Komamine, 1982). Harada *et al.* (1972) used mesophyll cells to study the effects of coumarin on the form and growth of cells. Rugman *et al.* (1982) investigated the effects of osmotic stress and abscisic acid on the division growth of mechanically isolated mesophyll cells of *Asparagus officinalis*. Beiderbeck and Hohl (1980) examined the influence of time-limited application of phytohormones on cell division of *Calystegia* mesophyll cells. Beiderbeck and Hohl (1981) propose the use of isolated leaf cells for studies in phytopathology. Whereas there are many papers on studies of metabolism using isolated leaf cells in short-term experiments (Colman *et al.*, 1979), there is so far only one investigation on cultured cells at different stages of development: freshly isolated, at extension growth, and forming cell clusters (Döhler and Kohlenbach, 1976). In *Asparagus officinalis* (Bui-Dang-Ha and Mackenzie, 1973) and in soybean (Schwenk *et al.*, 1981) mechanically isolated mesophyll cells were used to get isolated protoplasts.

REFERENCES

Albinger, G., and Beiderbeck, R. (1982). In-vitro Kultur von *Asparagus plumosus* ausgehend von isolierten Mesophyllzellen. *Abst. Tag., 100 Jahre Dtsch. Bot. Ges.* Abstr. No. 214.

Ball, E., and Joshi, P. C. (1965). Divisions in isolated cells of palisade parenchyma of *Arachis hypogaea*. *Nature (London)* **207,** 213–214.
Beiderbeck, R., and Hohl, R. (1980). Die Wirkungen zeitlich begrenzter Phytohormongaben auf isolierte Mesophyllzellen von *Calystegia sepium* L. *Biochem. Physiol. Pflanz.* **175,** 45–49.
Beiderbeck, R., and Hohl, R. (1981). Eine einfache Methode zur parabiotischen Kultur von Wirt- und Pathogenzellen. *Z. Planzenkr. Pflanzenschutz* **88,** 754–758.
Bhatt, P. H., and Mehta, A. R. (1974). Growth and differentiation in mechanically isolated mesophyll cells of *Ipomoea quamoclit*. *Can. J. Bot.* **52,** 2117–2118.
Bidney, D. L., and Shepard, J. F. (1980). Colony development from sweet potato petiole protoplasts and mesophyll cells. *Plant Sci. Lett.* **18,** 335–342.
Bui-Dang-Ha, D., and Mackenzie, I. A. (1973). The division of protoplasts from *Asparagus officinalis* L. and their growth and differentiation. *Protoplasma* **78,** 215–221.
Colman, B., Mawson, B. T., and Espie, G. S. (1979). The rapid isolation of photosynthetically active mesophyll cells from *Asparagus* cladophylls. *Can. J. Bot.* **57,** 1505–1510.
Dodds, J. H. (1980). The effect of 5-fluorodeoxyuridine and colchicine on tracheary element differentiation in isolated mesophyll cells of *Zinnia elegans* L. *Z. Pflanzenphysiol.* **99,** 283–285.
Döhler, G., and Kohlenbach, H. W. (1976). CO_2-Fixierung isolierter und sich teilender Mesophyllzellen von *Pharbitis purpurea*. *Z. Pflanzenphysiol.* **80,** 81–86.
Evans, P. K., and Cocking, E. C. (1974). The techniques of plant cell culture and somatic cell hybridization. *New Tech. Biophys. Cell Biol.* **2,** 127–158.
Fukuda, H., and Komamine, A. (1982). Lignin synthesis and its related enzymes as markers of tracheary-element differentiation in single cells from the mesophyll of *Zinnia elegans*. *Planta* **155,** 423–430.
Gnanam, A., and Kulandaivelu, G. (1969). Photosynthetic studies with leaf cell suspensions from higher plants. *Plant Physiol.* **44,** 1451–1456.
Haberlandt, G. (1902). Kulturversuche mit isolierten Pflanzenzellen. *Sitzungsber. Mat. Nat. Kl. Kais. Akad. Wiss. Wien.* **111,** 69–92.
Harada, H., Ohyama, K., and Cheruel, J. (1972). Effects of coumarin and other factors on the modification of form and growth of isolated mesophyll cells. *Z. Pflanzenphysiol.* **66,** 307–324.
Hsu, L.-C., Yeh, H. C., and Hung, W. L. (1975). Two convenient isolation methods for mesophyll cells used in plant cell culture. *Acta Bot. Sin.* **17,** 77–80.
Jullien, M. (1971). Régéneration de plantes entières à partir de cellules séparées des feuilles de *Nicotiana tabacum* diploides et haploides. *C. R. Hebd. Seances Acad. Sci, Ser. D* **273,** 1287–1290.
Jullien, M. (1974). La culture in vitro de cellules du tissu foliaire d'*Asparagus officinalis* L.: Obtention de souches à embryogénèse permanente et régéneration de plantes entières. *C. R. Hebd. Seances Acad. Sci., Ser. D.* **279,** 747–750.
Jullien, M. (1975). Contribution à l'étude des cultures in vitro de cellules séparées du tissu foliaire et de leurs potentialités organogènes chez quelques plantes supérieures, en particulier chez *Asparagus officinalis* L. Thesis, Univ. of Paris.
Jullien, M., and Rossini, L. (1977). L'obtention de cellules séparées à partir du tissu foliaire chez les plantes supérieures: Intérêt et potentialités d'une méthode mécanique. *Ann. Amelior. Plant.* **27,** 87–103.
Jullien, M., Rossini, L., and Guern, J. (1980). Some aspects of the induction of the first division and growth in cultures of mesophyll cells obtained from *Asparagus officinalis* L. *Proc. 5th, 1979 Int. Asparagus-Symp.* pp. 103–129.
Kohlenbach, H. W. (1959). Streckungs- und Teilungswachstum isolierter Mesophyllzellen von *Macleaya cordata* (Willd.) R. Br. *Naturwissenschaften* **46,** 116–117.
Kohlenbach, H. W. (1965). Über organisierte Bildungen aus *Macleaya cordata* Kallus, *Planta* **64,** 37–40.

Kohlenbach, H. W. (1966). Die Entwicklungspotenzen explantierter und isolierter Dauerzellen. I. Das Streckungs- und Teilungswachstum isolierter Mesophyllzellen von *Macleaya cordata*. *Z. Pflanzenphysiol.* **55,** 142–157.

Kohlenbach, H. W., and Schmidt, B. (1975). Cytodifferenzierung in Form einer direkten Umwandlung isolierter Mesophyllzellen zu Tracheiden. *Z. Pflanzenphysiol.* **75,** 369–374.

Kohlenbach, H. W., Meuser, M., and Schröter, H. (1971). Cytomorphological changes in explanted epidermal cells of *Rhoeo spathacea*. *In* "Morphogenesis in Plant Cell, Tissue and Organ Culture" (B. M. Johri and H. Y. Mohan Ram, eds.), pp. 23–24. Univ. of Delhi, Delhi, India.

Lang, H., and Kohlenbach, H. W. (1978). Development of isolated leaf parenchyma cells. *In* "Production of Natural Compounds by Cell Culture Methods" (A. W. Alfermann and E. Reinhard, eds.), pp. 274–283. Strahlen- und Umwelt Forschung, Munich.

Miksch, U., and Beiderbeck, R. (1976). Mechanische Isolierung von Einzelzellen aus Blättern höherer Pflanzen. *Biochem. Physiol. Pflanz.* **169,** 191–196.

Otsuki, Y., and Takebe, I. (1969). Isolation of intact mesophyll cells and their protoplasts from higher plants. *Plant Cell Physiol.* **10,** 917–921.

Rossini, L. (1969). Une nouvelle méthode de culture in vitro de cellules parenchymateuses séparées des feuilles de *Calystegia sepium* L. *C. R. Hebd. Seances Acad. Sci.* **268,** 683–685.

Rugman, E. E., Mackenzie, I. A., and Hall, J. F. (1982). The effects of abscisic acid and osmotic stress on division of isolated mesophyll cells. Poster demonstration at *Int. Conf. Plant Growth Subst., 11th, 1982.*

Schmidt, B. (1975). Untersuchungen zur Entwicklung isolierter Mesophyllzellen von *Zinnia elegans* und *Calystegia soldanella*. Thesis, Univ. of Frankfurt on Main, Federal Republic of Germany.

Schwenk, F. W. (1981). Callus formation from mechanically isolated mesophyll cells of soybean and sweet potato. *Plant Science Lett.* **23,** 147–151.

Schwenk, F. W., Pearson, C. A., and Roth, M. R. (1981). Soybean mesophyll protoplasts. *Plant Sci. Lett.* **23,** 153–155.

Takebe, I., Otsuki, Y., and Aoki, S. (1968). Isolation of tobacco mesophyll cells in intact and active state. *Plant Cell Physiol.* **9,** 115–124.

Takebe, I., Labib, G., and Melchers, G. (1971). Regeneration of whole plants from isolated mesophyll protoplasts of tobacco. *Naturwissenschaften* **58,** 318–320.

Usui, I., and Takebe, I. (1969). Division and growth of single mesophyll cells isolated enzymatically from tobacco leaves. *Dev. Growth Differ.* **11,** 143–151.

Wernicke, W., and Kohlenbach, H. W. (1976). Akkumulation von Chlorogensäure in isolierten Mesophyllzellen von *Ipomoea batatas* Poir. *Z. Pflanzenphysiol.* **77,** 464–470.

CHAPTER **26**

The Multiple-Drop-Array (MDA) Screening Technique

Christian T. Harms*

Department of Crop Science
Swiss Federal Institute of Technology (ETH) Zürich
Eschikon Experimental Station
Lindau-Eschikon, Switzerland

I. INTRODUCTION

The multiple-drop-array (MDA) technique has been developed by Potrykus and co-workers (1977a, 1979a; Harms *et al.*, 1979) as an experimental approach to the large-scale testing of culture media variations. The need for such a technique to screen systematically large numbers of multiple combinations of culture media constituents had grown from disappointing experiences with cereal mesophyll protoplasts which did not readily respond to the cultural regimes that had been shown to be successful with protoplast cultures of *Nicotiana* and *Petunia* species. Defining and optimizing the particular medium requirements was, therefore, considered as one of several approaches to the successful culture of protoplasts from recalcitrant species such as the cereals.

Such optimization requires the testing of numerous components in multiple combinations with other constituents of the culture medium. The culture methods employed so far had used only one factor combination per culture vessel, thus rendering such testing inefficient and limited by time, materials, space, and labor. The specifications of a more efficient large-

* Present address: CIBA-Geigy Biotechnology Institute, P.O. Box 12257, Research Triangle Park, North Carolina 27709.

ISBN 0-12-715001-3

scale screening technique had soon become clear, and the MDA technique then devised perfectly met the demands encountered in this task. These demands included that the technique in question should (1) use a minimum number of protoplasts per factor combination to allow a maximum number of combinations to be tested with a single population (preparation) of protoplasts; (2) be generally compatible with protoplast culture, allowing sterile handling and securing a high cell viability; (3) lend itself to an easy, rapid, and systematic testing of a large matrix of factor combinations including combinations of two, three or multiple factors.

II. PROCEDURE

The MDA screening technique uses hanging droplets of 40 μl as the experimental unit; each droplet represents just one combination of the factors to be tested. The droplets are arranged in a regular array of 7 × 7 drops on the lid of a 9- or 10-cm Petri dish. It is essential to use noncoated hydrophobic dishes (i.e., Greiner, Falcon, Petra Plastics) for optimal drop forming and positioning.

The following protocol (modified from Harms *et al.*, 1979) describes the testing of seven different medium components (factors A–G, i.e., seven different auxins) each in combination with four other factors (W–Z, i.e., four different cytokinins). Each of these factors is used in seven concentrations. The whole experiment, therefore, includes 4 × 7 Petri dishes, with 49 droplets each, to give a total of 1372 two-factor combinations. This experiment can be performed by one person within 5–6 hr plus the time necessary for media preparation, protoplast isolation, and culture evaluation. When using the technique with protoplasts, it is advisable to adopt an overnight isolation procedure (Harms and Potrykus, 1978).

The following procedures are performed aseptically in a laminar flow bench using sterilized materials and filter-sterilized media. Consult the flow diagram depicted in Fig. 1 for additional information on the various steps of the protocol.

1. Make up individually each of the components A–G and W–Z in basal culture medium (basal medium: containing all constituents that are kept constant in the experiment but lacking the components to be tested). Prepare a seven-step concentration series of each of the components to be tested, i.e., A_1–A_7, B_1–B_7, . . . Z_1–Z_7. The dilution series may be linear or logarithmic according to the purpose of the experiment. Prepare the concentrations of factors A–G two times higher and the concentration of factors W–Z four times higher than finally desired since they will be diluted to

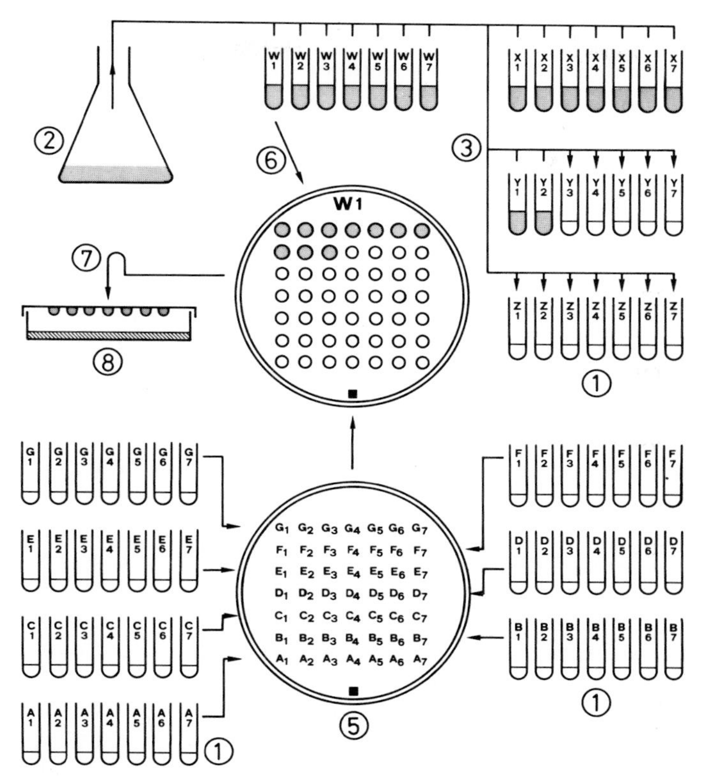

Fig. 1. Schematic flow diagram of the MDA screening technique. Circled numbers correspond to the various steps of the experimental protocol described in the text.

that extent in the course of the experiment. A minimum of 1 ml is required for each of the concentration steps. It is recommended that these preparations be made 1 or 2 days before the screening experiment in order not to spoil the screening by accidental contamination that may occur during medium preparation and dilution.

2. Isolate and purify protoplasts following an established protocol and suspend them in basal culture medium at a population density four times higher than desired for optimal culture response. If this is not known, it may be determined empirically in a previous experiment. A minimum of 28 ml of protoplast suspension is required.

3. Add 1 ml of the protoplast suspension to 1 ml of each of the concentration steps of the dilution series W_1–W_7, . . . Z_1–Z_7. Make sure that the protoplasts remain suspended until needed.

4. Cover the working area in the flow bench with blotting paper soaked with sterile water and reduce air flow to prevent evaporation from the droplets during the following steps of the procedure.

5. Mark the upper edge of the lids of 28 Petri dishes with pencil to avoid confusion about the lids' orientation once the lids are turned upside down. Place seven Petri dishes, lid down, on equal-sized template lids arranged on the wet blotting paper. The marks are now oriented down toward you. Premade Petri dish lids with marked positions of the drops' geometric array are recommended as a template which will greatly facilitate the correct placement of the droplets.

Twenty-microliter droplets of the dilution series A_1–A_7, . . . G_1–G_7 can now be arranged in a 7 × 7-drop array on the Petri dish lid. Each horizontal line contains the complete concentration series of just one factor, i.e., A_1–A_7 in the lowermost line (closest to the mark), continuing with B_1–B_7 in the next line and finishing with G_1–G_7 in the uppermost horizontal line. Each line begins with the highest concentration on the left and continues to the right of the lid with decreasing concentrations.

Start placing the droplets with concentration A_1 of the factor series A_1–A_7 at position A_1 (cf. Fig. 1); do the same with the next lid, and so on, until you finish with the whole series of seven lids. Then place a 20-μl droplet of concentration A_2 in position A_2 of all seven lids, continue with concentration A_3 in position A_3 of all seven lids, and so on.

With a little practice, accurate dosage and placement of the droplets are easily achieved using a graded 1-ml plastic pipette (Falcon) mounted on a piston pipette holder (Fortuna pipette) for smooth operation.

6. Add 20 μl of the protoplast suspension (made up with the concentration series of factors W, X, Y, and Z according to step 3 of the protocol) to each of the 7 × 7 droplets in one lid: Add suspension W_1 to all droplets of lid 1, use suspension W_2 for the second lid, and so on, finishing with suspension W_7 added to the droplets in lid 7. Indicate the added factor concentration on the lid for identification.

7. Insert the bottom part of the Petri dishes to the lids and put the seven dishes—still in an upside down position—in a stack. Hold the stack between the thumb and middle finger of both hands and quickly turn it to an upright position in which the drops are now hanging drops. This critical step requires some care, and practice beforehand is highly recommended for perfection. Note that all droplets will stay in their position if the centrifugal force acting on them when the dish is turned always presses them vertically onto the lid surface; they will slip away if this force acts from the side.

8. Add 20 ml sterile mannitol solution to the bottom of the dishes. This solution should have approximately 60% of the osmolality of the culture medium used in the hanging droplets.

9. Seal the dishes with Parafilm and incubate the MDA plates in moist plastic boxes (20 × 20 × 10 cm) under appropriate culture conditions.

10. Repeat instructions 6–9 with the next set of seven lids, this time adding the protoplast suspensions made up in the concentration series

X_1–X_7, each of them to all 49 droplets of one lid. Repeat this procedure for the remaining two sets of seven lids with the protoplast suspensions in series Y_1–Y_7 and Z_1–Z_7.

11. Evaluate the effects of the factor screening under the microscope at appropriate culture periods (2–14 days).

III. RESULTS AND CONCLUSIONS

An example of an MDA plate representing part of a multiple phytohormone screening with tobacco protoplasts is shown in Fig. 2 (Harms *et al.*, 1979), performed according to the above protocol. Although the protocol described here tests only two-factor combinations, the MDA technique was originally devised and instructions given for the screening of combinations of three or more variable factors (Potrykus *et al.*, 1979a).

Careful control of the osmotic conditions in the hanging droplets is a major factor contributing to the successful use of the MDA technique. The

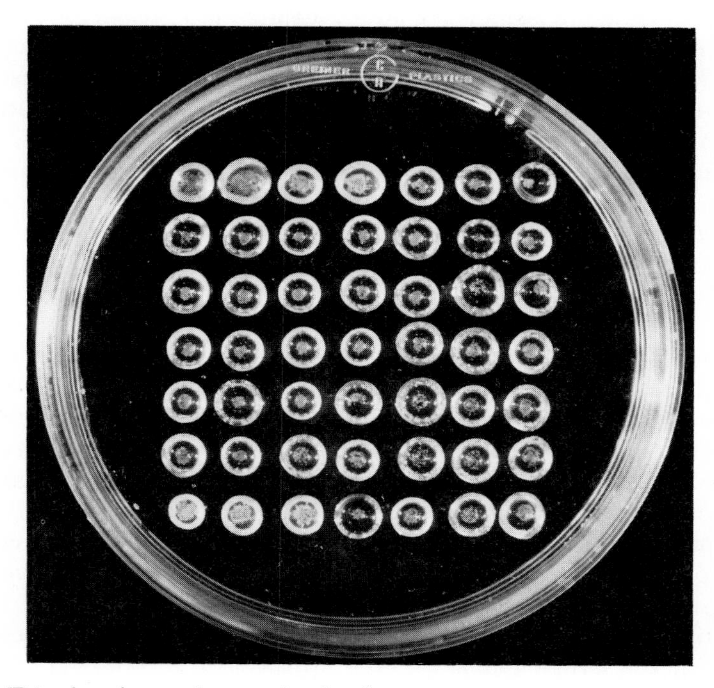

Fig. 2. MDA plate from a large-scale phytohormone screening experiment with tobacco mesophyll protoplasts.

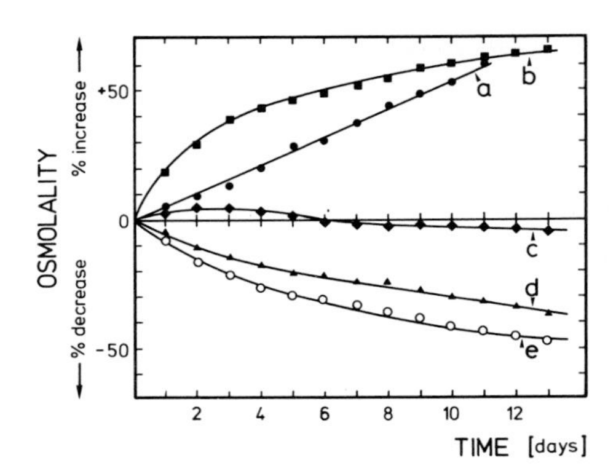

Fig. 3. Time course of the osmotic pressure in hanging 40-μl droplets of a 7 × 7 drop plate as affected by the absence (a) or presence (b–e) of sublayers (20 ml) of various osmolality added to MDA plates: (b) 0.6 *M* mannitol (660 mOs/kg H_20); (c) 0.4 *M* mannitol (440 mOs/kg H_2O); (d) 0.2 *M* mannitol (220 mOs/kg H_2O); (e) distilled water (0 mOs/kg H_2O). The MDA plates were incubated at 28 ± 1°C in moist plastic boxes. The initial osmotic pressure of the culture medium used in the hanging drops was 660 mOs/kg H_2O.

time course of the osmotic pressure has been extensively studied (Harms, 1977) in hanging drops as affected by either the absence or the presence of sublayers of different osmolality added to the MDA plates. It is obvious from Fig. 3 and many similar experiments that a sublayer having an osmolality 55–60% that of the culture medium (i.e., 0.4 *M* mannitol used with a culture medium of 660 mOs/kg H_2O; cf. protocol) serves best to maintain a constant osmotic pressure in the hanging droplets.

A division response exceeding 90% of the initial protoplasts clearly demonstrates that the hanging droplets as used in the MDA technique provide a culture device optimally suited for protoplast culture (Harms *et al.*, 1979). Medium conditions established in an MDA screening can easily be adopted to other culture procedures, such as liquid or soft agar plating in Petri dishes.

The culture responses can be easily inspected and evaluated since each droplet covers but a single field area in the microscope. The existence of approximately 600 protoplasts per droplet (corresponding to a protoplast population density of 1.5×10^4/ml) means that culture responses in the range of 1% can be detected and that the MDA technique tests 50–100 times more variations with a given population of cells than any other culture method. The MDA screening procedure has been extensively employed to define the culture conditions necessary for cereal mesophyll protoplasts to divide (Potrykus *et al.*, 1976), but despite testing approximately 100,000 factor combinations, these inductive conditions remain to be established (Harms, 1982).

In addition to protoplasts, the MDA technique can be used with various other single-cell systems including single cells, microspores, and unicellular algae (I. Potrykus, unpublished results). Factors which can be tested, singly or in complex mixtures, include phytohormones, mineral salts, vitamins, amino acids, carbon sources, nitrogen sources, osmolality or pH values, adsorbents, various extracts or conditioned medium, cell population density, and others. The phytohormone screening, as suggested in the above protocol, was chosen to illustrate just one of the many applications possible. A major feature of the MDA technique is its versatility and adaptability to the particular requirements posed by different experiments.

A culture medium particularly for cereal cultures has been designed from MDA media screenings (Potrykus *et al.*, 1976), and extensive phytohormone screening has led to a substantial improvement of protoplast cultures from tobacco (Harms *et al.*, 1979), maize stem (Potrykus *et al.*, 1977b), and maize cell suspension cultures (Potrykus *et al.*, 1979b). Although in these cases the main experimental emphasis had been put on detecting factor combinations effective in inducing cell division, the MDA screening technique also permits monitoring of the effects of media factors on developmental processes such as embryogenesis from single somatic cells or protoplasts (Lörz *et al.*, 1977; Harms *et al.*, 1979), meiosis, cytodifferentiation, and secondary metabolite biosynthesis (I. Potrykus, unpublished results).

In conclusion, the MDA technique has been developed into a powerful tool for the large-scale screening of multiple combinations of media factors which affect the cellular development in cultures of single cells. It can be anticipated that through its extensive application this versatile technique will continue to contribute to the improvement of the culture media conditions in single-cell systems.

REFERENCES

Harms, C. T. (1977). In vitro culture of corn and tobacco and development of a gradient system for the early selection of fused protoplasts. Ph.D. Thesis, Univ. of Heidelberg, Federal Republic of Germany.

Harms, C. T. (1982). Maize and cereal protoplasts—facts and perspectives. *In* "Maize for Biological Research" (W. F. Sheridan, ed.), pp. 373–384. Plant Mol. Biol. Assoc., Charlottesville, Virginia.

Harms, C. T., and Potrykus, I. (1978). Fractionation of plant protoplast types by iso-osmotic density gradient centrifugation. *Theor. Appl. Genet.* **53,** 57–63.

Harms, C. T., Lörz, H., and Potrykus, I. (1979). Multiple-drop-array (MDA) technique for the large-scale testing of culture media variations in hanging microdrop cultures of single cell systems. II. Determination of phytohormone combinations for optimal division response in *Nicotiana tabacum* protoplast cultures. *Plant Sci. Lett.* **14,** 237–244.

Lörz, H., Potrykus, I., and Thomas, E. (1977). Somatic embryogenesis from tobacco protoplasts. *Naturwissenschaften* **64,** 439–440.

Potrykus, I., Harms, C. T., and Lörz, H. (1976). Problems in culturing cereal protoplasts. *In* "Cell Genetics in Higher Plants" (D. Dudits, G. L. Farkas, and P. Maliga, eds.), pp. 129–140. Akadémiai Kiadó, Budapest.

Potrykus, I., Lörz, H., and Harms, C. T. (1977a). On some selected problems and results concerning culture and genetic modification of higher plant protoplasts. *In* "Plant Tissue Culture and its Biotechnological Application" (W. Barz, E. Reinhard, and M. H. Zenk, eds.), pp. 323–333. Springer-Verlag, Berlin and New York.

Potrykus, I., Harms, C. T., Lörz, H., and Thomas, E. (1977b). Callus formation from stem protoplasts of corn (*Zea mays* L.). *Mol. Gen. Genet.* **156,** 347–350.

Potrykus, I., Harms, C. T., and Lörz, H. (1979a). Multiple-drop-array (MDA) technique for the large-scale testing of culture media variations in hanging microdrop cultures of single cell systems. I. The technique. *Plant Sci. Lett.* **14,** 231–235.

Potrykus, I., Harms, C. T., and Lörz, H. (1979b). Callus formation from cell culture protoplasts of corn (*Zea mays* L.). *Theor. Appl. Genet.* **54,** 209–214.

CHAPTER **27**

Culture of Ovaries

T. S. Rangan

Phytogen
Pasadena, California

I. INTRODUCTION

In angiosperms, the twin processes of pollination and fertilization in addition to leading to the formation of the embryo and endosperm, contribute to the maturation and enlargement of the ovary into a fruit. When working with a whole plant, it is difficult to study the precise requirement for fruit development. Concomitant with other *in vitro* techniques, culture of ovaries offers a better elucidation of several aspects of fruit physiology such as fruit morphogenesis and physiological and biochemical changes. These aspects could be studied under precisely regulated environmental and nutritional conditions.

Ovaries from a number of species have been grown *in vitro* with variable degrees of success. Such studies have been mainly concerned with morphogenetic and physiological aspects of fruit and seed development (Johri and Guha, 1963; Nitsch, 1963a). Ovary culture was first attempted by LaRue (1942), who obtained limited growth of ovaries accompanied by rooting of pedicels in several species. Nitsch (1951) subsequently extended this technique to the study of fruit physiology, and similar results have been obtained by several investigators (Rangan, 1982).

CELL CULTURE AND SOMATIC CELL
GENETICS OF PLANTS, VOL. 1

ISBN 0-12-715001-3

II. PROCEDURE

Following pollination, whole flower buds are excised (2–15 days postpollination depending on the species), and the calyx, corolla, and stamen are removed. The ovaries are then surface sterilized in 5–10% calcium hypochlorite solution and rinsed three or four times with sterile distilled water. To obtain unpollinated ovaries, the flower buds are removed 24–48 hr prior to anthesis, and then the ovaries are excised and sterilized in the usual manner. Before culturing, the tip of the distal part of the pedicel is cut off (to remove the tissues exposed to calcium hypochlorite) and the ovary is implanted with the cut end inserted in the nutrient medium. Generally, the nutrient medium is rather simple, consisting of either Nitsch's (1951) or White's (1954) inorganic salt mixture; depending on the requirements, various growth substances such as auxins, cytokinins, and complex growth supplements (yeast extract, casein hydrolysate, coconut water) are added and the medium is gelled with agar. When liquid medium is to be employed, the ovary can be placed on a filter paper raft or float with the pedicel inserted through the filter paper and dipping into the medium. Cultures are incubated at 25 ± 2°C, 50–60% relative humidity, and under continuous or alternating periods (16:8 hr light:darkness schedule) of diffuse light.

The growth of the ovary *in vitro* is generally determined by measuring its length (from the base of the ovary to the tip of the stigma) and breadth (along the greatest diameter, generally in the middle of the fruit) by calipers. In species in which the seed set is satisfactory, the number of seeds formed could also be used as a parameter. Some investigators use the change in color of the ovary as the criterion for determining maturity.

III. RESULTS AND CONCLUSIONS

When cultured on an appropriate medium, the detached ovary enlarges and the pattern of growth follows the sigmoid type common among fruits. Ripening occurs in the culture vial, and tomatoes with the full red color and flavor of vine-ripened fruits have been grown from pollinated or auxin-treated ovaries (Nitsch, 1951). However, the size of mature fruits *in vitro* is usually smaller. Often the restricted space of the culture vial prevents the ovary from attaining its full size. To overcome this problem, Ito (1966) devised a partial sterile culture method for ovaries of *Dendrobium nobile.* In this method, only the surface-sterilized pedicel is inserted into the nutrient medium through an opening in the stopper, thus leaving the ovary free to grow outside the vial. In *Iberis amara* (Maheshwari and Lal, 1961), *Althaea rosea* (Chopra, 1962), and *Ranunculus sceleratus* (Sachar and Guha, 1962), it was possible to obtain fruits of natural or even larger size by adding suitable growth supplements to the medium and also by altering the stage of the ovary at culture.

The various species seem to have few requirements in common for growth *in vitro*. For example, ovaries of *Cucumis anguria* (Nitsch, 1951), *Lycopersicon esculentum* (Nitsch, 1951; Kano, 1962), *I. amara* (Maheshwari and Lal, 1961), and *Zephyranthes* sp. (Sachar and Kapoor, 1959) required only a basal medium with inorganic salts and sucrose for growth, although addition of an auxin in each case brought about a greater stimulation of growth. However, ovaries of *A. rosea* (Chopra, 1962), *Linaria maroccana* (Sachar and Baldev, 1958), and *R. sceleratus* (Sachar and Guha, 1962) did not grow satisfactorily even with the addition of complex growth substances such as casein hydrolysate or yeast extract. Addition of auxin, although stimulatory, often leads to parthenocarpic fruits or fruits with only a few viable seeds.

Sucrose as a source of carbon is essential, although a variety of other sources including maltose and lactose have been shown to be equally favorable (Leopold and Scott, 1952; Ito, 1966). Vitamins such as B_1 and B_6 or even a mixture of vitamins (Maheshwari and Lal, 1961) stimulate ovary growth. Vitamin E (tocopherol acetate) was shown to increase seed fertility in *D. nobile* (Ito, 1966).

The development of the ovary has long been associated with pollination and fertilization. In several apomictic species such as *Aerva tomentosa*, although there is no fertilization, pollination alone stimulates the growth of the ovary and ovule. In culture, pollinated and unpollinated ovaries of the same species respond very differently. The former grow to form miniature fruits; the latter fail to increase in size, although they remain alive for months (Nitsch, 1951). However, recent reports indicate that through unpollinated ovary culture the parthenogenetic development of the egg has been achieved in some species (Yang and Zhou, 1982). It has long been established that in many species the developing seeds are a rich source of auxin (Nitsch, 1952). Thus, in addition to pollen, which provides the initial stimulus, the developing seeds provide the necessary auxin for normal fruit growth. In species such as pear and apple, the size of the fruit is directly dependent on the number of seeds reaching maturity, and the asymmetrical fruits sometimes seen are due to the abortion of the ovules in one part of the ovary (Nitsch, 1952). Synthetic auxins such as 2-naphthoxyacetic acid (NOA) and 2,4-dichlorophenoxyacetic acid (2,4-D) can replace the stimulus provided by pollination. One can, therefore, conclude that auxins play a cardinal role in the growth of fruits.

Generally, in ovaries grown *in vitro*, the seed set is low and in several species the seeds fail to develop altogether. In *Allium cepa* (Guha and Johri, 1966), for example, the maximum seed set is only 30% as compared to about 60% in nature. The high sterility encountered in cultured ovaries is usually attributed to the degeneration of the endosperm, embryo, or entire ovules, thus indicating our inadequate knowledge of nutrition and growth. However, fertile seeds develop and even germinate *in situ* in several species (Nitsch, 1951; Leopold and Scott, 1952; Maheshwari and Lal, 1961; Sachar and Guha, 1962; Ito, 1966). The plantlets that develop can be transplanted to pots in the greenhouse after they have developed a satisfactory root system.

Several researchers studying the development of the ovary have found that other floral organs, especially the calyx or perianth, enhance growth and in some instances are necessary (Redei and Redei, 1955; LaCroix *et al.*, 1962; Bajaj, 1966; Guha and Johri, 1966). Nitsch's (1963b) studies seem to indicate that the calyx may be the source of some nitrogenous compounds indispensable for growth. Consequently, organic nitrogen supplementation of the medium may prove beneficial. This view is strengthened by the fact that in *Nigella sativa* Peterson (1973) found that media high in nitrogen were best in supporting the growth of the ovary. Later, investigating the effect of growth substances on development in *N. sativa*, he observed that gibberellic acid stimulated pistil growth but not ovule formation, whereas indoleacetic acid (IAA) and kinetin had no effect (Peterson, 1974). However, McHughen's (1982) studies suggest that no one type of floral organ may contribute exclusively to the growth or inhibition of the ovary.

The technique of ovary culture may perhaps also be utilized to obtain hybrids of normally incompatible species. This would involve culturing unpollinated pistils (excised from unopened buds) and pollinating them with pollen 24 hr later. Fertilization takes place in about 25% of the ovules, eventually resulting in mature seeds with viable hybrid embryos (Inomata, 1968).

In some species, especially the members of the family Umbelliferae, polyembryony has been observed in ovary cultures (Johri and Sehgal, 1966; Sehgal, 1972a,b). The embryo, which usually follows normal development, sometimes undergoes cleavage and budding, resulting in a polyembryonal mass. This polyembryonal mass with embryoids emerges by rupturing the pericarp and eventually forms plantlets. In other species such as *Haworthia turgida* (Majumdar, 1970), *Oryza sativa* (Nishii and Mitsuoka, 1969), *Citrus aurantifolia*, and *C. sinensis* (Mitra and Chaturvedi, 1972), callus and subsequently leafy shoots are formed from the ovary wall.

Although ovary culture is not employed as much today as it was 2 decades ago, it still offers a useful method to study some problems in fruit growth and physiology. Since the composition of the medium can be altered at will with no problems of application and translocation to the site of action, it is an ideal method for study of the effect of growth substances. The validity of such *in vitro* results, however, is dependent on the correlation existing between *in vitro* and *in vivo* observations. The ovary is a developmentally complex organ by itself and is an integral part of a more complex system. The developmental requirements and patterns appear to be species specific, and the regulatory factors that control the initiation and development of embryologically essential structures such as the placenta and ovule are not clearly understood. Perhaps pistil or ovary culture would provide a basis for further investigations of those regulatory mechanisms influencing various events that eventually lead to fruit and seed development.

REFERENCES

Bajaj, Y. P. S. (1966). Growth of *Hyoscyamus niger* ovaries in culture. *Phyton (Buenos Aires)* **23,** 57–62.

Chopra, R. N. (1962). Effect of some growth substances and calyx on fruit and seed development of *Althaea rosea* Cav. *In* "Plant Embryology—A Symposium", pp. 170–181. Council of Scientific and Industrial Research, New Delhi, India.

Guha, S., and Johri, B. M. (1966). In vitro development of ovary and ovule of *Allium cepa* L. *Phytomorphology* **16,** 353–364.

Inomata, N. (1968). In vitro culture of ovaries of *Brassica* hybrids between 2X and 4X. I. Culture medium. *Jpn. J. Breed.* **18,** 139–148.

Ito, I. (1966). In vitro culture of ovary in orchids (1). Effects of sugar, peptone and coconut milk upon the growth of ovary of *Dendrobium nobile. Sci. Rep. Kyoto Prefect. Univ., Agric.* **18,** 38–50.

Johri, B. M., and Guha, S. (1963). The technique of in vitro culture in the study of physiology of reproduction. *J. Indian Bot. Soc.* **42A,** 58–73.

Johri, B. M., and Sehgal, C. B. (1966). Growth response of ovaries of *Anethum, Foeniculum* and *Trachyspermum. Phytomorphology* **16,** 364–378.

Kano, K. (1962). In vitro culture of tomato ovaries. *J. Jpn. Soc. Hortic. Sci.* **31,** 197–206.

La Croix, L. J., Naylor, J. M., and Larter, E. N. (1962). Factors controlling embryo growth and development in barley (*Hordeum vulgare* L.). *Can. J. Bot.* **40,** 1515–1523.

LaRue, C. D. (1942). The rooting of flowers in sterile culture. *Bull. Torrey Bot. Club* **69,** 332–341.

Leopold, A. C., and Scott, F. I. (1952). Physiological factors in tomato fruit set. *Am. J. Bot.* **39,** 310–317.

McHughen, A. (1982). Some aspects of growth characteristics of tobacco pistils *in vitro. J. Exp. Bot.* **33,** 162–169.

Maheshwari, N., and Lal, M. (1961). In vitro culture of ovaries of *Iberis amara* L. *Phytomorphology* **11,** 17–23.

Majumdar, S. K. (1970). Production of plantlets from the ovary wall of *Haworthia turgida* var. *Palladifolia. Planta* **90,** 212–214.

Mitra, G. C., and Chaturvedi, H. C. (1972). Embryoids and complete plants from unpollinated ovaries and from ovules of in vivo-grown emasculated flower buds of *Citus* spp. *Bull. Torrey Bot. Club* **99,** 184–189.

Nishii, T., and Mitsuoka, S. (1969). Occurrence of various ploidy plants from anthers and ovary culture of rice plant. *Jpn. J. Genet.* **44,** 341–346.

Nitsch, J. P. (1951). Growth and development in vitro of excised ovaries. *Am. J. Bot.* **38,** 566–577.

Nitsch, J. P. (1952). Plant hormones in the development of fruits. *Q. Rev. Biol.* **27,** 33–59.

Nitsch, J. P. (1963a). Fruit development. *In* "Recent Advances in the Embryology of Angiosperms" (P. Maheshwari, ed.) pp. 361–394. Univ. of Delhi, Delhi, India.

Nitsch, J. P. (1963b). The in vitro culture of flowers and fruits. *In* "Plant Tissue and Organ Culture—A Symposium" (P. Maheshwari and N. S. Rangaswamy, eds.), pp. 198–214. Univ. of Delhi, Delhi, India.

Peterson, C. M. (1973). The nutritional requirement for ovule formation in excised pistils of *Nigella*. *Am. J. Bot.* **60,** 381–388.

Peterson, C. M. (1974). The effects of gibberellic acid and a growth retardant on ovule formation and growth of excised pistils of *Nigella* Ranunculaceae. *Am. J. Bot.* **61,** 693–698.

Rangan, T. S. (1982). Ovary, ovule and nucellus culture. *In* "Experimental Embryology of Vascular Plants" (B. M. Johri, ed.), pp. 105–129. Springer-Verlag, Berlin and New York.

Redei, G., and Redei, G. (1955). Rearing wheats from ovaries cultured in vitro. *Acta Bot. Acad. Sci. Hung.* **2,** 183–185.

Sachar, R. C., and Baldev, B. (1958). In vitro growth of ovaries of *Linaria maroccana* Hook. *Curr. Sci.* **27,** 104–105.

Sachar, R. C., and Guha, S. (1962). In vitro growth of achenes of *Ranunculus scleratus* L. *In* "Plant Embryology—A Symposium" pp. 244–253. Council of Scientific and Industrial Research, New Delhi, India.

Sachar, R. C., and Kapoor, M. (1959). Gibberellin in the induction of parthenocarpy in *Zephyrantes*. *Plant Physiol.* **34,** 168–170.

Sehgal, C. B. (1972a). In vitro induction of polyembryony in *Ammi majus* L. *Curr. Sci.* **41,** 263–264.

Sehgal, C. B. (1972b). Experimental induction of zygotic multiple embryos in *Coriandrum sativum* L. *Indian J. Exp. Biol.* **10,** 457–459.

White, P. R. (1954). "The Cultivation of Animal and Plant Cells." Ronald Press, New York.

Yang, H. Y., and Zhou, C. (1982). In vitro induction of haploid plants from unpollinated ovaries and ovules. *Theor. Appl. Genet.* **63,** 97–104.

CHAPTER **28**

Culture of Ovules

T. S. Rangan

Phytogen
Pasadena, California

I. INTRODUCTION

It is not too difficult to excise and culture angiosperm embryos at the heart-shaped stage or even earlier, but embryos at very early stages of development offer greater problems of isolation and culture. Very young embryos are liable to be mutilated during excision, and their nutritive requirements are more complex than those of mature embryos. In several groups of plants such as orchids, phanerogamic parasites, and some saprophytes, even the mature embryo is so minute (Rangaswamy, 1967) that its excision is extremely difficult. The difficulty of growing very young or minute embryos led to attempts to culture ovules. Culture of ovules is advantageous, as they can be excised even at the zygote stage and are thought to provide a "maternal environment" for the developing embryo.

Compared with studies on the embryo, reports of successful ovule culture are relatively few. White (1932) for the first time cultured the ovules of *Antirrhinum*. Subsequently, ovules of several species have been cultured, mostly with a view to understanding the factors regulating the development of the zygote through organized stages to a mature embryo (Maheshwari and Rangaswamy, 1965; Rangan, 1982).

CELL CULTURE AND SOMATIC CELL
GENETICS OF PLANTS, VOL. 1

ISBN 0-12-715001-3

II. PROCEDURE

In order to obtain ovules of uniform age, the flower buds are tagged to note the day of anthesis. Following pollination, the ovary (1–12 days postpollination) is collected and surface sterilized in 5–10% calcium hypochlorite and rinsed three or four times with sterile distilled water. In species such as *Abelmoschus*, in which the ovary wall is relatively thick, the ovaries can be sterilized by a quick dip in ethanol followed by a light flaming. The ovary is then cut open with a sterile scalpel, and the ovules are scooped out and placed as evenly as possible on the surface of the nutrient medium. To obtain unfertilized ovules, the ovary is collected 24–48 hr prior to anthesis and surface sterilized the usual way. For sterilizing the seeds of *Orobanche aegyptiaca*, Rangaswamy (1963) devised the following technique, which is ideal for sterilizing very minute and light seeds: A glass tube (about 7.5 × 1.5 cm) open at both ends was covered with a piece of muslin cloth at one end. The tube was kept vertical in a beaker containing a little chlorine water. The seeds were dusted on the muslin cloth and chlorine water was poured over them, so that the waste collected in the beaker. This technique proved very satisfactory in that it avoided the wastage of material, ensured safe sterilization by providing a chlorine-saturated environment, and enabled easy manipulation of seeds. The seeds following sterilization were removed with a fine sterilized brush and spread evenly on the medium.

The nutrient medium generally consists of either Nitsch's (1951) or White's (1954) inorganic salt mixture. Auxins, cytokinins, and other complex growth substances such as coconut water, casein hydrolysate, yeast extract, and plant extracts such as cucumber juice (Nakajima *et al.*, 1969) are added as required, and the medium is gelled with 0.6–0.8% agar. Cultures are incubated at 25 ± 2°C, 50–60% relative humidity, and under continuous or alternating (16:8 light:darkness) periods of diffuse light. The growth of the ovule *in vitro* is generally determined by comparison with that *in vivo*, as well as by the size and developmental stage of the embryo and endosperm.

III. RESULTS AND CONCLUSIONS

Withner (1943, 1959) employed ovule culture to shorten the time lag between pollination and maturation of orchid seeds, thereby hastening the production of seedlings. Subsequently, several researchers have shown that many nutritive supplements such as arginine and aspartic acid (Spoerl, 1948), tomato juice and prominogen, a protein hydrolysate (Vacin and Went, 1949), and peptone (Ito, 1961) significantly promoted the growth of orchid ovules.

N. Maheshwari (1958) demonstrated that it is possible to grow excised ovules (containing a zygote of four-celled proembryo) of *Papaver somniferum* to maturity *in vitro*. It has been reported that growth supplements such as kinetin (Maheshwari and Lal, 1961), coconut water and casamino acids (Kapoor, 1959), and casein hydrolysate (Bajaj, 1964) are favorable for the growth of ovules.

When ovules are cultured along with the placenta, the placental tissue has been found to have a beneficial effect (Chopra and Sabharwal, 1963; Pontovich and Sveshnikova, 1966). However, it is not clear whether this is due to increased surface absorption or to the possible presence of growth substances in the tissue.

The successful growth of ovules *in vitro* (i.e., the growth of the embryo in the ovule) is related to the age of the ovules (number of days postpollination) at culture. For example, in *Nicotiana tabacum*, ovules excised up to 9 days after pollination failed to grow, whereas those cultured 12 days after pollination grew to maturity (Siddiqi, 1964). The most responsive age for culture may vary from species to species.

It is well known that osmotic concentration of the nutrient medium has a significant effect on growth (Mauney, 1961; Maheshwari and Rangaswamy, 1965). Wakizuka and Nakajima (1974) found that ovules of *Petunia hybrida* cultured 4 days after pollination (containing the zygote and several endosperm nuclei) grew on a medium with 6% sucrose, whereas those excised 3 days after pollination (with the zygote and a primary endosperm nucleus) required 8% sucrose. The osmotic concentration can be increased by the addition of sucrose, mannitol, glycerol, and sodium chloride to the culture milieu. During the early stages of embryo development, the central vacuolar sap of the ovule has a high osmotic value, which gradually decreases with further growth of the embryo (Ryczkowski, 1962a,b). A study of the osmotic value of the ovule during different stages of its growth might prove helpful in understanding the requirement of excised ovules.

In many interspecific and intergeneric crosses, the hybrid embryo frequently aborts in the developing seed. Although the embryo rescue technique has been successful in growing the hybrid embryo to maturity, in many instances the embryo fails to develop beyond the heart- or torpedo-shaped stage. By resorting to ovule culture or a combination of ovule–embryo culture, it might be feasible to obtain the hybrid progeny (Gadwal *et al.*, 1968; Nitzsche and Hennig, 1976).

Although earlier attempts to culture unfertilized ovules proved to be in vain, several recent reports suggest that unfertilized ovule culture may prove to be a promising approach to obtain gynogenic haploids (Yang and Zhou, 1982). Likewise, a significant offshoot of research on ovule culture has been the technique of test tube pollination and fertilization, which has been effectively employed to raise interspecific and intergeneric hybrids (Rangaswamy, 1977; Zenkteler, 1970, see Chapters 32 and 33, this volume).

Research on ovule culture of phanerogamic parasites is comparatively recent. In species such as *O. aegyptiaca* and *Cistanche tubulosa,* the mature embryo is undifferentiated and lacks organization of the radicle, hypocotyl, cotyledons, and plumule. Ovule culture (or, perhaps more appropriately, seed culture) has facilitated our understanding of the embryo morphogenesis and host–parasite relationship of these obligate root parasites (Rangaswamy, 1963, 1967; Rangan and Rangaswamy, 1966).

Although the occurrence of embryos from the haploid components of the embryo sac has been recorded in more than 100 species, their potential application has long been neglected (Raghavan, 1976). The increasing success with culture of unfertilized ovules indicates its promising role in haploid breeding (Yang and Zhou, 1982). Although *in vitro* culture studies have significantly enhanced our knowledge of the reproductive cycle of angiosperms, little is known about the physiological and biochemical events involved in megasporogenesis and female gametophyte formation. Ovule culture may perhaps aid us in understanding more about what Heslop-Harrison (1980) considers to be "the forgotten generation—the angiosperm gametophyte."

REFERENCES

Bajaj, Y. P. S. (1964). Development of ovules of *Abelmoschus esculentus* var. Pusa sawani in vitro. *Proc. Natl. Inst. Sci. India, Part B* **30,** 175–185.

Chopra, R. N., and Sabharwal, P. S. (1963). In vitro culture of ovules of *Gynandropsis gynandra* (L.) Briq. and *Impatiens balsamina* L. *In* "Plant Tissue Culture—A Symposium" (P. Maheshwari and N. S. Rangaswamy, eds.), pp. 257–264. Univ. of Delhi, Delhi, India.

Gadwal, V. R., Joshi, A. B., and Iyer, R. D. (1968). Interspecific hybrids in *Abelmoschus* through ovule and embryo culture. *Indian J. Genet. Plant Breed.* **28,** 269–274.

Heslop-Harrison, J. (1980). The forgotten generation: Some thoughts on the genetics and physiology of angiosperm gametophyte. *In* "The Plant Genome" (D. R. Davies and D. A. Hopwood, eds.), pp. 1–14. John Innes Charity, Norwich, United Kingdom.

Ito, I. (1961). In vitro culture of ovary and seed in orchids. Ph.D. Thesis, Kyoto Prefect., Univ. of Kyoto.

Kapoor, M. (1959). Influence of growth substances on the ovules of *Zephyranthes. Phytomorphology* **9,** 313–315.

Maheshwari, N. (1958). In vitro culture of excised ovules of *Papaver somniferum. Science* **127,** 342.

Maheshwari, N., and Lal, M. (1961). In vitro culture of excised ovules of *Papaver somniferum* L. *Phytomorphology* **11,** 307–314.
Maheshwari, P., and Rangaswamy, N. S. (1965). Embryology in relation to physiology and genetics. *Adv. Bot. Res.* **2,** 219–321.
Mauney, J. R. (1961). The culture in vitro of cotton embryos. *Bot. Gaz. (Chicago)* **122,** 205–209.
Nakajima, T., Doyoma, Y., and Matsumoto, M. (1969). *In vitro* culture of excised ovules of white clover, *Trifolium repens* L. *Jpn. J. Breed.* **19,** 373–379.
Nitsch, J. P. (1951). Growth and development in vitro of excised ovaries. *Am. J. Bot.* **38,** 566–577.
Nitzsche, W., and Hennig, L. (1976). Fruchknotenkultur bei Gräsern. *Z. Pflanzenzuecht.* **77,** 80–82.
Pontovich, V. E., and Sveshnikova, I. N. (1966). Formation of *Papaver somniferum* L. embryos during cultivation of the ovules in vitro. *Fiziol. Rast. (Moscow)* **13,** 105–113.
Raghavan, V. (1976). "Experimental Embryogenesis in Vascular Plants." Academic Press, New York.
Rangan, T. S. (1982). Ovary, ovule and nucellus culture. *In* "Experimental Embryology of Vascular Plants" (B. M. Johri, ed.), pp. 105–129. Springer-Verlag, Berlin and New York.
Rangan, T. S., and Rangaswamy, N. S. (1966). Morphogenic investigations on parasitic angiosperms. I. *Cistanche tubulosa* Wight (family Orobanchaceae). *Can. J. Bot.* **46,** 263–266.
Rangaswamy, N. S. (1963). Studies on culturing seeds of *Orobanche aegyptiaca* Pers. *In* "Plant Tissue and Organ Culture—A Symposium" (P. Maheshwari and N. S. Rangaswamy, eds.), pp. 345–354. Univ. of Delhi, Delhi, India.
Rangaswamy, N. S. (1967). Morphogenesis of seed germination in angiosperms. *Phytomorphology* **17,** 477–487.
Rangaswamy, N. S. (1977). Application of in vitro pollination and in vitro fertilization. *In* "Applied and Fundamental Aspects of Plant Cell, Tissue, and Organ Cultures" (J. Reinert and Y. P. S. Bajaj, eds.), pp. 412–425. Springer-Verlag, Berlin and New York.
Ryczkwoski, M. (1962a). Changes in the osmotic value of the central vacuolar sap in developing ovules. (Dicotyledonous perennial plants). *Bull. Acad. Pol. Sci.* **10,** 371–374.
Ryczkowski, M. (1962b). Changes in the osmotic value of the sap from embryos, the central vacuole and the cellular endosperm during the development of ovules. *Bull. Acad. Pol. Sci.* **10,** 375–380.
Siddiqi, S. A. (1964). In vitro culture of ovules of *Nicotiana tabacum* L. var. N.P. 31. *Naturwissenschaften* **51,** 517.
Spoerl, E. (1948). Amino acids as sources of nitrogen for orchid embryos. *Am. J. Bot.* **35,** 88–95.
Vacin, E. F., and Went, F. W. (1949). Use of tomato juice in the asymbiotic germination of orchid seeds. *Bot. Gaz. (Chicago)* **111,** 175–183.
Wakizuka, T., and Nakajima, T. (1974). Effect of cultural condition on the in vitro development of ovules of *Petunia hybrida* Vilm. *Jpn. J. Breed.* **24,** 182–187.
White, P. R. (1932). Plant tissue culture: A preliminary report of results obtained in the culturing of certain plant meristems. *Arch. Exp. Zellforsch. Besonders Gewebezuecht.* **12,** 602–620.
White, P. R. (1954). "The Cultivation of Animal and Plant Cells." Ronald Press, New York.
Withner, C. L. (1943). Ovule culture: A new method for starting orchid seedlings. *Bull. Am. Orchid Soc.* **11,** 261–263.
Withner, C. L. (1959). "The Orchids. A Scientific Survey." Ronald Press, New York.
Yang, H. Y., and Zhou, C. (1982). In vitro induction of haploid plants from unpollinated ovaries and ovules. *Theor. Appl. Genet.* **63,** 97–104.
Zenkteler, M. (1970). Test-tube fertilization of ovules in *Melandrium album* Mill. with pollen grains of *Datura stramonium* L. *Experientia* **26,** 661–662.

CHAPTER **29**

Culture of Cotton Ovules

C. A. Beasley

Cooperative Agricultural Extension
University of California
El Centro, California

I. INTRODUCTION

Commercially produced lines of *Gossypium hirsutum* L. (cotton) yield two types of fibers: (1) lint fibers, primarily used in textiles, and (2) fuzz fibers, or "linters," much less profitably used in products such as seat cushion padding. Both types of fibers arise as elongations of epidermal cells of the immature seed and occupy similar sites on the surface of the cotton ovule. The extent of fuzz fibers varies from extremely dense to limited within *G. hirsutum*, and at least one cultivar of G. *barbadense* (57-4) has virtually no fuzz fibers (Beasley, 1979). Cotton cultivars also differ in the average length of their lint fibers, and in general, longer staple cottons are the most desirable. Differences also exist between species and cultivars in fiber strength, color, and sheen. Thus, the initiation and elongation of fiber cells and the deposition of cellulose as a secondary wall are under strict genetic regulation.

Despite this knowledge and the availability of a wide range of genetic material, few methods were available to study these important events.

CELL CULTURE AND SOMATIC CELL
GENETICS OF PLANTS, VOL. 1

ISBN 0-12-715001-3

Major efforts began in 1970 at the University of California, Riverside, with the initial support of Cotton Incorporated (grower funds), to develop the techniques for culturing and studying cotton ovules *in vitro*. In addition to providing answers to fuzz–lint fiber questions, it was hoped that *in vitro* manipulation of plant hormones would yield information on the relationships between fiber growth and the processes of fertilization and embryogenesis. This research generated more significant information on the latter than on fiber types, and several reports were issued as a result of continued funding from Cotton Incorporated and additional support from the National Science Foundation and the Cooperative Research Service of the United States Department of Agriculture.

II. PROCEDURES

A. Production of Parent Plants

Field-grown plants can serve as sources of cotton ovules; however, for many reasons (not the least of which are logistics and season-long production), greenhouse plants were used for the studies described herein. Similarly, many "upland" cottons were tested, but only work with the Acala varieties is described.

Carefully inspected, delinted seeds are overplanted in pots filled to a uniform head space with soil mix. About 2 weeks after planting, seedlings are reduced to one per pot, uniformity in size and vigor being the criteria for their retention. All seedlings are fertilized with a complete nutrient solution once a week and receive a KNO_3 supplemental nitrigation midweekly. In order to furnish a continuous supply of flowers for laboratory use, new sets of plants are started every 2 weeks, allowed to flower for about 140 days, and discarded. Obviously, the number of plants started, maintained, and discarded can be adjusted to the magnitude of laboratory need. Heaters and coolers are automatically activated to maintain an alternating temperature regime of about 30°C during the day and 20°C at night. Details of greenhouse cotton production have been presented elsewhere (Beasley, 1974, 1977a).

First bloom of cotton occurs on the seventh to ninth node about 60 days after planting. Anthesis (white bloom) begins about sunup, pollination rapidly occurs as anthers dry and dehisce, and most ovules are fertilized by the end of the first day postanthesis. On the day after anthesis (pink bloom), the closed, flaccid pink corolla begins to dry and in a few days abscises.

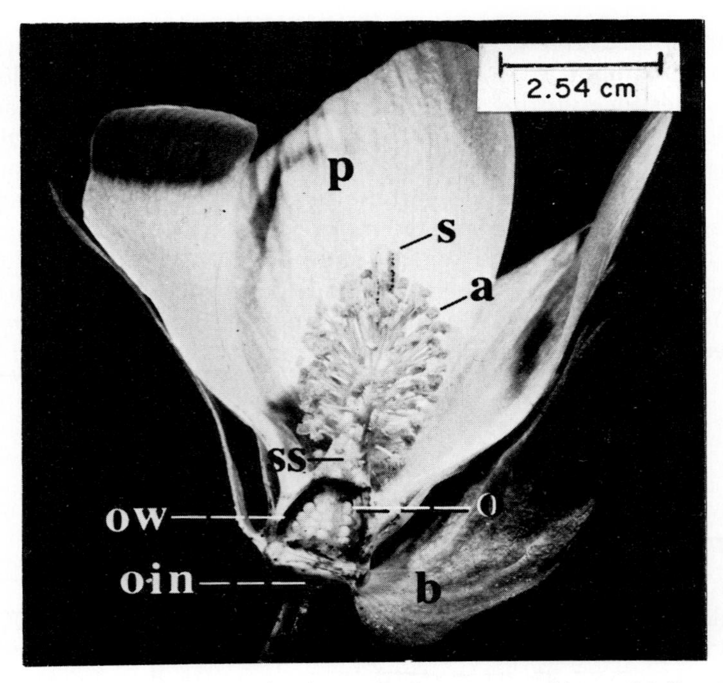

Fig. 1. Open flower on the morning of anthesis. Ovules are exposed by partial dissection. p, petal; b, involucral bract; s, stigma; ss, staminal sheath; a, anther; oin, outer involucral nectary; o, ovule; ow, ovary wall. (From Beasley, 1975.)

All flowers not needed for the laboratory on the day of anthesis or for transfer of ovules to culture a day or two later are removed from the plants, along with their associated flowering branches. Such daily removal forces progressive and uniform flowering on the main stem axis.

B. Collection and Handling of Plant Material

Figure 1 shows the components of the perfect (botanically speaking) cotton flower. If, for experimental purposes, it is desired to transfer preanthesis ovules or even nondehisced anthers to culture, floral bud measurements provide a good estimate of preanthesis developmental events (Beasley and Ting, 1974b). Unfertilized ovules are easily obtained by being transferred to culture on the morning of anthesis. An alternative procedure is to sever and remove the staminal sheath and stigma at the point just above the distal end of the ovary and permit the remaining floral parts to remain on the plant (e.g., for 2 days) until ovule transfer is desired (Beasley, 1973). Ovaries containing nonfertilized ovules abort 4–8 days pos-

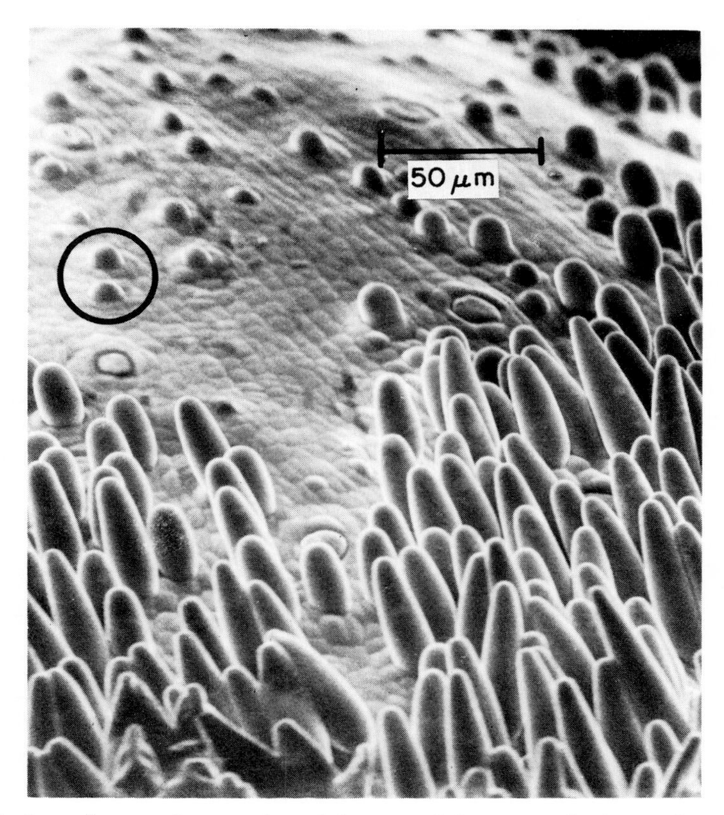

Fig. 2. Ovule surface on the morning of the second day postanthesis. At the top is the microplylar end, where fiber development is nil to limited. Circled two fiber initials illustrate the size and appearance of the longer fibers (foreground) covering the main surface of the ovule had the latter been photographed on the morning of anthesis.

tanthesis. If fertilized ovules are the object of study, they may be transferred (with assurance that all are fertilized) on the morning of the second day postanthesis (Beasley and Ting, 1973). Success in transfer without damage of ovules older than 2 days postanthesis diminishes with time. This is because very young fibers from adjacent ovules adhere to each other and the very thin primary walls are easily ruptured when ovules are separated from their attachment to the central placental column.

Figure 2 shows the approximate length of fiber one should expect over the major surface of the ovules on the second day postanthesis and illustrates the general size and appearance of epidermal cells as they begin their elongation phase on the day of anthesis. In Fig. 2 the long fibers would have appeared to be similar to those circled if the photograph had been taken 2 days earlier. Although the scanning electron microscope affords high-quality, three-dimensional viewing, one can easily assess the extent of early fiber development by the use of top/diagonal lighting and high-power dissecting or low-power compound microscopes. Fuller details of more structures have been presented elsewhere (Beasley, 1975).

Regardless of the time when ovules are to be used, preanthesis or post-anthesis, the procedure is as follows: First, petals, bracts, and sepals are removed. Then ovaries are soaked for 20 min in a 20% solution of bleach (6.0% NaOCl) containing 0.05% Tween-20 or other surfactant, subsequently rinsed with sterile distilled water, and opened with sterile instruments. An ovary is held at its base with tissue forceps while its distal tip is removed. A light incision is made along each locular suture line with a sharp (no. 11 blade) scalpel. With a little practice, one can soon gauge the depth of incision that is adequate to remove the distal tip and extend the blade through the ovary wall along the suture line, and yet not damage the underlying ovules. After an incision is made along the locular suture line, another is made at the base of the ovary, for each lock (carpel section), and the ovary walls are peeled off from base to distal end. Ovules are then prodded, by breaking each funiculus with a dissecting probe, onto a 1.5-cm spoon-shaped spatula. About 28–32 ovules are obtained from each ovary. The spatula is gently lowered below the surface of culture medium contained in a 125-ml culture vessel while the vessel is held at a 45° angle. The ovules float free, the spatula is removed, and the culture vessel is closed.

Transfer operations are performed in a sterile hood with a filtered, positive airflow system. Instruments may be dipped in 95% ethyl alcohol (ETOH) and flamed occasionally to assist in attaining complete asepsis. However, the practice of flaming the mouth of culture vessels prior to or after any operation is discouraged since ethylene and other inhibitory gaseous substances may be introduced via such a procedure (Beasley and Eaks, 1979). Additional details on these matters have been presented by Beasley (1974, 1977a), Beasley and Ting (1974a,b), and Beasley *et al.* (1975).

C. Nutrient Medium, Plant Growth Substances, and Culture Conditions

The composition of the basal medium and the effects of modification on ovule development and growth have been presented elsewhere (Beasley, 1977a,b; Beasley and Ting, 1973). With the exception of studies with boron (Birnbaum *et al.*, 1974, 1977), medium is prepared with glass-distilled water and autoclaved (121°C, 20 psi, 15 min) in 125-ml DeLong flasks closed with chrome Kaputs. Stock solutions of hormones are prepared by dissolving the chemicals in 1–2 ml dimethyl sulfoxide and making up to 1-liter volumes. Small aliquots are frozen in plastic vials and stored until needed. All additions to the basal nutrient medium are sterilized through 0.22-μm filters and buretted into the appropriate sterile treatment flasks prior to transferring ovules.

Except in studies specifically designed to investigate temperature– hormone–nutrient interactions (Beasley, 1977b), culture at constant 32°–34°C in darkness for 2 weeks is used as a standard. Reach-in laboratory incubators and walk-in growth chamber rooms have been used and are satisfactory. Small enclosures with limited air exchange are not recommended for housing culture vessels because of the possibility of imbalance in CO_2 and ethylene ratios (Beasley and Eaks, 1979).

D. Quantitative Determination of Fiber Production

It is a practical impossibility to count the number of fibers on cotton ovules and almost as difficult to determine fiber length on numerous ovules. As an estimation of fiber production *in vitro,* the stain–destain method was developed (Beasley *et al.*, 1974) and is described briefly: Twenty ovules with associated fibers from each culture vessel are (1) shaken in water to extend the fibers, (2) stained for 15 sec in 80 ml 0.018% toluidine blue O, (3) rinsed with running water for 60 sec, and (4) destained in 100 ml glacial acetic acid–ethanol–water (one part acid, nine parts 95% ETOH). Absorbency is determined after the ovules have been in the destaining solution for 1 hr, and that value is expressed in terms of Total Fiber Units. These arbitrary units have been correlated with fiber lengths but not with fiber numbers or extent of lint versus fuzz fibers. Nonetheless, they are useful in quantitatively determining the effects of a variety of treatments on fiber development from cultured ovules.

III. RESULTS AND CONCLUSIONS

Previously cited literature presents not only the procedures employed in ovule culture but also many of the results achieved over an 8-year period. This section summarizes some of the most salient of those results, without references, and suggests some areas for additional research.

On the day of anthesis, ovular epidermal cells that will elongate into lint fibers appear as small protuberances. *In vivo,* these cells continue to elongate through about 28 days postanthesis, and by 50–70 days postanthesis, the fiber has matured. Early in this program, we employed fertilized ovules cultured at 32–34°C. Such ovules produce fiber in response to a completely defined liquid medium containing no exogenous hormones. Fiber production from fertilized ovules is markedly promoted by gibberellic acid (GA_3) but not by indoleacetic acid (IAA).

Unfertilized ovules cultured on identical basal medium (no hormones added) produce no fiber. Such ovules do, however, produce fiber in response to IAA and/or GA_3. Of the two hormones, IAA produces the greater amount of fiber per ovule, and the effects of both hormones applied simultaneously are approximately additive and occasionally synergistic. Depending on the *in vitro* culture temperature, a variable number of unfertilized cotton ovules produce fiber in response to IAA. The response to IAA, for each ovule, is all or none. The percentage of ovules responding to 5.0 μM IAA ranges from 0 at 28°C to 100 at 34°C. The ability to respond at nonpermissive culture temperatures (i.e., 28–30°C) is markedly increased by including 2.5 mM NH_4NO_3 in the KNO_3-based medium or by increasing the concentration of IAA. The inclusion of NH_4^+ in the basal culture medium provides only an increased percentage of ovules producing fiber in the presence of IAA but does not increase fiber production per ovule. Temperature, like GA_3, increases both the number of ovules producing fiber in response to IAA and the amount of fiber produced per ovule.

Ovules acquire the capacity to respond (by producing fiber) to the *in vitro* application of IAA and GA_3 during the first few days preanthesis. Ovules placed in culture on the fourth day (or earlier) preanthesis produce no fiber in the presence of these hormones (Fig. 3). A few ovules respond to culture on the third day preanthesis, a widely variable percentage of ovules re-

Fig. 3. Ovules after 2 weeks of culture in basal medium plus 5.0 μM IAA and 0.5 μM GA_3, with culture beginning on the day before anthesis (left) or the fourth day preanthesis (right).

spond on the second day preanthesis, and essentially 100% respond on both the day before and the day of anthesis. The collective and intact tissues of the entire ovule are not essential to fiber elongation since fibers initiating growth from epidermal slices of ovules harvested on the day of anthesis continue elongation and develop normally in the presence of exogenous IAA and GA_3.

Among other things, the technique of cotton ovule culture offers potential for expanded studies of interspecific crosses (Stewart and Hsu, 1978); protoplast cultures from cells which are programmed to elongate only and never divide (Gould and Dugger, 1982); *in vitro* fertilization (Stewart, 1981); embryo development (Eid *et al.*, 1973; Shen *et al.*, 1978); and cell wall biosynthesis (Delmer, 1983; Dugger *et al.*, 1983).

REFERENCES

Beasley, C. A. (1973). Hormonal regulation of growth in unfertilized cotton ovules. *Science* **1979,** 1003–1005.

Beasley, C. A. (1974). Glasshouse production of cotton flowers, harvest procedures, methods of ovule transfer, and in vitro development of immature seed. *Cotton Grow. Rev.* **51,** 293–301.

Beasley, C. A. (1975). Developmental morphology of cotton flowers and seed, as seen with the scanning electron microscope. *Am. J. Bot.* **62,** 584–592.

Beasley, C. A. (1977a). Ovule culture: Fundamental and pragmatic research for the cotton industry. *In* "Applied and Fundamental Aspects of Plant Cell, Tissue, and Organ Culture" (J. Reinert and Y. P. S. Bajaj, eds.), pp. 160–178. Springer-Verlag, Berlin and New York.

Beasley, C. A. (1977b). A temperature dependent response to indoleacetic acid is altered by NH_4 in cultured cotton ovules. *Plant Physiol.* **59,** 203–209.

Beasley, C. A. (1979). Cellulose content in fibers of cottons which differ in their lengths and extent of fuzz. *Physiol. Plant.* **45,** 77–82.

Beasley, C. A., and Eaks, I. L. (1979). Ethylene from alcohol lamps and natural gas burners: Effects on cotton ovules cultured in vitro. *In Vitro* **15,** 263–269.

Beasley, C. A., and Ting, I. P. (1973). The effects of plant growth substances on in vitro fiber development from fertilized cotton ovules. *Am. J. Bot.* **60,** 130–139.

Beasley, C. A., and Ting, I. P. (1974a). Effects of plant growth substances on in vitro fiber development from unfertilized cotton ovules. *Am. J. Bot.* **61,** 188–194.

Beasley, C. A., and Ting, I. P. (1974b). Phytohormone effects on in vitro cotton seed development. *Plant Growth Subst., Proc. Int. Conf., 8th, 1973* pp. 907–914.

Beasley, C. A., Birnbaum, E. H., Dugger, W. M., and Ting, I. P. (1974). A quantitative procedure for estimation of growth of cotton fibers. *Stain Technol.* **49,** 85–92.

Beasley, C. A., Ting, I. P., Delmer, D. P., Linkins, A. E., and Birnbaum, E. H. (1975). Cotton ovule culture: A review of progress and a preview of potential. *In* "Tissue Culture and Plant Science 1974" (H. E. Street, ed.), pp. 169–192. Academic Press, New York.

Birnbaum, E. H., Beasley, C. A., and Dugger, W. M. (1974). Boron deficiency in unfertilized cotton (*Gossypium hirsutum*) ovules grown in vitro. *Plant Physiol.* **54,** 931–935.

Birnbaum, E. H., Dugger, W., M. and Beasley, C. A. (1977). Interaction of boron with components of nucelic acid metabolism in cotton ovules cultured in vitro. *Plant Physiol.* **59,** 1034–1038.

Delmer, D. P. (1983). Biosynthesis of cellulose. *Adv. Carbohydr. Chem. Biochem.* **41,** 105–153.

Dugger, W. M., Palmer, R. L., and Galloway, C. (1983). The incorporation of UDP[14c-glucose] by intact cotton fibers grown in vitro. *Plant Physiol., Suppl.* **72,** 61.

Eid, A. A. H., DeLange, E., and Waterkeyn, L. (1973). *In vitro* culture of fertilized cotton ovules. I. The growth of cotton embryos. *Cellule* **69,** 361–371.

Gould, J., and Dugger, W. M. (1982). Glucan biosynthesis by cotton ovule epidermal protoplasts. *Plant Physiol., Suppl.* **69,** 26.

Shan, T.-Y., Chang, S.-C., Yhn, Chi-C. (1978). The growth of fibers on excised cotton ovules and the formation of seedlings. *Acta Phytophysiol. Sin.* **4,** 183–187.

Stewart, J. M. (1981). *In vitro* fertilization and embryo rescue. *Environ. Exp. Bot.* **21,** 301–315.

Stewart, J. M., and Hsu, C. L. (1978). Hybridization of diploids and tetraploids through in ovulo embryo culture. *J. Hered.* **69,** 405–408.

CHAPTER **30**

Culture of Embryos

G. B. Collins
J. W. Grosser

Department of Agronomy
University of Kentucky
Lexington, Kentucky

I. INTRODUCTION

Since the turn of the century, when Hannig (1904) first cultured embryos of two cruciferous genera, *in vitro* zygotic embryo culture techniques have been developed with significant applications in plant breeding and horticulture, as well as in basic studies on embryo physiology and biochemistry. Many embryos do not survive *in vivo* or become dormant for long periods of time, and therefore successful *in vitro* embryo culture methods have become quite important. Mature embryos can usually be cultured on a simple medium of inorganic salts plus an energy source, but the culture of smaller embryos often requires the addition of different combinations of hormones, vitamins, amino acids, and possibly even natural endosperm extracts such as coconut milk or natural endosperm transplants. This chapter will review significant applications of embryo culture and discuss protocol and media requirements necessary for successful embryo culture.

CELL CULTURE AND SOMATIC CELL
GENETICS OF PLANTS, VOL. 1

ISBN 0-12-715001-3

II. APPLICATIONS

Raghavan (1976) has discussed evidence which suggests that embryos of inviable hybrids possess the potential for initiating development, but are inhibited from reaching adult size with normal differentiation. Endosperm development precedes and supports embryo development nutritionally, and endosperm failure has been implicated in numerous cases of embryo abortion. Endosperm failure generally results in abnormal embryo development and eventual starvation. There are other cases in which the hostile environment of the ovule (with tumor formation) leads to embryo abortion (Raghavan, 1976). In any event, isolation and culture of hybrid embryos prior to abortion may circumvent these strong postzygotic barriers to interspecific or intergeneric hybridization. The production of interspecific or intergeneric hybrids is the most conspicuous and impressive application of embryo rescue and culture, particularly for subsequent valuable gene transfer from wild species.

The first successful interspecific hybrid obtained via *in vitro* embryo rescue and culture was with *Linum perenne* and *L. austriacum* L. by Laibach in 1925 after conventional methods failed. Since that time, techniques and knowledge of media requirements have been expanded, making isolation and culture of hybrid embryos possible for crosses between numerous species (Table I).

Another significant application of embryo culture is the breakage of seed dormancy in order to shorten breeding cycles. Dormancy of seeds can be caused by numerous factors including endogenous inhibitors, specific light requirements, low temperatures, dry storage requirements, and embryo immaturity (Raghavan, 1977). These factors can be circumvented by embryo excision and culture. The classic example of this is with *Iris,* in which Randolph (1945) was able to shorten the breeding cycle from years to months.

Embryo culture can also be utilized in the production of monoploids. Kasha and Kao (1970) have developed a technique to produce barley monoploids. Interspecific crosses are made with *Hordeum bulbosum* as the pollen parent, and the resulting hybrid embryos are cultured but exhibit *H. bulbosum* chromosome elimination, resulting in monoploids of the female parent.

The development of embryo culture techniques has provided an excellent opportunity for studies in basic embryogenesis including the examination of embryo growth requirements, the effects of phytohormones, the nutrition and metabolism of embryos, and the effects of environmental conditions on zygotic embryogenesis. These topics have been reviewed by Norstog (1973) and Raghavan (1980).

TABLE I

Applied Examples of Embryo Culture[a]

Plant species	Purpose of embryo culture	Reference
Abelmoschus esculentus × *A. manihot* hybrid	Overcome inviability	Patil (1966)
A. esculentus × *A. moschatus*, *A. tulerculates* × *A. moschatus* hybrids	Overcome inviability	Gadwal *et al.* (1968)
Agropyron tsukushiense × *H. bulbosum* hybrid	Attempted polyhaploid production	Shigenobu and Sakamoto (1981)
Allium cepa × *A. fistolusum* hybrid	Study of embryo development	Dolezel *et al.* (1980)
Avena fatua	Overcome seed dormancy	Simpson (1965)
Brassica campestris × *B. oleracea* hybrid	Following ovule culture to overcome inviability	Inomata (1978a)
		Matsuzawa (1978)
	Study of media supplements	Inomata (1978b)
Brassica pekinensis × *B. oleracea* hybrid	Overcome inviability	Nishi *et al.* (1959)
Carthamus tinctorius × *C. lanatus* hybrid	Overcome inviability	Heaton and Klisiewicz (1981)
Cattleya, *Laelia* and other orchids	Induce embryo growth in the absence of symbiotic fungus	Knudson (1922)
Cerasus vulgaris × *C. tomentosa* hybrid	Overcome inviability	Kravtsov and Kas'yanova (1968)
Chrysanthemum boreale × *C. pacificum* hybrid	Overcome inviability	Kaneko (1957)
Colocasia esculentum, *C. antiquorum*	Overcome self-sterility of seeds	Abraham and Ramachandran (1960)
Corchorus capsularis × *C. olitorius* hybrid	Overcome inviability	Islam (1964)
Cucumis spp. (*C. metuliferus* and *C. melo*)	Study of self- and cross-compatibility relationships	Fassuliotus (1977)
Cucumis spp. interspecific hybrids	Overcome inviability	Kho *et al.* (1981)
Cucurbita pepo × *C. moschata* and reciprocal hybrids	Overcome inviability	Wall (1954)
Datura discolor × *D. stramonium* and other interspecific hybrids	Overcome inviability	Sanders (1950)
D. innoxia × Tree Datura (*Brugmansia*?)	Overcome inviability	Blakeslee and Satina (1944)
D. stramonium × *D. ceratocaula* and other interspecific hybrids	Overcome inviability	McLean (1946)

(*continued*)

TABLE I (*Continued*)

Plant species	Purpose of embryo culture	Reference
Drosophyllum lusitanicum	Overcome natural seed dormancy	Dore Swamy and Mohan Ram (1967)
Elaeis guineensis	Overcome natural seed dormancy	Rabéchault (1967)
Gossypium arboreum × *G. herbaceum* hybrid, *G. hirsutum* × *G. barbadense* hybrid	Overcome low seed set	Liang *et al.* (1978)
G. davidsonii × *G. sturtii* hybrid	Overcome inviability	Skovsted (1935)
G. hirsutum	Study of embryo development	Stewart and Hsu (1977)
G. hirsutum × *G. arboreum* and other interspecific hybrids	Overcome inviability	Beasley (1940)
Hordeum spp. (barley)	Study effects of irradiated food materials on tissues	Swaminathan *et al.* (1962)
	Study the possibility of selecting mutants with high specific amino acid content	Landa *et al.* (1980)
Hordeum brachyantherum × *H. vulgare* and other interspecific hybrids; *H. californicum* × *Secale cereale* and other intergeneric hybrids	Overcome inviability	Morrison *et al.* (1959)
H. jubatum × *Secale cereale* hybrid	Overcome inviability	Brink *et al.* (1944)
H. sativum × *H. bulbosum* hybrid	Overcome inviability	Konzak *et al.* (1951)
H. vulgare × *H. bulbosum* hybrid	Overcome inviability	Morrison *et al.* (1959); Davies (1960)
H. vulgare × (*H. compressum* × *H. pusillum*), *H. vulgare* × *H. hexopodium* hybrids	Overcome inviability	Schooler (1962)
Hordeum × *Triticum*	Overcome inviability	Kruse (1974)
Hordeum × *Agropyron*		
Hordeum × *Secale* intergeneric hybrids		
Impatiens hookeriana × *I. campanulata*	Overcome inviability	Arisumi (1980)
Iris sp.	Accelerate seed germination	Randolph and Cox (1943)
Iris (tall bearded) × *I. tectorum* (crested *Iris*) hybrid	Overcome inviability	Lenz (1954)
I. munzii × *I. sibirica* "Caesar's brother" hybrid	Overcome inviability	Lenz (1956)

I. pallida, *I. macrantha*, *I. pallida* × *I. chamaeris* hybrids	Overcome inviability	Werckmeister (1936)
I. pseudocorus × *I. versicolor* hybrid	Overcome inviability	Blakeslee and Satina (1944)
Lathyrus clymenum × *L. articulatus* hybrid	Overcome inviability	Pecket and Selim (1965)
Lilium auratum platyphyllum × *L. henryi*, *L. longiflorum* × *L. candidum* hybrids	Overcome inviability	Asano (1980)
L. henryi × *L. regale* hybrid	Overcome inviability	Skirm (1942)
L. longiflorum interspecific hybrids	Overcome inviability	Clark and Campbell (1978)
Lilium speciosum "album" × *L. auratum*, *L. speciosum* "Rubrum" × *L. auratum* hybrids	Overcome inviability	Emsweller *et al.* (1962)
Linum perenne × *L. austriacum* hybrid	Overcome inviability	Laibach (1925)
Lotus corniculatus × *L. filicaulis* and other interspecific hybrids	Overcome inviability	Grant *et al.* (1962)
L. pedunculatus × *L. tenuis* hybrid	Overcome inviability	Williams and DeLautour (1980); DeLautour *et al.* (1978)
L. tenuis × *L. corniculatus* hybrid	Overcome inviability	DeLautour *et al.* (1978)
Lycopersicon esculentum × *L. peruvianum* hybrid	Overcome inviability	Smith (1944); Choudhury (1955); Alexander (1956)
Malus sp. (weeping crabapple)	Accelerate seed germination	Nickell (1951)
Medicago sativa and unnamed interspecific hybrids	Overcome embryo abortion due to self-sterility and interspecific inviability	Fridriksson and Bolton (1963)
Melilotus officinalis × *M. alba* hybrid	Overcome inviability	Webster (1955); Schlosser-Szigat (1962)
Musa balbisiana	Overcome self-sterility of seeds	Cox *et al.* (1960)
Ornithopus sativus × *O. compressus* Pitman *O. pinnatus* × *O. sativus* hybrids	Overcome inviability	Williams and DeLautour (1980)
Oryza paraguaiensis × *O. brachyantha* and other interspecific hybrids	Overcome inviability	Li *et al.* (1961)
O. sativa × *O. minuta*, *O. sativa* × *O.* sp. (Paraguay) hybrids	Overcome inviability	Nakajima and Morishima (1958)
O. sativa × *O. officinalis* and other interspecific hybrids	Overcome inviability	Iyer and Govila (1964)
O. sativa × *O. schweinfurthiana* and other interspecific hybrids	Overcome inviability	Bouharmont (1961)

(*continued*)

TABLE I (*Continued*)

Plant species	Purpose of embryo culture	Reference
Phaseolus coccineus × *P. acutifolius*, *P. coccineus* × *P. vulgaris*, *P. vulgaris* × *P. acutifolius*, (*P. vulgaris* × *P. coccineus*) × *P. acutifolius* hybrids	Overcome inviability	Alvarez *et al.* (1981)
P. vulgaris × *P. acutifolius* hybrid	Overcome inviability	Honma (1955); Mok *et al.* (1978)
P. vulgaris × *P. lunatus* hybrid	Overcome inviability	Mok *et al.* (1978)
Phaseolus vulgaris	Study host–parasite relationship in phytopathology	Padmanabhan (1967)
Pinus lambertiana × *P. armandi*, *P. lambertiana* × *P. koraiensis* hybrids	Facilitate germination and overcome hybrid inviability	Stone and Duffield (1950)
Prunus avium (sweet cherry)	Overcome low seed viability	Tukey (1933)
P. cerasus (sour cherry), *P. persica* (plum), *Pyrus communis* (pear), *Malus domestica* (apple) (intervarietal hybrids)	Overcome low seed viability	Tukey (1934)
P. persica (plum)	Predict seed viability at planting	Tukey (1944)
Ribes nigrum × *Grossulaeia reclinata* hybrid	Overcome inviability	Kravtsov and Kas'yanova (1968)
Rosa sp. (Rose)	Accelerate seed germination	Asen (1948)
Solanum melongena cv. sonepat local × *S. khasianum* hybrid	Overcome inviability	Sharma *et al.* (1980)

S. nigrum × *S. luteum* hybrid	Overcome inviability	Jorgensen (1928)
Trifolium ambiguum × *T. hybridum* hybrid	Overcome inviability	Keim (1953); Evans (1962); Rupert and Richards (1979); Rupert and Evans (1980); Williams and DeLautour (1980)
T. ambiguum × *T. occidentale*, *T. ambiguum* × *T. montanum*, *T. repens* × *T. isthmocarpum* hybrids	Overcome inviability	Rupert and Evans (1980)
T. repens × *T. nigrescens* and other interspecific hybrids	Overcome inviability	Evans (1962)
T. ambiguum × *T. repens* hybrid	Overcome inviability	Williams (1978); Williams and DeLautour (1980)
T. repens × *T. uniflorum* hybrid	Overcome inviability	Williams and DeLautour (1980)
T. pratense × *T. sarosiense* hybrid	Overcome inviability	Phillips *et al.* (1982)
Tripsacum dactyloides × *Zea mays* hybrid	Overcome inviability	Farquharson (1957)
Triticum aestivum × *Aegilops* spp. hybrids	Overcome inviability	Cheuca *et al.* (1977)
T. durum × *Elymus arenarius* and other intergeneric hybrids	Overcome inviability	Ivanovskaya (1946, 1962)
T. durum abyssinicum × *Secale cereale* hybrid	Overcome inviability	Redei (1955)
Vigna umbellata × *V. angularis* hybrid	Overcome inviability	Ahn and Hartman (1978a)
V. radiata × *V. angularis* hybrid	Overcome inviability	Ahn and Hartman (1978b)
Zea mays	Test seed quality	Mukherje (1951)

[a] Updated and adapted with permission from Raghavan (1977).

Raghavan (1977) has discussed several examples in which embryo culture could be utilized in the testing of seed viability. These examples include work performed by Tukey (1944), who reported a correlation between the growth of excised embryos of nonafterripened peach seeds and the germination of afterripened seeds. Correlations such as this may be developed for other species. Mukherji (1951) has devised a method for testing *Zea mays* seed quality which utilizes embryo culture.

Other applications including the study of host pathogen interactions, overcoming self-sterility, and the germination of seeds of obligate parasites have been reviewed by Raghavan (1976, 1977).

III. GENERAL PROTOCOL AND DISCUSSION

The following protocol is based in general on the system developed in our laboratory for the isolation and culture of interspecific hybrid embryos within the genus *Trifolium*.

Successful utilization of embryo culture to produce interspecific or intergeneric hybrids can be greatly facilitated by preliminary experiments. Precrosses of available breeding material in a diallel fashion followed by checks for shriveled seed production can identify the best combinations of genotypes. Shriveled seed indicates endosperm failure, but at least some zygotic development. Williams and DeLautour (1980) and Phillips (1981) have demonstrated that considerable variation exists between cultivars and genotypes in hybrid embryos regarding *in vivo* size and ability to grow in culture. Efforts should therefore be concentrated on parental genotypes which give the best *in vivo* embryo development and the most efficient growth and development *in vitro* on the most efficient media. If possible, it is also important to use parental genotypes already identified as amenable to *in vitro* manipulation utilizing a well-defined tissue culture system, particularly when embryos tend to callus.

Many interspecific hybrids produced via embryo culture are sterile or of very low fertility. Doubling the chromosome number (usually with colchicine) can restore fertility in many cases when sterility is due to the lack of pairing of chromosomes during meiosis (Liang *et al.*, 1978; Heaton and Klisiewicz, 1981; Williams and DeLautour, 1981). Efficient chromosome doubling methods are not always available, and the use of tetraploid parents could possibly circumvent this problem. These methods cannot overcome genic sterility in hybrids.

In all cases, an understanding of the reproductive physiology of the parental species is fundamental. For example, attempted hybridizations with self-fertilizing species may require emasculations or bagging prior to anthesis in bisexual species. Examinations of zygote formation following pollination to determine optimal periods of *in situ* development of hybrid zygotes must also be performed. Phillips (1981) crossed *Trifolium sarosiense* × *T. pratense* in our laboratory. Florets were removed 11–23 days after pollination, using care to prevent mechanical ovule damage. Embryos were found to be more ontogenetically advanced as time progressed, but also a lower percentage of embryos survived over time. After 11 days, globular embryos were generally observed, but after 23 days only a few embryos were found, some of which were in the early torpedo stage. The best compromise between the maximum degree of development and the number of embryos must be determined (14–19 days in this case). This timing is critical for maximum success.

In some crosses in which the female parent is self-incompatible, rare selfs may be observed. However, these can usually be distinguished from hybrid embryos produced on the same flower by their advanced stage of development and their normal appearance.

Once the optimal parents are identified, appropriate crosses can be performed for subsequent embryo rescue. Florets are removed at the proper time. Either florets or ovaries can be sterilized in a sequence consisting of 2.5 min in 95% ethanol, 5 min in 2% sodium hypochlorite, and 5 min in sterile water. Ovules can then be removed from the ovaries. The tissue within the ovule, in which the embryo is embedded, is already sterile. The embryo can then be aseptically removed from the ovule with sharp pointed forceps and a sharp dissecting needle. Extreme care must be taken to prevent mechanical damage as well as desication of the embryo. The optimal point of incision into the ovule should be determined from preliminary investigations regarding the anatomy of the embryo sac and embryo location within the ovule. All embryo excisions and subsequent culture transfers should be performed in the sterile environment of a laminar flow hood. A dissecting microscope may be required for the excision of very small embryos.

Utilization of embryo culture to overcome seed dormancy requires a somewhat different procedure. Seeds which have hard coats are surface sterilized and soaked in water for a few hours to a few days. Sterile seeds are then split and the embryos excised (Yeung *et al.*, 1981).

In our *Trifolium* system, isolated embryos are placed directly on the first medium of the sequence. This medium has a high sucrose concentration to prevent precocious germination and a combination of hormones which supports the growth of heart-stage embryos without loss of organization (a moderate level of auxin and a low level of cytokinin). After 8–14 days, embryos cease to grow on this medium and must be transferred to a second medium. This medium has a normal sucrose concentration, a low level of auxin, and a moderate level of cytokinin, and allows for renewed embryo growth with direct shoot germination in many cases. Some embryos resume growth but suffer a loss of organization. Embryos which directly germinate shoots can be transferred to a third medium containing a low level of auxin with a high level of cytokinin, which promotes an increase in clonal shoot numbers. Shoots can then be rooted and eventually transferred to the greenhouse.

Embryos which exhibit renewed growth but a loss of organization on the second medium are transferred to a somatic embryogenesis induction medium. The production of hybrid plants from these embryos is generally less efficient.

Tissues surrounding the embryo provide a physical framework, as well as nutritional requirements which change as the embryo develops. There is a distinction regarding the dependence of the embryo on nutritional substances provided by the endosperm (or accessory cells of the embryo sac) between a heterotrophic phase and an autotrophic phase. The autotrophic phase apparently begins in the late heart stage (Raghavan, 1976). There is also a correlation between the developmental stage of the embryo in culture and media complexity, with less advanced embryos having more complex media requirements. Rescued embryos in the autotrophic phase are much more amenable to *in vitro* culture. The isolation and culture of globular and early heart-stage embryos is much more difficult, often requiring extensive media refinements which must be empirically defined. It is evident that the most important aspect of successful embryo culture is the definition of the media necessary to sustain growth and development of the embryo.

Another consideration regarding the culture of early staged embryos is the role of the suspensor. Monnier (1978) has pointed out that the suspensor must be intact for survival of early staged embryos. If the embryo is excised from the suspensor at an early stage, the wound results in embryo abortion.

Jensen (1977) has cultured very young barley embryos in the dark for 1–2 weeks at 18°C since light encourages precocious germination.

Monnier (1978) has engineered a clever device which eliminates the need for the sequential transfer from one medium to another. This device allows for the juxtaposition of two media of distinct composition. The first agar medium is liquefied by heating and then poured around a small central glass container in the center of a Petri dish. This medium makes up the external ring. Upon solidification of the first medium, the central container is removed and a second medium of unique composition is poured into the hole. Embryos are cultured on the second medium in the center of the Petri dish, and as a result of diffusion, they are gradually subjected to the action of a variable medium. Monnier was able to culture successfully globular embryos only 50μm long using this system.

IV. MEDIA REQUIREMENTS

A. Basic Requirements

1. Carbohydrate. The best energy source for embryo culture appears to be sucrose. High concentrations of sucrose (8–12.5%) are used which approximate the high osmotic potential of the intracellular environment of the young embryo sac. Raghavan (1976) believes that this condition of high osmolarity prevents precocious germination and switches cells from a state of elongation to one of division. He has also proposed the use of mannitol as an osmoticum. Norstog (1979) suggests that the condition of high osmolarity may prevent plasmolytic damage to very delicate proembryos in their early cleavage divisions.

2. Basal medium. In most cases, a standard basal plant growth medium with major salts and trace elements may be utilized. However, Monnier (1978) has pointed out that embryos are quite sensitive to the mineral solution used. Further, mineral solutions which promote growth are toxic, whereas nontoxic mineral solutions are not able to induce normal differentiation of embryos. The basal medium utilized may require empirical adjustments to produce a medium which promotes normal growth with minimum toxicity.

3. Nitrogen source. Hannig (1904) found asparagine to be a good source of reduced organic nitrogen for embryo culture. However, others have found glutamine to be a superior source of nitrogen (Paris *et al.*, 1953; Rijven, 1955; Raghavan, 1976; Mok *et al.*, 1978). Raghavan (1976) has reviewed inorganic nitrogen utilization in embryo culture. Limited data suggest that mechanisms of nitrogen utilization differ by species and are greatly affected by embryo maturity and culture conditions. NH_4NO_3 and KNO_3 are frequently used sources of inorganic nitrogen in embryo culture.

4. Hormones. In many cases, exogenously supplied hormones are not required for embryo culture. Monnier (1978) believes that embryos can be considered plants with their own endogenous hormones. However, there are several cases in which exogenously supplied hormones have greatly facilitated embryo culture. LaRue (1936) found low concentrations of indoleacetic acid (IAA) (0.05 μg/ml) helpful with embryos of several species. Raghavan and Torrey (1963) demonstrated that low concentrations of IAA and kinetin aided globular *Capsella* embryos. Phillips (1981) found that moderate levels of auxin with low levels of cytokinin aided the growth and survival of heart-shaped interspecific hybrid embryos within the genus *Trifolium*. It seems evident that for many species, natural endosperm may contain hormones which influence embryo growth and development, and in cases such as this where the endosperm fails, the culture media must

supply the necessary hormones. This subject has been reviewed by Norstog (1979) and Raghavan (1976, 1980).

B. Other Supplements

1. Vitamins. Commonly added vitamins are biotin, thiamine, pantothenic acid, nicotinic acid, ascorbic acid, inositol, and pyroxidine. Vitamins have not been proven to be essential for successful embryo culture.

2. Coconut milk. Coconut milk is a natural source of nutrients and was first used successfully by van Overbeek *et al.* (1941) with immature embryos of *Datura stramonium.* Since this time, coconut milk has been utilized in numerous instances. It appears to maintain embryos in a dormant state.

3. Seed or fruit diffusates. These substances, which are obtained from various plant parts, have been utilized in the culture of embryos from *Datura* and other species by Matsubara (1962, 1964).

4. Casein hydrolysate. This substance has been found to be effective in facilitating embryo culture by Ziebur *et al.* (1950) and Inomata (1978b). It may provide necessary amino acids and could also affect the osmolarity of the medium.

5. Amino acids. Amino acids are often added to embryo culture media. Jensen (1977) found that the addition of 14 different amino acids enhanced the growth and development of barley embryos.

6. Organic acids. Organic acids such as malic acid are often added to embryo culture media, but their role has not been well defined.

7. Transplanted nurse endosperm. The use of transplanted nurse endosperms with embryo culture has been discussed by Williams and DeLautour (1980). This technique is based on the insertion of a hybrid embryo into a healthy endosperm dissected from a normally developing ovule. The embryo and endosperm are then transferred together to the surface of a nutrient agar medium. Williams and DeLautour (1980) were able to obtain several interspecific hybrids within the genera *Trifolium, Lotus,* and *Ornithopus* utilizing nurse endosperm. Previously, Kruse (1974) utilized a similar technique to obtain several intergeneric hybrids involving *Hordeum, Triticum, Agropyron,* and *Secale.* Williams and DeLautour believed that the transplanted nurse endosperm improved growth and differentiation, improved recovery rates, supplied nutritional and hormonal requirements between the late globular and early heart stage, and played a physical role by providing a closer to normal *in vivo* microenvironment.

8. Agar. Experiments in our laboratory performed with the genus *Trifolium* indicate that the use of an agar-based culture medium better mimics *in situ* conditions than liquid media.

REFERENCES

Abraham, A., and Ramachandran, K. (1960). Growing *Colocasia* embryos in culture. *Curr. Sci.* **29,** 342–343.

Ahn, C. S., and Hartman, R. W. (1978a). Interspecific hybridization between rice bean *Vigna umbellata* and adzuki bean *Vigna angularis. J. Am. Soc. Hortic. Sci.* **103,** 435–438.

Ahn, C. S., and Hartman, R. W. (1978b). Interspecific hybridization between mung bean *Vigna radiata* and adzuki bean *Vigna angularis. J. Am. Soc. Hortic. Sci.* **103**(1), 3–6.

Alexander, L. J. (1956). Embryo culture of tomato interspecific hybrids. *Phytopathology* **46,** 6 (abstr.).

Alvarez, M. N., Ascher, P. D., and Davis, D. W. (1981). Interspecific hybridization in section Euphaseolus through embryo rescue. *HortScience* **16,** 541–543.

Arisumi, T. (1980). *In vitro* culture of embryos and ovules of certain incompatible selfs and crosses among *Impatiens* spp. *J. Am. Soc. Hortic. Sci.* **105,** 629–631.

Asano, Y. (1980). Studies on crosses between distantly related species of lilies. 5. Characteristics of newly obtained hybrids through embryo culture. *J. Jpn. Soc. Hortic. Sci.* **49,** 241–250.

Asen, S. (1948). Embryo culture of rose seeds. *Am. Rose Annu.* **33,** 151–152.

Beasley, J. O. (1940). Hybridization of American 26-chromosome and Asiatic 13-chromosome species of *Gossypium. J. Agric. Res.* **60,** 175–181.

Blakeslee, A. F., and Satina, S. (1944). New hybrids from incompatible crosses in *Datura* through culture of excised embryos on malt media. *Science* **99,** 331–334.

Bouharmont, J. (1961). Embryo culture of rice on sterile medium. *Euphytica* **10,** 283–293.

Brink, R. A., Cooper, D. C., and Ausherman, L. E. (1944). The antipodals in relation to abnormal endosperm behavior in *Hordeum jubatum* and *Secale cereale. J. Hered.* **35,** 67–75.

Cheuca, M. C., Cauderon, Y., and Tempe, J. (1977). *In vitro* embryo culture technique to obtain *Triticum aestivum* × *Aegilops* sp. hybrids. *Ann. Amelior. Plant.* **27,** 539–546.

Choudhury, B. (1955). Embryo culture technique. III. Growth of hybrid embryos (*Lycopersicon esculentum* × *Lycopersicon peruvianum*) in culture medium. *Indian J. Hortic.* **12,** 155–156.

Clark, D. R., and Campbell, R. J. (1978). Interspecific hybridization of *Lilium longiflorum* using stylar amputation pollinations stigmatic exudate and embryo culture. *HortScience* **13,** 350–351.

Cox, E. A., Stotzky, G., and Goors, R. D. (1960). *In vitro* culture of *Musa balbisiana* Colla embryos. *Nature (London)* **185,** 403–404.

Davies, R. D. (1960). The embryo culture of interspecific hybrids of *Hordeum. New Phytol.* **59,** 9–14.

DeLautour, G., Jones, W. T., and Ross, M. D. (1978). Production of interspecific hybrids in *Lotus* aided by endosperm transplants. *N. Z. J. Bot.* **16,** 61–68.

Dolezel, J., Novak, F. J., and Luzny, J. (1980). Embryo development and *in vitro* culture of *Allium cepa* and its interspecific hybrids. *Z. Pflanzenzuecht.* **85,** 177–184.

Dore Swamy, R., and Mohan Ram, H. Y. (1967). Cultivation of embryos of *Drosophyllum lusitanicum* link - an insectivorous plant. *Experientia* **23,** 675.

Emsweller, S. L., Asen, S., and Uhring, J. (1962). *Lilium speciosum* × *L. auratum. Am. Lily Soc.* **15,** 7–15.

Evans, A. M. (1962). Species hybridization in *Trifolium.* 1. Methods of overcoming species incompatability. *Euphytica* **11,** 164–176.

Farquharson, L. I. (1957). Hybridization of *Tripsacum* and *Zea. J. Hered.* **48,** 295–299.

Fassuliotis, G. (1977). Self fertilization of *Cucumis metuliferus* and its cross compatibility with *Cucumis melo. J. Am. Soc. Hortic. Sci.* **102,** 336–339.

Fridriksson, S., and Bolton, J. L. (1963). Preliminary report on the culture of alfalfa embryos. *Can. J. Bot.* **41,** 439–440.

Gadwal, V. R., Joshi, A. B., and Iyer, R. D. (1968). Interspecific hybrids in *Abelmoschus* through ovule and embryos culture. *Indian J. Genet. Plant Breed.* **28,** 269–274.

Grant, W. F., Bullen, M. R., and DeNettancourt, D. (1962). The cytogenetics of *Lotus.* I. Embryo-cultured interspecific diploid hybrids closely related to *L. corniculatus. Can. J. Genet. Cytol.* **4,** 105–128.

Hannig, E. (1904). Physiology of plant embryos. I. The culture of Cruciferous embryos outside the embryo sac. *Bot. Ztg.* **62,** 46–81.

Heaton, T. C., and Klisiewicz, J. M. (1981). A disease resistant safflower alloploid from *Carthamus tinctorius* × *Carthamus lanatus. Can. J. Plant Sci.* **61,** 219–224.

Honma, S. (1955). A technique for artificial culturing of bean embryos. *Proc. Am. Soc. Hortic. Sci.* **65,** 405–408.

Inomata, N. (1978a). Production of interspecific hybrids in *Brassica campestris* × *Brassica oleracea* sexual hybrid by culture *in vitro* of excised ovaries. Part 1. Development of excised ovaries in the crosses of various cultivars. *Jpn. J. Genet.* **53,** 161–174.

Inomata, N. (1978b). Production of interspecific hybrids between *Brassica campestris* and *Brassica oleracea* by culture *in vitro* of excised ovaries. Part 2. Effects of coconut-milk and casein hydrolysate on the development of excised ovaries. *Jpn. J. Genet.* **53,** 1–12.

Islam, A. S. (1964). A rare hybrid combination through application of hormone and embryo culture. *Nature (London)* **201,** 320.

Ivanovskaya, E. V. (1946). Hybrid embryos of cereals grown on artificial nutrient medium. *C. R. (Dokl.) Acad. Sci. URSS* **54,** 445–448.

Ivanovskaya, E. V. (1962). The method of raising embryos on an artificial nutrient medium and its application to wide hybridization. *In* "Wide Hybridization in Plants" (N. V. Tsitsin, ed.), p. 134. Israel Program for Scientific Translations, Jerusalem.

Iyer, R. D., and Govila, O. P. (1964). Embryo culture of interspecific hybrids in the genus *Oryza. Indian J. Genet. Plant Breed.* **24,** 116–121.

Jensen, C. J. (1977). 4. Monoploid production by chromosome elimination. *In* "Applied and Fundamental Aspects of Plant Cell, Tissue, and Organ Culture" (J. Reinert and Y. P. S. Bajaj, eds.), p. 299. Springer-Verlag, Berlin and New York.

Jorgensen, C. A. (1928). The experimental formation of heteroploid plants in the genus *Solanum. J. Genet.* **19,** 133–211.

Kaneko, K. (1957). Studies of the embryo culture on the interspecific hybridization of *Chrysanthemum. Jpn. J. Genet.* **32,** 300–305.

Kasha, K. J., and Kao, K. N. (1970). High frequency haploid production in barley (*Hordeum vulgare* L.). *Nature (London)* **225,** 874–876.

Keim, W. F. (1953). Interspecific hybridization in *Trifolium* using embryo culture techniques. *Agron. J.* **45,** 601–606.

Kho, Y. O., Den Nijs, A. P. M., and Franken, J. (1981). Interspecific hybridization in *Cucumis.* 2. The crossability of species and investigation of *in vivo* pollen tube growth and seed set. *Euphytica* **29,** 661–672.

Knudson, L. (1922). Nonsymbiotic germination of orchid seeds. *Bot. Gaz. (Chicago)* **73,** 1–25.

Konzak, C. F., Randolph, L. F., and Jensen, N. F. (1951). Embryo culture of barley species hybrids. Cytological studies of *Hordeum sativum* × *Hordeum bulbosum. J. Hered.* **42,** 125–134.

Kravtsov, P. V., and Kas'yanova, G. (1968). Culture of isolated embryos as a method for prevention of sterility in distant hybrids of fruit plants. *Fiziol. Rast. (Moscow),* **15,** 784–786.

Kruse, A. 1974. An *in vivo/vitro* embryo culture technique. *Hereditas* **77,** 219–224.

Laibach, F. (1925). Das Taubwerden von Bastardsmen und die kunsliche Aufzucht fruh absterbender Bastardembryonen. *Z. Bot.* **17,** 417–459.

Landa, Z., Novak, F. J., Opatrny, Z., Landova, B., and Petru, E. (1980). Use of explanted crops in genetics and plant selection. *Acad. Nakladatelstvi, Cesk. Akad. Ved. Rozpr. Cesk. Akad. Ved. Rada. Mat. Prir. Ved.* **90,** 5–62.

LaRue, C. D. (1936). The growth of plant embryos in culture. *Bull. Torrey Bot. Club* **63,** 365–382.

Lenz, L. W. (1954). The endosperm as a barrier to intersectional hybridization in *Iris*. *Aliso* **3,** 51–58.

Lenz, L. W. (1956). Development of the embryo sac, endosperm and embryo in *Iris munzii* and the hybrid *I. munzii* × *I. sibirica* "Caesars Brother." *Aliso* **3,** 329–343.

Li, H. W., Weng, T. S., Chen, C. C., and Wang, W. H. (1961). Cytogenetical studies of *Oryza sativa* L. and its related species. *Bot. Bull. Acad. Sin.* **2,** 79–86.

Liang, C. L., Sun, C. W., Liu, T. L., and Chiang, J. C. (1978). Studies on interspecific hybridization in cotton. *Sci. Sin. (Engl. Ed.)* **21,** 545–556.

McLean, S. W. (1946). Interspecific crosses involving *Datura ceratocoula* obtained by embryo dissection. *Am. J. Bot.* **33,** 630–638.

Matsubara, S. (1962). Studies on a growth promoting substance, "embryo factor," necessary for the culture of young embryos of *Datura tatula in vitro*. *Bot. Mag.* **75,** 10–18.

Matsubara, S. (1964). Effect of *Lupinus* growth factor on the *in vitro* growth of embryos of various plants and carrot root tissue. *Bot. Mag.* **77,** 403–411.

Matsuzawa, Y. (1978). Studies on the interspecific hybridization in genus *Brassica*. Part 1. Effects of temperature on the development of hybrid embryos and the improvement of crossability by ovary culture in interspecific cross *Brassica campestris mu Brassica oleracea*. *Jpn. J. Breed.* **28,** 186–196.

Mok, D. W. S., Mok, M. C., and Rabakoarihanta, A. (1978). Interspecific hybridization of *Phaseolus vulgaris* hybrid parent with *Phaseolus lunatus* hybrid parent and *Phaseolus acutifolius* hybrid parent. *Theor. Appl. Genet.* **52,** 209–215.

Monnier, M. (1978). 28. Culture of zygotic embryos. *In* "Frontiers of Plant Tissue Culture 1978" (T. A. Thorpe, ed.), p. 277. Univ. of Calgary Press, Calgary, Canada.

Morrison, J. W., Hannah, A. E., Loiselle, R., and Symko, S. (1959). Cytogenetic studies in the genus *Hordeum*. II. Interspecific and intergeneric crosses. *Can. J. Plant Sci.* **39,** 375–383.

Mukherji, D. K. (1951). Embryo culture as an aid to seed testing. *Proc. Natl. Inst. Sci. India, Part B* **17,** 253–259.

Nakajima, T., and Morishima, H. (1958). Studies on embryo culture in plants. II. Embryo culture of interspecific hybrids in *Oryza*. *Jpn. J. Breed.* **8,** 105–110.

Nickell, L. G. (1951). Embryo culture of weeping crabapple. *Proc. Am. Soc. Hortic. Sci.* **57,** 401–405.

Nishi, S., Kawata, J., and Toda, M. (1959). On the breeding of interspecific hybrids between two genomes, "c" and "a" of *Brassica* through the application of embryo culture techniques. *Jpn. J. Breed.* **8,** 215–222.

Norstog, K. (1973). New synthetic medium for the culture of barley embryos. *In Vitro* **8,** 307.

Norstog, K. (1979). 13. Embryo culture as a tool in the study of comparative and developmental morphology. *In* "Plant Cell and Tissue Culture: Principles and Applications" (W. R. Sharp, P. O. Larsen, E. F. Paddock, and V. Raghavan, eds.), p. 179. Ohio State Univ. Press, Columbus.

Padmanabhan, D. (1967). Effect of fusaric acid on *in vitro* culture of embryos of *Phaseolus vulgaris* L. *Curr. Sci.* **36,** 214–215.

Paris, D., Rietsema, J., Santina, S., and Blakeslee, A. F. (1953). Effects of amino acids, especially aspartic and glutamic acid and their amides, on the growth of *Datura stramonium* embryos *in vitro*. *Proc. Natl. Acad. Sci. U.S.A.* **39,** 1205–1212.

Patil, J. S. (1966). The culture *in vitro* of immature embryos of okra (*Abelmoschus esculentus*). *Plant Physiol.* **9,** 59–65.

Pecket, R. D., and Selim, A. R. A. A. (1965). Embryo culture in *Lathyrus*. *J. Exp. Bot.* **16,** 325–328.

Phillips, G. C. (1981). Hybridization of red clover with a perennial *Trifolium* species using *in vitro* embryo rescue. Ph.D. Dissertation, Univ. of Kentucky, Lexington.

Phillips, G. C., Collins, G. B., and Taylor, N. L. (1982). Interspecific hybridization of red clover *Trifolium pratense* cultivar Kenstar with *Trifolium sarosiense* using *in vitro* embryo rescue. *Theor. Appl. Genet.* **62,** 17–24.

Rabéchault, H. (1967). Relation entre le comportement des embryons de Palmier à huile (*Elaeis guineensis* Jacq.) en culture in vitro et la teneur en eau des graines. *C. R. Hebd. Seances Acad. Sci.* **264,** 276–279.

Raghavan, V. (1976). "Experimental Embryogenesis in Vascular Plants." Academic Press, New York.

Raghavan, V. (1977). 3. Applied aspects of embryo culture. *In* "Applied and Fundamental Aspects of Plant Cell, Tissue, and Organ Culture" (J. Reinert and Y. P. S. Bajaj, eds.), p. 375. Springer-Verlag, Berlin and New York.

Raghavan, V. (1980). Embryo Culture. *Int. Rev. Cytol., Suppl.* **11B,** 209–240.

Raghavan, V., and Torrey, J. G. (1963). Growth and morphogenesis of globular and older embryos of *Capsella* in culture. *Am. J. Bot.* **50,** 540–551.

Randolph, L. F. (1945). Embryo culture of *Iris* seed. *Bull. Am. Iris Soc.* **97,** 33–45.

Randolph, L. F., and Cox, L. G. (1943). Factors influencing the germination of *Iris* seed and the relation of inhibiting substances to embryo dormancy. *Proc. Am. Soc. Hortic. Sci.* **43,** 284–300.

Redei, G. (1955). *Triticum durum abyssinicum* × *Secale cereale* hybridek elöallitasa mesterséges embryo nevelés segitsegerel. *Novenytermeles* **4,** 365–367.

Rijven, A. H. G. C. (1955). Effects of glutamine, asparagine and other related compounds on the *in vitro* growth of embryos of *Capsella bursapastoris*. *Proc. K. Ned. Akad. Wet., Ser. C* **58,** 368–376.

Rupert, E. A., and Evans, P. T. (1980). Embryo development after interspecific cross-pollinations among species of *Trifolium* Section *Lotoidea*. *Agron. Abstr.* p. 68.

Rupert, E. A., and Richards, K. W. (1979). *Trifolium* species hybrids obtained from embryo-callus tissue cultures. *Agron. Abstr.* p. 75.

Sanders, M. E. (1950). Development of self and hybrid *Datura* embryos in artificial culture. *Am. J. Bot.* **37,** 6–15.

Schlosser-Szigat, G. (1962). Artbastardierung mit Hilfe der Embryokultur bei Steinklee (*Melilotus*). *Naturwissenschaften* **49,** 452–453.

Schooler, A. B. (1962). Technique of crossing wild and domestic barley species. *Bi-Mon. Bull., N. D. Agric. Exp. Stn.* **22,** 16–17.

Sharma, D. R., Chowdhury, J. B., Ahuja, U., and Dhankhar, B. S. (1980). Interspecific hybridization in genus *Solanum*. A cross between *Solanum melongena* cv. Sonepat local and *Solanum khasianum* through embryo culture. *Z. Pflanzenzuecht.* **85,** 248–253.

Shigenobu, T., and Sakamoto, S. (1981). Intergeneric hybridization between *Agropyron tsukushiense* and *Hordeum bulbosum* 4×. *Jpn. J. Genet.* **56,** 505–518.

Simpson, G. M. (1965). The role of giberellin in embryo dormancy. *Can. J. Bot.* **43,** 793–816.

Skirm, G. W. (1942). Embryo culture as an aid to plant breeding. *J. Hered.***33,** 211–215.

Skovsted, A. (1935). Cytological studies in cotton. III. A hybrid between *Gossypium davidsonii* Kell. and *G. sturtii* F. Muell. *J. Genet.* **30,** 397–405.

Smith, P. G. (1944). Embryo culture of a tomato species hybrid. *Proc. Am. Soc. Hortic. Sci.* **44,** 413–416.

Stewart, J. M., and Hsu, C. L. (1977). In ovulo embryo culture and seedling development of cotton *Gossypium hirsutum*. *Planta* **137**(2), 113–118.

Stone, E. C., and Duffield, J. W. (1950). Hybrids of sugar-pine embryo culture. *J. For.* **48,** 200–201.

Swaminathan, M. S., Chopra, V. S., and Bhaskaran, S. (1962). Cytological aberrations observed in barley embryos cultured in irradiated potato mash. *Radiat. Res.* **16,** 182–188.
Tukey, H. B. (1933). Artificial culture of sweet cherry embryos. *J. Hered.* **24,** 7–12.
Tukey, H. B. (1934). Artificial culture methods for isolated embryos of deciduous fruits. *Proc. Am. Soc. Hortic. Sci.* **32,** 313–322.
Tukey, H. B. (1944). The excised embryo method of testing the germinability of fruit seed with particular reference to peach seed. *Proc. Am. Soc. Hortic. Sci.* **45,** 211–219.
van Overbeek, J., Conklin, M. E., and Blakeslee, A. F. (1941). Factors in coconut milk essential for growth and development of very young *Datura* embryos. *Science* **94,** 350–351.
Wall, J. R. (1954). Interspecific hybrids of *Curcurbita* obtained by embryo culture. *Proc. Am. Soc. Hortic. Sci.* **63,** 427–430.
Webster, G. T. (1955). Interspecific hybridization of *Melilotus alba* × *M. officinalis* using embryo culture. *Agron. J.* **43,** 138–142.
Werckmeister, P. (1936). Über Herstellung und kunstliche Aufzucht von Bastarden der Gattung *Iris. Gartenbauwissenschaft* **10,** 500–520.
Williams, E. G. (1978). A hybrid between *Trifolium repens* and *T. ambiguum* obtained with the aid of embryo culture. *N. Z. J. Bot.* **16,** 499–506.
Williams, E. G., and DeLautour, G. (1980). The use of embryo culture with transplanted nurse endosperm for the production of interspecific hybrids in pasture legumes. *Bot. Gaz. (Chicago)* **141**(3), 252–257.
Williams, E. G., and DeLautour, G. (1981). Production of tetraploid hybrids between *Ornithopus pinnatus* and *Ornithopus sativus* using embryo culture. *N. Z. J. Bot.* **19,** 23–30.
Yeung, E. C., Thorpe, T. A., and Jensen, C. J. (1981). *In vitro* fertilization and embryo culture. *In* "Plant Tissue Culture: Methods and Applications in Agriculture" (T. A. Thorpe, ed.), p. 253. Academic Press, New York.
Ziebur, N. K., Brink, R. A., Graf, L. H., and Stahmann, M. A. (1950). The effect of casein hydrolysate on the growth *in vitro* of immature *Hordeum* embryos. *Am. J. Bot.* **37,** 144–148.

CHAPTER **31**

Culture of Endosperm

Sant S. Bhojwani

Department of Botany
University of Delhi
Delhi, India

I. INTRODUCTION

Except for members of the families Orchidaceae, Podostemaceae, and Trapaceae, all angiosperms form endosperm tissue. It may be consumed wholly by the developing embryo, or it may persist in mature seed as a massive tissue and store reserve food materials in diverse forms. The unique features of the endosperm tissue are as follows: (1) in over 81% of flowering plants it is triploid, being derived from the fusion product of three haploid nuclei (two from the female gametophyte and one from the male gametophyte), and (2) it is a homogeneous mass of parenchymatous tissue lacking any organogenic or vascular differentiation.

Following the discovery of syngamy and triple fusion, embryologists regarded endosperm as a second embryo modified to serve as the nutritive tissue for the zygotic embryo (LeMonnier, 1887; Sargant, 1900). Sargant suggested that in triple fusion "the third nucleus may have been introduced to maim the second embryo (endosperm) from the beginning and secure the survival of the first (zygotic embryo) without struggle." However, ploidy cannot be regarded as the cause of the formless nature of the endosperm tissue because the diverse destinies of the zygote and the pri-

ISBN 0-12-715001-3

mary endosperm nuclei are maintained even in the Onagraceae, in which the embryo and endosperm are both diploid.

Early interest in the culture of endosperm consisted of testing the growth and regeneration potentialities of this unique tissue (Narayanaswami, 1956). Lampe and Mills (1933) are credited for the pioneering attempts in this field. They cultured immature endosperm of maize and noted a slight proliferation of endosperm layers adjacent to the embryo. Extensive work on endosperm culture was started by LaRue and his co-workers at the University of Michigan. After many years of sustained work, LaRue (1949) raised, for the first time, continuously growing tissue cultures from immature endosperm of maize. Subsequently, many other workers examined maize endosperm cultures from diverse angles.

Since 1963 the Department of Botany at the University of Delhi has actively contributed to the field of endosperm culture. Maiden reports of cell divisions in mature endosperm and establishment of tissue cultures from it (Rangaswamy and Rao, 1963), and the demonstration of cellular totipotency of endosperm tissue (Johri and Bhojwani, 1965), were published from this school. Of late, the potential applications of endosperm culture for triploid production in relation to crop improvement have been recognized, and many workers, especially the Chinese, have made significant contributions in this direction (see Bhojwani and Razdan, 1983). During the last 5 years, differentiation of shoots or plantlets from the endosperm tissue of certain important crop plants has been reported (Table I).

This chapter is devoted chiefly to the technological aspects of endosperm culture. At the end, the applications of this technique in basic and applied areas of plant sciences are highlighted.

II. METHODOLOGY

A. Explant

The stage of seed development at which the endosperm can be used to initiate cultures depends on several considerations. In plants with nonendospermous seeds (e.g., apple and citrus), only immature seeds can provide the necessary explants. However, at the time of excision the endosperm must be cellular; free nuclear endosperm does not survive in culture (Mu *et al.*, 1977; Wang and Chang, 1978).

In cereals and grasses, where the mature endosperm represents a physiologically dead tissue except for a few outer aleurone layers, only imma-

TABLE I

Species Which Have Been Reported to Form Shoots or Plantlets from Endosperm Tissue[a]

Species	Reference
Actinidiaceae	
Actinidia chinensis	Gui *et al.* (1982)
Euphorbiaceae	
Codiaeum variegatum	Chikkannaih and Gayatri (1974)
Jatropha panduraefolia	Srivastava (1971)
Putranjiva roxburghii	Srivastava (1973)
Gramineae	
Oryza sativa	Nakano *et al.* (1975), Bajaj *et al.* (1980)
Loranthaceae	
Dendrophthoe falcata	Nag and Johri (1971)
Scurrula pulverulenta	Bhojwani and Johri (1970)
Taxillus vestitus	Nag and Johri (1971)
Rosaceae	
Prunus persica	Shu-quiong and Jia-qu (1980)
Pyrus malus	Mu *et al.* (1977)
Rutaceae	
Citrus grandis	Wang and Chang (1978)
Santalaceae	
Exocarpus cupressiformis	Johri and Bhojwani (1965)
Santalum album	Lakshmi Sita *et al.* (1980)

[a] Adapted from Bhojwani and Razdan (1983).

ture endosperm is amenable to culture (Narayanaswami, 1956; Tamaoki and Ullstrup, 1958; Varner, 1971). The most responsive age of the endosperm may vary with the plant: 8–10 days after pollination (DAP) in *Lolium perenne* (Norstog, 1956), 4–7 DAP in *Oryza sativa* (Nakano *et al.*, 1975), 8 DAP in *Triticum aestivum* and *Hordeum vulgare* (Sehgal, 1974), and 8–11 DAP in *Zea mays* (Straus and LaRue, 1954; Tamaoki and Ullstrup, 1958).

In dicotyledonous plants endosperm from immature (Mu *et al.*, 1977; Wang and Chang, 1978) as well as mature, dried seeds has been used to initiate cultures. In some plants, such as members of the Euphorbiaceae and Santalaceae, the association of the embryo is essential for the initial stages of dedifferentiation of the cells of mature endosperm. In such cases, the entire decoated seeds are used as the explant. Once endosperm cells have started dividing (Figs. 1, 2), the embryo/seedling is removed and the endosperm tissue is transferred to fresh medium for further growth (Fig. 3). If callusing of the embryo occurs concurrently with the proliferation of the endosperm (Fig. 2), care must be exercised to discard completely the tissue of embryo origin at the time of transfer of endosperm tissue. Com-

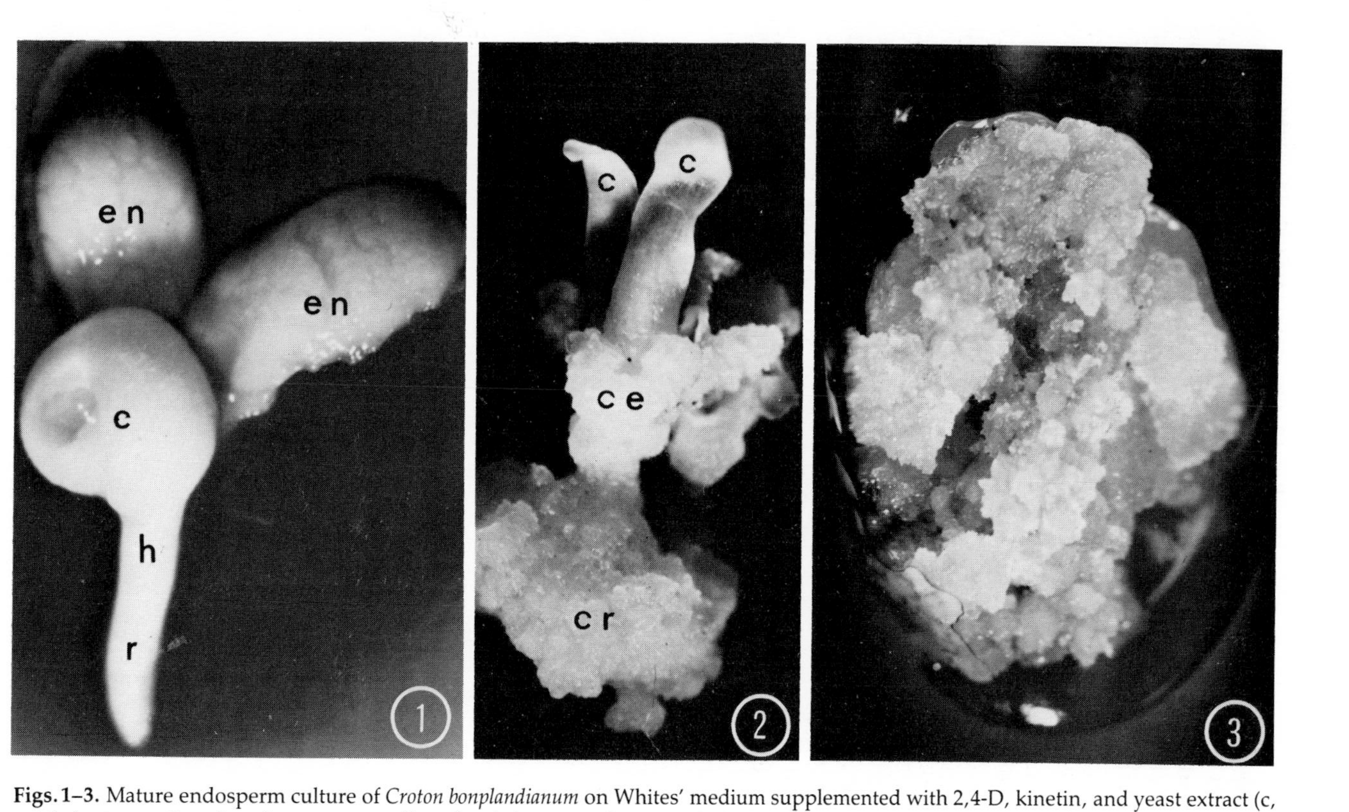

Figs. 1–3. Mature endosperm culture of *Croton bonplandianum* on Whites' medium supplemented with 2,4-D, kinetin, and yeast extract (c, cotyledon; ce, callused endosperm; cr, callused radicle; en, endosperm; h, hypocotyl; r, root). **Fig. 1.** Ten-day-old culture of decoated seeds showing a germinated embryo with enlarged, coiled cotyledons and an elongated hypocotyl; note the proliferation on the inner side of the endosperm. **Fig. 2.** Four-week-old culture showing enlarged cotyledons. The radicle as well as the endosperm has proliferated. **Fig. 3.** Profusely growing endosperm callus in a 4-week-old subculture. (From Bhojwani and Johri, 1970.)

plications arising due to the callusing of the embryo may be avoided by culturing excised endosperm pieces presoaked in GA_3 solution. The GA_3 treatment has been shown to replace the "embryo factor" (Srivastava, 1971; Johri and Srivastava, 1973). Brown *et al.* (1970) showed that soaking the entire seeds of castorbean in water for various periods before excising the endosperm pieces for culture also eliminates the need for the association of embryo during initial stages of culture of mature endosperm. A direct relationship occurs between the number of days the seeds are soaked in water and the percentage of cultures showing endosperm callusing (Fig. 4). Endosperm from immature seeds does not exhibit dependence on the embryo factor for proliferation (Mu *et al.*, 1977; Wang and Chang, 1978).

Preparation of the explant for the culture of mature endosperm is fairly simple. In plants having seeds with massive endosperm (Euphorbiaceae, Santalaceae), the decoated seeds are directly surface sterilized with a suitable disinfectant and, after two to three washings in sterile distilled water, planted on the medium. In the Loranthaceae the endosperm is surrounded by a sticky viscin layer which causes inconvenience in dissecting out the

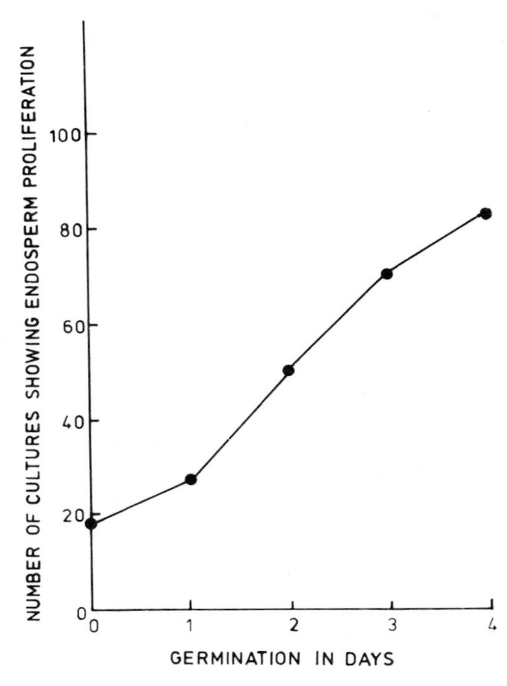

Fig. 4. Effect of soaking mature, dried castorbean seeds in water on the proliferation of endosperm; 0 day refers to seeds soaked for 24 hr. (Drawn from the data of Brown *et al.*, 1970).

endosperm. For preparing the explants of mature endosperm of these mistletoes, the fruits are surface sterilized with 90% alcohol and "seeds" (endosperm plus embryo) are dissected out under aseptic conditions. The fruits are held between the thumb and the forefinger, previously sterilized by rinsing in ethanol, and the fruit walls are removed with the help of a sterile needle. Frequent dipping of the fingertips in ethanol during the operation facilitates dissection.

For culture of immature endosperm, entire seeds or kernels are surface sterilized and the endosperm tissue is carefully excised under aseptic conditions using a stereoscopic microscope. In these cases, it is desirable to have a good knowledge of the structure of the seed. This is helpful in dissecting out endosperm explants completely free of embryonal tissue. Negligence in this regard could lead to contamination of the endosperm callus with the tissue of embryo origin. This could have been the cause of the sporadic differentiation of shoots in rice (Nakano *et al.*, 1975) and the origin of uniformly diploid plants in "endosperm" (actually seed) cultures of parsley (Masuda *et al.*, 1977).

To raise maize endosperm cultures, spikelets 10–12 DAP are brought to the laboratory. The glumes are removed and the immature ovaries surface sterilized. The tops of the kernels (micropylar end) are cut off with a sterile knife, and the exposed endosperm squeezed out and placed on the medium (Narayanaswami, 1956).

B. Medium

1. Callusing

The basal media used in endosperm culture studies are the popular White's Medium (White, 1943), its modified version (Rangaswamy, 1961), and MS medium (Murashige and Skoog, 1962). There are hardly any data on the comparative evaluation of one medium over the other. However, of late, MS medium has been used more frequently because of its completeness.

Best callusing of immature endosperm of graminaceous members occurs on a medium supplemented with yeast extract (400–2000 mg l^{-1}) with or without an auxin [1–2 mg l^{-1} 2,4-dichlorophenoxyacetic acid (2,4-D) or 1 mg l^{-1} indoleacetic acid (IAA)]. However, for the establishment of endosperm cultures of dicotyledonous plants, the presence of an auxin (2,4-D or IAA), a cytokinin (kinetin or benzylamino purine), and a rich source of organic nitrogen (yeast extract or casein hydrolysate) in the basal medium is essential.

Sucrose in the range of 2–4% has been found to be the most satisfactory source of carbohydrate for the growth of endosperm calli.

The optimum pH of the culture medium varies considerably with the species: 4.0 for *Asimina triloba* (Lampton, 1953), 5.6 for *Jatropha panduraefolia* and *Putranjiva roxburghii* (Srivastava, 1971, 1973), 5.0 for *Ricinus communis* (Johri and Srivastava, 1973), and 6.1 for *Zea mays* (Straus and LaRue, 1954).

2. Shoot Bud–Plantlet Differentiation

In loranthaceous and santalaceous parasites, shoot buds may differentiate directly from the peripheral cells of the endosperm (Figs. 5–7), or this may be preceded by a callusing phase. In all other plants, the latter pathway operates (Bhojwani and Razdan, 1983).

Generally, an exogenous cytokinin is essential for the differentiation of shoot buds from endosperm tissue. Whereas in *Scurrula pulverulenta* and *Taxillus vestitus* a cytokinin alone is able to induce bud formation, in *Dendrophthoe falcata* and *Leptomeria acida* cytokinin is effective only in the presence of a low concentration of an auxin, such as IAA and indolebutyric acid (IBA). Of the various cytokinins, 2-isopentenyl adenine has proved most effective for shoot bud formation from endosperm. In *Dendrophthoe* and *Leptomeria* casein hydrolysate promotes bud formation.

In some species, plantlet formation from endosperm callus may follow the embryogenic mode of development (*Citrus grandis*—Wang and Chang, 1978; *Pyrus malus*—Mu *et al.*, 1977; *Santalum album*—Lakshmi Sita *et al.*, 1980). Embryo differentiation occurs when the tissue is transferred from the callusing medium to a basal medium with or without gibberellin. Interestingly, in *Citrus* the concentration of the salts had to be raised to twice their level in the callusing medium in order to induce embryogenesis.

Injury to the endosperm enhances the frequency of shoot bud formation in *Taxillus vestitus* (Nag and Johri, 1971). Also, the position of the endosperm on the medium significantly affects the differentiation and distribution of buds on the explant. Half-split endosperm pieces planted with their cut surface in contact with the medium are more responsive than cultures with the cut surface of endosperm away from the medium. In these taxa the presence of the embryo significantly reduces the number of buds per culture and the number of cultures forming buds from endosperm. However, the subsequent development of the shoots is better in the presence of the embryo.

Full triploid plants of endosperm origin have been developed in *Actinidia chinensis, Citrus grandis, Putranjiva roxburghii, Pyrus malus,* and *Santalum album*.

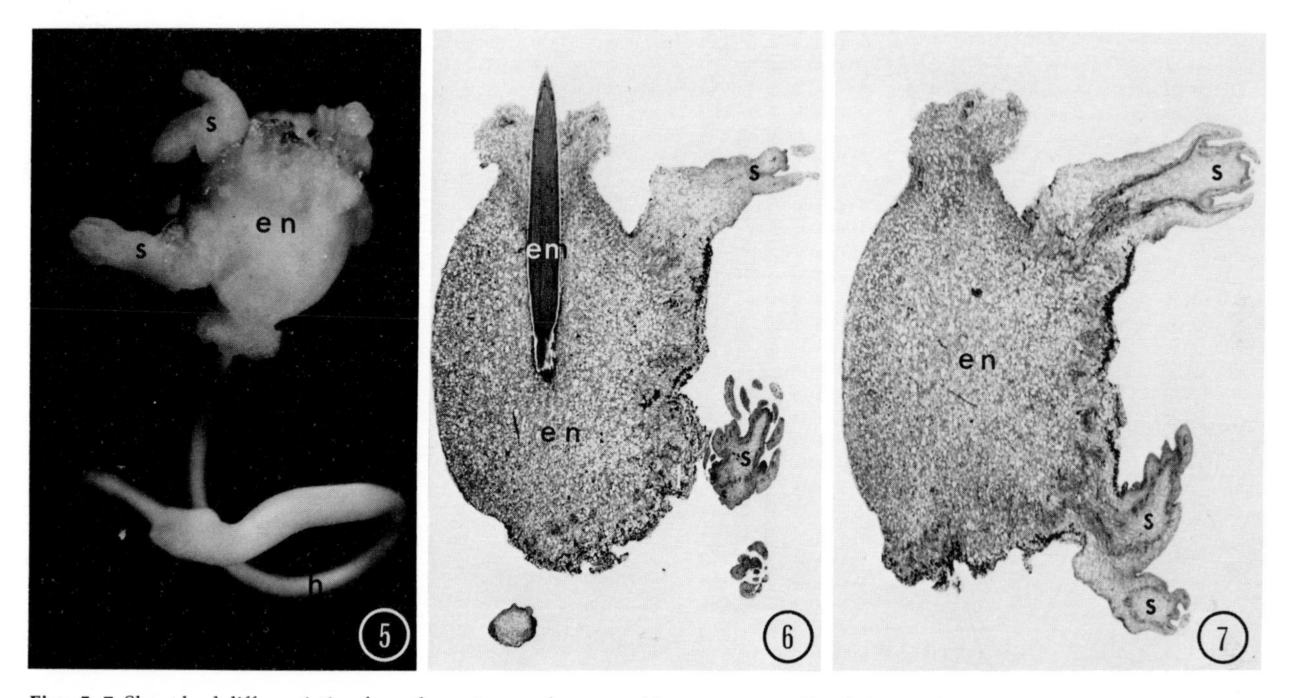

Figs. 5–7. Shoot bud differentiation from the mature endosperm of *Exocarpus cupressiformis* (em, embryo; en, endosperm; h, hypocotyl; s, shoot). **Fig. 5.** A 5-week-old "seed" (endosperm plus embryo) culture showing shoot buds on the endosperm and germinated embryo. **Figs. 6, 7.** Section of the seed shown in Fig. 5; note the origin of the shoots from the periphery of the endosperm. (From Johri and Bhojwani, 1965. Reprinted with permission from *Nature* **208**(5017), 1345–1347. Copyright © 1965 Macmillan Journals Limited.)

C. Storage Conditions

Not much is known about the optimum storage conditions for endosperm cultures. The available information is mainly with reference to castorbean and maize.

1. Light

For callusing the cultures are generally maintained in dark (Straus and LaRue, 1954; Wang and Chang, 1978) or diffuse light (Mu *et al.*, 1977), but for differentiation they are transferred to bright light (2000–4000 lx). *Ricinus* endosperm callus, however, grew better in continuous light than in dark (Srivastava, 1971).

2. Temperature

The optimum temperature for the growth of endosperm callus is reported to be around 25°C (Straus and LaRue, 1954; Srivastava, 1973). Even the differentiating cultures are usually maintained at this temperature.

III. APPLICATIONS

Since 1965 triploid shoots and plantlet formation from endosperm has been achieved in many species, including some important crop plants (see Table I). The list of species whose endosperm has yielded continuously growing tissue cultures is even longer (Johri and Bhojwani, 1977).

Where seed is not commercially important or seedlessness is desirable, as in some fruit trees (Wang and Chang, 1978), and where plants can be propagated vegetatively, the induction of triploidy may prove valuable for the improvement of crop plants. Some of the economically important plants whose triploids are already in commercial use include several varieties of apple, bananas, mulberry, sugar beets, and watermelons (Elliott, 1958). Triploid plants of *Petunia axillaris* raised from microspores were more vigorous and much more ornamental than their parental anther-donor diploids (Gupta, 1982). Similarly, triploid quaking aspen (*Populus tremuloides*) has more desirable pulp wood characteristics of interest to the forest industry than its diploids. Triploids are also useful for obtaining trisomic lines for genetic mapping (Gupta, 1982).

Triploid production by the conventional technique involves chromosome doubling followed by crossing the tetraploids with their diploids (Straub, 1973). This approach is not only laborious but in many cases may not be possible due to the high sterility of autotetraploids (Gupta, 1982; see also Esen and Soost, 1973). The demonstration of the totipotency of triploid cells of endosperm has created a new approach to the production of a large number of triploid plants in one step.

Endosperm tissue is a reservoir of stored food which is mobilized during seed development and germination. It therefore provides an excellent experimental system for studying the biosynthesis and metabolism of these natural products. Chu and Shannon (1975) have described corn endosperm callus as a useful model system for studying *in vivo* starch biosynthesis. Endosperm callus of coffee is shown to synthesize caffein. The level of the alkaloid increases by a factor of three after 2 weeks and by a factor of six after 4–5 weeks of culture.

REFERENCES

Bajaj, Y. P. S., Saini, S. S., and Bidani, M. (1980). Production of triploid plants from the immature and mature endosperm cultures of rice. *Theor. Appl. Genet.* **58,** 17–18.

Bhojwani, S. S., and Johri, B. M. (1970). Cytokinin-induced shoot bud differentiation in mature endosperm of *Scurrula pulverulenta*. *Z. Pflanzenphysiol.* **63,** 269–275.

Bhojwani, S. S., and Razdan, M. K. (1983). "Plant Tissue Culture: Theory and Practice." Elsevier, Amsterdam.

Brown, D. J., Canvin, D. T., and Zilkey, B. F. (1970). Growth and metabolism of *Ricinus communis* endosperm in tissue culture. *Can. J. Bot.* **48,** 2323–2331.

Chikkannaiah, P. S., and Gayatri, M. C. (1974). Organogenesis in endosperm tissue cultures of *Codiaeum variegatum* Blume. *Curr. Sci.* **43,** 23–24.

Chu, L. C., and Shannon, J. C. (1975). *In vitro* culture of maize endosperm—A model system for studying *in vitro* starch biosynthesis. *Crop Sci.* **15,** 814–819.

Elliott, F. C. (1958). "Plant Breeding and Cytogenetics." McGraw-Hill, New York.

Esen, A., and Soost, R. K. (1973). Seed development in citrus with special reference to 2x 4x crosses. *Am. J. Bot.* **60,** 448–462.

Gui, Y., Mu, X., and Xu, T. (1982). Studies on morphological differentiation of endosperm plantlets of Chinese gooseberry *in vitro*. *Acta Bot. Sin.* **24,** 216–221.

Gupta, P. P. (1982). Genesis of microspore-derived triploid petunias. *Theor. Appl. Genet.* **61,** 327–331.

Johri, B. M., and Bhojwani, S. S. (1965). Growth responses of mature endosperm in cultures. *Nature (London)* **208,** 1345–1347.

Johri, B. M., and Bhojwani, S. S. (1977). Triploid plants through endosperm culture. *In* "Applied and Fundamental Aspects of Plant Cell, Tissue, and Organ Culture" (J. Reinert and Y. P. S. Bajaj, eds.), pp. 398–411. Springer-Verlag, Berlin and New York.

Johri, B. M., and Srivastava, P. S. (1973). Morphogenesis in endosperm cultures. *Z. Pflanzenphysiol.* **70,** 285–304.

Lakshmi Sita, G., Raghava Ram, N. V., and Vaidyanathan, C. S. (1980). Triploid plants from endosperm cultures of sandalwood by experimental embryogenesis. *Plant Sci. Lett.* **20,** 63–69.

Lampe, L., and Mills, C. O. (1933). Growth and development of isolated endosperm and embryo of maize. *Abstr. Bot. Soc. Am.*

Lampton, R. K. (1953). Developmental and experimental morphology of the ovule and seed of *Asimina triloba.* Ph.D. Thesis, Univ. of Michigan, Ann Arbor.

LaRue, C. D. (1949). Cultures of the endosperm of maize. *Am. J. Bot.* **36,** 798 (abstr.).

LeMonnier, G. (1887). Sur la valeur morphologique de l'albumen chez les Angiosperms. *J. Bot. (Paris)* **1,** 140–142.

Masuda, K., Koda, Y., and Okazawa, Y. (1977). Callus formation and embryogenesis of endosperm tissues of parsley seed cultured on hormone-free medium. *Physiol. Plant.* **41,** 135–138.

Mu, S., Liu, S., Zhou, Y., Qian, N., Zhang, P., Xie, H., Zhang, F., and Yan, Z. (1977). Induction of callus from apple endosperm and differentiation of the endosperm plantlets. *Sci. Sin. (Engl. Transl.)* **20,** 370–375.

Murashige, T., and Skoog, F. (1962). A revised medium for rapid growth and bioassays with tobacco tissue cultures. *Physiol. Plant.* **15,** 473–497.

Nag, K. K., and Johri, B. M. (1971). Morphogenic studies on endosperm of some parasitic angiosperms. *Phytomorphology* **21,** 202–218.

Nakano, H., Tashiro, T., and Maeda, E. (1975). Plant differentiation in callus tissue induced from immature endosperm of *Oryza sativa* L. *Z. Pflanzenphysiol.* **76,** 444–449.

Narayanaswami, S. (1956). Plant endosperm and its culture *in vitro. Sci. Cult.* **22,** 132–136.

Norstog, K. J. (1956). Growth of rye-grass endosperm *in vitro. Bot. Gaz. (Chicago)* **117,** 253–259.

Rangaswamy, N. S. (1961). Experimental studies on female reproductive structures of *Citrus microcarpa* Bunge. *Phytomorphology* **11,** 109–127.

Rangaswamy, N. S., and Rao, P. S. (1963). Experimental studies on *Santalum album* L. Establishment of tissue culture of endosperm. *Phytomorphology* **13,** 450–454.

Sargant, E. (1900). Recent work on the results of fertilization in angiosperms. *Ann. Bot. (London)* **14,** 689–712.

Sehgal, C. B. (1974). Growth of barley and wheat endosperm in cultures. *Curr. Sci.* **43,** 38–40.

Shu-quiong, L., and Jia-qu, L. (1980). Callus induction and embryoid formation in endosperm cultures of *Prunus persica. Acta Bot. Sin.* **22,** 198–199.

Srivastava, P. S. (1971). *In vitro* induction of triploid roots and shoots from mature endosperm of *Jatropha panduraefolia. Z. Pflanzenphysiol.* **66,** 93–96.

Srivastava, P. S. (1973). Formation of triploid 'plantlets' in endosperm cultures of *Putranjiva roxburghii. Z. Pflanzenphysiol.* **69,** 270–273.

Straub, J. (1973). Die genetische variabilitat haplöider Petunien. *Z. Pflanzenzuecht.* **70,** 265–274.

Straus, J., and LaRue, C. D. (1954). Maize endosperm tissue grown *in vitro.* I. Culture requirements. *Am. J. Bot.* **41,** 687–694.

Tamaoki, T., and Ullstrup, A. J. (1958). Cultivation in vitro of excised endosperm and meristem tissue of corn. *Bull. Torrey Bot. Club* **85,** 260–272.

Varner, J. E. (1971). Digestion and metabolism in plants. *In* "Topics in the Study of Life," pp. 158–162. The Bio-Source Book, Harper, New York.

Wang, T., and Chang, C. (1978). Triploid citrus plantlet from endosperm culture. *In* "Proceedings of Symposium on Plant Tissue Culture," pp. 463–467. Science Press, Peking.

White, P. R. (1943). "A Handbook of Plant Tissue Culture." Jacques Cattell Press, Lancaster, Pennsylvania.

CHAPTER **32**

In Vitro Pollination and Fertilization

Maciej Zenkteler

Department of General Botany
Institute of Biology
Adam Mickiewicz University
Poznan, Poland

I. INTRODUCTION

In angiosperms the female gametophyte is situated deep inside the ovary. After the pollen grains germinate on the stigman, the pollen tubes grow through the stigma, style, and placenta and finally enter the embryo sacs, where the process of double fertilization occurs. The pathway of the pollen tube is one important, but not indispensable, factor in the formation of the zygote. Much more complicated events occur during and after syngamy, but our knowledge of this process is very meager. Angiosperms are mostly outbreeders, and therefore self-pollination is precluded. In nature, interspecific or intergeneric hybridization occurs very rarely; however, hybrid plants, as well as in some cases selfed plants, constitute valuable material in plant breeding. Any technique which would make possible either selfing or hybridization is of great value from both a theoretical and (especially) a practical point of view. The method of *in vitro* pollination and fertilization offers an opportunity to analyze these processes in completely controlled conditions. Besides, at present, in only a few cases is it possible to achieve hybrid embryos among plants that cannot cross by conventional methods *in vivo*. When the barriers hindering pollen tube growth lie in the stigma and style, then in plants which possess big pistils, part or all of the

CELL CULTURE AND SOMATIC CELL
GENETICS OF PLANTS, VOL. 1

ISBN 0-12-715001-3

style can be excised. Following this procedure, pollen grains can be placed on the cut surface of the ovary or transferred through the hole in the wall to the inside of the ovary. This technique, called *intraovarian pollination,* has been successfully applied to only a few species (*Papaver somniferum, Eschscholzia californica, Argemone mexicana,* and *A. ochroleuca*). A more complicated technique is the culture *in vitro* of ovules which have been excised together with the placenta and the direct pollination of these ovules. This technique was first successfully practiced by Maheshwari and Kanta (1964) at the University of Delhi with three members of the Papaveraceae and two of the Solanaceae. The placental pollination of ovules can be successfully applied to overcome self-incompatibility, for example, in *Petunia axillaris* (Rangaswamy and Shivanna, 1967, 1971). Another application of *in vitro* pollination is the development of hybrid embryos and in some cases plants (Zenkteler, 1980). *In vitro* pollination has also been used to obtain haploid plants, as shown in the case of *Mimulus luteus* when pollinated with *Torenia fournieri* (Hess and Wagner, 1974). The following procedure concerning pollination and fertilization of ovules cultured *in vitro* is based on the techniques employed by previous successful workers in this field.

II. THE EXPERIMENTAL MATERIAL

It is best to experiment with those species in which the ovaries are large and contain many ovules. The most successful results of pollination *in vitro* and development of seeds have been obtained with species belonging to the Solanaceae (*Nicotiana tabacum, N. alata, N. rustica, Petunia hybrida*), Papaveraceae (*Papaver somniferum, Eschscholzia californica, Argemone mexicana*), and Caryophyllaceae (*Melandrium album, M. rubrum, Agrostemma githago, Dianthus caryophyllus*). In all of these species the placentae are covered with several hundred ovules. Due to the large numbers, some ovules remain undamaged during the isolation of the whole placenta, or even of pieces, and therefore are able to develop further after pollination. Pollen grains of species belonging to Solanaceae, Caryophyllaceae, and Papaveraceae germinate easily on ovules, and pollen tubes usually grow abundantly all over the ovules and placenta. However, it can be rather difficult to achieve *in vitro* germination of pollen in other families, for example, the Cruciferae (Kameya *et al.*, 1966; Balatkova and Tupy, 1968; Guzowska, 1970). In the case of *Brassica oleracea,* if ovules, 1 day before pollination, are dipped in a 1% solution of calcium chloride, this treatment favors the growth of pollen tubes.

Before starting to experiment on test tube pollination, it is necessary to examine the following features: (1) the viability of ovules; (2) germination of pollen grains on ovules and the development of tubes (pollen tubes should develop normally, that is, not burst and grow quite long); (3) the entry of pollen tubes into the embryo sacs (Figs. 1c and 2a). In some cases it can be difficult to find tubes inside the micropyle when ovules, after separation from placenta, are squashed and stained prior to observation under the microscope. Therefore, in order to locate tubes inside the embryo sacs, it is sometimes necessary to cut sections of intact ovules before examining under the microscope (Fig. 2a).

III. DISINFECTION AND EXCISION OF THE OVULES ALONG WITH PLACENTA

Two or 3 days before the dehiscence of anthers, flower buds are emasculated and later bagged in order to prevent pollination. Ovules should be inoculated when, in the control material *in vivo,* anthers are at the stage of dehiscence. Anthers are dissected from buds just before opening and kept in sterile conditions until they burst. The whole pistils, or the ovaries alone, are sterilized. Sterilization can be varied according to the experimental material. Ovaries of some plants, especially those which grow in open air, should be sterilized longer than plants growing in a clean, well-kept greenhouse.

The procedure for sterilization is usually as follows: (1) Ovaries are dipped for about 1 min in 70% alcohol (when the ovaries are delicate, alcohol cannot be used). (2) After removal from alcohol, the ovaries are transferred into chlorine water or to another sterilizing solution (the concentrations of these solutions can be varied). It is, however, more suitable to sterilize for a longer period in a dilute solution rather than for a shorter period in a solution of higher concentration. When the experimental material is delicate, the period of sterilization and the concentration of the sterilizing solution must be carefully controlled. (3) Ovaries are washed thoroughly three or four times in sterile water. The ovary wall is then carefully peeled down with sterile scalpels, forceps, or needles, leaving behind the mass of bare ovules attached to the placenta. The whole placenta, or part of it when the placenta is well developed and covered with a large number of ovules, is placed on the medium. In some cases, for example, with *Dianthus caryophyllus* or *Melandrium album,* parts of the pedicel and shortened calyx can be left attached to the placenta transferred to the medium. The aseptically obtained pollen grains are now deposited on the ovules. The growth of pollen tubes attached to bare ovules is often inhibited by the presence of water on the surface of the ovules. If after the inoculation of ovules a film of

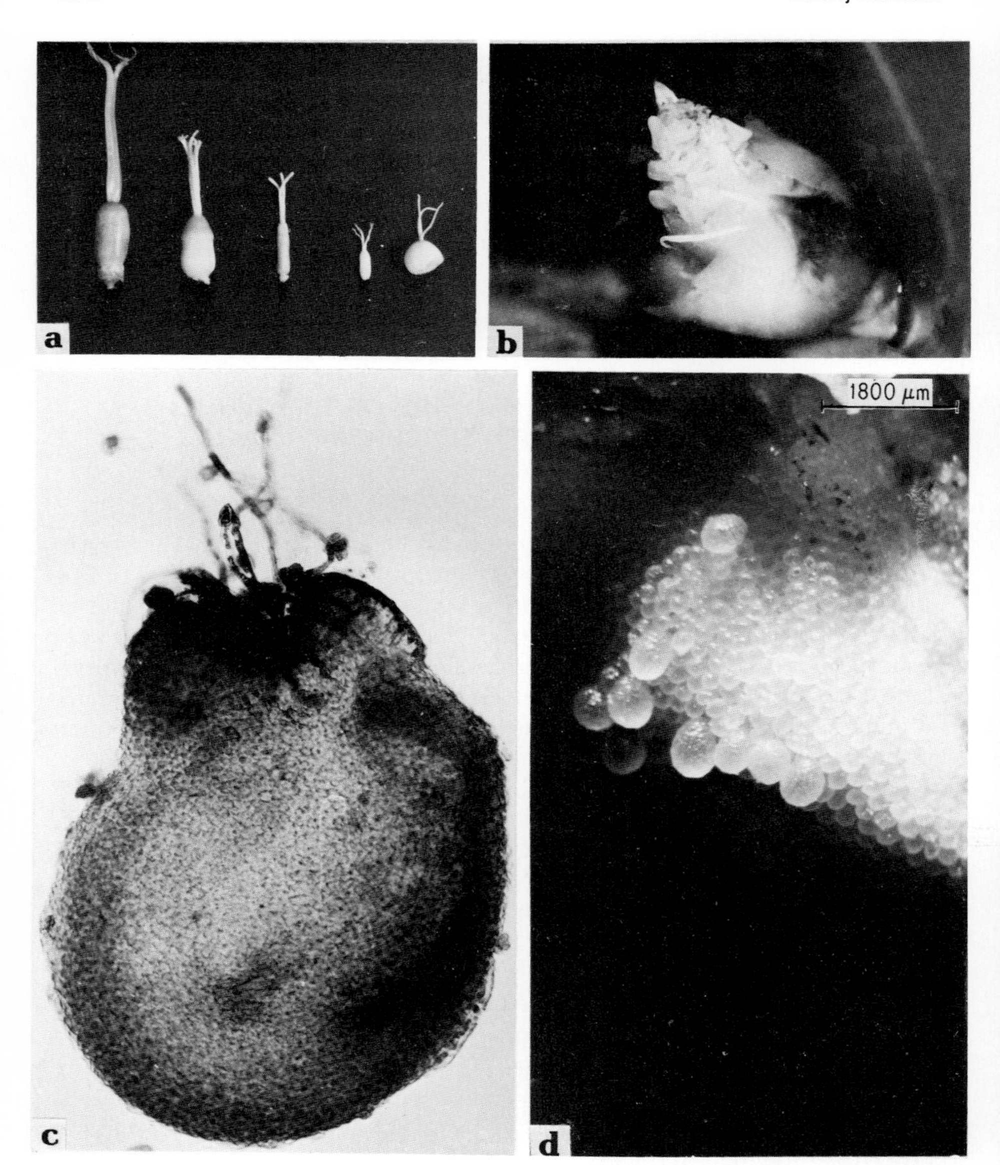

Fig. 1. (a) Styles of various species of the family Caryophyllaceae. After peeling of the ovary wall, placenta covered with many ovules can be successfully pollinated *in vitro*. (b) Developing ovules situated on the placenta of *Dianthus caryophyllus* 8 days after test tube pollination (66×). (c) Pollen tubes of *Zephyranthes lancasteri* entering the micropyle of ovules (280×). (d) Enlarged ovules of *Nicotiana tabacum* situated on the placenta 4 days after pollination *in vitro* with pollen grains of *Hyoscyamus niger*.

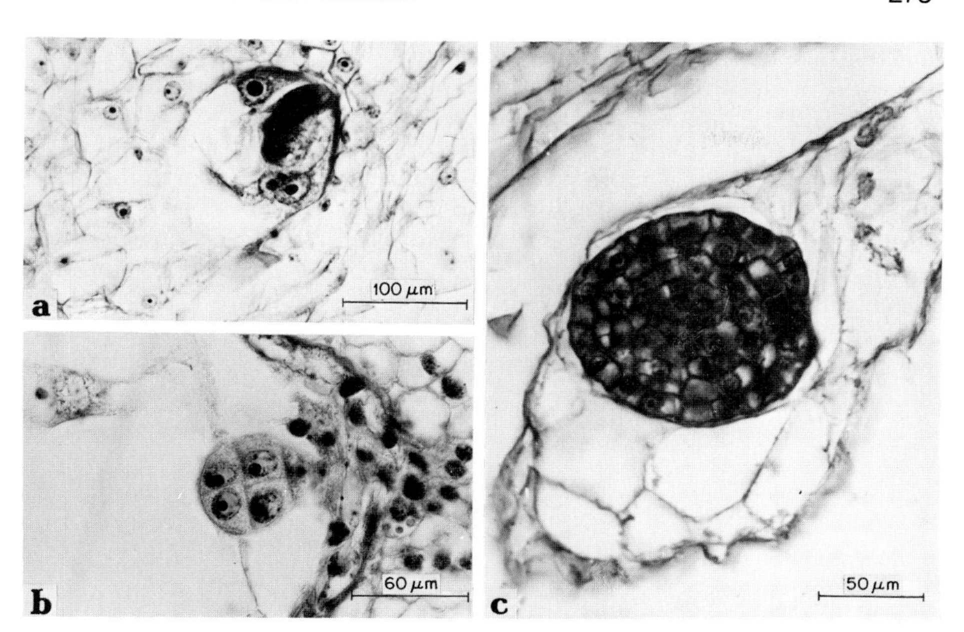

Fig. 2. (a) Pollen tube of *Silene schafta* in the micropyle of the embryo sac of *Melandrium album*. (b) Hybrid proembryo of *Nicotiana tabacum* × *Hyoscyamus niger* 4 days after test tube pollination. (c) Hybrid proembryo of *M. album* × *Cucubalus baccifer* 7 days after test tube pollination. The proembryo is surrounded by the degenerating endosperm.

water appears on their surface, they should be dried with filter paper. Later, the "dried" ovules can be covered by the pollen grains.

The pollinated ovules can be kept at 22–26°C. There is no precise information on the effects of temperature and light on test tube fertilization. Usually the first steps of this process occur at room temperature and without special lighting. Only later, during embryogenesis, should the ovaries be kept in more suitable conditions provided in a culture room. In some plants (*Nicotiana tabacum, Petunia hybrida, Melandrium album, Dianthus caryophyllus*) seeds containing mature embryos can germinate *in situ* and, within about 5 weeks following test tube pollination, many seedlings can be obtained from one explanted ovary.

The technical procedure for self-pollinating and cross-pollinating ovules is the same. The early growth of proembryos in both types of experimental material is more or less similar. Ovules which have been self-pollinated are usually kept on the placenta until the seeds develop. Cross-pollinated ovules, in contrast, can be kept on the placenta during the initial 6–8 days of culture without modification of the conditions of culture; various modifications must be applied when the small globular proembryos stop developing (Fig. 2c). In such cases enlarged ovules, or if possible globular proembryos, must be transferred to a fresh medium. Usually a liquid medium is used, and the cultures are agitated on a rotary shaker. An agar medium can also be used, but this is better for culturing ovules containing embryos whose development is more advanced (embryos with cotyledon initials). Only in such modified conditions is it possible to induce embryos to grow further, and consequently to obtain hybrid plants. There are no standard media which support the growth of hybrid proembryos of various species. Usually the media supplied do not support the good growth of hybrid proembryos or callus cells developed from the proembryos. Experiments concerning the culture of hybrid proembryos are difficult to perform, and at this time the results are very limited (Chapter 30, this volume).

IV. ANALYSIS OF OVULES POLLINATED *IN VITRO*

Usually several days after test tube pollination some ovules enlarge (Figs. 1b,d). This enlargement may show that the pollen tubes have entered the embryo sacs (Fig. 2a) and that the embryos are developing (Fig. 2b). However, the enlargement of ovules is not exclusive proof of the process of amphimixis. In some cases, especially when ovules are cross-pollinated with pollen belonging to other species or genera, only the endosperm develops. It must be remembered that the endosperm does not develop exclusively as a consequence of fertilization. The formation of this nourishing tissue can also be caused by the entrance of the pollen tube inside the embryo sac. In order to clarify this problem, it is important to analyze the pollinated ovules just after covering them with the pollen grains. Therefore, the pollinated material should be collected every few hours following pollination, fixed, embedded, and later cut with the microtome. Only the cytoembryological examination of the slides prepared from such material enables detailed, step-by-step study of the development of the endosperm and the embryo.

REFERENCES

Balatkova, V., and Tupy, J. (1968). Test-tube fertilization in *Nicotiana tabacum* by means of an artificial pollen tube culture. *Biol. Plant.* **10,** 266–270.

Guzowska, I. (1971). In vitro pollination of ovules and stigmas in several species. *Genet. Pol.* **12,** 261–265.

Hess, D., and Wagner, G. (1974). Induction of haploid parthenogenesis in *Mimulus luteus* by in vitro pollination with foreign pollen. *Z. Pflanzenphysiol.* **72,** 466–468.

Kameya, T., Hinata, K., and Mizushima, U. (1966). Fertilization in vitro of excised ovules treated with $CaCl_2$ in *Brassica oleracea* L. *Proc. Jpn. Acad.* **42,** 165–167.

Kanta, K., Rangaswamy, N. S., and Maheshwari, P. (1962). Test-tube fertilization in flowering plants. *Nature (London)* **194,** 1214–1217.

Linsmaier, E. M., and Skoog, F. (1965). Organic growth factor requirement of tobacco tissue culture. *Physiol. Plant.* **18,** 100–127.

Maheshwari, P., and Kanta, K. (1964). Control of fertilization. *In* "Pollen Physiology and Fertilization" (H. F. Linskens, ed.), pp. 187–193. Elsevier, Amsterdam.

Nitsch, J. P. (1951). Growth and development in vitro of excised ovaries. *Am. J. Bot.* **38,** 566–577.

Rangaswamy, N. S. (1961). Experimental studies on female reproductive structures of *Citrus microcarpa* Bunge. *Phytomorphology* **11,** 109–127.

Rangaswamy, N. S., and Shivanna, K. R. (1967). Induction of gametic compatibility and seed formation in axenic cultures of diploid self-incompatible species of *Petunia*. *Nature (London)* **16,** 937–939.

Rangaswamy, N. S., and Shivanna, K. R. (1971). Overcoming self-incompatibility in *Petunia axillaris* (Lam.) B.S.P. II. Placental pollination in vitro. *J. Indian Bot. Soc.* **50A,** 286–296.

Zenkteler, M. (1980). Intraovarian and in vitro pollination. *Int. Rev. Cytol., Suppl.* **11B,** 137–156.

CHAPTER **33**

In Vitro Pollination, Fertilization, and Development of Maize Kernels

Burle G. Gengenbach

Department of Agronomy and Plant Genetics
University of Minnesota
St. Paul, Minnesota

I. INTRODUCTION

Successful *in vitro* pollination and fertilization have been demonstrated primarily for various dicotyledonous plant species (see Chapter 32, this volume). Little experimentation of this type has been done with grass or cereal species, and reports of success are few. Zubkova and Sladky (1975) reported that they had pollinated excised maize (*Zea mays* L.) ovaries, but no embryo or seed development occurred. Details of the culture conditions and technique were not provided in their initial report.

Two reports (Sladky and Havel, 1976; Gengenbach, 1977a) subsequently demonstrated that unfertilized excised maize ovaries could be cultured on artificial media, pollinated *in vitro*, and the resultant caryopses recovered as mature kernels. The methods used in both studies were similar, and the frequency of mature kernels was about 5% in each case.

CELL CULTURE AND SOMATIC CELL
GENETICS OF PLANTS, VOL. 1

ISBN 0-12-715001-3

This technique provides the opportunity to study the processes of pollen germination and fertilization under conditions which are more easily controlled than is normally the case for maize. Various genetic manipulations may be made more feasible by overcoming certain biological barriers, such as self- or cross-incompatibility, or by reducing the distance the pollen tube must grow in order to reach the egg sac. Gamete selection experiments may also be possible upon modification of the basic experimental procedure. This basic procedure is directly applicable to studies of kernel development in which normal pollination and fertilization occur on the plant, after which the young kernels are cultured *in vitro* (Gengenbach, 1977a).

II. PROCEDURE

There is no single universal procedure for *in vitro* pollination and fertilization of maize. The protocol described here is primarily that published by Gengenbach (1977a) or as subsequently modified (Gengenbach, 1977b). Adaptations attributed to others are also cited and commented upon.

A. Source and Growth of Plants

Maize plants are grown in the conventional manner for field nurseries or in the greenhouse under conditions that optimize plant growth (i.e., high-intensity lighting, 15- to 16-hr photoperiod, 25–27°C). The female inflorescence (ear) is covered with a glassine bag prior to stigma (silk) emergence to prevent uncontrolled pollination. The date of silk emergence is recorded on the ear bag to serve as a reference of the physiological stage of development and to reduce variation among ears of the same genotype. Freshly shed pollen can be collected in bulk by covering the male inflorescence (tassel) with a paper bag for a short time or by removing freshly exserted, nondehisced anthers.

B. Dissection of the Ear

Ears are removed from the plant, keeping the peduncle (ear shank) attached; it becomes a convenient structure to hold during husk removal and ovary dissection. Protruding silks are removed with scissors or a scalpel, and the outer husks are removed by hand. Flame-sterilized instruments are used in all subsequent steps. Surface sterilization of the ear generally is not required, especially if plants can be grown in relatively disease- and insect-free environments (Gengenbach, 1977a,b; Bajaj, 1979; Raman *et al.*, 1980). In some plants the bacterium *Erwinia stewartii* may be present as an endophytic contaminant which is not eliminated by surface sterilization. When used, surface sterilization involves a brief rinse of the inner husks with 95% ethanol (Dhaliwal and King, 1978) or a 30-min treatment with 1% Famosept (Sladky and Havel, 1976). The inner husks are removed with forceps, taking care not to injure or remove the silks. The ear is then placed on a sterile surface, such as aluminum foil, in a laminar flow hood. Ovaries may be dissected in several arrangements such as a strip containing two rows of intact ovaries with silks attached. The silks are parted between two rows parallel to the axis, and a scalpel cut is made along this line about halfway into the pith tissue of the rachis (cob). This process is repeated on the other side of the two rows, and so on around the cob. The pith tissue subtending the ovaries is wedge-shaped and can be cut further to give segments with fewer ovaries. Ovaries near each end of the ear are usually discarded. Segments with more rows or circular pieces formed by two radial slices can also be used.

C. Pollination of Excised Ovaries

To prevent desiccation, ovaries should be placed as quickly as possible into a culture medium in a suitable container such as a 25 × 100-mm plastic Petri dish. Pollinations can be made outside the Petri dish by placing the silks over the rim of the dish so that they protrude when the lid is in place. This can be very useful if the pollen is not free of microbial contamination. Sladky and Havel (1976) and Raman *et al.* (1980) used a small dish containing media and ovaries inside a larger dish; pollinations were made in the enclosed larger dish but away from the culture media. When a substantial amount of pollen is available, the exposed silks can be dusted or the silks can be moved around in a container of pollen. In other situations, silks may be clipped very short or the tip of the ovary removed entirely to expose the nucellar tissue (Dhaliwal and King, 1978; Raman *et al.*, 1980). When this is done, pollination must be done in the same container as the media, and the pollen should be obtained from a microbe-free source such as a mature anther just before exsertion from the floret. Pollen can be teased from the anther with a sterile dissecting needle directly onto the silks or exposed nucellar tissue (Dhaliwal and King, 1978). Pollen pregerminated on a sterile pollen germination medium can be used by placing a small bit of agar medium containing pollen onto the silk (Raman *et al.*, 1980). After a 24- to 48-hr incubation period at 26 ± 1°C, the exposed silks are removed and the Petri dish is sealed with tape. Further incubation for kernel development is at 25–30°C in the dark.

D. Culture Media

Several media (White, 1943; Murashige and Skoog, 1962; Nitsch, 1969; Linsmaier and Skoog, 1965) and modifications thereof can be used. The media composition does not seem to affect the percentage of fertilization, as indicated by enlargement of the pollinated ovaries (Sladky and Havel, 1976; Gengenbach, 1977a; Bajaj, 1979; Raman *et al.*, 1980). Agar-solidified medium is also more convenient to use than liquid because of the support it provides for the ovaries and developing kernels. The medium listed in Table I uses as a nitrogen source an amino acid mixture based on the endosperm composition reported by Oaks and Beevers (1964). The nitrogen source does not affect fertilization frequency (Gengenbach, 1977a), but a source of reduced nitrogen is required for optimal kernel development and growth.

TABLE I

Culture Medium Containing a Complete Amino Acid Mixture as the Nitrogen Source

Component	μmoles/liter	Amino acid	mg/liter
KH_2PO_4	3750	Aspartate	500
$MgCl{\cdot}6H_2O$	1500	Threonine	198
$CaCl_2{\cdot}2H_2O$	1200	Serine	347
H_3BO_3	100	Glutamate	1761
$MnSO_4{\cdot}H_2O$	100	Proline	621
$ZnSO_4{\cdot}7H_2O$	10	Glycine	168
KI	5	Alanine	476
$NaMoO_4{\cdot}2H_2O$	1	Cysteine	160
$CuSO_4{\cdot}5H_2O$	0.1	Valine	330
$CoCl_2{\cdot}6H_2O$	0.1	Methionine	117
Fe-EDTA	50	Isoleucine	237
Thiamine-HCl	0.4 mg/liter	Leucine	1009
Folic acid	0.044 mg/liter	Tyrosine	262
Niacin	1.2 mg/liter	Phenylalanine	387
Sucrose	150 g/liter	Lysine	88
Agar	5.5 g/liter	Histidine	164
		Arginine	150
		Asparagine	300
		Glutamine	292
		Tryptophan	204

E. Culture of Developing Kernels

The preceding procedure is easily modified to facilitate studies of kernel development *in vitro*. The ears are pollinated on plants 3–5 days after silk emergence. Two to 5 days after pollination, the immature kernels with attached pith tissue are excised as described above. Cob segments may be cultured with all kernels intact; however, the uniformity and extent of kernel development generally are improved if kernels are removed from the cob in a ratio of at least five removed to one remaining. The kernels are removed (prevented from developing) by making a scalped cut through the kernel at the surface of the cob tissue before placing the cob segment into the culture medium. Petri dishes are sealed with masking tape and incubated in the desired conditions, usually in the dark at 28–30°C, for a specified time or until maturity at 35–45 days.

III. RESULTS AND DISCUSSION

A. Frequency of *in Vitro* Fertilization

Fertilization frequency as judged by enlargement of the ovary is influenced by the genotype of both the ovary source and the pollen source (Gengenbach, 1977b). Ovaries from single-cross hybrids show higher fertilization than ovaries from inbred lines, but there is a range in both groups and some inbreds are nearly as good as hybrids. Fertilization has been over 80% in some crosses. Ear characteristics such as rate of silk emergence differ considerably among genotypes; the structure of the ear may affect the ease of dissecting ovaries without injury and reducing the fertilization frequency. Ears with widely spaced, double ovary rows and soft cob tissues tend to perform best. The optimum ear age (number of days after silk emergence) also varies among genotypes. Fertilization frequency is not influenced by the number (1–10) of ovaries cultured together in one cob explant. The fertilization rate may not change significantly when pollinations are made 0–4 hr after ovary excision, but the rate declines after 24 hr.

Differences in fertilization frequency can be attributed to the pollen genotype even when freshly shed pollen is used. The time of pollen collection during the day may also influence the fertilization rate. Fertilization has been obtained with one germinated pollen grain per silk, although the use of 10 grains per silk increased the frequency from 4% to 55% (Raman *et al.*, 1980).

The fertilization frequency is determined primarily by factors contributing to the status of the plant material at the time of culture and pollination and, to a lesser extent, by media composition and culture environment. Although such tests have not been done, it is likely that *in vitro* fertilization would occur even if ovaries were cultured on water-agar medium, provided that the osmotic and humidity conditions prevented the silks from drying out.

B. Kernel Development

Development of maize kernels *in vitro* following fertilization either *in vitro* or on the plant is highly influenced by the culture media and environment. A source of reduced nitrogen such as glutamine (15 m*M*) or asparagine (15 m*M*), or the amino acid mixture cited in Table I, support growth much better than nitrate. From 5 to 50 m*M* NH_4Cl supports growth, but there may be toxicity problems at high concentrations. Sucrose, glucose, or fructose at 5–15% (w/v) is sufficient for the carbohydrate source. Other components of the medium cited in Table I have not been tested adequately to determine their contribution to kernel growth. The culture temperature regime also affects development significantly (Jones *et al.*, 1981).

The genotype and physiological status of the ear donor plant are both important in determining *in vitro* kernel growth after fertilization on the plant. Kernels from ears of a hybrid plant, regardless of the pollen genotype, generally grow well in culture. Inbred lines exhibit variable kernel growth *in vitro,* ranging from lines that do not grow to lines in which over 80% of the kernels reach near-normal growth, provided that the kernel number per cob segment was reduced initially. Ear-to-ear variation within a genotype that grows well indicates that microenvironmental factors affect the physiological status of the plant and ear, which in turn affects *in vitro* kernel development.

C. Uses

The availability of procedures for *in vitro* pollination and fertilization has led to several studies with specific objectives. Maize ovaries have been cultured and pollinated with *Sorghum bicolor* (Dhaliwal and King, 1978) or *Pennisetum typhoides* (Bajaj, 1979) pollen. Nucellar tissues enlarged, but no viable embryos were recovered. Potentially, male gametes may be selected by germinating pollen on selective media and transferring it to silks (Raman *et al.*, 1980) or by conducting the pollination and fertilization processes under controlled stress conditions such as high or low temperature.

Attempts to stimulate maize ovule development for recovery of maternal haploids have not yet succeeded, although success with other species has been reported (Yang and Zhou, 1982). Other potential genetic uses could include the rescue of embryo- or endosperm-defective mutants, mutagenic treatments of sperm nuclei during pollen tube growth in silks, and treatments that promote organelle transmission through the pollen.

Potential applications for physiological and kernel development studies are many and varied. Studies have been reported on (1) uptake of radiolabeled compounds into cultured kernels (Shimamoto and Nelson, 1981), (2) temperature effects on kernel development (Jones *et al.*, 1981), and (3) labeling of endosperm proteins with radioactive amino acids (Gengenbach *et al.*, 1981).

REFERENCES

Bajaj, Y. P. S. (1979). Test-tube fertilization and development of maize (*Zea mays* L.) plants. *Indian J. Exp. Biol.* **17,** 475–478.

Dhaliwal, S., and King, P. J. (1978). Direct pollination of *Zea mays* ovules in vitro with *Z. mays, Z. mexicana* and *Sorghum bicolor* pollen. *Theor. Appl. Genet.* **53,** 43–46.

Gengenbach, B. G. (1977a). Development of maize caryopses resulting from in-vitro pollination. *Planta* **134,** 91–93.

Gengenbach, B. G. (1977b). Genotypic influences on in vitro fertilization and kernel development of maize. *Crop Sci.* **17,** 489–492.

Gengenbach, B. G., Culley, D. E., and Rubenstein, I. (1981). Labelling corn endosperm proteins during development. *Agron. Abstr.* p. 85.

Jones, R. J., Gengenbach, B. G., and Cardwell, V. B. (1981). Temperature effects on in vitro kernel development of maize. *Crop Sci.* **21,** 761–766.

Linsmaier, E. M., and Skoog, F. (1965). Organic growth factor requirements for tobacco tissue cultures. *Physiol. Plant.* **18,** 100–127.

Murashige, T., and Skoog, F. (1962). A revised medium for rapid growth and bioassays with tobacco cultures. *Physiol. Plant.* **15,** 473–498.

Nitsch, J. P. (1969). Experimental androgenesis in *Nicotiana*. *Phytomorphology* **10,** 389–404.

Oaks, A., and Beevers, H. (1964). The requirements for organic nitrogen in *Zea mays* embryos. *Plant Physiol.* **39,** 37–43.

Raman, K., Walden, D. B., and Greyson, R. I. (1980). Fertilization in *Zea mays* by cultured gametophytes. *J. Hered.* **71,** 311–314.

Shimamoto, K., and Nelson, O. E. (1981). Movement of ^{14}C-compounds from maternal tissue into maize seeds grown *in vitro*. *Plant Physiol.* **67,** 429–432.

Sladky, Z., and Havel, L. (1976). The study of the conditions for the fertilization *in vitro* in maize. *Biol. Plant.* **18,** 469–472.

White, P. R. (1943). "A Handbook of Plant Tissue Culture." Jacques Cattell Press, Lancaster, Pennsylvania.

Yang, H. Y., and Zhou, C. (1982). In vitro induction of haploid plants from unpollinated ovaries and ovules. *Theor. Appl. Genet.* **63,** 97–104.

Zubkova, M., and Sladky, Z. (1975). The possibility of obtaining seeds following placental pollination *in vitro*. *Biol. Plant.* **17,** 276–280.

CHAPTER **34**

Anther Culture of *Nicotiana tabacum*

N. Sunderland

John Innes Institute
Norwich, England

I. INTRODUCTION

Nicotiana tabacum was the second species after *Datura innoxia* (Guha and Maheshwari, 1964) to be used successfully in anther culture as a means of diverting immature pollen grains from their normal gametophytic role to develop as sporophytes possessing the gametic number of chromosomes. The first reports on *N. tabacum* (tobacco) came from the laboratory of J. P. Nitsch, who, until his untimely death in 1972, developed appropriate culture conditions for obtaining mature plants in large numbers from the anthers and laid down morphological criteria for recognition of flower buds at a responsive anther stage (Bourgin and Nitsch, 1967; Nitsch *et al.*, 1968; Nitsch and Nitsch, 1969; Nitsch, 1971, 1972). Other scientists (Nakata and Tanaka, 1968; Bernard, 1971; Sunderland and Wicks, 1969, 1971) also recognized the value of developing the technique on a more economically

CELL CULTURE AND SOMATIC CELL
GENETICS OF PLANTS, VOL. 1

ISBN 0-12-715001-3

acceptable species than *D. innoxia* and one which was playing a leading role specially in the culture of other parts of the plant. The last authors in particular sought to resolve discrepancies in relation to the initial patterns of division and to devise a reliable cytological procedure for standardizing anthers at a uniform stage for culture.

Nicotiana tabacum has proved to be one of the most responsive species in anther culture and undoubtedly the easiest to use. Single flower buds are readily harvested without damage to younger buds. Anthers are large enough to be dissected out without optical aids. The same plant produces flower buds in a regular sequence over many weeks, first from the main inflorescence and then from laterals. Provided unused flowers are removed to prevent seeding, one plant will suffice for a whole series of experiments.

II. PROCEDURE I

This is essentially the procedure that emerged in the early 1970s for use with agar media and for culture of anthers direct from the plant.

A. Sterilization of Buds and Removal of Anthers

Flower buds should be harvested as the corolla is just emerging from the calyx (Nitsch and Nitsch, 1969). To kill surface contaminants, immerse the excised buds for 10–15 min in a solution of sodium hypochlorite (0.1% available chlorine) containing a drop of Tween-20 or similar detergent as a wetting agent. Rinse the buds in sterile water and put them in a sterile Petri dish. It is useful to group buds in an age sequence of increasing corolla length (Sunderland and Wicks, 1971). Remove the calyx by means of flamed forceps, preferably in a stream of sterile air, and measure the length of the corolla from base to tip. Slit open the corolla and remove the five stamens, placing them in the Petri dish. Gently separate each anther from its filament. The two pollen sacs comprising the anther may sometimes fall apart, but this is of little consequence unless the pollen sacs are otherwise damaged. Badly damaged anthers are best rejected, as are any that inadvertently come into contact with the sterilant.

B. Staging of Anthers

In general, tobacco anthers are responsive over a corolla length range of 15–25 mm, but the developmental stages relative to this range vary with the genotype and, more particularly, with the conditions under which plants are grown. For the most consistent results, tobacco anthers are best cultured as the microspores (unicellular) pass through the first mitotic division, which leads to the formation of bicellular pollen grains (Sunderland and Wicks, 1969). To assess the developmental stage, put one test anther from each bud on a microscope slide and gently crush it in a drop of acetocarmine stain prepared according to Sunderland and Wicks (1971). Remove the anther debris, apply a coverslip, and leave for 15–30 min. Dividing microspores can easily be discerned in the light microscope. In practice, because neither the microspores nor the five anthers in a bud are entirely synchronous, cultures can be initiated when the test anther shows the spores to be ready to divide (stage 3; Sunderland, 1974), to be in division (stage 4), or to have just completed the division (stage 5). At stage 4, undivided as well as dividing and divided spores will be present. At stage 3, the single nucleus will be seen to occupy much of the spore volume. The vacuole is relatively small, and there is little cytoplasmic staining. By contrast, at stage 5, cytoplasmic staining will be more intense and the small generative cell will be separated from the larger vegetative (tube) cell by a hemispherical wall.

Feulgen reagent does not stain the hemispherical wall, and because the chromatin of the generative nucleus is more condensed than that of the vegetative nucleus, the one stains much more densely than the other. The vegetative nucleus is so faintly stained that it is easily overlooked. With Feulgen staining alone, anthers at stage 5 may be confused with anthers at stage 3.

C. Agar Cultures

Place the anthers of appropriate stage on the medium so that the two pollen sacs are in contact with it and the dehiscence furrows exposed to the culture atmosphere (Sunderland, 1973). For the anthers of one bud, a 50-mm disposable plastic Petri dish is recommended (preferably the deep variety made by Sterilin, 50 × 18 mm) containing 5 ml medium. Seal the dishes with Parafilm or its equivalent and incubate them in darkness at 25–28°C. If screw-capped vessels are used, there is a critical relationship between the size of the vessel, the volume of medium, and the number of anthers cultured (Dunwell, 1979). The N_6 medium described by Chu (1978) is recommended, with MS medium (Murashige and Skoog, 1962) or H medium (Nitsch, 1972) as alternatives. With quantities expressed in milligrams per liter, N_6 medium consists of $(NH_4)_2SO_4$, 463; KNO_3, 2830; KH_2PO_4, 400; $MgSO_4 \cdot 7H_2O$, 185; $CaCl_2 \cdot 2H_2O$, 166; $MnSO_4 \cdot 4H_2O$, 4.4, $ZnSO_4 \cdot 7H_2O$, 1.5; H_3BO_3, 1.6; KI, 0.8; glycine, 2.0; thiamine HCl, 0.5; nicotinic acid, 0.5; sucrose, 20,000 or 30,000; and agar, 8,000. To this add a solution of FeEDTA freshly prepared from $FeSO_4 \cdot 7H_2O$, 27.9, and Na_2EDTA, 37.2. Adjust the pH to 5.8 before autoclaving the medium. Do not include hormones. Both hormones and activated charcoal have been used (Anagnostakis, 1974). Charcoal may be beneficial in adsorbing inhibitors from the agar (Kohlenbach and Wernicke, 1978), but it will also adsorb hormones.

Bipolar embryoids can usually be seen emerging from the anthers after 21–28 days. As soon as embryoids are discernible, transfer the cultures to light. If growth rooms are available, a temperature of 20–25°C is recommended, with a 16-hr day and Grolux light [total radiation of at least 1 klx and a photosynthetically active radiation (PAR) value of about 40 microeinsteins m^{-2} sec^{-1}]. With these conditions, embryoids will soon develop green leaflets and roots.

D. Transfer of Plantlets to Soil

Plantlets should first be transferred to an agar maintenance medium (half-strength N_6, MS, or H). Begin this transfer not later than 35–42 days after initiation of the cultures, while the plantlets are free of each other. Plantlets become increasingly difficult to tease apart without damage the longer they are left in the original culture vessel. Transfer them singly into 100-ml containers with 20 ml of medium and illuminate them as before. After several weeks, when the plantlets have developed a good root system and several leaves, transplant them into compost or a mixture of peat and sand. Leave agar attached to the roots rather than risk breaking roots by removing it. Freshly potted plants will benefit from a few days in a mist propagator. If mist is not available, cover each pot with a polyethylene hood.

III. PROCEDURE II

This modified procedure involves the use of liquid media (float culture) and anthers specially prepared for culture (Sunderland and Roberts, 1977). It gives much improved and more easily quantified culture yields than procedure I. Pollen grains are released into the medium and develop free of the anther tissues.

A. Preparation of Anthers for Culture (Stress Pretreatment)

Buds are harvested as described above, but before sterilization they are subjected to a period of temperature stress. The pollen is switched into embryogenesis by the pretreatment instead of during culture, as in procedure I (Sunderland, 1978, 1979). The stress pretreatment is an extension of the "cold shock" devised by C. Nitsch (1974) for culture of isolated tobacco pollen.

Put the excised buds in a polyethylene bag, Petri dish, or other container that can easily be sealed. Water need not be included; if it is included, ensure that it is physically separated from the buds. Store the sealed containers in darkness, preferably at 7–9°C (Sunderland and Roberts, 1979). These temperatures are easily attained in a refrigerator by choosing the appropriate shelf. With storage of the excised buds at 7°C for 14–21 days, embryo yields were enhanced 10- to 15-fold (Sunderland, 1978). Higher temperatures are also effective and shorten the pretreatment. Pretreatments for 7 days at 14°C, 4 days at 20°C, or 2 days at 25°C all proved effective (Sunderland, 1982), but it is emphasized that timing becomes progressively more critical the higher the temperature. To make the most of the pretreatment, determine systematically the best combination of time and temperature for the genotype and particular plant growth conditions being used.

B. Staging of Anthers

Experience has shown that the pretreatment is more effective when buds are pretreated directly from the plant. If sterilized and opened before pretreatment, buds deteriorate more rapidly due to the presence of surface water and necrosis of damaged tissues. Sacrifice a few buds beforehand to determine the corolla lengths most appropriate to the mitotic anther stage (stage 4), as described above. Restrict pretreatment to buds within that range. Errors arising from the use of slightly less uniform buds will be adequately compensated for by the pretreatment.

C. Float Cultures

Plastic Petri dishes (50 × 18 mm), as specified above, are ideal. Place the anthers on the surface of the medium (5 ml liquid N_6), where they will float without support due to the waxy cuticle and occlusion of air by epidermal hairs. Seal and incubate the dishes as in procedure I. Do not shake the dishes. Anthers inadvertently sticking on the side or base of the dish when cultures are examined can easily be dislodged onto the surface of the medium by one or two sharp taps of the dish against the laboratory bench. With such float cultures, up to 40 anthers per 5 ml of medium may be used without incurring serious competitive inhibition (N. Sunderland, unpublished).

For cultures of free pollen, remove the anthers from the dish after a few days and transfer them to another dish of medium. Pollen shed into the medium will form embryoids in the absence of the anthers. For a series of pollen cultures from the same batch of anthers, repeat the transfer several times (Sunderland and Roberts, 1977).

With Procedure II, embryoids become discernible after about 14 days. The medium becomes rapidly depleted when many thousands of embryoids are formed. The embryoids may not develop further than the torpedo stage. In this event, add more medium or transfer the embryoids to fresh medium until plantlets are obtained that are large enough for transfer to soil.

IV. WHICH GENOTYPE TO USE?

Embryoid yields vary from one genotype to another (Hlasnikova, 1977). With the use of procedure II and anthers restricted to the 15- to 20-mm corolla range, the 5 most productive genotypes among 24 tested by the author were White Burley Groen (subgroup pallescens), Coulo (tabacum), Tamasesti 50 (attenuata), Virginia Gold B (havanensis), and Cabot (sagittata), giving yields ranging from 800 to 100 embryoids per anther with a pretreatment of 5 days at 14°C. The five least responsive genotypes were Banat (macrophylla), Dubek (lingua), Amersfoorter (attenuata), Kentucky (havanensis), and Hicks (undulata), yields being less than 10 embryoids per anther. Mean yields ranging from 100 to 10 were obtained with White Burley (pallescens), Virginia Bright Leaf (undulata), Kapolnai (macrophylla), Samsun (lingua), Djubek (tabacum), Havana (tabacum), White Burley Geel (havanensis), Herzegovina (pallescens), Gold Dollar (attenuata), Kerti (serotina), Samsun Dere (macrophylla), Virginia Gold Leaf (tabacum), Xanthi (lingua), and Burley (undulata). The taxonomic nomenclature is that used by the Central Institute for Genetics and Research on Cultivated Plants, Gatersleben.

As a generalization, White Burley cultivars are among the most responsive in anther culture.

V. CULTIVATION OF TOBACCO PLANTS FOR ANTHER CULTURE

Most tobaccos can be cultivated throughout the year in a glasshouse provided with supplementary illumination in winter. Germinate seeds in a seedpan and plant resulting seedlings singly in 3½-in. pots in a rich compost. Water pots sparingly, and feed the plants occasionally. Conditions leading to quick flowering should be fostered. In the author's experience, conditions giving luxuriant vegetative growth are adverse to high-yielding cultures. Overfeeding of plants is particularly undesirable during winter. It leads to weak, spindly plants and poorly responding anthers (Sunderland, 1978). If space allows and batches of plants sown at regular intervals can be accommodated, leave the plants in 3½-in. pots and feed them once every 28 days. If space is at a premium and few plants can be accommodated, transplant them into 5-in. pots when a rosette of leaves has been formed. As soon as the main inflorescence has stopped flowering, cut it off together with some of the upper internodes and repot the plant in a larger container to promote further branching. Repeat the process, if necessary, after the next sequence of flowering. Remove all unused flower buds regularly to maintain a steady rate of flower production.

Cultivation of plants in controlled environments is desirable for long-term experimentation demanding a constant supply of uniform material. Cultivate seedlings as above and keep them at 20–25°C, preferably under sodium lamps at an intensity of 15–20 klx. With a White Burley cultivar at 20°C and an intensity of 15 klx, short days (8 hr) resulted in more responsive anthers than long days (16 hr) (procedure I) (Dunwell, 1976). Anthers of the same cultivar from plants grown at 25°C in 16-hr days (20 klx) responded similarly to anthers from glasshouse plants grown in summer (procedure II) (Sunderland, 1984). To economize on the use of controlled conditions, keep stocks of plants in a glasshouse and transfer them to controlled conditions as the inflorescence axis starts to develop.

Insect-proof glasshouses are recommended. Every precaution must be taken to control insect pests such as white fly, red spider, and sciarids.

VI. APPLICATIONS OF TOBACCO ANTHER CULTURE

Procedure I is ideal for classroom demonstrations, undergraduate exercises, and the rapid production of haploid and doubled haploid plants on a commercial scale. Chromosome doubling techniques are described in detail by Jensen (1974).

Several doubled haploid lines selected for improved yield, earlier flowering, increased disease resistance, and, in some instances, better smoking qualities, have been developed and released to the industry (Nakamura *et al.*, 1974; Atanassov *et al.*, 1978; Institute of Tobacco, Shantung and Institute of Botany, Peking, 1974; Institute of Tobacco Research, Chinese Academy of Agricultural Sciences, 1974, 1978). Both haploids and doubled haploids are being subjected to close genetic evaluation (Collins and Legg,

1980). In many cases, doubled haploids gave uniform progeny on selfing, remained genetically stable over several generations, and revealed the same adaptability to the environment as did standard commercial inbreds (Oka *et al.*, 1977). However, reductions in vigor and size were also observed (Burk and Matzinger, 1976; Arcia *et al.*, 1978). The extent to which such variation might be exploitable has still to be assessed.

Procedure I has also proved valuable in (1) production of aneuploids from monosomic tobaccos (Mattingley and Collins, 1974); (2) production of chlorophyll mutants from plants derived from mutagenized seeds (Vagera *et al.*, 1976); (3) irradiation and chemical mutagenesis of anthers, plantlets, or mature plants (Nitsch *et al.*, 1969; Devreux and Laneri, 1974); and (4) fundamental studies of the mechanism of induction (Sunderland and Dunwell, 1977).

The greatest impact of tobacco anther culture in the future will probably be as a source of haploids for the development of cell suspension, callus, and protoplast culture systems. Such systems are being used for the isolation of mutant cells (Carlson, 1970, 1973; Maliga *et al.*, 1973; Berlyn and Zelitch, 1975; Muller and Grafe, 1978; Radin and Carlson, 1978). Regeneration of plants expressing the mutation has been achieved in some, although not all, cases. Protoplast systems are ideal for organelle transference and DNA manipulation. With the development of free pollen grain cultures (procedure II), mutant selection becomes feasible at the pollen grain level, as does the microinjection of individual pollen grains.

REFERENCES

Anagnostakis, S. L. (1974). Haploid plants from anthers of tobacco—enhancement with charcoal. *Planta* **115,** 281–283.

Arcia, M. A., Wernsman, E. A., and Burk, L. G. (1978). Performance of anther-derived haploids and their conventionally inbred parent as lines, in F1 hybrids, and in F2 generations. *Crop Sci.* **18,** 413–418.

Atanassov, A., Pamukov, I., Kunev, K., and Nedeltcheva, S. (1978). Results from the application of haploids in tobacco breeding. *Coresta Inf. Bull. Int. Tob. Sci. Symp., 1978* p. 82.

Berlyn, M. D., and Zelitch, I. (1975). Photoauxotrophic growth and photosynthesis in tobacco callus cells. *Plant Physiol.* **56,** 752–756.

Bernard, S. (1971). Développement d'embryons haploides à partir d'anthères cultivées *in vitro.* Etude cytologique comparée chez le Tabac et la Pétunia. *Rev. Cytol. Biol. Veg.* **34,** 165–188.

Bourgin, J. P., and Nitsch, J. P. (1967). Obtention de *Nicotiana* haploides à partir d'étamines cultivées *in vitro. Ann. Physiol. Veg.* **9,** 377–382.

Burk, L. G., and Matzinger, D. F. (1976). Variation among anther-derived doubled haploids from an inbred line of tobacco. *J. Hered.* **67,** 381–384.

Carlson, P. S. (1970). Induction and isolation of auxotrophic mutants in somatic cell cultures of *Nicotiana tabacum*. *Science* **168,** 487–489.

Carlson, P. S. (1973). Methionine sulfoximine-resistant mutants of tobacco. *Science* **180,** 1366–1368.

Chu, C. C. (1978). The N_6 medium and its application to anther culture of cereal crops. *In* "Proceedings of Symposium on Plant Tissue Culture," pp. 43–50. Science Press, Peking.

Collins, G. B., and Legg, P. D. (1980). Recent advances in the genetic application of haploidy in *Nicotiana*. *In* "The Plant Genome" (D. R. Davies and D. A. Hopwood, eds.), pp. 197–213. John Innes Charity, Norwich, England.

Devreux, M., and Laneri, V. (1974). Anther culture, haploid plant, isogenic line and breeding researches in *Nicotiana tabacum* L. *Polyploidy Induced Mutat. Plant Breed. Eucarpia/FAO/IAEA Conf. Proc., 1972* pp. 101–107.

Dunwell, J. M. (1976). A comparative study of environmental and developmental factors which influence embryo induction and growth in cultured anthers of *Nicotiana tabacum*. *Environ. Exp. Bot.* **16,** 109–118.

Dunwell, J. M. (1979). Anther culture in *Nicotiana tabacum:* The role of the culture vessel atmosphere in pollen-embryo induction and growth. *J. Exp. Bot.* **30,** 419–428.

Guha, S., and Maheshwari, S. C. (1964). *In vitro* production of embryos from anthers of *Datura*. *Nature (London)* **204,** 497.

Hlasnikova, A. (1977). Androgenesis *in vitro* evaluated from the aspects of genetics. *Z. Pflanzenzuecht.* **78,** 44–56.

Institute of Tobacco, Shantung and Institute of Botany, Peking (1974). Success of breeding new tobacco cultivar "Tan-Yuh no. 1." *Acta Bot. Sin.* **16,** 300–303.

Institute of Tobacco Research, Chinese Academy of Agricultural Sciences (1974). The evaluation of the progenies of the pollen plants of tobacco. *Acta Genet. Sin.* **1,** 26–29.

Institute of Tobacco Research, Chinese Academy of Agricultural Sciences (1978). A preliminary study on the heredity and vitality of the progenies of tobacco pollen plants. *In* "Proceedings of Symposium on Plant Tissue Culture," pp. 223–225. Science Press, Peking.

Jensen, C. J. (1974). Chromosome doubling techniques in haploids. *In* "Haploids in Higher Plants: Advances and Potential" (K. J. Kasha, ed.), pp. 153–190. Univ. of Guelph Press, Guelph, Ontario, Canada.

Kohlenbach, H. W., and Wernicke, W. (1978). Investigations on the inhibitory effect of agar and the function of active carbon in anther culture. *Z. Pflanzenphysiol.* **86,** 463–472.

Maliga, P., Sz-Breznovits, A., and Marton, L. (1973). Streptomycin-resistant plants from callus culture of haploid tobacco. *Nature (London)* **244,** 29–30.

Mattingly, C. F., and Collins, G. B. (1974). The use of anther-derived haploids in *Nicotiana*. III. Isolation of nullisomics from monosomic lines. *Chromosoma* **46,** 29–36.

Muller, A. J., and Grafe, R. (1978). Isolation and characterization of cell lines of *Nicotiana tabacum* lacking nitrate reductase. *Mol. Gen. Genet.* **161,** 67–76.

Murashige, T., and Skoog, F. (1962). A revised medium for rapid growth and bioassays with tobacco tissue cultures. *Physiol. Plant.* **15,** 473–497.

Nakamura, A., Yamada, T., Kadotani, N., Itagaki, R., and Oka, M. (1974). Studies on the haploid method of breeding in tobacco. *SABRAO J.* **6,** 107–131.

Nakata, K., and Tanaka, M. (1968). Differentiation of embryoids from developing germ cells in anther culture of tobacco. *Jpn. J. Genet.* **43,** 65–71.

Nitsch, C. (1974). Pollen culture—a new technique for mass production of haploid and homozygous plants. *In* "Haploids in Higher Plants: Advances and Potential" (K. J. Kasha, ed.), pp. 123–135. Univ. of Guelph Press, Guelph, Ontario, Canada.

Nitsch, J. P. (1971). La production *in vitro* d'embryons haploides: Résultats et perspectives. *Colloq. Int. C.N.R.S.* **193,** 281–294.

Nitsch, J. P. (1972). Haploid plants from pollen. *Z. Pflanzenzuecht.* **67,** 3–18.
Nitsch, J. P., and Nitsch, C. (1969). Haploid plants from pollen grains. *Science* **163,** 85–87.
Nitsch, J. P., Nitsch, C., and Hamon, S. (1968). Réalisation expérimentale de l'androgénèse chez divers *Nicotiana. C.R. Seances Soc. Biol. Ses Fil.* **162,** 369–372.
Nitsch, J. P., Nitsch, C., and Péreau-Leroy, P. (1969). Obtention de mutants à partir de *Nicotiana* haploides issus de grains de pollen. *C.R. Hebd. Seances Acad. Sci.* **269,** 1650–1652.
Oka, M., Nakamura, A., and Yamada, T. (1977). Adaptability to cultural environments of doubled haploid lines of flue-cured tobacco. *SABRAO J.* **9,** 111–116.
Radin, D. N., and Carlson, P. S. (1978). Herbicide-resistant tobacco mutants selected *in situ* and recovered via regeneration from cell culture. *Genet. Res.* **32,** 85–89.
Sunderland, N. (1973). Pollen and anther culture. *In* "Plant Tissue and Cell Culture" (H. E. Street, ed.), pp. 161–190. Univ of California Press, Berkeley.
Sunderland, N. (1974). Anther culture as a means of haploid induction. *In* "Haploids in Higher Plants: Advances and Potential" (K. J. Kasha, ed.), pp. 91–122. Univ. of Guelph Press, Guelph, Ontario, Canada.
Sunderland, N. (1979). Comparative studies of anther and pollen culture. *In* "Plant Cell and Tissue Culture: Principles and Applications" (W. R. Sharp, P. O. Larsen, E. F. Paddock, and V. Raghavan, eds.), pp. 203–219. Ohio State Univ. Press, Columbus.
Sunderland, N. (1978). Strategies in the improvement of yields in anther culture. *In* "Proceedings of Symposium on Plant Tissue Culture," pp. 65–85. Science Press, Peking.
Sunderland, N. (1982). Induction of growth in the culture of pollen. *In* "Differentiation In Vitro" (M. M. Yeoman and D. E. S. Truman, eds.), pp. 1–24. Cambridge Univ. Press, London and New York.
Sunderland, N. (1984). Haplogenesis and improvement of vegetatively-propagated plants. *In* "Propagation of Vegatatively-Propagated Plants." Academic Press, New York.
Sunderland, N., and Dunwell, J. M. (1977). Anther and pollen culture. *In* "Plant Tissue and Cell Culture" (H. E. Street, ed.), 2nd ed., pp. 223–265. Univ. of California Press, Berkeley.
Sunderland, N., and Roberts, M. (1977). New approach to pollen culture. *Nature (London)* **270,** 236–238.
Sunderland, N., and Roberts, M. (1979). Cold-pretreatment of excised flower buds in float culture of tobacco anthers. *Ann. Bot. (London)* [N.S.] **43,** 405–414.
Sunderland, N., and Wicks, F. M. (1969). Cultivation of haploid plants from tobacco pollen. *Nature (London)* **224,** 1227–1229.
Sunderland, N., and Wicks, F. M. (1971). Embryoid formation in pollen grains of *Nicotiana tabacum. J. Exp. Bot.* **22,** 213–226.
Vagera, J. F., Novak, F. J., and Vyskot, B. (1976). Anther cultures of *Nicotiana tabacum* L. mutants. *Theor. Appl. Genet.* **47,** 109–114.

CHAPTER **35**

Anther Culture of *Solanum tuberosum*

G. Wenzel
B. Foroughi-Wehr

Institute for Resistance Genetics
Federal Biological Research Center for Agriculture and Forestry
Grünbach, Federal Republic of Germany

I. INTRODUCTION

One of the greatest obstacles in potato breeding is the tetraploid nature of *Solanum tuberosum* ($2n = 4x = 48$). With the steadily increasing number of breeding purposes and the need to concentrate more on quantitatively than on qualitatively inherited characters, the populations of seedlings needed for the successful production of a new variety is growing to such an extent that the classic approaches of $4x$ selection breeding are no longer feasible. Since dihaploids ($2n = 2x = 24$) can be produced parthenogenetically (Hougas and Peloquin, 1957), the idea arose that breeding should be converted from the tetraploid to the dihaploid level (Hougas and Peloquin, 1958; Chase, 1963). By the use of *S. phureja,* the dihaploid induction frequency could be increased to such an extent that the procedure becomes applicable. It could be even further increased using marker genes

CELL CULTURE AND SOMATIC CELL
GENETICS OF PLANTS, VOL. 1

ISBN 0-12-715001-3

TABLE I

Comparative Efficiency of Different Procedures for Haploid Induction in *Solanum tuberosum*

Ploidy	Parthenogenesis					Androgenesis			
	Number of:					Number of:			
2x → 1x; 4x → 2x	Fertilized flowers (100%)	Seeds	Haploid plants	Percent	Reference	Anthers plated	Plants regenerated	Percent	Reference
4x → 2x	49,004	35,547	222	0.5	Hougas *et al.* (1964)	50	1	2	Irikura (1975b)
4x → 2x	14,631	54,024	1,978	14	Frandsen (1967)	2,700	16	0.6	Wenzel *et al.* (1982)
4x → 2x	3,938	6,854	1,824	46	Wenzel *et al.* (1983)	8,750	52	0.6	Mix (1982)
2x → 1x	55	6,374	5	9	van Breukelen *et al.* (1975)	195	6	3	Foroughi-Wehr *et al.* (1977)
2x → 1x	7,965	262,648	5	0.6	Wenzel (1979)	54,931	2,317	4	Wenzel *et al.* (1980)

(Hermsen and Verdenius, 1973). In addition to the parthenogenetic approach, trials were undertaken to reduce the ploidy level androgenetically (Dunwell and Sunderland, 1973; Irikura, 1975a).

The production of dihaploids is, however, only the first step in ploidy level reduction of *S. tuberosum*. By both the parthenogenetic (van Breukelen *et al.*, 1975) and the androgenetic technique, a second chromosome reduction cycle is possible, resulting in monohaploid potatoes ($2n = x = 12$; Foroughi-Wehr *et al.*, 1977; Wenzel and Sopory, 1978; Sopory *et al.*, 1978). This second reduction enables the formation of homozygous diploid and tetraploid potatoes by one or two endoreduplications. Monohaploids are, consequently, a central part of a breeding scheme by which the efficiency of potato breeding could be increased (Wenzel *et al.*, 1979), and they may be used further to produce homozygous material for tetraploid hybrid seed production (Wenzel *et al.*, 1982).

II. PARTHENOGENESIS OR ANDROGENESIS

In Table I the success rates using parthenogenetic and androgenetic techniques obtained in several laboratories are summarized. It is obvious that parthenogenesis has its highest potential in the production of dihaploids from tetraploids; for monohaploid production, androgenesis is more efficient, although van Breukelen *et al.* (1975) could select from the clone Gineke 69G609, by pollination with the *S. phureja* clones IvP 35 or IvP 48, high amounts of parthenogenetic monohaploids. As androgenesis is superior only in monohaploid production, we will concentrate on methods for microspore culture of dihaploid potatoes. The androgenetic procedures for dihaploid extraction are nearly identical to the monohaploid techniques. Irikura (1975a) concluded from comparative experiments that Nitsch and Nitsch (1969) medium is better for the haploid extraction from polyploid *Solanum* species (including *S. tuberosum*), whereas Murashige and Skoog (MS) (1962) medium is better in the reduction of diploid *Solanum* species. In relation to tetraploid reduction, Mix (1982) reported that auxin concentration (1–2 mg/liter) and the presence of zeatin (0.1 mg/liter) are more critical for tetraploids than for dihaploids.

III. PRECULTURE OF ANTHER DONOR PLANTS

Most critical is the physiological state of the anther donor plants. They should be routinely grafted onto tomato stocks to guarantee vigorous flowering that is not retarded by tuberization (Fig. 1). The hybrid tomato

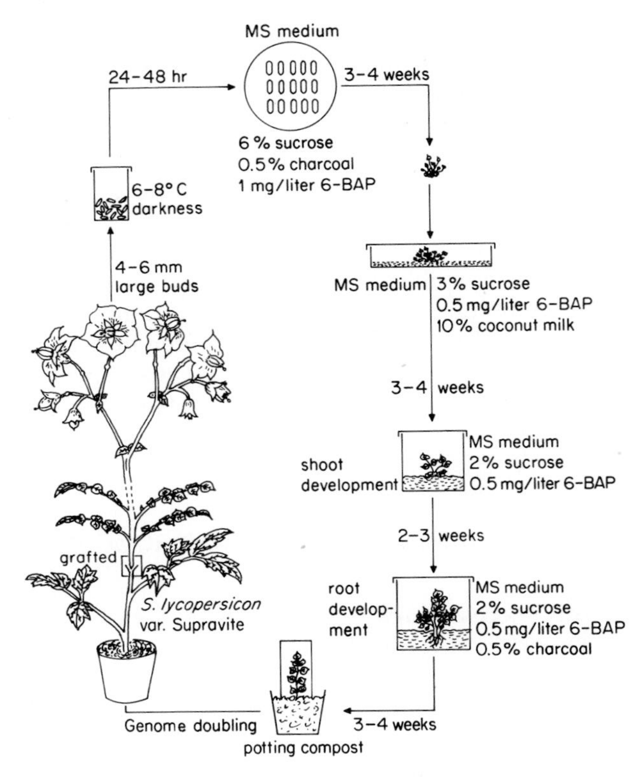

Fig. 1. Procedure for anther culture in *Solanum tuberosum*.

variety *Supravite,* resistant to tobacco mosaic virus, has turned out to be very useful. The plants should be kept in an aphid protected greenhouse at temperatures not exceeding 18°C in a 15-hr light regime, extended during the short day season by OSRAM HQI lamps with an intensity designed to deliver at least 10,000 lx at the growing point of the plants. The distance between the lamps and the plants should be a minimum of 1.0 m for 400-W lamps and 1.5 m for 1000-W lamps to prevent damage of the microspores by heat. From such vigorous material, pathogen-free, well-nourished buds are harvested.

IV. DEVELOPMENTAL STAGE OF MICROSPORES

Best results in microspore development have been obtained using anthers with microspores in the late uninucleate stage, free from starch. The length of the anther should be 3–4 mm, corresponding to a bud size of 4–6 mm. The bud size depends to some extent on its position in the inflorescence. Early flowers have large buds and relatively small anthers, whereas late buds possess relatively large anthers in small buds. Furthermore, the bud size varies in relation to the genotype. Therefore, occasionally some anthers should be fixed in Carnoy's solution and stained with acetocarmine to make sure that the late uninucleate stage is present. The rather thick exine in potato prevents the use of rapid staining procedures.

V. *IN VITRO* CULTURE PROCEDURE

After being picked, the buds are maintained for 48 hr in a dry test tube kept in the dark, in a refrigerator, at 6–9°C. Then the buds are surface sterilized in 70% aqueous ethanol with 0.1% Tween-20 added, and washed three times with sterile tap water. Subsequently, the anthers are dissected in a clean bench area. The anthers of three buds are plated per a 6-cm glass Petri dish. The choice of glass or plastic dishes depends predominantly on the specific culture room conditions; normally, the condensation water is more disturbing in the plastic than in the glass dishes. As long as no condensation is formed, the quality of the vessels is not critical.

VI. CULTURE MEDIA

In general, the basic media of Nitsch and Nitsch (1969), Murashige and Skoog (1962), or Linsmaier and Skoog (1965), or the potato extract medium (Anonymous, 1976) have been used for induction of microspore development. The potato extract medium was used in potato only when no success could be obtained on a defined medium. It is not easy to obtain a constant response with this undefined compound. The critical compounds in standard media are the sucrose concentration, ranging from 3 to 9% with an optimum at 6%, and the phytohormones. The addition of 3–5 g/liter charcoal was beneficial (Table II).

The addition of auxins is not essential, but normally 1 mg/liter indoleacetic acid (IAA) is recommended. More critical is the cytokinin. In most cases, 6-benzylaminopurine (6-BAP) at a concentration of 1 mg/liter is most inductive, although some recalcitrant genotypes need zeatin even for the first induction (0.1–0.5 mg/liter). Zeatin can be replaced, however, normally by 6-BAP, as soon as other culture parameters have been better coordinated.

TABLE II

Medium Composition for the Different Developmental Steps during Potato Anther Culture

	Basic medium	IAA (mg/liters)	BAP (mg/liter)	Zeatin (mg/liter)	Sucrose (g/liter)	Active carbon (g/liter)	Coconut milk (%)
Anthers	MS	1	1.0	—	60	5	—
Embryoids	MS	—	0.5	—	30	—	10
Callus	MS	—	—	0.3	30	—	10
Plantlets	MS	—	0.5	—	20	5	—
Shoots	B5	—	0.5	—	20	5	(5)

The Petri dishes are incubated in a culture room at 20–25°C with a 15-hr light regime of 5000 lx (a 1:1 mixture of a white fluorescent tube with OSRAM Natura was the most favorable). Within 6–8 weeks, macroscopic structures become visible which can be transferred to a morphogenetic medium. The ratio of embryoids to calli is under genetic control primarily, but is also influenced physiologically, for example, by the season.

The morphogenetic medium consists again of the basic compounds of MS medium. At this time, 0.3 mg/liter zeatin is recommended for most types (for genotypes with a high regeneration capacity, it may be replaced by 0.5 mg/liter 6-BAP), sucrose can be reduced stepwise from 30 to 20 g/liter and charcoal is not essential. For embryoids, this medium is sufficient, whereas for calli, 10% coconut milk should be added. Organogenesis from calli may start even after 1 year of subculture. During this time, calli should be transferred to fresh medium every 4–6 weeks. From microspores, nonmorphogenetic callus has not yet been obtained. Young plantlets are normally propagated by shoot tip culture on B5 medium (Gamborg *et al.*, 1968) containing only 0.5 mg/liter 6-BAP and supplemented with 5 g/liter charcoal if root development is desired. For transfer to the greenhouse, *in vitro*-formed roots are normally not used, but cuttings are transferred and rooted in the soil. It is important to keep the humidity high during this transfer process.

VII. THE PLOIDY LEVEL

As soon as normal leaflets are developed, the ploidy level may be determined by using the rapid procedure of counting plastids in the stomata (Frandsen, 1968). Monohaploids have 5–8, dihaploids 10–15, and tetraploids 18–24 chloroplasts per guard cell. Only in doubtful cases do the chromosomes in the root tips have to be counted. Normally, more than two-thirds of all regenerated plants have doubled their ploidy during *in vitro* culture.

VIII. BACK TO THE GREENHOUSE

Most monohaploids are very fragile and have to be looked after carefully. A routine grafting onto tomatoes is recommended, especially during the short-day season. For tuberization, dihaploids should be grafted on top of monohaploids. The spontaneous doubled plants do not need any specific precaution.

IX. CHROMOSOME DOUBLING

Chromosome doubling of monohaploids is not an easy procedure. In our own experiments, doubling failed with all known procedures, particularly colchicine applied either via roots or via cotton wool plugs through axillary buds, or by auxin concentrations in a secondary *in vitro* circle. J. Straub (personal communication) has now succeeded in doubling the same genotypes by a repeated colchicine treatment of axillary buds with cotton wool plugs over a period of 14 days.

Doubling of dihaploids to tetraploids is easily achieved using 0.5% colchicine solution applied to axillary buds, which are first dissected. Application may also be performed using a syringe. However, chimeral formation happens quite often. Fewer chimeras are formed by *in vitro* doubling. Jacobsen (1981) found a high percentage of doubled dihaploids without a single chimera in 129 plants tested.

X. PROBLEMS

It should be mentioned, finally, that results have shown a strong influence of the genotype on the success rate. Some clones respond actively in anther culture, whereas others do not. As most of the easily responding anthers do not possess valuable characters, it is recommended that they be combined prior to anther culture with well-responding clones by breeding.

Once the sexual hybrids have been screened for the presence of the desired character, they can be satisfactorily used in haploid induction. We believe that this approach is easier than the elaboration of specific media for specific genotypes. Especially since each microspore of an F_1 hybrid anther donor plant possesses a different genome, specific media would not be too helpful. On the other hand, one must also take into account a specific selection among the total microspore population using genotypes with good regeneration capacity and a simple medium.

In potato many genotypes possess a high proportion of unreduced microspores. Our results show that from such diplandroids many spontaneous doubled plants are regenerated. These diploids may result from unreduced gametes which may even have a selective advantage over the normal reduced spores. In addition, plantlets from reduced spores double quite early, normally while still in culture. The homozygous nature of naturally doubled plants has to be confirmed carefully by crossing experiments or marker analysis.

XI. CONCLUSIONS

Microspore regeneration within the anthers in culture is possible in potato at both the tetraploid and dihaploid ploidy levels. The greatest success has been obtained starting from the dihaploid level and paying attention to the regeneration capacity of the clones. The production of androgenetic plants with these methods was sufficiently large to allow combinations to be made with plants reared by other unconventional and classic potato breeding procedures, and the first useful clones to be produced in the field (Wenzel and Uhrig, 1981).

REFERENCES

Anonymous (1976). Simplified medium for anther culture study of *Oryza sativa* subspecies Shien. *Acta Genet. Sin.* **3,** 169–170.

Chase, S. S. (1963). Analytic breeding in *Solanum tuberosum* L.—A scheme utilizing parthenotes and other diploid stocks. *Can. J. Genet. Cytol.* **5,** 359–363.

Dunwell, J. M., and Sunderland, N. (1973). Anther culture of *Solanum tuberosum* L. *Euphytica* **22,** 317–323.

Foroughi-Wehr, B., Wilson, H. M., Mix, G., and Gaul, H. (1977). Monohaploid plants from anthers of a dihaploid genotype of *Solanum tuberosum* L. *Euphytica* **26,** 361–367.

Frandsen, N. O. (1967). Haploidproduktion aus einem Kartoffelzuchtmaterial mit intensiver Wildarteinkreuzung. *Zuechter* **37,** 120–134.

Frandsen, N. O. (1968). Die Plastidenzahl als Merkmal bei der Kartoffel. *Theor. Appl. Genet.* **38,** 153–167.

Gamborg, O. L., Miller, R. A., and Ojima, K. (1968). Nutrient requirements of suspension culture of soybean root cells. *Exp. Cell Res.* **50,** 151–158.

Hermsen, J. G. T., and Verdenius, J. (1973). Selection from *Solanum tuberosum* group *phureja* of genotypes combining high frequency haploid induction with homozygosity for embryo-spot. *Euphytica* **22,** 244–259.

Hougas, R. W., and Peloquin, S. J. (1957). A haploid plant of the potato variety Katahdin. *Nature (London)* **180,** 1202–1210.

Hougas, R. W., and Peloquin, S. J. (1958). The potential of potato haploids in breeding and genetic research. *Am. Potato J.* **35,** 701–707.

Hougas, R. W., Peloquin, S. J., and Gabert, A. C. (1964). Effect of seed-parent and pollinator on frequency of haploids in *Solanum tuberosum. Crop Sci.* **4,** 593–595.

Irikura, Y. (1975a). Induction of haploid plants by anther culture in tuber-bearing species and interspecific hybrids of *Solanum. Potato Res.* **18,** 133–140.

Irikura, Y. (1975b). Cytogenetic studies on the haploid plants of tuber-bearing *Solanum* species. I. Induction of haploid plants of tuber-bearing *Solanums. Res. Bull. Hokkaido Natl. Agric. Exp. Stn.* **112,** 1–67.

Jacobsen, E. (1981). Polyploidization in leaf callus tissue and in regenerated plants of dihaploid potato (*Solanum tuberosum*). *Plant Cell, Tissue Organ Cult.* **1,** 77–84.

Linsmaier, E. M., and Skoog, F. (1965). Organic growth factor requirements of tobacco tissue cultures. *Physiol. Plant.* **18,** 100–127.

Mix, G. (1982). Dihaploide Pflanzen aus *Solanum tuberosum* Antheren. *Landbauforsch. Voelkenrode* **32,** 34–36.

Murashige, T., and Skoog, F. (1962). A revised medium for rapid growth and bio assays with tobacco tissue culture. *Physiol. Plant.* **15,** 473–497.

Nitsch, J. P., and Nitsch, C. (1969). Haploid plants from pollen grains. *Science* **163,** 85–87.

Sopory, S. K., Jacobsen, E., and Wenzel, G. (1978). Production of monohaploid embryoids and plantlets in cultured anthers of *Solanum tuberosum. Plant Sci. Lett.* **12,** 47–54.

van Breukelen, E. W., Ramanna, M. S., and Hermsen, J. G. T. (1975). Monohaploids (n = x = 12) from autotetraploids *Solanum tuberosum* (2x = 4x = 48) through two successive cycles of female parthenogenesis. *Euphytica* **24,** 567–574.

Wenzel, G. (1979). Neue Wege in der Kartoffelzüchtung I. + II. Vom Sortenklon zu Haploiden. *Kartoffelbau* **30,** 126–129.

Wenzel, G., and Sopory, S. K. (1978). Production and utilisation of dihaploid or monohaploid potatoes. *In* "Production of Natural Compounds by Cell Culture Methods" (A. W. Alfermann and E. Reinhard, eds.), pp. 303–305. Munich, Federal Republic of Germany.

Wenzel, G., and Uhrig, H. (1981). Breeding for nematode and virus-resistance in potato via anther culture. *Theor. Appl. Genet.* **59,** 333–340.

Wenzel, G., Schieder, O., Przewozny, T., Sopory, S. K., and Melchers, G. (1979). Comparison of single cell culture derived *Solanum tuberosum* L. plants and a model for their application in breeding. *Theor. Appl. Genet.* **55,** 49–55.

Wenzel, G., Bapat, V. A., and Uhrig, H. (1982). New strategy to tackle breeding problems of potato. *In* "Plant Cell Culture in Crop Improvement" (S. K. Sen and K. L. Giles, eds.), pp. 337–349. Plenum, New York.

Wenzel, G., Uhrig, H., and Burgermeister, W. (1983). Potato—a first crop improvement by the application of microbiological techniques? *In* "Genetic Manipulation Impact on Man and Society." ICSU Press (in press).

CHAPTER **36**

Anther Culture of *Brassica*

W. A. Keller

Genetic Engineering Section
Ottawa Research Station, Ontario Region
Agriculture Canada
Ottawa, Ontario, Canada

I. INTRODUCTION

The utilization of anther culture techniques for haploid production in the genus *Brassica* was first reported in *B. oleracea* by Kameya and Hinata in 1970. Further published information on this subject appeared in 1975 when anther culture procedures were employed to obtain plants from microspores in *B. napus* (Thomas and Wenzel, 1975) and *B. campestris* (Keller *et al.*, 1975). Since then, significant progress in producing and characterizing anther-derived haploids has been reported with *B. napus* (Keller and Armstrong, 1978; Renard and Dosba, 1980; Lichter, 1981), *B. campestris* (Keller and Armstrong, 1979), *B. oleracea* (Keller and Armstrong, 1981,

CELL CULTURE AND SOMATIC CELL
GENETICS OF PLANTS, VOL. 1

ISBN 0-12-715001-3

1983), *B. hirta* (Klimaszewska and Keller, 1983), and *B. juncea* (George and Rao, 1982). In *Brassica,* the developmental pathway leading to haploid regeneration generally involves the induction of microspore embryogenesis. However, the proliferation of microspore-derived callus and subsequent regeneration of plants through organogenesis have been observed in some cases, including the original report by Kameya and Hinata (1970). In this chapter, procedures for reliable haploid production in several *Brassica* species will be outlined, and potential applications of haploids in this genus will be discussed.

II. PROCEDURE

A. Growth of Donor Plants

Brassica species can be grown in 12-cm-diameter round pots in a soil–peat–sand mixture (1:2:1) containing 0.8% (w/v) slow-release fertilizer (e.g., Osmocote). A weekly nutrient supplementation of 20N–20P–20K is sufficient for active growth.

The physiological condition of the donor plants will significantly influence anther culture success. The most suitable donors are grown in climate-controlled chambers, although greenhouse- and field-grown material can be successfully used. A 20–15°C day–night cycle is satisfactory, although a 15–10°C cycle is preferable in *B. napus* (Keller and Stringam, 1978). A 16-hr photoperiod is generally employed; however, continuous light has also been used for *B. campestris* (Keller *et al.*, 1983). Light intensity is an important factor. Superior donors are obtained under 3000 foot candles (f-c), in comparison to 750 or 1500 f-c (Keller *et al.*, 1983). The light source consists of fluorescent tubes and incandescent bulbs, which contribute 84 and 16% of the intensity, respectively.

B. Anther Staging and Bud Selection

In *Brassica,* as in many other species, the uninucleate microspore stage is the most responsive in culture, although it appears that embryogenesis is induced at the mitotic stage as well. Accurate staging can be achieved by fixing anthers in Carnoy's 6:3:1 (ethanol:acetic acid:chloroform) solution and staining with propiono-carmine. Rapid staging can be achieved by preparing fresh squashes in lactopropionic orcein or 0.05% toluidine blue in citric acid buffer (pH 4.0).

An effective external staging system based on petal length and anther appearance has been developed for *Brassica* spp. (Keller *et al.*, 1975, 1983). Buds in which petal length varies from one-fourth to slightly more than one-half of the length of the anther contain pollen at the optimal developmental stage. The anthers in these buds have a yellowish-green color. It is advisable to check closely and modify external staging procedures to suit new genotypes or changes in donor growth conditions.

The most suitable material for anther culture is the main inflorescence taken just prior to flower emergence. However, flowering inflorescences or axillary buds may also be used if the donor material is valuable. In cases of limited supply, donor plants may also be cut back and anthers cultured again after a period of regrowth.

C. Bud Pretreatment

Pretreatment of *B. juncea* inflorescences at 10°C for 2–15 days prior to anther culture is reported to enhance microspore embryogenesis (George and Rao, 1982); however, a low-temperature pretreatment (5°C for 7 days) is inhibitory to *B. campestris* (Keller *et al.*, 1983). In *B. campestris* and *B. napus*, specific bud pretreatments are generally not employed. In *B. oleracea* var. *italica* cv. Green Mountain, pretreatment of excised inflorescences (in water) at 45°C for 1 hr followed by 40°C for 3 hr prior to anther culture, stimulates embryogenesis (Keller and Armstrong, 1983). Subjection of buds to reduced atmospheric pressure (500 mm Hg for 0.5 hr) increases the yield of microspore-derived embryos in *B. hirta* (Klimaszewska and Keller, 1983).

D. Surface Sterilization and Anther Extraction

Excised buds are placed in tissue capsules or polyester screen sacks and surface sterilized by a 10-min immersion in 7% (w/v) calcium hypochlorite. They are then rinsed three times and finally floated on sterile distilled water. Buds should be used within 4 hr of surface sterilization.

Manipulation of buds and extraction of anthers are achieved with the use of two finely pointed surgical forceps (e.g., no. 5). The forceps are sterilized by immersion in 70% ethanol, followed by a double rinse in sterile demineralized water.

Buds are removed from water in groups of three to six and placed on the stage of a dissecting microscope which has been wiped with 70% ethanol. Two forceps are used to remove carefully sepals and petals. Anthers are folded away from the pistil with the use of a forceps arm. The forceps are then placed under the anther, with the filament between the arms, and are gently raised to remove the anther. Anthers which are completely free of filament tissue appear to undergo higher frequencies of embryogenesis. Anthers should be placed on the medium so that the outer flat side is in contact with the medium. Care must be taken not to puncture or crush the anther locule, as damaged anthers rarely respond *in vitro*.

E. Culture Medium

The culture medium is derived from B5 medium (Gamborg *et al.*, 1968) with the following modifications: 100 g/liter sucrose, 750 mg/liter $CaCl_2 \cdot 2 H_2O$, 40 mg/liter sequestrene 330 Fe (Ciba-Geigy), 800 mg/liter glutamine, 100 mg/liter serine, and 0.1 mg/liter of 2,4-dichlorophenoxyacetic acid (2,4-D) and naphthaleneacetic acid (NAA). The pH level is adjusted to 5.8 with 0.1 *N* KOH, and the medium is sterilized by autoclaving for 20 min at 1.1 kg/cm^2.

Brassica anthers may be cultured on solid or liquid medium. When solid medium is used, six anthers are plated in 60 × 15-mm sterile disposable Petri dishes containing 5–6 ml of medium solidified with 0.8% agar. In most cases, increased embryo yields are obtained with liquid medium. A thin liquid layer, approximately 3.5 ml per 60 × 15-mm dish, is used, and 18 anthers are placed on it. The anthers float on the medium and usually aggregate in the center of the dish. The Petri dishes are carefully sealed with Parafilm prior to incubation.

F. Incubation of Anthers

An initial period of culture at elevated temperature prior to maintenance at 25°C greatly enhances embryo yield. In most cases, the elevated temperature treatment is 35°C for 1–3 days. The optimal exposure period may vary for different genotypes. In the case of *B. napus* cv. Tower, a 14-day, 30°C treatment is the most effective (Keller and Armstrong, 1978); however, other genotypes respond better to a short (1- to 2-day) treatment at 35°C. *Brassica* anthers are routinely incubated in darkness.

The emergence of whitish-yellow embryos is detected as early as 15 days after culture and usually does not continue beyond 35 days. A range of developmental stages from globular to fully differentiated mature embryos can be identified.

G. Plant Regeneration

Plantlets will not develop from embryos in the anther culture medium, and an embryo culture procedure is required. The embryo culture medium is derived from the anther culture medium by the reduction of sucrose to 2% and by elimination of the growth regulators and amino acid supplements. Solid medium is required for embryo development even if the anthers are cultured on liquid. It is essential to culture the embryos soon after emergence, as prolonged maintenance on the anther culture medium will lead to senescence. Embryo culture conditions include maintenance of 1–10 embryos per 60 × 15-mm Petri dish (5–6 ml of solid medium) at 25°C under continuous fluorescent light (150 ft-c).

The frequency of embryo survival is dependent on the developmental stage. Most globular embryos do not survive, whereas the majority of bipolar embryos survive and turn green. A gradual sequential reduction of the sucrose level in the embryo culture medium may aid embryo survival. For example, survival efficiency of anther-derived *B. napus* embryos is higher with 6% than with 2% sucrose (Keller and Armstrong, 1978).

Only a small fraction (1–10%) of cultured embryos develop directly into plantlets. The majority develop in an abnormal manner, with elongated or swollen hypocotyls and massive cotyledons. In some cases, especially in *B. napus,* large numbers of secondary embryos develop on the hypocotyls of the anther-derived embryos (Thomas *et al.*, 1976; Keller and Armstrong, 1977).

Plantlets can be obtained from abnormal embryos through organogenesis. Shoots are regenerated by cutting the embryos into three to four explants and culturing on a Murashige and Skoog (MS) medium (1962) modified as follows: inclusion of B_5 vitamins (Gamborg *et al.*, 1968) rather than MS vitamins, substitution of 40 mg/liter Fe-EDTA (ethylenedinitrilo tetracetic acid, iron derivative, sodium salt; Baker Chemical Co.) for $FeSO_4 \cdot 7H_2O + Na_2EDTA$, reduction of sucrose to 2%, addition of 5×10^{-6} *M* benzylaminopurine (BAP) and 10^{-7} *M* NAA or 10^{-6} *M* indoleacetic acid (IAA), and inclusion of 0.8% agar (Keller and Armstrong, 1977). Shoot regeneration occurs during the first 3 weeks of culture. In cases in which normal shoot meristems do not appear, a second round of explant culture on the same medium is recommended. Shoots can be successfully rooted on the embryo culture medium described above. Incubation conditions for shoot induction are the same as those for embryo culture. Plantlets with active root systems can be directly transferred to peat pellets (e.g., Jiffy-7) and placed in a mist chamber for 1–2 weeks to facilitate hardening. They can be potted and maintained in a greenhouse or planted in the field.

H. Cytological Characterization of Regenerants

Chromosome numbers of anther-derived *Brassica* plants can be determined by cytological analysis of pollen mother cells (at anaphase I or metaphase II). Clusters of buds are fixed in Carnoy's 6:3:1 solution containing a trace of ferric chloride and subsequently stained with acetocarmine or propionocarmine.

Root tips can also be utilized to determine ploidy level. This involves pretreatment in 0.05% colchicine for 2, 12, or 24 hr, fixation in acetic ethanol containing a trace of ferric chloride, hydrolysis in 1 *N* HCl for 5 min at room temperature, and squashing in acetic-orcein.

Although cytological confirmation is required, flowering populations of anther-derived *Brassica* plants can be quickly screened for haploids, as they have small, sterile flowers with shrunken anthers.

I. Chromosome Doubling

For most practical applications of haploids, chromosome doubling must be achieved in order to restore fertility. The most common method of doubling is through colchicine treatment (Wenzel *et al.*, 1977), although tissue culture methods involving shoot organogenesis may be used in some cases (Keller and Stringam, 1978, and Section III,A, this chapter).

III. RESULTS AND CONCLUSIONS

A. Yield of Microspore-Derived Plants and Frequency of Haploids

If prescribed conditions for establishing anther cultures are established, large numbers of microspore-derived embryos can be obtained. Embryo yield is genotype dependent, and in *B. napus* 300–6000 embryos can be produced from 1000 cultured anthers. Embryo yields in the other species are somewhat lower, ranging from 50 to 300 embryos per 1000 anthers. Under conditions routinely employed, a 40–50% efficiency of plant recovery from cultured embryos can be expected.

The frequency of haploids is genotype dependent and may include 20–80% of the regenerants. The majority of nonhaploid plants (>90%) are diploids, although a few triploids, tetraploids, and hexaploids have also been detected. Morphological analysis of anther-derived diploids has demonstrated homozygosity, indicating that these plants originated from haploid microspores, possibly through a process of endomitosis and/or nuclear fusion (Keller *et al.*, 1975). The employment of elevated culture temperatures apparently reduces the frequency of spontaneous doubling (Keller and Armstrong, 1978). On the other hand, induction of shoot organogenesis from abnormal embryo explants favors diploidization (Keller and Armstrong, 1981).

B. Utilization of *Brassica* Haploids

Haploids can be used to select effectively desirable genetic recombinants from F_1 hybrids. In *Brassica,* some preliminary encouraging studies on selection have been reported (Wenzel *et al.*, 1977; Hoffmann *et al.*, 1982; Keller *et al.*, 1983). Further progress should be expected, as a number of plant breeders are evaluating doubled haploids in their breeding programs.

Haploid production is an effective method for generating homozygous lines for use in hybrid cultivar development in cross-pollinated crops. This approach may be especially applicable for *B. oleracea* crops of high cash value.

Haploids can also serve as the source of explants for haploid tissue cultures or as a direct source of cells for use in mutant isolation. Plants have been regenerated from haploid *B. napus* callus (Stringam, 1979; Sacristan, 1981), protoplasts (Kohlenbach *et al.*, 1982), and secondary embryos (Loh and Ingram, 1982). Haploid *B. napus* tissue cultures have been used in preliminary studies on the selection of disease-resistant variants (Sacristan and Hoffmann, 1979; Sacristan, 1982).

Finally, *Brassica* haploids may be used in chromosome pairing studies to gain further information on genome evolution (Armstrong and Keller, 1981, 1982).

REFERENCES

Armstrong, K. C., and Keller, W. A. (1981). Chromosome pairing in haploids of *Brassica campestris*. *Theor. Appl. Genet.* **59,** 49–52.

Armstrong, K. C., and Keller, W. A. (1982). Chromosome pairing in haploids of *Brassica oleracea*. *Can. J. Genet. Cytol.* **24,** 735–739.

Gamborg, O. L., Miller, R. A., and Ojima, K. (1968). Nutrient requirements of suspension cultures of soybean root cells. *Exp. Cell Res.* **50,** 151–158.

George, L., and Rao, P. S. (1982). *In vitro* induction of pollen embryos and plantlets in *Brassica juncea* through anther culture. *Plant Sci. Lett.* **26,** 111–116.

Hoffmann, F., Thomas, E., and Wenzel, G. (1982). Anther culture as a breeding tool in rape. *Theor. Appl. Genet.* **61,** 225–232.

Kameya, T., and Hinata, K. (1970). Induction of haploid plants from pollen grains of *Brassica. Jpn. J. Breed.* **20,** 82–87.

Keller, W. A., and Armstrong, K. C. (1977). Embryogenesis and plant regeneration in *Brassica napus* anther cultures. *Can. J. Bot.* **55,** 1383–1388.

Keller, W. A., and Armstrong, K. C. (1978). High frequency production of microspore-derived plants from *Brassica napus* anther cultures. *Z. Pflanzenzuecht.* **80,** 100–108.

Keller, W. A., and Armstrong, K. C. (1979). Stimulation of embryogenesis and haploid production in *Brassica campestris* anther cultures by elevated temperature treatments. *Theor. Appl. Genet.* **55,** 65–67.

Keller, W. A., and Armstrong, K. C. (1981). Production of anther-derived dihaploid plants in autotetraploid marrowstem kale (*Brassica oleracea* var. Acephala). *Can. J. Genet. Cytol.* **23,** 259–265.

Keller, W. A., and Armstrong, K. C. (1983). Production of haploids via anther culture in *Brassica oleracea* var. italica. *Euphytica* **32,** 151–159.

Keller, W. A., and Stringam, G. R. (1978). Production and utilization of microspore-derived haploid plants. *In* "Frontiers of Plant Tissue Culture 1978" (T. A. Thorpe, ed.), pp. 113–122. Univ. of Calgary Press, Calgary, Alberta, Canada.

Keller, W. A., Rajhathy, T., and Lacapra, J. (1975). *In vitro* production of plants from pollen in *Brassica campestris. Can. J. Genet. Cytol.* **17,** 655–666.

Keller, W. A., Armstrong, K. C., and de la Roche, A. I. (1983). The production and utilization of microspore-derived haploids in *Brassica* crops. *In* "Plant Cell Culture in Crop Improvement" (S. K. Sen and K. L. Giles, eds.), pp. 169–183. Plenum, New York.

Klimaszewska, K., and Keller, W. A. (1983). The production of haploids from *Brassica hirta* Moench (*Sinapis alba* L.) anther cultures. *Z. Pflanzenphysiol.* **109,** 235–241.

Kohlenbach, H. W., Wenzel, G., and Hoffmann, F. (1982). Regeneration of *Brassica napus* plantlets in cultures from isolated protoplasts of haploid stem embryos as compared with leaf protoplasts. *Z. Pflanzenphysiol.* **105,** 131–142.

Lichter, R. (1981). Anther culture of *Brassica napus* in a liquid culture medium. *Z. Pflanzenphysiol.* **103,** 229–237.

Loh, C. S., and Ingram, D. S. (1982). Production of haploid plants from anther cultures and secondary embryoids of winter oilseed rape, *Brassica napus* ssp. *oleifera. New Phytol.* **91,** 507–516.

Murashige, T., and Skoog, F. (1962). A revised medium for rapid growth and bioassays with tobacco tissue cultures. *Physiol. Plant.* **15,** 473–497.

Renard, M., and Dosba, F. (1980). Etude de l'haploidie chez le Colza (*Brassica napus* L. var. *oleifera* Metzger). *Ann. Amelior. Plant.* **30,** 191–209.

Sacristan, M. D. (1981). Regeneration of plants from long-term callus cultures of haploid *Brassica napus. Z. Pflanzenzuecht.* **86,** 248–253.

Sacristan, M. D. (1982). Resistance responses to *Phoma lingam* of plants regenerated from selected cell and embryogenic cultures of haploid *Brassica napus. Theor. Appl. Genet.* **61,** 193–200.

Sacristan, M. D., and Hoffman, F. (1979). Direct infection of embryogenic tissue cultures of haploid *Brassica napus* with resting spores of *Plasmodiophora brassicae. Theor. Appl. Genet.* **54,** 129–132.

Stringam, G. R. (1979). Regeneration in leaf-callus cultures of haploid rapeseed. *Z. Pflanzenphysiol.* **92,** 459–462.
Thomas, E., and Wenzel, G. (1975). Embryogenesis from microspores of *Brassica napus*. *Z. Pflanzenzuecht.* **74,** 77–81.
Thomas, E., Hoffmann, F., Potrykus, I., and Wenzel, G. (1976). Protoplast regeneration and stem embryogenesis of haploid androgenetic rape. *Mol. Gen. Genet.* **145,** 245–247.
Wenzel, G., Hoffmann, F., and Thomas, E. (1977). Anther culture as a breeding tool in rape. I. Ploidy level and phenotype of androgenetic plants. *Z. Pflanzenzuecht.* **78,** 149–155.

CHAPTER **37**

Anther Culture of Cereals and Grasses

G. Wenzel
B. Foroughi-Wehr

Institute for Resistance Genetics
Federal Biological Research Center for Agriculture and Forestry
Grünbach, Federal Republic of Germany

I. INTRODUCTION

The last 15 years of research into the production of haploids in cereals—the world's most important crops—have aimed at producing them at satisfactory frequencies. As with most other plants, it is possible to induce haploids either from unfertilized egg cells (parthenogenesis) or from microspores (androgenesis). The parthenogenetic approach is an older method, and examples exist of its successful application (Nitzsche and Wenzel, 1977). In maize, by the use of color markers, haploid seeds can be selected in reasonable amounts (Chase, 1974). Using the *ig* gene in maize (Kermicle, 1969) or the *hap* gene in barley (Hagberg and Hagberg, 1980), haploids can be produced parthenogenetically without using the tissue culture step. Since the *Hordeum bulbosum* technique works for *H. vulgare*, functioning via chromosome elimination, substantial progress in barley breeding using such haploids has been made (Kasha and Reinbergs, 1980). The cultivar

CELL CULTURE AND SOMATIC CELL
GENETICS OF PLANTS, VOL. 1

ISBN 0-12-715001-3

TABLE I

Early Experiments on Anther Culture of Cereals

Variety	First anther derived plant in:	References
Oryza spp.	1968	Niizeki and Oono
Hordeum vulgare L.	1971	Clapham
	1973	
Triticum aestivum L.	1973	Ouyang *et al.*, Wang *et al.*, Picard and Buyser
Triticale	1973	Wang *et al.*
Secale cereale L.	1974	Wenzel and Thomas
	1975	
Zea mays L.	1975	Anonymous

Mingo was produced by this technique and licensed 5 years after the first cross was made. So far, however, parthenogenetic techniques are working satisfactorily only in barley and to some extent in maize. Androgenesis (anther culture) is more universal, and its successful use opens a broader genetic spectrum per cross when the immense number of microspores is compared with the limited number of egg cells. During the last few years, progress has been made in improving the anther culture techniques in cereals to such an extent that rice (Hu and Shao, 1981), wheat (Henry *et al.*, 1982), barley (Foroughi-Wehr *et al.*, 1982), and maize (Hu and Shao, 1981; Ting *et al.*, 1981) lines descending from microspores are now being checked in official yield trials by seed boards for licensing. This may create the impression that all problems have been solved, but the anther culture procedure is still very empirical, and results cannot yet be generalized. Most research groups use their specific media, varieties, and even tricks.

In this chapter, we have tried to extract general principles in order to specify in a somewhat subjective way the most promising approach and to cite papers dealing predominantly with methods used. Table I summarizes the early papers on haploid induction in cereals. For further information, see the reviews of Nitzsche and Wenzel (1977) and Vasil (1980).

II. GENOTYPE

Success in anther culture is predominantly dependent on the genotype of the anther donor material. In the first experiments using rice, Niizeki and Oono (1968) found significant differences in the frequency of callus formation; only 2 of the 10 varieties tested produced plants. In anther culture of two maize hybrids, in one (*Dan-San 91*) 4.1% and in the other

(*King Hwang 13*) 0.4% of the plated anthers produced callus. Similar differences are reported for rye (Wenzel *et al.*, 1977) and wheat (Picard and DeBuyser, 1973).

When 19 spring barley varieties were checked, the variety *Dissa* expressed the highest level of tissue culture ability (Foroughi-Wehr *et al.*, 1976). Good tissue culture ability is equivalent to a good regeneration capacity under given culture conditions. Probably culture conditions could be optimized for each genotype, as proposed by Dunwell (1981). We believe, however, that it is cheaper to broaden the genetic bases for tissue culture ability by selection and combination breeding. The spring barley variety *Dissa* was consequently crossed with the top varieties *Aramir* and *Mutina*. *Aramir* has a poor and *Mutina* a medium reaction in anther culture. The F_1 progeny of these crosses were used as anther donors. The results are given in Fig. 1. In both cases, the F_1 is intermediate between both parents, making it feasible that tissue culture ability is heritable. Similar results were reported by Wenzel *et al.* (1977) for rye and by Bullock *et al.* (1982) for wheat. In wheat, the good regeneration capacity of the variety *Centurk* could be transferred in reciprocal crosses to the F_1 progeny. The cytoplasmic effects were negligible.

III. PHYSIOLOGICAL STATUS OF THE DONOR PLANTS

The physiological status of the plants at the time of anther excision strongly influences the sporophytic potential of microspores in the anther. The response in culture is predominantly influenced by the different growth conditions during various seasons. As microscopic examination of

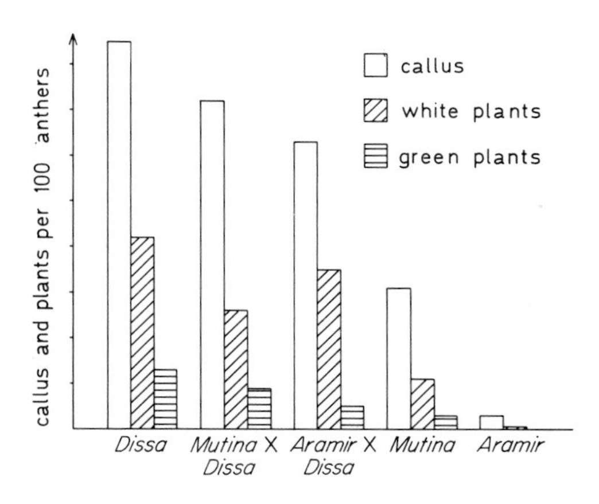

Fig. 1. Influence of the genotype of varieties and hybrids in barley anther culture.

the microspores did not show any morphological or developmental variation, the environment probably influences the endogenous state of the anther donor plant. Critical environmental factors are light intensity, photoperiod, temperature, nutrition, and concentration of CO_2. As these factors also interact, they affect the donor plant in a complex way. For rye we found it beneficial to increase the CO_2 concentration during the natural short-day season. This can easily be achieved by opening a CO_2 cylinder for an adequate time interval. As plant growth conditions cannot be controlled in the field, we use only barley, rye, and wheat material grown under semicontrolled conditions, although plants grown outdoors during natural growing season are more responsive than greenhouse-grown material. Anthers taken from plants grown under about 20,000 lx (the emission spectra of HQI lamps are most suitable) at the growing point of the plants produced macroscopic structures in much higher frequencies than anthers from donors under lower light intensity. Also, the temperature during plant growth was critical. A constant temperature of 18–20°C, coupled with low light intensities (less than 10,000 lx), caused, for example in barley, a reduced rate of microspore–callus formation (Foroughi-Wehr and Mix, 1979). In rye, temperatures over 25°C were disadvantageous, as the microspores tended to burst in isolated pollen culture or during pollen tube germination. By checking the pollen tube germination of fully developed pollen grains, one can be deduce the quality of the microspores. For most cereals, an aqueous solution of 2–3% sucrose, 10–20 mg/liter H_3BO_4, and 200 mg/liter $CaCl_2$ works satisfactorily. Finally, it should be mentioned that pest control procedures may have a detrimental effect on microspore development.

IV. EFFECT OF POLLEN STAGE AND POLLEN DEVELOPMENT

Another critical factor to be considered is the selection of anthers at an appropriate stage of pollen development. Anthers with microspores ranging from tetrad to the binucleate stage (Fig. 2a) are most responsive. But as soon as starch deposition has begun, no sporophytic development and subsequently no macroscopic structure formation (i.e., calli and embryoids) occur. In all cereals investigated, the early or mid-uninucleate stage of microspore development (stages 1 and 2, respectively; *after* Sunderland, 1974) was found to give the best results: *Oryza sativa* (Guha *et al.*, 1970; Niizeki and Oono, 1971; Wang *et al.*, 1974); *H. vulgare* (Clapham, 1971; Gaul *et al.*, 1976); *Triticum aestivum* (Ouyang *et al.*, 1973; Wang *et al.*, 1973); *Zea*

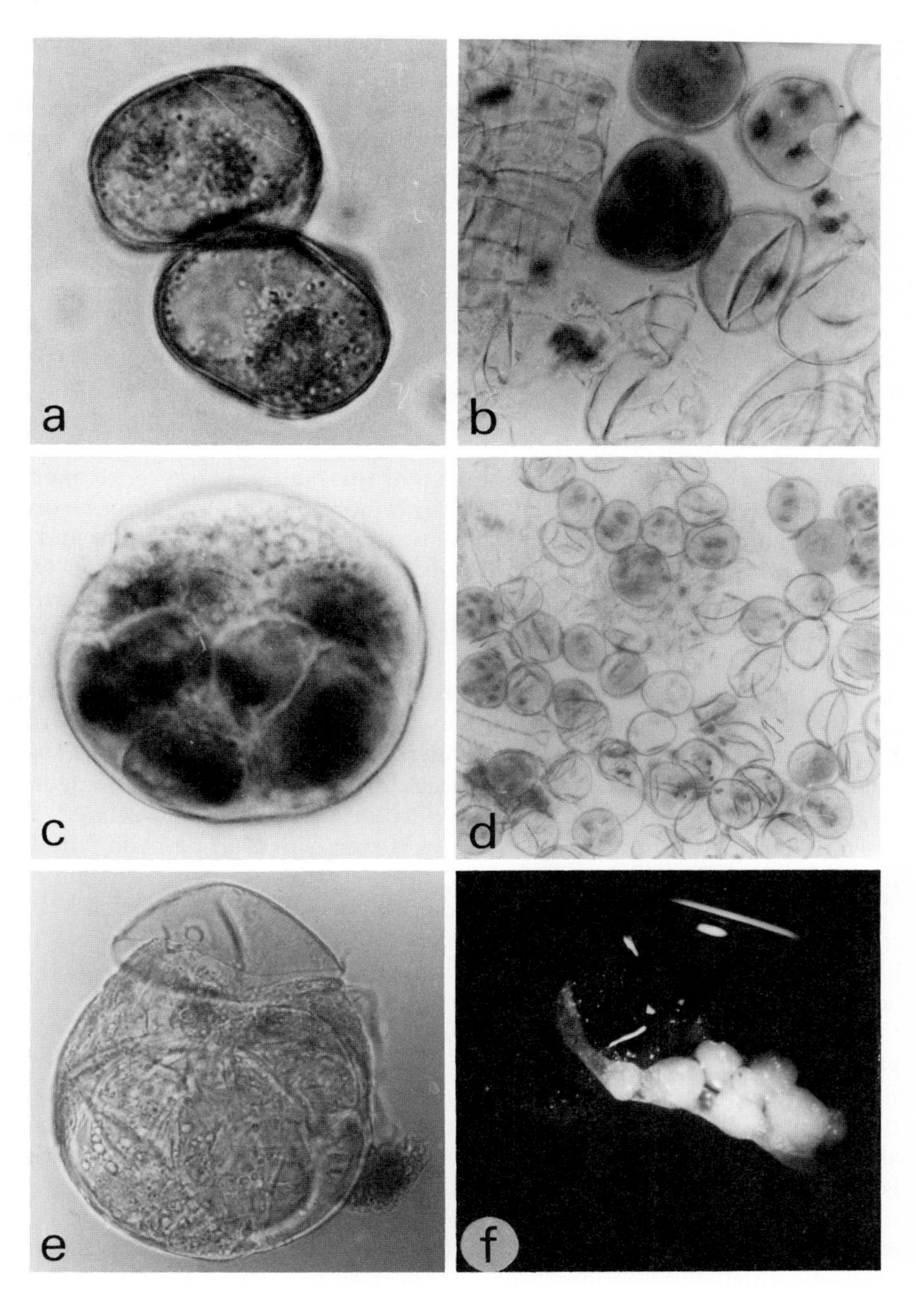

Fig. 2. Microspore development in *Hordeum vulgare.* (a) Ideal developmental stage for starting anther culture; the upper spore has just two nuclei. (b–d) Dividing microspores within the anther 10 days after plating. (e) Globular embryoid formation; the exine is still attached. (f) Macroscopic structure formation 4 weeks after anther plating.

mays (Miao *et al.*, 1978); Triticale (Sozinov *et al.*, 1981); and *Secale cereale* (Wenzel and Thomas, 1974; Thomas and Wenzel, 1975). The pollen stage in the anther was determined with acetocarmine squashes. These checks can be used as markers for the estimation of microspore development in the remaining florets from the same variety during the same season. Not all anthers from one spike are at the same stage of development. Pollen in florets in the middle of the inflorescence is further developed than that in apical and basal florets. In rye, when central florets show the first microspore mitosis, then anthers from the lower and the upper third of the ear are in the correct developmental stage for stimulating sporophytic development.

Growth of somatic tissue (filament, connective, or anther wall) instead of microspores is also under the influence of the medium. In rye, filaments tend to grow preferentially on media with more than 2 mg/liter of auxins. Similar results are obtained from old anthers; from young anthers and at too high cytokinin concentrations, preferential proliferation of anther wall and connective tissues occurs. In barley, however, even under a wide range of conditions, the formation of somatic callus was never observed (Fig. 3c).

The induction of uninucleate microspores to form sporophytic embryoids or calli has been studied in detail by Sunderland in *H. vulgare* (Sunderland and Dunwell, 1977; Sunderland *et al.*, 1979; Sunderland and Evans, 1980). It was found that the three different pathways of androgenetic microspore development elaborated in *Datura* (Sunderland *et al.*, 1974) are transferable to the cereals. Here, however, along all pathways multicellular microspores develop, even if the vegetative or the generative nucleus starts to divide first (Sunderland and Evans, 1980). During establishment of an anther culture procedure, it is recommended that some anthers be sacrificed after 2–3 weeks of culture. These should be squashed in acetocarmine and checked microscopically.

In rye, responsive anthers do not turn brown and flaccid during the first weeks; they stay firm and white (on average, 10% of the plated anthers). Early sporophytic development can most likely be observed in white anthers. Figure 2 documents the development in barley. The white anthers should not become covered by a film of the liquid medium. Its prevention is important for later callus culture. After further divisions (Fig. 2b–e), macroscopic structures become visible (Fig. 2f). Depending on the genotype, callus or embryoids may be formed. In rice and barley, callus is formed exclusively, whereas in other cereals embryoids quite often rupture the anther wall. In rye, embryoids show secondary callusing under the influence of the auxin necessary for the induction of microspore development.

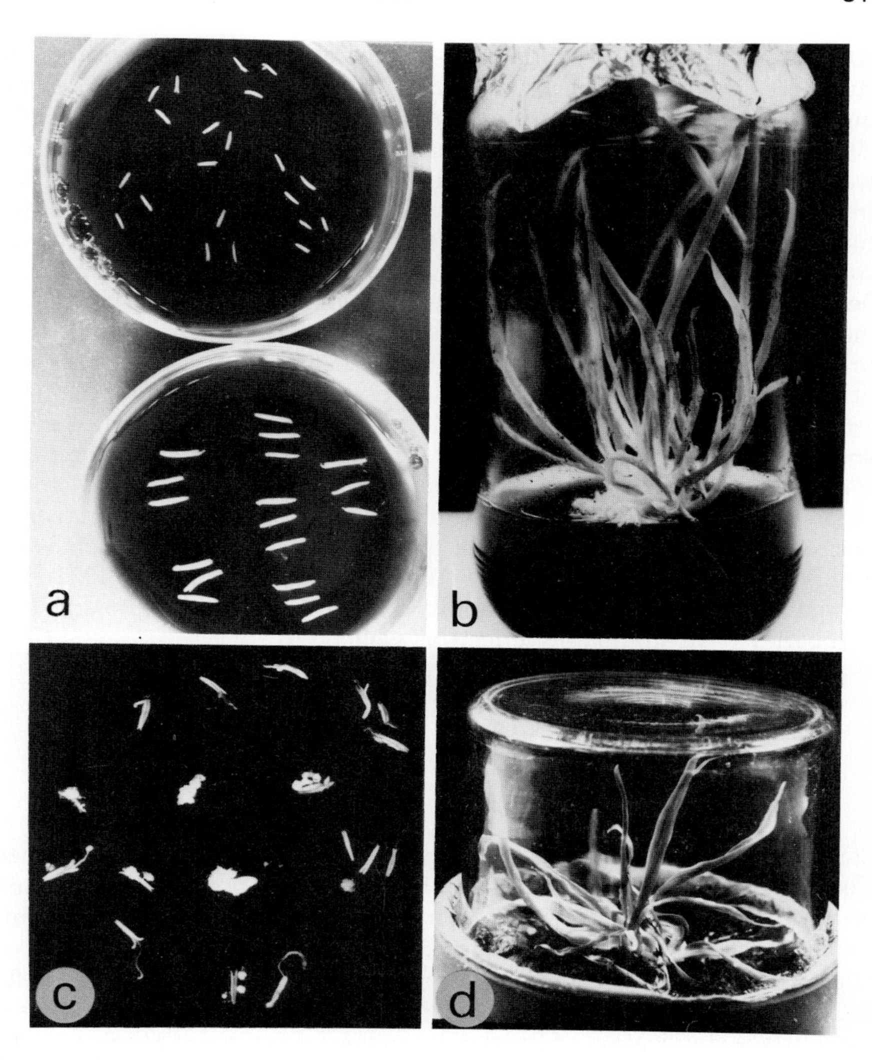

Fig. 3. Anther culture in cereals. (a) Six-centimeter Petri dishes with barley (upper) and rye anthers. (b) Young barley plantlet on rooting medium. (c) Macroscopic structure formation from barley anthers. (d) Rye plantlet transferred to potting compost.

V. PRETREATMENT OF ANTHERS

Certain treatments given to the whole spike or to the anthers can have a positive effect on the development of the microspore. The most effective technique used in anther culture is cold treatment. Sunderland *et al.* (1981)

recommend for barley a period of 21–25 days at 4°C or 14–21 days at 7°C for cut tillers. In rye a cold treatment of 6°C for a period of 3–15 days, depending on the plant genotype, is optimal (Wenzel *et al.*, 1977). Wang *et al.* (1974) increased the amount of callus formation in rice when the raceme was pretreated at 10°C for 48 hr, but other investigators showed the same frequency of callus and plantlet formation without cold pretreatment. Actual mechanisms producing this effect are unknown. As a result of cold storage, weak or nonviable anthers and microspores are killed; they then become dark brown in color. Thus, the spikes are enriched in vigorous material. At 6°C, increased numbers of microspores are arrested during the first mitosis, as starch production is blocked. This increases the possibility of subsequent development. It is also possible that the cold pretreatment retards aging of the anther wall, allowing a higher proportion of microspores to change their developmental pattern from gametophytic to sporophytic. Cold pretreatment may also provide an unspecific shock, resulting in the establishment of endogenous cellular conditions which favor the desired development. This is quite probable, as other shocks such as cutting (Wilson, 1977), spraying with ethrel (Bennett and Hughes, 1972), centrifugation (Nitsch, 1974), and irradiation with 1 krad in rye (Stolarz, 1974) and barley have stimulatory effects as well.

VI. CULTURE PROCEDURE

Spikes are harvested at the uninucleate stage of microspore development close to the first microspore mitosis. From one well-nourished, vigorously growing plant, up to 10 spikes can be harvested. Their developmental sequence has no effect on the *in vitro* reaction as long as the ears are normally developed. Inflorescences of most cereals are covered by the flag leaf at this early developmental stage and keep the florets completely axenic. Surface sterilization of the flag leaf allows subsequent axenic isolation of the anthers at the clean bench. If the spike bearing the correct microspores has already emerged from its flag leaf, sterilization is more difficult. In order to obtain microspores in the correct stage for culture and to avoid contamination of emergent spikes, the spikes still within the flag leaves can be harvested, sterilized, transferred for several days to culture medium lacking growth hormones, and covered with a sterilized plastic bag. Several possibilities exist for plating of anthers. Most commonly the anthers are plated on solid agar media in Petri dishes. However, larger vessels, such as 50-ml Erlenmeyer flasks, are preferable. For barley, Kao (1981) and Sunderland *et al.* (1979) proposed liquid culture media. Normally the media are solidified by 6–10 g/liter agar; Nitsch *et al.* (1982) recommend the use of agarose in a concentration of 0.8 g/liter for maize anther culture. It is usual to plate 10–20 anthers per 6-cm Petri dish, about 50 anthers per 10 ml of liquid medium, or 20 anthers per 50-ml Erlenmeyer flask. The use of glass or plastic dishes is predominantly dependent on the amount of condensation water formed. Plastic dishes require better temperature-controlled culture rooms.

Wilson (1977) developed a technique that avoids the laborious excision of anthers from florets by culturing the whole inflorescences in liquid culture media. Although reproducible results have been obtained by others (Sunderland, 1980), the method cannot withstand the comparison with the rate of green plantlet production of normal anther culture.

VII. CULTURE MEDIA

It is difficult to draw a conclusion as to which medium is most suitable for different cereal species. As species or even genotypes may demand different nutritional conditions, no general recommendation can be given. Beyond that, the conditions needed for sporophytic microspore induction, for callus or embryoid formation, and for organogenesis normally also differ markedly. The importance of certain ingredients of the culture medium, particularly minerals, organic additives, carbohydrates, hormones, and unidentified growth adjuvants, will therefore be discussed only briefly.

The media used most successfully for the induction of sporophytic microspore development are summarized in Table II. Addition of hormones to the basal medium seems to be necessary, but recent investigations indicate that they are not essential in these monocotyledonous species. Miao *et al.* (1978) reported the development of embryoids in anther cultures of *Z. mays* in the absence of growth hormones, although their addition raised the response considerably. When the response of cereal cells and tissues to different phytohormone concentrations or combinations with that of dicots is compared, the poor reaction in the former is striking. It reflects the relatively high amount of phytohormone autotrophy of these plants.

Several amino acids, vitamins, and inositol stimulate androgenesis. A beneficial effect of glutamine has been noted in *Z. mays* (Ku *et al.*, 1978), *T. aestivum* (De Buyser and Henry, 1980), and Triticale (Sozinov *et al.*, 1981). In wheat, glutamine can replace potato extract in the medium. Glutathione is effective in rye anther culture (Wenzel *et al.*, 1977). Inositol is beneficial especially for the growth of multicellular microspores in barley anther cul-

TABLE II

Media Composition for Cereal Anther Culture

Variety	Basal medium	Additional components	mg/liter
Triticum aestivum	Potato 2 medium (Chuang *et al.*, 1978) Potato 1 medium (Anonymous, 1976)	Glutamine (Henry and De Buyser, 1981) (Schaeffer *et al.*, 1979)	200
Triticale	B5 (Gamborg, 1970)	2,4-D	2.0
		Proline	100–160
		Oxyproline	100–160
		Traumatic acid	5–50
		IBA	100
		Glutamine	100
		Sucrose	30,000
		(Sozinov *et al.*, 1981)	
Secale cereale	N6 (Chu, 1978)	Glutathione	30
		Ascorbic acid	10
		2,4-D	2.0
		Kinetin	0.5
		Glucose	50,000
		(Wenzel *et al.*, 1977)	
Hordeum vulgare	Potato 2 medium N6 (liquid and solid)	Inositol	1,000
		Kinetin	0.5
		2,4-D	1.5
		Sucrose	90,000
		(Sunderland *et al.*, 1981; Xu and Sunderland, 1981)	
	MS (Murashige and Skoog, 1962; Linsmaier and Skoog, 1965) without $CoCl_2 \cdot 6H_2O$, $FeSO_4 \cdot 7H_2O$	Inositol	100
		Thiamine	0.4
		IAA	1.0
		BAP	1.0
		Fe NaEDTA	40
		Sucrose	60,000
		(Foroughi-Wehr *et al.*, 1976)	
Zea mays	N6	Casein hydrolysate	500
		L-Proline	100
		TIBA	0.1
		Sucrose	60,000
		(Nitsch *et al.*, 1982)	
	Yü-pei (Ku *et al.*, 1978)	Glycine	7.7
		Thiamine	0.25
		Nicotinic acid	1.3

TABLE II (*Continued*)

Variety	Basal medium	Additional components	mg/liter
		TIBA	0.1
		Casein hydrolysate	500
		Sucrose	120,000
		(Genovesi and Collins, 1982)	
Oryza spp.	MS (half concentration)	Kinetin	2.0
		NAA	4.0
		Sucrose	60,000–90,000
		(Chen, 1978; Chen *et al.*, 1982)	
	Miller (Miller, 1963) N6	2,4-D	2.0
		(Yin *et al.*, 1976)	

ture (Xu and Sunderland, 1981), whereas it is detrimental in maize callus induction (Brettel *et al.*, 1981). In barley we found it beneficial to use filter-sterilized instead of autoclaved media, especially for plant regeneration. This is probably not only due to heat unstable compounds such as indoleacetic acid (IAA) in the medium. Sucrose is obligatory for all tissue culture steps (table sugar is equally satisfactory). An increased concentration of sucrose (6–15 g/liter) yields better results in most of the cereals.

Quite often, undefined compounds are added successfully to culture media. The most common are coconut milk or deproteinized coconut water, pretreated at 60°C for 30 min at concentrations of 5–20%. More recently, the potato extract medium (Anonymous, 1975) caused substantial progress in cereal anther culture. Originally it was developed for wheat, but it also yielded good results in rye (Wenzel *et al.*, 1977). The composition of potato extract medium 1 is as follows: 20% potato extract, 9% sucrose, 1.5–2.0 mg/liter 2,4-dichlorophenoxyacetic acid (2,4-D), 0.5 mg/liter kinetin, 10^{-4} mg/liter FeEDTA, and 0.7–0.8% agar. In preparing the extract, 200 g fresh potato tubers were cleaned and the sprouted eyes removed. The tubers are cut into 1-cm^3 pieces, placed in 400–600 ml distilled water, and boiled for 30 min. After the macerate is squeezed through two layers of cheesecloth, the liquid extract is collected and mixed with the other compounds to make 1 liter of medium. Potatoes not stored too long give better results than fresh harvested or old tubers. Among the German varieties, *Culpa* and *Clivia* were better than several others tested. The potato extract purchased from several companies was comparable to extracts prepared from nonoptimal varieties. Optimal extract preparations may be stored frozen at −18°C without loosing their activity.

Chuang *et al.* (1978) reported greater success using a new potato extract medium 2, a combination of a salt and a potato medium. Its composition is as follows: 10% aqueous potato extract, 10^{-4} mg/liter FeEDTA, 1 mg/liter thiamine HCl, 1.5 mg/liter 2,4-D, 0.5 mg/liter kinetin, 1000 mg/liter KNO_3, 100 mg/liter $(NH_4)_2SO_4$, 100 mg/liter $Ca(NO_3)_2 \cdot 4H_2O$, 125 mg/ liter $MgSO_4 \cdot 7H_2O$, 200 mg/liter KH_2PO_4, 35 mg/liter KCl, 10% sucrose, and 0.6% agar. Despite success with these media, we think it is better to work with defined media; occasionaly lower success rates are balanced by greater reproducibility.

Activated charcoal may enhance the percentage of androgenetic anthers in some cereals (maize: Miao *et al.*, 1978; rye: Wenzel *et al.*, 1977). The reason is probably its absorbance of inhibitors, produced either by the anther or by interactions in or with the medium.

Anther culture of most cereals is performed at temperature of 20–26°C. Darkness is normally more effective than any illumination. After induction, macroscopic structures are transplanted to a regeneration medium similar in composition to the induction medium but generally with reduced sucrose and auxin concentrations. The cultures should be kept at a 14-hr light regime of 4000–6000 lx. After organogenesis or morphogenesis has started, a further transfer to media containing charcoal is recommended, as in dark medium roots develop more vigorously (Fig. 3b). After plantlet formation, they are transferred to the greenhouse and potting compost; during the first 2 weeks low temperatures (8–15°C) and high humidity are suggested (Fig. 3d).

VIII. PLOIDY LEVEL AND GENOME DOUBLING

For estimation of ploidy level, root tips are incubated for 5 hr in a solidified aqueous solution of bromonaphthaline and fixed in a 3:1 alcohol:acetic acid mixture. After hydrolysis in 1 *N* HCl at 60°C for 10 min, root tips are stained with Feulgen and treated for 2 hr with 2% pectinase. For a rapid check of ploidy level, it is often sufficient to count the number of nucleoli per nucleus after overnight incubation of leaves with orcein. Cells of haploid plants contain only one nucleolus, whereas diploids in up to 70% of nuclei contain two nucleoli (Reitberger, 1977). This procedure can be used before rooting and transfer to soil.

In anther culture of cereals, the majority of the plants developed are diploid (ca. 70% in barley and rye); the rest are haploid or tetraploid. Various factors affect this doubling of the ploidy level of the regenerants: pollen stage at the time of inoculation, hormones, or certain epigenetic factors. For example, Amssa *et al.* (1980) suggest that endoreduplication or endomitosis in pollen of *T. aestivum* is enhanced by a cold pretreatment.

Haploid plants are doubled by colchicine, as described by Jensen (1974). This occurs without problems in wheat, rice, and barley, whereas in rye the procedure must often be repeated. In maize, colchicine does not work satisfactorily (Genovesi and Collins, 1982).

IX. THE ALBINO PROBLEM

The biggest problem in cereal anther culture is the formation of large numbers of albino plantlets (up to 90%). Electron microscopic observations in albino rice plants revealed the presence of proplastids and the complete absence of ribosomes in the plastids. This indicates that the protein required for transformation to grana lamellae is not produced. The deficiency in plastid DNA of albinos is further underlined by the observed absence of fraction I protein (Sun *et al.*, 1979; Wang *et al.*, 1978). The development of albinos may also be due to mutations or expression of recessive genes (Wang *et al.*, 1973; Clapham, 1977; Wang *et al.*, 1978). As the rate of albino production in androgenetic haploids is higher than that in parthenogenetic ones (e.g., in rice the percentage of green plants in ovary culture was 89%, compared to 36% in anther culture; Yang and Zhou, 1982), it was suggested that insufficient numbers of proplastids are present in the microspore.

X. CONCLUSION

Microspore culture is a most promising approach for the induction of haploids at frequencies useful for applied breeding programs. Although a great deal of progress has been made, however, convincing proof, that is, a general, simple, reproducible method applicable to a wide range of cereals, is still lacking. On the other hand, single-cell regeneration from microspores works, whereas cereal protoplast regeneration is still a problem (Chapter 45, this volume). In all cases, it is important to monitor accurately the developmental stage of the microspore. The response sporophytic development might be increased by subjecting the anthers or flowers to shock treatments. Cytokinins are normally not essential in the media, whereas 2,4-D is the most effective auxin. High sucrose concentrations have a striking effect on regeneration frequency. Two further general points should be stressed. One is the importance of the genotype. We strongly recommend testing new material first for their tissue culture and regeneration capacity. This does not necessarily have to be done with microspores; rates of callusing of stem nodes, embryos, or other callusing tissues of those genotypes is sufficient. Further broadening of that character can subsequently be achieved by combination breeding and selection. The second and probably most critical point in anther culture is the preculture of the anther donor material. The best genotypes and optimal culture conditions will fail when the plants are badly cared for. McComb (1979) gave space in her report of a research trip to Chinese anther culture laboratories to a large picture of a gardener. He was proud to be 82 years old. He always transferred plants from agar medium to soil and had, according to his long experience, 75% success. The nature of such experience cannot be described in words, but it is a crucial one.

REFERENCES

Amssa, M., De Buyser, J., and Henry, Y. (1980). Origine des plantes diploides obtenues par culture *in vitro* d'anthères de Blè tendre (*Triticum aestivum* L.): Influence du prétraitement au froid et de la culture *in vitro* sur le doublement. *C. R. Hebd. Seances Acad. Sci.* **290,** 1095–1097.

Anonymous (1975). Primary study on induction of plants of *Zea mays. Acta Genet. Sin.* **2,** 138–142.

Anonymous (1976). A sharp increase of the frequency of pollen plant induction in wheat with potato medium. *Acta Genet. Sin.* **3,** 30–31.

Bennett, M. D., and Hughes, W. G. (1972). Additional mitosis in wheat pollen induced by ethrel. *Nature (London)* **240,** 566–568.

Brettel, R. J. S., Thomas, E., and Wernicke, W. (1981). Production of haploid maize plants by anther culture. *Maydica* **26,** 101–111.

Bullock, W. P., Baenziger, P. S., Schaeffer, G. W., and Bottino, P. J. (1982). Anther culture of wheat (*Triticum aestivum* L.) F_1's and their reciprocal crosses. *Theor. Appl. Genet.* **62,** 155–159.

Chase, S. S. (1974). Breeding diploid species. *In* "Haploids in Higher Plants" (K. J. Kasha, ed.), pp. 211–230. Univ. of Guelph, Guelph, Ontario, Canada.

Chen, C.-C. (1978). Effects of sucrose concentration on plant production in anther culture of rice. *Crop Sci.* **18,** 905–906.

Chen, C.-M., Chen, C.-C., and Lin, M.-H. (1982). Genetic analysis of anther-derived plants of rice. *J. Hered.* **73,** 49–52.

Chu, C.-C. (1978). The N6 medium and its applications to anther culture of cereal crops. *In* "Proceedings of Symposium on Plant Tissue Culture," pp. 43–45. Science Press, Peking.

Chuang, C.-C., Ouyang, T.-W., Chia, H., Chou, S. M., and Ching, C.-K. (1978). A set of potato media for wheat anther culture. *In* "Proceedings of Symposium on Plant Tissue Culture," pp. 51–56. Science Press, Peking.

Clapham, D. (1971). *In vitro* developement of callus from the pollen of *Lolium* and *Hordeum. Z. Pflanzenzuecht.* **65,** 285–292.

Clapham, D. (1973). Haploid *Hordeum* plants from anthers *in vitro*. *Z. Pflanzenzuecht.* **69,** 142–155.

Clapham, D. (1977). Haploid induction in cereals. *In* "Applied and Fundamental Aspects of Plant Cell Tissue and Organ Culture" (J. Reinert and Y. P. S. Bajaj, eds.), pp. 279–298. Springer-Verlag, Berlin and New York.

De Buyser, J., and Henry, Y. (1980). Comparaison de différents milieux utilisés en culture d'anthères *in vitro* chez le Blé tendre. *Can. J. Bot.* **58,** 997–1000.

Dunwell, J. M. (1981). Influence of genotype and environment on growth of barley embryos *in vitro*. *Ann. Bot. (London)* [N.S.] **48,** 535–542.

Foroughi-Wehr, B., and Mix, G. (1979). In vitro response of *Hordeum vulgare* L. anthers cultered from plants grown under different environments. *Environ. Exp. Bot.* **19,** 303–309.

Foroughi-Wehr, B., Mix, G., Gaul, H., and Wilson, H. M. (1976). Plant production from cultered anthers of *Hordeum vulgare* L. *Z. Pflanzenzuecht.* **77,** 198–204.

Foroughi-Wehr, B., Friedt, W., and Wenzel, G. (1982). On the genetic improvement of androgenetic haploid formation in Hordeum vulgare L. *Theor. Appl. Genet.* **62,** 233–239.

Gamborg, O. L. (1970). The effects of amino acids and ammonium on the growth of plant cells in suspension culture. *Plant Physiol.* **45,** 372–375.

Gaul, H., Mix, G., Foroughi-Wehr, B., and Okamoto, M. (1976). Pollen grain developement of *Hordeum vulgare*. *Z. Pflanzenzuecht.* **76,** 77–85.

Genovesi, A. D., and Collins, G. O. (1982). *In vitro* production of haploid plants of corn via anther culture. *Crop Sci.* **22,** 1137–1144.

Guha, S., Iyer, R. B., Gupta, N., and Swaminathan, M. S. (1970). Totipotency of gametic cells and the production of haploids in rice. *Curr. Sci.* **39,** 174–176.

Hagberg, A., and Hagberg, G. (1980). High frequency of spontaneous haploids in the progeny of an inducted mutation in barley. *Hereditas* **93,** 341–343.

Henry, Y., and De Buyser, J. (1981). Float culture of wheat. *Theor. Appl. Genet.* **60,** 77–79.

Henry, Y., De Buyser, J., Amssa, M., and Taleb, G. (1982). In vitro androgenesis in winter wheat breeding. *Proc. Int. Congr. Plant Tissue Cell Cult. 5th 1982* p. 569.

Hu, H., and Shao, Q. (1981). Advances in plant cell and tissue culture in China. *Adv. Agron.* **34,** 1–11.

Jensen, C. J. (1974). Chromosome doubling techniques in haploids. *In* "Haploids in Higher Plants: Advances and Potential" (K. J. Kasha, ed.), pp. 153–190. Univ. of Guelph, Guelph, Ontario, Canada.

Kao, K. N. (1981). Plant formation from barley anther culture with Ficoll media. *Z. Pflanzenphysiol.* **103,** 437–443.

Kasha, K. S., and Reinbergs, E. (1980). Achievements with haploids in barley research and breeding. *In* "The Plant Genome" (D. R. Davies and D. A. Hopwood, eds.), pp. 215–230. John Innes Charity, Norwich, England.

Kermicle, J. L. (1969). Androgenesis conditioned by a mutation in maize. *Science* **166,** 1322–1424.

Ku, M.-K., Cheng, W.-C., Kuo, L.-C., Kuan, Y.-L., An, H. P., and Huang, C.-H. (1978). Induction factors and morpho-cytological characteristics of pollen-derived plants in maize (*Zea mays*). *In* "Proceedings of Symposium on Plant Tissue Culture," pp. 35–42. Science Press, Peking.

Linsmaier, E. M., and Skoog, F. (1965). Organic growth factor requirements of tobacco tissue cultures. *Physiol. Plant.* **18,** 100–127.

McComb, J. A. (1979). Use of tissue culture, particularly anther culture for plant breeding and propagation in China. *J. Aust. Inst. Agric. Sci.* pp. 187–190.

Miao, S.-H., Kuo, C.-S., Kwei, Y.-J., Sun, A.-J., Ku, S.-Y., Lu, W.-L., Wang, Y.-Y., Chen, M.-L., Wu, M.-K., and Hong, L. (1978). Induction of pollen plants of maize and observations on their progeny. *In* "Proceedings of Symposium on Plant Tissue Culture," pp. 23–33. Science Press, Peking.

Miller, C. O. (1963). Kinetin and kinetin like compounds. *In* "Moderne Methoden der Pflanzenanalyse" (H. F. Linskens and M. V. Tracey, eds.), Vol. 6, pp. 194–202. Springer-Verlag, Berlin and New York.

Murashige, T., and Skoog, F. (1962). A revised medium for rapid growth and bio assays with tobacco tissue cultures. *Physiol. Plant.* **15,** 473–497.

Niizeki, H., and Oono, K. (1968). Induction of haploid rice plant from anther culture. *Proc. Jpn. Acad.* **44,** 554–557.

Niizeki, H., and Oono, K. (1971). Rice plants obtained by anther culture. *Colloq. Int. C.N.R.S.* **193,** 251–257.

Nitsch, C. (1974). La culture de pollen isolé sur milieu synthétique. *C. R. Hebd. Seances Acad. Sci., Ser. D* **278,** 1031–1034.

Nitsch, C., Andersen, J., Godard, A., Neuffer, M. F., and Sheridan, H. F. (1982). Production of haploid plants of *Zea mays* and *Pennisetum* through androgenesis. *In* "Variability in Plants Regenerated from Tissue Culture" (E. D. Earle and Y. Demarly, eds.), pp. 69–91. Praeger Press, New York.

Nitzsche, W., and Wenzel, G. (1977). "Haploids in Plant Breeding," Fortschr. Pflanzenzuecht., Vol. 8. Parey, Berlin.

Ouyang, T.-W., Hu, H., Chuang, C.-C., and Tseng, C.-C. (1973). Induction of pollen plants from anthers of *Triticum aestivum* cultured *in vitro. Sci. Sin. (Engl. Transl.)* **16,** 79–95.

Picard, E., and De Buyser, J. (1973). De plantules haploids de *Triticum aestivum* L. a partir de culture d'anthères *in vitro. C. R. Hebd. Seances Acad. Sci., Ser. D* **277,** 1463–1466.

Reitberger, A. (1977). Methodische Untersuchungen zur Ploidiebestimmung an Ruhekernen. II Gramineen. *Z. Pflanzenzuecht.* **79,** 14–25.

Schaeffer, G. W., Baenziger, P. S., and Worley, S. (1979). Haploid plant development from anthers and in vitro embryo culture of wheat. *Crop Sci.* **19,** 697–702.

Sozinova, A., Luksansuk, S., and Ignatova, S. (1981). Anther cultivation and induction of haploid plants in Triticale. *Z. Pflanzenzuecht.* **86,** 272–285.

Stolarz, A. (1974). The induction of androgenesis in pollen grains of *Secale cereale* L. Strzekecińskie Jare in *in vitro* conditions. *Hodowla Rosl., Aklim. Nasienn.* **18,** 217–220.

Sun, C. S., Wu, S. C., Wang, C. C., and Chu, C. C. (1979). The deficiency of soluble proteins and plastid ribosomal RNA in the albino pollen plantlets of rice. *Theor. Appl. Genet.* **55,** 193–197.

Sunderland, N. (1974). Anther culture as a means of haploid induction. *In* "Haploids in Higher Plants: Applications and Potential" (K. J. Kasha, ed.), pp. 91–122. Univ. of Guelph, Guelph, Ontario, Canada.

Sunderland, N. (1980). Anther and pollen culture 1974–1979. *In* "The Plant Genome" (D. R. Davies and D. A. Hopwood, eds.), pp. 171–183. John Innes Charity, Norwich, England.

Sunderland, N., and Dunwell, J. M. (1977). Anther and pollen culture. *In* "Plant Tissue and Cell Culture" (H. E. Street, ed.), 2nd ed., pp. 223–264. Univ. of California Press, Berkeley.

Sunderland, N., and Evans, L. J. (1980). Multicellular pollen formation in cultered barley anthers. II. The A-pathway, B-pathway and C-pathway. *J. Exp. Bot.* **31,** 501–514.

Sunderland, N., Collins, G. B., and Dunwell, J. M. (1974). The role of nuclear fusion in pollen embryogenesis of *Datura innoxia* Mill. *Planta* **117,** 227–241.

Sunderland, N., Roberts, M., Evans, L. J., and Wildon, D. C. (1979). Multicellular pollen formation in cultered barley anthers. I. Independent division of the generative and vegetative cells. *J. Exp. Bot.* **30,** 1133–1144.

Sunderland, N., Xu, Z. H., and Huan, B. (1981). Recent advances in barley anther culture. *Barley Genet. 4, Proc. Int. Symp., 4th, 1981* pp. 599–703.

Thomas, E., and Wenzel, G. (1975). Embryogenesis from microspores of rye. *Naturwissenschaften* **62,** 40–41.

Ting, Y. C., Yu, M., and Zheng, W.-Z. (1981). Improved anther culture of maize (*Zea mays*). *Plant Sci. Lett.* **23,** 139–145.

Vasil, I. K. (1980). Androgenetic haploids. *Int. Rev. Cytol. Suppl.* **11B,** 195–217.

Wang, C.-C., Chu, C.-C., Sun, C.-S., Wu, S.-H., Yin, K.-C., and Hsü, C. (1973). The androgenesis in wheat (*Triticum aestivum*) anthers cultered in vitro. *Sci. Sin. (Engl. Transl.)* **16,** 218–222.

Wang, C.-C., Sun, C.-S., and Chu, Z.-C. (1974). On the conditions for the induction of rice pollen plantlets and certain factors affecting the frequency of induction. *Acta Bot. Sin.* **16,** 43–54.

Wang, C. C., Sun, C.-S., Chu, C.-C., and Wu, S.-C. (1978). Studies on the albino pollen plantlets of rice. *In* "Proceedings of Symposium on Plant Tissue Culture," pp. 149–160. Science Press, Peking.

Wenzel, G., and Thomas, E. (1974). Observations on the growth in culture of anthers of *Secale cereale*. *Z. Pflanzenzuecht.* **72,** 89–94.

Wenzel, G., Hoffmann, F., and Thomas, E. (1977). Increased induction and chromosome doubling of androgenetic haploid rye. *Theor. Appl. Genet.* **51,** 81–86.

Wilson, H. M. (1977) Culture of whole barley spikes stimulates high frequencies pollen calluses in individual anthers. *Plant Sci. Lett.* **9,** 233–238.

Xu, Z. H., and Sunderland, N. (1981). Glutamine, inositol and conditioning factors in the production of barley pollen callus *in vitro*. *Plant Sci. Lett.* **23,** 161–168.

Yang, H. Y., and Zhou C. (1982). In vitro induction of haploid plants from unpollinated ovaries and ovules. *Theor. Appl. Genet.* **63,** 97–104.

Yin, K.-C., Hsu, C., Chu, C.-Y., Pi, F.-Y., Wang, S.-T., Lin, T.-Y., Chu, C.-C., Wang, C.-C., and Sun, C.-S. (1976). A study of the new cultivar of rice raised by haploid breeding method. *Sci. Sin. (Engl. Transl.)* **19,** 227–242.

CHAPTER **38**

Isolation and Culture of Protoplasts: Tobacco

Itaru Takebe
Toshiyuki Nagata

Department of Biology
Nagoya University
Chikusa-ku, Nagoya, Japan

I. INTRODUCTION

Soon after the new era of plant protoplast research was opened by Cocking (1960), we started a project to use isolated protoplasts from tobacco leaves as a host cell system for plant viruses (Chapter 56, this volume). By testing a number of commercial preparations of cell wall lytic enzymes and by examining the conditions suitable for digestion of leaf materials without impairing cell viability, we developed a method for isolating tobacco mesophyll protoplasts in quantities sufficient for biochemical experimentation (Takebe *et al.*, 1968). This was the first report of the isolation of protoplasts

CELL CULTURE AND SOMATIC CELL
GENETICS OF PLANTS, VOL. 1

ISBN 0-12-715001-3

from tobacco and of the use of mesophyll tissues as the source of protoplasts.

Our method consisted of enzyme treatments in two steps. Tobacco leaves (lower epidermis stripped off) were first macerated with pectinase to release consecutively spongy and palisade parenchyma cells. The cells thus isolated were then treated with cellulase to convert them into protoplasts. The isolated tobacco mesophyll protoplasts supported vigorous virus multiplication (Takebe and Otsuki, 1969), thus demonstrating their high metabolic activity.

Although no measure was taken in the original method to exclude microorganisms, an aseptic version of the method was developed later to allow a long-term culture of the isolated protoplasts (Nagata and Takebe, 1970). We then studied the developmental potential of the isolated protoplasts, and defined the conditions under which tobacco mesophyll protoplasts form cell walls and undergo divisions (Nagata and Takebe, 1970), give rise to callus-like colonies through sustained division in an agar medium (Nagata and Takebe, 1971), and eventually regenerate whole plants via organogenesis (Takebe *et al.*, 1971; Nagata and Takebe, 1971). These represented the first instances of the induction of cell division and subsequent plant development from isolated plant protoplasts.

This chapter describes the latest version of our procedures for the aseptic isolation and culture of protoplasts from the palisade tissue of tobacco (*Nicotiana tabacum* L.). Isolation and culture of protoplasts from other *Nicotiana* species will be only briefly discussed.

II. ISOLATION AND CULTURE OF TOBACCO MESOPHYLL PROTOPLASTS

A. Materials

1. Plants

Cultivars *Xanthi* nc and *Xanthi* of *N. tabacum* L. are suitable, although other cultivars may also be used. Seedlings about 5 cm in size are planted in 15-cm pots containing a 2:1 mixture of peat moss and vermiculite in the lower and upper halves, with a layer of about 15 g of chemical fertilizer (consisting of a 2:3:1 mixture of ammonium nitrate, calcium superphosphate, and potassium chloride) in between. After 2 weeks of growth, about 3 g of chemical fertilizer is added to the surface of the soil. Plants are grown in a greenhouse with a daytime (16-hr) temperature of 28°C and a night-time (8-hr) temperature of 20°C. During the short-day seasons, supplemental illumination is provided from white fluorescent lamps to maintain the long-day conditions. Just fully expanding leaves of 50- to 60-day-old plants are used as the source of protoplasts.

2. Enzyme Solutions

Enzyme solution for maceration of leaves contains 0.1% Pectolyase Y-23 (Nagata and Ishii, 1979; Seishin Pharmaceutical Co., Koamicho, Nihonbashi, Chuo-ku, Tokyo, Japan) and 0.5% potassium dextran sulfate (sulfur content 17%, degree of polymerization 3.5; Meito Sangyo Co., Muromachi, Nihonbashi, Chuo-ku, Tokyo, Japan) dissolved in 0.6 *M* mannitol solution, with the pH adjusted to 5.8 with 0.1 *N* HCl. Cellulase solution contains 1% Cellulase Onozuka RS (Yakult Honsha Co., Nishinomiya Office, Shingikan-cho, Nishinomiya, Japan) dissolved in 0.6 *M* mannitol solution, pH 5.2. The enzyme solutions are sterilized by filtration through a membrane filter with a pore size of 0.22 μm. The solutions may be stored frozen.

B. Isolation Procedure

1. Surface Sterilization of Leaves

Excised leaves are rinsed in a solution of household detergent for vegetable washing and are surface sterilized by being dipped in a 2% solution of sodium hypochlorite for 20–30 min, followed by rinsing in sterilized distilled water. All subsequent operations are performed under aseptic conditions.

2. Removal of Lower Epidermis

The lower epidermis of sterilized leaves is peeled off using sterilized forceps and plastic gloves. Peeling of the epidermis is much easier with leaves which have slightly reduced turgor. Peeled leaves are cut into pieces of about 4 cm^2 using a razor blade.

3. Maceration of Leaf Pieces

Leaf pieces of about 2 g fresh weight are soaked in 20 ml maceration solution in a screw-capped 100-ml Erlenmeyer flask. The flask is placed in a vacuum desiccator and evacuated for about 2 min using a rotary pump. Air is then introduced into the desiccator through a glass fiber filter. The leaf pieces turn dark green as they are infiltrated with enzyme solution.

The flask containing the leaf pieces is shaken at a frequency of 120 excursions/min and a stroke of 4 cm in a water bath of 25°C. Incubation for about 2 min releases most of the spongy parenchyma cells into solution, as well as damaged cells from the cut surfaces. The maceration solution containing these cells is removed using a Pasteur-type pipette and is replaced by 20 ml fresh maceration solution. Further incubation for 15–20 min with shaking results in complete maceration of the mesophyll tissues. The maceration solution containing isolated palisade cells is filtered through a sheet of nylon bolting cloth (mesh opening 200 μm) to remove the undigested upper epidermis and veins, and the filtrate is centrifuged at 100 *g* for 2 min to collect palisade cells.

4. Cellulase Treatment

The isolated palisade cells are suspended in 10 ml cellulase solution, transferred into a screw-capped 50-ml Erlenmeyer flask, and incubated in a water bath at 30°C with occasional swirling. All the living cells turn spherical within 20 min of incubation. The complete removal of cell walls may be ascertained by staining a drop of sample with Calcofluor White ST (Nagata and Takebe, 1970), now available from Calbiochem-Behring (La Jolla, California) as Bioglo. The protoplasts are collected by centrifugation at 100 *g* for 1–2 min, and washed twice by suspension in 10 ml mannitol solution and centrifugation. The speed and duration of centrifugation are adjusted so that small numbers of protoplasts remain in the supernatant solution with damaged cells and cell debris. The latter are partly removed in this way because they sediment more slowly than intact protoplasts.

C. Procedure for Plating

The isolated protoplasts are suspended in the medium (Table I) to a cell population density of about 1×10^4/ml, as determined using a hemocytometer, and the suspension is distributed into 50-ml Erlenmeyer flasks in 10-ml portions. Meanwhile, the same medium containing 1.2% Bacto agar (Difco) is melted, distributed into test tubes in 10-ml portions, and kept in a water bath at 43°C. The melted agar medium in the test tube is poured into the Erlenmeyer flask containing the protoplast suspension, and the contents are mixed quickly by being sucked into and out of a 10-ml wide-mouth pipette. The mixture is then quickly distributed in 5-ml portions into 6-cm Falcon plastic dishes. After the medium solidifies, the dish-

TABLE I

Medium for Culturing Tobacco Mesophyll Protoplasts[a]

Mineral salts (mg/liter)		Organic constituents	
NH_4NO_3	825	Sucrose	10 g/liter
KNO_3	950	meso-Inositol	100 mg/liter
$CaCl_2 \cdot 2H_2O$	220	Thiamine HCl	1 mg/liter
$MgSO_4 \cdot 7H_2O$	1233	1-NAA	1 mg/liter
KH_2PO_4	680	6-BAP	1 mg/liter
Na_2-EDTA	37.3	D-Mannitol	109.3 g/liter
$FeSO_4 \cdot 7H_2O$	27.8		
$MnSO_4 \cdot 4H_2O$	22.3		
$ZnSO_4 \cdot 4H_2O$	8.6		
H_3BO_3	6.2		
KI	0.83		
$NaMoO_4 \cdot 2H_2O$	0.25		
$CuSO_4 \cdot 5H_2O$	0.025		
$CoSO_4 \cdot 7H_2O$	0.030		

[a] The pH is adjusted to 5.8 before autoclaving. The medium is the same as that of Nagata and Takebe (1971), except that the concentration of NAA is reduced to 1 mg/liter and that of mannitol to 0.6 *M*.

es are sealed with a strip of Parafilm, and are placed in a culture chamber at 28°C. Continuous illumination is provided at 2000–3000 lx from white fluorescent lamps.

D. Comments

The yield of viable and active protoplasts is influenced considerably by the quality of source leaves. "Good" leaves are fresh light green in color and have relatively thin lamina. Overfertilized plants give dark green leaves with a thick lamina, and underfertilized plants have yellowish senescing leaves. Such leaves yield unstable protoplasts which degenerate during enzyme treatment or subsequent culture.

Mesophyll cells may vary in the osmotic property and the digestability of their walls according to the cultivar, the method of plant growth, the season, and other factors. Therefore, some of the parameters in the procedure described above may have to be varied to meet the requirement of particular materials. For example, a higher or lower mannitol concentration may be appropriate in some cases. The proper mannitol concentration is one that renders the solution slightly hypertonic; the mesophyll cells released into the maceration solution should be slightly plasmolyzed.

Pectolyase Y-23 in the maceration solution and Cellulase Onozuka RS in the cellulase solution may be substituted by Macerozyme R-10 (0.5%; Yakult Honsha Co., Nishinomiya Office) and by Cellulase Onozuka R-10 (2%; Yakult Honsha Co., Nishinomiya Office), respectively. Incubation for a somewhat longer time is necessary with the latter enzyme preparations. Dextran sulfate in the maceration solution helps to isolate the majority of mesophyll cells in an intact state (Takebe *et al.*, 1968).

Plating efficiency (percentage of protoplasts forming colonies) depends strongly on the population density of protoplasts (Nagata and Takebe, 1971). Few if any colonies are formed when protoplasts are plated at densities below 1×10^3/ml. Colony formation by plated protoplasts is also influenced by the light intensity during culture. A lower light intensity than specified (800 lx) gives poor plating efficiency, whereas illumination at a higher light intensity than specified (5000 lx) does not improve colony formation (Nagata and Takebe, 1971). Enzmann-Becker (1973) recommends culture of tobacco mesophyll protoplasts under dim light (400 lx) for the first 2 days and thereafter at 3000 lx.

Tobacco mesophyll protoplasts can also be cultured in the liquid medium specified in Table I. However, the single-cell origin of the resulting colonies is not guaranteed in this type of culture, since protoplasts settle down to the bottom and come into contact with each other. On the other hand, the culture in a liquid medium has the advantage that the composition of the medium can be changed at will by adding a solution or by substituting a new medium. For example, the growth of colonies derived from protoplasts can be accelerated by gradually reducing the osmoticum concentration, since the hypertonic condition required to start the culture of protoplasts is unfavorable for the growth of normal cells.

III. RESULTS AND IMPLICATIONS

The procedure described above yields about 10^7 protoplasts of palisade cells from 1 g fresh weight tobacco leaves. The protoplasts embedded in the agar medium form walls and initiate division within 3 days of culture. The cells derived from protoplasts subsequently undergo sustained division and give rise to visible colonies within 3 weeks. The cells now assume the appearance of callus cells, as chloroplasts degenerate during the course of repeated cell divisions. The colonies further grow to a size of 0.5–1.0 mm after 6 weeks of culture. Usually about 60% of plated protoplasts develop to this stage.

The colonies can be transferred to the agar medium of Table I (mannitol omitted) to establish clonal cell lines of single-cell origin. On the other hand, shoot formation can be induced in the colonies by transferring them to suitable differentiation medium, for example, the medium of Table I (mannitol omitted), in which naphthaleneacetic acid (NAA) and benzylaminopurine (BAP) are substituted for by 3-indoleacetic acid (IAA) (4 mg/liter) and kinetin (2.56 mg/liter), respectively. After initial growth as callus masses, they differentiate many shoots, which can be further transplanted to White's basal medium to induce root formation (Nagata and Takebe, 1971). The plantlets thus formed can be planted in soil when they are about 7 cm tall. The plants grow to maturity, flower, and set seeds.

The development of the procedures for isolation (Takebe *et al.*, 1968) and culture (Nagata and Takebe, 1970, 1971) of tobacco mesophyll protoplasts had important implications for developmental biology and somatic cell genetics of plants. First, large-scale isolation of single somatic plant cells became possible using the mesophyll tissues as a source material. Second, high developmental competence was demonstrated and direct evidence for the totipotency of mesophyll cells was provided. Third, high-efficiency plating of protoplasts opened a way to the application of the techniques of microbial genetics to somatic plant cells. Finally, plant regeneration from protoplasts indicated the possibility of linking genetic manipulation of protoplasts to plant improvement.

IV. MODIFICATIONS AND VARIATIONS

A. Materials

Mesophyll protoplasts have also been isolated from haploid tobacco plants and were shown to develop into flowering plants (Ohyama and Nitsch, 1972). Binding (1975) used tobacco shoots cultured *in vitro* as the source of mesophyll protoplasts. Leaves of the shoot cultures were cut into slices and digested with a mixture of pectinase and cellulase. Shoot cultures have the advantages of not requiring a greenhouse and of supplying uniform materials. It is also reported that protoplasts from shoot cultures reproducibly show high plating efficiency.

Tobacco protoplasts were also obtained from suspension cultures by treatment with a mixture of pectinase and cellulase (Uchimiya and Murashige, 1974). However, only 30% of the cells were converted into protoplasts, indicating that the walls of suspension-cultured cells are tougher than those of mesophyll cells. The protoplasts from tobacco suspension cultures formed walls and resumed cell division when returned to cell culture medium (Uchimiya and Murashige, 1976). Recently, a procedure was developed using Pectolyase Y-23 and Cellulase Onozuka RS to convert essentially all the cells in a tobacco suspension culture into protoplasts (Nagata *et al.*, 1981).

B. Isolation Procedure

Mesophyll protoplasts can be isolated directly from peeled tobacco leaves by treatment with a mixture of pectinase and cellulase (Power and Cocking, 1970). The "one-step" procedure is widely used because it is simple and has the merit of not requiring the use of dextran sulfate. On the other hand, the one-step procedure yields heterogeneous populations consisting of spongy and palisade protoplasts. In addition, such preparations usually contain significant numbers of subprotoplasts, as well as multinucleate protoplasts formed as a result of "spontaneous fusion" (Withers and Cocking, 1972). The incidence of multinucleate protoplasts may be reduced by preplasmolyzing peeled leaves before enzyme treatment (Frearson *et al.*, 1973).

C. Method of Culture

Since colony formation by tobacco mesophyll protoplasts requires a population density above 10^3/ml, very large numbers of colonies are formed in each plate, sometimes causing difficulty in separating individual colonies. Raveh *et al.* (1973) used a feeder technique to plate tobacco protoplasts at low population densities (Chapter 24, this volume). Protoplasts from diploid and haploid tobacco were plated on a feeder layer containing tobacco mesophyll protoplasts irradiated with x-rays. Colony formation occurred in 10–40% of the protoplasts plated at a density of 5–50/ml.

Attempts to define a synthetic medium which could support colony formation by tobacco mesophyll protoplasts at low population densities have been unsuccessful. Caboche (1980) found that NAA at 3 mg/liter, which is necessary for the initiation of division in tobacco mesophyll protoplasts, is toxic to protoplasts at low population densities. However, once the first division was completed at a high population density, the cells derived from protoplasts could be diluted and cultured at low densities, provided that the NAA concentration is lowered to 0.1 mg/liter. A plating efficiency of 30–40% was attained this way at a population density as low as 1–10/ml, with the cells derived from haploid tobacco protoplasts.

Gleba (1978) applied the microdroplet culture technique to tobacco mesophyll protoplasts (Chapter 46, this volume). Diploid and haploid protoplasts were suspended in a liquid medium at a population density of 24 $\times 10^3$/ml so that each microdroplet of 0.25–0.5 μl contained one protoplast. The microdroplets were cultured in the wells of Cuprak dishes (Costar). Using a medium supplemented with coconut milk, casein hydrolysate, and glutamine, 5–40% of the individually cultured haploid protoplasts developed into colonies, from which whole plants could be regenerated.

V. OTHER *NICOTIANA* SPECIES

Table II lists other *Nicotiana* species from which mesophyll protoplasts have been isolated and cultured successfully. Whole plants have been recovered from protoplasts in all the species except *N. glutinosa*.

The one-step procedure was used to isolate protoplasts from the species in Table II. Conditions of enzyme treatment of leaves and culture of isolated protoplasts differ in detail from one species to another. The very high plating efficiencies reported for haploid protoplasts of *N. sylvestris* and *N. plumbaginifolia* are apparently due to the use of conditioned medium and the addition of organic acids, respectively.

TABLE II

***Nicotiana* Species in Which Colony Formation by Mesophyll Protoplasts Have Been Reported**

Section and species	Ploidy	Source material[a]	Plating efficiency	Medium[b]	Reference
Tomentosae					
N. otophora	$2n = 24$	Pl	39	L/S	Banks and Evans (1976)
N. glutinosa	$2n = 24$	Pl	20–50	L	Bourgin *et al.* (1979)
Paniculatae					
N. paniculata	$2n = 24$	Pl	5	L	Bourgin *et al.* (1979)
N. glauca	$2n = 24$	Pl	5–40	L	Bourgin *et al.* (1979)
N. knightiana	$2n = 24$	Pl	0.1–1.0	L	Maliga *et al.* (1978)
N. rustica	$2n = 48$	Sc	6–40	L/S	Gill *et al.* (1979)
Alatae					

TABLE II (*Continued*)

Section and species	Ploidy	Source material[a]	Plating efficiency	Medium[b]	Reference
N. alata	$2n = 18$	Pl	1–5	L	Bourgin *et al.* (1979)
	$n = 9$	Sc	60	L/S	Bourgin and Missonier (1978)
N. langsdorffii	$2n = 18$	Pl	5–40	L	Bourgin *et al.* (1979)
N. longiflora	$2n = 20$	Pl	5–10	L	Bourgin *et al.* (1979)
N. plumbaginifolia	$2n = 20$	Pl	20–60	L	Bourgin *et al.* (1979)
	$n = 10$	Pl	22–63	L	Negrutiu (1981)
	$n = 10$	Sc	26–100	L	Negrutiu (1981)
N. sylvestris	$2n = 24$	Pl	60–90	L	Nagy and Maliga (1976)
	$2n = 24$	Sc	90–100	L	Negrutiu and Mousseau (1980)
	$n = 12$	Sc	100	L/S	Durand (1979)
Repandae					
N. repanda	$2n = 48$	Pl	80–90	L	Evans (1979)
N. nesophila	$2n = 48$	Pl	—	L	Evans (1979)
N. stocktonii	$2n = 48$	Pl	—	L	Evans (1979)
Suaveolentes					
N. suaveolens	$2n = 32$	Pl	10–40	L	Bourgin *et al.* (1979)
N. debneyi	$2n = 48$	Pl	50	L/S	Scowcroft and Larkin (1980)
Accuminata					
N. accuminata	$2n = 24$	Pl	1	L	Bourgin *et al.* (1979)

[a] Pl, plants; Sc, shoot culture.
[b] L, liquid medium; S, solid medium.

REFERENCES

Banks, M. S., and Evans, P. K. (1976). A comparison of the isolation and culture of mesophyll protoplasts from several *Nicotiana* species and their hybrids. *Plant Sci. Lett.* **7,** 409–416.

Binding, H. (1975). Reproducively high plating efficiencies of isolated mesophyll protoplasts from shoot cultures of tobacco. *Physiol. Plant.* **35,** 225–227.

Bourgin, J. P., and Missonier, C. (1978). Culture de protoplastes de mesophylle de *Nicotiana alata* Link et Otto haploide. *Z. Pflanzenphysiol.* **87,** 55–64.

Bourgin, J. P., Chupeau, E., and Missonier, C. (1979). Plant regeneration from mesophyll protoplasts of several *Nicotiana* species. *Physiol. Plant.* **45,** 288–292.

Caboche, M. (1980). Nutritional requirements of protoplast-derived haploid tobacco cells grown at low cell densities in liquid medium. *Planta* **149,** 7–18.

Cocking, E. C. (1960). A method for the isolation of plant protoplasts and vacuoles. *Nature (London)* **187,** 927–929.

Durand, J. (1979). High and reproducible plating efficiencies of protoplasts isolated from *in vitro* grown haploid *Nicotiana sylvestris* Spegaz. et Comes. *Z. Pflanzenphysiol.* **93,** 283–295.

Enzmann-Becker, G. (1973). Plating efficiency of protoplasts of tobacco in different light conditions. *Z. Naturforsch., C: Biochem., Biophys., Biol., Virol.* **28C,** 470–471.

Evans, D. A. (1979). Chromosome stability of plants regenerated from mesophyll protoplasts of *Nicotiana* species. *Z. Pflanzenphysiol.* **95,** 459–463.

Frearson, E. M., Power, J. B., and Cocking, E. C. (1973). The isolation, culture and regeneration of *Petunia* leaf protoplasts. *Dev. Biol.* **33,** 130–137.

Gill, R., Rashid, A., and Maheshwari, S. C. (1979). Isolation of mesophyll protoplasts of *Nicotiana rustica* and their regeneration into plants flowering *in vitro. Physiol. Plant.* **47,** 7–10.

Gleba, Y. Y. (1978). Microdroplet culture: Tobacco plants from single mesophyll protoplasts. *Naturwissenschaften* **65,** 158–159.

Maliga, P., Kiss, Z. R., Nagy, A. H., and Lazar, G. (1978). Genetic instability in somatic hybrids of *Nicotiana tabacum* and *Nicotiana knightiana. Mol. Gen. Genet.* **163,** 145–151.

Nagata, T., and Ishii, S. (1979). A rapid method for isolation of mesophyll protoplasts. *Can. J. Bot.* **57,** 1820–1823.

Nagata, T., and Takebe, I. (1970). Cell wall regeneration and cell division in isolated tobacco mesophyll protoplasts. *Planta* **92,** 301–308.

Nagata, T., and Takebe, I. (1971). Plating of isolated tobacco mesophyll protoplasts on agar medium. *Planta* **99,** 12–20.

Nagata, T., Okada, K., Takebe, I., and Matsui, C. (1981). Delivery of tobacco mosaic virus RNA into plant protoplasts mediated by reverse-phase evaporation vesicles (liposomes). *Mol. Gen. Genet.* **184,** 161–165.

Nagy, J. I., and Maliga, P. (1976). Callus induction and plant regeneration from mesophyll protoplasts of *Nicotiana sylvestris. Z. Pflanzenphysiol.* **78,** 453–455.

Negrutiu, I. (1981). Improved conditions for large-scale culture, mutagenesis, and selection of haploid protoplasts of *Nicotiana plumbaginifolia. Z. Pflanzenphysiol.* **104,** 431–442.

Negrutiu, I., and Mousseau, J. (1980). Protoplast culture from *in vitro* grown plants of *Nicotiana sylvestris* Spegg. and Comes. *Z. Pflanzenphysiol.* **100,** 373–376.

Ohyama, K., and Nitsch, J. P. (1972). Flowering haploid plants obtained from protoplasts of tobacco leaves. *Plant Cell Physiol.* **13,** 229–236.

Power, J. B., and Cocking, E. C. (1970). Isolation of leaf protoplasts: Macromolecule uptake and growth substance response. *J. Exp. Bot.* **21,** 64–70.

Raveh, D., Huberman, E., and Galun, E. (1973). In vitro culture of tobacco protoplasts: Use of feeder techniques to support division of cells plated at low densities. *In Vitro* **9,** 216–222.

Scowcroft, W. R., and Larkin, P. J. (1980). Isolation, culture and plant regeneration from protoplasts of *Nicotiana debneyi. Aust. J. Plant Physiol.* **7,** 635–644.

Takebe, I., and Otsuki, Y. (1969). Infection of tobacco mesophyll protoplasts by tobacco mosaic virus. *Proc. Natl. Acad. Sci. U.S.A.* **64,** 843–848.

Takebe, I., Otsuki, Y., and Aoki, S. (1968). Isolation of tobacco mesophyll cells in intact and active state. *Plant Cell Physiol.* **9,** 115–124.

Takebe, I., Labib, G., and Melchers, G. (1971). Regeneration of whole plants from isolated mesophyll protoplasts of tobacco. *Naturwissenschaften* **58,** 318–320.

Uchimiya, H., and Murashige, T. (1974). Evaluation of parameters in the isolation of viable protoplasts from cultured tobacco cells. *Plant Physiol.* **54,** 936–944.
Uchimiya, H., and Murashige, T. (1976). Influence of the nutrient medium on the recovery of dividing cells from tobacco protoplasts. *Plant Physiol.* **57,** 424–429.
Withers, L. A., and Cocking, E. C. (1972). Fine-structural studies on spontaneous and induced fusion of higher plant protoplasts. *J. Cell Sci.* **11,** 59–75.

CHAPTER **39**

Isolation and Culture of Protoplasts: *Petunia*

Horst Binding
Gabriela Krumbiegel-Schroeren

Botanical Institute and Botanical Garden
Biology Center
Christian Albrechts University
Kiel, Federal Republic of Germany

I. INTRODUCTION

The genus *Petunia,* consisting of about 14 species, belongs to the family Solanaceae. *Petunia hybrida,* in particular, is one of the best-investigated plant species. This is the consequence of a number of advantageous properties: it is a beloved ornamental plant; it has high breeding qualities because of its high genotypic variability; it is easily grown in open air, greenhouses, and tissue culture; pollination is well controlled experimentally; sexual reproduction is high; interesting genetic properties are established, for example, multiple alleles of sexual self-incompatibility; and chromosomes are low in number ($n = 7$) and large enough for detailed cytogenetic investigations. *Petunia* is favored as a model plant genus for molecular

CELL CULTURE AND SOMATIC CELL
GENETICS OF PLANTS, VOL. 1

ISBN 0-12-715001-3

biology by the Plant Molecular Biology Association. For reviews on *Petunia,* see the monograph edited by Sink (1984).

Petunia hybrida was the third species of higher plants in which protoplast regeneration was established (Frearson *et al.*, 1973; Donn *et al.*, 1973; Durand *et al.*, 1973; Vasil and Vasil, 1974), and was the first in which monohaploid plants were grown from isolated protoplasts (Binding, 1974a,c). Furthermore, protoplast regeneration was investigated in great detail, revealing results and experiences which were useful for achieving regeneration in other genera of the Solanaceae and other taxa of the class Magnoliatae. Hence *Petunia* proved to be an excellent model plant—next to tobacco—in the field of protoplast regeneration.

The first reports on the isolation and culture of *Petunia* protoplasts appeared in 1972 (Hess and Potrykus, 1972; Potrykus and Durand, 1972). Detailed investigations on plant regeneration have been published by Frearson *et al.* (1973), Donn *et al.* (1973), Durand *et al.* (1973), Binding (1974c), and Vasil and Vasil (1974). The following presentation of the methodology is mainly based on the last four papers. Some more recent findings, especially in the course of investigations summarized by Binding *et al.* (1981), are included. References will also be made to a number of other protocols which have been used for *Petunia* protoplast isolation and regeneration and which gave good protoplast yields and high plating efficiencies. Differences between *Petunia* species in their ability to regenerate have been observed. However, the demands and developmental patterns appeared to be basically so similar that the results referred mainly to the genus as a whole.

The procedures used, and the behavior of *Petunia* protoplasts in culture, are very similar to those of tobacco, *Datura,* and potato (Chapters 38, 40 and 43, this volume). The protocols for *Petunia* were originally modified from those for tobacco. These modifications have also been found to be useful for tobacco (Binding, 1975). *Datura* protoplasts have been cultured based on the protocols for *Petunia* (Schieder, 1975). The procedure which had been elaborated for potato (Binding *et al.*, 1978) has been found to be highly efficient for *Petunia* (Binding *et al.*, 1981).

II. PROTOPLAST ISOLATION

A. Plant Material

Protoplasts have been isolated from various genotypes of *P. hybrida* (Frearson *et al.*, 1973; Vasil and Vasil, 1974; Izhar and Power, 1977), as well as from *P. parodii* (Hayward and Power, 1975), *P. inflata, P. violocea, P. axillaris* (Power *et al.*, 1976), and *P. parviflora* (Sink and Power, 1977). *Pe-*

tunia hybrida, P. parodii, and *P. axillaris* have also been investigated recently by Dulieu *et al.* (1983).

1. Mesophyll

Homogeneous protoplast suspensions are prepared from leaves of young plants grown in the greenhouse or the growth chamber. The yields and stability of the protoplasts are highly dependent on the conditions of the plants. This was first found in mosses (Binding, 1966) and has been most extensively investigated in tobacco. Light intensity and water supply seem to play a decisive role in the cultivation of *Petunia.* These factors are not under precise control in the greenhouse.

Leaves are harvested at the end of the dark period to obtain optimal quantities of viable protoplasts. They are decontaminated by any of the established procedures of surface sterilization. These procedures are not sufficient for old plants (Durand *et al.,* 1973) and after propagation by cuttings due to systemic contamination. Antibiotics are used in the following steps of preparation to prevent propagation of the microorganisms (Power *et al.,* 1976).

Leaves are allowed to wilt after sterilization. The lower epidermis is then easily peeled off, or the cuticule is broken by carborundum and/or a strong brush. The prepared leaf pieces are floated on the wall-degrading enzyme solutions. Preincubation in plasmolytica is usually not needed in *Petunia* (Frearson *et al.,* 1973; Binding, 1974c).

Problems in controlling the environment and microbial infection are greatly reduced when leaves of *in vitro* cultures are used. It is therefore not surprising that leaves from shoot cultures have been widely used for *Petunia* protoplast preparation (Binding, 1974c), as well as for other taxa in which shoot cultures have been grown (Binding, 1975; Chapter 40, this volume). Peeling of the epidermis of shoot culture leaves is difficult and time-consuming. The leaves are therefore usually sliced by a sharp razor blade into strips or small pieces (1–2 mm) in a droplet of plasmolyticum or enzyme solution. The slices are suspended in the enzyme solution and submitted to a brief treatment by a weak vacuum for infiltration.

2. Apices

Protoplasts of the apical region, including leaf primordia, proved to be superior to leaves for protoplast isolation in a number of plant species (Binding *et al.,* 1982). In *Petunia,* similar results have been obtained with apical and mesophyll protoplasts of shoot cultures, indicating that the developmental stages of the tissue have little or no influence on the yield of viable protoplasts. However, in the case of adult plants, shoot tips were more advantageous than leaves. They were less contaminated by micro-organisms and less sensitive to environmental conditions. Consequently, reproducibly high yields of protoplasts were obtained irrespective of light intensity, season, and other factors. Apical material is sliced after surface sterilization.

3. Callus and Cell Suspension Cultures

There are some reports on protoplasts from callus cultures (Binding, 1974b,c; Vasil and Vasil, 1974; Power *et al.*, 1980) and cell suspension cultures (Cocking *et al.*, 1977; Power *et al.*, 1979). Plant regeneration was obtained by Vasil and Vasil (1974).

B. Preparation of Protoplasts

Cell walls are digested by enzyme preparations [0.5% macerozyme and 3% Cellulase Onozuka R-10 or Meicelase] osmotically stabilized by 0.5 M mannitol or sucrose. Addition of calcium ions [5 μM $CaCl_2$ or $Ca(NO_3)_2$] and dextran sulfate (1–2%) is useful. The ratio of amount of plant material to enzyme solution has some influence on the yield of protoplasts. In the case of mesophyll protoplasts from shoot cultures, 1 g leaflets per 100 ml was found to be adequate. Some authors prefer agitated incubation on reciprocal shakers; others float peeled leaves on enzyme solutions in Petri dishes or suspend sliced material in rather thin layers of liquid in Petri dishes (e.g., 12 ml in a 10-cm dish). Incubation in light (1–2 klx) seemed to be superior to incubation in the dark. Incubation at 25–30°C is stopped after about 5 hr or carried overnight. The resulting protoplast suspensions are poured through steel or nylon sieves. Sieves measuring 40 μM are used for apical meristem protoplasts, and those of 80–100 μM for mesophyll protoplasts. Most types of protoplasts are then spinned down by centrifugation at 50–100 g. When sucrose is used as the osmotic stabilizer in the enzyme solution, protoplasts with large vacuoles (e.g., from leaf mesophyll or from mutant white tissue) float. Floating is useful for removal of debris which sediments in sucrose solutions. Most mutant white protoplasts do not sediment even in enzyme solution with mannitol. Therefore, ionic solutions (0.25 M NaCl or sea water diluted to a respective osmolality; Binding, 1974c) are added before centrifugation. Pelleted protoplasts are taken up in solutions of lower density, preferentially mannitol, NaCl, or sea water. Occasionally, solutions are used which contain the inorganic components of culture media at the original or one-tenth concentrations or salts of other formulations. Ionic plasmolytica (NaCl or sea water) are highly appropriate for the removal of remnants of broken protoplasts. Hence, floating and sedimentation are equally good procedures for the purification of protoplast suspensions.

Protoplasts may be taken in culture after just one step of washing. It has been usual and is still preferred in several laboratories to repeat the washing procedures three to five times. However, no further increase in plating efficiency was found in our experiments. In fact, the total number of protoplasts isolated decreased with repeated washings.

III. CULTURE OF PROTOPLASTS

Nearly all established culture techniques for plant protoplasts have been utilized in *Petunia*.

A. Culture Media

The culture media DPD (Durand *et al.*, 1973), FPC (Frearson *et al.*, 1973), and V-47 (Binding, 1974a,c) have been elaborated especially for *P. hybrida* (Table I). V-47 has been developed from the DPD medium to obtain increased plating efficiencies in haploid and diploid genotypes. The V-KM medium has recently been preferred in our laboratory; this medium is a combination of the rich organic components of the 8p medium (Kao and Michayluk, 1975) with the salts of V-47. It was originally developed for *Solanum dulcamara* (Binding and Nehls, 1977), but has been found useful for a great number of species (Binding *et al.*, 1980, 1981, 1982). The medium contains 2.5 μM 6-benzylaminopurine (6-BAP), 5 μM naphthaleneacetic acid (NAA), and 0.5 μM 2,4-dichlorophenoxyacetic acid (2,4-D).

The V-KM medium has several advantages over the other media. The mitotic activities are higher, incomplete cell divisions are rarer, and adventitious shoots are more reliably and more rapidly formed after transfer to low osmotic media.

Initial protoplast culture is carried out preferentially in liquid media. Agar often affects the osmotic stability. Agarose, on the other hand, is highly appropriate for embedding nacked protoplasts (Shillito *et al.*, 1983).

B. Physical Culture Conditions

Protoplasts are suspended in culture media at population densities of 0.7–10.0 $\times$ 10^4 cells/ml. The suspensions are usually incubated in Petri

TABLE I

Protoplast Culture Media for *Petunia*

A. Mineral salts (mg/liter)

Major salts	DPD[a]	V-47[b]	FPC[c]
NH_4NO_3	270	280	412.5
KNO_3	1480	1400	525
$CaCl_2 \times 2H_2O$	570	735	850
$MgSO_4 \times 7H_2O$	340	556	739
KH_2PO_4	80	68	353.6
Na_2EDTA	37.3	37.3	18.75
$FeSO_4 \times 7H_2O$	27.8	27.8	13.9

Minor salts	DPD	V-47	FPC
$MnSO_4 \times 4H_2O$	7.2	6.7	11.2
$ZnSO_4 \times 7H_2O$	1.5	1.4	4.3
KI	0.25	0.25	0.49
H_3BO_3	2.0	1.9	3.1
$Na_2MoO_42H_2O$	0.1	0.12	0.125
$CuSo_4 \times 5H_2O$	0.015	0.013	0.025
$CoCL_2 \times 6H_2O$	0.01	0.012	—
$CoSO_4 \times 7H_2O$	—	—	0.015

B. Organic constituents (mg/liter)

Vitamins	DPD	V-47	FPC
Thiamine HCl	4.0	4.0	1.0
Pyridoxine HCl	0.7	0.7	0.5
Nicotinic acid	4.0	4.0	5.0
Biotin	0.04	0.04	0.05
Folic acid	0.4	0.4	0.5
Meso-inositol	100.0	100.0	100.0
amino acid			
Glycine	1.4	1.4	1.0

Hormones	DPD	V-47	FPC
6-BAP	0.4	0.4	1.0
NAA	—	1.1	2.0
2,4-D	1.3	—	—
Sugars			
Sucrose	17,100	17,100	10,000
Mannitol	55,000	91,000	130,000

[a] DPD: Durand *et al.* (1973).
[b] V-47: Binding (1974c); modified from DPD.
[c] FPC: Frearson *et al.* (1973).

dishes, forming layers of 1–2 mm. This condition is established by gently shaking 0.7 ml protoplast suspension in a dish 3 cm in diameter (1.8 ml in 5 cm and 6 ml in 10 cm, respectively). Culture in droplets of 20–100 μl has also been reported. The dishes are sealed with Parafilm. Growth chambers adjusted to temperatures of 25–29°C are convenient for incubation. Continuous illumination is provided by cool white fluorescent tubes at about 2 klx.

IV. CULTURE OF PROTOPLAST REGENERANTS

Cell wall regeneration starts soon after incubation of isolated protoplasts in culture media. First bicellular regenerants are formed after 5 days in mesophyll preparations (Frearson *et al.*, 1973; Durand *et al.*, 1973); protoplasts from shoot apical meristems start mitotic activities after 30 hr. This has also been observed in mesophyll protoplasts from shoot culture leaves.

A. Cell Propagation

The regenerants are grown in the initial culture for 10–14 days when the protoplasts are plated at low densities (around 10^4/ml) and when mitotic activity starts late (e.g., in leaf mesophyll protoplasts) or when the growth rates are rather low (true in some species and genotypes, as well as in relatively simple culture media such as V-47).

If high cell populations are provided by high plating densities (around 5–10 × 10^4 cells ml) and by rapid proliferation (e.g., in V-KM medium), it is convenient to carry out several steps of dilutions. These procedures are especially important. They are highly useful for plant regeneration in *P. hybrida* and *P.* Mitchell. They were originally devised for culture of dihaploid potato protoplasts (Binding *et al.*, 1978) and have been successfully applied to a number of other protoplast systems (Binding and Nehls, 1980; Binding *et al.*, 1980, 1981). Protoplasts of shoot cultures are plated in V-KM medium at densities of about 5 × 10^4/ml. The suspensions are diluted by addition of equal volumes of semisolid V-KM medium (containing 0.25% agar). Media with this agar concentration can be pipetted at room temperature; when more agar is added, the media should be kept at 40–45°C (Nagata and Takebe, 1971). Half of the diluted suspension is transferred to a new dish. Next, dilution is carried out at about the sixth day of culture. The diluted suspensions may be poured into larger Petri dishes when a little more fresh soft agar media are used than the volume of the suspensions. The timing of the next step of dilution depends on the growth rates of the regenerated cell clusters. At about the ninth day, it is usually convenient to dilute again and layer the suspensions on top of solid agar V-KM medium. The colonies are allowed to grow in this condition to sizes of about 1–2 mm in diameter before being transferred to low osmotic media. Even though this transfer is usually well tolerated by the colonies, several authors prefer to reduce the osmolarities of the media stepwise. It is useful to keep the cultures for about 1 week at 1 klx and a 16-hr day and then to transfer them to higher light intensities.

B. Shoot Induction

Shoot formation sets in usually on low osmotic culture media. However, it has also been observed during the growth on solid agar V-KM medium. A low osmotic medium well suited for *P. hybrida* and *P.* Mitchell was the B5C medium, which resembles B5 medium, (Gamborg *et al.*, 1968), enriched by 5% coconut milk and 2.5 μM 6-BAP. Shoot primordia were observed as early as 7 days after reduction of the osmolarity and 5 weeks after protoplast isolation. Up to 80% of the calli produced shoots within 3 weeks. Transfer of calli larger than 2 mm after prolonged culture on V-KM medium usually reduced organogenesis. The same is true when coconut milk omitted or an auxin is added (NAA, 2,4-D). Other *Petunia* species have been regenerated with auxins in addition to cytokinin: about 11–23 μM indoleacetic acid (IAA) for *P. parviflora* (Sink and Power, 1977) and *P. parodii* (Power *et al.*, 1976), respectively, and 0.2 μM NAA for *P. axillaris* (Power *et al.*, 1976).

C. Root Induction

As in many other plant genera, reduction of the amounts of nutrients, and of cytokinins as well as initially high concentrations of auxins, are appropriate means for the induction of roots. B5, MS (Murashige and Skoog, 1962), or NT media are mostly used. Hormones are either omitted totally, or IAA is preferentially added at various concentrations. Details on the procedures of root induction of shoots and on the transfer to soil are given in Chapter 6, this volume.

V. CONCLUSIONS

Protoplast isolation and regeneration are well established in *Petunia*. *Petunia* is therefore exceptionally well suited for experiments which rely upon isolated protoplasts. However, there are still some restrictions which must be overcome by improvement of the applied technologies, above all with respect to the following situations: Plant regeneration is sufficient in a number of genotypes of *P. hybrida*, in *P. parodii*, and *Petunia* Mitchell, but was less efficient, for instance, in *P. hybrida* cultivar Commanche (Power *et al.*, 1976). Reduction of 2,4-D, establishment of high growth rates, and early induction of organogenesis are helpful in getting rather low incidence of polyploid and aneuploid regenerants. However, further reduction of cytogenetic variability is desirable.

REFERENCES

Binding, H. (1966). Regeneration and Verschmelzung nackter Laubmoosprotoplasten. *Z. Pflanzenphysiol.* **55,** 305–321.

Binding, H. (1974a). Cell cluster formation by leaf protoplasts from axenic cultures of haploid *Petunia hybrida* L. *Plant Sci. Lett.* **2,** 185–188.

Binding, H. (1974b). Fusionsversuche mit isolierten Protoplasten von Petunia hybrida L. Z. *Pflanzenphysiol.* **72,** 422–426.

Binding; H. (1974c). Regeneration von haploiden und diploiden Pflanzen aus Protoplasten von *Petunia hybrida* L. Z. *Pflanzenphysiol.* **74,** 327–356 [translated to English by F. de Bruijn, *Plant Mol. Biol. Newsl.* **1,** 77–95 (1980)].

Binding, H. (1975). Reproducibly high plating efficiencies of isolated mesophyll protoplasts from shoot cultures of tobacco. *Physiol. Plant.* **35,** 225–227.

Binding, H., and Nehls, R. (1977). Regeneration of isolated protoplasts to plants in *Solanum dulcamara* L. Z. *Pflanzenphysiol.* **85,** 279–280.

Binding, H., and Nehls, R. (1980). Protoplast regeneration to plants in *Senecio vulgaris* L. Z. *Pflanzenphysiol.* **99,** 183–185.

Binding, H., Nehls, R., Schieder, O., Sopory, S. K., and Wenzel, G. (1978). Regeneration of mesophyll protoplasts isolated from dihaploid clones of *Solanum tuberosum. Physiol Plant.* **43,** 52–54.

Binding, H., Nehls, R., and Kock, R. (1980). Versuche zur Protoplastenregeneration dikotyler Pflanzen unterschiedlicher systematischer Zugehörigkeit. *Ber. Dtsch. Bot. Ges.* **93,** 667–671.

Binding, H., Nehls, R., Kock, R., Finger, J., and Mordhorst, G. (1981). Comparative studies on protoplast regeneration in herbaceous species of the dicotyledoneae class. Z. *Pflanzenphysiol.* **101,** 119–130.

Binding, H., Nehls, R., and Jörgensen, J. (1982). Protoplast regeneration in higher plants. In "Plant Tissue Culture 1982" (A. Fujiwara, ed.), pp. 575–578. Maruzen, Tokyo.

Cocking, E. C., George, D., Price-Jones, M. J., and Power, J. B. (1977). Selection procedures for the production of inter-species somatic hybrids of *Petunia hybrida* and *Petunia parodii.* II. Albino complementation selection. *Plant Sci. Lett.* **10,** 7–12.

Donn, G., Hess, D., and Potrykus, I. (1973). Wachstum und Differenzierung in aus isolierten Protoplasten von *Petunia hybrida* entstandenem Kallus. *Z. Pflanzenphysiol.* **69,** 423–437.

Dulieu, H. L., Bruneau, R., and Pelletier, A. (1983). Heritable differences in in vitro regenerability in *Petunia,* at the protoplast and at the seedling stage. *In* "Protoplasts 1983" (I. Potrykus, C. T. Harms, A. Hinnen, R. Hütter, P. J. King, and R. D. Shillito, eds.), pp. 236–237. Birkhäuser, Basel.

Durand, J., Potrykus, I., and Donn, G. (1973). Plantes issues de protoplastes de Pétunia. Z. *Pflanzenphysiol.* **69,** 26–34.

Frearson, E. M., Power, J. B., and Cocking, E. C. (1973). The isolation, culture and regeneration of *Petunia* leaf protoplasts. *Dev. Biol.* **33,** 130–137.

Gamborg, O. L., Miller, R. A., and Ojima, K. (1968). Nutrient requirements of suspension cultures of soybean root cells. *Exp. Cell Res.* **50,** 151–158.

Hayward, C., and Power, J. B. (1975). Plant production from leaf protoplasts of *Petunia parodii. Plant Sci. Lett.* **4,** 407–410.

Hess, D., and Potrykus, I. (1972). Teilung isolierter Protoplasten von *Petunia hybrida. Naturwissenschaften* **59,** 273–274.

Izhar, S., and Power, J. B. (1977). Genetical studies with *Petunia* leaf protoplasts. 1. Genetic variation to specific growth hormones and possible genetic control on stages of protoplast development in culture. *Plant Sci. Lett.* **8,** 375–383.

Kao, K. N., and Michayluk, M. R. (1975). Nutritional requirements for growth of *Vicia hajastana* cells and protoplasts at a very low population density in liquid media. *Planta* **126,** 105–110.

Murashige, F., and Skoog, F. (1962). A revised medium for rapid growth and bioassays with tobacco tissue culture. *Physiol. Plant.* **15,** 473–497.

Nagata, T., and Takebe, I. (1971). Plating of isolated tobacco mesophyll protoplasts on agar medium. *Planta* **99,** 12–20.

Potrykus, I., and Durand, J. (1972). Callus formation from single protoplasts of *Petunia*. *Nature (London), New Biol.* **37,** 286–287.
Power, J. B., Frearson, E. M., George, D., Evans, P. K., Berry, S. F., Hayward, C., and Cocking, E. C. (1976). The isolation, culture and regeneration of leaf protoplasts in the genus *Petunia*. *Plant Sci. Lett.* **7,** 51–55.
Power, J. B., Berry, S. F., Chapman, J. V., and Cocking, E. C. (1979). Somatic hybrids between unilateral cross-incompatible *Petunia* species. *Theor. Appl. Genet.* **55,** 97–99.
Power, J. B., Berry, S. F., Chapman, J. V., and Cocking, E. C. (1980). Somatic hybridization of sexually incompatible *Petunias: Petunia parodii, Petunia parviflora*. *Theor. Appl. Genet.* **57,** 1–4.
Schieder, O. (1975). Regeneration von haploiden und diploiden *Datura innoxia* Mill. Mesophyllprotoplasten zu Pflanzen. *Z. Pflanzenphysiol.* **76,** 462–466.
Shillito, R. D., Paszkowski, I., and Potrykus, I. (1983). Culture in agarose improves protoplast plating and proliferation, and permits division in otherwise unresponsive systems. *In* "Protoplasts 1983" (I. Potrykus, C. T. Harms, A. Hinnen, R. Hütter, P. J. King, and R. D. Shillito, eds.), pp. 266–267. Birkhäuser, Basel.
Sink, K. C. (1984). "*Petunia* Monograph." Springer-Verlag, Berlin and New York.
Sink, K. C., and Power J. B. (1977). The isolation, culture and regeneration of leaf protoplasts of *Petunia parviflora* Juss. *Plant Sci. Lett.* **10,** 335–340.
Vasil, V., and Vasil, I. K. (1974). Regeneration of tobacco and *Petunia* plants from protoplasts and culture of corn protoplasts. *In Vitro* **10,** 83–96.

CHAPTER **40**

Isolation and Culture of Protoplasts: *Datura*

O. Schieder

Max Planck Institute for Plant Breeding Research
Cologne, Federal Republic of Germany

I. INTRODUCTION

The genus *Datura* is of some pharmaceutical importance because of alkaloid production. One of the main alkaloids synthesized in several *Datura* species is scopolamine. The herbaceous species *D. innoxia* and the tree species *D. sanguinea* serve predominantly as the sources for scopolamine. Additionally, some *Datura* species, including *D. innoxia*, are suitable as model plants for somatic cell genetics using diploid ($2n = 24$) or haploid ($n = 12$) protoplasts or cell suspensions. Diploid protoplasts of *D. innoxia* have been used for successful intraspecific (Schieder, 1977a), interspecific (Schieder, 1978, 1980, 1982), and intergeneric (Krumbiegel and Schieder, 1979) somatic hybridization experiments. Moreover, asymmetric somatic hybrids of *D. innoxia* containing only a few chromosomes from another species could also be obtained (Krumbiegel and Schieder, 1981; Gupta *et al.*, 1984). Finally, haploid protoplasts of *D. innoxia* have been used in mutation experiments leading to albino mutants (Schieder, 1976; Krumbiegel, 1979).

The success of the experiments described above demonstrates the great

CELL CULTURE AND SOMATIC CELL
GENETICS OF PLANTS, VOL. 1

ISBN 0-12-715001-3

usefulness of *Datura* in somatic cell genetics. The following description of the isolation and culture of *Datura* protoplasts guarantees their regeneration to complete flowering plants and is useful for both the haploid and diploid protoplasts of the herbaceous species *D. innoxia, D. metel,* and *D. meteloides* (Schieder, 1975, 1977b; Furner *et al.*, 1978).

II. PROTOPLAST ISOLATION

A. The Source of Protoplasts

In general, leaf material from plants grown in the greenhouse or in the field can be used as the source for protoplasts. However, with such material, some disadvantages exist. First, the osmotic pressure in the mesophyll cells can differ due to the weather conditions and the humidity, thereby altering the yield of protoplasts at each isolation. Second, the necessary sterilization of such leaf material produces considerable debris. Third, it is evident, not only from our experience but from that of others as well, that protoplasts isolated from greenhouse or field leaf material do not regenerate as well as protoplasts derived from aseptically grown shoot cultures *in vitro* (Binding, 1975). Therefore, aseptic shoot cultures as the source for protoplasts are preferred.

For establishing aseptic shoot cultures, in many cases one can start with sterilized seeds sown on agar plates (Binding, 1975). However, this procedure for *Datura* is not practical because *Datura* seeds normally germinate successively over a period of several weeks (Conklin, 1976). Moreover, *in vitro* germination is very poor. A more efficient method is to start aseptic shoot cultures with internodes or shoot tips from young plants grown in the greenhouse.

The following procedure for establishing shoot cultures is necessary. Take young shoots and remove all large leaves, including their petioles. Cut out all internodes (10–20 mm), including the shoot tip. Every internode contains dormant axillary shoot buds. For sterilization, transfer the internodes into a solution of 0.1% $HgCl_2$ and 0.1% Na-dodecyl sulfate. After a 7-min treatment, wash the internodes three to five times with sterilized tap water, and transfer them into Petri dishes containing B5 agar medium (Gamborg *et al.*, 1968) supplemented with 6-benzylaminopurine (6-BAP) (0.5 or 1.0 mg/liter). Cultivation follows in a culture room at 2000–5000 1x, 16 hr/day, at 25–28°C. After 1–2 weeks under these conditions, the axillary buds form shoots and produce after a longer period of culture numerous adventitious shoots which can be removed with a razor blade and used for propagation of the culture. These shoots have relatively small leaves which can be used as the source of protoplasts. However, rooted shoots with large leaves can be obtained upon transfer into larger culture vessels containing B5 medium without any phytohormones. Subculturing every 14 days given the best leaf material for protoplast isolation.

B. Isolation Procedure

1. Mesophyll Protoplasts

Cut about 1 g of the larger leaves, place them in a sterilized glass Petri dish (100 mm in diameter), and add some droplets of a 0.3-*M* mannitol solution to prevent desiccation. Cut the leaves with a razor blade into pieces 3–5 mm^2 in diameter and add 10 ml of the 0.3-*M* mannitol solution. Shake and remove the mannitol solution with a pipette. This procedure cleans the preparation from released chloroplasts and other broken material. With forceps the shredded leaf material can be transferred into a flask containing 50 ml of the enzyme solution for digesting the cell walls. The enzyme solution can be prepared as follows:

1% Cellulase Onozuka R-10
0.2% Macerozyme R-10 (both Kinki Yakult Co., Nishinomiya, Japan)
0.6 *M* mannitol
pH 5.5, adjusted with 1 *M* KOH

This solution should have an osmotic pressure of about 730 mOsm and must be filter sterilized.

The flask containing the enzyme solution and the shredded leaf material can be shaken gently in a water bath shaker at about 27°C. However, higher yields of protoplasts can be obtained by rotating the flask on a roller at 3 rpm at the same temperature in either a culture room or an incubator (Schieder, 1977b).

After 4–5 hr of incubation, filter the enzyme solution containing the protoplasts and undigested leaf material through a steel sieve (pore size 100 μm). Dilute in a ratio of 1:1 the protoplast containing enzyme solution with 80% filter sterilized sea water (ca. 730 mOsm) and centrifuge for 3 min at 100 *g* to sediment the protoplasts.

We are using sea water from the North Sea (>800 mOsm); however, the osmolality differs throughout the year and has to be determined by an osmometer. The advantages of sea water are that it is an excellent osmotic stabilizer and is easy to prepare. Moreover, protoplasts derived from albino mutants sediment in sea water much better than in a mannitol solution. However, other salt solutions, for example, a culture medium containing NaCl instead of mannitol, may also be useful.

The sedimented protoplasts contaminated with chloroplasts and undigested cell aggregates have to be resuspended with 80% seawater and centrifuged again. For separating the protoplasts from chloroplasts and cell aggregates, protoplast floatation with sucrose is helpful (Davey *et al.*, 1974). Resuspend the pellet with a 0.6-*M* sucrose solution and centrifuge for 10 min at 100 *g*. Under this treatment the protoplasts continue to float, whereas most of the chloroplasts and cell aggregates sediment. Using a pipette, remove the protoplast containing supernatant, dilute it in a ratio of 1:3 with sea water, and centrifuge again. Resuspend the sedimented protoplasts with regeneration medium V-47 (ca. 730 mOsm), according to Binding (1974), containing 1.5 mg/liter α-naphthaleneacetic acid (α-NAA) and 0.4 mg/liter 6-benzylaminopurine (6-BAP), giving a final density of 10^4–10^5 protoplasts/ml.

2. Suspension Protoplasts

The cells of suspension cultures (Chapters 44 and 45, this volume), in general are somewhat different and need a more severe enzyme treatment than mesophyll cells. Furner *et al.* (1978) used, for a predominantly haploid cell suspension of *D. innoxia*, the following enzyme solution:

1.5% Driselase (Kyowa Hakko, Kogyo Co., Tokyo, Japan)
0.5% pectinase (Sigma)
3 m*M* 2 (*N*-morpholino) ethane sulfonic acid
7 m*M* $CaH_4(PO_4)_2 \cdot H_2O$
0.5 *M* mannitol
pH 5.8, filter sterilized

The cells were incubated with this enzyme solution for 8 hr on a gyrotory shaker at room temperature.

We are using another enzyme solution described as follows:

1% Cellulase Onozuka R-10
0.2% Macerozyme R-10
0.05% Pectolyase (Seishin Pharmaceutical Co., Ltd., Chiba-ken, Japan)
0.6 *M* mannitol
pH 5.5, adjusted with 1 *M* KOH

For protoplast isolation, centrifuge 50 ml of a cell suspension from *D. innoxia* and resuspend the sedimented cells in the enzyme solution in a roller flask. After incubation on a roller for 15 hr, prepare the protoplasts exactly as described for the mesophyll protoplasts.

III. PROTOPLAST CULTURE

For cultivation, pipette either 0.7 ml of the protoplast suspension into a 40-mm-diameter plastic Petri plate or 2 ml into a 60-mm-diameter plate and seal with Parafilm. Incubate under permanent illumination of about 1500 lx at 25–27°C. After 2–3 days the first protoplasts start to divide, and 14 days later small cell colonies of 10–20 cells are formed. At this time, for separating the colonies and for enhancing their growth it is advantageous to dilute the cultures 1:1 with V-47 medium (ca. 630 mOsm) containing 0.4% agar, giving a final agar concentration of 0.2%. The soft agar should not have a temperature higher than 40°C. After a further 2–3 weeks, the colonies are visible macroscopically and can be transferred onto B5 agar medium supplemented with 6-BAP (1 mg/liter). In the beginning, it is important to seal the dishes with Parafilm, which, however, should be removed after about 10 days, because redifferentiation of the colonies works much better in unsealed dishes. The culture can be kept in a culture room illuminated 16 hr/day with 2000–5000 lx at 27°C.

The cell colonies will become green after 3–4 weeks of culture. After about 4–5 weeks, pick up the single colonies and transfer them onto freshly prepared B5 agar medium. Here they will soon begin to produce leaves and small shoots. Following two to three subcultures on the same medium, larger shoots can be transferred without any phytohormones onto B5 medium, where they produce roots. Rooted shoots can be transferred into soil in the greenhouse. The potted plantlets have to be kept for about 10 days in a moist chamber for acclimatization.

IV. FINAL REMARKS

With the regeneration procedure described above, it is relatively easy to regenerate the protoplasts from *D. innoxia*, *D. meteloides*, and *D. metel*, whereas *D. ferox*, *D. stramonium*, *D. discolor*, and all the tree *Datura* species do not regenerate (Schieder, 1977b, 1980). However, the regeneration capacity can differ strongly depending upon the genotype used. For example, mesophyll protoplasts of 25 haploid lines of *D. innoxia*, which were derived via anther culture from a diploid line from which up to 70% of the mesophyll protoplasts start divisions, were tested for their division capacity. Some of them showed division rates similar to those of the parental line, but others showed a lower rate or no divisisions at all (Schieder, unpublished). Moreover, from two diploid lines of *D. meteloides*, one line yields mesophyll protoplasts with a plating efficiency of about 50%, whereas from the other line no divisions can be obtained. Therefore, before starting somatic cell genetics with *Datura*, it is important to screen for lines showing a high protoplast division rate. It seems that the genotype, at least for the genus *Datura*, is more important with respect to protoplast regeneration than the culture medium. We cannot exclude the possibility that for the other *Datura* species in which the regeneration experiments were unsuccessful, genotypes can be found that will give a more positive result.

REFERENCES

Binding, H. (1974). Regeneration von haploiden und diploiden Pflanzen aus Protoplasten von *Petunia hybrida* L. *Z. Pflanzenphysiol.* **74,** 327–356.

Binding, H. (1975). Reproducibly high plating efficiencies of isolated mesophyll protoplasts from shoot cultures of tobacco. *Physiol. Plant.* **35,** 225–227.

Conklin, M. E. (1976). "Genetic and Biochemical Aspects of the Development of *Datura*." Karger, Basel.

Davey, M. R., Bush, E., and Power, J. B. (1974). Cultural studies of a dividing legume leaf protoplast system. *Plant Sci. Lett.* **3,** 127–133.

Furner, I. J., King, J., and Gamborg, O. L. (1978). Plant regeneration from protoplasts isolated from a predominantly haploid suspension culture of *Datura innoxia* (Mill.). *Plant Sci. Lett.* **11,** 169–176.

Gamborg, O. L., Miller, R. A., and Ojima, K. (1968). Nutrient requirements of suspension cultures of soybean root cells. *Exp. Cell Res.* **50,** 151–158.

Gupta, P. P., Shaw, D. S., and Schieder, O. (1984). Intergeneric nuclear gene transfer between somatically and sexually incompatible plants through asymmetric protoplast fusion. *Mol. Gen. Genet.* (in press).

Krumbiegel, G. (1979). Response of haploid and diploid protoplasts from *Datura innoxia* Mill. and *Petunia hybrida* L. to treatment with X-rays and a chemical mutagen. *Environ. Exp. Bot.* **19,** 99–103.

Krumbiegel, G., and Schieder, O. (1979). Selection of somatic hybrids after fusion of protoplasts from *Datura innoxia* Mill. and *Atropa belladonna* L. *Planta* **145,** 371–375.

Krumbiegel, G., and Schieder, O. (1981). Comparison of somatic and sexual incompatibility between *Datura innoxia* and *Atropa belladonna*. *Planta* **153,** 465–470.

Schieder, O. (1975). Regeneration von haploiden und diploiden *Datura innoxia* Mill. Mesophyll-Protoplasten zu Pflanzen. *Z. Pflanzenphysiol.* **76,** 462–466.

Schieder, O. (1976). Isolation of mutants with altered pigments after irradiating haploid protoplasts from *Datura innoxia* Mill. with X-rays. *Mol. Gen. Genet.* **149,** 251–254.

Schieder, O. (1977a). Hybridization experiments with protoplasts from chlorophyll deficient mutants of some solanaceous species. *Planta* **137,** 253–257.

Schieder, O. (1977b). Attempts in regeneration of mesophyll protoplasts of haploid and diploid wild type lines, and those of chlorophyll deficient strains from different *Solanaceae*. *Z. Pflanzenphysiol.* **84,** 275–281.

Schieder, O. (1978). Somatic hybrids of *Datura innoxia* Mill. + *Datura discolor* Bernh. and *Datura innoxia* Mill. + *Datura stramonium* L. var. *tatula* L. I. Selection and characterization. *Mol. Gen. Genet.* **162,** 113–119.

Schieder, O. (1980). Somatic hybrids between a herbaceous and two tree *Datura* species. *Z. Pflanzenphysiol.* **98,** 119–127.

Schieder, O. (1982). Somatic hybridization: A new method for plant improvement. *In* "Plant Improvement and Somatic Cell Genetics" (I. K. Vasil, W. R. Scowcroft, and K. J. Frey, eds.), pp. 239–253. Academic Press, New York.

CHAPTER **41**

Isolation and Culture of Protoplasts: *Brassica*

Hellmut R. Schenck

Institut für Pflanzenbau und Pflanzenzüchtung
Georg August University
Göttingen, Federal Republic of Germany

Franz Hoffmann

Department of Developmental and Cell Biology
University of California
Irvine, California

I. INTRODUCTION

Within the tribe Brassiceae the genus *Brassica* is the most important from an economic point of view (for more general information, see the monographs by Vaughan *et al.*, 1976; Tsunoda *et al.*, 1980). It includes the three diploid species *B. campestris* (oil turnip, turnip rape), *B. nigra* (black mustard), and *B. oleracea* (cabbage), and their corresponding amphidiploids *B. carinata* (ethiopian mustard), *B. juncea* (black or Indian mustard), and *B. napus* (rapeseed). *Brassica campestris* and, above all, *B. napus* are the most

ISBN 0-12-715001-3

important oil seed crops of the temperate climate zone. *Brassica oleracea*, with numerous subspecies, is a major vegetable crop for human consumption. Furthermore, *B. napus* is one of the very few important crop plants which can be successfully established in tissue culture and probably one of the first examples of crop improvement by a combination of classical plant breeding methods and modern *in vitro* techniques (Hoffmann, 1980).

Brassica protoplasts have been used in fusion experiments for the successful creation of hybrid cells (Kartha *et al.*, 1974a; Hoffmann *et al.*, 1980) and *Arabidobrassica* plantlets (Gleba and Hoffmann, 1980; Hoffmann and Adachi, 1981). Furthermore, the first fusion-derived synthetic *B. napus* plants, obtained by somatic hybridization of *B. campestris* with *B. oleracea*, already exist (Schenck and Röbbelen, 1982). Such synthetic amphidiploids represent a promising challenge for breeders because of the narrow genetic diversity of the existing material. The use of *Brassica* protoplasts for the mass propagation of highly valued individual plants or for the production and selection of mutants (intentionally induced as well as accidentally regenerated so-called somaclonal variants) is limited due to the poor rate of plant regeneration from protoplast-derived callus. The first regeneration of plants from mesophyll protoplasts of amphidiploid *B. napus* was reported by Kartha *et al.* (1974b), and that of amphihaploid rapeseed by Thomas *et al.* (1976). Meanwhile, additional reports have been published concerning *Brassica* species (Table I).

Brassica protoplasts used in studies on cauliflower mosaic virus (CaMV) infection (e.g., Howell and Hull, 1978) were mostly isolated from "turnips," a subspecies of *B. rapa*. However, the tissue culture success with some of these turnips seems to be limited. We therefore suggest the utilization of *B. napus* cultures and protoplasts for this kind of work, preferably the Canadian spring type variety "Tower."

II. PROTOPLAST ISOLATION

A. Protoplast Isolation from Mesophyll of Potted Plants

Leaves of 4- to 6-week-old greenhouse-grown plants should have a size of 5–8 cm. After washing in tap water, the leaves are sterilized in 2% NaOCl for 20 min. One drop of a detergent (e.g., Tween-80) per 200 ml solution should be added to reduce the surface tension of the water. The leaves are rinsed three to five times with sterile water for 5 min. All of the following operations are carried out under aseptic conditions. Leaves are

TABLE I

Callus, Root, and Plant Regeneration from Protoplasts of Different *Brassica* Species

Species	Origin of protoplasts	Development obtained			Reference
		Callus	Roots	Plants	
B. napus					
Amphidiploid	Mesophyll	+	+	+	Kartha *et al.* (1974b)
	Roots	+	+	+	Xu *et al.* (1982)
Amphihaploid	Mesophyll	+	+	+	Thomas *et al.* (1976)
	Stem embryos	+	+	+	Kohlenbach *et al.* (1982a)
	Mesophyll	+	+	+	Li and Kohlenbach (1982)
B. oleracea					
Marrow stem Kale	Mesophyll	+	+	−	Gatenby and Cocking (1977)
Kohlrabi	Mesophyll	+	+	−	Schenck and Hoffmann (1979)
Fodder kale	Mesophyll	+	+	+	Schenck (1981)
"Cabbage"	Roots	+	+	+	Xu *et al.* (1982)
Broccoli[a]	Mesophyll	+	+	+	Schenck and Röbbelen (1982)
Head cabbage[a]	Mesophyll	+	+	+	Schenck and Röbbelen (1982)
B. campestris	Mesophyll	+	+	−	Schenck and Hoffmann (1979)
	Roots	+	+	−	Xu *et al.* (1982)
B. nigra	Mesophyll	+	−	−	Schenck and Hoffmann (1979)
B. rapa	Mesophyll	+	+	−	Ulrich *et al.* (1980)
B. alba (*Sinapis alba*)	Roots	+	−	−	Xu *et al.* (1982)

[a]Obtained in protoplast fusion experiments from nonhybrid cells.

cut into small strips of 0.5–1.0 mm with a scalpel or razor blade. Removal of the lower epidermis is not required but is sometimes suggested. A preplasmolysis of the cells is frequently described. The leaf strips can also be brought directly into the final mannitol concentration (0.4–0.7 *M* depending on the material used). A preculture of the whole leaves is described by Gatenby and Cocking (1977). In this case, the lower epidermis is removed, and the leaves are placed abaxial surface down on an agar medium containing inorganic salts, vitamins, sucrose, and phytohormones, resulting in more readily dividing protoplasts of *B. oleracea*.

Two general procedures are recommended for the enzyme treatment. In the rapid procedure, a more concentrated enzyme solution is added. The tissue pieces are incubated in a water bath shaker, usually at 27–30°C for 2–3 hr. To save the expensive enzymes, overnight incubations are carried out using about one-fourth of the enzyme concentration. The incubation time is 14–17 hr at room temperature. Table II summarizes the different enzyme treatments described for Braciceae. Some authors add supplementary compounds to the enzyme solution such as antibiotics—for instance,

400 μg/ml ampicillin, 10 μg/ml tetracycline, and 10 μg/ml gentamicin—to avoid bacterial contamination (Gatenby and Cocking, 1977), nutrient salts, or calcium salts for membrane stabilization (Kartha *et al.*, 1974b).

After the enzyme treatment, the released protoplasts are filtered through 50- to 100-μm sieves (stainless steel or nylon). Following centrifugation at 30–100 *g* (depending on cell characteristics and incubation medium) for about 3 min, the pellet is washed twice with mannitol (0.4–0.7 *M*). Floating protoplasts can be obtained in sucrose-containing washing solution and then pelleted again in high-salt-containing solution (for details, see Section II,C). After supernatant mannitol is added, the final culture medium is added and the protoplasts are cultured in Petri dishes at a plating density of 2.5×10^4–5×10^5 cells/ml (Table III). Cultures are placed in the dark or under dim light for a few days or several weeks before being exposed to stronger illumination (Table III).

B. Protoplast Isolation from Mesophyll of Axenically Grown Plants

To achieve more constant growth conditions and to avoid stress due to the sterilization procedure, it is useful to use axenically grown plants and tissues as the protoplast source. Additionally, interior (not intracellular) infections (not eliminated by surface sterilization of greenhouse-grown plants) can be avoided. *Brassica* seeds can be sterilized in 5% NaOCl (for 30 min) or 0.1% $HgCl_2$ (30 min) containing 1 drop of Tween-80 or other nontoxic detergent per 200 ml. For germination the seeds are sown in Petri dishes on 0.8% water-agar. The germinated seedlings are subsequently transferred to larger culture vessels containing any common medium without any hormones. Leaves 3–5 cm in length can be used for protoplast isolation.

C. Protoplast Isolation from Stem Embryo Cultures

Stem embryo cultures in rapeseed, first described by Thomas *et al.* (1976) as possessing high embryogenic potential, were used by Kohlenbach *et al.* (1982a) as plant material for protoplast isolation. To achieve clean suspensions without larger debris, protoplasts are purified with a flotation method. Following the method of Potrykus *et al.* (1977) for stem protoplasts of corn, the protoplasts were centrifuged in a mixture of 1 volume of suspension (containing 0.55 *M* mannitol) plus 2 volumes of 0.24 *M* $CaCl_2$ at 50 *g*. The sediment was resuspended in 1 volume of culture medium plus 2

TABLE II

Enzyme Mixtures Used for the Isolation of Protoplasts from *Brassica* Species

Species	Origin of protoplasts	Enzyme mixture		Incubation		Reference
				Time	Temperature (°C)	
B. napus	Mesophyll (epidermis removed)	Cellulase Onozuka P-1500[a]	0.5%	2 hr	19–21	Kartha *et al.* (1974b)
		Hemicellulase[b]	0.5%			
		—After 2 hr, transferred to—				
		Driselase[c]	0.5%	3 hr	19–21	
		Hemicellulase	0.5%			
		Sorbitol	0.25 *M*			
		Mannitol	0.25 *M*			
		pH	6.2			
B. napus	Mesophyll	PATE[d]	0.1%	2.5 hr	29	Thomas *et al.* (1976)
		Cellulase Onozuka R-10[e]	0.5%			
		Mannitol	0.5 *M*			
		pH	7.0			
B. napus	Mesophyll	PATE	0.025%	14 hr	28	Li and Kohlenbach (1982)
		Cellulase Onozuka R-10	0.125%			
		Mannitol	0.55 *M*			
		pH	7.0			
B. napus	Stem embryos	PATE	0.2%	16 hr	28	Kohlenbach *et al.* (1982a)
		Cellulase Onozuka R-10	0.75%			
		Mannitol	0.55 *M*			
		pH	7.0			
B. napus	Stem embryos	PATE	0.2%	16 hr	28	Kohlenbach *et al.* (1982a)
		Driselase	0.3%			
		Cellulase Onozuka R-10	0.3%			
		Mannitol	0.65 *M*			
		pH	7.6			

B. oleracea	Mesophyll (epidermis removed)	Macerozyme[e]	0.4%	14 hr	27	Gatenby and Cocking (1977)
		Driselase	0.5%			
		Meicelase[f]	4.0%			
		Mannitol	0.6 *M*			
		pH	5.8			
B. campestris	Mesophyll	PATE	0.1%	2 hr	30	Schenck and Hoffmann (1979)
B. nigra		Cellulase Onozuka R-10	0.5%			
B. oleracea		Mannitol	0.55 *M*			
		pH	7.0			
B. alba	Root	Rhozyme[b]	2.0%	16 hr	25	Xu *et al.* (1982)
B. campestris		Meicelase	4.0%			
B. napus		Macerozyme	0.3%			
b. oleracea		Mannitol	0.7 *M*			
		pH	5.6			
B. campestris	Mesophyll	Pectolyase Y 23[g]	0.1%	2.5 hr	30	Schenck and Röbbelen (1982) (modified)
B. oleracea		Cellulase Onozuka RS[e]	0.8%			
		Mannitol	0.55%			
		pH	5.6			
B. campestris	Mesophyll	Pectolyase Y 23	0.025%	16 hr	20	Schenck and Röbbelen (1982) (modified)
B. oleracea		Cellulase Onozuka RS	0.2%			
		Mannitol	0.55 *M*			
		pH	5.6			
B. rapa	Mesophyll (epidermis removed)	Macerozyme	0.1%	3.5 hr	27	Ulrich *et al.* (1980)
		Cellulase Onozuka R-10	1.0%			
		Mannitol	0.4 *M*			
		pH	5.4			

[a] All Japan.
[b] Rohm and Haas.
[c] Kyowa Hakko Kogyo.
[d] Hoechst.
[e] Kinki Yakult.
[f] Meiji Seika Kaisha.
[g] Kikkoman Shoyn.

TABLE III

Survey of Culture Conditions for *Brassica* Protoplasts[a]

Species	Tissue[b]	Initial medium + supplements	Titer	Temp. (°C)	Light conditions: Dark	Light conditions: Weak	Light conditions: Light
napus	M	B5 medium Sorbitol 45.5 g/liter Mannitol 45.5 Glucose 2.5 B-Ribose 125 mg/liter *N*-Z-amine 875 $CaCl_2 \cdot 2H_2O$ 875 2,4-D 0.5 NAA 0.18 BAP 0.22 15 days	2–2.5 × 10^5	26	—	100 lx first 2 weeks	5000 lx after 2 weeks
napus	M	N medium Sucrose 10 g/liter Glucose 10 Caseine Hydrolysate 100 mg/liter Mannitol 0.5 *M* 2,4-D 0.25 NAA 0.5 BAP 0.1 30 days	5 × 10^4– 5 × 10^5	25	First 2 days	—	2000 lx after 2 days
napus	SES	N9 medium Sucrose 10 g/liter Glucose 10 Caseine Hydrolysate 100 mg/liter Mannitol 0.55–0.65 *M* 2,4-D 1.0 ZEAribosid 0.5 4 weeks	2.5 × 10^5	25	—	—	1500–2000 lx after 3–4 weeks
napus	M	N medium Sucrose 10 g/liter Glucose 10 Caseine Hydrolysate 100 mg/liter Mannitol 0.55 *M* 2,4-D 0.5 NAA 0.5 BAP 0.5 Or alternatively: 2,4-D 1.0 NAA 0.5 BAP 1.0 2 weeks	4 × 10^5	25	2 weeks	—	1500–2000 lx after 2 weeks
napus *oleracea* *alba*	R	KM8p/KM8 2:1 Glucose 68.4 g/liter 2,4-D 0.13 mg/liter NAA 1.0	2.5–3 × 10^4	23	14 days	700 lx after 2 weeks	—

Addition of fresh medium	Transfer I	Transfer II	Transfer III	Reference
B5 medium instead of sorbitol and mannitol; 2% sucrose volume: 10% of initial ml used per Petri dish Weekly for first 2 weeks	MS agar + B5 vitamins Coconut milk 10% Casamino acid 0.1% (vitamin free) 5–10 mm	MS medium GA_3 0.035 mg/liter BAP 1.1	—	Kartha *et al.* (1974b)
Fresh medium lacking mannitol, volume: 30% of initial ml used per Petri dish Weekly	N agar Sucrose 20 g/liter Caseine Hydrolysate 100 mg/liter 2,4-D 0.1 BAP 0.25 1 mm	MS medium lacking hormones	—	Thomas *et al.* (1976)
After 3–5 days, replacement of the initial medium by fresh medium, protoplasts resuspended in only one-half of the initial medium	MS11 medium Sucrose 10 g/liter Or alternatively: MS13 medium Sucrose 20 g/liter Glutamine 500 mg/liter Inositol 500 Both media liquid; edamine omitted 4 weeks	PCW agar, i.e., potato extract, 20% Sucrose 30 g/liter 2,4-D 1.75 mg/liter KIN 0.5 Fe-EDTA 100 μM 4 weeks	MS14 agar Sucrose 35 g/liter Galactose 2 Glutamine 800 mg IAA 0.5 BAP 0.05 2–6 weeks	Kohlenbach *et al.* (1982a)
Fresh medium mannitol 0.4 *M* volume: one-sixth of initial ml used per Petri dish After 2 weeks	MS13 as listed above Mannitol 0.4 *M* 2,4-D 0.2 mg/liter KIN 3.0 2–3 weeks	MS13 as listed above Mannitol 0.4 *M* 2,4-D 0.01 mg/liter BAP 1.0 3–4 weeks	MS ABA 0.05 mg GA_3 1–3 or hormone free	Li and Kohlenbach (1982)
Addition of 25% (v/v) K8 medium 2× in 7- to 10-day intervals, third	B5 agar (B5-2) sucrose 3% 2,4-D 1 mg/liter NAA 1 BAP 1	MS agar (MS-D3) IAA 2.0 mg/liter BAP 1.0 Or alternatively: NAA 0.05	—	Xu *et al.* (1982)

(continued)

TABLE III (*Continued*)

Species	Tissue[b]	Initial medium + supplements			Titer	Temp. (°C)	Light conditions: Dark	Weak	Light
		ZEA	0.5						
			30–45 days						
rapa	M	KM8p			1 × 10^5	24	—	—	1000 lx first 4 weeks; 12,000 lx after 4 weeks
		Glucose	68.4	g/liter					
		2,4-D	0.2	mg/liter					
		NAA	1.0						
		ZEA	0.5						
			4 weeks						
oleracea	M	B5 medium			2.25 × 10^5	25	—	100 lx first 7 weeks	4500 lx after 7 weeks
		Sucrose	10	g/liter					
		Glucose	5						
		N-Z-amine	150	mg/liter					
		Mannitol	0.44	*M*					
		2,4-D	1.0						
		NAA	0.5						
		BAP	0.5						
			3 weeks						
campestr *oleracea*	M	S1 medium			8 × 10^4 8 × 10^5	25	First 2 weeks	500 lx after 2 weeks	2000–3000 lx after 3 weeks
		Sucrose	20	g/liter					
		Inositol	100	mg/liter					
		fe(II)-K-tartrate	2.5						
		$MnSO_4 \cdot H_2O$	1.0						
		Na_2MoO_4	0.5	μg/liter					
		2,4-D	0.2						
		NAA	1.0						
		BAP	0.5						
		Mannitol	0.55	*M*					
campestr. *oleracea* *nigra*	M	GK2 medium			8 × 10^4 8 × 10^5	25	First 2 weeks	500 lx after 2 weeks	2000–3000 lx after 3 weeks
		Glucose	0.55	*M*					
		2,4-D	0.25	mg/liter					
		NAA	0.5						
		BAP	0.1						

[a]Notes: B5 medium: Gamborg *et al.* (1968); KM medium: Kao and Michayluk (1975); LS medium: Linsmaier and Skoog (1965); MS medium: Murashige and Skoog (1962); N medium: Nitsch (1969).
[b]Abbreviations: M, mesophyll; SES, stem embryo system; R, roots.

Addition of fresh medium	Transfer I	Transfer II	Transfer III	Reference
time B5 +sucrose 3% +CPA 1.0 mg/liter +BAP 1.0	Or alternatively: NAA 0.02 mg/liter BAP 2.0 2–3 weeks	BAP 0.5 2–3 weeks		
Fresh medium volume: 60% of initial ml used per Petri dish	MS solid or liquid IAA 4.0 mg/liter KIN 2.56 4 weeks	—	—	Ulrich *et al.* (1980)
1 ml medium, containing colonies placed onto 2 ml B5 medium with mannitol (0.27 *M*) agar (0.6%) Colonies settle onto the agar 2 weeks	Transfer of the settled colonies on agar blocks to solidified culture medium with 0.1 *M* mannitol 2 weeks	Same procedure as transfer I medium without mannitol 2 weeks	Agar medium MS salts B5 vitamins sucrose 3% mg/liter N-Z-amine 150 NAA 0.02 BAP 2 2 weeks	Gatenby and Cocking (1977)
Replacement of old medium by fresh medium Weekly for first 2 weeks	S1 agar 2,4-D 0.02 mg/liter NAA 1.0 BAP 0.5 lacking mannitol 3–4 weeks	LS agar BAP 0.2 mg/liter lacking mannitol 3–4 weeks	—	Schenck and Röbbelen (1982) H. R. Schenck (unpublished research 1983)
Replacement of old medium by fresh medium Weekly for first 2 weeks	LS agar Sucrose 20 g/liter Inositol 100 mg/liter GA_3 1.0 NAA 1.0 BAP 0.5 Or alternatively: NAA 0.1 ZEA 2.0 without mannitol	—	—	Schenck and Hoffmann (1979)

volumes of 0.24 *M* $CaCl_2$ and centrifuged again. This pellet was suspended in 1 volume of culture medium plus 5 volumes of 0.54 *M* sucrose and centrifuged at 200 *g*. The undamaged floating protoplasts were pipetted off and centrifuged twice more in a solution containing culture medium and $CaCl_2$ (1:2) at 50 *g*; then the pelleted protoplasts were suspended in culture medium.

Stem embryo protoplasts may be an alternative source of leaf material in other plant species as well. The possibility of inducing stem embryos seems to be a general feature. It is reported, for instance, in *Ranunculus* (Konar *et al.*, 1972), *Datura* (Geier and Kohlenbach, 1973), and *Solanum* (Sopory *et al.*, 1978). In rapeseed stem embryo systems can be obtained best via androgenetic embryogenesis (Thomas *et al.*, 1976).

D. Protoplast Isolation from Roots

Xu *et al.* (1982) described the isolation of *Brassica* protoplasts from roots. One-centimeter-long apical root pieces were excised from sterile agar-germinated seedlings. Those of *B. alba* were cut transversely into 0.5- to 1.0-mm segments, whereas the thinner roots of *B. campestris, B. napus,* and *B. oleracea* were left intact. Root material was plasmolyzed and enzymatically treated, and protoplasts were released by gently squeezing the root segments. Yields were, on average, 1.5×10^6 protoplasts per 200 root explants (ca. 1 g fresh weight).

III. PROTOPLAST CULTURE

The most frequently used culture media contain the phytohormones 2,4-dichlorophenoxyacetic acid (2,4-D) (0.2–1.0 mg/liter), naphthaleneacetic acid (NAA) (0.5–1.0 mg/liter), and benzylaminopurine (BAP) (0.1–0.5 mg/liter) during the initial culture period. All authors describe subsequent washings of protoplasts or the introduction of fresh medium with constant or slightly reduced levels of osmotica. These adjustments appear to be necessary prerequisites for further development. *Brassica* protoplasts apparently release toxic substances into the medium, as suggested by Schenck and Hoffmann (1979) and by Ulrich *et al.* (1980). Addition of 0.05% activated charcoal obviously adsorbs some of these substances, yielding a 10 times higher microcolony formation (Kohlenbach *et al.*, 1982a). The release of brown particles by the cells directly or precipitated as a result of interactions between cell products and media components seems to be correlated with high polyphenol oxidase (PPO) activity. These particles can completely obliterate the protoplasts, as well as small colonies. Therefore, selection of individual plants with low PPO activity is favorable, especially in *B. campestris* and *B. oleracea* (Schenck and Hoffmann, 1979).

An initial dark period is almost always necessary. This is followed by a phase under weak light and a final stronger light phase (Table III).

Haploid protoplasts of stem embryos of *B. napus* can also differentiate directly into tracheary elements without intervening cell divisions (Kohlenbach *et al.*, 1982b). Up to 600 such elements per milliliter can be obtained (initial density 2×10^5 protoplasts per milliliter) 4–55 days after protoplast isolation. Preliminary investigations show that the formation of tracheary elements may be influenced by the hormone concentrations and the nature of the osmotic stabilizer.

IV. PLANT REGENERATION

As in most other plant species under investigation, plant regeneration from protoplast-derived callus remains a major problem for *Brassica*. However, considerable progress has recently been made. Shoot formation in the early publications was an extremely rare event. Those experiments were performed with mesophyll protoplasts (Kartha *et al.*, 1974b; Thomas *et al.*, 1976). Therefore, alternative sources for *Brassica* protoplasts were looked for and were found in stem embryo systems (Kohlenbach *et al.*, 1982a) and roots (Xu *et al.*, 1982). A limited but reproducible number of regenerated *B. napus* plants can be obtained via somatic embryogenesis from cultures of isolated protoplasts from stem embryos. The procedure consists of a sequence of culture steps, including a medium with potato extract. On this medium, calli with embryogenic nodules arise which may release their embryogenic capacity when they are transferred to a medium with reduced auxin. Defined plantlets and/or new-stem embryos arise from up to one-third of the cultures. Following this procedure, it is possible to obtain three to six stem embryo systems per isolation experiment (starting with 2×10^6 protoplasts isolated from 6 g stem embryo material). The method works reliably but is not yet suited for application (e.g., mutagenesis), being too indirect and demanding too much time from the initiation of protoplast cultures and the regeneration of plants (35 weeks). Xu *et al.* (1982) report the regeneration of one shoot per 6–10 root protoplast-derived calli in *B. napus* and *B. oleracea*.

Progress has also been achieved by Li and Kohlenbach (1982). Somatic embryos and plantlets have been obtained from mesophyll protoplasts isolated from androgenetic rapeseed. Proembryos were induced in a medium supplemented with an auxin and a cytokinin at comparatively high concentrations. They developed the mature embryos when the auxin was reduced or omitted. The regeneration occurs directly, without an intervening callus phase. The method yields about 1000 normal plantlets per 10^6 cultured protoplasts, and the regeneration cycle requires only about 4 months. This system should be suitable for testing in applied experiments. Furthermore, the authors suggest that this material might become a system for analyzing somatic embryogenesis *ab initio*. In *B. napus* each embryo comes from undertermined mesophyll cells. Analysis of the initiation of embryogenesis in such cells may produce a general procedure for other species.

REFERENCES

Gamborg, O. L., Miller, R. A., and Ojima, K. (1968). Nutrient requirements of suspension cultures of soybean root cells. *Exp. Cell Res.* **50,** 151–158.

Gatenby, A. A., and Cocking, E. C. (1977). Callus formation from protoplasts of marrow stem kale. *Plant Sci. Lett.* **8,** 275–280.

Geier, T., and Kahlenbach, H. W. (1973). Entwicklung von Embryonen und embryogenem Kallus aus Pollenkörnern von *Datura meteloides* und *Datura innoxia*. *Protoplasma* **78,** 381–396.

Gleba, Y. Y., and Hoffmann, F. (1980). "Arabidobrassica": A novel plant obtained by protoplast fusion. *Planta* **149,** 112–117.

Hoffmann, F. (1980). Pflanzliche Zellkulturtechniken als Züchtungsschritt am Beispiel Raps. *Naturwissenschaften* **67,** 301–306.

Hoffmann, F., and Adachi, T. (1981). "Arabidobrassica": Chromosomal recombination and morphogenesis in asymmetric intergeneric hybrid cells. *Planta* **153,** 586–593.

Hoffmann, F., Schenck, H., Kohlenbach, H. W., and Gleba, Y. Y. (1980). Regeneration and fusion of protoplasts from important crop plants of the *Brassiceae*. *In* "Advances in Protoplast Research" (L. Ferenczy, G. L. Farkas, and G. Lazar, eds.), pp. 287–292. Akadémiai Kiadó, Budapest.

Howell, S. H., and Hull, R. (1978). Replication of cauliflower mosaic virus and transcription of its genome in turnip leaf protoplasts. *Virology* **86,** 468–481.

Kao, K. N., and Michayluk, M. R. (1975). Nutritional requirements for growth of *Vicia hajastana* cells and protoplasts at a very low population density in liquid media. *Planta* **126,** 105–110.

Kartha, K. K., Gamborg, O. L., Constabel, F., and Kao, K. N. (1974a). Fusion of rape seed and soybean protoplasts and subsequent division of heterokaryocytes. *Can. J. Bot.* **52,** 2435–2436.

Kartha, K. K., Michayluk, M. R., Kao, K. N., Gamborg, O. L., and Constabel, F. (1974b). Callus formation and plant regeneration from mesophyll protoplasts of rape plants (*Brassica napus* cv. Zephyr). *Plant Sci. Lett.* **3,** 265–271.

Kohlenbach, H. W., Wenzel, G., and Hoffmann, F. (1982a). Regeneration of *Brassica napus* plantlets in cultures from isolated protoplasts of haploid stem embryos as compared with leaf protoplasts. *Z. Pflanzenphysiol.* **105,** 131–142.

Kohlenbach, H. W., Körber, M., and Li, L. (1982b). Cytodifferentiation of protoplasts isolated from a stem embryo system of *Brassica napus* to tracheary elements. *Z. Pflanzenphysiol.* **107,** 367–371.

Konar, R. N., Thomas, E., and Street, H. E. (1972). Origin and structure of embryoids arising from epidermal cells of the stem of *Ranunculus scelerathus. J. Cell Sci.* **11,** 77–93.

Li, L., and Kohlenbach, H. W. (1982). Somatic embryogenesis in quite a direct way in cultures of mesophyll protoplasts of *Brassica napus. Plant Cell Rep.* **1,** 209–211.

Linsmaier, E. M., and Skoog, F. (1965). Organic growth factor requirements of tobacco tissue cultures *Physiol. Plant.* **18,** 100–127.

Murashige, T., and Skoog, F. (1962). A revised medium for rapid growth and bioassays with tobacco tissue cultures. *Physiol. Plant.* **15,** 473–479.

Nitsch, J. P. (1969). Experimental androgenesis in *Nicotiana. Phytomorphology* **10,** 389–404.

Potrykus, I., Harms, C. T., Lörz, H., and Thomas, E. (1977). Callus formation from stem protoplasts of corn (*Zea mays*). *Mol. Gen. Genet.* **156,** 347–350.

Schenck, H. R. (1981). Plant regeneration from protoplasts of *Brassica oleracea. Cruciferae Newsl.* **6,** 23–24.

Schenck, H. R., and Hoffmann, F. (1979). Callus and root regeneration from mesophyll protoplasts of basic *Brassica* species: *B. campestris, B. oleracea* and *B. nigra. Z. Pflanzenzuecht.* **82,** 354–360.

Schenck, H. R., and Röbbelen, G. (1982). Somatic hybrids by fusion of protoplasts from *Brassica oleracea* and *B. campestris. Z. Pflanzenzuecht.* **89,** 278–288.

Sopory, S. K., Jacobsen, E., and Wenzel, G. (1978). Production of monohaploid embryoids and plantlets in cultured anthers of *Solanum tuberosum. Plant Sci. Lett.* **12,** 47–54.

Thomas, E., Hoffmann, F., Potrykus, I., and Wenzel, G. (1976). Protoplast regeneration and stem embryogenesis of haploid androgenetic rape. *Mol. Gen. Genet.* **145,** 245–248.

Tsunoda, S., Hinata, K., and Gömez-Campo, C., eds. (1980). "*Brassica* Crops and Wild Allies." Jpn. Sci. Soc. Press, Tokyo.

Ulrich, T. H., Chowdhury, J. B., and Widholm, J. W. (1980). Callus and root formation from mesophyll protoplasts of *Brassica rapa. Plant Sci. Lett.* **19,** 347–354.

Vaughan, J. G., MacLeod, A. J., and Jones, B. M. G., eds. (1976). "The Biology and Chemistry of Cruciferae." Academic Press, New York.

Xu, Z., Davey, M. R., and Cocking, E. C. (1982). Plant regeneration from root protoplasts of *Brassica. Plant Sci. Lett.* **24,** 117–121.

CHAPTER **42**

Isolation and Culture of Protoplasts: Tomato

Elias A. Shahin

ARCO Plant Cell Research Institute
Dublin, California

I. INTRODUCTION

In recent years, considerable progress has been made toward the development of protoplast technology in a wide variety of plants. The suitability of protoplasts for fusion experiments, as well as uptake of DNA and cell organelles, have made them an important tool for genetic engineering of crops. However, plant regeneration from protoplasts has been achieved only in a limited number of species, mostly from the Solanacae (Vasil and Vasil, 1980). Nevertheless, plant regeneration from tomato protoplasts remains unattainable and unpredictable (F. J. Zapata *et al.*, 1977; F. J. Zapata *et al.*, 1981) even though plants could be easily regenerated from leafexplants (Behki and Lesley, 1976).

CELL CULTURE AND SOMATIC CELL
GENETICS OF PLANTS, VOL. 1

ISBN 0-12-715001-3

Fig. 1. Plant regeneration from tomato (*Lycopersicon esculentum* cultivar VF-36) mesophyll protoplasts. (A) Freshly isolated protoplast (420X). (B,C) Cell growth and first cell divisions (224X). (D) Cluster of regenerated cells after 10 days of culture in TM-2 liquid medium (70X). (E) Adventitious shoots forming on mini-calli 21 days after transfer to TM-4 medium. (F) Protoplast-derived plant growing in pot.

Tomato is an agriculturally important plant which has almost every characteristic required of a model plant system: extensive genetic mapping, availability of genetically marked lines, vigorous growth, short generation time, and high fertility. One drawback is the lack of a protocol to regenerate reproducibly a large number of plants from isolated tomato protoplasts. In this chapter, we describe the establishment of such a system utilizing *in vitro* preconditioned plants as a source for protoplast isolation.

II. PROCEDURE

A. *In Vitro* Preconditioning of Tomato Plants prior to Protoplast Isolation

1. Seed Sterilization and Germination

Transfer about 200 tomato seeds to 20 ml of 70% ethanol in a 50-ml beaker and place it on a magnetic stirrer. After 2 min, discard the ethanol and add a 20% solution of a commercial bleach (1.05% Na-hypochlorite). Stir for 15–20 min, and then discard the floating seeds as well as the solution. Rinse the remaining seeds (four to five times) with sterile distilled water. Thereafter, place the seeds on a double layer of presterilized filter paper (Whatman no. 9) in a glass Petri dish and add 5 ml sterile distilled water. Seal the Petri dish with Parafilm, and place it in a dark incubator at 28°C.

2. *In Vitro* Establishment of Plants

When seeds have germinated (3–5 days), cut off the roots and a portion 2–5 mm) of the hypocotyl. Transfer the excised shoot into a Plant Con (four to eight per container) containing 200 ml TM-1 medium (Table I). Place the containers in an incubator with the following light and temperature regime: a constant temperature of 25°C, a 16/8-hr photoperiod, and a light intensity of 4800 lx.

3. Dark Treatment

When plants have grown enough to reach the lid of the Plant Con (3–4 weeks), or have expanded leaves with a sizable stem, remove the plant containers from the incubator and place them in total darkness at room temperature (25°C) for 48 hr; 10- to 12-day-old cotyledons are an attractive

TABLE I

Constituents and Concentrations of Tomato Culture Media[a]

Constituent	TM-1	TM-2	TM-3	TM-4	TM-5
Macronutrients					
KH_2PO_4	—	170	170	—	—
$CaCl_2 \cdot 2H_2O$	150	440	440	150	75
KNO_3	2530	1500	1500	1900	1265
NH_4NO_3	320	—	—	320	160
$NH_4H_2PO_4$	230	—	—	230	115
$(NH_4)_2SO_4$	134	—	—	134	67
$MgSO_4 \cdot 7H_2O$	250	370	370	247	125
Micronutrients					
Potassium iodide	0.83	0.83	0.83	0.83	0.83
H_3BO_3	6.20	6.20	6.20	6.20	6.20
$MnSO_4 \cdot 4H_2O$	22.30	22.30	22.30	22.30	22.30
$ZnSO_4 \cdot 7H_2O$	8.60	8.60	8.60	8.60	8.60
$Na_2MoO_4 \cdot 2H_2O$	0.25	0.25	0.25	0.25	0.25
$CuSO_4 \cdot 5H_2O$	0.025	0.025	0.025	0.025	0.025
$CoCl_2 \cdot 6H_2O$	0.025	0.025	0.025	0.025	0.025
$FeSO_4 \cdot 7H_2O$	13.90	13.90	13.90	13.90	13.90
$Na_2 \cdot EDTA$	18.50	18.50	18.50	18.50	18.50
Vitamins					
Nicotinic acid	2.50	2.50	5	5	5
Thiamine·HCl	10	10	0.50	0.50	0.50
Pyridoxine HCl	1	1	0.50	0.50	0.50
Folic acid	0.50	0.50	0.50	0.50	0.50
Biotin	0.05	0.05	0.05	0.05	0.05
D-Ca-pantothenate	0.50	0.50	—	—	—
Choline chloride	0.10	0.10	0.10	0.10	0.10
Glycine	0.50	0.50	2.50	2.50	2.50
Casein hydrolysate	50	150	100	—	—
L-Cysteine	1	1	—	—	—
Malic acid	10	10	—	—	—
Ascorbic acid	0.50	0.50	—	—	—
Adenine sulfate	—	40	40	—	—
L-Glutamine	—	100	100	—	—
Myo-inositol	100	4600	100	100	100
Riboflavin	0.25	0.25	—	—	—
Others					
Sucrose	30.0 g	68.40 g	50.0 g	30.0 g	30.0 g
Mannitol	—	4.56 g	—	—	—
Xylitol	—	3.80 g	—	—	—
Sorbitol	—	4.56 g	—	—	—
MES	—	97.60	97.60	97.60	—
Noble agar	6.0 g	—	7.0 g	7.0 g	9.0 g
Hormones					
NAA	—	1	—	—	—

(continued)

TABLE I (*Continued*)

Constituent	TM-1	TM-2	TM-3	TM-4	TM-5
Zeatin riboside	—	0.5	—	1	—
2,4-D	—	—	0.20	—	—
BAP	—	—	0.50	—	—
GA_3[b]	—	—	—	0.20	—
IBA	—	—	—	—	0.1
pH	5.80	5.60	5.80	5.80	5.80

[a] Culture media were brought to pH 5.60–5.80 with 0.1 *N* KOH after addition of all ingredients, except agar, and then sterilized by autoclaving for 15 min at 115°C. TM-2 medium was filter sterilized by using Nalgene filter. The unit of measure for all media is milligrams per liter except as noted.

[b] GA_3 applied after autoclaving.

source of protoplasts because of the ease of producing the material. This treatment helps reduce starch accumulation and favors osmotic adjustment of the cells upon isolation.

4. Pre-Enzyme Cold Temperature Treatment (PET)

Remove the leaves or cotyledon from the stem, and use a sharp pair of scissors to cut fine strips (3–5 mm). Place the tissues (0.5–1.0 g) in a 125-ml side arm Erylenmyer flask that contains 30 ml PET solution (Table II). Keep it at 10°C for 12–14 hr. The hypocotyl and stem should be cut first longitudinally, followed by thin transverse sections. This treatment will increase protoplast yield and division.

B. Enzyme Digestion

Discard the PET solution and add the enzyme solution (Table II) at a rate of 25 ml/0.5g tissue. Seal the flask with a rubber stopper no. 5, and then vacuum the infiltrate for 1 min. Place the flask in a water bath shaker at 60–75 rpm at 28–29°C. Incubation varies with the source of protoplasts: 4 hr are needed for complete digestion of cotyledon and leaf tissue, whereas 14–16 hr are needed for stem tissue. Use enzyme solution TSE-1 for leaves, TSE-2 for both cotyledon and leaf tissue, and TSE-3 for stem and hypocotyl tissue (Table II).

C. Protoplast Purification

Pour the digested tissues into a funnel containing two layers of cheesecloth and collect protoplasts in Babcock bottles (Kimble, Co., serial 1000).

TABLE II

Components of Solutions for Preconditioning and Isolation of Tomato Protoplasts[a]

Constituent	PET	TSE-1	TSE-2	TSE-3	TWS
Macronutrients					
KH_2PO_4	42	85	85	85	85
$CaCl \cdot 2H_2O$	110	220	220	220	220
KNO_3	375	750	750	750	750
$MgSO_4 \cdot 7H_2O$	92.50	185	185	185	185
Vitamins					
Nicotinic acid	2.50	2.50	2.50	2.50	2.50
Thiamine·HCl	10.00	10.00	10.00	10.00	10.00
Pyridoxine HCl	1.00	1.00	1.00	1.00	1.00
Folic acid	0.50	0.50	0.50	0.50	0.50
Biotin	0.05	0.05	0.05	0.05	0.05
D-Ca-pantothenate	0.50	0.50	0.50	0.50	0.50
Choline chloride	0.10	0.10	0.10	0.10	0.10
Ascorbic acid	0.50	0.50	0.50	0.50	0.50
Myo-inositol	100.00	100.00	100.00	100.00	100.00
Casein hydrolysate	50.00	50.00	50.00	50.00	50.00
L-Cysteine	1.00	1.00	1.00	1.00	1.00
Malic acid	10.00	10.00	10.00	10.00	10.00
Glycine	0.50	0.50	0.50	0.50	0.50
Riboflavin	0.25	0.25	0.25	0.25	0.25
Enzymes					
Cellulysin[b]	—	0.75%	1.50%	—	—
Macerozyme[b]	—	0.10%	0.30%	1.0%	—
Cellulase R-10[c]	—	—	—	2.0%	—
Others					
Sucrose	102.60 g	102.60 g	102.60 g	102.60 g	102.60 g
PVP-10	—	1.00%	1.00%	1.00%	—
MES	97.60	97.60	97.60	97.60	97.60
2,4-D	1.00	—	—	—	—
BAP	0.50	—	—	—	—
pH	5.80	5.60	5.60	5.60	5.80
Sterilize	Autoclave	——Filter sterilize——			Autoclave

[a] The unit of measure for all media is milligrams per liter except as noted.
[b] Calbiochem-Behring Corp., La Jolla, California.
[c] Yakult Biochemicals Co., Nishinomiya, Japan.

Allow to stand for 30 min in the dark, and then centrifuge for 10 min at 1000 rpm in an International Model HN-S centrifuge. During centrifugation, the viable protoplasts float to the top, whereas the debris settles to the bottom or remains suspended. Collect the protoplasts by using a sterile Pasteur pipette, and then resuspend in a tomato washing solution (TWS) (Table II) in a Babcock bottle. Allow it to stand for 15 min, and then centrifuge for 5 min at 1000 rpm.

D. Osmolarity Adjustment of the Protoplasts

Collect the clean floating protoplasts using a sterile Pasteur pipette and place them in a small volume (0.5–2.0 ml) of the plating medium minus hormones TM-2 (Table I). The top of the Babcock bottle is graduated from 1 to 8 units, with 10 divisions per unit. The protoplast yield can be calculated on the basis of their banding in correspondence with those divisions. Under our conditions, each division is calibrated to have 350,000–400,000 protoplasts. However, this figure will vary considerably with the size of the protoplasts, centrifugation speed and time, and osmolarity. Therefore, it is necessary to use a hemocytometer to confirm the number of protoplasts. Adjust the protoplast density to give a final figure of 500,000/ml. Allow the protoplasts to stand for at least 1 hr before plating.

E. Plating and Culture of the Protoplasts

Culture the protoplasts in a thin layer (2 ml) of TM-2 medium in 60 × 15-mm plastic Petri dishes. The hormonal components of TM-2 medium are not critical to protoplast survival and division; a wide range of auxins and cytokinins can be used in addition to naphthaleneacetic acid (NAA) and zeatin riboside. Protoplasts divide and form cell colonies (Fig. 1A,B,C) in the basal TM-2 medium that is supplemented with 2,4-dichlorophenoxyacetic acid/benzylaminopurine (2,4-D/BAP), 2,4-D/zeatin riboside, *para*-chlorophenoxyacetic acid (*p*-CPA)/zeatin riboside, NAA/BAP, NAA/kinetin,2,4-D/kinetin, and *p*-CPA/BAP combinations in the range of 0.5–1.0 mg/liter.

Pour the culture medium first in the Petri dish, and then gently add 0.2 ml of the protoplast solution. Rotate the plates very gently to allow even distribution of the protoplasts, and then seal them with Parafilm. Incubate at 25°C in diffused light (500 lx, 16 hr/day) in a covered plastic box humidified by moist blotting paper in a beaker with 2% $CuSO_4$ solution. Be sure to place the box on a leveled shelf to minimize the uneven distribution of the protoplasts.

F. Mini-Callus Formation

Ten days or more after culturing the protoplasts, multicellular colonies will be visible (Fig. 1D). At this stage, use a sterile Pasteur pipette to transfer colonies onto the agar surface of TM-3 medium (Table I). Allow the colonies to spread evenly onto the surface of the agar. Use about 0.5 ml aliquot per plate and dilute when necessary with 0.2 ml TM-2 medium. Do not allow the cell colonies to be covered with the liquid medium. For this reason, remove the excess solution by using a sterile Pasteur pipette or by keeping the plate uncovered for 10–15 min in the hood.

Use a sterile scapel (Razor no. 15) to remove the mini-calli that will stay attached to the bottom or sides of the Petri dishes. Seal the plates with Parafilm, place them in a clear plastic box, and incubate under the same environmental conditions as described in Section II,E.

G. Shoot Induction

Six days later, white mini-calli (0.5–2.0 mm) will be apparent. Carefully transfer the mini-calli (10 per plate) onto the surface of TM-4 medium (Table I). Use a scalpel (Razor no. 15) to lift the tiny calli from the side in contact with the agar. Be sure to clean the mini-callus from any attached agar. Seal the plates with Parafilm and place them in a clear plastic box. Incubate them under a 16-hr/day light period of 4800 lx light intensity and 25°C constant temperature. Shoots will be observed after 9 days or longer, but this varies with the cultivar. A wide range of auxin/cytokinin can be used in the basal TM-4 medium other than GA_3 and zeatin riboside; indoleacetic acid (IAA)/kinetin, GA_3/2iP, IAA + GA_3/BAP, IAA/2iP, GA_3/kinetin, IAA + GA_3/kinetin, and IAA + GA_3/zeatin riboside. However, each of these combinations is genotype specific.

H. Root Initiation

When shoots reach 1 cm or more in length (Fig. 1E), dissect them from the callus and transfer them to TM-5 medium (Table I). Make sure that the shoot is cut at a 60° angle to the stem. Incubate them under the same conditions as described in Section II,G. Shoots that are not fully developed, leaf buds, and meristematic domes have to be subcultured on fresh TM-4 medium to allow extensive growth before being transferred to rooting medium. By following this step, all the existing shoots can be recovered, and new multiple shoots will emerge from the organized callus. In some cases, the shoots will develop callus around the base of the stem, which apparently retards growth. To avoid this, cut off the callus and subculture the stem onto fresh TM-5 medium. Usually allow 3 weeks for root formation before proceeding with any of the previously mentioned steps.

I. Plant Formation

Once the rooted plantlets are well developed (2–3 weeks after root initiation), transfer them to 4-in. pots containing a mixture of soil:sand (1:1). Cover the shoots with plastic cups to avoid dehydration and place the pots in a growth chamber of 16-hr/day light (7000 lx) and 25°C constant temperature. Fertilize weekly with a diluted solution of 20 N:20 P:20 K fertilizer (1 g/liter). Once the plants attain a certain size (above 15 cm), they can be moved to the greenhouse (Fig. 1F).

III. RESULTS AND DISCUSSION

Our results (Table III) demonstrate the feasibility of regenerating tomato protoplasts by utilizing the protocol described above and summarized in

TABLE III

Summarized Response of Protoplast Culture of Different Tomato Cultivars[a]

Cultivar	Source of protoplasts	Division	Colony formation	Callus formation	Shoot formation	Root formation	Whole plant
Red cherry	Leaf	+[b]	+	+	+	+	+
Cocktail cherry	Leaf	+	+	+	+	+	+
VFNT-cherry	Leaf	+	+	+	+	+	+
VF-36	Leaf	+	+	+	+	+	+
VF-36	Stem	+	+	+	+	+	+
Manapal	Leaf	+	+	+	+	+	+
Manapal	Cotyledon	+	+	+	+	+	+
Manapal	Stem	+	+	+	+	+	+
Floradade	Leaf	+	+	+	+	+	+
UC-82	Leaf	+	+	+	+	+	+
Red Ace–VF type	Cotyledon	+	+	+	+	+	+
Roma	Leaf	+	+	+	+	+	+
Beefsteak	Leaf	+	+	+	+	+	+
San Marzano	Leaf	+	+	+	+	+	+

[a] Protoplasts isolated and cultured from different cultivars under identical experimental conditions as given in this chapter.
[b] + = positive response.

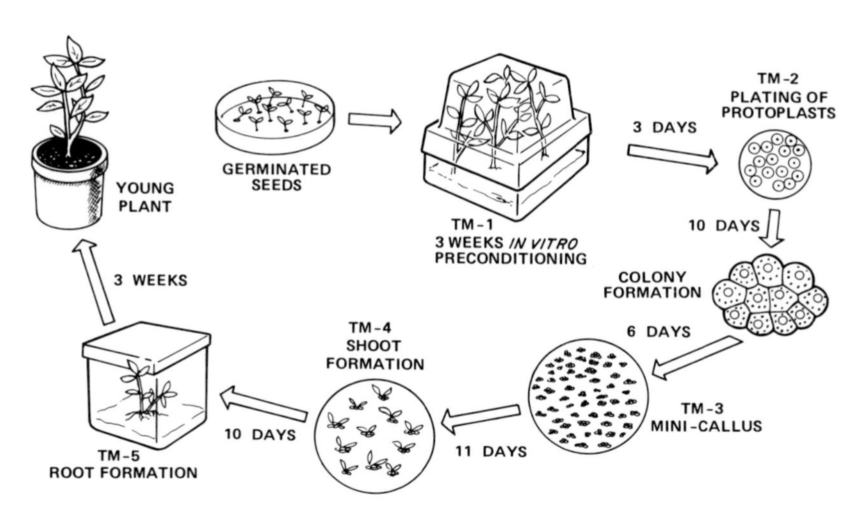

Fig. 2. Schematic sequence for *in vitro* preconditioning, isolation, culture, and plant regeneration from protoplasts of cultivated tomato.

Fig. 2. Protoplasts from all commercial tomato cultivars tested (eleven cultivars) have proliferated to mini-calli and subsequently to shoots without any modification of the media or the protocol used. Some differences were observed in the plating efficiency and the frequency of shoot induction, but these could be optimized by minor modification of the growth regulators in the basal TM-2 and TM-4 media (Shahin, unpublished data). As for the shooting medium, the combination of GA_3 and zeatin riboside in basal medium TM-4 was very effective in shoot induction in all tested cultivars. On the other hand, other auxin/cytokinin combinations (i.e., IAA/kinetin, GA_3/2iP) were effective only on a specific genotype, while failing to induce shoots on others (E. A. Shahin, unpublished data). Therefore, it is very likely that protoplast-derived calli from other tomato cultivars will develop shoots on GA_3/zeatin riboside-supplemented TM-4 medium or a slightly modified medium.

Whole plants regenerated from seven cultivars were grown in the greenhouse. No somaclonal variations were observed. Most of the plants have flowered and set viable seeds by self-pollination. Presently, the first selfed progeny (TP-1 generation) are being planted in the field to screen for any phenotypic variations under normal conditions of cultivation. It is possible that this tomato protoplast system could prove useful in parasexual hybridization and transformation studies in *Lycopersicon* species.

REFERENCES

Behki, R. M., and Lesley, S. M. (1976). *In vitro* plant regeneration from leaf explants of *Lycopersicon esculentum* (tomato). *Can. J. Bot.* **54,** 2409–2414.

Vasil, I. K., and Vasil, V. (1980). Isolation and culture of protoplasts. *Int. Rev. Cytol., Suppl.* **11B,** 1–20.

Zapata, F. J., Evans, P. K., Power, J. B., and Cocking, E. C. (1977). The effect of temperature on the division of leaf protoplasts of *Lycopersicon esculentum* and *Lycopersicon peruvianum.* *Plant. Sci. Lett.* **8,** 119–124.

Zapata, F. J., Sink, C. K., and Cocking, E. C. (1981). Callus formation from leaf mesophyll protoplasts of three *Lycopersicon* species: *L. esculentum* c.v. Walter, *L. pimpinellifolium* and *L. hirsutum, F. glabratum.* *Plant Sci. Lett.* **23,** 41–46.

CHAPTER **43**

Isolation, Culture, and Regeneration of Potato Leaf Protoplasts from Plants Preconditioned *in Vitro*

Elias A. Shahin

ARCO Plant Cell Research Institute
Dublin, California

I. INTRODUCTION

Considerable interest in culture of potato protoplasts was developed after Shepard and co-workers exploited the natural and/or induced genetic variability associated with regenerating plants via protoplast culture (Secor and Shepard, 1980; Shepard *et al.*, 1980; Shepard, 1980, 1981). The protocol for the isolation, culture, and regeneration from mesophyll protoplasts was described earlier (Shepard and Totten, 1977). Elsewhere, other researchers have described different protocols for regenerating plants from leaf protoplasts of dihapoid *Solanum tuberosum* and *S. chacoense* (Binding *et al.*, 1978;

CELL CULTURE AND SOMATIC CELL
GENETICS OF PLANTS, VOL. 1

ISBN 0-12-715001-3

Butenko *et al.*, 1977). However, these techniques have not succeeded, in our and other laboratories (Shepard, 1982), with U.S. and Canadian cultivars. On the other hand, the techniques employed by Shepard *et al.* (1980) were applicable on both North American and European cultivars (E. A. Shahin, unpublished data; Gun and Shepard, 1981). These protocols are labor intensive, require the growth of plants in growth chambers prior to protoplast isolation, and utilize complex plating medium (reservoir medium). We have modified the procedures by employing more defined media and have developed controlled conditions for growing plants *in vitro*.

Presented here are our operative protocols (which may not be optimal, equal to, or superior to others); they are very reliable and reproducible, and offer the following advantages: (1) preculture of the plants can be optimized, and consequently the physiological conditions are much more constant than they are in growth chamber plants; (2) no surface sterilization is needed; (3) less time is needed for enzymatic digestion because of the more tender leaves; (4) establishment of such cultures is less time-consuming; (5) *in vitro* materials are available; and (6) continuous use of virus-free materials is assured.

II. PROCEDURE

A. *In Vitro* Preconditioning of Potato Plants

1. Plant Materials

Leaves from tetraploid potatoes ($2n = 4x = 48$), dihaploid *S. tuberosum* ($2n = 2x = 24$), and wild species *S. chacoense* ($2n = 2x = 24$) have been chosen to illustrate all the principles involved in regenerating plants from protoplasts. In tetraploid and dihaploid potatoes, only virus-free plants should be used as starting materials. Virus-free clones can be easily obtained by applying the meristem culture technique developed by Stace-Smith and Mellor (1968). On the other hand, true seeds are used for establishing starting materials from *S. chacoenese*.

2. *In Vitro* Establishment of Axillary Bud Cultures

All of the following operations should be performed aseptically, preferably in a laminar flow hood. For *S. chacoense*, germinate the true seeds in sterile distilled water by following standard procedures (i.e., 10 min in 2% NaOCl, rinse three times in sterile water). Ten days later, transfer the seedlings to Nalgene jars (250 ml), each containing 150 ml modified

TABLE I

Stock Solutions for Preparing Media for Culturing Potato Leaf Protoplasts

Constituent	Per 1000 ml
MS major salts (10X)[a]	
KNO_3	19.00 g
NH_4NO_3	16.50 g
$CaCl_2 \cdot 2H_2O$	4.40 g
$MgSO_4 \cdot 7H_2O$	3.70 g
KH_2PO_4	1.70 g
Modified MS major salts (M-MS) (10X)[b]	
KNO_3	19.00 g
$CaCl_2 \cdot 2H_2O$	4.40 g
$MgSO_4 \cdot 7H_2O$	3.70 g
KH_2PO_4	1.70 g
Micronutrients (100X)	
KI	83.00 mg
$ZnSO_4 \cdot 7H_2O$	920.00 mg
H_3BO_3	620.00 mg
$MnSO_4 \cdot 4H_2O$	2.23 g
$Na_2MoO_4 \cdot 2H_2O$	25.00 mg
$CuSO_4 \cdot 5H_2O$	2.50 mg
$CoCl \cdot 6H_2O$	2.50 mg
Vitamins (200X)[c]	
Myo-inositol	20.00 g
Thiamine·HCl	100.00 mg
Pyridoxine·HCl	100.00 mg
Nicotinic acid	1.00 g
Glycine	500.00 mg
Biotin	10.00 mg
Folic acid	100.00 mg
Fe/EDTA (20X)	
$Na_2 \cdot EDTA$	373.00 mg
$FeSO_4 \cdot 7H_2O$	278.00 mg

[a] Murashige and Skoog (1962).
[b] Shepard and Totten (1977).
[c] Modified from Nitsch and Nitsch (1969).

Murashige and Skoog (MS) medium supplemented with kinetin (KN) (Table II) for normal growth and expansion of the leaves. Maintain the cultures at 6000 lx light intensity at a constant temperature of 25°C and a 16/8-hr light and dark cycle. Similarly, for tetraploid and dihaploid potatoes, culture stem nodal segments from virus-free plantlets. Once rooted, plantlets continue to grow without transfer until the culture jar is crowded with roots and shoots. These plantlets can be maintained indefinitely by transferring shoot tips or nodal segments (2–4 cm) to fresh culture jars at intervals of about 3–4 months. However, we recommend that only 3- to 4-

TABLE II

Components of Medium and Solutions for *in Vitro* Preconditioning and Isolation of Potato Leaf Protoplasts

Constituent	PM (amount/liter)	ES-1 (amount/100 ml)	ES-2 (amount/100 ml)	WS (amount/1000 ml)
MS major salts (10X)[a]	100.0 ml	–	–	–
M-MS major salts (10X)[a]	–	5.0 ml	5.0 ml	5.0 ml
Fe/EDTA (100X)[a]	10.0 ml	1.0 ml	1.0 ml	–
Vitamins (200X)[a]	5.0 ml	0.5 ml	0.5 ml	5.0 ml
Micronutrients (100X)[a]	10.0 ml	1.0 ml	1.0 ml	10.0 ml
Casein hydrolysate	–	5.0 ml	5.0 mg	10.0 mg
MES	97.0 mg	9.7 mg	9.7 mg	97.0 mg
Cellulase R-10[b]	–	0.5%	–	–
Macerozyme R-10[b]	–	0.1%	–	–
Cellulysin[c]	–	–	2.0%	–
Macerase[c]	–	–	0.25%	–
PVP-10	–	1.0 g	1.0 g	–
Sucrose	30.0 g	10.27 g	10.27 g	102.7 g
Kinetin	0.2 mg	–	–	–
Agar (Difco)	7.0 g	–	–	–
pH	5.8	5.65	5.65	5.8
Autoclave for 15 min	+	–	–	+
Filter sterilize	–	+	+	–

[a] Use the corresponding stock solutions presented in Table I.
[b] Yakult Biochemicals Co., Nishinomiya, Japan.
[c] Calbiochem-Behring Corp., La Jolla, California.

week-old plantlets be used as a source of protoplasts. Older plant materials often produce low frequency of viable protoplasts and are more difficult to digest enzymatically.

3. Dark Treatment

When the plantlets develop expanded leaves, remove the jars from the incubator and place them in darkness at 25°C for 2 days. This type of pretreatment is beneficial, although not essential; it increases protoplast yield and rate of growth significantly.

B. Enzyme Digestion

Remove the leaves from the stem, and then gently brush the lower epidermis with a nylon brush to enhance the infiltration of the enzyme solution. Cut the leaves into small pieces (0.5–1.0 cm) and then place them in a 125-ml side arm Erlenmyer flask that contains the enzyme solution (Table II). Enzymes supplied by the Yakult Biochemical Co. (Japan) or the Calbiochem-Behring Corp. (La Jolla, California) are recommended for this particular protocol. Enzyme solution ES-1 is designed for tetraploid materials, whereas ES-2 is for dihaploid *S. tuberosum* and wild species. Use 1 g of leaf tissues per 25 ml of enzyme solution. Seal the flask with a rubber stopper no. 5, and then vacuum the infiltrate for 1 min. Place the flask in a water bath shaker at 60 rpm at 28–29°C. Incubation time varies with the age of the source plants; 4 hr are needed for complete digestion of leaf materials obatined from 4- to 6-week-old plants, whereas 8 hr are needed for older plants.

C. Protoplast Purification

Pour the digested leaf tissues into a funnel containing two layers of cheesecloth and collect protoplasts in Babcock bottles (Kimble 516). Centrifuge the bottles for 10 min at 1000 rpm (in an International Model HN-S centrifuge). The protoplasts will float to the top of the enzyme solution, whereas the debris settles to the bottom or remains suspended. Collect the protoplasts by using sterile Pasteur pipettes, and then resuspend them in a washing solution (WS) (Table II) in Babcock bottles. Centrifuge for 5 min at 1000 rpm, and then collect the protoplasts as before. This washing procedure may be repeated (at least twice) to ensure adequate cleaning of the protoplasts from enzymes and cell debris.

D. Osmolarity Adjustment of Protoplasts

At the end of the final wash, collect the floating protoplasts by Pasteur pipettes and place them in a small volume (1 ml) of the liquid plating medium but without growth regulators (Table III). Confirm their number by using a hemocytomer and then dilute with the liquid plating medium to give a final density of 500,000 protoplasts per milliliter. Allow the protoplasts to remain in this medium for at least 1 hr before plating.

E. Plating and Culture of Protoplasts

The protoplasts can be cultured in either liquid or semisolid medium (Table III). For plating in liquid medium, culture the protoplasts in a thin layer (2 ml) of liquid medium in 60 × 15-mm plastic Petri dishes. Pour the

TABLE III

Constituents of Protoplast-Plating Media

Constituent	Liquid medium (100 ml)	Semisolid medium[a]	
M-MS major salts (10X)	10.00 ml	12.50 ml	Part A (100 ml)
Sucrose 0.2 *M*	6.84 g	8.56 g	
Mannitol 0.025 *M*	456.00 mg	570.00 mg	
Xylitol 0.025 *M*	380.00 mg	450.00 mg	
Myo-inositol 0.025 *M*	450.00 mg	562.00 mg	
Sorbitol 0.025 *M*	456.00 mg	570.00 mg	
MES	9.70 mg	10.00 mg	
Micronutrients (100X)	1.00 ml	5.00 ml	Part B (25 ml)
Vitamins (200X)	0.50 ml	2.50 ml	
Fe/EDTA (20X)	5.00 ml	5.00 ml	
Casein hydrolysate	10.00 mg	25.00 mg	
Naphthaleneacetic acid (NAA)	0.10 mg	1.00 mg	
6-Benzylaminopurine (6-BAP)	0.05 mg	0.25 mg	
Agarose (Typ IV, Sigma)	–	1.10 g	
pH	5.65	5.80	
Filter sterilize	+	–	
Autoclave for 15 min	–	+	

[a] The semisolid medium is composed of two major separate components which have to be prepared and sterilized separately. Part A solution should have a pH adjusted to 5.65 and filter sterilized. After autoclaving part B, take out 5 ml and add it to 20 ml of part A. Mix well, and then allow to cool at room temperature before adding the protoplasts.

medium in the Petri dish, and then gently pipette 0.2 ml of the protoplast solution. Similarly, the protoplasts can be cultured in a thin layer (3 ml) of the agarose-embedded medium in 60 × 15-mm plastic Petri dishes. Gently pipette the protoplasts into the semisolid medium and mix by rotating the test tube. Apply 3 ml of the semisolid medium–protoplast mixture into the Petri dish, and then rotate the dish gently to allow even distribution of the protoplasts. Seal the dishes with Parafilm and place them in a covered plastic box humidified by a moist blotting paper in a beaker with 2% $CuSO_4$ solution. Incubate under continuous light, 500–800 lx, and 24°C constant temperature. Protoplasts should be plated, in both types of media, at a minimum density of 30,000 and not more than 120,000 per milliliter of medium. Protoplasts will not survive if the plating density is too low, whereas at a very high plating density, cell proliferation will cease. The protoplast's first division can be observed within 4–5 days after plating. However, the period of sustained division and formation of cell colonies varies among cultivars.

F. Callus Formation

After 10–14 days, the cell colonies will be ready for transfer to solid agar medium (Table IV) for further growth. Use a spatula to lift the agarose-embedded medium and place it on the surface of the solid agar medium. Also, it is possible to use a sterile Pasteur pipette with a wide opening. Gently spread the suspension over the plate with a sterile bent glass rod. Seal the plates with Parafilm, and incubate under continuous light of 4000 lx at 24°C. Ten days later, small white colonies will be visible to the naked eye, and in some cases they will appear as clusters of cells. Use a scalpel (with razor no. 15) to transfer the individual colonies (30–40 colonies per plate) onto the surface of fresh medium. Incubate in the same conditions as

TABLE IV

Media Constituents for Callus Formation, Shoot Induction, and Root Initiation[a]

Constituent	Callus formation	Shoot induction[b] (4X)	Shoot induction[b] (2X)	Root initiation
M-MS major salts (10X)	100.00 ml	100.00 ml	–	–
NH_4NO_3	300.00 mg	300.00 mg	–	–
NH_4Cl	100.00 mg	100.00 mg	–	–
MS major salts (10X)	–	–	100.00 ml	100.00 ml
Micronutrients (100X)	10.00 ml	10.00 ml	10.00 ml	10.00 ml
Fe/EDTA (20X)	50.00 ml	50.00 ml	50.00 ml	50.00 ml
Vitamins (200X)	5.00 ml	5.00 ml	5.00 ml	5.00 ml
Sucrose	2.50 g	2.50 g	30.00 g	30.00 g
Mannitol	54.20 g	54.20 g	–	–
Adenine sulfate	40.00 mg	80.00 mg	–	–
Casein hydrolysate	400.00 mg	80.00 mg	–	–
MES	97.00 mg	97.00 mg	97.00 mg	97.00 mg
NAA	0.10 mg	–	0.02 mg	–
6-BAP	0.50 mg	–	–	–
Indoleacetic acid (IAA)[c]	–	0.10 mg	–	–
Zeatin	–	0.50 mg	2.00 mg	–
GA_3[c]	–	–	0.20 mg	–
IBA	–	–	–	0.10 mg
Noble agar (purified)	9.00 g	15.00 g	15.00 g	–
Difco agar	–	–	–	6.50 g
pH	5.8	5.8	5.8	5.8
Autoclave for 15 min	+	+	+	+
Pour into Petri dishes	15 × 100 mm	20 × 100 mm	20 × 100 mm	Nalgene jars

[a] All amounts are per liter.
[b] 4X = shoot induction medium for tetraploid potato; 2X = shoot induction medium for dihaploid potato.
[c] Add these hormones after autoclaving.

before. As cell colonies proliferate on the medium, their color will change from white-beige (at 12 days) to light yellow (22 days), and then to green (30 days). At this time, they will be ready for transfer to shoot medium.

G. Shoot Induction

Carefully transfer the dark green calli (20–30 per plate) onto the surface of shoot medium (Table IV). Special attention should be given to avoid causing injury to the callus while it is being transferred. Injury will result in callus proliferation and no shoot induction. For this reason, gently lift the callus from the side that is touching the agar and then place it on the surface of the shoot medium. Use the appropriate medium for each protoplast-derived callus based on the ploidy level of the source materials. Incubate the plates at 24°C constant temperature and a 16-hr photoperiod. Shoot emergence should be apparent within 4 weeks. However, this varies from one cultivar to another.

H. Root Initiation

Potato protoplast-derived shoots root extremely well in modified MS medium supplemented with IBA (Table IV). Transfer the shoot along with the callus and place it on the agar; then incubate at 26°C under 8000 lx and a 16-hr photoperiod. In some cultivars, shoots can root easily on almost any commonly used basal medium without the presence of any auxins; however, high light intensity is critical.

I. Plant Formation

When a root system has developed (3–4 weeks), apply running tap water to loosen the agar. After removing the medium from the roots, place the plantlet in a soil:vermiculite (1:1) mixture. The shoot should be initially kept in very humid conditions (underneath a plastic cup) for a week for acclimatization. The plantlets should be maintained under high light intensity (>10,000 lx) in a moist chamber (>80% relative humidity) and a 14-hr photoperiod. Fertilize weekly by using Peter's solution 20:20:20, 1 g/liter. Once the plants attain a certain height (15 cm), they can be moved to the greenhouse.

III. RESULTS AND DISCUSSION

The regeneration potential of potato leaf protoplasts isolated from different genotypes and species under identical experimental conditions has been presented. The results given in Table V demonstrate that it is feasible using simple techniques and defined media, to regenerate plants from protoplasts isolated from *in vitro* preconditioned plantlets of tetraploid and dihaploid potatoes as well as from *S, chacoense*. Interestingly, a great deal of phenotypic variation occurred among plants, such as changes in stamen morphology, tuber color and shape, and stem wing. Field evaluation is needed to identify fully important agronomic traits such as yield and disease resistance.

Although this protocol was effective, without alteration, against six different genotypes (Table V), the media are not universal. Therefore, it should not be assumed that it can be applied to other potato cultivars without modification. In this case, the principal limitation is the researcher, who must know how and when to manipulate the various steps or media in order to obtain optimum results. For instance, attention should be given to the hormonal composition and concentration of the plating and shoot induction media. Equally important is the sucrose and/or mannitol concentration in the shoot induction medium. Also, it may be necessary to transfer the protoplast-derived calli to the shoot induction medium at an earlier stage to increase the frequency of shoot regeneration.

TABLE V

Potato Mesophyll Protoplast Isolation, Plating Efficiency, and Shoot Formation from Different Genotypes

Potato accession	Ploidy	Protoplasts (yield/g)	Plating density	Plating efficiency (%)[a]	Shoots[b]/callus (%)
A503-72	4X	6.8×10^6	60,000/ml	36	261/2500 (10.44)
Wn 708-27	4X	4.1×10^6	60,000/ml	18	1042/1250 (83.36)
NDD47-1	4X	5.9×10^6	60,000/ml	42	1838/2100 (87)
USDA-41956	4X	3.7×10^6	60,000/ml	20	314/985 (31.87)
PI 16194	2X	2.7×10^6	70,000/ml	26	256/1510 (16.95)
Solanum chacoense	2X	1.2×10^6	50,000/ml	53	126/850 (14.82)

[a] Plating efficiency represents the percentages of plated protoplasts that produced callus.
[b] Multiple shoots (three to four per callus).

REFERENCES

Binding, H., Nehls, R., Schieder, O., Sopory, S. K., and Wenzel, G. (1978). Regeneration of mesophyll protoplasts isolated from dihaploid clones of *Solanum tuberosum*. *Physiol. Plant.* **43,** 52–54.

Butenko, R. G., Kuchko, A. A., Vitenko, V. A., and Avetisov, V. A. (1977). Obtaining and cultivation of leaf mesophyll protoplasts of *Solanum tuberosum* L. and *Solanum chacoense* Bitt. *Sov. Plant Physiol. (Engl. Transl.)* **24,** 660–664.

Gun, R. E., and Shepard, J. F. (1981). Regeneration of plants from mesophyll-derived protoplasts of British potato (*Solanum tuberosum* L.) cultivars. *Plant Sci. Lett.* **22,** 97–101.

Murashige, T., and Skoog, F. (1962). A revised medium for rapid growth and bioassay with tobacco tissue cultures. *Physiol. Plant.* **15,** 473–497.

Nitsch, J. P., and Nitsch, C. (1969). Haploid plants from pollen grains. *Science* **163,** 85–87.

Secor, G., and Shepard, J. F. (1980). Variability of protoplast-derived potato clones. *Crop Sci.* **21,** 102–105.

Shepard, J. F. (1980). Mutant selection and plant regeneration from potato mesophyll protoplasts. *In* "Genetic Improvement of Crops: Emergent Techniques" (I. Rubenstein, B. Gengenbach, R. L. Phillips, and C. E. Green, eds.), pp. 185–219. Univ. of Minnesota Press, Minneapolis.

Shepard, J. F. (1981). Protoplasts as sources of disease resistance in plants. *Annu. Rev. Phytopathol.* **19,** 145–166.

Shepard, J. F. (1982). Cultivar dependent cultural refinements in potato protoplast regeneration. *Plant Sci. Lett.* **26,** 127–132.

Shepard, J. F., and Totten, R. E. (1977). Mesophyll cell protoplasts of potato: Isolation proliferation and plant regeneration. *Plant Physiol.* **60,** 313–316.

Shepard, J. F., Bidney, D., and Shahin, E. (1980). Potato protoplasts in crop improvement. *Science* **208,** 17–24.

Stace-Smith, R., and Mellor, F. C. (1968). Eradication of potato viruses X and S by thermotherapy and axillary bud culture. *Phytopathology* **58,** 199–203.

CHAPTER 44

Isolation and Culture of Protoplasts from Carrot Cell Suspension Cultures

Denes Dudits

Institute of Genetics
Biological Research Center
Hungarian Academy of Sciences
Szeged, Hungary

I. INTRODUCTION

As shown by early studies on plant tissue and cell cultures, carrot (*Daucus carota*) cells can be cultured in a liquid medium (Steward and Shantz, 1956) and their totipotency expressed through somatic embryo differentiation (Steward *et al.*, 1958; Reinert, 1959). Because of the high morphogenic potential of carrot somatic cells, carrot cell lines offer an appropriate experimental system for analysis of plant cell differentiation as well as protoplast-mediated genetic manipulation of plants (Sung and Dudits, 1981). In the early 1970s, when protoplast research became an intensively studied field of plant sciences, carrot cell suspensions were successfully used in the isolation of protoplasts (Chupeau and Morel, 1970; Hellmann and Reinert, 1971; Butenko and Ivantsov, 1973; Wallin and Eriksson, 1973). Studies on plant redifferentiation from protoplast-derived calli were first published by two research groups (Grambow *et al.*, 1972; Kameya and Uchimiya, 1972). Later modifications in the hormone com-

CELL CULTURE AND SOMATIC CELL
GENETICS OF PLANTS, VOL. 1

ISBN 0-12-715001-3

position of the culture medium made possible the early induction of embryogenesis and formation of somatic embryos without an intermediate callus stage (Dudits *et al.*, 1976).

Several isolation and culture procedures have been published by different laboratories working with carrot protoplasts (in addition to those previously mentioned: Gosch *et al.*, 1975; Harms *et al.*, 1981; Kameya *et al.*, 1981). Frequently, the efficiency of different modifications cannot be compared due to the lack of quantitative data. Occasionally, there is a need to develop a special medium to inhibit the division of a fusion partner (Dudits *et al.*, 1977) or to fulfill the additional requirements of an albino mutant exhibiting low division frequency in protoplasts (Dudits *et al.*, 1980a).

A protocol which has been used in our laboratory to culture protoplasts from cell suspensions of various genotypes, variants of wild and domesticated carrot, is described here.

II. ISOLATION OF CARROT PROTOPLASTS FROM CELL SUSPENSION CULTURES

Usually cell suspensions are used as protoplast sources, and few if any attempts have been made to induce cell division and colony formation from carrot leaf protoplasts. Among the factors determining the efficiency of culture methods, the most significant is the type and condition of the cell line; therefore, we present a protocol to establish and maintain a carrot cell suspension.

As far as the genotype is concerned, both the wild and domesticated carrots can form fine suspension cultures. However, the morphogenic potential was found to be more stable in wild carrot cultures. Varieties of domesticated carrots differ significantly in synchrony of embryo formation and retention of morphogenic capability. The use of a newly established cell line—not older than 1 year—is suggested if plant regeneration is a basic requirement in experiments with carrot protoplasts. With prolonged time in culture, the selection of actively dividing cell populations may result in a higher frequency of protoplast division with a simultaneous decline or loss of morphogenic capacity.

Primary callus tissues can be initiated from hypocotyl or leaf segments of young seedlings that are grown on a hormone-free culture medium (see below), supplemented with agar (0.8% w/v). After surface sterilization with mercurous chloride (0.1% w/v), the seeds are germinated under sterile conditions. Pieces (1.0–1.5 cm long) of hypocotyl or leaf tissues are placed on the surface of an agar basic culture medium with the following composition in milligrams per liter (after Murashige and Skoog, 1962; Gamborg and Eveleigh, 1968): NH_4NO_3, 1650; KNO_3, 1900; $MgSO_4 \cdot 7H_2O$, 370; KH_2PO_4, 170; $FeSO_4 7 \cdot H_2O$, 27.8; Na_2EDTA, 37.3; $CaCl_2 \cdot 2H_2O$, 440; potassium iodide (KI), 0.83; $MnSO_4 \cdot 4H_2O$, 22.3; H_3BO_3, 6.2; $ZnSO_4 \cdot 7H_2O$, 8.6; $Na_2MoO_4 \cdot 2H_2O$, 0.25; $CuSO_4 \cdot 5H_2O$, 0.025; $CoCl_2 \cdot 6H_2O$, 0.025; nicotinic acid, 1; thiamine HCl, 10; pyridoxine HCl, 1; inositol, 100; casein hydrolysate, 250; sucrose, 20,000; 2,4-dichlorophenoxyacetic acid (2,4-D), 0.5; pH: 5.6.

After incubation at room temperature in the dark for 4–6 weeks, calli are transferred into a liquid medium of the same composition without agar (Chapters 17 and 18, this volume). In order to ensure the formation of a fine suspension, 5–10 callus pieces should be placed into 20–30 ml of a culture medium in a 100-ml flask. These cultures are shaken on a gyrotory shaker at 25°C at a speed that swirls medium within the flask. During the first period of culture, the medium is changed at 7- to 10-day intervals. Later the cells are subcultured twice a week. When removing the old medium, the large, nondividing cells should be discarded from the culture. Separation of cell clumps can be achieved by choosing a proper sedimentation time. This enrichment for small aggregates, consisting of actively dividing cells, will significantly improve the division frequency in protoplast cultures.

In general, protoplasts are isolated from 2-day-old subcultures that are incubated with an equal volume of an enzyme solution (Dudits *et al.*, 1976): 2% Onozuka R-10 (Yakult Biochemicals Co., Japan) or 2% Cellulysin (Calbiochem-Behring Corp., La Jolla, California); 1% Pectinase (Sigma); 0.5% Driselase (Kyowa Hakko Kogyo, Tokyo, Japan) and 0.5% Rhozyme (Rohm and Haas, Philadelphia, U.S.A.) in 0.35 *M* sorbitol; 0.35 *M* mannitol; 3 m*M* 2-(*N*-morpholino) ethane-sulfonic acid (MES); 6 m*M* $CaCl_2 \cdot 2H_2O$; 0.7 m*M* $NaH_2PO_4 \cdot H_2O$ at pH 5.6. The mixture is incubated on a gyrotory shaker (50 rpm) at 25°C overnight. After digestion the material is filtered through a 44-μm stainless steel sieve, washed two times by centrifugation at 800 rpm, and suspended in 10 ml washing solution: 0.4 *M* glucose; 3 m*M* $CaCl_2 \cdot 2H_2O$; 0.7 m*M* NaH_2PO_4 at pH 5.6. At this stage, the protoplasts are ready to be used in fusion or uptake experiments.

III. CULTURE OF PROTOPLASTS

The protoplasts are washed once with a culture medium and cultured as a suspension in plastic Petri dishes at a density of 10^5–5×10^6 protoplasts per milliliter. The protoplast culture medium consists of the basic cell culture medium supplemented with an osmotic stabilizer, amino acids, and plant hormones. For carrot protoplasts the following culture media can be suggested:

Medium A: basic cell culture medium supplemented with glucose (68400 mg/liter). The hormone can be 2,4-D (0.5 mg/liter) alone or 2,4-D (0.5 mg/liter) + naphthaleneacetic acid (NAA) (0.19 mg/liter) + zeatin (0.11 mg/liter). pH, 5.6.

Medium B: casein hydrolysate-free basic cell culture medium supplemented with the following (milligrams per liter): glucose, 68,400; L-glycine, 10; L-glutamine, 100; L-tryptophan, 10; L-cysteine, 10; L-methionine, 5; choline chloride, 10: pH: 5.6. The combinations of hormone components are the same as in medium A.

The protoplast culture media are filter sterilized. After 10–14 days from initiation of the cultures, the protoplasts can be gradually fed with a fresh cell culture medium. The frequency of colony formation is highly variable according to the genotype and the condition of the cell line. In medium A, $4–9 \times 10^4$ colonies can be generated from $2–5 \times 10^5$ protoplasts isolated from cell suspensions of wild carrot established by Z. R. Sung. Colony formation is less efficient (60%) in synthetic medium B in comparison with medium A. The multicellular colonies can be plated onto an agar-solidified medium with 0.5 mg/liter 2,4-D and grown as callus tissues.

IV. INDUCTION OF SOMATIC EMBRYO DIFFERENTIATION AND PLANT REGENERATION

The protoplast-derived calli have to be transferred into a hormone-free medium in order to exploit the embryogenic potential of the original cell line. In attempting to accelerate embryo formation, the 10- to 14-day-old protoplast cultures can be fed with a hormone-free medium, or the protoplasts can be cultured in a medium supplemented with NAA (0.18 mg/liter) and zeatin (0.11 mg/liter). In the latter case, embryo development can be seen in the original protoplast culture (Dudits *et al.*, 1976). Zeatin exhibits a positive effect on embryo formation.

Somatic embryos and calli developed from the protoplast suspension can be transferred into cotton support cultures (Dudits *et al.*, 1977). The advantage of the cotton support culture method is that the medium can be changed without transplanting the tissues. During embryogenesis, the medium should be changed several times in order to remove the hormone residues inhibiting embryo formation. The redifferentiated plantlets are also cultured on cotton support wetted by a hormone-free cell culture medium. The transfer of these plantlets from culture into soil can be safely accomplished by placing the roots with the cotton into the soil after washing out the medium. In this way, the root system is not disturbed and a higher percentage of the plants will survive the change in their growth environment.

V. RESULTS AND CONCLUSION

As shown by the results of studies on culture techniques, mature carrot plants can be regenerated from protoplasts. Therefore, the carrot can become involved extensively in somatic hybridization experiments. Interspecific (Dudits *et al.*, 1977) as well as intergeneric (Dudits *et al.*, 1979, 1980a) hybrid plants have been produced by fusion of carrot protoplasts. Since somatic hybridization studies—with very few exceptions—are restricted to species belonging to the family Solanaceae (Schieder and Vasil, 1980), analysis of somatic incompatibility between carrot and other Umbellifereae species (Dudits *et al.*, 1980b) may help to outline a general view of the potential use of somatic cell hybridization. Fusion between carrot protoplasts has been successfully used for genetic analysis of mutant traits (Harms *et al.*, 1981; Kameya *et al.*, 1981; Lazar *et al.*, 1981).

Carrot protoplasts have served as recipients in several DNA uptake experiments (Ohyama *et al.*, 1972; Fernandez *et al.*, 1978; Rollo *et al.*, 1981; Matthews and Cress, 1981; Uchimiya and Harada, 1981). Considering the high morphogenic potential of carrot somatic cells, the carrot can be an experimental material in somatic cell genetic studies of higher plants.

ACKNOWLEDGMENTS

The author thanks Richard Ferrao for his critical reading of the English manuscript.

REFERENCES

Butenko, R. G., and Ivantsov, A. (1973). The isolation of protoplasts from different strains of carrot and tobacco tissue cultures. *Colloq. Int. C.N.R.S.* **212,** 79–84.

Chupeau, M. Y., and Morel, G. (1970). Obtention de protoplastes de plantes supérieur a partir de tissus cultives in vitro. *Hebd. Seances Acad. Sci., Ser. D* **270,** 2659–2662.

Dudits, D., Kao, K. N., Constabel, F., and Gamborg, O. L. (1976). Embryogenesis and formation of tetraploid and hexaploid plants from carrot protoplasts. *Can. J. Bot.* **54,** 1063–1067.

Dudits, D., Hadlaczky, Gy., Levi, E., Fejer, O., Haydu, Zs., and Lazar, G. (1977). Somatic hybridization of *Daucus carota* and *D. capillifolius* by protoplast fusion. *Theor. Appl. Genet.* **51,** 127–132.

Dudits, D., Hadlaczky, Gy., Bajszar, Gy., Koncz, Cs., Lazar, G., and Horvath, G. (1979). Plant regeneration from intergeneric cell hybrids, *Plant Sci. Lett.* **15,** 101–112.

Dudits, D., Fejer, O., Hadlaczky, Gy., Koncz, Cs., Lazar, G., and Horvath, G. (1980a). Intergeneric gene transfer mediated by plant protoplast fusion. *Mol. Gen. Genet.* **179,** 283–288.

Dudits, D., Hadlaczky, Gy., Lazar, G., and Haydu, Zs. (1980b). Increase in genetic variability through somatic cell hybridization of distantly related plant species. *In* "Plant Cell Cultures: Results and Perspectives" (F. Sala, B. Parisi, R. Cella, and O. Ciferri, eds.), pp. 207–214. Elsevier/North-Holland, Amsterdam.

Fernandez, S. M., Lurquin, P. L., and Kado, C. (1978). Incorporation and maintenance of recombinant-DNA plasmid vehicles pBR 313 and pCR1 in plant protoplasts. *FEBS Lett.* **87,** 277–282.

Gamborg, O. L., and Eveleigh, D. E. (1968). Culture methods and detection of glucanases in suspension cultures of wheat and barley. *Can. J. Biochem.* **46,** 417–421.

Gosch, G., Bajaj, Y. P. S., and Reinert, J. (1975). Isolation, culture and fusion studies on protoplasts from different species. *Protoplasma* **85,** 327–336.

Grambow, H. J., Kao, K. N., Miller, R. A., and Gamborg, O. L. (1972). Cell division and plant development from protoplasts of carrot cell suspension cultures. *Planta* **103,** 348–355.

Harms, C. T., Potrykus, I., and Widholm, J. M. (1981). Complementation and dominant expression of amino acid analogue resistance markers in somatic hybrid clones from *Daucus carota* after protoplast fusion. *Z. Pflanzenphysiol.* **101,** 377–390.

Hellmann, S., and Reinert, J. (1971). Protoplasten aus Zellkulturen von *Daucus carota. Protoplasma* **72,** 479–484.

Kameya, T., and Uchimiya, H. (1972). Embryos derived from isolated protoplasts of carrot. *Planta* **103,** 356–360.

Kameya, T., Horn, M. E., and Widholm, J. M. (1981). Hybrid shoot formation from fused *Daucus carota* and *D. capillifolius* protoplasts. *Z. Pflanzenphysiol.* **104,** 459–466.

Lazar, G. B., Dudits, D., and Sung, Z. R. (1981). Expression of cycloheximide resistance in carrot somatic hybrids and their segregants. *Genetics* **98,** 347–356.

Matthews, B. F., and Cress, D. E. (1981). Liposome-mediated delivery of DNA to carrot protoplasts. *Planta* **153,** 90–94.

Murashige, T., and Skoog, F. (1962). A revised medium for rapid growth and bioassays with tobacco tissue culture. *Physiol. Plant.* **166,** 473–497.

Ohyama, K., Gamborg, O. L., and Miller, R. A. (1972). Uptake of exogenous DNA by plant protoplasts. *Can. J. Bot.* **50,** 2077–2080.

Reinert, J. (1959). Uber die Kontrolle der Morphogenese und die Induktion von Adventurembryonen an Gewebekuturen aus Karotten. *Planta* **53,** 318–333.

Rollo, F., Galli, M., and Paresi, B. (1981). Liposome-mediated transfer of DNA to carrot protoplasts: A biochemical and autoradiographic analysis. *Plant Sci. Lett.* **20,** 347–354.

Schieder, O., and Vasil, I. K. (1980). Protoplast fusion and somatic hybridization. *Int. Rev. Cytol., Suppl.* **11B,** 21–26.

Steward, F. C., and Shantz, E. M. (1956). The chemical induction of growth in plant tissue cultures. *In* "The Chemistry and Mode of Action of Plant Growth Substances" (R. L. Wain and F. Wightman, eds.), p. 8. Butterworth, London.

Steward, F. C., Mapes, M. O., and Mears, K. (1958). Growth and organized development of cultured cells. II. Organization in cultures grown from freely suspended cells. *Am. J. Bot.* **45,** 705–708.

Sung, Z. R., and Dudits, D. (1981). Carrot somatic cell genetics. *In* "Genetic Engineering in the Plant Sciences" (N. J. Panopoulos, ed.), pp. 11–37. Praeger, New York.

Uchimiya, H., and Harada, H. (1981). Transfer of liposome-sequestering plasmid DNA into *Daucus carota* protoplasts. *Plant Physiol.* **68,** 1027–1030.

Wallin, A., and Eriksson, T. (1973). Protoplast cultures from cell suspensions of *Daucus carota. Physiol. Plant.* **28,** 33–39.

CHAPTER **45**

Isolation and Culture of Embryogenic Protoplasts of Cereals and Grasses

Vimla Vasil
Indra K. Vasil

Department of Botany
University of Florida
Gainesville, Florida

I. INTRODUCTION

The successful regeneration of plants from cultured protoplasts is a key requirement in most of the current schemes for the genetic modification of plants by the techniques of cell and/or molecular biology. Mesophyll protoplasts isolated from leaves of many dicotyledonous species have proved very useful for this purpose (Chapters 38–43, this volume). However, even after prolonged and extensive efforts, it is not yet possible to obtain sustained and reproducible divisions in mesophyll protoplasts of the family Gramineae, which contains many economically important species (Potrykus, 1980; I. K. Vasil and Vasil, 1980; Vasil, 1983a,b). Nevertheless, sustained divisions leading to callus formation have been obtained in protoplasts isolated from cell cultures of a number of gramineous species

CELL CULTURE AND SOMATIC CELL
GENETICS OF PLANTS, VOL. 1

ISBN 0-12-715001-3

(Maretzki and Nickell, 1973; Koblitz, 1976; Deka and Sen, 1976; Cai *et al.*, 1978; Nemet and Dudits, 1977; Vasil and Vasil, 1979; Potrykus *et al.*, 1979; Brar *et al.*, 1980; Chourey and Zurawski, 1981; Jones and Dale, 1982). No shoots, somatic embryos, or plants were recovered in any of the above instances. V. Vasil and Vasil (1980) described the formation of somatic embryos and plantlets from protoplasts isolated from an embryogenic cell suspension culture of *Pennisetum americanum*. We have progressively improved upon these early results and have obtained somatic embryos and green plants from protoplasts isolated from embryogenic suspensions of *Panicum maximum* (Lu *et al.*, 1981) and *Pennisetum purpureum* (Vasil *et al.*, 1983). The procedures used in these studies are described.

II. ISOLATION OF PROTOPLASTS

A. Embryogenic Cell Suspension Culture

The cell suspension cultures used for the isolation of protoplasts are derived from embryogenic calli obtained from cultured immature embryos and young inflorescences or leaves (see Chapter 18, this volume, for details). These finely dispersed and highly friable suspensions, lacking any meristems, grow rapidly and consist almost entirely of unorganized groups of embryogenic cells. The cells are small and richly cytoplasmic, and contain many small starch grains which mask the nucleus. The suspensions are subcultured at 4- to 5-day intervals, and 3- to 4-day-old suspensions are used for the isolation of protoplasts. Appropriate dilution ratios during subculture and the duration of each subculture must be determined for each cell line to obtain optimal results (Chapter 18, this volume).

B. Enzyme Solution and Osmoticum

The solutions used for the isolation of protoplasts contain one or more of the following cellulolytic enzymes, along with a plasmolyzing agent which also controls the osmoticum and a membrane stabilizer: Driselase, Onozuka Cellulase R-10, Pectinase, Pectolyase, and Rhozyme. The amount and combination of enzymes used are determined by the nature of the cell groups in the suspension, the rate of suspension growth, and the phase of growth when the cells are harvested for the preparation of protoplasts.

For 3- to 4-day-old predominantly embryogenic suspensions of *Panicum maximum* and *Pennisetum americanum* the following enzymes give the best results: Onozuka Cellulase R-10 (2.0–2.5%), Rhozyme (0.1–0.5%), Driselase (0.1–0.5%), Pectinase (0.1–0.5%), and/or Pectolyase (0.1–0.5%). Cell suspensions of *Pennisetum purpureum* are more sensitive to the enzymes, particularly Rhozyme and Driselase. However, excellent protoplast preparations and yields can be obtained with Onozuka Cellulase R-10 alone.

Mannitol alone or a mixture of sorbitol and mannitol is suitable as an osmotic stabilizer. Sucrose does not appear to be suitable, and is not necessary for flotation since very little debris is present after filtration and washing of protoplasts. A combined molarity of 0.3–0.5 *M* is sufficient for the plasmolysis of cells.

The enzymes as well as the osmotic stabilizer are dissolved either in Murashige and Skoog's (MS) (1962) basal medium, without any growth substances, or in distilled water to which the following are added to maintain the integrity of the plasmalemma: 7 m*M* $CaCl_2 \cdot 2H_2O$, 0.7 m*M* $NaH_2PO_4 \cdot H_2O$, and 3 m*M* MES buffer adjusted to pH 5.6. The enzyme mixture is first centrifuged for 5 min at 7500 *g* to remove undissolved coarse debris, and then sterilized by passing through a Millipore filter (0.45 μm).

C. Nutrient Medium

Kao and Michayluck's (1975) modified nutrient medium (V. Vasil and Vasil 1980) is suitable for both washing and culture of protoplasts, although sustained divisions can also be obtained in MS medium. Glucose (0.3–0.5 *M*) is included in the medium, both as a source of carbon and as an osmotic stabilizer, until after the regeneration of cell walls. The basal medium is supplemented with 2,4-dichlorophenoxyacetic acid (2,4-D) (0.25–1.0 mg/liter), alone or in combination with 6-benzylaminopurine (6-BAP) (0.1–0.25 mg/liter) or zeatin (0.1–0.5 mg/liter). The pH of the medium is adjusted to 5.6, and it is sterilized by autoclaving.

D. Isolation of Protoplasts

The Erlenmeyer flask containing the cell suspension is removed from the gyrotory shaker and allowed to stand for a few seconds. Approximately 12–16 ml of the clear medium on the surface is removed by a pipette. Next, about 6 ml of the suspension from the upper part of the culture is removed and mixed with 50 ml of the sterilized enzyme solution. This deliberate selection of a specific part of the suspension is important since it allows discarding both of the 12–16 ml supernatant which contains cellular debris and any enlarged or senescing cells, as well as the lower portion of the culture which contains larger cell groups. The 6 ml of suspension used for protoplast isolation contains mostly very small groups of cells. The protoplast–enzyme mixture is placed in Erlenmeyer flasks (50 ml) or Petri dishes (100 × 15 mm), which are sealed with Parafilm and placed on a gyrotory shaker (50–60 rpm) at 25–26°C. Within 5–6 hr 90–95% of the cells have formed protoplasts. Further incubation is not necessary since protoplast yields are not improved significantly.

The enzyme–protoplast mixture is first passed through a single or double layer of Miracloth, and then successively through 100-, 50-, and 25-μm stainless steel filters to remove undigested cells and cellular debris. The protoplasts are collected from the enzyme solution by centrifugation at 100 *g* for 3 min. The protoplasts settle in a loose pellet which is gently resuspended in fresh nutrient medium. The process is repeated three times to remove the enzyme solution. After the final washing, the protoplasts are resuspended at a density of 1–5 × 10^5/ml.

III. CULTURE OF PROTOPLASTS

The washed protoplasts are cultured either in liquid medium or in agar. Four or 5 small drops (0.05–0.1 ml) are placed in Falcon Petri dishes (50 × 9 mm), or a very shallow layer (0.2–0.3 ml) is evenly spread in a Costar Petri dish (35 × 10 mm) at a density of 1–4 × 10^5/ml. Protoplasts can also be mixed with an equal volume of 0.5–0.6% cooled Agarose (FMC Corp., Rockland, Maine) and plated in a shallow layer in Falcon or Costar Petri dishes. The dishes are sealed with Parafilm and incubated in diffuse light or in dark at 27°C.

Fresh nutrient medium with a reduced concentration of glucose (0.15–0.2 *M*) is added in small volumes (0.03–0.15 ml) following the formation of cell walls and the initiation of cell divisions (7–10 days). Later, fresh medium is added at weekly intervals or as needed until cell colonies visible to the unaided eye have been formed. Finally, protoplast-derived cell colonies and some somatic embryos are transferred to various agar media for further development and differentiation of somatic embryos and plantlets.

A majority of the freshly isolated protoplasts are rather small in size (10–25 μm), are richly cytoplasmic, and contain an abundance of small starch grains. Some vacuolated protoplasts derived from a few nonembryogenic cells as well as large (more than 30 μm) multinucleate protoplasts formed by spontaneous fusion are also seen. The embryogenic protoplasts have a tendency to stick together in masses that float to the surface of the liquid medium. In agar media they are stable and evenly dispersed without clumping. Within 2 days a majority of the protoplasts form cell walls, and in 5–10 days two- to four-celled structures are seen.

The plating efficiency of cultured protoplasts varies considerably from culture to culture. This is caused by differences in plating densities and the manner and extent of protoplast dispersal in the medium. Conservatively speaking, 25–35% of the protoplasts that survive 24 hr after washing and culture divide and form cell colonies of 16 or more cells in 2–3 weeks. Colonies of various sizes result because of asynchrony of cell division in the cultures. The protoplast-derived cell colonies are remarkably similar to the cell groups present in cell suspension cultures from which the protoplasts were initially isolated. Cell colonies are transferred to various agar media for further development and differentiation of somatic embryos and plantlets (for details, see Lu *et al.*, 1981; Vasil *et al.*, 1983).

IV. IMPORTANT PARAMETERS

The nature and condition of the suspensions used for the isolation of protoplasts are the most important factors because they affect protoplast yield, the duration of enzyme treatment necessary to obtain protoplasts, and the extent of spontaneous fusion. This can be achieved by careful monitoring of the suspension and making changes in subculture schedules, the temperature at which the suspensions are maintained, and selection of suitable portions of the suspension after it has been allowed to settle momentarily. Careful and delicate handling of the protoplasts during washing and centrifugation is also important. Low plating density results in poor plating efficiency. The even distribution of protoplasts during culture and plating, the use of agar medium at nearly room temperature, and a low final concentration (0.25–0.3%) of agar are all helpful in obtaining optimum plating efficiencies.

V. CONCLUSIONS

Protoplasts isolated from embryogenic cell suspension cultures are currently the only source of totipotent protoplasts of cereal and grass species (I. K. Vasil and Vasil, 1980; V. Vasil and Vasil, 1980; Lu *et al.*, 1981; Vasil *et al.*, 1983; Vasil, 1983a,b). The initial problems of very low plating efficiencies have been largely resolved. Plantlets derived from the protoplasts have not yet been successfully transferred to soil and grown to maturity. This remaining problem must be resolved for the further use of this system in somatic cell genetics studies.

REFERENCES

Brar, D. S., Rambold, D., Constabel, F., and Gamborg, O. L. (1980). Isolation, fusion and culture of *Sorghum* and corn protoplasts. *Z. Pflanzenphysiol.* **96,** 269–275.

Cai, Q., Quain, Y., Zhou, Y., and Wu, S. (1978). A further study on the isolation and culture of rice (*Oryza sativa* L.) protoplasts. *Acta Bot. Sin.* **20,** 97–102.

Chourey, P. S., and Zurawski, D. B. (1981). Callus formation from protoplasts of maize cell culture. *Theor. Appl. Genet.* **59,** 341–344.

Deka, P. C., and Sen, S. K. (1976). Differentiation in calli originated from isolated protoplasts of rice (*Oryza sativa* L.) through plating technique. *Mol. Gen. Genet.* **145,** 239–243.

Jones, M. G. K., and Dale, P. J. (1982). Reproducible regeneration of callus from suspension culture protoplasts of the grass *Lolium multiflorum*. *Z. Pflanzenphysiol.* **105,** 267–274.

Kao, K. N., and Michayluk, M. R. (1975). Nutritional requirements for growth of *Vicia hajastana* cells and protoplasts at a very low population density in liquid media. *Planta* **126,** 105–110.

Koblitz, H. (1976). Isolation and cultivation of protoplasts from callus cultures of barley. *Biochem. Physiol. Pflanz.* **170,** 287–293.

Lu, C., Vasil, V., and Vasil, I. K. (1981). Isolation and culture of protoplasts of *Panicum maximum* Jacq. (Guinea grass): Somatic embryogenesis and plantlet formation. *Z. Pflanzenphysiol.* **104,** 311–318.

Maretzki, A., and Nickell, L. G. (1973). Formation of protoplasts from sugarcane cell suspensions and the regeneration of cell cultures from protoplasts. *Colloq. Int. C.N.R.S.* **212,** 51–63.

Murashige, T., and Skoog, F. (1962). A revised medium for rapid growth and bioassays with tobacco tissue cultures. *Physiol. Plant.* **15,** 473–497.

Nemet, G., and Dudits, D. (1977). Potential of protoplast, cell and tissue culture research in cereal research. *In* "Use of Tissue Cultures in Plant Breeding" (F. J. Novak, ed.), pp. 145–163. Czech. Acad. Sci., Inst. Exp. Bot., Prague.

Potrykus, I. (1980). The old problem of protoplast culture: Cereals. *In* "Advances in Protoplast Research" (L. Ferenczy, G. L. Farkas, and G. Lazar, eds.), pp. 243–254. Akadémiai Kiadó, Budapest.

Potrykus, I., Harms, C. T., and Lorz, H. (1979). Callus formation from cell culture protoplasts of corn (*Zea mays* L.). *Theor. Appl. Genet.* **54,** 209–214.

Vasil, I. K. (1983a). Regeneration of plants from single cells of cereals and grasses. *In* "Genetic Engineering in Eukaryotes" (P. Lurquin and A. Kleinhofs, eds.), pp. 233–252. Plenum, New York.

Vasil, I. K. (1983b). Isolation and culture of protoplasts of grasses. *Int. Rev. Cytol., Suppl.* **14,** 79–88.

Vasil, I. K., and Vasil, V. (1980). Isolation and culture of protoplasts. *Int. Rev. Cytol., Suppl.* **11B,** 1–19.

Vasil, V., and Vasil, I. K. (1979). Isolation and culture of cereal protoplasts. I. Callus formation from pearl millet (*Pennisetum americanum*) protoplasts. *Z. Pflanzenphysiol.* **92,** 379–383.

Vasil, V., and Vasil, V. (1980). Isolation and culture of cereal protoplasts. II. Embryogenesis and plantlet formation from protoplasts of *Pennisetum americanum*. *Theor. Appl. Genet.* **56,** 97–99.

Vasil, V., Wang, D., and Vasil, I. K. (1983). Plant regeneration from protoplasts of *Pennisetum purpureum* Schum. (Napier grass). *Z. Pflanzenphysiol.* **111,** 233–239.

CHAPTER **46**

Mechanical Isolation and Single-Cell Culture of Isolated Protoplasts and Somatic Hybrid Cells

Y. Y. Gleba
V. A. Sidorov

Department of Cytophysiology and Plant Cell Engineering
Institute of Botany
Academy of Sciences of the Ukrainian Soviet Socialist Republic
Kiev, Union of Soviet Socialist Republics

Franz Hoffmann

Department of Developmental and Cell Biology
University of California
Irvine, California

I. INTRODUCTION

The feasibility of *in vitro* single-cell culture of higher plants was demonstrated several decades ago (Muir *et al.*, 1958; Steward and Shantz, 1955; Nickell, 1956). The technique for isolation and cloning of individual cells was further developed by Bergmann (1959, 1960). This allowed observations of division patterns of isolated cells under the microscope. However, the totipotency of a single higher plant cell was not strictly confirmed until Vasil and Hildebrandt (1965) observed the regeneration of flowering plants

CELL CULTURE AND SOMATIC CELL
GENETICS OF PLANTS, VOL. 1

ISBN 0-12-715001-3

which had irrefutably originated from one tobacco cell in isolation from others. Progress in plant protoplast techniques provided further development of single-cell culture methods. Mechanical isolation of heterokaryocytes and their culture introduced by Kao (1977) was improved for single protoplast culture, with subsequent regeneration of whole plants (Gleba, 1978).

The procedures described below outline the general approach for culture of single protoplasts as well as for somatic hybrid cells. A detailed protocol describes the procedure for hybrid cells (Section II) but is also applicable to the culture of single protoplasts. Some additional remarks on single protoplast culture are given in Section III. The protocol is based on visual identification and subsequent mechanical isolation of fusion products. Cells isolated in this way are then cultured individually (cloned), or are cultured together with other cells (hybrids, mutants, etc.), as proposed by Kao (1977). Modifications of these methods have been used successfully in studies by Gleba and Hoffmann (1978, 1980), Menczel *et al.* (1978), Gleba and Berlin (1979), Wetter and Kao (1980), Sidorov *et al.* (1981), Gleba *et al.* (1982), Chien *et al.* (1982), Skarzhinskaya *et al.* (1982), and Hein *et al.* (1983).

Further experiments, including more than 20 intra- and interspecific combinations in *Nicotiana* (Gleba and Evans), as well as the intertribal combination between *Brachycome* and *Crepis* (Hahne and Hoffmann) and the interfamiliar combinations between *Pisum* and *Nicotiana*, and *Vicia* and *Nicotiana* (Gleba, Okolot and Momot), are currently in progress. Active application of the method became possible only after the solution of a number of methodological problems. The first of these is the difficulty of identifying heterocellular fusion products. To identify these products accurately, morphologically differing cells must be used as parental forms. In all of the successful experiments, leaf mesophyll protoplasts and protoplasts from callus cultures were the two parental cell sources. Under these conditions, the hybrid fusion products are usually still identifiable several days after fusion. Evidently, the problem of identification might also be solved for morphologically uniform parental cells by means of vital staining (Galbraith and Galbraith, 1979; Galbraith and Mauch, 1980; see also Chapter 50, this volume). However, using this method, the successful production of a dividing somatic hybrid cell has not yet been reported.

The second problem related to mechanical isolation is the need to condition the secluded fusion product (i.e., to create culture conditions suitable for cell divisions and colony formation). The most elegant approach consists of providing conditions for the cultivation of one cell. It may be attained by using enriched nutritive medium (Kao, 1977), or by reducing the volume of the culture medium (microdroplet culture; Gleba, 1978) and postponing isolation until the fusion product resumes cell divisions. Other ways are joint culturing of several dozen heterokaryocytes in the same droplet (Menczel *et al.*, 1978) or growing fusion products in the presence of mutant cells, which can be selectively eliminated at later stages of culturing (Hein *et al.*, 1983).

Third, and finally, the method of manual mechanical isolation appeared to be problematic for some investigators. However, in most cases there is no necessity to resort to complicated micromanipulation apparatus, proposed, for example, by Patnaik *et al.* (1982). We found after 6 years of experience that the easiest and most convenient approach for the isolation of single cells was manual isolation using standard semiautomatic micropipettes.

II. ISOLATION AND CULTURE OF HYBRID CELLS

The isolation and fusion of protoplasts may be performed by any existing methods, but we strongly recommend the fusion procedure described by Menczel *et al.* (1981). This involves treatment with polyethylene glycol (PEG) solution, followed by treatment with an alkaline solution containing calcium and dimethylsulfoxide. The use of this procedure permits fusion between callus-derived and mesophyll protoplasts with high frequency. The treatment does not result in interfering cell agglutination. During the final dilution with the culture medium, in which a small stream of medium is directed from the pipette to the cells, large amounts of floating hybrid cells are usually observed. As a rule, a cell mixture with a ratio of 1:3 callus:mesophyll protoplasts is used in our experiments. After fusion the protoplasts are cultured for 3–6 days in a manner similar to that of other screening procedures. Under our conditions, the best results for *Nicotiana* hybrids were obtained on modified Nagata and Takebe (1971) medium containing 0.4 *M* mannitol, 20 ml/liter coconut milk, 300 mg/liter casein hydrolysate, and 100 mg/liter glutamine (Gleba, 1978), or on 8p medium (Kao and Michayluk, 1975). Cultures were placed under diffuse light at room temperature. The optimal time for the isolation of the fusion products has to be determined individually for each species combination. The main criteria are that (1) hybrid products are still distinguishable and (2) cell divisions in the fusion products are as advanced as possible.

As a rule, the identification of heteroplasmic fusion products (i.e., hybrid cells between mesophyll and callus protoplasts) immediately after fusion poses no difficulty. The mesophyll protoplasts contain the green chloroplasts, and most of their cell volume is occupied by the central vacuole. The callus protoplasts contain more cytoplasm and bear visible cytoplasmic strands. They are also often deprived of visible large organelles and contain colorless plastids. Commonly, the heteroplasmic fusion products have the morphology of callus cells but possess green chloroplasts. In the process of cultivation, these distinctions gradually disappear. After 6–8 days, identification often becomes very difficult. However, during the first week it is still possible to identify fusion products as a result of the following peculiarities of the cultured hybrid cells: (1) they contain a mixture of chloroplasts and colorless plastids that can be unambiguously revealed under inverted microscopes with good optics; (2) they are considerably larger than mesophyll cells, and begin dividing much earlier than the former; (3) in many cases, the fusion treatment result in destruction of almost all of the nonfused mesophyll cells; and (4) in other cases the mesophyll protoplasts, under the conditions used, do not divide (Fig. 1).

In most cases, the optimal period for isolation is 3–6 days of initial cultivation. Fusion products which have undergone at least one cell division can be removed. Before removal, the cultures are diluted by 3–4 volumes of fresh nutrient medium of the same composition. Removal of fusion products is performed by using 1- to 2-μl pipettes with plastic tips (e.g., Eppendorf, Hamburg, Federal Republic of Germany). The tips are sterilized by autoclaving. The remaining parts of the pipette need not be sterilized. Dishes containing cell suspensions are placed under an inverted microscope in a sterile flow cabinet (i.e., inside a laminar flow unit; alternatively, if the cabinet has vibrations which hamper the microscopy, placement should be on a separate table near the outlet of the laminar flow unit). The most convenient equipment for removal of heterokaryocytes by hand is microscopes with low-situated (15 cm over the table surface) specimen tables. These permit stabilization of the hand holding the pipette. Systems similar to the Diavert (Zeiss, Wetzlar, Federal Republic of Germany) can be used. The identification of heterokaryocytes is performed optimally at a magnification of 320–400×. For removal, the lowest (40–100×) enlargement is used. The pipette tip is introduced into the field of view, and the cell is taken up into the tip. Cells with medium are then transferred into the wells in the inner chambers of Cuprak dishes (Costar, Cambridge, Massachusetts). The outside chamber of the dish is filled with 3 ml of the nutrient medium diluted to half-concentration with water. The osmotic value of the solution in the outside chamber is therefore half of that in the microdroplets. After the isolation is complete, the contents of every well are examined and recorded.

Other important practical aspects of the procedure are as follows:

1. Sterility. This is much less of a problem than in macrocultures. The chances of infecting volumes as small as 0.5–2.0 μl are theoretically two to three orders of magnitude lower than those of milliliters; this has been observed in practice.

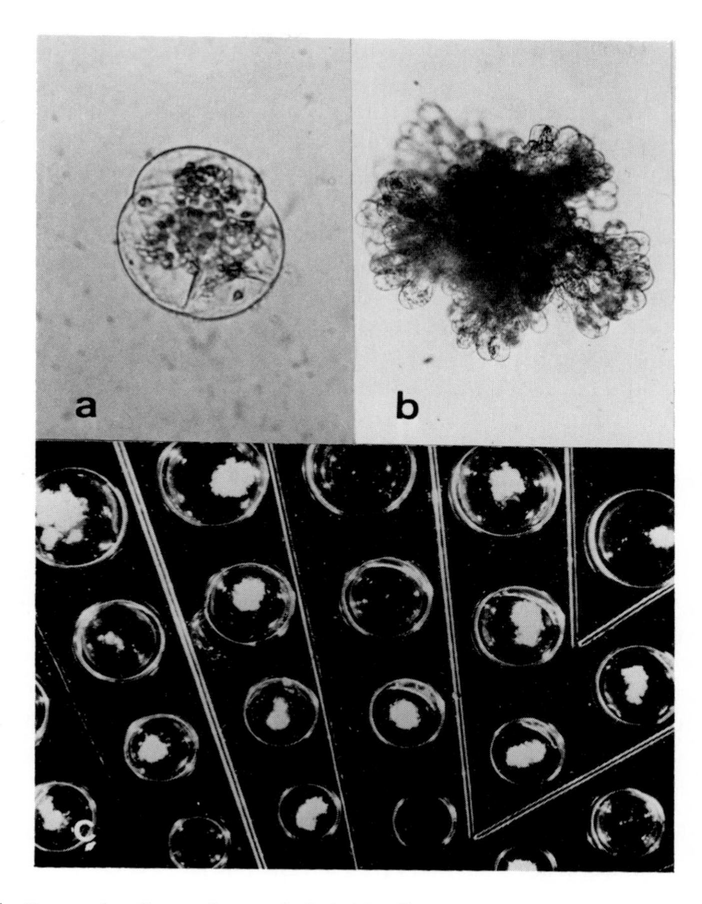

Fig. 1. Selection and culture of somatic hybrid cells. (a) Mechanically isolated hybrid cell derived from fused protoplasts of *Brassica* (mesophyll) and *Arabidopsis* (callus). (b) Cell colony developed from a single hybrid cell. (c) Small hybrid calli cultured in microdroplets in the smaller wells of the inner chamber of Cuprak dishes.

2. Humidity. Droplets measuring ~1 μl dry very quickly if precautions are not taken to prevent this. Drying during the process of cell transfer to Cuprak dish results in drastically increased osmotic pressures. This is often a main cause of cell death. Therefore, the following should be done: (1) Outer chambers of dishes have to be prefilled several hours prior to transfer. (2) All procedures connected with the opening of dishes have to be shortened. Not more than 30–40 wells in the same dish should be filled, otherwise the wells filled first will dry. (3) Minimal airflow speed should be used in the cabinet; usually it can be much lower than sterility standards require (see above).

If the conditions for cultivation are correctly selected, after 1–3 days in microdroplets the cells usually recover from the shock caused by the changes in osmotic pressure (evaporation) and mechanical stress during transfer. They acquire turgor and pass to active divisions. After 10–14 days of microculture (in darkness at 27°C), the formation of colonies consisting of several dozen cells takes place. At the same time, the "growth" of microdroplets is also observed due to "pumping" by evaporation and water condensation from the external chamber. Thus, the final concentration of the osmoticum in microdroplets attains roughly the value of the outer chamber. The colonies formed during 14 days of cultivation are isolated by using Pasteur pipettes and are transferred into one droplet (50 μl each) in 6-cm Petri dishes (one colony per dish!). Each dish contains seven such droplets of nutrient medium. These are identical to the ones used earlier, but have lower (0.2 *M*) osmoticum. When the colonies have increased sufficiently in size (after about 10–14 days), the droplets are fused together by shaking the dish. After 3–4 weeks, it is necessary to add 0.3–0.5 ml of fresh medium. The colonies formed during 4–6 weeks of culture (2–3 mm in diameter) can subsequently be grown by using standard techniques or planted onto regeneration medium, for example, MS medium (Murashige and Skoog, 1962) containing 1 mg/liter 6-benzylaminopurine (6-BAP) as the sole hormone. The culture is carried out under 2–3 klx continuous lighting. Plants are rooted on MS medium containing 0.1 mg/liter 1-naphthaleneacetic acid (1-NAA). In *Nicotiana* the entire cycle from protoplasts to rooted hybrid plants takes 3–4 months.

III. ISOLATION AND CULTURE OF NONHYBRID CELLS

Before individual protoplasts are isolated mechanically, suspensions of freshly prepared protoplasts should be diluted with culture medium to 200–1000 protoplasts per milliliter. Modified NT medium (Nagata and Takebe, 1971) containing 0.4 *M* mannitol, 20 mg/liter coconut milk, 300 mg/liter casein hydrolysate, and 100 mg/liter glutamine can be used for microdroplet culture of tobacco single mesophyll protoplasts (Gleba, 1978). Preculture of protoplasts for 1 or 2 days at high densities prior to transfer into microculture can considerably increase the survival of single cells.

In 2–3 weeks, colonies from single cells can be subcultured as described for hybrid cell lines. After another 2–3 weeks, individual colonies can be transferred onto regeneration medium. The plating efficiency of a single tobacco protoplast varies from 5 to 40% (Gleba, 1978).

IV. DISCUSSION

The most important advantages of the method of mechanical isolation and individual cloning of fusion products over all other techniques of somatic hybrid screening are as follows:

1. The selection of hybrid cells (namely, heterokaryocytes) is also the procedure for cloning. Thus, in analyzing the progeny of fusion products, one deals with clones arising (in each case) from single hybridization events. The application of most other selection procedures provides no guarantee that all hybrids isolated in one experiment are products from different hybridization events. The cloning gives perspectives for strict genetic analysis of the diversity of recombinants arising from somatic cell fusion products.
2. Screening of single cells involves essentially work with individuals, not populations, in contrast to any other procedure for hybrid selection (that is, based on the methodology of microbial genetics dealing with the rare genetic events in large populations). Consequently, the analysis and interpretation of the outcome of hybridization experiments are greatly simplified in most cases. Rare genetic events, such as mutation at the moment of cell fusion, may be neglected. For example, any variegated plant obtained from protoplast fusion of a plastome chlorophyll-deficient mutant with the wild type might arise *a priori,* not only due to hybridization but possibly also as a result of a new mutation or a chimera formation containing original cells of both parents (Gleba *et al.*, 1978). However, cloning of single fusion products can exclude any ambiguities. As another example, the rigorous interpretation of the reported experiments on protoplast fusions with cytoplasts (Maliga *et al.*, 1982) is greatly hampered, since the population of cytoplasts contains considerable contamination of protoplasts and nucleoplasts. The only correct answer to the question of cybrid origin in such experiments is the elimination of statistical uncertainty by application of the individual culture of single fusion products between a protoplast and a cytoplast.
3. Mechanical isolation of hybrids is not necessarily connected with the use of mutants, or the application of differential culture conditions, or the utilization of inactivating agents. This implies that (a) in many cases, the application of this procedure is time-saving and permits simplification of the experiments (the production of mutants and the elaboration of conditions for "physiological" selection are unnecessary) and (b) the procedure does not involve obvious selection pressure against any hybrid forms. Therefore, the spectrum of genetic diversity of the progeny (upon mechanical cloning) can surpass that arising in experiments with other selection techniques, in which elimination of some of the recombinants may occur.

For example, the method permits the identification of segregants for nuclei (cybrids), a category of recombinant forms which is automatically rejected by most other selection procedures.

4. Finally, mechanical isolation provides the investigator with certain psychological advantages. As soon as the isolated fusion products commence dividing in individual culture (usually 8–10 days after hybridization), the experimenter knows that the hybrid is present. In contrast, the use of most other selection procedures involves a period of uncertainty which may last for several months.

However, despite all the advantages described here, disadvantages are also inherent in this method of hybrid selection. For example, mechanical cloning cannot be applied in all cases, either because the identification of heterokaryons is impossible or because the suitable conditions for growing heterokaryons in individual culture are not known. In most cases, mechanical isolation necessitates the use of cells from callus tissues or suspension cultures. This is another very important shortcoming of the method. Since tissue culture cells are often aneuploid, they are genetically and physiologically heterogeneous. If the investigator has the simple aim of producing somatic hybrids, then some of the restrictions mentioned above might be overcome by using mechanical isolation without application of individual culturing (Menczel *et al.*, 1978; Hein *et al.*, 1983). Finally, although no strict data have been obtained, it seems probable that the use of the method of mechanical isolation and individual cloning is connected with a certain selective pressure. As a rule, the most actively dividing fusion products are isolated, and there is a tendency for screening of the larger (multinuclear) fusion products (in which the "hybridity" is more fully expressed). Furthermore, part of the heterokaryocyte population perishes during the early stages of individual culture.

REFERENCES

Bergmann, L. (1959). A new technique for isolating and cloning single cells of higher plants. *Nature (London)* **184,** 648–649.

Bergmann, L. (1960). Growth and division of single cells of higher plants in vitro. *J. Gen. Physiol.* **43,** 841–851.

Chien, Y., Kao, K. N., and Wetter, L. R. (1982). Chromosomal and isozyme studies of *Nicotiana tabacum-Glycine max* hybrid cell lines. *Theor. Appl. Genet.* **62,** 301–304.

Galbraith, D. W., and Galbraith, J. E. C. (1979). A method for identification of fusion of plant protoplasts derived from tissue cultures. *Z. Pflanzenphysiol.* **93,** 149–158.

Galbraith, D. W., and Mauch, T. J. (1980). Identification of fusion of plant protoplasts. II. Conditions for the reproducible fluorescence labelling of protoplasts derived from mesophyll tissue. *Z. Pflanzenphysiol.* **98,** 129–140.

Gleba, Y. Y. (1978). Microdroplet culture—tobacco plants from single mesophyll protoplasts. *Naturwissenschaften* **65,** 158–159.

Gleba, Y. Y., and Berlin, J. (1979). Somatic hybridization by protoplast fusion in *Nicotiana:* Fate of nuclear genetic determinants. *Abstr., Int. Protoplast Symp., 5th, 1979* p. 73.

Gleba, Y. Y., and Hoffmann, F. (1978). Hybrid cell lines *Arabidopsis thaliana* + *Brassica campestris:* No evidence for specific chromosome elimination. *Mol. Gen. Genet.* **165,** 257–264.

Gleba, Y. Y., and Hoffmann, F. (1980). "Arabidobrassica": A novel plant obtained by protoplast fusion. *Planta* **149,** 112–117.

Gleba, Y. Y., Piven, N. M., Komarnitsky, I. K., and Sytnik, K. M. (1978). Parasexual cytoplasmic hybrids (cybrids) *Nicotiana tabacum* + *Nicotiana debneyi* obtained by protoplast fusion. *Dokl. Akad. Nauk SSSR* **240,** 1223–1226.

Gleba, Y. Y., Momot, V. P., Cherep, N. N., and Skarzynskaya, M. V. (1982). Intertribal hybrid cell lines of *Atropa belladonna* (X) *Nicotiana chinensis* obtained by cloning individual protoplast fusion products. *Theor. Appl. Genet.* **62,** 75–79.

Hein, T., Przewozny, T., and Schieder, O. (1983). Culture and selection of somatic hybrids using an auxotrophic cell line. *Theor. Appl. Genet.* **64,** 119–122.

Kao, K. N. (1977). Chromosomal behaviour in somatic hybrids of soybean—*Nicotiana glauca. Mol. Gen. Genet.* **150,** 225–230.

Kao, K. N., and Michayluk, M. R. (1975). Nutritional requirements for growth of *Vicia hajastana* cells and protoplasts at a very low population density in liquid media. *Planta* **126,** 105–110.

Maliga, P., Lörz, H., Lázár, G., and Nagy, F. (1982). Cytoplast-protoplast fusion for interspecific chloroplast transfer in *Nicotiana. Mol. Gen. Genet.* **185,** 211–215.

Menczel, L., Lázár, G., and Maliga, P. (1978). Isolation of somatic hybrids by cloning *Nicotiana* heterokaryons in nurse culture. *Planta* **143,** 29–32.

Menczel, L., Nagy, F., Kiss, Z. R., and Maliga, P. (1981). Streptomycin resistant and sensitive somatic hybrids of *Nicotiana tabacum* + *Nicotiana knightiana:* Correlation of resistance to *N. tabacum* plastids. *Theor. Appl. Genet.* **59,** 191–195.

Muir, W. H., Hildebrandt, H. C., and Riker, A. J. (1958). The preparation, isolation and growth in culture of single cells from higher plants. *Am. J. Bot.* **45,** 589–597.

Murashige, T., and Skoog, F. (1962). A revised medium for rapid growth and bio-assays with tobacco tissue cultures. *Physiol. Plant.* **15,** 473–497.

Nagata, T., and Takebe, I. (1971). Plating of isolated tobacco mesophyll protoplasts on agar medium. *Planta* **99,** 12–20.

Nickell, L. G. (1956). The continuous submerged cultivation of plant tissues as single cells. *Proc. Natl. Acad. Sci. U.S.A.* **42,** 848–850.

Patnaik, G., Cocking, E. C., Hamill, J., and Pental, D. (1982). A simple procedure for the manual isolation and identification of plant heterokaryons. *Plant Sci. Lett.* **24,** 105–110.

Sidorov, V. A., Menczel, L., Nagy, F., and Maliga, P. (1981). Chloroplast transfer in *Nicotiana* based on metabolic complementation between irradiated and iodoacetate treated protoplasts. *Planta* **152,** 341–345.

Skarzhinskaya, M. V., Cherep, N. N., and Gleba, Y. Y. (1982). Somatic hybridization and production of cell lines for potato + tobacco plants. *Tsitol. Genet.* **16,** 42–48.

Steward, F. C., and Shantz, E. M. (1955). The chemical induction of growth in plant tissue cultures. I. Methods of tissue culture and the analysis of growth. *In* "The Chemistry and Mode of Action of Plant Growth Substances" (R. L. Wain and F. Wightman, eds.), pp. 165–203. Butterworth, London.

Vasil, V., and Hildebrandt, A. C. (1965). Differentiation of tobacco plants from single, isolated cells in microcultures. *Science* **150,** 889–892.

Wetter, L. R., and Kao, K. N. (1980). Chromosome and isoenzyme studies on cells derived from protoplast fusion of *Nicotiana glauca* with *Glycine max-Nicotiana glauca* cell hybrids. *Theor. Appl. Genet.* **57,** 273–276.

CHAPTER **47**

Fusion of Protoplasts by Polyethylene Glycol (PEG)*

F. Constabel

Plant Biotechnology Institute
National Research Council
Saskatoon, Saskatchewan, Canada

I. INTRODUCTION

Production of somatic cell hybrids and certain investigations of the structure and function of cell membranes require protoplast fusion. A reliable method with polyethylene glycol (PEG) as an agent to induce protoplasts to adhere to each other and fuse has received worldwide acceptance. Since its inception (Kao and Michayluk, 1974; Wallin *et al.*, 1974), PEG-mediated fusion has replaced the less productive method employing sodium nitrate (Power *et al.*, 1970). It has been complemented by a method which leads to protoplast fusion on its own, that is, fusion by buffered calcium chloride and mannitol at pH 10.5 and 37°C (Keller and Melchers, 1973). The method and merits of fusing protoplasts by means of an electric field are described in Chapter 48, this volume. Before beginning a fusion experiment, one must consider (1) the kind and quality of protoplasts, the fusion partners, (2) solutions which induce protoplast adhesion and fusion, and (3) the detection and culture of fusion products.

*NRCC No. 22971.

CELL CULTURE AND SOMATIC CELL
GENETICS OF PLANTS, VOL. 1

ISBN 0-12-715001-3

II. PROTOPLASTS AS FUSION PARTNERS

Protoplasts generally are obtained from cell cultures derived from a wide variety of species and from primary, soft, and easily accessible parenchyma of annuals and mesophytic perennials, most commonly leaf tissue (see Chapters 38–43, this volume). Crucial for successful fusion is a complete absence of cell wall residue or newly formed cellulose fibrils. The optimum state of protoplasts for fusion may be achieved by keeping protoplasts in media containing cellulase until preparations for fusion are completed and by using protoplasts washed free of enzymes without delay.

The success of fusion can be determined only if fusion products are detectable. The selection of protoplasts for fusion is therefore limited to those which provide a marker. Some systems which allow for the detection of hybrid cells as a result of protoplast fusion are described in Chapter 49, this volume. Here, concern is directed to markers which can be recognized in live or in fixed and processed protoplasts subsequent to completion of the fusion process. Chloroplasts and pigments or stains stored in vacuoles are particularly useful markers (Potrykus, 1972). Clear protoplasts, which often result from cells grown *in vitro,* and mesophyll or petal protoplasts are most suitable partners (Kao *et al.*, 1974), as are protoplasts which have been differentially stained with conventional (Nagy *et al.*, 1977) and fluorescent dyes (Galbraith and Mauch, 1980; Patnaik *et al.*, 1982). Also, protoplasts with nuclei that differ in size, amount of heterochromatin, or radioactive label have successfully been used (Keller *et al.*, 1973). In rare cases, the formation of precipitate as a reaction product of two different cell saps indicates the presence of fusion products (Constabel *et al.*, 1980).

Given somatic cell hybridization as the objective of protoplast fusion, the viability of fusion products is of prime concern. To that end, protoplasts obtained with rapidly growing cell suspensions and with young leaves would form most suitable combinations. Furthermore, analysis of fusion products by Feulgen microspectrophotometry and autoradiography has revealed that whereas fusion products of all cell cycle combinations occur, protoplasts of certain cycle phases participate in fusion more frequently than expected, and there is a slight predominance of "like-with-like" cycle combinations (Ashmore and Gould, 1982).

Protoplasts isolated for a fusion experiment while showing a marker and appearing most viable may not all be uninucleate, which is desirable for a hybridization experiment. Some protoplasts may be binucleated and multinucleated due to spontaneous fusion (Withers and Cocking, 1972). Avoidance of this situation would require slight plasmolysis by means of 0.5–0.7 *M* sorbitol or mannitol prior to incubation of cells with wall-digesting enzymes. Rupture of plasmodesmata caused by plasmolysis will reduce the chance of spontaneous fusion during wall removal and protoplast isolation.

III. ADHESION AND FUSION

When one is aiming at a high rate of fusion, both partners should be present and mixed at a ratio of approximately 1 and at a density of 5×10^4–2×10^5 protoplasts per milliliter. Determination of the protoplast density is usually performed by using a hemocytometer.

PEG is not the only agent that facilitates fusion. Apart from $NaNO_3$ and high Ca^{2+}/pH, dextran sulfate and gelatin (Kameya, 1975), and polyvinyl alcohol (Nagata, 1978) have successfully been tested. Dimethyl sulfoxide and concanavalin A were found to enhance PEG-mediated protoplast fusion (Norwood *et al.*, 1976; Glimelius *et al.*, 1978). PEG is commercially available in a wide range of molecular weights. Experience has shown PEG 1540 (MW 1300–1600), PEG 4000 (MW 3000–3700), and PEG 6000 (MW 6000– 7500) to be the most effective polymers. The final concentration of PEG in solutions for fusion is adjusted to 25–33% (w/v). Being a rather bulky substance, PEG may more easily be added to a certain volume of water, as required in formulas by Kao (1977), than dissolved first and brought to volume later, as required in formulas by Berry and Power (1980). The quality and concentration of PEG display their effect on protoplasts instantly. Adhesion among protoplasts, and between protoplasts and glass or plastic surfaces, and distortions strongly increase with higher concentrations and with higher molecular weights of PEG. Overoptimal conditions will lead to mass agglutination among protoplasts and lethal distortion. Underoptimal conditions will result in some adhesion but no fusion. Obviously, favorable conditions have to be found by trial.

Employing PEG for protoplast fusion is simple and effective, but may have a drawback when used in high concentration. Mercer *et al.* (1979) reported on the cytotoxicity of PEG with mammalian cells and recommended that 41% PEG 6000 be used in combination with 250 μg/ml phytohemagglutinin for the fusion of quiescent syrian hamster cell line BHK 21/13 (comparable data for plant protoplasts not available). Furthermore, Honda *et al.* (1981) shed some doubt on PEG as a fusogenic agent. By precipitation with ether and/or by dialysis in water, the fusion activity of PEG 6000 commercial grade disappeared completely, whereas the adhesion activity was retained. The authors concluded that PEG 6000 contains at least two components, one of which has the activity for cell adhesion, the other for perturbation of the phospholipid bilayer of adhering membranes. The former is the PEG 6000 itself; the latter is considered to be the catalyst of polymerization of ethylene oxide and/or an antioxidant for PEG.

Solutions for fusion require bivalent cations, up to 10 m*M* Ca^{2+}, in order to enhance adhesion by PEG (Wallin *et al.*, 1974).

While exposed to PEG, protoplasts adhere tightly. They have been observed to fuse at this time. Generally, however, fusion is the result of the disturbance which is generated when, after 10–30 min of incubation, PEG is washed out by very gently added elution medium. This medium may simply be a solution of sorbitol (0.5–0.7 *M*), the protoplast culture medium, or, more effectively, a high pH/Ca^{2+} solution. Alkaline conditions and the presence of calcium ions, although favoring fusion, may substantially decrease the viability of fusion products (Binding, 1974). After 10–15 min of elution or exposure to high pH, the protoplasts are finally washed and cultured as droplets or as a thinly layered suspension in protoplast culture medium.

IV. FUSION PRODUCTS

When one inspects by microscope the protoplasts washed free of PEG and suspended in culture medium, the fusion products should be detectable due to markers. They will be partly clear and partly green when protoplasts of cells cultured *in vitro* and of leaves have been employed. Within a few hours, the cytoplasm of fusion partners mix and make the detection of fusion products more difficult. From 10 to 20% (up to 40%) of all living protoplasts may be fusion products. Fixation and nuclear staining will reveal the nature of these fusion products. Only a fraction of them will display one nucleus of each kind, an A + B combination. Often, several protoplasts participate in the formation of one fusion product, furnishing multinucleated cells (Constabel *et al.*, 1975). This problem may be reduced by adjusting the quality and concentration of the PEG employed.

The culture of fusion products is easiest when in suspension with nonfused protoplasts. The conditions are the same as those for protoplast culture: dim light or darkness, 25–28°C, and protection from evaporation. After a week or longer, a relatively small amount of culture medium of reduced osmolality may be added to the drops or layer of suspension with fusion products, thereby enhancing wall formation and mitosis. When kept in Petri dishes, cultures can conveniently be checked by the use of an inverted microscope. For a discussion of the isolation of single fusion products, see Chapter 46, this volume.

Once fusion products have formed cell walls and start to divide, the suspension may be plated. This procedure permits recognition of individual products and monitoring of their development. Vigorous and viable fusion products may be plated before wall formation.

V. PROTOCOL FOR THE FUSION OF JACKBEAN AND PERIWINKLE PROTOPLASTS (J. W. Kirkpatrick)

1. Material
   a. Well-established and rapidly growing cell cultures of jackbean (*Canavalia ensiformis* L.).
   b. Young but fully expanded leaves of periwinkle plants [*Catharanthus roseus* (L.) G. Don] grown in a growth room under 4000 lx of fluorescent and incandescent light, for 16/24 hr, at about 24°C and kept in the dark or in dim light for 30 hr prior to protoplast isolation.
2. Protoplast isolation as in Constabel (1982)
3. Fusion procedure (cf. Figs. 1–5)
   a. Solutions for protoplast fusion.
      i. Prefusion solution for parental protoplasts (solution 1): Dissolve 9.0 g sorbitol, 1.0 g glucose, 50 mg $CaH_4(PO_4)_2 \cdot H_2O$ in water adjusted to pH 5.8 with KOH and 100 ml. Sterilize by filtration.
      ii. PEG solution (solution 2): Dissolve 50 g PEG 1540 (pharmaceutical grade), 150 mg $CaCl_2 \cdot 2H_2O$, 10 mg KH_2PO_4, and 1 g glucose in 100 ml water adjusted to pH 5.8 with KOH. Sterilize by filtration.
      iii. Elution and culture medium after Kao (1982), slightly modified (solution 3):

Mineral salts (mg)			
NH_4NO_3	250	Sequestrene 330 Fe	28
KNO_3	2500	KI	0.75
$CaCl_2 \cdot 2H_2O$	600	H_3BO_3	3
$MgSO_4 \cdot 7H_2O$	250	$MnSO_4 \cdot 2H_2O$	10
$(NH_4)_2SO_4$	134	$ZnSO_4 \cdot 7H_2O$	2
$NaH_2PO_4 \cdot H_2O$	150	$NaMoO_4 \cdot 2H_2O$	0.25
K_2CO_3	75	$CuSO_4 \cdot 5H_2O$	0.025
		$CoCl_2 \cdot 6H_2O$	0.025

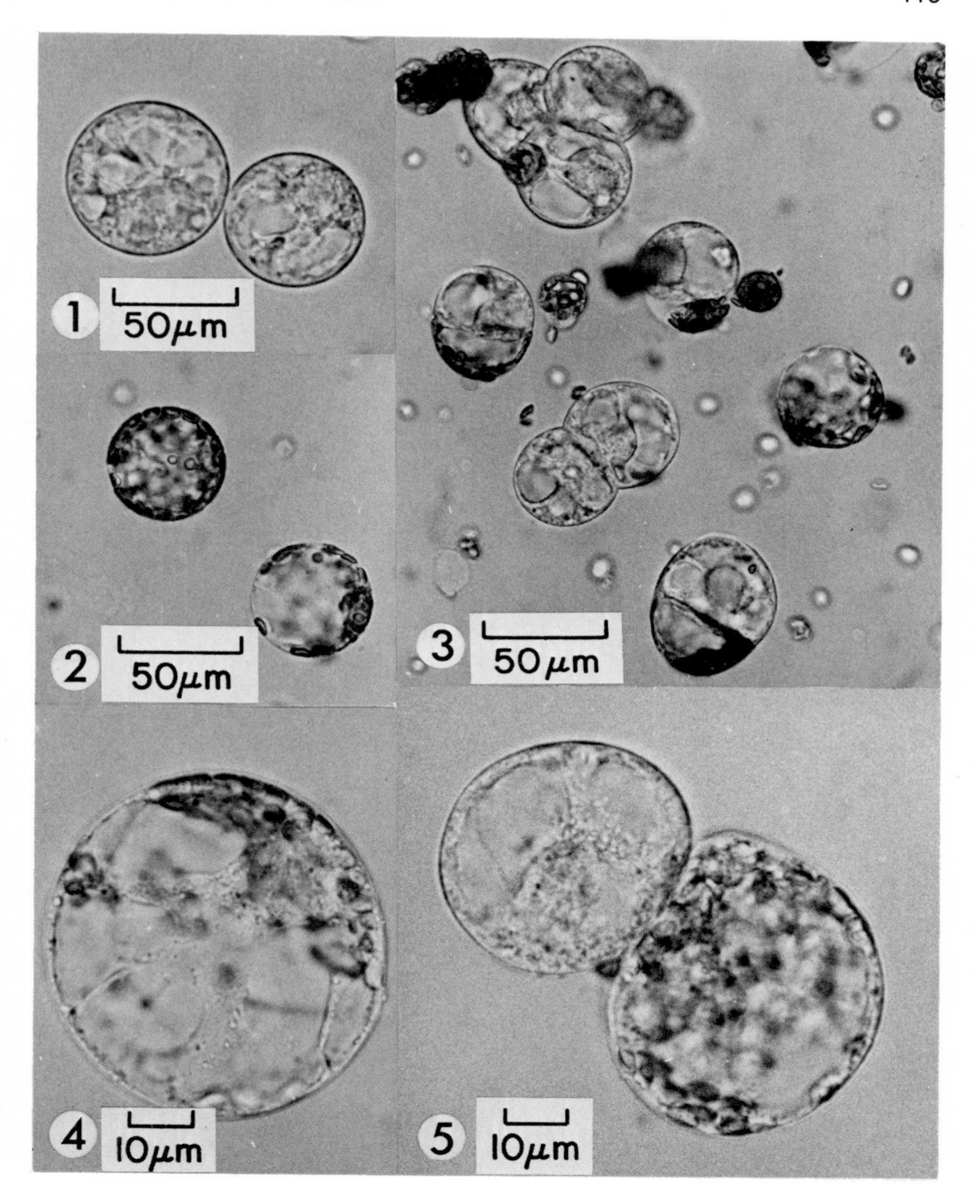

Figs. 1–5. PEG-mediated fusion of jackbean cell culture and periwinkle leaf protoplasts. **Fig. 1.** Jackbean protoplasts, clear and rich in cytoplasm. **Fig. 2.** Periwinkle protoplasts containing chloroplasts. **Fig. 3.** Agglutination of protoplasts in pairs and triplets. **Fig. 4.** Fusion product. **Fig. 5.** Nonfused protoplasts.

Vitamins (mg)			
Inositol	100	Choline chloride	0.5
Thiamine·HCl	10	Folic acid	0.2
Nicotinamide	1	Riboflavin	0.1
Pyridoxin·HCl	1	*p*-Aminobenzoic acid	0.01
Ascorbic acid	1	Biotin	0.01
Calcium pantothenate	0.5		
Organic acids (mg)			
Sodium pyruvate	5	Malic acid	10
Citric acid	10	Fumaric acid	10
Sugars (mg)			
Glucose	73,000	Ribose	125
Fructose	125	Xylose	125
Hormones (mg)			
2,4-Dichlorophenoxy-acetic acid (2,4-D)	1.0	Zeatin	0.5
1-Naphthaleneacetic acid (1-NAA)	0.5		
Miscellaneous			
L-Glutamine	580 mg		
Coconut water	20 ml		

Adjust to pH 5.8 with KOH, 1000 ml water.

iv. High pH/Ca^{2+} solution (solution 4):
(a) Dissolve 9.1 g glucose and 0.75 g glycine in 100 ml water adjusted to pH 10.5 with NaOH.
(b) Dissolve 9.1 g glucose and 1.47 g $CaCl_2 \cdot 2H_2O$ in 100 ml water. Mix the two solutions 1:1. Sterilize by filtration.

b. Fusion.
i. Suspend protoplasts in solution 1, adjust density to 5×10^4 to 2×10^5/ml, and mix equal volumes (1 ml) of the suspensions in a centrifuge tube.
ii. Dispense droplets of ca. 0.2-ml protoplast suspension in 60 × 15-mm plastic Petri dishes, 6 drops per dish. When after 2–5 min the protoplasts have settled, slowly add 2 drops of 0.2 ml each of solution 2 at the edge of the protoplast layer.
iii. Incubate the preparation at room temperature (ca. 24°C) for 10–20 min.
iv. Gently add several drops of 0.2–0.5 ml of solution 3 or solution 4 over a period of 15 min. Continue to add solution 3 until the Petri dish is flooded.

v. Tilt Petri dishes and collect the liquid in a centrifuge tube. Wash the protoplasts adhering to the Petri dish one more time with solution 3 and collect the liquid as before. Finally, culture the protoplasts in the Petri dish with a thin layer of solution 3; centrifuge the collected liquid for 3 min at 100 *g*, resuspend the sediment in a small amount of solution 3, and spread the suspension as a thin layer in a Petri dish.

vi. Upon inspection under an inverted microscope, more than 10% of all viable cells should display jackbean–periwinkle fusion.

4. Culture of fusion products: Culture the fusion products together with the unfused protoplasts. Seal the Petri dishes with Parafilm and incubate the preparations in diffuse, cool fluorescent dim light (ca. 0.6 W m^{-2}) in a plastic box humidified with moist blotting paper.

Several protocols for the fusion of protoplasts by means of PEG have been published (Berry and Power, 1980; Kao, 1982) and should be consulted if the rate of fusion and survival of fusion products need to be improved.

REFERENCES

Ashmore, S. E., and Gould, A. R. (1982). Protoplast fusion and cell cycle. *Plant Cell Rep.* **1,** 225–229.

Berry, S. F., and Power, J. B. (1980). Fusion of *Petunia* protoplasts. *Plant Mol. Biol. Newsl.* **1,** 23–25.

Binding, H. (1974). Fusion experiments with isolated protoplasts of *Petunia hybrida* L. *Z. Pflanzenphysiol.* **72,** 422–426.

Constabel, F. (1982). Isolation and culture of plant protoplasts. *In* "Plant Tissue Culture Methods" (L. R. Wetter and F. Constabel, eds.), 2nd rev. ed., NRCC No. 19876, pp. 38–48. National Research Council of Canada, Ottawa.

Constabel, F., Dudits, D., Gamborg, O. L., and Kao, K. N. (1975). Nuclear fusion in intergeneric heterokaryons. *Can. J. Bot.* **53,** 2092–2095.

Constabel, F., Koblitz, H., Kirkpatrick, J. W., and Rambold, S. (1980). Fusion of cell sap vacuoles subsequent to protoplast fusion. *Can. J. Bot.* **58,** 1032–1034.

Galbraith, D. W., and Mauch, T. J. (1980). Identification of fusion of plant protoplasts. II. Conditions for the reproducible fluorescence labelling of protoplasts derived from mesophyll tissue. *Z. Pflanzenphysiol.* **98,** 129–140.

Glimelius, K., Wallin, A., and Eriksson, T. (1978). Concanavalin A improves the polyethylene glycol method for fusing plant protoplasts. *Physiol. Plant.* **44,** 92–96.

Honda, K., Maeda, Y., Sasakawa, S., Ohno, H., and Tsuchida, E. (1981). The components contained in PEG of commercial grade (PEG 6000) as cell fusogen. *Biochem. Biophys. Res. Commun.* **101,** 165–171.

Kameya, T. (1975). Induction of hybrids through somatic cell fusion with dextran sulfate and gelatin. *Jpn. J. Genet.* **50,** 235–246.

Kao, K. N. (1977). Chromosomal behavior in somatic hybrids of soybean-*Nicotiana glauca*. *Mol. Gen. Genet.* **150,** 225–230.

Kao, K. N. (1982). Plant protoplast fusion and isolation of heterokaryocytes. *In* "Plant Tissue Culture Methods" (L. R. Wetter and F. Constabel, eds.), 2nd rev. ed., NRCC No. 19876, pp. 49–56. National Research Council of Canada, Ottawa.

Kao, K. N., and Michayluk, M. R. (1974). A method for high-frequency intergeneric fusion of plant protoplasts. *Planta* **115,** 335–367.

Kao, K. N., Constabel, F., Michayluk, M. R., and Gamborg, O. L. (1974). Plant protoplast fusion and growth of intergeneric hybrid cells. *Planta* **120,** 215–227.

Keller, W. A., and Melchers, G. (1973). The effect of high pH and calcium on tobacco leaf protoplast fusion. *Z. Naturforsch., C: Biochem., Biophys. Biol., Virol.* **28C,** 737–741.

Keller, W. A., Harvey, B. L., Kao, K. N., Miller, R. A., and Gamborg, O. L. (1973). Determination of the frequency of interspecific protoplast fusion by differential staining. *Colloq. Int. C.N.R.S.* **212,** 455–463.

Mercer, W. E., and Schlegel, R. A. (1979). Phytohemagglutinin enhancement of cell fusion reduces polyethylene glycol cytotoxicity. *Exp. Cell Res.* **120,** 417–421.

Nagata, T. (1978). A novel cell-fusion method of protoplasts by polyvinyl alcohol. *Naturwissenschaften* **65,** 263–264.

Nagy, J. I., Paszkowski, J., and Joo, F. (1977). Identification of parental nuclei in *Nicotiana* heterokaryons. *Z. Pflanzenphysiol.* **84,** 295–301.

Norwood, T. H., Zeigler, C. J., and Martin, G. M. (1976). DMSO enhances polyethylene glycol-mediated somatic cell fusion. *Somatic Cell Genet.* **2,** 263–270.

Patnaik, G., Cocking, E. C., Hamill, J., and Pental, D. (1982). A simple procedure for the manual isolation and identification of plant heterokaryons. *Plant Sci. Lett.* **24,** 105–110.

Potrykus, I. (1972). Fusion of differentiated protoplasts. *Phytomorphology* **22,** 91–96.

Power, J. B., Cummins, S. E., and Cocking, E. C. (1970). Fusion of isolated plant protoplasts. *Nature (London)* **225,** 1016–1018.

Wallin, A., Glimelius, K., and Eriksson, T. (1974). The induction of aggregation and fusion of *Daucus carota* protoplasts by polyethylene glycol. *Z. Pflanzenphysiol.* **74,** 64–80.

Withers, L. A., and Cocking, E. C. (1972). Fine-structural studies on spontaneous and induced fusion of higher plant protoplasts. *J. Cell Sci.* **11,** 59–75.

CHAPTER **48**

Fusion of Protoplasts by Dextran and Electrical Stimulus

Toshiaki Kameya

Institute for Agricultural Research
Tohoku University
Sendai, Japan

I. INTRODUCTION

Several fusion methods have been employed in somatic cell hybridization with plant protoplasts. Keller and Melchers (1973) found that incubation of protoplasts at high temperatures in a strongly alkaline environment with high concentrations of Ca^{2+} resulted in high fusion frequencies. Ever since Kao and Michayluk (1974) and Wallin *et al.* (1974) demonstrated that polyethylene glycol (PEG) induced protoplast fusion, PEG has been used for cell fusion (Chapter 47, this volume). Recently, Zimmermann and Scheurich (1981) reported high-frequency fusion of plant protoplasts by electric fields. Also, Kameya (1975, 1979, 1982; Kameya *et al.*, 1981) found that high molecular weight dextrans in the presence of high concentrations of inorganic salts cause protoplast aggregation and fusion, which are enhanced by NaOH or by electrical treatment. In this chapter a combination fusion method using dextran and an electrical stimulus is described.

CELL CULTURE AND SOMATIC CELL
GENETICS OF PLANTS, VOL. 1

ISBN 0-12-715001-3

II. PROCEDURE

A. Special Equipment

1. Function generator or sliding transformer
2. Electrode of platinum (thickness, 0.1 mm; width, 1 cm^2)
3. Glass chamber for electrical treatment (Fig. 1)

B. Materials and Reagents

1. Protoplasts. Protoplasts are isolated from leaves, roots, or cultured cells by various methods (Chapters 38–45, this volume). They must be washed with a solution of 0.55 *M* mannitol (pH 5.5) by centrifugation (80 *g*, 5 min).
2. Solutions

a. Fusion medium (pH 5.5)	Amount (g/100 ml water)
Dextran (MW 500,000; Sigma)	15
NaCl	6
Sterilize by autoclaving.	

b. NaOH–NaCl solution	Amount (per 100 ml)
NaOH	0.1–0.01 *N*
NaCl	8 g
Filter sterilize (pore size, 0.45 μm).	

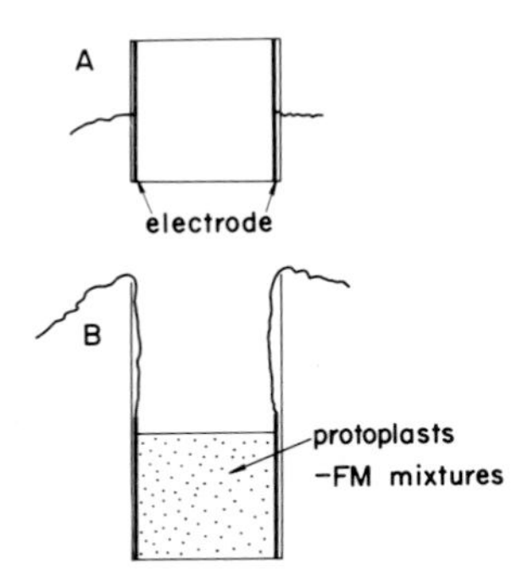

Fig. 1. Schematic diagram of a chamber for electrical treatment. (A) Top view; (B) side view. The distance between the electrodes (platinum) is 1 cm, connected to a function generator or sliding transformer.

c. Elution medium (pH 5.5)	Amount (g/100 ml)
Mannitol	5
$CaCl_2 \cdot 2H_2O$	2

Sterilize by autoclaving.

C. Procedure for Fusion by NaOH Treatment (Fig. 2)

1. Place 2 drops (0.4 ml) of fusion medium in a Petri dish.
2. Place 1 drop (0.2 ml) of protoplast suspension (10^5/ml) containing two kinds of cells on fusion medium.
3. Add 2 drops of fusion medium to the protoplast suspension and mix by gently shaking the Petri dish.
4. Incubate for 10–20 min at 30°C.
5. Observe the protoplast adhesion through an inverted microscope.
6. Pipette 1 drop of the NaOH–NaCl solution onto the mixture with a Pasteur pipette and incubate for 10–20 min at 30°C.
7. Add 5 ml elution medium to the mixture. From 5 to 10 min later, mix by gently shaking the Petri dish and incubate at 5°C for 1 hr.
8. Collect protoplasts in a centrifuge tube with disposable pipettes and centrifuge at 80 *g* for 3–5 min.
9. Remove the supernatant with a Pasteur pipette.

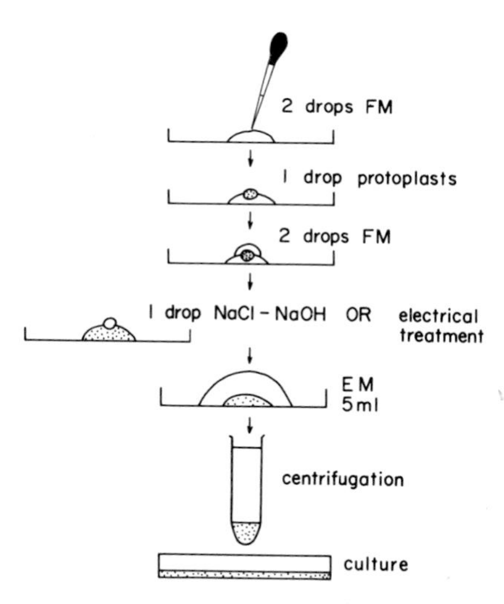

Fig. 2. Procedures for fusion using the dextran method. FM, fusion medium. EM, elution medium.

10. Resuspend the protoplasts in 5 ml elution medium and centrifuge at 80 *g* for 3–5 min.
11. Remove the supernatant with a Pasteur pipette.
12. Suspend the protoplasts in culture medium for culture.

D. Procedure for Fusion by Electrical Treatment

1. Place 2 drops (0.4 ml) fusion medium in a chamber (Fig. 1).
2. Place 1 drop (0.2 ml) protoplast suspension (10^5/ml) containing two kinds of cells on fusion medium.
3. Add 2 drops of fusion medium to the protoplast suspension and mix them by gently shaking the chamber.
4. Incubate for 10–20 min at room temperature.
5. Observe the protoplast adhesion through an inverted microscope.
6. Apply alternating current (50 or 60 Hz, 30–40 V) for 1 sec and incubate for 10–20 min at room temperature.
7. Add 5 ml elution medium to the mixture. From 5 to 10 minutes later, mix by gently shaking the chamber at room temperature for 1 hr.
8. Collect protoplasts in a centrifuge tube with disposable pipettes and centrifuge at 80 *g* for 3–5 min.
9. Remove the supernatant with a Pasteur pipette.
10. Resuspend the protoplasts in 5 ml elution medium and centrifuge at 80 *g* for 3–5 min.
11. Remove the supernatant with a Pasteur pipette.
12. Suspend the protoplasts in culture medium for culture.

III. RESULTS AND CONCLUSIONS

The procedures described above were effective for fusing protoplasts of *Nicotiana tabacum, Petunia hybrida, Brassica pekinensis,* and *Daucus carota.* When the dextran method is used, fusion frequencies and the ratio of surviving cells varied among different kinds of protoplasts. Sometimes leaf protoplasts from plants cultivated under high humidity conditions were sensitive to high pH and high electric field intensity. Therefore, it may be necessary to check the optimum concentrations of NaOH in the NaOH–NaCl solution and the intensity of the electric field.

High NaOH concentration (0.1 *N*) and high electric field intensity (50 V/cm) could cause high fusion frequency. In this case, as the number of surviving cells may decrease, it is better to use a increased amount (two to four times) of fusion medium.

A long electrical treatment and use of direct current induce lysis of protoplasts because of electrolysis of NaCl in fusion medium. The treatment should be performed within 1 sec using alternating current. A large number of protoplasts could be treated by using a big chamber with electrodes.

REFERENCES

Kameya, T. (1975). Induction of hybrids through somatic cell fusion with dextran sulfate and gelatin. *Jpn. J. Genet.* **50,** 235–246.

Kameya, T. (1979). Studies on plant cell fusion: Effect of dextran and pronase E on fusion. *Cytologia* **44,** 449–456.

Kameya, T. (1982). The method for fusion with dextran. *Proc. Int. Congr. Plant Tissue Cell Cult., Plant Tissue Cult., 1982* pp. 613–614.

Kameya, T., Horn, M. E., and Widholm, J. (1981). Hybrid shoot formation from fused *Daucus carota* and *D. capillifolius* protoplasts. *Z. Pflanzenphysiol.* **104,** 459–466.

Kao, K. N., and Michayluk, M. R. (1974). A method for high frequency inter-generic plant protoplast fusion. *Planta* **115,** 355–367.

Keller, W. A., and Melchers, G. (1973). The effect of high pH and calcium on tobacco leaf protoplast fusion. *Z. Naturforsch., C: Biochem., Biophys., Biol., Virol.* **28C,** 737–741.

Wallin, A., Glimelius, K., and Eriksson, T. (1974). The induction of aggregation and fusion of *Daucus carota* protoplasts by polyethylene glycol. *Z. Pflanzenphysiol.* **74,** 64–84.

Zimmermann, U., and Scheurich, P. (1981). High frequency fusion of plant protoplasts by electric fields. *Planta* **151,** 26–32.

CHAPTER 49

Inactivation of Protoplasts before Fusion to Facilitate Selective Recovery of Fusion-Derived Clones

Laszlo Menczel*

Institute of Plant Physiology
Biological Research Center
Hungarian Academy of Sciences
Szeged, Hungary

I. INTRODUCTION

In fusion experiments it is necessary to distinguish clones established by the division of parental cells from those which are obtained from fused cells. Normally, the various colony types are distinguished by using genetic traits. In many higher plant systems, however, suitable genetic traits are not available. In this chapter, nongenetic methods which make feasible the selective recovery of colonies derived from fused protoplasts are described. These methods were first established using mammalian cell lines (Harris, 1972; Wright, 1978).

The methods are based on inactivation of protoplasts before fusion (by iodoacetic acid treatment or lethal irradiation), which renders them unable

*Present address: Department of Botany, University of Florida, Gainesville, Florida 32611.

ISBN 0-12-715001-3

to divide and therefore prevents the formation of parental-type clones. The inactivated protoplasts, however, are reactivated in the fusion products by their fusion partner, which is either not inactivated, or inactivated using a different treatment. Details of the applications of inactivation methods will be discussed in Section IV.

II. MATERIALS AND EQUIPMENT

Cobalt-60 source capable of providing a minimum dose rate of 0.05 Gy/sec
Low-speed bench centrifuge
Controlled temperature water bath
Sterile polycarbonate centrifuge tubes (Nalgene)
2-Iodoacetic acid
Miscellaneous equipment (Pasteur pipettes, etc.) to handle protoplasts under sterile conditions
W5 solution (125 m*M* $CaCl_2$ + 155 m*M* NaCl + 5 m*M* KCl + 5 m*M* glucose, pH 5.6)

III. INACTIVATION PROCEDURES

A. Iodoacetate Treatment

1. Prepare protoplasts in 0.4 M osmoticum.
2. Wash protoplasts with W5 twice by resuspending and centrifuging at 50 *g* for 3 min.
3. Prepare the inactivating solution by dissolving 18.6 mg iodoacetic acid in 10 ml W5 (10 m*M*; the pH should be adjusted to 5.6 with NaOH).
4. Resuspend protoplasts in inactivating solution and incubate at 20°C for 20 min (shaking occasionally).
5. Wash protoplasts as in step 2 at least twice.

Inactivated protoplasts should show no signs of damage immediately after treatment, but should shrink and turn brown within 24 hr when cultured. The conditions indicated above are suitable for the treatment of various *Nicotiana* protoplasts. In other systems it may be necessary to modify the concentration and/or the time of treatment to achieve complete inactivation without immediate damage. Protoplast density during treatment should not exceed 10^5/ml since inactivation depends on protoplast density.

B. Cobalt-60 Irradiation

1. Prepare protoplasts in 0.4 M osmoticum.
2. Wash protoplasts twice in W5 by resuspending and centrifuging at 50 *g* for 3 min.
3. Resuspend protoplasts in a polycarbonate centrifuge tube in W5 at a density of about 10^5/ml.
4. Place the tube into ^{60}Co equipment for a period calculated to give the desired dose. Protoplasts should be irradiated using supralethal doses delivered at a rate of 0.05–0.1 Gy/sec.
5. Wash protoplasts once in W5, as in step 2.

The irradiated protoplasts should show no signs of damage. When cultured, they form a cell wall and remain alive for a few days; a small proportion (depending on the dose applied) may undergo the first division. A dose of 50 Gy or higher is suitable for the inactivation of various *Nicotiana* protoplasts.

IV. APPLICATIONS

The use of inactivation in fusion experiments is based on the reactivation of inactivated cells in the fusion products. In the original experiment on mammalian cells (Wright, 1978), reactivation occurred through metabolic complementation (mutual reactivation) of two fusion partners inactivated by different treatments (iodoacetic acid and diethylpyrocarbonate). This "double inactivation" approach did not yield viable fusion products in *Nicotiana* when used in its original form (P. Medgyesy, personal communication). A modified form of the double inactivation scheme, iodoacetic acid treatment of one parent in combination with irradiation of the other, has been successfully applied to obtain fusion-derived clones in *Nicotiana* (Sidorov *et al.*, 1981). However, due to the effects of irradiation on the nuclear genetic material of the irradiated partner (see below), this method is suitable for the production of cybrids rather than hybrids. On the other hand, inactivation of one parent in combination with a genetic marker has been used successfully to produce both hybrids and cybrids (Medgyesy *et al.*, 1980; Menczel *et al.*, 1982). Recognition of fusion-derived clones in this case is based on the appearance of a trait originally carried by the inactivated parent (marker rescue, Fig. 1.)

Irradiation and iodoacetate treatment have different effects on the genetic constitution of fusion-derived clones. Iodoacetate treatment apparently leaves intact the genetic material of inactivated cells and yields a high proportion of hybrids among the clones (Medgyesy *et al.*, 1980; Menczel *et al.*, 1982). On the other hand, irradiation results in the loss of irradiated nuclei from a significant proportion of fusion products (up to 80%, depending on the dose). This leads to the formation of segregant clones carrying the nucleus of the untreated parent and cytoplasmic elements from both parents (Zelcer *et al.*, 1978; Aviv and Galun, 1980; Menczel *et al.*, 1982, 1983). Irradiation is used, therefore, in experiments aimed at transferring cytoplasmic factors between cultivars or species. In such experiments the marker exploited for the recognition of fusion-derived clones should be located in chloroplasts or mitochondria.

The high doses of irradiation necessary to prevent cell division certainly induce several mutations in the nuclear and organellar DNA of irradiated cells. In cytoplasm transfer experiments (the main application of inactivation by irradiation), only those clones which have no contribution of nu-

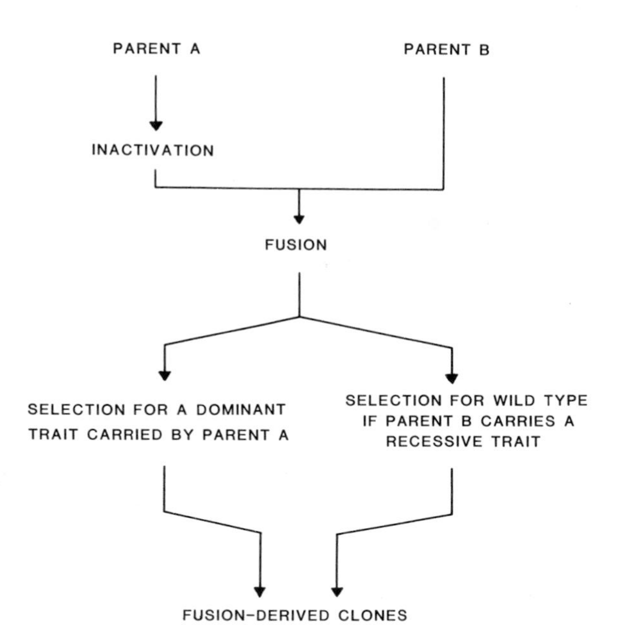

Fig. 1. Marker rescue from inactivated protoplasts to produce fusion-derived clones.

clear genetic material from the irradiated parent should be used. Thus, in such experiments, only mutations induced in chloroplasts and mitochondria have to be considered. Carryover of organelle mutations from the irradiated parent does not seem to be frequent. No orgenelle mutations were detected in over 100 plants regenerated from 57 clones in a chloroplast transfer experiment using irradiation (Menczel *et al.*, 1982).

Iodoacetic acid treatment apparently has no effect on the genetic material of fusion-derived clones. This has been shown by the regeneration of a number of morphologically normal and fertile hybrid and cybrid *Nicotiana* plants from clones produced by the use of iodoacetic acid treatment (Medgyesy *et al.*, 1980).

If exposure of protoplasts to mutagenic irradiation or toxic chemicals interferes with subsequent analysis in any way, cloning of isolated fusion products described in Chapter 46, this volume, should be used for the production of fusion-derived clones.

REFERENCES

Aviv, D., and Galun, E. (1980). Restoration of fertility in cytoplasmic male sterile (CMS) *Nicotiana sylvestris* by fusion with X-irradiated *N. tabacum* protoplast. *Theor. Appl. Genet.* **58,** 121–127.

Harris, M. (1972). Effect of X-irradiation of one partner on hybrid frequency in fusion between Chinese hamster cells. *J. Cell. Physiol.* **80,** 119–128.

Medgyesy, P., Menczel, L., and Maliga, P. (1980). The use of cytoplasmic streptomycin resistance: Chloroplast transfer from *Nicotiana tabacum* into *Nicotiana sylvestris,* and isolation of their somatic hybrids. *Mol. Gen. Genet.* **179,** 693–698.

Menczel, L., Galiba, G., Nagy, F., and Maliga, P. (1982). Effect of radiation dosage on efficiency of chloroplast transfer by protoplast fusion in *Nicotiana. Genetics* **100,** 487–495.

Menczel, L., Nagy, F., Lázár, G., and Maliga, P. (1983). Transfer of cytoplasmic male sterility by selection for streptomycin resistance after protoplast fusion in *Nicotiana. Mol. Gen. Genet.* **189,** 365–369.

Sidorov, V. A., Menczel, L., Nagy, F., and Maliga, P. (1981). Chloroplast transfer in *Nicotiana* based on metabolic complementation between irradiated and iodoacetate treated protoplasts. *Planta* **152,** 341–345.

Wright, W. E. (1978). The isolation of heterokaryons and hybrids by a selective system using irreversible biochemical inhibitors. *Exp. Cell Res.* **112,** 395–407.

Zelcer, A., Aviv, D., and Galun, E. (1978). Interspecific transfer of cytoplasmic male sterility by fusion between protoplasts of normal *Nicotiana sylvestris* and X-ray irradiated protoplasts of male sterile *N. tabacum. Z. Pflanzenphysiol.* **90,** 397–407.

CHAPTER **50**

Selection of Somatic Hybrid Cells by Fluorescence-Activated Cell Sorting

David W. Galbraith

School of Biological Sciences
University of Nebraska at Lincoln
Lincoln, Nebraska

I. INTRODUCTION

Somatic hybridization offers potential for the development of novel plant species of agronomic importance. The combination of techniques of protoplast fusion and protoplast regeneration into complete plants enables the possibility of transfer of genetic information between plants that are sexually incompatible. A major prerequisite for successful somatic hybridization is a method for the identification and selection of hybrids, since fusion treatments generally produce low numbers of hybrids within an unfused parental background. Various selection schemes have been developed. In most cases, these methods are tailored to the specific parental combination, hence, for example, the use of complementation of chlorophyll- or nitrate reductase-deficient mutants, or of specific growth media require-

CELL CULTURE AND SOMATIC CELL
GENETICS OF PLANTS, VOL. 1

ISBN 0-12-715001-3

ments. These methods are clearly not appropriate in somatic hybridization experiments for which the choice of parental species is unrestricted.

An alternative procedure for the identification and selection of fused protoplasts is the use of fluorescence markers. The principle involves the separate introduction of exogenous, distinguishable fluorescent dyes into the two parental populations of protoplasts. Following fusion the heterokaryons can be distinguished from the parentals by the presence of both dyes. Selection can then be performed either by micromanipulation or by electronic sorting using the commercially available fluorescence-activated cell sorter.

My interest in the potential of fluorescence as a selection technique was initiated by a report by Keller *et al.* (1977), in which two lipophilic, differently colored fluorescent dyes were employed to measure the process of animal cell fusion. Incorporation of the fluorescent dyes involved a preincubation period, a method which seemed appropriate for use with plant suspension cultures. In our study (Galbraith and Galbraith, 1979), we showed that protoplasts prepared from tissue cultures preincubated for 12–15 hr in the presence of octadecanoyl aminofluorescein (F18) or octadecyl rhodamine B (R18) became brightly labeled. We further demonstrated that the labels were nontoxic and that following fusion treatments three classes of fluorescent protoplasts could be distinguished. Two classes represented parental protoplasts in that they contained a single fluorescent dye. The third class of protoplasts contained both dyes and therefore represented heterokaryons.

We realized that the process of preincubation of tissue cultures with the lipophilic dyes would not be appropriate for the preparation of fluorescent protoplasts from intact plant organs. This is because the presence of external barriers such as the cuticle prevents access of the dyes to the protoplasts contained within the tissues. We therefore developed an alternative labeling procedure appropriate for leaf protoplasts (Galbraith and Mauch, 1980), which involves the incubation of the leaf tissue in a mixture of polysaccharidases and low levels of one of the two fluorescent labeling reagents, fluorescein isothiocyanate (FITC) and rhodamine isothiocyanate (RITC). During the period of protoplast release, they became stained by the fluorescent dyes. We demonstrated that protoplast viability is not affected by this treatment, that the labels do not leak between two protoplast populations separately labeled and cocultured, and finally, that induction of fusion gives rise to a subset of protoplasts that contain both labels.

The selection of the heterokaryons based upon fluorescence can theoretically be achieved either by micromanipulation or by fluorescence-activated cell sorting. The former procedure is time-consuming and results in the selection of only a few potential hybrids (Patniak *et al.*, 1982). The latter can be exceptionally rapid. However, there are still technical barriers to be overcome, many of which relate to the mechanical design of cell sorters, which are tailored for use with animal cells in general and lymphoid cells in particular. Nevertheless considerable advances in the use of cell sorting for the selection of plant protoplasts have recently been made (Galbraith and Harkins, 1982; Redenbaugh *et al.*, 1982). We are optimistic that this process will become the one of choice in the future selection of somatic hybrids.

The process of electronic fluorescence-activated selection of somatic hybrids can be divided into four major stages. First, there must be a reproducible introduction of fluorescence into the two parental protoplast populations. It must be possible to distinguish the two protoplast populations by fluorescence analysis using the cell sorter. The labeling techniques should be nontoxic and should not prevent the regeneration of the protoplasts into plants. Second, techniques must be available for the reproducible induction of fusion between the parental protoplasts. Such techniques preferably should give rise to dispersed populations of protoplasts rather than to clumps. Third, conditions should be available under which cell sorting can be carried out under viable conditions without contamination by microorganisms. Finally, conditions for the regeneration of callus from protoplasts and induction of organogenesis from callus are required. In this chapter, I focus on those aspects directly applicable to the processes of fluorescence labeling and cell sorting that are not covered in detail elsewhere in this volume.

II. PROCEDURES

A. Fluorescence Labeling of Protoplasts

1. Protoplasts Derived from Suspension Cultures

The following method for the labeling of suspension cultures of *Nicotiana tabacum* and *Glycine max* relies on the partitioning of lipophilic fluorescent dyes into the membranes of the suspension cultures during a preincubation period. The two dyes, F18 and R18, can be synthesized according to the method of Keller *et al.* (1977) or can be purchased (Molecular Probes, Inc., Junction City, Oregon).

The cells to be labeled (25 ml cell suspension containing a packed cell volume of approximately 5-ml cells) are transferred under sterile conditions into 150-ml Erlenmeyer flasks. To these are added 0.020-ml aliquots of a 5-mg/ml solution of F18 or R18 dissolved in ethanol. The flasks are maintained under normal growth conditions for at least 12 hr to enable partitioning of the fluorescent dye into the cell membranes (Galbraith and Galbraith, 1979).

An alternative labeling technique involves staining of the suspension culture cells with the nucleophilic fluorescent dyes FITC and RITC. Aliquots (25 ml of cell suspension, as above) are transferred into sterile 150-ml Erlenmeyer flasks. To these are added 0.02-ml aliquots of solution of FITC or RITC (5 mg/ml) freshly dissolved in ethanol, acetone, or dimethylsulfoxide. The flasks are maintained under regular growth conditions for 15–18 hr. In contrast to the lipophilic dyes, FITC and RITC at high levels are toxic to cells. For each suspension culture system, I recommend the use of standard growth measurements (Galbraith and Galbraith, 1979) as a function of addition of increasing concentrations of the fluorescent dyes to establish these limits of toxicity.

For both labeling protocols, after the period of preincubation the tissue culture cells are transferred into sterile 50-ml conical centrifuge tubes and collected by centrifugation at 33 *g* for 3 min. Prior to preparation of protoplasts, the cells are washed by sequential suspension and centrifugation in 30 ml growth medium. Protoplast preparation can then be carried out using procedures appropriate for the suspension culture under study.

2. Protoplasts Derived from Leaf Tissues

Since leaf tissues comprise complex three-dimensional structures enclosed by barrier layers that are relatively impermeable to aqueous solutions, it is not possible to employ preincubation treatments for the introduction of fluorescent dyes into the tissues. This alternative method can be used for the production of fluorescent protoplasts derived from leaf tissue. It relies on the observation that addition of nucleophilic fluorescent dyes directly to the enzyme solution used for protoplast production results in the appearance of viable, labeled protoplasts. This procedure, originally devised for leaf protoplasts of *Nicotiana* (Galbraith and Mauch, 1980), has been reproducibly applied by other workers for the labeling of leaf protoplasts of *Euphorbia lathyrus* (Redenbaugh *et al.*, 1982) and suspension protoplasts of *Petunia hybrida* (Patniak *et al.*, 1982).

The procedure, described here for *N. tabacum* var. Xanthi leaf protoplasts is as follows. Fully expanded leaves from young plants grown under standard greenhouse conditions are washed in tap water for 15–30 min. The leaves are sterilized by sequential immersion in 70% ethanol (30 sec) and 33% Clorox (15 min). All further operations are carried out under sterile conditions in a laminar flow transfer hood. The residual Clorox is removed by two sterile water rinses. The leaves are trimmed to remove the veins, midrib, and all bleached tissue areas and are transferred into osmoticum (solution 1; see Appendix). The leaves (approximately 0.75 *g* fresh weight) are sliced into 30 × 2-mm pieces and transferred into 10-ml polysaccharidase solution (solution 2; see Appendix) contained in 250-ml Erlenmeyer flasks. To each flask is immediately added 5–25 μl of a 5 mg/ml solution in ethanol or acetone of FITC, RITC, or tetramethylrhodamine isothiocyanate (TRITC). The flasks are evacuated for 90 sec using a vacuum aspirator. Incubation is carried out for a period of 15–18 hr at 25°C with reciprocal shaking at 20 excursions per minute. Following incubation the protoplasts are separated from large debris by filtration through six layers of cheesecloth. The protoplasts are pelleted by centrifugation at 100 *g* for 5 min. The pellet is resuspended in 20 ml of a 25% (w/v) sucrose solution (solution 3; see Appendix). The suspension is transferred into a 50-ml polycarbonate centrifuge tube, is overlaid with 20 ml osmoticum, and is centrifuged at 100 *g* for 5 min. The intact protoplasts float to the interface, from which they are recovered by aspiration using a Pasteur pipette. The protoplasts are washed twice in osmoticum prior to fusion or culturing.

3. Observation of Protoplasts by Fluorescent Microscopy

For routine observations of the labeled protoplasts, we employ a Zeiss Standard microscope equipped with epi-fluorescence optics and a 50-W mercury arc light source. For fluorescein the filter combinations are exciter filter KP490, dichroic mirror FT510, and barrier filters LP520 and KP560. This combination provides broad excitation in the range of 400–490 nm, beam splitting at 510 nm, and detection of fluorescence emission in the range 520–560 nm. For rhodamine the filter combinations are exciter filters KP600 and BP546/12, dichroic mirror FT580, and barrier filters LP590 and BG18. This combination provides excitation at 546 nm with elimination of background red light from the mercury source, beam splitting at 580 nm, and detection of fluorescence emission in the range 590–690 nm. We routinely record photographs using Kodacolor ASA 400 slide film for which the exposures require approximately 20–60 sec.

B. Analysis of Fluorescent Protoplasts by Flow Cytometry

Several commercial flow cytometer cell sorters are available. The following description, which is specific for the Coulter Electronics EPICS V, is nevertheless applicable in general terms to the other available machines. The operation of the flow cytometer involves the rapid analysis of the degree of fluorescence associated with each of the protoplasts within the labeled populations. This is achieved by passage of the protoplasts within a sheath fluid through a flow cell containing a jeweled orifice 100 μm in diameter (Fig. 1). The resultant stream intersects a focused laser beam. As the protoplasts pass through the focus, they absorb light, which is subsequently reemitted in the form of fluorescence. The fluorescence signals comprise a rapid series of pulses. These are separated according to emission wavelength, using specific optical filters and a dichroic mirror located orthogonally to the laser light path and the fluid flow direction, and are converted into corresponding DC voltages by means of two photomultipliers. The use of the two filter sets and photomultipliers makes it possible to measure the fluorescence of two different dyes (e.g., FITC and RITC) associated with single protoplasts in labeled populations. The voltage pulses are subsequently amplified and are converted into binary equivalents by means of two independent analog-to-digital converters. The binary information is stored in the form of 256-channel one-dimensional, or 64 × 64-channel two-dimensional histograms in the memory of a microprocessor.

For the analysis of protoplasts labeled with FITC and RITC, the flow

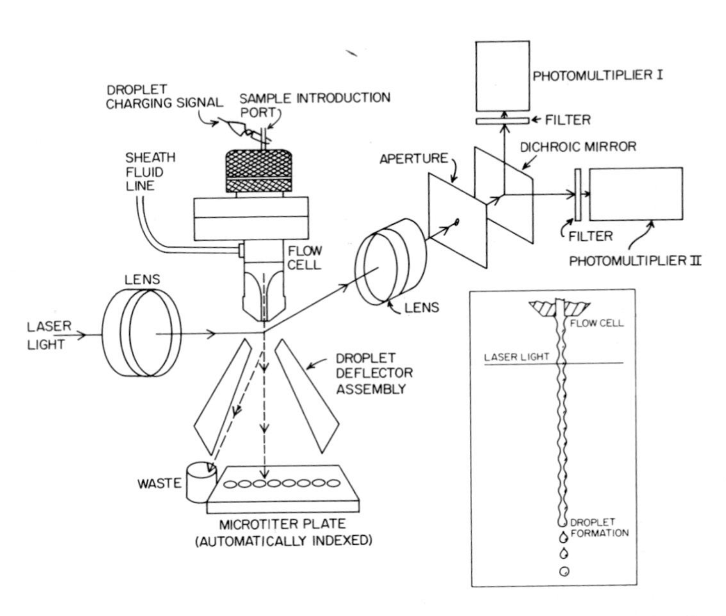

Fig. 1. Operation of the cell sorter in schematic form. The fluorescence of individual protoplasts passing through the focus of the laser beam is analyzed by two photomultipliers, one for green (fluorescein) fluorescence and one for red (rhodamine) fluorescence. A piezoelectric crystal in the body of the flow cell causes the fluid streams to break into droplets. The droplets containing heterokaryons are automatically collected in a sterile microtiter plate.

cytometer is operated at a wavelength of 514 nm with a laser output of 400 mW using barrier filters LP540 and BG38, dichroic mirror FT560, green photomultiplier filter SP560, and red photomultiplier filter LP590. The subtraction module, a device that permits subtraction of a portion of each of the signals of the green and red channels from the corresponding opposite channel, is adjusted to amplification settings that eliminate "cross talk" between the two channels. This is achieved by passage through the flow cytometer of protoplasts singly labeled with either FITC or RITC. The gain settings are then adjusted such that FITC-labeled and RITC-labeled protoplasts appear along orthogonal axes in two-dimensional histograms. In these histograms, the x axis corresponds to green (FITC) fluorescence, the y axis to red (RITC) fluorescence, and the z axis to protoplast number. Histograms are routinely accumulated to a total of 10,000 protoplasts.

For analysis of endogenous chlorophyll autofluorescence within leaf protoplasts, the flow cytometer is operated at a laser wavelength of 457 nm with an output of 200 mW and barrier filter LP510. One-dimensional histograms, of which the abscissa corresponds to the logarithm of red fluorescence, are accumulated to a total of 10,000 protoplasts.

C. Induction of Protoplast Fusion

In experiments involving the fusion of fluorescently labeled protoplast populations we have employed a variety of fusion techniques, most of which are based upon the use of polyethylene glycol (Galbraith and Galbraith, 1979; Galbraith and Mauch, 1980; Galbraith and Harkins, 1982). Fusion techniques are addressed elsewhere in this volume in detail (Chapters 47 and 48). We anticipate that all such techniques, including the use of electrical fusion, will be applicable to protoplasts that have been subjected to fluorescent labeling.

D. Protoplast Sorting under Conditions That Maintain Viability

Cell sorting based upon fluorescence involves the operation of the microprocessor in conjunction with a piezoelectric crystal and a droplet charging mechanism (Fig. 1). Those protoplasts to be sorted are defined by means of a series of electronic cursors which can be placed upon the displays of the histograms that have been previously accumulated. The protoplasts that, in passing through the focus of the laser beam, generate the desired fluorescence signals (e.g., those producing both a green and a red fluorescence signal, hence representing heterokaryons) are instantaneously identified by the microprocessor, this information being used for the generation of the sorting signal. The sorting procedure relies on the conversion of the fluid stream into a series of precisely defined droplets at a fixed distance below the laser intersection point. This is achieved by the oscillation of a piezoelectric crystal which is attached to the body of the flow cell at a fixed frequency in the range of 10–40 kHz. A sorting DC voltage pulse is applied to the fluid stream after a fixed time delay following identification of the desired protoplast. The time delay is determined empirically so that the sorting pulse is applied when the desired protoplast is at the point of droplet formation. The application of the voltage pulse to the fluid stream leaves a small residual charge on the surface of the droplets that are formed. These droplets then enter a fixed electrostatic field between two deflection plate assemblies. The charged droplets are deflected and can be collected in separate containers. The sorting process is designed for operation at rates of up to 10^4 positive events per second.

We have modified the basic principles outlined above for use with plant protoplasts. We operate the cell sorter in conjunction with the Coulter Electronics Autoclone. This device attaches below the deflection plate assembly and comprises an electrically driven platform that accommodates a standard 96-well sterile plastic tissue culture–microtiter plate (Costar, type 3596). The droplet-charging command circuit is modified so that all droplets that do *not* contain the desired protoplasts are deflected to waste. Those droplets containing the desired heterokaryons are allowed to pass undeflected into the first well of the microtiter plate. The number of these positive events is automatically monitored. Based upon our suggestion, Coulter can now provide integrated circuits that allow the quantitation of up to 1000 positive events per well, after which point the plate is automatically indexed. This process brings the second well of the microtiter plate into register for collection of a second series of protoplasts. The sequence is repeated until the plate has been filled. The use of microtiter plates drastically reduces problems relating to the maintenance of sterility within the protoplast cultures, since any contamination will be restricted to the individual wells of the plate.

For the sorting of protoplasts under conditions in which viability is maintained, we employ the following procedure. Leaf protoplasts from *N. tabacum* are prepared and purified as described above. The final protoplast pellet is resuspended in 5 ml hypertonic growth medium (solution 5; see Appendix) which has been precooled to 4°C. The suspension is transferred into a sterile sample-introduction tube for sorting. The cell sorter is presterilized by autoclaving the sheath fluid filter and all connecting tubing and by flushing the sample line and flow cell with a 10% solution of Clorox. Residual bleach is removed by backflushing with filter-sterilized sheath

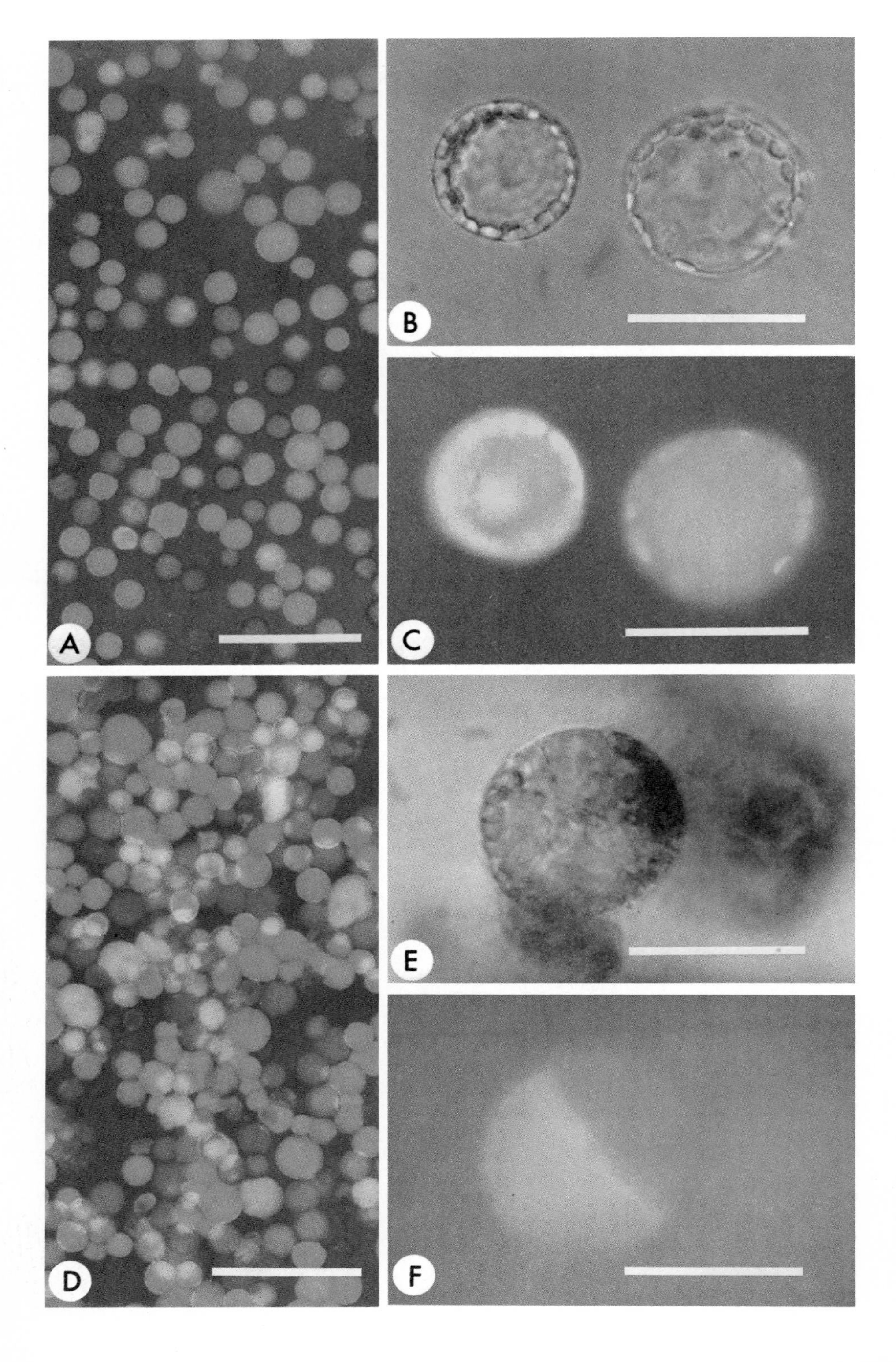
A
B
C
D
E
F

fluid (solution 4; Appendix) for 10 min. The sort delay parameter is determined during sterilization by the sorting of fluorescent microspheres (Coulter Electronics "Fullbright") that have been added to the bleach solution in the sample introduction tube. For this the piezoelectric crystal is maintained at 14.5 kHz, which gives rise to a delay parameter of 10 and an overall efficiency of sorting of the microspheres of 100%. Sorting of protoplasts is carried out at 4°C by placing the sheath fluid container in an ice water bath. This bath possesses a recirculation pump which chills the jacketing around the sample introduction tube. Sorting is carried out into 96-well microtiter plates prefilled with 0.150 ml culture medium (solution 4; see Appendix). Repeated passages allow the collection of rows within a single plate containing 1000–8000 protoplasts per well in stages of 1000. The plates are then sealed with Parafilm and incubated in a sealed, humid box at 26°C in darkness. After 12 days the cell colonies contained within individual wells of the tissue culture plates are transferred into 6 ml cell culture medium (solution 6; see Appendix) contained in 5-cm-diameter plastic Petri dishes.

III. RESULTS AND CONCLUSIONS

A. Fluorescence Labeling of Protoplasts

The introduction of exogenous fluorescent dyes into protoplasts provides a means for the identification of heterokaryons produced by protoplast fusion. The method is unrestricted to protoplast type, requiring only that the protoplasts remain viable and reproducibly fluorescent after staining and during the fusion and selection stages. These criteria are satisfied by both of the labeling techniques described in Section II. Experiments in which the increase in packed cell volume of suspension cultures labeled with the lipophilic dyes F18 and R18 was monitored indicated no

---

Fig. 2. Populations of *Nicotiana* leaf protoplasts separately labeled with FITC and RITC. (A) Parental protoplast populations under episcopic illumination, exhibiting green (FITC) and red (RITC) fluorescence (175×). (B) As for (A), at greater magnification, under bright field illumination (1130×). (C) As for (B), under episcopic illumination. Fluorescence is distributed throughout the protoplasts (1130×). (D) Protoplast populations after completion of polyethylene glycol-induced fusion. Heterokaryons appear yellow (175×). (E) An intermediate stage in fusion, prior to elution of the polyethylene glycol solution, under bright field illumination (1130×). (F) As for (E), under episcopic illumination (1130×). In (A) and (D), the bars represent 150 μm. In (B), (D), (E), and (F), the bars represent 30 μm.

alterations in this parameter as compared to controls (Galbraith and Galbraith, 1979). Protoplast production from these cultures proceeded normally. Although in these experiments protoplast regeneration was not monitored, in current work we have been maintaining suspension cultures of *Nicotiana* species permanently stained with F18 and R18. Protoplasts prepared from these cultures regenerate cell walls and produce callus that can regenerate into normal fertile plants. The lipophilic dyes can be observed to be localized in the membranes of the suspension culture cells and protoplasts, including the plasma membrane, the tonoplast, and the nuclear membrane. Fluorescence is particularly intense with R18. It may therefore be possible to employ photobleaching of R18, using a concentrated light source focused through a standard microscope onto a small area of the plasma membrane, tonoplast (around cytoplasmic strands), or nuclear membrane, to determine membrane microviscosities.

The fluorescence exhibited by cells or protoplasts labeled with FITC, RITC, or TRITC is evenly distributed throughout the cytoplasm and vacuole (Fig. 2; Galbraith and Mauch, 1980). This is consistent with the premise that these dyes react with available molecules, such as proteins and amino acids, which are not localized within specific organelles. We have shown that the process of labeling can be carried out under conditions in which the protoplasts become intensely fluorescent without loss of viability (Galbraith and Mauch, 1980). This was indicated by measurements in culture of protoplast protein synthesis and volume expansion and by the recovery of normal plants from labeled protoplasts. The results above have all been independently confirmed (Patniak *et al.*, 1982; Redenbaugh *et al.*, 1982), which suggests that the degree of substitution required for recognition of fluorescence within the protoplasts is not sufficient to disrupt metabolic pathways such that cell death is induced. The process of labeling does result in the retention of the fluorescent dyes for considerable periods of time during culture and during fusion treatments (Galbraith and Mauch, 1980). This means that the labeling process is suitable for the proposed selection of heterokaryons following protoplast fusion.

B. Flow Cytometric Analysis of Protoplasts

A prerequisite for the sorting of heterokaryons is that the parental protoplast populations, separately labeled with the fluorescein and rhodamine dyes, be differentiated by the cell sorter. Analysis of these protoplast popu-

Fig. 3. Two-dimensional contour projections of the fluorescence of *Nicotiana* leaf protoplasts. (A) FITC-labeled protoplasts. (B) RITC-labeled protoplasts. (C) Protoplasts subjected to fusion with polyethylene glycol.

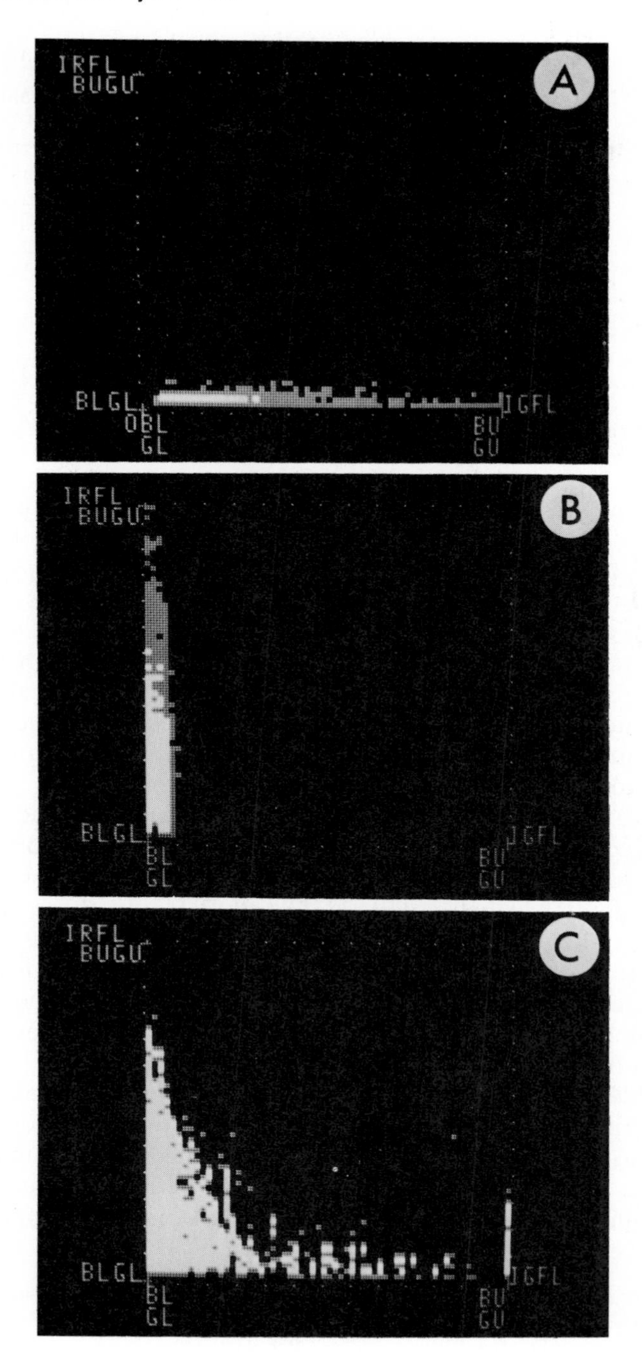
A
IRFL
BUGU
BLGL
IGFL
BL
GL
BU
GU
B
IRFL
BUGU
BLGL
IGFL
BL
GL
BU
GU
C
IRFL
BUGU
BLGL
IGFL
BL
GL
BU
GU

TABLE I

Specificity of Discrimination of Fluorescence within Leaf Protoplasts Using the Cell Sorter

Description		Proportion of sorted protoplasts exhibiting fluorescence corresponding to:		
		FITC	RITC	Both
Mixture of singly and doubly labeled leaf protoplasts of *Nicotiana tabacum*		16	0	84
Fused *N. tabacum* and *N. nesophila* leaf protoplasts	(i)	0	22	78
	(ii)	0	0	100
Fused *N. tabacum* and *N. stocktonii* leaf protoplasts		0	0	100

lations by the cell sorter differs in several respects from that using fluorescence microscopy. Most cell sorters contain a single laser light source. Although the laser is tunable to several lines in the visible spectrum, only one line can be selected for any one experiment. This means that analysis of protoplast fluorescence can only be performed using a single wavelength of excitation of fluorescence. Hence the machine must be adjusted using filter combinations and subtraction module settings to allow optimal and equal detection of fluorescence emission of both fluorescein and rhodamine. Furthermore, under certain circumstances, signals due to chlorophyll autofluorescence must be selectively eliminated.

It is nevertheless possible to establish conditions in which FITC- and RITC-labeled protoplasts can be distinguished by the cell sorter. Figure 3 illustrates contour projections of two-dimensional histograms of leaf protoplasts separately labeled with FITC and RITC. In these the origin is located to the lower left-hand corner of the projections, the x axis corresponding to green (FITC) fluorescence and the y axis to red (rhodamine) fluorescence. In Fig. 3C, we illustrate the appearance of populations of protoplasts separately labeled with FITC and RITC that have been subjected to a fusion treatment.

By establishing electronic sorting windows that exclude all protoplasts falling within the parental zones in these projections (approximately channels 1–10 in both dimensions), it is possible to collect the putative heterokaryons. In four independent experiments, we have sorted those protoplasts falling outside the parental zones and have examined them by fluorescence microscopy for the presence of FITC and RITC fluorescence.

Fig. 4. Growth in culture of sorted *N. tabacum* leaf protoplasts (A) after 1 day, (B) after 2 days, (C) after 5 days, and (D) after 12 days. (200×). Bar represents 150 μm.

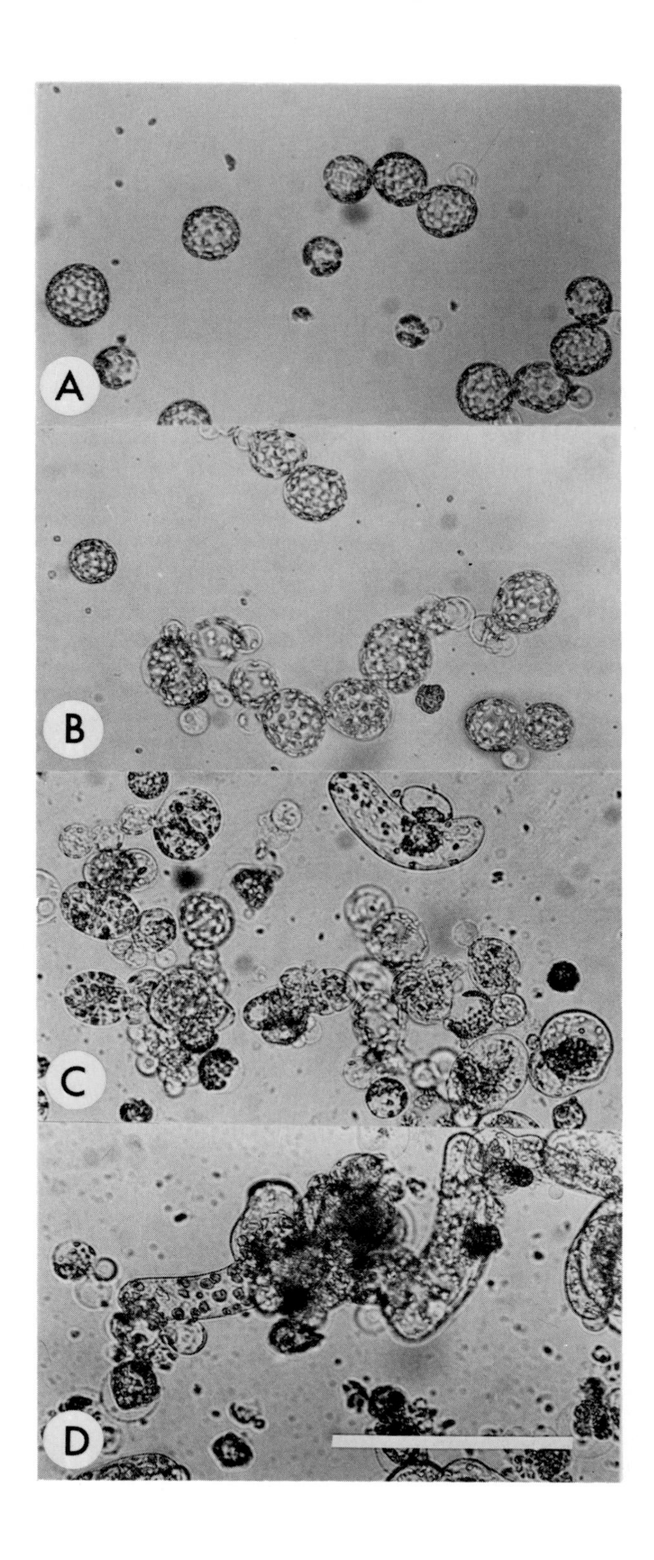
A
B
C
D

The data, presented in Table I, indicate that cell sorting can produce populations of heterokaryons uncontaminated by parental protoplasts.

C. Cell Culture of Viable Sorted Protoplasts

A major concern is that the protoplasts remain viable and sterile during the process of sorting of the heterokaryons. We have extensively investigated this using leaf protoplasts from *N. tabacum.* For these studies the protoplasts were sorted based on their endogenous chlorophyll autofluorescence. Three parameters were measured. The first was the degree of protoplast breakage induced by the sort process. This was calculated by dividing the number of protoplasts actually recovered in the microtiter plate wells (as counted using light microscopy) by the number detected and sorted by the cell sorter. The second parameter was the percentage of intact protoplasts that were viable, as determined by fluorescein diacetate staining. The third parameter was the plating efficiency, determined as the percentage of the viable protoplasts that initiated cell division and formed calli. Using the conditions listed in Section II,D, we routinely obtain a 46–50% efficiency of recovery of intact protoplasts. The process of cell sorting does not reduce the viability of these protoplasts as determined by fluorescein diacetate staining, and the corresponding plating efficiency approaches 100%. During culture we find that approximately 5% of the wells containing sorted protoplasts display contamination. The appearance of typical examples of sorted protoplasts after various periods of time in culture is given in Fig. 4.

At the present time, we have not yet selected somatic hybrid plants by the use of the techniques given above. However, we believe that we have developed conditions for the successful operation of all of the stages required for such selection. Furthermore, some of these stages will be of importance in other aspects of plant tissue culture. For example, in *Catharanthus roseus* tissue cultures, a correlation between endogenous cellular fluorescence and the corresponding production of alkaloids has been demonstrated (Deus and Zenk, 1982). The rapid selection of cell cultures of high endogenous fluorescence by the use of flow analysis, viable sorting, and culture of protoplasts would clearly be of great advantage in this system.

APPENDIX

Solutions Employed in Cell Sorting of Protoplasts

1. Osmoticum for preparation of tobacco leaf protoplasts
   130 g mannitol
   TO salts and hormones (Chupeau *et al.*, 1978) to 1 liter.

APPENDIX (*Continued*)

Adjust pH to 5.5 with 0.1 *M* NaOH.
The final solution is filtered through a 0.22-μm filter prior to use in order to remove any insoluble material.

2. Enzyme solution used in the preparation of tobacco leaf protoplasts
   1.0 g Cellulysin
   0.5 g Driselase
   0.2 g Macerase
   Dissolve to a final volume of 1 liter with solution 1. This enzyme solution is clarified by centrifugation at 10,000 *g* for 10 min and is sterilized by passage through a Millipore Millex-GS type SLGS0250S disposable ultrafiltration unit.
3. Sucrose solution used in protoplast purification
   250 g sucrose
   Dissolve in an aqueous solution of TO salts and hormones (Chupeau *et al.*, 1978) to a final volume of 1 liter. This solution should be sterilized only by ultrafiltration.
4. Protoplast growth medium
   KOM growth medium (Galbraith and Mauch, 1980) is based on that of Kao (1977), modified to contain 68.4 g/liter glucose and 50.5 g/liter mannitol. This solution is used as the sheath fluid for protoplast sorting.
5. Hypertonic growth medium
   This comprises solution 4, containing an additional 50.5 g/liter mannitol.
6. Cell culture medium
   This comprises medium AG as described by Caboche (1980).

REFERENCES

Caboche, M. (1980). Nutritional requirements of protoplast-derived haploid tobacco cells grown at low cell densities in liquid medium. *Planta* **149,** 7–18.

Chupeau, Y., Missonier, C., Hommel, M.-C., and Goujard, J. (1978). Somatic hybrids of plants by fusion of protoplasts. *Mol. Gen. Genet.* **165,** 239–245.

Deus, B., and Zenk, M. H. (1982). Exploitation of plant cells for the production of natural compounds. *Biotechnol. Bioeng.* **24,** 1965–1974.

Galbraith, D. W., and Galbraith, J. E. C. (1979). A method for identification of fusion of plant protoplasts derived from tissue cultures. *Z. Pflanzenphysiol.* **93,** 149–158.

Galbraith, D. W., and Harkins, K. R. (1982). Cell sorting as a means of isolating somatic hybrids. *In* "Plant Tissue Culture, 1982" (A. Fujiwara, ed.), pp. 617–618, Maruzen, Tokyo.

Galbraith, D. W., and Mauch, T. J. (1980). Identification of fusion of plant protoplasts. II. Conditions for the reproducible fluorescence labelling of protoplasts derives from mesophyll tissue. *Z. Pflanzenphysiol.* **98,** 129–140.

Kao, K. N. (1977). Chromosomal behavior in somatic hybrids of soybean—*Nicotiana glauca. Mol. Gen. Genet.* **150,** 225–230.

Keller, P. M., Person, S., and Snipes, W. (1977). A fluorescence enhancement assay of cell fusion. *J. Cell Sci.* **28,** 167–177.

Patniak, G., Cocking, E. C., Hamill, J., and Pental, D. (1982). A simple procedure for the manual isolation and identification of plant heterokaryons. *Plant Sci. Lett.* **24,** 105–110.

Redenbaugh, K., Ruzin, S., Bartholomew, J., and Bassham, J. A. (1982). Characterization and separation of plant protoplasts via flow cytometry and cell sorting. *Z. Pflanzenphysiol.* **107,** 65–80.

CHAPTER 51

Enucleation of Protoplasts: Preparation of Cytoplasts and Miniprotoplasts

Horst Lörz

Max Planck Institute for Plant Breeding Research
Cologne, Federal Republic of Germany

I. INTRODUCTION

In somatic hybridization by protoplast fusion the immediate fusion product is a binucleated structure, a heterokaryon with mixed cytoplasm consisting of a combination of plastidom and chondriom of both parents. The analysis of these hybridization products with respect to nuclear–cytoplasmic interaction is rather complicated. To avoid some of the problems associated with the combination of both nuclear genomes and mixed cytoplasmic material, subprotoplasts (protoplast fragments) can be used to replace one or even both of the fusion partners. Research on fragmentation of plant cells and protoplasts was greatly stimulated by the successful work in somatic animal cell genetics using enucleated cells (cytoplasms) and nucleated minicells (Ege, 1980).

Somatic combination of separate parts of the plant cell genome is of special interest in plant cell biology and plant breeding. Experiments using isolated cell organelles such as nuclei (Lörz and Potrykus, 1978) chloroplasts (Potrykus, 1973; Bonnett and Eriksson, 1974), or mitochondria have been described and have also been successful with respect to physical uptake of the organelles into protoplasts. However, convincing evidence

ISBN 0-12-715001-3

for integration, expression, and multiplication of transferred organelles is still lacking (Potrykus and Lörz, 1976; Giles, 1978). New combinations of nuclear and cytoplasmic genetic information have been achieved by protoplast fusion and subsequent selection and analysis based on suitable markers (Medgyesy *et al.*, 1980). A step toward a more defined combination of nuclear and cytoplasmic material was achieved by inactivating the nuclear genome of one of the fusion partners by irradiation, thus producing "physiological" subprotoplasts (Zelcer *et al.*, 1978; Sidorov *et al.*, 1981). Subprotoplasts are also formed naturally, for example, in the ripening pericarp of some Solanaceae species (Binding, 1976), or spontaneously by "budding" of protoplasts in culture, or during plasmolysis of elongated cells. The last can be collected as a distinct fraction from protoplast preparations (Bradley, 1978; Archer *et al.*, 1982). A more general and experimentally feasible approach for the preparation of subprotoplasts, however, is the fragmentation of isolated protoplasts into miniprotoplasts and enucleated cytoplasts (Wallin *et al.*, 1978; Lörz *et al.*, 1981; Bradley, 1983).

II. PRINCIPLE OF THE PROCEDURE

The fragmentation of protoplasts is achieved by centrifugal forces during centrifugation. Different specific densities of the cellular components (nuclei versus cytoplasmic material) allow the enucleation of protoplasts in isoosmotic density gradients (Lörz *et al.*, 1981). Additional exposure of isolated protoplasts to cytochalasin B in combination with centrifugation was also found to be beneficial for enucleation (Wallin *et al.*, 1978; Bracha and Sher, 1981). Due to the centrifugal forces during centrifugation, protoplasts are dramatically deformed and "drawnout" (Fig. 1). Cellular material of high density, for example, the nucleus, is oriented toward the bottom of the centrifuge tube, whereas the less dense components, such as vacuoles, are oriented toward the top of the tube. After prolonged centrifugation the nuclei, still surrounded by some cytoplasmic material, are finally pinched off, giving rise to enucleated protoplasts. Nucleated miniprotoplasts and enucleated cytoplasts again become spherical after removal from the gradient.

III. TECHNIQUE

Many different types of protoplasts have been used for enucleation. The composition of the gradient, speed of centrifugation, and treatment with

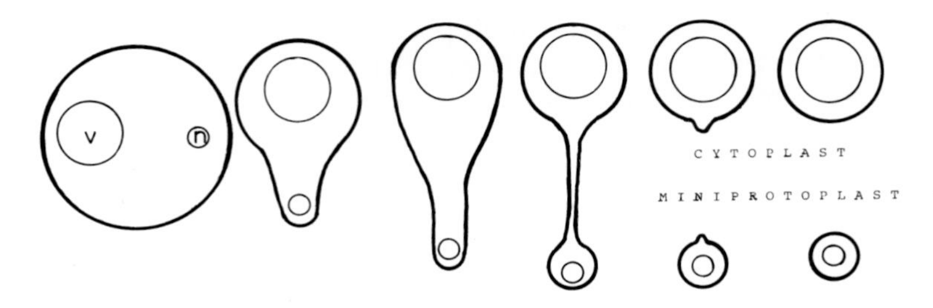

Fig. 1. Fragmentation of protoplasts during centrifugation. The diagrammatic presentation is based on microscopic observations of centrifuged and fragmented protoplasts removed from different areas of iso-osmotic density gradients and samples taken after different times of centrifugation (n, nucleus; v, vacuole).

cytochalasin B have to be worked out individually and adapted for the specific type of protoplast used for enucleation. The following methodological instructions, therefore, are given as a framework within which to develop a procedure for individual needs.

Suitable components for establishing gradients for protoplast centrifugation are inorganic salts, sugars, and modified silica gels such as Percoll (Harms and Potrykus, 1978; Lörz *et al.*, 1981; Lesney *et al.*, 1983). Table I describes a gradient composition including cytochalasin B, according to Wallin *et al.* (1978; Table 1A) and an iso-osmotic density gradient without cytochalasin B after Lörz *et al.* (1981, Table 1B). The discontinuous gradients are formed by gently layering the different solutions, without mixing, starting from layer I. Centrifugation is performed in a swinging-bucket rotor with 10-ml tubes, each loaded with 2–5 × 10^5 protoplasts. Depending on the material used in the experiment, centrifugation speed is between 20,000 and 40,000 *g* for 15–20 min at 37°C (Wallin *et al.*, 1978; Bracha and Sher, 1981) or 45–90 min at 12°C (Lörz *et al.*, 1981). After centrifugation,

TABLE I

Gradients for Enucleation of Protoplasts[a]

Gradient A		Gradient B
	IV	Protoplasts in 0.22 *M* $CaCl_2$
Protoplasts in culture medium	III	0.5 *M* mannitol, 5% Percoll
1.5 *M* sorbitol, 50 μg/ml cytochalasin B	II	0.48 *M* mannitol, 20% Percoll
Saturated sucrose	I	0.45 *M* mannitol, 50% Percoll

[a]Gradient A with cytochalasin B is centrifuged at 37°C for 15 min at 20,000 *g* and used for mesophyll protoplasts of *Pisum sativum* and suspension protoplasts of *Nicotiana sylvestris* or at 40,000 *g* with protoplasts of suspension cultures of *Daucus carota* and *N. tabacum* (Wallin *et al.*, 1978). The iso-osmotic density gradient B is also centrifuged at 40,000 *g*, but for 60 min at 12°C, and used for protoplasts isolated from cultured cells of *Hyoscyamus muticus, N. tabacum,* and *Zea mays* (Lörz *et al.*, 1981).

enucleated cytoplasts are located in the top fraction of the gradient and nucleated miniprotoplasts form a band between layers I and II. Both subprotoplast fractions are collected by Pasteur pipettes and resuspended in medium.

IV. RESULTS

The success in enucleation of protoplasts is greatly influenced by the quality of the protoplast preparation, for example, the homogeneity and stability of the protoplasts. Of the original protoplasts placed on a gradient, 90–95% are recovered as nucleated miniprotoplasts and about 60–65% as enucleated cytoplasts. Microscopic analysis of the cytoplast fraction and staining (as control) for remaining nucleated protoplasts gives evidence that the proportion of cytoplasts is very high (up to 99%). Miniprotoplasts derived from cultured cells are reduced in size to about 15% of the original protoplast volume. Determination of the soluble protein in the subprotoplasts shows about one-third of the original content with miniprotoplasts and two-thirds with enucleated cytoplasts. Cytoplasts are rather fragile structures and are metabolically less active than nucleated protoplasts (Lörz *et al.*, 1981). Most important, however, miniprotoplasts and enucleated protoplasts are suitable for fusion experiments, and cytoplasts are especially useful experimental tools for transfer of chloroplasts and mitochondria (Wallin *et al.*, 1979; Bracha and Sher, 1981; Maliga *et al.*, 1982).

V. CONCLUSION

Enucleation of protoplasts by centrifugation and treatment with cytochalasin B has become a useful technique in plant cell biology. The standardization of the procedure is hampered and limited by the different types of protoplasts isolated from different plant species and used for subprotoplast preparation.

In general, protoplasts without green chloroplasts isolated from cell suspension or callus are more suitable for enucleation than mesophyll protoplasts. Fully developed chloroplasts are very similar in specific density to nuclei, and therefore fragmentation of mesophyll protoplasts often leads to a "heavy" subprotoplast containing a nucleus and chloroplasts and a "light" fragment consisting of the large vacuole and some residual cytoplasm only (Lörz *et al.*, 1976). Removing the vacuole and preparing nonvacuolated miniprotoplasts is of special interest for microinjection studies. Highly vacuolated cells are easily damaged during injection. In contrast, nonvacuolated miniprotoplasts are more stable for this type of manipulation, and the smaller the recipient cell the better is the chance to aim directly into the nucleus (Steinbiss and Stabel, 1983).

In addition to their use for biochemical and physiological experiments, subprotoplasts are of increasing importance in somatic plant cell genetics. These areas include reconstitution of cells in homologous and heterologous systems, fusion experiments with miniprotoplasts for somatic hybridization, and transfer of cytoplasmic genetic information by fusion with enucleated cytoplasts.

REFERENCES

Archer, E. K., Landgren, C. R., and Bonnett, H. T. (1982). Cytoplast formation and enrichment from mesophyll tissues of *Nicotiana* ssp. *Plant Sci. Lett.* **25,** 175–185.

Binding, H. (1976). Somatic hybridization experiments in Solanaceous species. *Mol. Gen. Genet.* **155,** 171–175.

Bonnett, H. T., and Eriksson, T. (1974). Transfer of algal chloroplasts into protoplasts of higher plants. *Planta* **120,** 71–79.

Bracha, M., and Sher, N. (1981). Fusion of enucleated protoplasts with nucleated miniprotoplasts in onion (*Allium cepa* L.). *Plant Sci. Lett.* **23,** 95–101.

Bradley, P. M. (1978). Production of enucleated plant protoplasts of *Allium cepa. Plant Sci. Lett.* **13,** 287–290.

Bradley, P. M. (1983). The production of higher plant subprotoplasts. *Plant Mol. Biol. Rep.* **1,** 117–123.

Ege, T. (1980). Fusion of cell fragments as a method in cell genetics. *In* "Transfer of Cell Constituents into Eukaryotic Cells" (J. E. Celis *et al.*, eds.), pp. 201–233. Plenum, New York.

Giles, K. L. (1978). The uptake of organelles and microorganisms by plant protoplasts: Old ideas but new horizons. *In* "Frontiers of Plant Tissue Culture 1978" (T. A. Thorpe, ed.), pp. 67–79. Univ. of Calgary Press, Calgary, Alberta, Canada.

Harms, C. T., and Potrykus, I. (1978). Fractionation of plant protoplast types by iso-osmotic density gradient centrifugation. *Theor. Appl. Genet.* **53,** 57–63.

Lesney, M. S., Callow, P. C., and Sink, K. C. (1983). A simplified method for bulk production of cytoplasts from suspension-culture derived protoplasts of *Solanum nigrum* (L). *In* "Protoplasts 1983" (I. Potrykus, C. T. Harms, A. Hinnen, R. Hütter, P. J. King, and R. D. Shillito, eds.), pp. 116–117. Birkhäuser, Basel.

Lörz, H., and Potrykus, I. (1978). Investigation on the transfer of isolated nuclei into plant protoplasts. *Theor. Appl. Genet.* **53,** 251–256.

Lörz, H., and Potrykus, I. (1980). Isolation of subprotoplasts for genetic manipulation studies. *In* "Advances in Protoplast Research" (L. Ferenczy, G. L. Farkas, and G. Lázár, eds.), pp. 377–382. Akadémiai Kiadó, Budapest.

Lörz, H., Harms, C. T., and Potrykus, I. (1976). Isolation of "vacuoplasts" from protoplasts of higher plants. *Biochem. Physiol. Pflanz.* **169,** 617–620.

Lörz, H., Paszkowsky, I., Dierks-Ventling, C., and Potrykus, I. (1981). Isolation and characterization of cytoplasts and miniprotoplasts derived from protoplasts of cultured cells. *Physiol. Plant.* **53,** 385–391.

Maliga, P., Lörz, H., Lázár, G., and Nagy, F. (1982). Cytoplast-protoplast fusion for interspecific chloroplast transfer in *Nicotiana. Mol. Gen. Genet.* **185,** 211–215.

Medgyesy, P., Menczel, L., and Maliga, P. (1980). The use of cytoplasmic streptomycin resistance: Chloroplast transfer from *Nicotiana tabacum* into *Nicotiana sylvestris,* and isolation of their somatic hybrids. *Mol. Gen. Genet.* **179,** 693–698.

Potrykus, I. (1973). Transplantation of chloroplasts into protoplasts of *Petunia. Z. Pflanzenphysiol.* **70,** 364–366.

Potrykus, I., and Lörz, H. (1976). Organelle transfer into isolated protoplasts. *In* "Cell Genetics in Higher Plants" (D. Dudits, G. L. Farkas, and P. Maliga, eds.), pp. 183–190. Akadémiai Kiadó, Budapest.

Sidorov, V. A., Menczel, L., Nagy, F., and Maliga, P. (1981). Chloroplast transfer in *Nicotiana* based on metabolic complementation between irradiated and iodoacetate treated protoplasts. *Planta* **152,** 341–345.

Steinbiss, H. H., and Stabel, P. (1983). Protoplast derived tobacco cells can survive capillary microinjection of the fluorescent dye lucifer yellow. *Protoplasma* **116,** 223–227.

Wallin, A., Glimelius, K., and Eriksson, T. (1978). Enucleation of plant protoplasts by cytochalasin B. *Z. Pflanzenphysiol.* **87,** 333–340.

Wallin, A., Glimelius, K., and Eriksson, T. (1979). Formation of hybrid cells by transfer of nuclei via fusion of miniprotoplasts from cell lines of nitrate reductase deficient tobacco. *Z. Pflanzenphysiol.* **91,** 89–94.

Zelcer, A., Aviv, D., and Galun, E. (1978). Interspecific transfer of cytoplasmic male sterility by fusion between protoplasts of normal *Nicotiana sylvestris* and X-ray irradiated protoplasts of male-sterile *N. tabacum. Z. Pflanzenphysiol.* **90,** 397–407.

CHAPTER **52**

Isolation of Organelles: Nuclei

L. Willmitzer

Max Planck Institute for Plant Breeding Research
Cologne, Federal Republic of Germany

I. INTRODUCTION

The technique described in this chapter has been used in our laboratory routinely during the last 4 years for the isolation of highly purified, physiologically active nuclei from plant tissue cultures and leaves of tobacco. This method was developed because of our interest in a comparison of normal and transformed (crown gall) plant cells on the basis of their respective chromosomal proteins (histones and nonhistone proteins), different nuclear enzymatic activities, and chromatin structure. Therefore, the initial aim was to develop a method which allowed the isolation of highly purified nuclei which had retained their endogenous enzyme activities as well as their chromatin structure. In addition, as we were particularly interested in tumorous (crown gall) cell lines, the method was first developed for callus and suspension cultures.

Isolation of pure nuclei from tissue culture cells has proven to be rather difficult, mainly for the following reasons:

CELL CULTURE AND SOMATIC CELL
GENETICS OF PLANTS, VOL. 1

ISBN 0-12-715001-3

1. There is a very low concentration of nuclei in tissue culture plant cells (e.g., 4 kg of tobacco plant suspension culture are needed to produce the same number of nuclei in the starting material as in 100 g of rat liver).
2. Nuclei from plant tissue cultures have a strong tendency to form large aggregates with cell wall fragments, especially in the presence of divalent cations.
3. High amounts of phenolic compounds and phenol oxidases are located in the vacuole and are liberated during disruption of the tissue, leading to denaturation of proteins and nucleic acids (Loomis, 1974).

The main modifications introduced into the method described below compared to previously described methods are as follows:

1. There is a brief (30- to 60-min) incubation of the cells in cellulase/pectinase. Although during this treatment tissue-cultured tobacco cells do not change, as judged by microscopic criteria, this treatment is necessary in order to open the cells mechanically in a gentle and efficient way. In addition, we found that the aggregation problem (see above) is reduced by this treatment.
2. Throughout the isolation procedure, the nuclei are stabilized by the polyamines spermine/spermidine (first introduced by Hewish and Bourgoyne, 1973) instead of divalent cations. This has the additional advantage that the chelating agent EDTA can be introduced into the medium, resulting in an inhibition of DNases as well as the Cu^{2+}-dependent phenol oxidases (Anderson, 1968).
3. The osmotically inactive density gradient medium Percoll is used instead of osmotically active sucrose and glycerol media.
4. Only filtration and sedimentation/flotation steps are included in the method, thus allowing the processing of large amounts of material (1–2 kg) within a reasonable time (3–5 hr).

II. PROCEDURE

A. Composition of the Buffer

Buffer A: 0.7 *M* mannitol
10 m*M* MES
5 m*M* EDTA
0.1% bovine serum albumin (BSA)

0.2 mM phenylmethylsulfonyl fluoride (PMSF)
pH 5.8

Buffer B: Buffer A supplemented with 150 mg purified pectinase and 300 mg cellulase in a total of 3 liters. In order to dissolve the lyophilized enzymes, the pH has to be raised to 11 for 2 min and then returned to 5.8 (PMSF is then added).

Buffer C: 0.25 M sucrose
10 mM NaCl
10 mM MES, pH 6.0
5 mM EDTA
0.15 mM spermine
0.5 mM spermidine
20 mM mercaptoethanol
0.2 mM PMSF
0.6% Nonidet P40
0.1% BSA

Buffer D: 30 g of 5X concentrated buffer C (the concentration of Nonidet P40 is kept to 0.6%) added to 116 g Percoll with continuous stirring, pH adjusted to 6.0 by 1 N HCl (do this carefully, as Percoll may gelate irreversibly upon back-titration).

Buffer E: 6 g of 5X concentrated buffer C added to 45 g Percoll, pH adjusted to 6.0.

B. Purification of Cellulase and Pectinase

Cellulase and pectinase, purchased from Serva Feinbiochemica (Heidelberg, Federal Republic of Germany) (numbers 16420 and 31660, respectively), were purified before use. Fifty grams of cellulase or pectinase were dissolved in 250 ml 0.5 M acetic acid. The undissolved particles were pelleted by centrifugation for 15 min at 3000 g, and the clear supernatant was applied to a Biogel P6 column (5 × 90 cm) which was eluted at 80 ml/hr with 0.5 M acetic acid. The high molecular weight peak was collected and lyophilized, yielding 1 g of cellulase and 0.5 g of pectinase, respectively.

C. Experimental Protocol

All steps, unless stated otherwise, are performed at 4°C. Swinging bucket rotors are used for all centrifugation steps in order to minimize mechanical stress on the nuclei.

Step 1: Wash cells (200 g fresh weight) from late log phase with buffer A (1 liter) on wide-mesh sieve cloth (130 μm) or on commercial curtain cloth.

Step 2: Incubate in 1 liter of buffer A at 25°C for 15 min.

Step 3: After filtration and resuspension in 1 liter of buffer B, incubate the cells at 25°C for 30–60 min.

Step 4: Wash again with 1 liter of buffer C.

Step 5: Resuspend in 0.3 liter of buffer C and homogenize for 4 × 15 sec using the Ultra-Turrax T45 (Janke & Kunkel, Staufen, Federal Republic of Germany) at full speed.

Step 6: Filter the suspension repeatedly on sieve cloth (Schweizer Siedengazefabrik, Zurich, Switzerland, P series) with decreasing mesh (130, 80, and 20 μm).

Step 7: Centrifuge filtrate at 2000 *g* (brakes off) for 5 min.

Step 8: Resuspend the pellet (continuing the nuclei) in buffer C, adjusting a volume equivalent to 45 g. Add 60 g of butter D. Mix the two solutions in an Erlenmeyer flask by manual shaking.

Step 9: Filter the suspension through a 20-μm-sieve cloth.

Step 10: Centrifuge filtrate at 1000 *g* (brakes off) for 5 min.

Step 11: Resuspend the pellet containing nuclei and contaminations in 2–5 ml of buffer C. After shaking in an Erlenmeyer flask (to abolish density differences) centrifuge supernatant at 1000 *g* for 5 min. Repeat three to five times until the amount of nuclei in sediment is small compared to the contaminating material.

Step 12: Spin collected pellets down at 1000 *g* for 5 min.

Step 13: Resuspend pellets by repeated sucking with a Pasteur pipette. Add 20 ml buffer E to the final pellet.

Step 14: Centrifuge suspension at 5000 *g* (brakes off) for 5 min.

Step 15: Take off the floating layer containing the nuclei. Dilute with 20 ml of buffer C. Spin down nuclei at 1000 *g* for 5 min.

Step 16: Wash nuclei two to three times in buffer C (1000 *g*) for 3 min.

D. Comments on the Different Steps in the Purification Procedure

Step 3: The time of incubation and/or the amount of cellulase/pectinase contained in buffer B has to be adjusted for every cell line. The recipe given above was used by us for suspension culture cells of tobacco in the middle to late log phase. For older tissue or callus cultures the incubation time was prolonged to 2–4 hr. When leaf tissue was used, the leaves were macerated before incubation in buffer B. The same was done with big callus clumps.

Step 5: To prevent foaming, a few drops of *n*-octanol should be added.

Step 6: To allow rapid filtration, the sievers must be in close contact with the funnel; any air bubbles between the sieve and the funnel will retard this process. Normally, each step takes less than 5 min. During filtration through the 20-μm sieve, the material should not be pressed; otherwise unbroken cells will enter the filtrate.

Step 7: We used a Beckman J6 centrifuge, as it allows the processing of up to 6 liters in one step.

Step 8: The exact ratio of homogenate to buffer D has to be determined for every cell line. A ratio between 6:7 and 6:10 was found to be suitable for all cell cultures used in our work. To determine the optimal ratio, one should check a series of mixtures in the above range. The important criterion is that no cell wall debris should be sedimented by the subsequent centrifugation, as it is nearly impossible, in our experience, to get rid of this debris at a later stage.

Step 10: We used a Minifuge 2 (Heraeus-Christ G.m.b.H., Osterode, Federal Republic of Germany), which accelerates and decelerates very rapidly.

Step 11: This step increases the yield, as many nuclei are trapped in empty cell wall residues and thus will not be pelleted.

Step 13: The ratio of sediment to buffer E added must not be higher than 1:20 (v/v); otherwise the density of the mixture will be too low and the nuclei will not float.

III. RESULTS AND CONCLUSION

The procedure described has been used routinely for the isolation of nuclei from a number of different tobacco suspension culture lines as well as leaves. The nuclei obtained are highly purified. Particles other than nuclei (mostly cell wall residues) contributed not more than 5% of the total particles. The nuclei have retained their morphological characteristics, as evidenced by phase contrast and electron microscopy; high molecular weight DNA can be extracted from the nuclei; the chromatin (nucleosomal) arrangement is retained; and extracted histones display the expected gel-electrophoretic pattern (Willmitzer and Wagner, 1981). Nuclei isolated according to this procedure have been demonstrated to contain the following enzymatic activities: RNA polymerase I, II, and III (Willmitzer *et al.*, 1981a), DNA polymerase (Meier, 1981), histone-acetylase (Arfmann and Haase, 1981), poly-ADP-ribosylatinon activity (Willmitzer, 1979; Böcker and

Szopa, 1982), and protein kinase (Erdmann *et al.*, 1982; Arfmann and Willmitzer, 1982).

The endogenous activity of the RNA polymerases allows the analysis of "run-off" transcripts in isolated nuclei (Willmitzer *et al.*, 1981b). However, as no precautions are taken during the isolation procedure, the RNA synthesized in isolated nuclei is partly degraded. A similar result has been reported for run-off transcripts synthesized in isolated nuclei from carrot (Calza *et al.*, 1982). The only possible additive which, to my knowledge, would not interfere with RNA polymerase activity is the placental ribonuclease inhibitor, which, however, is too expensive to be used in mass isolations. Thus, for the isolation of undegraded nuclear RNA, the procedure described is, in our experience, not suitable (cf., however, Willmitzer *et al.*, 1981b).

The procedure described above has been used most extensively with tissue cultures and leaves from tobacco. However, it has been shown to be applicable, with some modifications (e.g., time of incubation with cellulase/pectinase), to tissue cultures of potato (H.-P. Mühlbach, personal communication) as well as tissue cultures of parsley (K. Hahlbrock, personal communication).

REFERENCES

Anderson, J. W. (1968). Extraction of enzymes and subcellular organelles from plant tissue. *Phytochemistry* **7,** 1973–1988.

Arfmann, H.-A., and Haase, E. (1981). Effect of sodium butyrate on the modification of histones in cell cultures of *Nicotiana tabacum*. *Plant Sci. Lett.* **21,** 317–324.

Arfmann, H.-A., and Willmitzer, L. (1982). Endogenous protein kinase activity in tobacco nuclei. Comparison of transformed and nontransformed cell cultures of *Nicotiana tabacum*. *Plant Sci. Lett.* **26,** 31–38.

Böcker, M., and Szopa, J. (1982). Properties of ADP-ribosylation is isolated nuclei from *Nicotiana tabacum* 1. cell cultures. *Z. Pflanzenphysiol.* **108,** 113–124.

Calza, R. E., Oelke, S. M., and Lurquin, P. F. (1982). Transcription in nuclei isolated from carrot protoplasts: Effects of exogenous DNA. *FEBS Lett.* **143,** 109–114.

Erdmann, H., Böcker, M., and Wagner, K. G. (1982). Two protein kinases from nuclei of cultured tobacco cells with properties similar to the cyclic nucleotide-independent enzymes (NI and NII) from animal tissue. *FEBS Lett.* **137,** 245–248.

Hewish, D. R., and Bourgoyne, L. A. (1973). Chromatin substructure. The digestion of chromatin DNA at regularly spaced sites by a nuclear deoxyribonuclease. *Biochem. Biophys. Res. Commun.* **52,** 504–510.

Loomis, W. D. (1974). Overcoming problems of phenolics and guinones in the isolation of plant enzymes and organelles. *In* "Methods in Enzymology" (S. Fleischer and L. Packer, eds.), Vol. 31, Part A, pp. 528–544. Academic Press, New York.

Meier, R. (1981). Diploma thesis, Technical Univ., Braunschweig, Federal Republic of Germany.

Willmitzer, L. (1979). Demonstration of *in vitro* covalent modification of chromosomal proteins by poly (ADP) ribosylation in plant nuclei. *FEBS Lett.* **108,** 13–16.

Willmitzer, L., and Wagner, K. G. (1981). Isolation of nuclei from tissue-cultured plant cells. *Exp. Cell Res.* **135,** 69–77.

Willmitzer, L., Schmalenbach, W., and Schell, J. (1981a). Transcription of T-DNA in octopine and nopaline crown gall tumors is inhibited by low concentrations of α-amanitin. *Nucleic Acids Res.* **9,** 4801–4812.

Willmitzer, L., Otten, L., Simons, G., Schmalenbach, W., Schröder, J., Schröder, G., Van Montagu, M., De Vos, G., and Schell, J. (1981b). Nuclear and polysomal transcripts of T-DNA in octopine crown gall suspension and callus cultures. *Mol. Gen. Genet.* **182,** 255–262.

CHAPTER **53**

Isolation of Organelles: Chromosomes

Gyula Hadlaczky

Institute of Genetics
Biological Research Center
Hungarian Academy of Sciences
Szeged, Hungary

I. INTRODUCTION

The overwhelming majority of our knowledge on the biochemistry of eukaryotic chromosomes came from studies of isolated mammalian chromosomes. Also, the use of isolated chromosomes provided essential information about their structural organization (for a review, see Bostock and Sumner, 1978). By means of isolated chromosomes, a promising new scheme for genetic manipulation called *chromosome-mediated gene transfer* has been developed (Klobutcher and Ruddle, 1981). The contributions of the plant material to such studies has been hindered, until recently, by the lack of reliable procedures for mass isolation of plant chromosomes. Recent developments in protoplast and cell culture of plants may soon change this situation. Although considerable efforts have been made in this field

CELL CULTURE AND SOMATIC CELL
GENETICS OF PLANTS, VOL. 1

ISBN 0-12-715001-3

(Malmberg and Griesbach, 1980; Szabados *et al.*, 1981; Griesbach *et al.*, 1982), the isolation of plant chromosomes is still not well developed.

We have developed a procedure for mass isolation of plant chromosomes, in milligram quantities, from protoplasts. Plant chromosomes isolated by this method exhibited excellent preservation of morphology, and the purity of the chromosomes has made them suitable for structural and biochemical studies (Hadlaczky *et al.*, 1982, 1983). The procedure described here has been tested for both monocotyledonous (wheat) and dicotyledonous (poppy) plant cell lines, as well as for mammalian (human, hamster, and mouse) cell lines with satisfactory results. However, considering the heterogeneity of different plant cell lines, the optimal procedure may vary somewhat from one to another. This should be kept in mind if one wishes to apply the procedure for different cell lines or with disparate culture conditions.

II. PROCEDURE

A. Culture of Cell Suspension

Wheat (*Triticum monococcum*) cells (Gamborg and Eveleigh, 1968) and poppy (*Papaver somniferum*) cells are maintained in C8 I medium (Dudits *et al.*, 1977) and in Medium I (Kao *et al.*, 1974), respectively. Suspensions are cultured in continuous light (3000 lx) on a rotary shaker (120 rpm) at 25°C and subcultured at 2-day intervals. In both suspensions, 2 ml of settled cells in 100 ml culture medium is optimal for producing exponentially growing cell populations.

B. Cell Synchronization

Synchronization of wheat and poppy cells for metaphase is made by combined hydroxyurea-colchicine (Eriksson, 1966; Szabados *et al.*, 1981) and colchicine treatment, respectively. From 1-day-old wheat suspensions, 5 ml of settled cells are harvested and resuspended in 100 ml fresh medium containing 2.5 m*M* hydroxyurea (Sigma). After 24-hr incubation, hydroxyurea is removed by three subsequent washes with fresh medium, and cells are incubated in fresh medium supplemented with 0.05% colchicine (Serva, Heidelberg, Federal Republic of Germany) for another 11 hr on a shaker (180 rpm) in the dark.

Poppy cells are blocked for metaphase with simple colchicine treatment. After 1 day of subculturing, 5 ml of settled cells are transferred to 100 ml of fresh medium containing 0.1% colchicine and incubated for 12 hr in the dark on a shaker. With wheat and poppy cells, the above procedures give a reproducible, partial, mitotic synchrony. The yield of metaphase cells is usually around 30%.

C. Isolation of Protoplasts

Colchicine-treated cells are harvested by centrifugation (100 *g* for 5 min), and the pelleted cells are resuspended in the supernatant medium in a ratio of 1:1. Cell suspension is mixed with an equal volume of enzyme solution containing 0.05 and 0.1% of colchicine for wheat and poppy cells, respectively. The enzyme solution contains 5 m*M* $MgCl_2$, 2 m*M* $CaCl_2$, 3 m*M* 2-(*N*-morpholino) ethane sulfonic acid, 170 m*M* mannitol, 250 m*M* glucose, 6% Cellulase (Onozuka R-10), 2% Rhozyme (Rohm and Haas), 2% Pectinase (Serva), and 5% Driselase (Fluka) at pH 5.6. In many instances, enzymes (especially Driselase) contain insoluble material. Therefore, it is advisable to centrifuge (at 4000 rpm for 10 min) the enzyme solution before adjusting the pH.

After a 2- to 3-hr incubation of cells on a shaker (at 50 rpm) in the dark, protoplasts are collected by centrifugation (100 *g* for 3 min), and pelleted protoplasts are washed free from enzymes with a solution containing 100 m*M* glycine, 2.5 m*M* $CaCl_2$, and 5% glucose at pH 6.0. Protoplasts are pelleted again by centrifugation (100 *g* for 3 min). The above procedure leads to only a modest (2–4%) decrease in the number of mitotic cells.

D. Rupture of Protoplasts

All of the following operations of the chromosome isolation procedure are performed on ice or at 0–4°C, except where especially noted, and siliconized glass tubes are used. Two milliliters of pelleted protoplasts are resuspended in 100 ml of hypotonic glycine–hexylene glycol buffer (GH buffer) containing 100 m*M* glycine, 1% hexylene glycol at pH 8.4–8.6 adjusted by a saturated solution of $Ca(OH)_2$. For wheat protoplasts, the buffer is supplemented with 2.5% glucose.

After incubation of protoplasts for 10 min at room temperature, the suspension is chilled down in ice water and Triton X-100 (Sigma) detergent is added (from a 10% stock solution) to a final concentration of 0.1%. Repeated pipetting of suspension to new tubes by plastic Pasteur pipettes causes complete disruption of protoplasts, and chromosomes are liberated into the suspension.

E. Purification of Chromosomes

Suspension of ruptured protoplasts is centrifuged at 1000 *g* for 20 min. The pellet containing chromosomes, nuclei, and cellular debris is resuspended in GHT buffer (GH buffer supplemented with 0.1% Triton X-100) containing 1 m*M* phenylmethyl sulfonyl fluorid (PMSF, Sigma) and 1% isopropyl alcohol, and then centrifuged at 200 *g* for 10 min. If necessary, this differential centrifugation is repeated until the nuclei and cellular debris from the chromosome suspension are totally eliminated. The supernatant containing only chromosomes and slow-sedimenting contaminants is mixed with a 1 *M* sucrose solution in a ratio of 1:1 and layered onto the top of a 1 *M* sucrose solution made up in GHT buffer and centrifuged at 1000 *g* for 20 min. This step is repeated, usually two or three times, until the chromosome suspension is of satisfactory purity.

F. Light Microscopy

Steps in the isolation process are checked by phase-contrast microscopy and by Giemsa staining. For stained preparations, a drop of sample is dried onto a slide, immersed in 0.1 *N* HCl for 30 sec, rinsed with distilled water, and stained with a 2% solution of Giemsa stain in Sörensen's phosphate buffer (pH 6.8) for 2–5 min at room temperature. Alternatively, to examine the final chromosome suspension, Feulgen staining is made. A drop of concentrated chromosome suspension is fixed onto the slide with acetic-alcohol (1:3), hydrolyzed in 1 *N* HCl at 60°C for 10 min, and stained with Schiff's reagent for 1 hr at room temperature.

G. Electron Microscopy

For whole-mount preparations, a concentrated chromosome suspension (10–15 μl in GHT buffer) is spread on a hypophase made from double glass-distilled water. Samples are picked up on Formvar-coated 150-mesh copper grids and dehydrated through 30, 50, 70, 90, and 100% ethanol, a 1:1 mixture of ethanol and amyl acetate, and 100% amyl acetate. The grids are dried by a critical point dryer, using carbon dioxide, and then coated

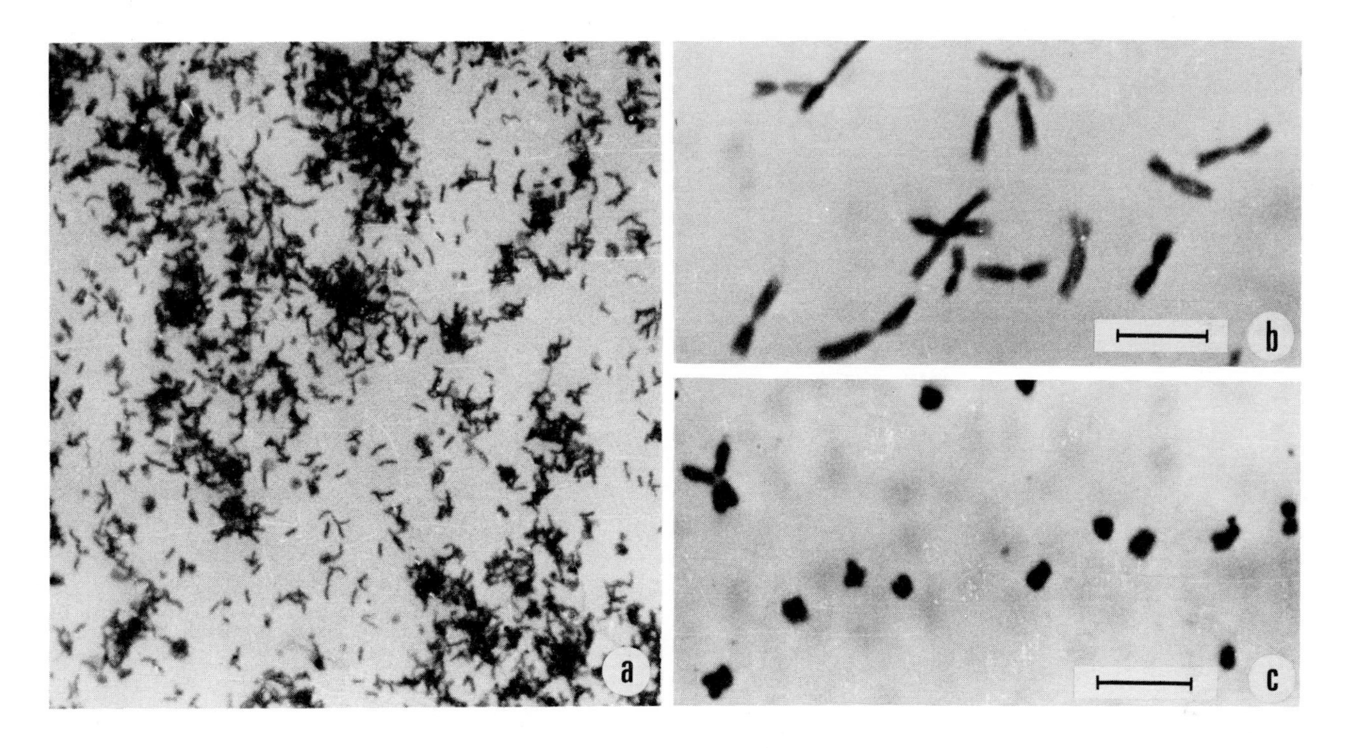

Fig. 1. Light microscopy of isolated plant chromosomes with Feulgen staining. (a) Isolated wheat chromosomes at a 2 mg/ml DNA concentration. Morphology of (b) isolated wheat and (c) poppy chromosomes. Bars represent 10 μm. (From Hadlaczky *et al.*, 1983.)

with carbon in a vacuum evaporator. Preparations are examined by transmission electron microscopy.

Preparations for scanning electron microscopy are fixed onto a coverslip with 2.5% glutaraldehyde, dehydrated with ethanol and acetone and dried by critical point drying, as described above. Coverslips carrying dry samples are cut into small pieces with a diamond, mounted on the preparation holder with a conductive paint, and coated with gold.

H. Storage of Isolated Chromosomes

Isolated chromosomes in GHT buffer containing 1 m*M* phenylmethylsulfonyl fluoride (PMSF) are stable for days when stored in a refrigerator. After 48 hr of storage, no marked degradation of the chromosomal proteins and DNA occurs (Hadlaczky *et al.*, 1983).

III. RESULTS AND DISCUSSION

By the above procedure, fairly pure wheat and poppy chromosomes can be isolated in milligram quantities (Fig. 1a). Isolated plant chromosomes show excellent preservation of morphology with light microscopy (Fig. 1b, c); their ultrastructure also seems to be intact (Fig. 2a–c). The purity of the chromosome suspension, as judged by phase-contrast microscopy and stained preparations, is satisfactory. In a successful isolation, the amount of visible contaminants (nucleoli in wheat; starch granules and protein aggregates in poppy chromosomes) must not be higher than 5%.

Nonhistone proteins and histones of pure isolated chromosomes show rather constant electrophoretic patterns (Fig. 3).

Problems and Perspectives

1. The primary requisite for a successful chromosome isolation is a cell population enriched in metaphase cells. In our experience, the minimal yield of the metaphase cells required for chromosome isolation is about 20%. Although chromosomes can be isolated from cultures having a lower

Fig. 2. Whole-mount electron micrographs of isolated plant chromosomes. (a) Isolated wheat chromosome spread from GHT buffer. (b) Isolated poppy chromosome spread from GHT buffer. (c) Scanning electron micrograph of isolated wheat chromosome. Bars represent 1 μm. [(a) From Hadlaczky *et al.*, 1983; (b,c) from Hadlaczky *et al.*, 1982.]

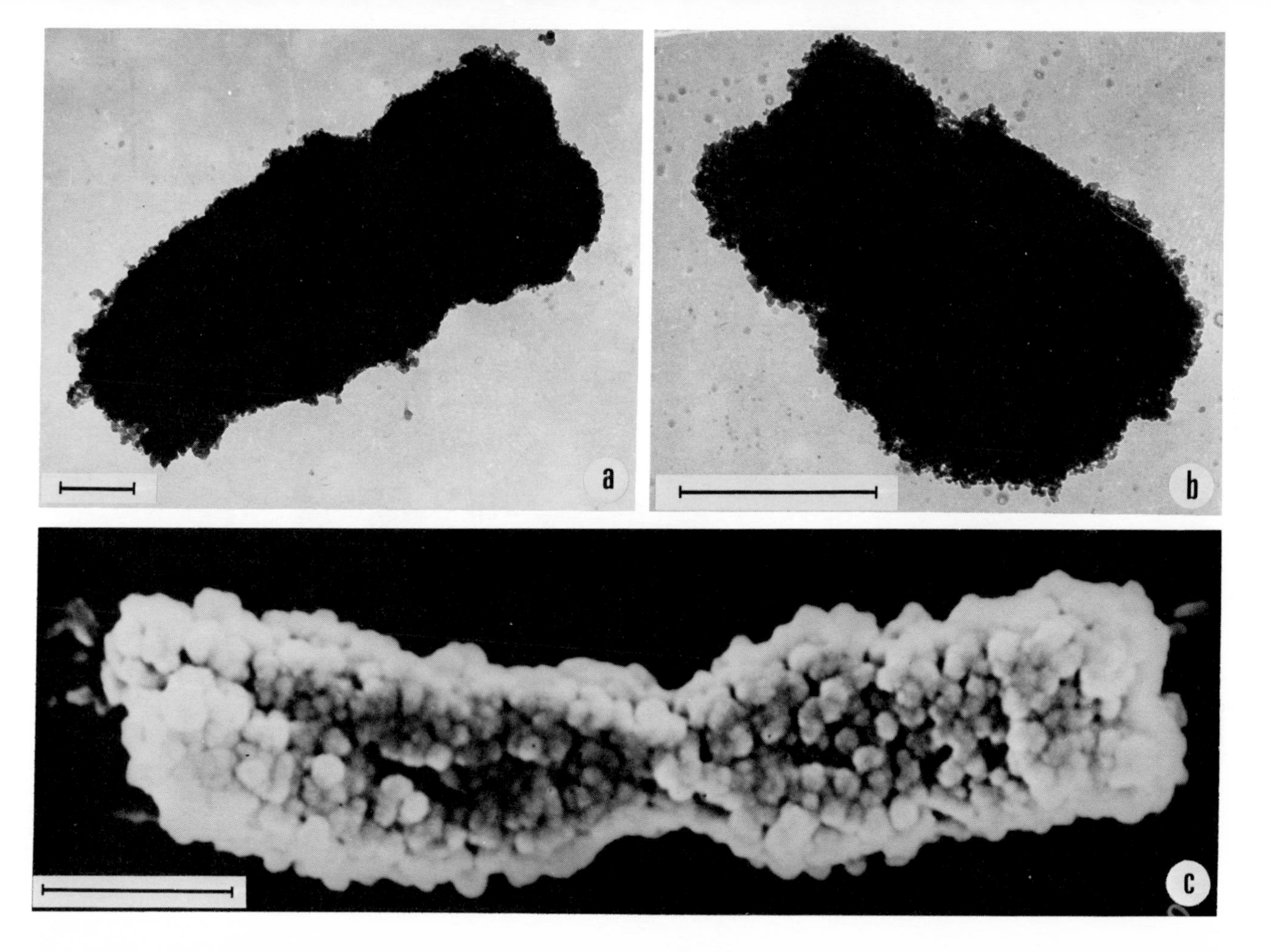
a
b
c

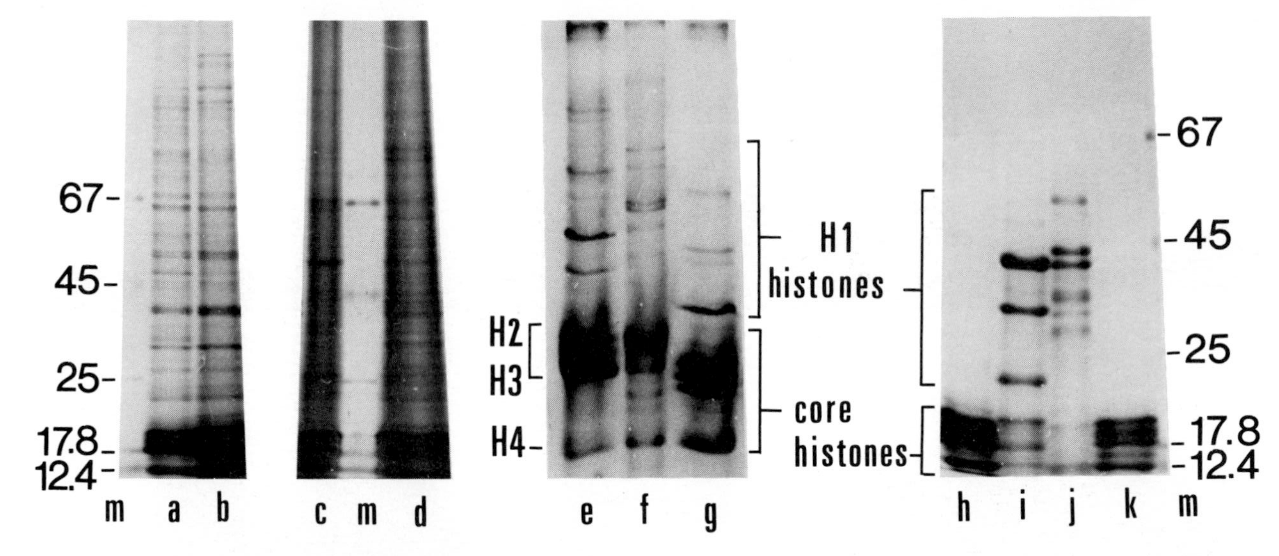

Fig. 3. Polyacrylamide gel electrophoresis of chromosomal proteins. Proteins of (a) isolated wheat nuclei, (b) isolated wheat chromosomes, (c) isolated poppy chromosomes, and (d) isolated poppy nuclei on SDS gel. Gel electrophoresis of histones extracted with sulfuric acid: (e) histones of wheat, (f) histones of poppy, (g) histones of Chinese hamster, which served as a control for sulfuric acid extraction (acid urea gel), (h) core histones of wheat extracted with sulfuric acid after H1 depletion, (i) presumptive H1 histones of wheat and (j) poppy extracted with perchloric acid, (k) core histones of poppy extracted with sulfuric acid after H1 depletion, and (m) molecular weight markers (SDS gel). (From Hadlaczky *et al.*, 1983.)

degree of synchrony, their purity leaves much to be desired. Using more effective synchronizing agents such as aphidicoline (Nagata *et al.*, 1982) and improvements in culture conditions may bring about significant increases in the effectiveness of chromosome isolation.

2. A crucial step in the isolation procedure is a rapid but effective enzyme treatment of cells to produce mitotic protoplasts. Inadequate digestion of the cell wall and the presence of cell wall residues can seriously contaminate isolated chromosomes. On the other hand, longer enzymatic treatment may cause an increased stickiness and disintegration of chromosomes. This step of the isolation procedure may also be improved by more vigorous enzyme combinations (for an example, see Hasezawa *et al.*, 1981).

3. The purification of chromosomes requires the removal of cytoplasmic contaminants and slow-sedimentating cellular debris, and the separation of chromosomes from nuclei and from undisrupted cells. Since the removal of cytoplasmic contaminants has the greatest influence on the purity of the final chromosome suspension, a few comments are in order. The cytoplasmic protein content of the ruptured protoplast suspension must be low enough to avoid protein aggregation. In our hands, the density of 2 ml pelleted protoplasts in 100 ml buffer meets this requirement. Separation of chromosomes from the cytoplasmic material at the earliest stage of the isolation procedure is also an important point. The viscosity of burst protoplast suspensions may vary from one cell line to another, and chromosomes of different species may show different sedimentations. Therefore, one should find the optimal parameters of centrifugation for each cell line.

4. The problem of determining the purity of the isolated chromosome suspensions must be mentioned briefly. Hearst and Botchan (1970) stated that the purity of the chromosomes after isolation cannot be determined with any degree of accuracy at present. This is still true. The purity of the isolated chromosomes can be judged by the ratio of visible contaminants and by the constancy of the gross chemical composition of chromosomes.

Finally, it is clear that isolated plant chromosomes will become an important source in chromosome research and genetic manipulation. For this reason, even the smallest progress is of great importance in a challenging field. The procedure described here may be only a prototype for others; however, it may serve as a starting point in future studies.

REFERENCES

Bostock, C. J., and Sumner, A. T. (1978). "The Eukaryotic Chromosome." Elsevier/North Holland Biomedical Press, Amsterdam.

Dudits, D., Hadlaczky, G., Lévi, Ě., Fejér, O., Haydu, Z., and Lázár, G. (1977). Somatic hybridization of *Daucus carota* and *D. capillifolius* by protoplast fusion. *Theor. Appl. Genet.* **51,** 127–132.

Eriksson, T. (1966). Partial synchronization of cell division in suspension cultures of *Haplopappus gracilis. Physiol. Plant.* **19,** 900–910.

Gamborg, O. L., and Eveleigh, D. E. (1968). Culture methods and detection of glycanases in suspension cultures of wheat and barley. *Can. J. Biochem.* **46,** 417–421.

Griesbach, R. J., Malmberg, R. L., and Carlson, P. S. (1982). An improved technique for the isolation of higher plant chromosomes. *Plant Sci. Lett.* **24,** 55–60.

Hadlaczky, G., Praznovszky, T., and Bisztray, G. (1982). Structure of isolated protein-depleted chromosomes of plants. *Chromosoma* **86,** 643–659.

Hadlaczky, G., Bisztray, G., Praznovszky, T., and Dudits, D. (1983). Mass isolation of plant chromosomes and nuclei. *Planta* **157,** 278–285.

Hasezawa, S., Nagata, T., and Syono, K. (1981). Transformation of Vinca protoplasts mediated by Agrobacterium spheroplasts. *Mol. Gen. Genet.* **182,** 206–210.

Hearst, J. E., and Botchan, M. (1970). The eukaryotic chromosome. *Annu. Rev. Biochem.* **39,** 151–182.

Kao, K. N., Constabel, F., Michayluk, M. R., and Gamborg, O. L. (1974). Plant protoplast fusion and growth of intergeneric hybrid cells. *Planta* **120,** 215–227.

Klobutcher, L. A., and Ruddle, F. H. (1981). Chromosome mediated gene transfer. *Annu. Rev. Biochem.* **50,** 533–554.

Malmberg, R. L., and Griesbach, R. J. (1980). The isolation of mitotic and meiotic chromosomes from plant protoplasts. *Plant Sci. Lett.* **17,** 141–147.

Nagata, T., Okada, I., and Takebe, I. (1982). Mitotic protoplasts and their infection with Tobacco Mosaic Virus RNA encapsulated in liposomes. *Plant Cell Rep.* **1,** 250–252.

Szabados, L., Hadlaczky, G., and Dudits, D. (1981). Uptake of isolated chromosomes by plant protoplasts. *Planta* **151,** 141–145.

CHAPTER **54**

Isolation of Organelles: Chloroplasts

J. Kobza

Department of Botany
Washington State University
Pullman, Washington

G. E. Edwards

Department of Botany and Institute of Biological Chemistry
Washington State University
Pullman, Washington

I. INTRODUCTION

In studies on somatic cell genetics, there is interest in transplantation of chloroplasts and their extrachromosomal genetic information into protoplasts. From a fundamental point of view, such studies are of importance in understanding developmental biology and how the development of the chloroplast is controlled through nuclear versus chloroplastic DNA. From a practical point of view, such procedures are of interest in considering the potential for improving photosynthetic efficiency within or between spe-

ISBN 0-12-715001-3

cies. In higher plants, there is a range of differences among chloroplasts including size, degree of grana stacking, ability to accumulate starch, and the capacity of noncyclic versus cyclic and pseudocyclic photophosphorylation. There is variation in photochemical components (e.g., chlorophyll content, reaction centers, and electron transport carriers) and in the photochemical capacity relative to the carbon assimilation capacity of the stroma (e.g., in shade-type versus sun-type plants). Finally, among C_3, C_4, and Crassulacean acid metabolism (CAM) plants there are qualitative differences in the proteins of carbon assimilation located in the stroma, and those of the chloroplast envelopes which are involved in metabolite transport.

A first step in developing the transfer of chloroplasts from one species into protoplasts of another is to isolate pure, intact, functional chloroplasts. The purpose of this chapter is to describe the available techniques for chloroplast isolation.

A. General Background

Whenever plant tissue is macerated in an attempt to isolate chloroplasts, the resulting brei is a complex mixture of cell fragments and organelles. In such a brei, chloroplasts themselves are a heterogeneous group, ranging from broken chloroplasts, devoid of their envelopes and stroma, to completely intact chloroplasts with an intact envelope and a full complement of stromal components. It is this latter group (designated type A by Hall, 1976) which is the desired product of any isolation procedure that attempts to procure intact chloroplasts. These chloroplasts closely resemble *in vivo* chloroplasts in appearance and in having high photosynthetic rates (50–250 μmoles CO_2 fixed per milligram of chlorophyll per hour).

Until the early 1970s, the only established means for isolation of chloroplasts was to disrupt the plant cell wall by mechanically grinding the leaf material. This was a limitation, since most species are very resistant to mechanical grinding. However, in the 1970s, the procedures for isolation of protoplasts from various species became well established. It was found that chloroplasts could be efficiently isolated from protoplasts by mild lysis of the plasmalemma (Gutierrez *et al.*, 1975). Since then, the list of species from which intact, functional chloroplasts can be isolated has grown dramatically, but the full potential has not been realized. One obvious measure of retention of function of the isolated chloroplast is its ability to carry on photosynthetic carbon assimilation. A number of isolation procedures have been developed which maximize this function.

B. Mechanical versus Enzymatic Isolation

The two procedures utilized for isolation of chloroplasts, mechanical grinding of the leaf material and the isolation and rupture of protoplasts, have their advantages and disadvantages. Mechanical grinding procedures work well with only a few C_3 species (spinach, pea shoots, and a few others: Robinson *et al.*, 1979; Walker, 1980; Leegood and Walker, 1979; Robinson, 1983). However, mechanical grinding is not as time-consuming as protoplast isolation, can be less expensive to perform, and eliminates subjection of the chloroplasts to inhibitory agents which may be present in some protoplast isolation enzyme mixtures.

Protoplast isolation allows a much wider range of species from which intact chloroplasts can be isolated, including C_4 and CAM plants (Table I). However, among the leaf materials of species examined, not all are susceptible to digestion by the commonly used commercial cellulase (which usually contains cellulase- and hemicellulase-degrading enzymes) and pec-

TABLE I

Examples of Species from Different Photosynthetic Groups Which May Be Used for Isolating Chloroplasts from Protoplasts

C_3 mesophyll protoplasts	
Monocots	Dicots
Wheat	Spinach
Barley	Peas
(Very consistent)	(Problems with some dicots in digestion and yields)
C_4 mesophyll protoplasts	
Monocots	Dicots
Maize	*Amaranthus graecizans*
Sorghum	*Portulaca grandiflora*
Crabgrass	(Limited success; many species resistant)
(Very consistent)	
C_4 bundle sheath protoplasts	
Monocots	Dicots
Panicum miliaceum	*Amaranthus graecizans*
Urochloa mosambicensis	*Portulaca grandiflora*
(Digestion and stability problems with many species, particularly with older tissue)	
CAM mesophyll protoplasts	
Sedum praealtum	
Mesembryanthemum crystallinum	
(Limited success with other species which have problems in yield and protoplast stability)	

tinase (which contains depolymerases of pectin and/or pectate). Also, isolation of protoplasts is a more difficult and time-consuming process, and the yields are often relatively low. Nonetheless, protoplast isolation is an excellent procedure by which intact chloroplasts can be isolated from many species.

For a limited number of C_4 species, mesophyll and bundle sheath protoplasts have been isolated and separated according to differences in size and density. Subsequently, intact chloroplasts were isolated from the protoplasts (Table I; Edwards *et al.*, 1979). It has generally not been possible to isolate mechanically intact mesophyll and bundle sheath chloroplasts of C_4 plants. However, chloroplasts have been isolated mechanically from *Panicum maximum* and separated in a Ludox gradient on the basis that the bundle sheath chloroplasts were high in starch, and therefore were of a higher density than mesophyll chloroplasts (Walbot, 1977).

Mesophyll protoplasts have been isolated from a number of CAM plants. Intact chloroplasts have been isolated from protoplasts of the CAM species *Sedum praealtum* and *Mesembryanthemum crystallinum* (Spalding and Edwards, 1980; Piazza *et al.*, 1982; Winter *et al.*, 1982). Also, intact chloroplasts have been isolated by mechanical procedures from the inducible CAM plant *M. crystallinum* and purified on a sucrose gradient, which is an exceptional case among CAM species (J. G. Foster and G. E. Edwards, unpublished). In most CAM plants, grinding procedures generally fail to produce intact and functional chloroplasts due to the high acidity (from malic acid in vacuoles) and phenols. Even with *M. crystallinum,* high rates of photosynthesis have been obtained only with chloroplasts isolated from protoplasts (Demmig and Winter, 1983; Monson *et al.*, 1983). Finally, protoplast isolation produces a uniform stock of protoplasts which can serve as a chloroplast supply for multiple experiments over an extended period (e.g., for 1 or 2 days). This is a distinct advantage due to the greater lability of a chloroplast suspension (Rathnam and Edwards, 1976).

II. PROCEDURES

A. Mechanical Isolation of Chloroplasts

1. Plant Material

In mechanical isolation, the choice of plant material is extremely important to ensure adequate success in isolating chloroplasts. The plant material should be soft, pliable, and easily macerated, which generally excludes the use of most monocots. However, chloroplasts capable of high rates of photosynthesis have been isolated from 6- to 8-day-old wheat seedlings (Leegood and Walker, 1979).

Tissue with high concentrations of inhibitory compounds, such as phenols or tannins, should be avoided, as should tissue which has a high activity of degradative enzymes such as lipolytic enzymes or proteinases. By the use of protective agents (Galliard, 1974; Loomis, 1974), it is possible to reduce the effects of these compounds or degradative enzymes on the chloroplasts once they have been released by homogenization of the tissue. However, these additions complicate the procedures and are not always effective.

Since chloroplasts which have large starch granules tend to rupture upon centrifugation, it is desirable to experiment on plants low in starch content. Tissue with a low starch content can be achieved by using plants grown under a photoperiod which minimizes starch accumulation. Surprisingly, recent studies indicate that several species store high amounts of starch in the chloroplasts when the plants are grown under a short photoperiod but not under a long photoperiod (Chatterton and Silvius, 1981). Spinach and, to a lesser extent, young pea shoots have become the plants of choice for most investigators when isolating chloroplasts mechanically. This is due to the limitations described above and to the extensive knowledge of the properties of chloroplasts from these species (see Robinson *et al.*, 1979; Walker, 1980, for growth conditions).

2. Isolation Procedure

There are a number of articles and reviews covering mechanical procedures for isolating intact chloroplasts (Nobel, 1974; Robinson *et al.*, 1979; Walker, 1980). Table II lists isolation media which have been used to isolate intact and photosynthetically competent chloroplasts from several species. The grinding medium should be a semifrozen slurry, and all isolation procedures should be carried out between 0 and 4°C.

The leaf material is preilluminated for 20–30 min before harvesting. This increases the yield and reduces the photosynthetic induction period (Walker, 1976). After the leaf material (60–80 g) is deribbed of major veins (i.e., in spinach), it is washed and ground in 200–250 ml of the grinding medium using either a precooled mortar and pestle or a blender (Waring or Polytron). The grinding time should not exceed 5 sec if a Waring-type blender is used or 3 sec if a Polytron type is used. Because the Polytron gives high yields, 4–6 mg chlorophyll as intact chloroplasts from 80 g of leaf material, it is the preferred instrument (Walker, 1980).

The homogenate from such a grind should be rapidly filtered through two layers of muslin or Miracloth and eight layers of cotton wool. This filtration step, which removes cellular debris such as unbroken cells, cell walls, and some broken chloroplasts, is faster and more beneficial than low-speed centrifugation. In fact, it is necessary that the chloroplasts be

TABLE II

Grinding Media Used in Mechanical Isolation Techniques for Obtaining Photosynthetically Functional Chloroplasts of C_3 Plants

Spinach[a]	Pea shoots[b]	Wheat seedlings[c]
0.33 *M* sorbitol	0.3 *M* glucose	0.33 m*M* sorbitol
10 m*M* $Na_4P_2O_7$	50 m*M* Na_2HPO_4	20 m*M* HEPES[d]
5 m*M* $MgCl_2$	50 m*M* KH_2PO_4	10 m*M* $NaHCO_3$
2 m*M* Na isoascorbate	5 m*M* $MgCl_2$	10 m*M* EDTA
pH 6.5	0.1% NaCl	5 m*M* $MgCl_2$
	0.2% Na isoascorbate	0.1% BSA
	0.1% bovine serum albumin (BSA)	0.2% Na isoascorbate
	pH 6.5	pH 7.6

[a] Walker (1971).
[b] Walker (1964).
[c] Leegood and Walker (1979).
[d] *N*-(2-hydroxyethyl)piperazine–*N*–2–ethanesulfonic acid.

removed from the homogenate as rapidly as possible; otherwise the chloroplasts may lose photosynthetic activity and intactness (Walker, 1964).

Following filtration, the homogenate is centrifuged to pellet the intact chloroplasts. This is accomplished by centrifugation in a swing-out head at 2000 *g* for 30–40 sec, or by acceleration to a higher *g* force (e.g., 5000–6000 *g*) and quick deceleration by hand braking such that the entire run does not exceed 90 sec. The centrifuge tubes should be smooth since scratched surfaces can cause the chloroplasts to rupture.

The supernatant is decanted, and the pellets are rinsed with small aliquots of grinding or resuspension medium to remove the upper layer of the pellet, which is largely broken chloroplasts. The pellet is resuspended by gentle shaking or use of a cotton swab (not by vortexing). Suspension in a minimal volume of medium will keep the chloroplasts concentrated and more stable. Further washing of the chloroplasts, to help remove other organelles, cellular debris, and inhibitory substances, may be performed by diluting the chloroplasts and repeating the centrifugation and resuspension steps (for further information on resuspension and assay media for measuring photosynthesis of isolated chloroplasts, see Robinson *et al.*, 1979; Walker, 1980).

B. Chloroplasts from Protoplasts

With the rapid development of methodology for protoplast isolation from leaf tissue over the last 15 years (Chapters 38–43, this volume), protoplasts have become a very useful tool to the plant scientist. For more descriptive information on development of isolation techniques, see Ruesink (1980), Edwards *et al.* (1978a), and Edwards and Walker (1983).

1. Plant Material

In general, young, actively growing leaf material or cell suspensions produce the highest yields of protoplasts (Huber and Edwards, 1975; Hughes *et al.*, 1978; Eriksson *et al.*, 1978). Isolating chloroplasts from protoplasts also affords the investigator some flexibility in the choice of plant material. For example, Leegood and Walker (1979) have isolated protoplasts, and subsequently chloroplasts with high photosynthetic rates, from flag leaves of *Triticum aestivum*. They were not able to isolate functional chloroplasts mechanically from the same plant material.

2. Protoplast Isolation

Depending on the quantity and type of leaf material utilized, tissue can be prepared for enzymatic maceration by slicing it into 0.5- to 1.0-mm segments with a sharp razor blade, by abrading the epidermis with a brush or carborundum, or by peeling off the epidermis. Monocots are generally sliced into segments perpendicular to the veins, whereas with dicots various methods are used depending on the species. Once prepared, the leaf material is added to the digestion medium, which consists of the macerating enzymes, an osmoticum, buffer, and $CaCl_2$ (Table III). The leaf tissue is incubated for 2–4 hr at 25–30°C, after which the digestion medium is decanted. The tissue is then swirled gently in an isolation medium to release the protoplasts. This suspension is filtered (e.g., through a 1-mm and a 400-μm net) to remove large debris. The filtered suspension and the digestion medium are centrifuged at 50–100 *g* for 5 min to sediment the protoplasts. The protoplasts can be further purified by flotation on a

TABLE III

Typical Constituents of Protoplast Isolation Media

Digestion medium	Protoplast storage medium
0.5 *M* sorbitol	0.5 *M* sorbitol
5.0 m*M* MES[a]	5.0 m*M* MES
1.0 m*M* $CaCl_2$	1.0 m*M* $CaCl_2$
0.05% BSA	pH 6
2.0% cellulase	
0.3% pectinase	
pH 5–6	

[a] 2-(*N*-morpholino)ethanesulfonic acid.

sucrose–sorbitol step gradient (see Edwards *et al.*, 1978a, 1979) or by the use of a liquid–liquid two-phase system composed of polyethylene glycol and dextran (Kanai and Edwards, 1973). When using a sucrose gradient overlayed with sorbitol, dextran can be added to the sucrose, if necessary, to increase the density of the medium and allow the protoplasts to float to the sucrose–sorbitol interphase (Edwards *et al.*, 1979). If protoplasts are kept at low temperatures, they are stable in sorbitol and $CaCl_2$ at a relatively low pH (Table III), whereas at room temperature they are more stable in a sucrose or all-salt medium (Edwards *et al.*, 1976).

3. Chloroplast Isolation from Protoplasts

Once protoplasts are isolated and purified, intact chloroplasts can be isolated by gentle rupture of the plasmalemma. The protoplast suspension (or an aliquot thereof) is centrifuged again at low speed for 2 min to pellet the protoplasts. The pellet is resuspended in the desired breakage medium (Table IV), and the protoplasts are ruptured by forcing them through a 20-μm nylon net fixed to the end of a plastic syringe, by passing the suspension through a syringe needle or by the use of a Yeda press under mild pressures (65–75 psi) (Gutierrez *et al.*, 1975). If breakage of the protoplasts is incomplete, the pressure used with the Yeda press is increased or a nylon net is selected which has a smaller pore size. The use of EDTA in the break medium tends to destabilize the plasmalemma, making breakage easier. It is necessary to use minimal force in breaking protoplasts to release chloroplasts which are about 70–100% intact, but at the same time to exert adequate force to break all the protoplasts, thus preventing chloroplasts from being trapped in resealed plasma membranes, as observed

TABLE IV

Example of Media Which May Be Used for Isolation of Chloroplasts from Protoplasts

Breakage and resuspension medium	Media for assay of photosynthetic activity
0.3 *M* sorbitol 25 m*M* HEPES-KOH 10 m*M* EDTA[a] pH 7.5	+ 10 m*M* $NaHCO_3$ + 0.1–0.3 m*M* KH_2PO_4[b]

[a] EDTA may aid in protoplast breakage and is essential for obtaining photosynthetic activity with some species (Edwards *et al.*, 1978b). However, other media may be more suitable for longer-term chloroplast stability (e.g., including a divalent cation such as Mg^{2+}) or for inducing the uptake of chloroplasts into protoplasts (see Chapter 57, this volume).

[b] With chloroplasts of different species, the Pi optimum is somewhat variable and thus needs to be determined experimentally.

with *S. praealtum* (Piazza *et al.*, 1982), or the aggregation of chloroplasts, as occassionally seen in some species (e.g., with celery; Rumpho *et al.*, 1983). Yields of chloroplasts can be routinely determined by measuring the chlorophyll by the method of Arnon (1949) or Wintermans and De Mots (1965). To determine that the chloroplasts are functional, CO_2 fixation can be assayed with C_3 chloroplasts in the presence of bicarbonate and inorganic phosphate (Pi) (Table IV). Other photosynthetic functions can be measured with C_4 mesophyll chloroplasts (see Edwards and Walker, 1983).

C. Further Purification of Chloroplasts Isolated Mechanically or from Protoplasts

As already described for chloroplasts isolated either mechanically or from protoplasts, low-speed centrifugation can be used to obtain intact chloroplasts. These preparations will have low contamination by mitochondria, peroxisomes (2–10% depending on the degree of washing), and nuclei. If the chloroplast preparation is to be used for somatic cell genetics, complete elimination of other organelles, particularly other DNA-containing organelles (i.e., nuclei and mitochondria), is desirable. Various gradients can be used to purify further the chloroplasts, although not all of these will result in chloroplasts which remain functional, in particular the isopynic-type sucrose gradients, which have a high osmolarity. Silica sols (e.g., Ludox) and, more recently, PVP-coated silica sols (Percoll), have been used successfully in density gradients to isolate pure, functional chloroplasts (see Price *et al.*, 1979; Takebe *et al.*, 1979; Winter *et al.*, 1982; Robinson, 1983). However, Ludox has the disadvantage of requiring further purification (Mogenthalar *et al.*, 1974), and in some cases Percoll needs to be dialyzed before use in order to obtain functional chloroplasts on the gradients (Stitt and Heldt, 1981).

D. Chloroplast Intactness

The Hill oxidant ferricyanide is excluded from type A chloroplasts. Thus, the relative rate of uncoupled ferricyanide reduction by the chloroplast suspension before and after osmotic shock provides a good indication of the percentage of intactness of chloroplast preparations (Lilley *et al.*, 1975). This procedure may slightly overestimate the degree of intactness of the chloroplasts since a few chloroplasts apparently rupture, lose their stromal contents, and then reseal. However, this method has become widely used due to its simplicity and reproducibility. Ferricyanide reduction may be followed polarographically with an oxygen electrode or spectrophotometrically by the change in absorbance at 410 nm.

An alternative method for determining chloroplast intactness when chloroplasts are isolated from protoplasts is to use a chloroplast marker enzyme, such as NADP-triose phosphate dehydrogenase. For example, following the isolation of chloroplasts from protoplasts, the enzyme can be assayed in the supernatant and in the chloroplast pellet. The activity of the enzyme in the chloroplast pellet, divided by the total activity of the enzyme in both fractions, all multiplied by 100, gives the percentage of intactness of the chloroplasts in the original chloroplast suspension.

III. CONCLUSIONS

Originally, intact chloroplasts were isolated to study their photosynthetic function and determine if they were capable of autonomous CO_2 fixation. The ability to isolate chloroplasts from protoplasts has allowed a number of studies with various species, including intracellular compartmentation of enzymes, metabolite transport, metabolic activity, and the isolation and study of the properties of chloroplast envelopes. Intact chloroplasts can now be obtained from a variety of plants for such studies or for transplantation of these organelles into protoplasts.

ACKNOWLEDGMENTS

Results from our laboratory presented in this chapter were supported by grants from the National Science Foundation and the USDA Competitive Grants Program.

REFERENCES

Arnon, D. I. (1949). Copper enzymes in isolated chloroplasts. Polyphenoloxidase in *Beta vulgaris*. *Plant Physiol.* **24,** 1–15.

Chatterton, N. J., and Silvius, J. E. (1981). Photosynthate partitioning into starch in soybean leaves. II. Irradiance level and daily photosynthetic period duration effects. *Plant Physiol.* **67,** 257–260.

Demmig, B., and Winter, K. (1983). Photosynthetic characteristics of chloroplasts isolated from *Mesembryanthemum crystallinum* L., a halophilic plant capable of Crassulacean acid metabolism. *Planta* **159,** 66–76.

Edwards, G. E., and Walker, D. A. (1983). "C_3, C_4: Mechanisms and Cellular and Environmental Regulation of Photosynthesis." Blackwell, Oxford.

Edwards, G. E., Huber, S. C., and Gutierrez, M. (1976). Photosynthetic properties of plant protoplasts. *In* "Microbial and Plant Protoplasts" (J. F. Peberdy, A. H. Rose, H. J. Rogers, and E. C. Cocking, eds.), pp. 299–332. Academic Press, New York.

Edwards, G. E., Robinson, S. P., Tyler, N. J. C., and Walker, D. A. (1978a). Photosynthesis by isolated protoplasts, protoplast extracts, and chloroplasts of wheat. *Plant Physiol.* **62,** 313–317.

Edwards, G. E., Robinson, S. P., Tyler, N. J. C., and Walker, D. A. (1978b). A requirement for chelation in obtaining functional chloroplasts of sunflower and wheat. *Arch. Biochem. Biophys.* **190,** 412–423.

Edwards, G. E., Lilley, R. McC., Craig, S., and Hatch, M. D. (1979). Isolation of intact and functional chloroplasts from mesophyll and bundle sheath protoplasts of the C_4 plant *Panicum miliaceum. Plant Physiol.* **63,** 821–827.

Eriksson, T., Glimelius, K., and Wallin, A. (1978). Protoplast isolation, cultivation and development. *In* "Frontiers of Plant Tissue Culture 1978" (T. A. Thorpe, ed.), pp. 131–139. Univ. of Calgary Press, Calgary, Alberta, Canada.

Galliard, T. (1974). Techniques for overcoming problems of lipolytic enzymes and lipooxygenases in the preparation of plant organelles. *In* "Methods in Enzymology" (S. Fleischer and L. Packer, eds.), Vol. 31, Part A, pp. 528–544. Academic Press, New York.

Gutierrez, M., Huber, S. C., Ku, S. B., and Edwards, G. E. (1975). Intracellular carbon metabolism in mesophyll cells of C_4 plants. *Proc. Int. Congr. Photosynth. Res., 3rd, 1974* pp. 1219–1230.

Hall, D. O. (1976). The coupling of photophosphorylation to electron transport in isolated chloroplasts. *In* "The Intact Chloroplast" (J. Barber, ed.), pp. 135–170. Elsevier/North-Holland Biomedical Press, Amsterdam.

Huber, S. C., and Edwards, G. E. (1975). An evaluation of some parameters required for the enzymatic isolation of cells and protoplasts with CO_2 fixation capacity from C_3 and C_4 grasses. *Physiol. Plant.* **35,** 203–209.

Hughes, B. G., White, F. G., and Smith, M. A. (1978). Effect of plant growth, isolation and purification conditions on barley protoplast yield. *Biochem. Physiol. Pflanz.* **172,** 67–77.

Kanai, R., and Edwards, G. E. (1973). Separation of mesophyll protoplasts and bundle sheath cells from maize leaves for photosynthetic studies. *Plant Physiol.* **51,** 1133–1137.

Leegood, R. C., and Walker, D. A. (1979). Isolation of protoplasts and chloroplasts from flag leaves of *Triticum aestivum* L. *Plant Physiol.* **63,** 1212–1214.

Lilley, R. McC., Fitzgerald, M. P., Rienits, K. G., and Walker, D. A. (1975). Criteria of intactness and the photosynthetic activity of spinach chloroplast preparations. *New Phytol.* **75,** 1–10.

Loomis, W. D. (1974). Overcoming problems of phenolics and quinones in the isolation of plant enzymes and organelles. *In* "Methods in Enzymology" (S. Fleischer and L. Packer, eds.), Vol. 31, Part A, pp. 528–544. Academic Press, New York.

Mogenthaler, J. J., Price, C. A., Robinson, J. M., and Gibbs, M. (1974). Photosynthetic activity of spinach chloroplasts after isopycnic centrifugation in gradients of silica sol. *Plant Physiol.* **54,** 532–534.

Monson, R. K., Rumpho, M. E., and Edwards, G. E. (1983). The influence of inorganic phosphate on photosynthesis in intact chloroplasts from *Mesembryanthemum crystallinum* L. plants exhibiting C_3 photosynthesis or Crassulacean acid metabolism. *Planta* **159,** 97–104.

Nobel, P. S. (1974). Rapid isolation techniques for chloroplasts. *In* "Methods in Enzymology" (S. Fleischer and L. Packer, eds.), Vol. 31, Part A, pp. 600–606. Academic Press, New York.

Piazza, G. J., Smith, M. G., and Gibbs, M. (1982). Characterization of the formation and distribution of photosynthetic products by *Sedum praealtum* chloroplasts. *Plant Physiol.* **70,** 1748–1758.

Price, C. A., Bartolf, M., Ortiz, W., and Reardon, E. M. (1979). Isolation of chloroplasts in Silica-sol gradients. *In* "Plant Organelles" (E. Reid, ed.), pp. 35–46. Wiley, New York.

Rathnam, C. K. M., and Edwards, G. E. (1976). Protoplasts as a tool for isolating functional chloroplasts from leaves. *Plant Cell Physiol.* **17,** 177–186.

Robinson, S. P. (1983). Isolation of intact chloroplasts with high CO_2 fixation capacity from sugarbeet leaves containing calcium oxalate. *Photosynth. Res.* **4,** 281–287.

Robinson, S. P., Edwards, G. E., and Walker, D. A. (1979). Established methods for the isolation of intact chloroplasts. *In* "Plant Organelles" (E. Reid, ed.), pp. 13–24. Wiley, New York.

Rumpho, M. E., Edwards, G. E., and Loescher, W. H. (1983). A pathway for photosynthetic carbon flow to mannitol in celery leaves: Activity and localization of key enzymes. *Plant Physiol.* **73,** 869–873.

Ruesink, A. (1980). Protoplasts of plant cells. *In* "Methods in Enzymology" (A. San Pietro, ed.), Vol. 69, pp. 69–84. Academic Press, New York.

Spalding, M. H., and Edwards, G. E. (1980). Photosynthesis in isolated chloroplasts of the Crassulacean acid metabolism plant *Sedum praealtum*. *Plant Physiol.* **65,** 1044–1048.

Stitt, M., and Heldt, H. W. (1981). Physiological rates of starch breakdown in isolated intact spinach chloroplasts. *Plant Physiol.* **68,** 755–761.

Takebe, T., Mishimura, M., and Akazawa, T. (1979). Isolation of intact chloroplasts from spinach leaf by centrifugation in gradients of modified silica "Percoll." *Agric. Biol. Chem.* **43**(10), 2137–2142.

Walbot, V. (1977). Use of silica-sol step gradients to prepare bundle sheath and mesophyll chloroplasts from *Panicum maximum*. *Plant Physiol.* **60,** 102–108.

Walker, D. A. (1964). Improved rates of carbon dioxide fixation by illuminated chloroplasts. *Biochem. J.* **92,** 22c.

Walker, D. A. (1971). Chloroplast (and Grana): Aqueous (including high carbon fixation ability). *In* "Methods in Enzymology" (A. San Pietro, ed.), Vol. 23, pp. 211–220. Academic Press, New York.

Walker, D. A. (1976). CO_2 fixation by intact chloroplasts: Photosynthetic induction and its relation to transport phenomena and control mechanisms. *In* "The Intact Chloroplast" (J. Barber, ed.), pp. 235–278. Elsevier/North-Holland Biomedical Press, Amsterdam.

Walker, D. A. (1980). Preparation of higher plant chloroplasts. *In* "Methods in Enzymology" (A. San Pietro, ed.), Vol. 69, pp. 94–104. Academic Press, New York.

Winter, K., Foster, J., Edwards, G. E., and Holtum, J. A. M. (1982). Intracellular localization of enzyme of carbon metabolism in *Mesembryanthemum crystallinum* exhibiting C_3 photosynthetic characteristics or Crassulacean acid metabolism. *Plant Physiol.* **69,** 300–307.

Wintermans, J. F. G. M., and De Mots, A. (1965). Spectrophotometric characteristics of chlorophyll and their pheophytins in ethanol. *Biochim. Biophys. Acta* **109,** 448–453.

CHAPTER **55**

Liposome Preparation and Incubation with Plant Protoplasts

Robert T. Fraley

Corporate Research Laboratories
Monsanto Company
St. Louis, Missouri

I. INTRODUCTION

Liposomes, or phospholipid vesicles, are closed vesicular structures which are formed spontaneously upon hydration of phospholipids with aqueous buffers. They are characterized by one or more phospholipid bilayers (or lamellae) which surround an aqueous interior; liposomes prepared by various methods can range from ~200Å to over 10 μm in diameter (Szoka and Papahadjopoulos, 1980; Papahadjopoulos *et al.*, 1981). The phospholipid composition of the bilayer can be altered to vary both the physical and the chemical properties of the liposome preparation (Szoka and Papahadjopoulos, 1980).

CELL CULTURE AND SOMATIC CELL
GENETICS OF PLANTS, VOL. 1

ISBN 0-12-715001-3

Historical Background

Since their discovery (Bangham *et al.*, 1965), liposomes have been studied extensively as model membranes (Papahadjopoulos and Kimelberg, 1974; Gennis and Jonas, 1977; Seelig, 1976) and as pharmacological carriers (for review, see Kimelberg and Mayhew, 1978). Detailed studies on their interaction with cultured animal cells (Tyrrell *et al.*, 1976; Pagano and Weinstein, 1978; Finkelstein and Weissmann, 1978) demonstrated that liposomes could become associated with the cell surface and actually fuse with the plasma membrane or be taken up by cells by endocytosis. Lipids or proteins embedded in the liposome membrane or small molecules entrapped in the aqueous interior of the vesicles could also be introduced into cells by these uptake mechanisms. Following the development of improved methods for producing large unilammelar vesicles (for review, see Szoka and Papahadjopoulos, 1980; Deamer and Uster, 1980), it was demonstrated that macromolecules, including nucleic acids, could be encapsulated in liposomes and that these could be introduced into animal cells in a functional form (Wilson *et al.*, 1979; Fraley *et al.*, 1980, 1981; for review, see Fraley and Paphadjopoulos, 1982). As the efficiency of the liposome delivery method was improved, it became apparent that this technique might be useful for genetic studies in a variety of cell systems, including plant cells.

A number of studies established that incubation of liposomes with plant protoplasts resulted in their association with plant cells (Cassells, 1978; Lurquin, 1979; Lurquin and Sheehy, 1982; Matthews *et al.*, 1979; for review, see Fraley and Papahadjopoulos, 1982; Fraley, 1984). More recently, it has been demonstrated by several laboratories that plant viral RNAs encapsulated in liposomes can be used to infect protoplasts at high efficiency (Fraley *et al.*, 1982; Fraley, 1983; Nagata *et al.*, 1981; Watanabe *et al.*, 1982; Rollo and Hull, 1982; Christen and Lurquin, 1983). It is likely that this method will also have application to DNA delivery experiments in studies on stable plant cell transformation or in short-term transient expression assays.

In this chapter, the most recent methods for liposome preparation, purification, and incubation with plant cells will be discussed in detail; special emphasis will be given to the encapsulation of nucleic acids in liposomes.

II. LIPOSOME PREPARATION

A. Lipids

Equimolar mixtures of bovine brain phosphatidyl serine (PS) and cholesterol (Chol) have been shown to date to be the most effective lipid composition for liposome-mediated delivery experiments with a variety of mammalian and plant cell types (Fraley and Papahadjopoulos, 1982). The higher intracellular delivery observed with PS-Chol vesicles has been found to result from their (1) low cell toxicity, (2) high binding to cells, (3) low level of cell-induced leakage, and (4) ability to respond to polyalcohol uptake treatments.

These lipids may be obtained from several commercial sources (Avanti Polar Lipids, Supelco, etc.); their purity should be checked by thin-layer chromatography in at least two solvent systems (Kates, 1972). PS should be converted into its sodium form by washing with EDTA and NaCl (Papahadjopoulos and Miller, 1967). Chol should be recrystallized from methanol to remove oxidized contaminants which exhibit high cell toxicity (Chen *et al.*, 1974). Following quantitation, these lipids can be stored in chloroform, under N_2 or Ar at $-70°C$ for several years without significant degradation. Sealed glass ampules, containing 5- or 20-μmol aliquots of PS and Chol, provide for convenient storage.

B. Sample Preparation and Buffers

Protein contaminants in RNA or DNA samples can interfere with liposome preparation and reduce encapsulation efficiency. The viscosity of nucleic acid samples at high concentrations (>2–3 mg/ml) can also reduce encapsulation efficiency. Divalent metals interact with the polar head group of PS and interfere with vesicle preparation, so EDTA should be present in the sample buffer. The presence of monovalent ions (~50 m*M*) reduces ionic interactions between phospholipids and DNA and facilitates liposome preparation. Since large unilamellar vesicles are osmotically sensitive, the sample buffer osmolarity should be adjusted to be equivalent to that of the medium or buffer used for protoplast incubations. The following sample buffer has been found to be optimal for most applications: 10 m*M* Tris, 50 m*M* KCl, 0.4 *M* mannitol, 0.1 m*M* EDTA, pH 7.0.

C. Preparation of Liposomes by the Reverse Phase Evaporation Technique

Although several methods are currently available for preparing liposomes (Szoka and Papahadjopoulos, 1980; Deamer and Uster, 1980), none combine the versatility, efficiency, and ease of preparation found in the reverse phase evaporation method (Szoka and Papahadjopoulos, 1978). In addition, this method was specifically developed for the encapsulation of large macromolecules such as RNA or DNA.

In this method, the phospholipid is first dissolved in an organic solvent, usually diethyl ether, although other solvents or mixtures can be used. A sample buffer solution containing RNA or DNA is added, and an emulsion is formed by mixing or by brief (2- to 3-sec) sonication. The ratios of solvent:buffer:lipid are critical for obtaining high levels (50%) of encapsulation. The brief period of sonication used to form the emulsion has been shown not to result in the degradation of DNA (Fraley *et al.*, 1980) or RNA molecules (Fraley *et al.*, 1982). Other solvent systems can be used which spontaneously form emulsions without the need for sonication (Darszon *et al.*, 1979; Rollo and Hull, 1982); however, these exhibit reduced values for maximal encapsulation.

Following formation of the emulsion, the excess solvent phase is slowly removed by rotary evaporation and a gel is formed. After continued evaporation and mechanical disruption, this gel phase will convert into a typical translucent liposome suspension.

Standard Procedure

1. PS and Chol (5 μmol each), dissolved in chloroform, are transferred to a sterile 10 × 100-mm glass tube (sample tube) with a Teflon-lined screw cap. The sample tube is inserted into a conventional evaporation tube, and the chloroform is evaporated (T = 25–30°C) in a standard rotary evaporator (~500 mm Hg for 20 min).
2. Diethyl ether (0.6 ml) is added to dissolve the dry lipid film. (The ether is extracted with H_2O before use to remove water-soluble peroxides.)
3. Sample buffer (0.175 ml) containing the nucleic acid to be encapsulated and a small amount (0.1 μCi) of ^3H-poly A (or another labeled nucleic acid) to permit quantitation is added to the sample tube.
4. An emulsion is formed by sonicating the sample tube containing the two-phase mixture in a bath-type sonicator (Laboratory Supplies Co. Inc., Hicksville, New York, #G112SP1) for 2–3 sec.
5. The sample tube is immediately transferred to the evaporation tube, a vacuum of 300–400 mm Hg is established, and the bulk of the ether is removed in 5–10 min by rotary evaporation. During this period, a viscous gel-like phase forms, which can be disrupted by frequent vortexing. The vacuum can be steadily increased to ~650 mm Hg, during which time a translucent liposome suspension will form. Any residual ether can be removed by extended evaporation, but care should be taken to minimize evaporation of the aqueous sample.

This completes the liposome preparation; the sample may be stored under N_2 or Ar at 4°C for several weeks. Quantitation of vesicle lipid concentration and encapsulation efficiency can be made following separation of vesicles from unencapsulated nucleic acids (described below).

D. Liposome Purification

Since many of the methods (i.e., centrifugation, molecular sieve chromatography, dialysis) which have been routinely used to separate liposomes from small molecules are not applicable or convenient for separations from nucleic acids, a new procedure has been developed for this purpose (Fraley *et al.*, 1980). This method, which utilizes discontinuous polymer gradients to separate liposomes from unencapsulated RNA, DNA, or protein molecules on the basis of their differing densities, also has the advantage of being rapid and adaptable for use with multiple samples. The method can easily be scaled up or down with sample size, is performed under sterile conditions, and allows the vesicles to be collected in a concentrated form. The polymers used (ficoll, metrizamide, etc.) have no effect on vesicle properties and are nontoxic to plant cells, so the resulting liposome preparation can be used directly in cell incubations.

Standard Procedure

1. The liposome preparation (0.175 ml) is transferred under sterile conditions from the sample tube used for vesicle preparation to a centrifuge tube (SW50.1 tubes are of convenient size; they may be sterilized by immersion in 70% ethanol). The sample tube is rinsed with 0.25 ml of sample buffer; this step is repeated twice to ensure good vesicle recovery.
2. One milliliter of a solution of 25% ficoll (prepared in the above sample buffer) is added to the liposome preparation and mixed gently by vortexing. Two milliliters of a solution of 10% ficoll (prepared in the above sample buffer) is gently layered over the sample, followed by addition of a buffer layer (0.5–1.0 ml). The ficoll solutions can be sterilized by filtration or autoclaving; care should be taken to check the pH (7.0) of the final polymer solution.
3. Centrifugation is carried out at 30,000 rpm for 30–40 min. The liposome band will appear at the buffer/10% ficoll interface and can be readily collected by gradient fractionation or with a pipette. Control experiments using "empty" liposomes and exogenously added DNA have established the reliability of this separation method (Papahadjopoulos *et al.*, 1981).
4. The phospholipid concentration (Bartlett, 1959) and radioactivity of the purified vesicle preparations are measured to permit calculation of encapsulation efficiency and to determine the volume of the sample to be

used in subsequent cell incubations. After correction for lipid loss (typically 10–20%), the encapsulation efficiency is usually 40–50% of the initial sample.

III. CONDITIONS FOR LIPOSOME–PROTOPLAST INCUBATION

Protoplasts used in liposome studies have been prepared from a variety of plant species including carrot (Matthews *et al.*, 1979), tobacco (Fraley *et al.*, 1982), petunia (Fraley, 1983), cowpea (Lurquin, 1979), and turnip (Rollo and Hull, 1982), using relatively standard enzymatic isolation methods (Chapters 38–45, this volume). Complete removal of the cell wall is essential for maximum uptake (Nagata *et al.*, 1981; Watanabe *et al.*, 1982), and contamination by cell debris should be minimized since this increases cell aggregation during incubation with liposomes.

Usually protoplasts are resuspended in isotonic buffer or medium (10^5–10^6/ml), and aliquots of a liposome preparation are added. The efficiency of liposome delivery to protoplasts has been shown to be dependent on several parameters (Fraley, 1983), including (1) vesicle composition, (2) encapsulation efficiency, (3) buffer composition and divalent metal on concentration, (4) incubation time, and (5) postincubation treatment with polyalcohols. (The optimal incubation conditions determined to date are described below.) At the end of the incubation period, the protoplasts are usually collected and washed by repeated centrifugation at low speeds. Polymers such as ficoll can be included in these washes to maximize the separation between liposomes and protoplasts (Fraley *et al.*, 1982). The cells are then analyzed for the uptake of labeled (radioactive, fluorescent, etc.) markers which were included in the liposome preparation, or resuspended in culture medium and incubated for longer periods for transient expression or transformation experiments. Usually only a slight reduction (10–20%) in cell viability is observed with lipid concentrations of 50–100 nmol/10^6 protoplasts (Fraley *et al.*, 1982).

Standard Conditions

1. Protoplasts are resuspended in incubation buffer (5 m*M* Tris, 0.5 *M* mannitol, 5 m*M* $CaCl_2$, pH 7.05) at a cell density of 10^6/ml. Then 0.5 ml of cells is transferred to a 15-ml conical centrifuge tube (Corning, 25310).

2. From 50 to 100 nmol of vesicle lipid in a volume of 10–50 μl is added directly to the protoplast sample. The tube is gently mixed, and incubation is carried out for 5 min at 30°C.

3. A 4.5-ml solution of polyethylene glycol (PEG; 15% wt/v, made up in the above incubation buffer) is added to the tube; the contents are mixed gently by inversion, and the incubation is continued for an additional 25 min. A variety of polyalcohols have been tested for their effectiveness in stimulating liposome uptake by protoplasts (Fraley, 1983); PEGs ranging in molecular weight from 6000 to 10,000 are most effective).

4. A 10-ml solution of incubation buffer is added to dilute the viscous PEG solution, and the tube is centrifuged 100 *g* for 10 min. The protoplasts are resuspended in culture medium or incubation buffer and washed twice by centrifugation (100 *g* for 5 min).

Protoplast recovery after the incubations and washes usually ranges between 50 and 75%. The protoplasts can be cultured using standard techniques; there is no effect of the liposome treatment and incubation conditions on the long-term viability of the cells.

IV. CONCLUDING REMARKS

Liposome-mediated delivery is a promising new technique for introducing macromolecules into plant protoplasts. The method appears to be widely applicable for use with cells prepared from a variety of plant species, and it can be expected to facilitate studies in plant virology and molecular biology. Further refinement of this procedure, along with the construction of suitable plant vectors and selectable markers, should allow for the development of efficient systems for plant cell transformation.

REFERENCES

Bangham, A., Standish, M., and Watkins, J. (1965). Diffusion of univalent ions across the lamellae of swollen phospholipids. *J. Mol. Biol.* **13,** 238–252.

Bartlett, G. (1959). Phosphorus assay in column chromatography. *J. Biol. Chem.* **234,** 466–468.

Cassells, A. (1978). Uptake of charged lipid vesicles by isolated tomato protoplasts. *Nature (London)* **275,** 780.

Chen, H., Kandutsch, A., and Waymouth, C. (1974). Inhibition of cell growth by oxygenated derivatives of cholesterol. *Nature (London)* **251,** 419–421.

Christen, A., and Lurquin, P. (1983). Infection of cowpea mesophyll protoplasts with cowpea chlorotic mottle virus (CCMV) RNA encapsulated in large liposomes. *Plant Cell Rep.* **2,** 43–46.

Darszon, A., Vandenberg, C., Ellisman, M., and Montal, M. (1979). Incorporation of membrane proteins into large single bilayer vesicles; application to rhodopsin. *J. Cell Biol.* **81,** 446–454.

Deamer, D., and Uster, P. (1980). Liposome preparation methods and monitoring liposome fusion. *In* "Introduction of Macromolecules into Visible Mammalian Cells" (R. Baserga, C. Croce, and G. Roueza, eds.), pp. 205–220. Liss, New York.

Finkelstein, M., and Weissmann, G. (1978). The introduction of enzymes into cells by means of liposomes. *J. Lipid Res.* **19,** 289–303.

Fraley, R. (1983). Liposome-mediated delivery of tobacco mosaic virus RNA into petunia protoplasts; improved conditions for liposome: Protoplast incubations. *Plant Mol. Biol.* **2,** 5–14.

Fraley, R. (1984). Liposomes as carriers for introducing nucleic acids into plant protoplasts. *In* "Plant Genetic Engineering" (D. E. Cress, ed.). Dekker, New York (in press).

Fraley, R. T., and Papahadjopoulos, D. (1982). Liposomes: The development of a new carrier system for introducing nucleic acids into plant and animal cells. *Curr. Top. Microbiol. Immunol.* **96,** 171–191.

Fraley, R., Subramani, S., Berg, P., and Papahadjopoulos, D. (1980). Introduction of liposome-encapsulated SV40 DNA into cells. *J. Biol. Chem.* **255,** 10,431–10,435.

Fraley, R., Straubinger, R., Rule, G., Springer, L., and Papahadjopoulos, D. (1981). Liposome-mediated delivery of DNA to cells: Enhanced efficiency of delivery related to lipid composition and incubation conditions. *Biochemistry* **20,** 6978–6987.

Fraley, R., Dellaporta, S., and Papahadjopoulos, D. (1982). Liposome mediated delivery of tobacco mosaic virus RNA into tobacco protoplasts: A sensitive assay for monitoring liposome-protoplast interactions. *Proc. Natl. Acad. Sci. U.S.A.* **79,** 1859–1863.

Gennis, R., and Jones, A. (1977). Protein-lipid interactions. *Annu. Rev. Biophys. Bioeng.* **6,** 195–238.

Kates, M. (1972). Techniques in lipidology. *Lab. Tech. Biochem. Mol. Biol.* **3.**

Kimelberg, H., and Mayhew, E. (1978). Properties and biological effects of liposomes and their uses in pharmacology and toxicology. *CRC Crit. Rev. Toxicol.* **6,** 25–79.

Lurquin, P. (1979). Entrapment of plasmid DNA by liposomes and their interactions with plant protoplasts. *Nucleic Acids Res.* **6,** 3773–3784.

Lurquin, P., and Sheehy, R. (1982). Binding of large liposomes to plant protoplasts and delivery of encapsulated DNA. *Plant Sci. Lett.* **25,** 133–146.

Matthews, B., Dray, S., Widhold, J., and Ostro, M. (1979). Liposome-mediated transfer of bacterial RNA into carrot protoplasts. *Planta* **145,** 37–44.

Nagata, T., Okada, K., Takebe, I., and Matsui, C. (1981). Delivery of tobacco mosaic virus RNA into plant protoplasts mediated by reverse-phase evaporation vesicles. *Mol. Gen. Genet.* **184,** 161–165.

Pagano, R., and Weinstein, J. (1978). Interaction of liposomes with mammalian cells. *Annu. Rev. Biophys. Bioeng.* **7,** 435–468.

Papahadjopoulos, D., and Kimelberg, H. (1974). Phospholid vesicles (liposomes) as models for biological membranes: Their properties and interactions with cholestrol and proteins. *In* "Progress in Surface Science" (S. G. Davidson, ed.), pp. 141–232. Permagon, Oxford.

Papahadjopoulos, D., and Miller, W. (1967). Phospholid model membranes. I. Structural characteristics of hydrated liquid crystals. *Biochim. Biophys. Acta* **135,** 624–638.

Papahadjopoulos, D., Fraley, R., Heath, T., and Straubinger, R. (1981). Liposomes: Recent advances in methodology for introducing macromolecules into eukaryotic cells. *In* "Techniques in Cellular Physiology" (P. F. Baker, ed.), pp. 1–18. Elsevier/North-Holland, New York.

Rollo, F., and Hull, C. (1982). Liposome-mediated infection of turnip protoplasts with turnip rosette virus and RNA. *Gen. Virol.* **60,** 359–363.

Seelig, J. (1976). ^{31}P nuclear magnetic resonance and the headgroup structure of phospholids in membranes. *Biochim. Biophys. Acta* **515,** 105–140.

Szoka, F., and Papahadjopoulos, D. (1978). Procedure for preparing liposomes with large intestinal aqueous space and high capture by reverse-phase evaporation. *Proc. Natl. Acad. Sci. U.S.A.* **75,** 145–149.

Szoka, F., and Papahadjopoulos, D. (1980). Comparative properties and methods of preparation of lipid vesicles (liposomes). *Annu. Rev. Biophys. Bioeng.* **9,** 467–508.

Tyrrell, D., Heath, T., Colley, C., and Ryman, B. (1976). New aspects of liposomes. *Biochim. Biophys. Acta* **457,** 259–302.

Watanabe, Y., Ohno, T., and Okada, Y. (1982). Virus multiplication in tobacco protoplasts inoculated with tobacco mosaic virus RNA encapsulated in large unilamellar vesicle liposomes. *Virology* **120,** 478–480.

Wilson, T., Papahadjopoulos, D., and Taber, R. (1979). The introduction of poliovirus RNA into cells via lipid vesicles (liposomes). *Cell* **17,** 77–84.

CHAPTER **56**

Inoculation of Protoplasts with Plant Viruses

Itaru Takebe

Department of Biology
Nagoya University
Chikusa-ku, Nagoya, Japan

I. INTRODUCTION

Infection by plant viruses is initiated by virus particles which enter host cells through wounds in the cell walls. This is a very inefficient process and gives rise to infection in only an extremely small fraction (less than 1 in 10,000) of host cells, even when the tissues are abraded to produce wounds. The low efficiency of infection has been one of the major causes of technical difficulty in studying plant virus multiplication.

An idea that protoplasts, cells in a wall-less form, may be inoculated without wounding and possibly with higher efficiency led us to develop a method by which metabolically active protoplasts are isolated in large amounts from the mesophyll tissues of tobacco (Takebe *et al.*, 1968). We subsequently showed that a brief incubation of these protoplasts with tobacco mosaic virus (TMV) under defined conditions results in the infection

CELL CULTURE AND SOMATIC CELL
GENETICS OF PLANTS, VOL. 1

ISBN 0-12-715001-3

of up to 30% of protoplasts (Takebe and Otsuki, 1969). A critical factor for the infection of protoplasts was the addition of poly-L-ornithine, a macromolecular polycation. The procedure for inoculating tobacco mesophyll protoplasts with TMV was subsequently refined to such an extent that nearly all of the protoplasts are infected (Otsuki *et al.*, 1972).

The successful inoculation of tobacco mesophyll protoplasts with TMV provided a system of single leaf cells in which infection occurs synchronously in a majority of cells, a system far more suitable than the conventional tissue materials for the study of virus multiplication in host cells. Similar systems have since been established for many other viruses using protoplasts from a few dicot and monocot species.

In this chapter, I shall describe our procedure for inoculating tobacco mesophyll protoplasts with TMV in its latest version. Procedures for other viruses or protoplasts will be discussed only briefly, and inoculation of protoplasts with isolated viral nucleic acids will not be dealt with (for a review of this subject, see Mühlbach, 1982, and Takebe, 1983). Methods for using protoplasts for plant virus studies have been reviewed (Sarkar, 1977). There are several review articles that provide a more general discussion of the protoplast systems in plant virology (Takebe, 1975, 1977, 1983; Mühlbach, 1982; Harrison and Mayo, 1983).

II. PROCEDURE FOR INOCULATING TOBACCO MESOPHYLL PROTOPLASTS WITH TMV

A. Protoplasts and Virus

Nicotiana tabacum cv. *Xanthi* and *Xanthi* nc are commonly used as the source of mesophyll protoplasts, although other varieties of tobacco may be used as well. The method for isolating protoplasts from tobacco leaves is described in Chapter 38, this volume. In many cases, it is not necessary to use aseptic protoplasts because inoculated protoplasts are usually cultured for only 2 days; protoplasts isolated from unsterilized leaves may be inoculated under nonsterile conditions. However, it is recommended that inoculated protoplasts be cultured in sterilized medium and flasks to minimize the growth of contaminating microorganisms, because the antibiotics added to the medium (Table I) do not completely prevent microbial growth.

The two-step isolation procedure described in Chapter 38 permits separation of palisade from spongy parenchyma cells, and protoplasts from palisade cells are used for inoculation with virus. In contrast, the one-step

TABLE I

Medium for Culturing Inoculated Tobacco Mesophyll Protoplasts

KH_2PO_4	0.2 m*M*	2,4-Dichlorophenoxyacetic acid	1 μg/ml
KNO_3	1 m*M*	D-Mannitol	0.6 *M*
$MgSO_4$	1 m*M*	Cephaloridine[a]	300 μg/ml
$CaCl_2$	10 m*M*	Mycostatin[a]	10 μg/ml
KI	1 μ*M*	pH	5.4
$CuSO_4$	0.01 μ*M*		

[a] Added after autoclaving of the medium.

isolation procedure (Kassanis and White, 1974) yields a mixture of spongy and palisade protoplasts which also contains significant numbers of subprotoplasts. The heterogeneous preparation gives lower levels of infection than palisade protoplasts obtained by the two-step procedure. In general, protoplasts containing as few damaged cells or as little debris as possible should be used, because such "impurities" absorb virus and reduce the effective concentration of inoculum.

The quality of inoculum virus is also important; virus preparations with sufficiently high purity and infectivity should be used to ensure high levels of infection. A common strain of TMV as propagated and purified according to the method of Otsuki *et al.* (1977) produces over 100 necrotic lesions on half leaves of *N. tabacum* cv. Xanthi nc at a concentration of 0.02 μg/ml and is satisfactory for infection of protoplasts. A concentrated solution of purified TMV can be stored frozen without loss of infectivity, but repeated thawing and freezing result in a gradual decrease in infectivity.

B. Solutions to Be Prepared

1. A 0.6 *M* solution of D-mannitol
2. Solution of purified TMV freshly diluted with 0.6 *M* mannitol to a concentration of 10 μg/ml
3. Solution of poly-L-ornithine (molecular weight over 100,000) freshly diluted with 0.6 *M* mannitol to a concentration of 10 μg/ml
4. A 0.2 *M* citrate buffer, pH 5.2
5. Protoplast culture medium (Table I)
6. Phosphate-buffered saline (PBS) (0.1 *M* phosphate buffer, pH 7.0, containing 0.85% NaCl)
7. Solution of TMV antibody conjugated with fluorescein isothiocyanate (FITC) and appropriately diluted with PBS (for preparation of fluorescent antibody, see Otsuki and Takebe, 1969)

8. Mayer's albumin (1:1 mixture of egg white and glycerin)
9. 10% glycerin solution in PBS

C. Inoculation Procedure

1. Prepare the inoculum solution in a 50-ml centrifuge tube by adding to 5 ml mannitol solution 1 ml citrate buffer, 2 ml poly-L-ornithine solution, and 2 ml TMV solution.
2. Incubate the inoculum solution for 10 min in a water bath at 25°C.
3. Meanwhile, centrifuge a 10-ml suspension of freshly isolated protoplasts (4×10^6 cells) at about 100 *g* for 2 min using a small centrifuge tube (17 × 105 mm), and remove the supernatant solution.
4. Quickly resuspend the pelleted protoplasts with 10 ml mannitol solution and immediately pour them into the preincubated inoculum solution.
5. Incubate the mixture for 10 min at 25°C with occasional swirling.
6. Spin down and carefully resuspended the protoplasts with 10 ml mannitol solution, and transfer them into a small centrifuge tube (17 × 105 mm).
7. Wash the protoplasts twice by centrifugation and resuspension in 10 ml mannitol solution.
8. Resuspend the washed protoplasts in 10 ml protoplast culture medium and transfer them to a 100-ml conical flask with a cotton plug. Incubate the flask at 28°C under illumination with white fluorescent lamps (ca. 3000 lx).

D. Determination of the Level of Infection

1. After 24 hr of culture, harvest a 0.5-ml sample of protoplasts into a small centrifuge tube and dilute with 5 ml mannitol solution. Collect the protoplasts by centrifugation and resuspend them with a drop of mannitol solution.
2. Using a Pasteur pipette, place a minute drop of the thick suspension of protoplasts on a glass slide smeared with Mayer's albumin.
3. Allow the protoplasts to settle down onto the surface of the slide (about 5 min), and then dry the drop completely with warm air from a hair drier.
4. Fix the protoplasts by dipping the slide into 95% ethanol for 10 min (chlorophyll is extracted by this treatment).
5. Wash the specimen by dipping the slide for 10 min in PBS.

6. Using a piece of filter paper, blot both surfaces of the slide except the area where the protoplasts are attached.

7. Cover the protoplasts with a drop of diluted fluorescent antibody solution, and incubate the slide for 1 hr at 37°C in a moist chamber.

8. Wash the specimen by dipping the slide for 10 min in PBS.

9. Blot the slide as in step 6 and cover the protoplasts with a drop of glycerin solution. Place a coverslip over the protoplasts and remove the excess glycerin solution with a piece of filter paper.

10. Examine the specimen under a fluorescence microscope using the filter combination for FITC immunofluorescence (e.g., exciter filters BG-12 and 20IF-490, dichroic mirror DM 500, and barrier filters O-515 and O-530 for an epi-illumination-type microscope).

11. Score over 300 protoplasts and determine the percentage of protoplasts containing materials with yellowish-green specific fluorescence (Fig. 1).

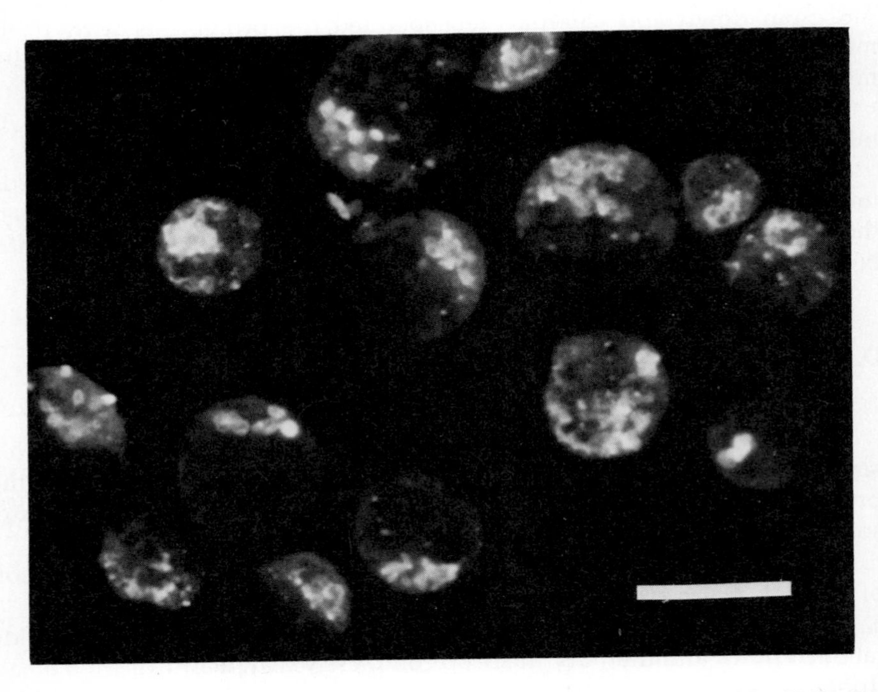

Fig. 1. Fluorescence micrograph of tobacco mesophyll protoplasts infected by TMV. Protoplasts were cultured for 24 hr after inoculation and were then stained with fluorescent antibody to virus. Virus masses in protoplasts fluoresce in yellowish-green. Bar = 50 μm.

E. Comments

The addition of poly-L-ornithine to inoculum solution is essential for infection of tobacco mesophyll protoplasts by TMV, and the inoculum virus has to be preincubated with the polycation to attain maximal infection. Poly-L-ornithine with a molecular weight below 100,000 is not only less effective but also detrimental to protoplasts. Some variation with respect to effectiveness and toxicity has been noticed among different lots of poly-L-ornithine with similar specification. Poly-D-lysine or poly-L-lysine with a molecular weight over 100,000 may be used in place of poly-L-ornithine.

Citrate buffer can be substituted for by phosphate buffer (0.5 *M*, pH 5.8). With phosphate buffer, maximal levels of infection are obtained with an inoculum virus concentration one-tenth that necessary for inoculation with citrate buffer, provided that the final population density of the protoplasts does not exceed 2×10^5/ml. At a higher population density the level of infection decreases markedly, whereas infection in citrate buffer is relatively insensitive to variations in the population density of protoplasts (Takebe, 1977).

For unknown reasons, protoplasts remaining for some time in suspension are less susceptible to infection than freshly resuspended protoplasts. It is therefore recommended that the protoplasts be spun down while the inoculum virus is preincubated with poly-L-ornithine, and they be added to the inoculum solution as soon as they are resuspended.

Protoplasts are particularly vulnerable after inoculation. Resuspension of inoculated protoplasts should therefore be performed as gently as possible. A small drop of mannitol solution is added first to the pelleted protoplasts, and the centrifuge tube is rotated slowly. When the protoplasts are dispersed into a thick suspension, they are diluted with an appropriate volume of mannitol solution. Damaged protoplasts tend to sediment more slowly than intact ones. By finely adjusting the speed and duration of centrifugation, it is possible to pellet intact protoplasts while leaving damaged ones in the supernatant solution. Damaged protoplasts may be partly removed in this way during the washing process.

The medium of Table I contains neither a carbon source nor vitamins. TMV multiplies actively in this medium when the inoculated protoplasts are cultured under illumination. Richer media such as those which support cell wall formation and cell division (Nagata and Takebe, 1971) do not improve TMV multiplication and enhance the growth of contaminating microorganisms.

F. Other Combinations of Virus and Protoplasts

Inoculation of protoplasts has been extended to a number of viruses, including a DNA virus (Table II). Since the procedures for inoculating these viruses are essentially a modification or variation of that for TMV and tobacco mesophyll protoplasts, details of the procedures for particular viruses are left to the respective literatures, and the kinds of modification and variation will be briefly discussed here.

Most plant viruses have lower infectivity than TMV, so that higher virus concentrations are usually needed to infect protoplasts. In such cases, it is often necessary to use poly-L-ornithine also at higher concentrations (Otsuki *et al.*, 1974).

Although poly-L-ornithine or another polycation is essential for the infection of protoplasts by most viruses, there are exceptions. Thus, brome mosaic virus can be inoculated without poly-L-ornithine to mesophyll protoplasts of tobacco (Motoyoshi *et al.*, 1974) and barley (Furusawa and Okuno, 1978). This virus has an exceptionally high isoelectric point and should hence be positively charged at the inoculation pH. On the other hand, protoplasts from cowpea primary leaves are infected in the absence of poly-L-ornithine by cowpea mosaic virus (Hibi *et al.*, 1975) and cucumber mosaic virus (Koike *et al.*, 1977). Both of these viruses require polycation for infection of tobacco mesophyll protoplasts (Huber *et al.*, 1977; Otsuki and Takebe, 1973). A cell electrophoresis study showed that cowpea protoplasts have less negative surface charge than tobacco protoplasts (Nagata and Melchers, 1978). These observations suggest that the major role of polycation is to assist electrostatic adsorption of negatively charged virus to the surface of protoplasts.

In many cases, the infection of protoplasts is favored by a lower pH. However, the pH optimum for infection may vary according to the virus, and may also be influenced by the kind of buffer used (Mayo and Roberts, 1979). For example, Tris-HCl at pH 8.0 was found to be superior to phosphate at pH 6.7 for the inoculation of tomato mesophyll protoplasts with TMV (Motoyoshi and Oshima, 1976).

Infection of barley mesophyll protoplasts by brome mosaic virus is enhanced by raising the mannitol concentration from 0.5 *M* to 0.85 *M* at the time of inoculation (Okuno and Furusawa, 1978). The so-called direct methods of inoculation are sometimes used (Motoyoshi *et al.*, 1973; Kubo *et al.*, 1975). In these methods, protoplasts are directly suspended in virus/poly-L-ornithine solution rather than first being suspended in mannitol solution and then mixed with the inoculum solution.

TABLE II

Some Plant Viruses Inoculated to Protoplasts[a]

Virus	Source of protoplasts[b]	Reference[c]
A. RNA viruses		
Tobacco mosaic	Tobacco	This article
	Tobacco cell culture	Kikkawa *et al.* (1982)
	Tomato	Motoyoshi and Oshima (1976)
Tobacco rattle	Tobacco	Kubo *et al.* (1975)
Potato X	Tobacco	Otsuki *et al.* (1974)
Turnip yellow mosaic	Chinese cabbage	Renaudin *et al.* (1975)
Cowpea mosaic	Cowpea	Hibi *et al.* (1975)
	Soybean cell culture	Jarvis and Murakishi (1980)
Cucumber mosaic	Tobacco	Otsuki and Takebe (1973)
	Cowpea	Koike *et al.* (1977)
	Cucumber	Maule *et al.* (1980)
Brome mosaic	Barley	Furusawa and Okuno (1978)
	Radish	Furusawa and Okuno (1978)
	Tobacco	Motoyoshi *et al.* (1974)
Alfalfa mosaic	Cowpea	Nassuth *et al.* (1981)
Raspberry ringspot	Tobacco	Barker and Harrison (1977)
Southern bean mosaic	Soybean cell culture	Jarvis and Murakishi (1980)
Potato leaf roll	Tobacco	Takanami and Kubo (1979)
	Potato	Barker and Harrison (1982)
B. DNA virus		
Cauliflower mosaic	Turnip	Furusawa *et al.* (1980)

[a] For a more comprehensive list, see Mühlbach (1982) and Takebe (1983).
[b] Mesophyll tissues unless otherwise mentioned.
[c] Literature containing the latest inoculation procedure.

Recently, protoplasts from suspension-cultured plant cells were also found to be suitable for inoculation with viruses. Infection of protoplasts from soybean suspension culture by legume viruses was stimulated by Ca^{2+} added to the inoculum solution (Jarvis and Murakishi, 1980). However, a similar effect of Ca^{2+} was not seen with protoplasts from tobacco suspension culture and TMV, and the conditions that were optimal for the infection of these protoplasts were very similar to those for tobacco mesophyll protoplasts (Kikkawa *et al.*, 1982).

Finally, it should be mentioned that a procedure using polyethylene glycol, an agent inducing protoplast fusion, was introduced for inoculation of tomato mesophyll protoplasts with TMV (Cassels and Barlass, 1978) and was later applied to tobacco mesophyll protoplasts (Cassels and Cocker, 1980).

III. RESULTS AND CONCLUSIONS

The procedure described above consistently gives rise to TMV infection in over 90% of tobacco mesophyll protoplasts if necessary precautions are taken. Virus multiplies in the protoplasts at a rate and to an extent comparable to those in the cells of intact leaves (Takebe, 1975). The procedure also formed a basis for the inoculation of many other plant viruses to protoplasts of a few species (Table II).

The successful inoculation of protoplasts with plant viruses provided an *in vitro* system of single host cells in which viruses multiply synchronously in a majority of cells under well-defined and easily controllable conditions. Because of their unique features, protoplasts are now widely used to study various aspects of virus infection and multiplication, ranging from viral RNA replication and translation to expression of resistance genes. Indeed, advanced studies of many problems in plant virology will not be feasible without protoplast systems. A discussion of the particular uses of protoplasts in virus research is beyond the scope of this chapter; readers are referred to relevant review articles (Takebe, 1977, 1983; Mühlbach, 1982; Harrison and Mayo, 1983).

REFERENCES

Barker, H., and Harrison, B. D. (1977). Infection of tobacco mesophyll protoplasts with raspberry ringspot virus alone and together with tobacco rattle virus. *J. Gen. Virol.* **35,** 125–133.

Barker, H., and Harrison, B. D. (1982). Infection of potato mesophyll protoplasts with five plant viruses. *Plant Cell Rep.* **1,** 247–249.

Cassels, A. C., and Barlass, M. (1978). The initiation of TMV infection in isolated protoplasts by polyethylene glycol. *Virology* **87,** 459–462.

Cassels, A. C., and Cocker, F. M. (1980). TMV inoculation of tobacco protoplasts in the presence of protoplast fusion agents. *Z. Naturforsch., C: Biosci.* **35C,** 1057–1061.

Furusawa, I., and Okuno, T. (1978). Infection with BMV of mesophyll protoplasts isolated from five plant species. *J. Gen. Virol.* **40,** 489–491.

Furusawa, I., Yamaoka, N., Okuno, T., Yamamoto, M., Kohno, M., and Kunoh, H. (1980). Infection of turnip protoplasts with cauliflower mosaic virus. *J. Gen. Virol.* **48,** 431–435.

Harrison, B. D., and Mayo, M. A. (1983). The use of protoplasts in plant virus research. *In* "Use of Tissue Culture and Protoplasts in Plant Pathology" (J. P. Helgeson and B. J. Deverall, eds.), pp. 69–137. Academic Press, New York.

Hibi, T., Rezelman, G., and van Kammen, A. (1975). Infection of cowpea mesophyll protoplasts with cowpea mosaic virus. *Virology* **64,** 308–318.

Huber, R., Rezelman, G., Hibi, T., and van Kammen, A. (1977). Cowpea mosaic virus infection of protoplasts from Samsun tobacco leaves. *J. Gen. Virol.* **34,** 315–323.

Jarvis, N. P., and Murakishi, H. H. (1980). Infection of protoplasts from soybean cell culture with southern bean mosaic and cowpea mosaic viruses. *J. Gen. Virol.* **48,** 365–376.

Kassanis, B., and White, R. F. (1974). A simplified method of obtaining tobacco protoplasts for infection with tobacco mosaic virus. *J. Gen. Virol.* **24,** 447–452.

Kikkawa, H., Nagata, T., Matsui, C., and Takebe, I. (1982). Infection of protoplasts from tobacco suspension cultures by tobacco mosaic virus. *J. Gen. Virol.* **63,** 451–456.

Koike, M., Hibi, T., and Yora, K. (1977). Infection of cowpea mesophyll protoplasts with cucumber mosaic virus. *Virology* **83,** 413–416.

Kubo, S., Harrison, B. D., Robinson, D. J., and Mayo, M. A. (1975). Tobacco rattle virus in tobacco mesophyll protoplasts: Infection and virus multiplication. *J. Gen. Virol.* **27,** 293–304.

Maule, A. J., Boulton, M. I., and Wood, K. R. (1980). An improved method for the infection of cucumber leaf protoplasts with cucumber mosaic virus. *Phytopathol. Z.* **97,** 118–126.

Mayo, M. A., and Roberts, I. M. (1979). Some effects of buffers on the infectivity and appearance of virus inocula used for tobacco protoplasts. *J. Gen. Virol.* **44,** 691–698.

Motoyoshi, F., and Oshima, N. (1976). The ue of Tris-HCl buffer for inoculation of tomato protoplasts with tobacco mosaic virus. *J. Gen. Virol.* **32,** 311–314.

Motoyoshi, F., Bancroft, J. B., Watts, J. W., and Burgess, J. (1973). The infection of tobacco protoplasts with cowpea chlorotic mottle virus and its RNA. *J. Gen. Virol.* **20,** 177–193.

Motoyoshi, F., Bancroft, J. B., and Watts, J. W. (1974). The infection of tobacco protoplasts with a variant of brome mosaic virus. *J. Gen. Virol.* **25,** 31–36.

Mühlbach, H. P. (1982). Plant cell cultures and protoplasts in plant virus research. *Curr. Top. Microbiol. Immunol.* **99,** 81–129.

Nagata, T., and Melchers, G. (1978). Surface charge of protoplasts and their significance in cell-cell interaction. *Planta* **142,** 235–238.

Nagata, T., and Takebe, I. (1971). Plating of isolated tobacco mesophyll protoplasts on agar medium. *Planta* **99,** 12–20.

Nassuth, A., Alblas, F., and Bol, J. F. (1981). Localization of genetic information involved in the replication of alfalfa mosaic virus. *J. Gen. Virol.* **53,** 207–214.

Okuno, T., and Furusawa, I. (1978). The use of osmotic shock for the inoculation of barley protoplasts with brome mosaic virus. *J. Gen. Virol.* **39,** 187–190.

Otsuki, Y., and Takebe, I. (1969). Fluorescent antibody staining of tobacco mosaic virus antigen in tobacco mesophyll protoplasts. *Virology* **38,** 497–499.

Otsuki, Y., and Takebe, I. (1973). Infection of tobacco mesophyll protoplasts by cucumber mosaic virus. *Virology* **52,** 433–438.

Otsuki, Y., Takebe, I., Honda, Y., and Matsui, C. (1972). Ultrastructure of infection of tobacco mesophyll protoplasts by tobacco mosaic virus. *Virology* **49,** 188–194.

Otsuki, Y., Takebe, I., Honda, Y., Kajita, S., and Matsui, C. (1974). Infection of tobacco mesophyll protoplasts by potato virus X. *J. Gen. Virol.* **22,** 375–385.

Otsuki, Y., Takebe, I., Ohno, T., Fukuda, M., and Okada, Y. (1977). Reconstitution of tobacco mosaic virus rods occurs bidirectionally from an internal initiation region: Demonstration by electron microscopic serology. *Proc. Natl. Acad. Sci. U.S.A.* **74,** 1913–1917.

Renaudin, J., Bove, J. M., Otsuki, Y., and Takebe, I. (1975). Infection of Brassica leaf protoplasts by turnip yellow mosaic virus. *Mol. Gen. Genet.* **141,** 59–68.

Sarkar, S. (1977). Use of protoplasts for plant virus studies. *In* "Methods in Virology" (K. Maramorosch and H. Koprowski, eds.), Vol. 6, pp. 435–456. Academic Press, New York.

Takanami, Y., and Kubo, S. (1979). Enzyme-assisted purification of two phloem-limited plant viruses: Tobacco necrotic dwarf and potato leaf roll. *J. Gen. Virol.* **44,** 153–159.

Takebe, I. (1975). The use of protoplasts in plant virology. *Annu. Rev. Phytopathol.* **13,** 105–125.

Takebe, I. (1977). Protoplasts in the study of plant virus replication. *In* "Comprehensive Virology" (H. Fraenkel-Conrat and R. R. Wagner, eds.), Vol. 11, pp. 237–283. Plenum, New York.

Takebe, I. (1983). Protoplasts in plant virus research. *Int. Rev. Cytol., Suppl.* **16,** 89–111.

Takebe, I., and Otsuki, Y. (1969). Infection of tobacco mesophyll protoplasts by tobacco mosaic virus. *Proc. Natl. Acad. Sci. U.S.A.* **64,** 843–848.

Takebe, I., Otsuki, Y., and Aoki, S. (1968). Isolation of tobacco mesophyll cells in intact and active state. *Plant Cell Physiol.* **9,** 115–124.

CHAPTER 57

Uptake of Organelles

Anita Wallin

Department of Plant Physiology
Swedish University of Agricultural Sciences
Uppsala, Sweden

I. INTRODUCTION

Over the last few years, considerable attention has been paid to the possibility of creating genetic variation in plant cells by organelle transplantation. As the degree and nature of the dependence of chloroplasts and mitochondria on the nucleus are not known fully, organelle transplantation allows the investigation of the ability of a foreign nucleus to code for gene products of transplanted organelles.

The first report of a successful transfer of chloroplasts (Carlson, 1973) was therefore received with great interest. Carlson transferred isolated tobacco chloroplasts into white protoplasts from a variegated albino mutant of *Nicotiana tabacum.* Plants were regenerated with normal functional plastids. However, the absence of evidence for uptake and survival of

CELL CULTURE AND SOMATIC CELL
GENETICS OF PLANTS, VOL. 1

ISBN 0-12-715001-3

chloroplasts and the use of genetic markers as the only proof of uptake have made this work controversial. Later work (Uchimiya and Wildman, 1979) performed with a similar material did not support the view that genetic information from foreign organelles was translated into polypeptides in leaves derived from the receptor protoplasts either in Carlson's work or in similar experiments by Kung *et al.* (1975).

Another approach is to prove incorporation in connection with organelle uptake based on microscopic or physiological evidence. Isolated nuclei of *Petunia hybrida* stained with fluorescent ethidium bromide were transferred to protoplasts of white petunia petals (Potrykus and Hoffmann, 1973). Potrykus (1973) also incorporated green petunia chloroplasts into albino protoplasts isolated from a *P. hybrida* plant with extrachromosomally controlled variegation using lysozyme or incubation of chloroplasts in the enzyme solution during protoplast isolation. A higher frequency of incorporated chloroplasts was obtained by the use of polyethylene glycol (PEG) to induce uptake (Bonnett and Eriksson, 1974). Clear evidence for real uptake was shown by light and electron microscopic studies of the incorporation of algal chloroplasts of *Vaucheria dichotoma* into carrot protoplasts (Bonnett, 1976), and of spinach chloroplasts into fungal protoplasts (Vasil and Giles, 1975). PEG has also been used for incorporation of nuclei (Lörz and Potrykus, 1978). The mechanism of PEG-induced uptake has been studied for algal chloroplasts (Bonnett, 1976; Bonnett and Banks, 1977) and for petunia chloroplasts (Davey *et al.*, 1976). In 1980, Giles introduced liposomes as chloroplast carriers for uptake. Spinach chloroplasts were incorporated at high frequency per protoplast, and photosynthetic activity could also be registered. There is no clear evidence for replication of any incorporated organelles, but receptor protoplasts containing transferred chloroplasts have been observed to divide (Bonnett *et al.*, 1980).

Several reviews on the problems and possibilities of organelle uptake were published earlier by Potrykus (1975), Giles (1977, 1978), Cocking (1977), Bonnett *et al.* (1980), and Chaleff (1981). This chapter will consider the experimental procedure used, especially for chloroplast uptake, where most progress has been made.

II. SELECTION AND PREPARATION OF PROTOPLASTS SUITABLE AS RECEPTORS FOR ORGANELLE TRANSMISSION

Different types of protoplasts, from higher plants as well as from fungi, have been used as receptor protoplasts. These have all lacked functional chloroplasts, so as to eliminate competition with the incorporated plastids. In experiments studying uptake procedures or short-term behavior of incorporated chloroplasts, plant cells have been selected that under special culture conditions remain colorless, for example, suspension cultures of carrot (Bonnett and Eriksson, 1974) and wild carrot (Fowke *et al.*, 1979) grown in diffuse light, or other types of cell that are colorless, such as crown gall tissue (Davey and Power, 1975).

In most cases, mutants with a deficiency in chloroplast DNA have been utilized. Protoplasts have been isolated from pure white leaves or *in vitro* cultured cells induced from white parts of plants showing extrachromosomally inherited variegation (Carlson, 1973; Potrykus, 1973; Kung *et al.*, 1975; Landgren and Bonnett, 1979).

Freshly isolated protoplasts without residual cell walls should be purified and kept in a solution without debris when used for uptake. Debris can damage protoplasts during uptake and even prevent incorporation of organelles by forming clods. Another important feature of the protoplasts is that they are able to grow and differentiate into intact plants. In order to facilitate observations of incorporated organelles in individual cells, it is suitable to culture the protoplasts in small drops or agar layers.

III. SELECTION OF CHLOROPLASTS

A. Chloroplast Donors

Up to now, the most promising experiments on organelle transmission have been done with algal chloroplasts. Chloroplasts of green algae such as *Vaucheria* are small and surrounded by a rigescent integument protecting them during isolation and uptake (Bonnett, 1976). The choice of algal chloroplasts is also based on the fact that plant cells with high probability are developed from an endosymbiotic association of ancestral algae and heterotrophic organisms. Solanaceous species and spinach have been chosen as organelle donors among higher plants since their organelle inheritance is well known and there are standard methods for isolation of active chloroplasts. The initial treatment of the plants used for isolation of chloroplasts varies. Plants of *Nicotiana excelsior* were placed in the dark 2–3 days before preparation in order to minimize the quantity of starch grains and broken cells (Landgren and Bonnett, 1979). However, experience in a number of laboratories indicates that rapidly growing and actively photosynthesizing plants yield the most active chloroplasts (Chapter 54, this volume). Proplastids have been suggested as being more suitable than mature plastids, as they are relatively undifferentiated and therefore perhaps more easily adapted to the host cell. Further advantages with proplastids are their small size and active division when taken from tissue close to the meristem (Boffey *et al.*, 1979). These conditions may facilitate their uptake and survival in the receptor cell.

Of special interest as chloroplast donors are mutants resistant to streptomycin and kanamycin (Maliga *et al.*, 1975; Dix *et al.*, 1977), since it has been shown that the mutation has occurred in the chloroplast DNA. By introduction of these organelles into a drug-sensitive receptor cell, it would be possible to select transformed cells by culture in the presence of the drug. Another trait available for chloroplast identification is chloroplast-encoded herbicide resistance (Arntzen and Duesing, 1983).

B. Quality of Isolated Chloroplasts

Functional chloroplasts are an absolute necessity for successful plastid transfer. The environmental conditions and the treatment during isolation and uptake are probably the most critical steps for survival of the incorporated organelles. Chloroplasts have been isolated mainly by two different methods, mechanical dispersion of cells and rupture of protoplasts. In order to free chloroplasts from other cellular organelles and extrachloroplastic enzymes, and with their photosynthetic function intact, one has to select a suitable purification method. So far, chloroplasts for transmission have been isolated by differential centrifugation in media with salts and sugar alcohols. For the details of chloroplast isolation and suitable media for isolation, see Chapter 54, this volume.

Microscopic and biochemical methods have been used to estimate the quality of the isolated chloroplasts. Chloroplasts should routinely be examined in the light microscope before the uptake procedure is done. Chloroplast quality can be roughly determined by checking that the chloroplasts have a bright opaque appearance in the phase-contrast microscope and that there is no abnormal clumping or breakage of the organelles. However, highly reflecting chloroplasts with a bright opaque appearance do not necessarily have an intact, double-membraned envelope (Davey *et al.*, 1976). To make sure that the chloroplasts are biochemically active, a measurement of photosynthetic activity is recommended. The oxygen evolution is easily tested with an oxygen electrode.

IV. PROCEDURE FOR ORGANELLE TRANSPLANTATION

Transmission of foreign organelles to protoplasts is made by mixing isolated organelles and protoplasts in a medium suitable for maintaining their integrity. In order to increase the frequency of uptake, agents such as lysozyme or PEG can be added. Recently, liposomes have been introduced as vehicles for plastid transplantation.

In the first published method for transmission of organelles, isolated nuclei and protoplasts were layered like a sandwich in a centrifuge tube suspended in hypotonic mannitol solution with addition of 0.03% lysozyme (Potrykus and Hoffman, 1973). About 0.5% of the receptor-protoplasts contained incorporated nuclei after slow centrifugation. A similar technique was used for plastids, but Potrykus (1973) found that a more efficient procedure was to incubate single cells in digestive enzyme solution with isotonic concentrations of sodium nitrate together with the chloroplasts. The yield of protoplasts containing chloroplasts was up to 0.5%.

Better results were obtained when PEG was used for adhesion and uptake of chloroplasts. PEG facilitated chloroplast uptake in protoplasts of higher plants and of fungi, independently of chloroplast origin. Bonnett and Banks (1977) used *Codium* chloroplasts and white carrot protoplasts in a model system to test favorable conditions in order to promote uptake frequency and to conserve protoplast viability. They suspended organelles and protoplasts in a protoplast culture medium at room temperature and added PEG. Maximum uptake of chloroplasts was obtained at a protoplast density of about 10^6/ml, a chloroplast density of about 10^8/ml, and with PEG (MW 4000) treatment at a final concentration of 28% (w/w) for a period of 10 min, followed by slow dilution and washing. Under these conditions, normally about 20%, and sometimes up to 50%, of the protoplasts contained chloroplasts. Using chloroplasts of higher plants Landgren and Bonnett (1979) found it necessary to decrease the density of chloroplasts when transferring them into albino leaf protoplasts in order to avoid destruction of the receptor protoplasts. Otherwise the experimental procedure was mainly in accordance with those developed for algal chloroplasts. About 10% of the protoplasts contained chloroplasts after PEG treatment.

Total disruption of incorporated chloroplasts of higher plants was obtained using PEG (MW 6000) in a final concentration of about 28% (w/v) at 2–4°C (Davey *et al.*, 1976). Effects of PEG treatment on mitochondria showed that concentrations of up to 10% (w/v) (MW 4000) had no effect on the functional and structural integrity, but concentrations normally used to induce organelle uptake extensively altered the organelles (Benbadis and de Virville, 1982). Dextran in the medium even prevents PEG-induced uptake (Landgren and Bonnett, 1979). When comparing PEG treatments, one should be conscious that different authors use various PEGs. Differences in molecular weight, manufacture, and contamination, as well as methods of expressing the concentrations, make a direct comparison between treatments difficult. A PEG that is active and harmless for fusion is often most suitable for organelle uptake as well.

After PEG treatment, only a few plastids are normally incorporated per protoplast. To improve the frequency of uptake and to avoid detrimental effects of PEG, Giles (1980) introduced fusion by chloroplast-charged liposomes and protoplasts. Liposomes were made using a positively charged liposomekit containing lecithin, sterylamine, and cholesterol mixed in the ratio 10:3:7 (see also Chapter 55, this volume). The lipids, dissolved in cool chloroform, were placed in a round-bottomed flask. After evaporation, a film of lipids appeared on the glass surface. The bottle with the chloroplasts in isolation medium was warmed to 25°C. After gentle shaking, groups of chloroplasts were concentrated by centrifugation and washed. Up to 85% of *Datura innoxia* protoplasts contained 6–20 chloroplasts following treatment with the lipochloroplasts. After uptake the plastids remained in groups in the new protoplast.

V. MICROSCOPIC AND BIOCHEMICAL METHODS TO PROVE UPTAKE AND SURVIVAL OF ISOLATED ORGANELLES

There are reliable microscopic and biochemical methods to prove that organelles are incorporated after induced uptake. But the fate of organelles after uptake has been incompletely analyzed. First, one has to prove that the organelles have biological activity directly after uptake. Such activity would show at least a short-time survival. Second, one has to prove that the organelles can be expressed and replicated in their new environment, which can be done first when the recipient cells have been cultured for some time or after regeneration of plants.

A. Uptake

Uptake is most easily observed in the light microscope. In order to prove that organelles are incorporated, the receptor protoplasts are rolled between the slide and the coverslip. By focusing through the protoplast from different positions, it is possible to observe where the organelle is situated. Additional cytological evidence can sometimes be obtained by visible changes in the internal structure of the recipient protoplast such as a curving vacuole. Fluorescence microscopy can reveal chloroplasts that are difficult to observe in bright field microscopy, as chloroplasts are strongly fluorescent. Ultrastructural analysis with the electron microscope is an alternative to light microscopy to check that organelles really are inside the protoplasts. It is, of course, necessary that organelles in the recipient protoplast have features that distinguish them from the incorporated organelles. But electron microscopy can also show if the organelles are intact after incorporation and allow the study of the membranes at uptake. Suitable methods for ultrastructural analysis have been published by Fowke (1975) and by Davey *et al.* (1976).

Additional evidence for uptake can be obtained by biochemical tests. A dye test with nitro blue tetrazolium chloride was introduced by Bonnett and Banks (1977). The dye was reduced to an insoluble blue precipitate in the presence of photosynthetic electron transport. All isolated chloroplasts outside the protoplast turned blue after exposure to light, but chloroplasts inside the plasmalemma remained green, as the dye could not penetrate the protoplast.

A complication in analyzing organelle uptake is the frequent adhesion of organelles to the outside of the protoplasts. These organelles are difficult to distinguish from incorporated organelles by light microscopy. Such chloroplasts can also be misleading when measuring the photosynthetic activity of incorporated plastids.

B. Short-Time Survival

Biological activity present in the organelles, but not in the recipient protoplast, would be suitable for proving that the organelles are functional. Giles (1980) determined the net photosynthesis by measuring oxygen evolution per milligram of chlorophyll in induced plastids using an oxygen electrode. These chloroplasts showed some photosynthetic activity over 24 hr. Prerequisites for this method are that transferred plastids are in an active photosynthesizing state at isolation, that the number of plastids is fairly high, and that the measurement takes place soon after uptake. After a few days of culture of mesophyll protoplasts, the chloroplasts have a tendency to dedifferentiate (Rennie *et al.*, 1980) and the photosynthetic activity is fairly low. Transplanted plastids are probably also degraded.

C. Long-Time Survival

Genetic markers can be used to prove uptake in receptor protoplasts by analysis of tissue or plants regenerated from such protoplasts. Most commonly the isoelectric focusing pattern of the enzyme ribulose-1,5-biphosphate carboxylase (RuBPcase) has been determined, since it is species specific. This protein consists of two kinds of subunits differing in size. The large subunit is coded by the chloroplast DNA and the small subunit by the nuclear DNA. Expression of the large subunit of the enzyme RuBPcase, coded by the incorporated chloroplasts, would be evidence for replication and function of the transferred chloroplasts (Carlson, 1973; Kung *et al.*, 1975; Uchimiya and Wildman, 1979).

To facilitate analysis, it is suitable to select chloroplasts for transmission whose electrophoretic mobility of RuBPcase clearly deviates from the chloroplasts of the receptor cells. However, when working with genetic markers, one has to remember that they cannot be used uncritically (see Chaleff, 1981; p. 127).

In the future, analysis of organelle DNA using restriction enzymes would be a still more sensitive method for detecting transferred extrachromosomal DNA (Belliard *et al.*, 1979). Another organelle character that could be used to distinguish mitochondrial types is cytoplasmic male sterility and resistance to phytotoxins (Galun, 1982). In order to exclude misinterpretations, it is best to combine genetic characters with some type of evidence for biological activity in connection with the uptake procedure.

VI. CONCLUSIONS

Incorporation of organelles is an artificial method for mixing organelles of different genetic traits, allowing the analysis of recombination and segregation of genetic information as well as of incompatibility responses. New combinations of cytoplasmic markers are seldom obtained in agricultural plants, as organelles in most such plants are transmitted almost invariably via the female parent. Incorporation of isolated organelles permits transfer of a single type of organelle, which is a simplification compared to fusion by protoplasts or subprotoplasts (cf. Chapter 49, this volume). With this method, it will be possible to obtain more knowledge of plant functions that require information from both nucleus and organelles, such as pollen formation, carbon dioxide fixation, and toxin resistance. This would be valuable in plant breeding. Isolated organelles might also be used as vehicles for transmission of foreign genes to plants.

How should one overcome the remaining obstacles to the replication and functioning of incorporated organelles? Unsatisfactory handling during the incorporation of organelles has to be improved before one can ascribe the failures entirely to physiological, genetic, or statistical causes.

More attention ought to be given also to the selection and treatment of the plastids, especially from higher plants. Up to now, only mature plastids have been tested after isolation in media favoring photosynthetic activity in isolated organelles. Experiments should also be performed with plastids in different stages of development cultured under various environmental conditions. Alternative methods for measuring the physiological activity of incorporated chloroplasts can be introduced, for example, autoradiographic determination of $^{14}CO_2$ fixation or use of 3-(3,4-dichlorophenol)-1,1-dimethylurea (DCMU) to determine the variable fluorescence of the plastids.

The organelles must be better protected during isolation and uptake. Modern techniques using silica gel gradients (Price *et al.*, 1979) and two-phase systems including dextran and PEG (Larsson and Andersson, 1979) ought to be tested for isolation of organelles (see Chapter 54, this volume). Organelles isolated by these methods have proved to be clean, intact, and highly active. The media for mixing protoplasts and chloroplasts before PEG addition are often designed for the protoplasts and are used at room temperature. These conditions do not necessarily favor the survival of naked chloroplasts. PEG has detrimental effects on the mitochondria and chloroplasts of higher plants and should not be used at high concentrations for these organelles. Liposomes, as vectors for organelle transmission, seem promising (Giles, 1983) but require further study, including study with the electron microscope. A great advantage of this method is the high frequency of incorporated plastids per cell. This can also be done if the organelles are injected by micromanipulation—time-consuming work but presumably without negative effects on the organelles.

REFERENCES

Arntzen, C. J., and Duesing, J. H. (1983). Chloroplast-encoded herbicide resistance. *In* "Advances in Gene Technology: Molecular Genetics of Plants and Animals" (F. Ahmad, K. Downey, J. Schulz, and R. W. Voellmy, eds.), pp. 23–24.

Belliard, G., Vedel, F., and Pelletier, G. (1979). Mitochondrial recombination in cytoplasmic hybrids of *Nicotiana tabacum*. *Nature (London)* **281,** 401–403.

Benbadis, A., and de Virville, J. D. (1982). Effects of polyethylene glycol treatment used for protoplast fusion and organelle transplantation on the functional and structural integrity of mitochondria isolated from spinach leaves. *Plant Sci. Lett.* **26,** 257–264.

Boffey, S. A., Ellis, J. R., Selldén, G., and Leech, R. M. (1979). Chloroplast division and DNA synthesis in light-grown wheat leaves. *Plant Physiol.* **64,** 502–505.

Bonnett, H. T. (1976). On the mechanism of the uptake of *Vaucheria* chloroplasts by carrot protoplasts treated by polyethylene glycol. *Planta* **131,** 229–233.

Bonnett, H. T., and Banks, M. S. (1977). Chloroplast incorporation, survival, and replication in foreign cytoplasm. *Int. Cell Biol., Pap. Int. Congr., 1st, 1976* pp. 225–231.

Bonnett, H. T., and Eriksson, T. (1974). Transfer of algal chloroplasts into protoplasts of higher plants. *Planta* **120,** 71–79.

Bonnett, H. T., Wallin, A., and Glimelius, K. (1980). The analysis of organelle behavior and genetics following protoplast fusion or organelle incorporation. *In* "Genetic Improvement of Crops: Emergent Techniques" (I. W. Rubenstein, B. Gengenbach, R. L. Phillips, and C. E. Green, eds.), pp. 137–152. Univ. of Minnesota Press, Minneapolis.

Carlson, P. S. (1973). Towards a parasexual cycle in higher plants. *In* "Protoplastes et fusion de cellules somatiques végetales" (J. Tempé, ed.), pp. 497–505. Inst. Natl. Rech. Agron., Paris.

Chaleff, R. S. (1981). "Genetics of Higher Plants. Applications of Cell Culture." Cambridge Univ. Press, London and New York.

Cocking, E. C. (1977). Uptake of foreign genetic material by plant protoplasts. *Int. Rev. Cytobiol.* **148,** 323–343.

Davey, M. R., and Power, J. B. (1975). Polyethylene glycol-induced uptake of micro-organisms into higher plant protoplasts: An ultrastructural study. *Plant Sci. Lett.* **5,** 269–274.

Davey, M. R., Frearson, E. M., and Power, J. B. (1976). Polyethylene glycol-induced transplantation of chloroplasts into protoplasts: An ultrastructural assessment. *Plant Sci. Lett.* **7,** 7–16.

Dix, P. J., Joo, F., and Maliga, P. (1977). A cell line of *Nicotiana sylvestris* with resistance to kanamycin and streptomycin. *Mol. Gen. Genet.* **157,** 285–290.

Fowke, L. C. (1975). Electron microscopy of protoplasts. *In* "Plant Tissue Culture Methods" (O. L. Gamborg and L. R. Wetter, eds.), pp. 72–82. National Research Council of Canada, Saskatoon.

Fowke, L. C., Gresshoff, P. M., and Marchant, H. J. (1979). Transfer of organelles of the alga *Chlamydomonas reinhardii* into carrot cells by protoplast fusion. *Planta* **144,** 341–347.

Galun, E. (1982). Somatic cell fusion for inducing cytoplasmic exchange: A new biological system for cytoplasmic genetics in higher plants. *In* "Plant Improvement and Somatic Cell Genetics" (I. K. Vasil, W. R. Scowcroft, and K. J. Frey, eds.), pp. 205–219. Academic Press, New York.

Giles, K. L. (1977). Chloroplast uptake and genetic complementation. *In* "Plant Cell, Tissue, and Organ Culture" (J. Reinert and Y. P. S. Bajaj, eds.), pp. 536–577. Springer-Verlag, Berlin and New York.

Giles, K. L. (1978). The uptake of organelles and microorganisms by plant protoplasts: Old ideas but new horizons. *In* "Frontiers of Plant Tissue Culture 1978" (T. A. Thorpe, ed.), pp. 67–74. Univ. of Calgary Press, Calgary, Alberta, Canada.

Giles, K. L. (1980). The maintainance of isolated chloroplasts *in vitro* and *in planta*. *In* "Endocytobiology: Endosymbiosis and Cell Biology" (W. Schwemmler and H. E. A. Schenk, eds.), Vol. 1, pp. 673–684. de Gruyter, Berlin.

Giles, K. L. (1983). Mechanisms of uptake into plant protoplasts. *In* "Plant Cell Culture in Crop Improvement" (S. K. Sen and K. L. Giles, eds.), pp. 227–235. Plenum, New York.

Kung, S. D., Gray, J. C., and Wildman, S. G. (1975). Polypeptide composition of fraction 1 protein from parasexual hybrid plants in the genus *Nicotiana*. *Science* **31,** 353–355.

Landgren, C. R., and Bonnett, H. T. (1979). The culture of albino protoplasts treated with polyethylene glycol to induce chloroplast incorporation. *Plant Sci. Lett.* **16,** 15–22.

Larsson, C., and Andersson, B. (1979). Two-phase methods for chloroplasts, chloroplast elements and mitochondria. *Methodol. Surv.* **9,** 35–46.

Lörz, H., and Potrykus, I. (1978). Investigations of the transfer of isolated nuclei into plant protoplasts. *Theor. Appl. Genet.* **53,** 251–266.

Maliga, P., Sz-Breznovits, A., and Márton, L. (1975). Non-Mendelian streptomycin-resistant tobacco mutant with altered chloroplasts and mitochondria. *Nature (London)* **255,** 401–402.

Potrykus, I. (1973). Transplantation of chloroplasts into protoplasts of *Petunia*. *Z. Pflanzenphysiol.* **70,** 364–366.

Potrykus, I. (1975). Uptake of cell organelles into isolated protoplasts. *In* "Modification of the Information Content of Plant Cells" (R. Markham, D. R. Davies, D. A. Hopwood, and R. W. Horne, eds.), pp. 169–179. Elsevier/North-Holland, New York.

Potrykus, I., and Hoffmann, F. (1973). Transplantation of nuclei into protoplasts of higher plants. *Z. Pflanzenphysiol.* **69,** 287–289.
Price, C. A., Bartolf, M., Oritz, W., and Reardon, E. M. (1979). Isolation of chloroplasts in silica-sol gradients. *Methodol. Surv.* **9,** 25–33.
Rennie, P. J., Weber, G., Constabel, F., and Fowke, L. C. (1980). Dedifferentiation of chloroplasts in interspecific and homospecific protoplast fusion products. *Protoplasma* **103,** 253–262.
Uchimiya, H., and Wildman, S. G. (1979). Nontranslation of foreign genetic information for fraction 1 protein under circumstances favorable for direct transfer of *Nicotiana gossei* isolated chloroplasts into *N. tabacum* protoplasts. *In Vitro* **15,** 463–468.
Vasil, I. K., and Giles, K. L. (1975). Induced transfer of higher plant chloroplasts into fungal protoplasts. *Science* **190,** 680.

CHAPTER **58**

Transformation of Tobacco Cells by Coculture with *Agrobacterium tumefaciens*

László Márton

Institute of Plant Physiology
Biological Research Center
Hungarian Academy of Sciences
Szeged, Hungary

I. INTRODUCTION

Interest in genetic engineering of plant cells focused attention on the only known natural transformation system of plants, the tumor induction by a soil bacterium, *Agrobacterium tumefaciens*, on dicotyledonous plants (Roberts, 1982; Kahl and Schell, 1982).

In addition to obtaining a fairly good understanding of the genetics of *A. tumefaciens* and its Ti-plasmid, an effort has been made to develop a single-cell crown gall transformation system instead of the traditional wound infection procedure of whole plants, seedlings, or different parts of a plant. The advantage of a single-cell transformation system lies not only in the controlled conditions but also in the possibility of obtaining a large number of simultaneously transformed cells (cell lines) derived from individual transformation events, which can be used in comparative studies (Wullems *et al.*, 1981a; Ooms *et al.*, 1982).

In contrast to tumor formation on plants, crown gall transformation of cultured cells or protoplasts requires the selection of transformants. Trans-

CELL CULTURE AND SOMATIC CELL
GENETICS OF PLANTS, VOL. 1

ISBN 0-12-715001-3

formed cells differ from normal ones in their ability to grow on hormone-free medium in culture and in the production of opines which are not found in normal cells. Using the hormone autotrophic growth of transformed cells in their selection, a successful attempt has been made to work out crown gall transformation of tobacco protoplast cultures by coculture with intact bacteria (Marton *et al.*, 1979), by bacterial protoplasts (Hasezawa *et al.*, 1981), and by Ti-plasmid DNA (Wullems *et al.*, 1980; Davey *et al.*, 1980; Krens *et al.*, 1982). Selection based on opine production has not been successful so far, although preliminary results were published (Van Slogteren *et al.*, 1982).

In this chapter, we present a recent version of the procedure of crown gall transformation of protoplast cultures by coculture with the Ach-5 strain of *A. tumefaciens* which has been successfully used with *Nicotiana tabacum* and *N. plumbaginifolia* in our laboratory.

II. PROCEDURES

A. Protoplast Isolation and Culture

Of a variety of protoplast isolation procedures and culture conditions, we present only one which is used in our laboratory routinely to isolate and culture protoplasts from leaves, calli, and suspension cultures of different *Nicotiana* species. This is basically a one-step isolation procedure using a mixture of Onozuka R-10 Cellulase and Macerozyme enzyme preparations and sucrose as osmoticum during digestion and washing (Nagy and Maliga, 1976).

1. Plant Material

Nicotiana tabacum (e.g., cv. Petit Havana SR1) sterile plantlets derived from sterilized seeds or cuttings of different origin are kept in 500-ml commercial jars containing 80 ml solidified (0.7% agar) MS medium (Murashige and Skoog, 1962) without hormones and vitamins. Subculture the plants for 4–6 weeks in order to maintain proper quality of the plant material for protoplast isolation.

2. Protoplast Isolation

1. Collect 1.0–1.5 g leaves from the sterile plant cultures mentioned above.

2. Mount them on each oth in a 10-cm Petri dish, keep them together by gentle pressure with the tip of a forceps, and slice them with a razor blade or sharp surgical knife into 2- to 3-mm slices.

3. Add 10 ml of K_3 medium (Nagy and Maliga, 1976) containing 0.4 M sucrose, 0.1 mg/liter naphthaleneacetic acid (NAA), and 0.2 mg/liter kinetin as hormones.

4. Allow the leaf tissue to plasmolyze for 1 hr under ordinary laboratory conditions.

5. Replace the medium with 10 ml enzyme solution (2% Cellulase and 0.5% Macerozyme R-10 Onozuka are dissolved in K_3 medium, pH 5.8, which is adjusted after the enzymes are dissolved) and incubate the tissue at 23°C overnight or at least for 8–10 hr without shaking.

6. At the end of the incubation period, swirl the mixture to release the protoplasts.

7. Filter the preparation through a nylon or stainless steel sieve (60- to 90-μm pore size) to get rid of the macrodebris.

8. Centrifuge the filtrate in screw-cap tubes at ambient temperature for 5 min at 500 rpm. The protoplasts will float and the debris will pellet.

9. Collect the layer of protoplasts from the surface by a Pasteur pipette and resuspend them in K_3 medium (repeat the last two steps twice more).

10. Culture the protoplasts in K_3 medium at a density of about 10^5/ml (In *N. plumbaginifolia* protoplast cultures, a different hormone combination was found to be effective: NAA, 1 mg/liter, and benzyladenine, 0.1 mg/liter).

11. During the first 48 hr, keep the cultures in dim light, and then at 500–1000 lx in the culture room at 27°C.

B. Transformation by Coculture

For efficient transformation, the age of protoplast cultures was found to be important (Márton *et al.*, 1979; Wullems *et al.*, 1981a; Krens *et al.*, 1982). The optimal age of the protoplast culture is about 72 hr after isolation, but it can vary depending on the culture conditions and the plant species used (e.g., freshly isolated protoplasts of *N. plumbaginifolia* could be transformed at a comparable frequency, 10^{-2}–10^{-3}; Czakó *et al.*, 1984).

1. Preparation of *Agrobacterium* Suspension

1. Grow bacterial culture in YTG medium up to about 1.0 A at 660 nm (dilute an overnight culture five times and culture up to this absorbance the next morning). YTG medium consists of yeast extract, 1 g/liter; Bacto tryp-

tone, 10 g/liter; glucose, 10 g/liter; NaCl, 1 g/liter; $MgSO_4 \cdot 7H_2O$, 0.25 g/liter; and $CaCl_2 \cdot 2H_2O$, 0.25 g/liter.

2. Spin the bacteria down (10 min at 5000 rpm) and resuspend them in K_3 medium up to a density of about 10^9/ml (determine the number of bacteria on the basis of a calibration curve at A_{660}).

2. Coculture of Bacterial and Plant Cells

1. Add a few drops of bacterial suspension to the protoplast culture, to a final density of 10^7/ml, to obtain a ratio of about 1:100 protoplasts to bacterial cells.
2. Coculture them under culture room conditions for 32–48 hr. Depending on the bacterial strain used in the coculture, a stronger or weaker aggregation of the cells is seen within a few hours (Márton *et al.*, 1979).
3. At the end of the coculture period, add Cellulase R-10 enzyme solution up to 0.5% to the culture from a filter-sterilized stock solution (5%) and allow it to stand for 1–4 hr with occasional gentle shaking.
4. Stop the enzymatic treatment when the culture contains only single cells.
5. Remove the majority of bacteria by repeated washing. First, centrifuge the cultures at 250 rpm, collect the cells from the surface and the bottom, and resuspend them in W5 solution (NaCl, 9 g/liter; $CaCl_2 \cdot 2H_2O$, 18.4 g/liter; KCl, 0.4 g/liter; glucose, 1 g/liter; pH 5.8). Spin them down and discard the supernatant. Repeat washing in W5 solution at least three times.
6. Finally, resuspend and culture the cells in K_3 medium containing antibiotics such as carbenicillin (200 mg/liter) and vancomycin (250 mg/liter). (On the possible use of different antibiotics in plant cell cultures, see Pollock *et al.*, 1983.) The cell density in the cultures will be about $3–6 \times 10^4$/ml (the decrease is due to the enzymatic digestion and washings).
7. After 10–14 days, dilute the cultures threefold by adding fresh medium (with antibiotics, as above) and incubate them for an additional 10 days. [In *N. plumbaginifolia* cultures the gradual decrease in sucrose concentration from 0.4 *M* to 0.2 *M* during these 20–24 days was found to be important for selection (Czakó *et al.*, 1984).]

3. Selection of Transformants

1. Collect the colonies by centrifugation. Wash them three times in hormone-free K_3 medium.
2. Finally, resuspend the colonies in hormone-free K_3 medium (containing 0.3 *M* sucrose, 500 mg/liter; vancomycin and 0.5% agar) at 35°C, and

plate them into a Petri dish (20 ml medium into a 10-cm dish). The density of the colonies can be 200–500/ml depending on their size.

3. Keep the colonies for 4–8 weeks under culture room conditions. A few weeks after the plating into selective medium, certain changes can be observed under the microscope. Cells in most of the colonies undergo considerable enlargement and elongation (0.5–1.0 mm) but do not divide. The cells of transformed colonies under the same hormone-free conditions are small, round, and dividing.

4. Pick the obviously overgrowing colonies from the selective plates (Fig. 1A,B) and inoculate them onto solid K_3 medium again, but with no antibiotics and a sucrose concentration of 0.2 *M*. The size of the colonies to be picked is at least 2 mm, and from them about 50 can be cultured in a 10-cm Petri dish.

5. Subculture the growing colonies on hormone-free MS medium after 3 weeks and maintain them as cell lines. The clones that grow continuously on hormone-free medium are putative transformants.

4. Regeneration of Plants

Plant regeneration can be obtained from the transformed clones spontaneously or after subculture(s) on MS medium containing 2 mg/liter benzyladenine. Transformed shoots usually grow without apical dominance, and roots do not develop from them.

Culture the regenerated plants as normal shoot culture (Section II,A,1).

5. Classification of Transformants

The putative transformants should be further tested for the presence of opines and/or the integrated "T" DNA sequences in their nuclear DNA. The synthesis of octopine or nopaline is usually tested by detecting octopine synthase (OS) or nopaline synthase (NOS) enzyme activity by the rapid, simple, and very sensitive assay of Otten and Schilperoort (1978). The presence of agropine and mannopine is detected directly in the tissue extracts after paper electrophoresis and special staining, as described by Tempé *et al.* (1980) and Tepfer and Tempé (1981). Since the production of opines is encoded by the integrated sequence of Ti-plasmids in the plant cells, their presence is reliable evidence of transformation. Integrated Ti-plasmid sequences may also be found in opine-negative clones by DNA–

Fig. 1. Selection of transformants on hormone-free K_3 medium. The cell wall regenerating protoplasts were cocultured with (A) nonpathogenic and (B) pathogenic strains of *Agrobacterium tumefaciens.*

A

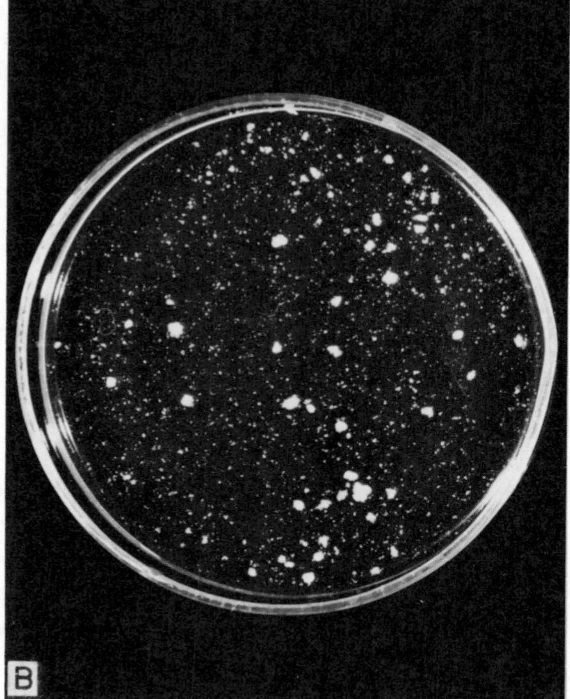
B

DNA hybridization techniques, as shown by Tomashow *et al.* (1980) and Ooms *et al.* (1982). Special application of DNA–DNA hybridization by Southern-type blotting (Southern, 1975) was developed by Thomashow *et al.* (1980) to detect and map integrated Ti-plasmid sequences in plant tumors.

III. RESULTS AND CONCLUSIONS

The coculture technique has become a procedure of general use in the molecular biology of the crown gall transformation of plant cells, resulting in a better understanding of the mechanisms involved in the transformation process (Márton *et al.*, 1979; Wullems *et al.*, 1981a; Ooms *et al.*, 1982). It was used with different strains of *Agrobacterium* and plant species (Wullems *et al.*, 1981a; Hasezawa *et al.*, 1981; Czakó *et al.*, 1984). Coculture-derived transformants were used in experiments showing transmission of tumor markers into sexual progeny (Wullems *et al.*, 1981b).

The high transformation frequency in cocultures and the selection at the level of cultured cells made possible significant progress in the field of plant cell genetic engineering. Foreign genes, other than those of the integrating fragment of Ti-plasmids, can be transferred into plant cells with the help of bacterial transformation machinery, as first shown by Hernalsteens *et al.* (1980) in *in vivo* tumors. Recently, preliminary information has been released about new achievements in plant cell genetic engineering using the coculture techniques. Ti-plasmids carrying chimeric resistance genes were expressed in plants, thus conferring drug resistance on the plant cells in culture (Caplan *et al.*, 1983). The possibility of selection based on drug resistance of transformed plant cells allows the elimination of those genes from the Ti-plasmids that cause the tumorous growth of transformants after integration.

REFERENCES

Caplan, A., Herrera-Estrella, L., Inzé, D., Van Haute, E., Van Montagu, M., Schell, J., and Zambryski, P. (1983). Introduction of genetic material into plant cells. *Science* **222,** 815–821.

Czakó, M., Maliga, P., and Márton, L. (1984). Agropine and mannopine or mannopine alone is present in cell lines transformed by co-culture of *Nicotiana plumbaginifolia* protoplasts with an octopine-type *Agrobacterium tumefaciens* strain (in preparation).

Davey, M. R., Cocking, E. C., Freeman, H., Pearce, N., and Tudor, I. (1980). Transformation of *Petunia* protoplasts by isolated *Agrobacterium* plasmids. *Plant Sci. Lett.* **18,** 307–313.

Hasezawa, S., Nagata, T., and Syono, K. (1981). Transformation of *Vinca* protoplasts mediated by *Agrobacterium tumefaciens* spheroplasts. *Mol. Gen. Genet.* **182,** 206–210.

Hernalsteens, J. P., Van Vliet, F., De Beuckeleer, M., Depicker, A., Engler, G., Holsters, M., Van Montagu, M., and Schell, J. (1980). The *Agrobacterium tumefaciens* Ti plasmid as a host vector system for introducing foreign DNA in plant cells. *Nature (London)* **287,** 654–656.

Kahl, G., and Schell, J. S. (1982). "Molecular Biology of Plant Tumours." Academic Press, New York.

Krens, F. A., Molendijk, L., Wullems, G. J., and Schilperoort, R. A. (1982). *In vitro* transformation of plant protoplasts with Ti plasmid DNA. *Nature (London)* **298,** 72–74.

Márton, L., Wullems, G. J., Molendijk, L., and Schilperoort, R. A. (1979). *In vitro* transformation of cultured cells from *Nicotiana tabacum* by *Agrobacterium tumefaciens. Nature (London)* **277,** 129–131.

Murashige, T., and Skoog, F. (1962). A revised medium for rapid growth and bioassay with tobacco tissue cultures. *Physiol. Plant.* **15,** 472–497.

Nagy, J. I., and Maliga, P. (1976). Callus induction and plant regeneration from *Nicotiana sylvestris* mesophyll protoplasts of. *Z. Pflanzenphysiol.* **78,** 453–455.

Ooms, G., Bakker, A., Molendijk, L., Wullems, G. J., Gordon, M. P., Nester, E. W., and Schilperoort, R. A. (1982). T-DNA organization in homogeneous and heterogeneous octopine-type crown gall tissues of *Nicotiana tabacum. Cell* **30,** 589–597.

Otten, L. A. B. M., and Schilperoort, R. A. (1978). A rapid microscale method for the detection of lysopine and nopaline dehydrogenase. *Biochim. Biophys. Acta* **527,** 497–500.

Pollock, K., Barfield, D. G., and Schields, R. (1983). The toxicity of antibiotics to plant cell culture. *Plant Cell Rep.* **2,** 36–39.

Roberts, W. P. (1982). The molecular basis of crown gall induction. *Int. Rev. Cytol.* **80,** 63–92.

Southern, E. M. (1975). Detection of specific sequences among DNA fragments separated by gel electrophoresis. *J. Mol. Biol.* **98,** 503–517.

Tempé, J., Guyon, P., Petit, A., Ellis, J. G., Tate, M. E., and Kerr, A. (1980). Preparation et propriétés de nouveaux substrats cataboliques pour deux types de plasmides oncogenes d'*Agrobacterium tumefaciens. C.R. Hebd. Seances Acad. Sci., Ser. D* **290,** 1173–1176.

Tepfer, D. A., and Tempé, J. (1981). Production d'agropine par des racines formées sous l'action d'*Agrobacterium* rhizogenes souche A4. *C.R. Hebd. Seances Acad. Sci., Ser. C* **292,** 153–156.

Thomashow, M. W., Nutter, R., Montoya, A. L., Gordon, M. P., and Nester, E. W. (1980). Integration and organization of Ti plasmid sequences in crown gall tumors. *Cell* **19,** 729–739.

Van Slogteren, G. M. S., Hooykaas, P. J. J., Planque, K., and De Groot, B. (1982). The Lysopine-dehydrogenase gene used as a marker for selection of octopine crown gall cells. *Plant Mol. Biol.* **1,** 133–142.

Wullems, G. J., Márton, L., Molendijk, L., Krens, F., Ooms, G., Wurzer-Figurelli, L., and Schilperoort, R. A. (1980). Genetic modification of plant cells by transformation and somatic hybridization. *In* "Advances in Protoplast Research" (L. Ferenczy, G. L. Farkas, and G. Lázár, eds.), pp. 407–422. Akadémiai Kiadó, Budapest.

Wullems, G. J., Molendijk, L., Ooms, G., and Schilperoort, R. A. (1981a). Differential expression of crown gall tumor markers in transformants obtained after *in vitro Agrobacterium tumefaciens*-induced transformation of cell wall regenerating protoplasts derived from *Nicotiana tabacum. Proc. Natl. Acad. Sci. U.S.A.* **78,** 4344–4348.

Wullems, G. J., Molendijk, L., Ooms, G., and Schilperoort, R. A. (1981b). Retention of tumor markers in F1 progeny plants from *in vitro* induced octopine and nopaline tumor tissues. *Cell* **24,** 719–727.

CHAPTER **59**

Ti-Plasmid DNA Uptake and Expression by Protoplasts of *Nicotiana tabacum*

F. A. Krens
R. A. Schilperoort

Department of Plant Molecular Biology
State University of Leiden
Leiden, The Netherlands

I. INTRODUCTION

Genetic manipulation of plants is a powerful tool in the fundamental research on the molecular basis of processes involved in cell development and differentiation. Differential gene expression is one of the interesting features in cell specialization. In order to understand the regulation of gene expression during these processes, one must have information on gene structure and its organization in the genome. With recombinant DNA technology, even plant genes can now be isolated and studied in detail. Such well-defined genes can be manipulated in their sequences and inserted again into the host plant DNA; afterward, the effect of both sequence alteration and chromosome position can be investigated during cell growth

CELL CULTURE AND SOMATIC CELL
GENETICS OF PLANTS, VOL. 1

ISBN 0-12-715001-3

and differentiation. Most of the genetic background of the host remains unchanged except for the desired trait introduced.

Another goal of genetic manipulation of plants has a more applied character. It concerns the introduction of only a few desired genes in an otherwise unchanged crop for plant improvement or the genetic manipulation of cells for plant biotechnology, that is, the industrial production of valuable compounds which are of chemical or pharmaceutical interest. To accomplish genetic engineering of cells, well-equipped plant vectors are needed. The vectors must carry sites in which genes can be stably inserted and should have selectable markers for both bacteria and plant cells. In addition, they should have the capacity to replicate in bacteria, and after introduction into plant cells, they should either show autonomous replication or should have special sequences which allow the integration of the desired genes. In order to be expressed, introduced genes must harbor regulatory sequences, or such sequences present in the vector must be connected to them. To guarantee stable transmission to progeny plants, newly introduced genes have to survive meiosis, which is best accomplished by integration of these genes into chromosomes.

Ti-plasmids and Ri-plasmids of *Agrobacterium tumefaciens* and *A. rhizogenes*, respectively, meet many of these requirements. They provide a natural system for genetic engineering of plants. Genes on a special area of Ti- and Ri-plasmids, called T-DNA, become integrated and expressed in the nuclear DNA of the host upon the induction of crown gall tumors and the hairy root disease by the respective bacterial agents (Chilton *et al.*, 1980, 1982; Willmitzer *et al.*, 1980, 1982a,b; Gelvin *et al.*, 1982; Murai and Kemp, 1982). Certain phenotypic tumorous traits of plants regenerated from transformed tissues are transmitted to progeny plants (Wullems *et al.*, 1981) in a mendelian fashion (Otten *et al.*, 1981). It has also been found that the T-DNA present in transformed plants remains unchanged after sexual propagation (De Greve *et al.*, 1982; Memelink *et al.*, 1983). By transposon mutagenesis, evidence has been obtained for the presence on T-DNA of sequences involved in cell growth and regeneration. On separate T-DNA loci, genes have been identified that give rise to phytohormone-like activities (Ooms *et al.*, 1981; Garfinkel *et al.*, 1981; Leemans *et al.*, 1982). Due to these genes, transformed tissues have acquired the capacity to grow on phytohormone-free medium. This is a good trait for selection of cells after transformation of plant protoplasts with Ti-plasmid DNA. As an additional marker, the presence of lysopine dehydrogenase (LpDH), the enzyme responsible for the synthesis of tumor-specific compounds such as octopine in crown gall tissues, can be used. However, its presence is not a condition for the tumorous character of crown gall cells.

Plant protoplasts have already been shown to be suitable for transformation at the cellular level by *A. tumefaciens* using the cocultivation method developed in our laboratory (Márton *et al.*, 1979; Chapter 58, this volume). Monocotyledons, however, are not susceptible to *A. tumefaciens*, and therefore the cocultivation method cannot be directly applied to them. This was one of the reasons for developing an *in vitro* DNA transformation procedure. Many methods for DNA uptake by plant protoplasts based on the results from research on the uptake of viral RNA, on somatic cell fusion, and on DNA transformation of animal cells have been tried in our laboratory and by others (Wullems *et al.*, 1980a; Davey *et al.*, 1980; Dellaporta and Fraley, 1981; Krens *et al.*, 1982, 1983; Draper *et al.*, 1982). In our hands, only one procedure has led to a reproducible method for the stable transformation of tobacco protoplasts with Ti-plasmid DNA. A stepwise scheme is presented in Fig. 1.

1 ml (5×10^5) protoplasts in K_3 medium 0.5 ml 40% (w/v) PEG 6000 in F medium

↘ ↙

Mix and add 10 μg Ti-plasmid DNA
and 50 μg calf thymus DNA

↓

Mix and incubate for 30 min at 26°C with occasional shaking
5 × [add 2 ml F medium, mix, and incubate for 5 min at 21°C]

↓

Centrifuge at 600 rpm

↓

Remove supernatant and resuspend protoplasts in 10 ml K_3 medium containing
0.4 *M* sucrose, 250 μg/ml carbenicillin, and phytohormones

↓

Plate and leave in the dark at 26°C for 24 hr

↓

Place in 2000 lx for 12 hr/day and culture for approximately 2 weeks

↓

Add fresh K_3 medium containing 0.4 *M* sucrose and phytohormones, and culture for approximately 2 weeks

↓

Dilute the suspension 1:10 with K_3 medium containing 0.3 *M* sucrose, hormones, and 0.3% agar, and culture for approximately 4 weeks

↓

Take colonies and place them on K_3-H containing 0.2 *M* sucrose
and 0.5% agar; culture for approximately 4 weeks

↓

Take surviving colonies and put them on fresh K_3-H, 0.2 *M* sucrose,
0.5% agar, and culture for approximately 4 weeks

↓

Putative transformed calli are placed on LS medium and tested for LpDH activity

Fig. 1. A stepwise scheme of the *in vitro* transformation procedure.

II. ISOLATION PROCEDURES

Starting materials in our DNA transformation experiments are tobacco leaf protoplasts and octopine-type Ti-plasmid DNA of *A. tumefaciens*. Procedures for their isolation are described in the next sections.

A. Plant Protoplasts

The development of an *in vitro* DNA transformation procedure for plant cells requires a well-established protoplast system with high plating efficiency, that is, good viability, cell division, and growth of protoplasts in order to have a chance to select transformants even if transformation frequencies are low. It must be recalled that the transformation procedure undoubtedly will cause cell death to a certain extent. Therefore, we used a well-defined protoplast system that has proven to be very suitable for various purposes during many years in our laboratory. Also, regeneration can be accomplished easily by varying the hormone levels in the medium.

Shoots of *Nicotiana tabacum* var. Petit Havanna SR_1 are cultured axenically on T medium (Nitsch and Nitsch, 1969) and subcultured every 4 weeks. For protoplast isolation, 3-week-old leaves are taken from shoots by using a sterile scalpel. The main vein is removed, and the remaining parts are placed in a sterile plastic Petri dish with their undersurface directed downward. Enzyme solution is added, while care is taken to submerge completely all the leaf halves. Incubation is performed at room temperature for about 18 hr in the dark. The medium used in further handling is K_3 medium supplemented with phytohormones. It consists of the following salts (in milligrams/liter): $NaH_2PO_4{\cdot}H_2O$, 150; $CaHPO_4{\cdot}2H_2O$, 63; $CaCl_2{\cdot}2H_2O$, 900; KNO_3, 2500; NH_4NO_3, 250; $(NH_4)_2SO_4$, 134; $MgSO_4{\cdot}7H_2O$, 250; Na_2EDTA, 37.3; $FeSO_4{\cdot}7H_2O$, 27.8; KI, 0.75; H_3BO_3, 3.0; $MnSO_4{\cdot}H_2O$, 10; $ZnSO_4{\cdot}7H_2O$, 2.0; $Na_2MoO_4{\cdot}2H_2O$, 0.25; $CuSO_4{\cdot}5H_2O$, 0.025; $CoCl_2{\cdot}6H_2O$, 0.025; organic substances (in milligrams/liter): myo-inositol, 100; thiamine·HCl, 10; nicotinic acid, 1.0; pyridoxine·HCl, 1.0; sucrose, 136,800; xylose, 250; phytohormones (in milligrams/liter): naphthaleneacetic acid (NAA), 0.1; kinetin, 0.2. The medium is adjusted with NaOH to pH 5.6 and autoclaved prior to use.

The enzymes used in protoplast isolation are cellulase (Onozuka R-10), 0.2 g/15 ml, and Macerozyme (Onozuka R-10), 0.07 g/15 ml. Usually 1.5 g leaf material is incubated with 15 ml filter-sterilized enzyme solution. Protoplast yield varies from 5 to 8 million. After the 18-hr incubation period, the Petri dish is shaken gently by hand to free the protoplasts from their matrix. The protoplast suspension is then filtered twice through a set of steel filters, mesh size approximately 400 and 100 μm, to remove large pieces of debris. Subsequently the protoplasts are washed and further purified in K_3 medium by centrifugation at 600 rpm. This is done twice. The K_3 medium at this stage contains 0.4 *M* sucrose, which makes the protoplasts float. After gentle centrifugation they form a band at the surface, which allows effective separation from residual debris and the final elimination of medium containing residues and enzymes. The elimination is achieved by putting a 15-cm-long, sterile glass capillary (Hirschmann, Eberstadt, Federal Republic of Germany), connected to a peristaltic pump by a silicon tube, through the protoplast layer and by removing the medium and residues below the protoplast layer by suction. The protoplasts are counted and then suspended at the appropriate density in K_3 medium. All handlings are performed aseptically in a laminar flow cabinet.

B. Ti-Plasmid Isolation

The octopine-type *A. tumefaciens* strain LBA 4001 used in these studies is streaked on a solid medium in order to obtain single colonies. TY medium solidified with agar (tryptone, 5 g/liter; yeast extract, 3 g/liter; agar, 18 g/liter) is used. For plasmid isolation, one colony is suspended in 10 ml liquid TY medium in a 100-ml Erlenmeyer flask and cultured at 29°C with continuous shaking for 24 hr. The culture is then added to a 3-liter Erlenmeyer flask containing 1 liter of TY medium and further incubated under the same conditions for 18 hr. In general, 2 liters of culture are used for plasmid isolation. The bacteria are harvested by centrifugation for 20 min at 7000 rpm at 4°C in 500-ml buckets. The supernatant is discarded, and the bacterial pellets are suspended in 1 liter of T–E buffer (Tris, 50 m*M*; EDTA, 20 m*M*; adjusted to pH 8.0 with HCl). The suspensions are collected in one glass beaker, and the bacteria are lysed with pronase and 1% (w/v) SDS for 30 min at 37°C (pronase solution: 240 mg in 38 ml T-E buffer preincubated for 1.5 hr at 37°C in order to destroy any contaminating DNase). After lysis, when the lysate appears to be completely clear, the pH is raised to 12.2 by the dropwise addition of 3 *N* NaOH under constant gentle stirring, and incubation is continued for an additional 10 min. Under these conditions, chromosomal DNA but not Ti-plasmid DNA will denature. Thereafter the pH is lowered to about 8.6 by adding dropwise 2 *M* Tris, pH 7.0, and the suspension is brought to 1 *M* NaCl. During a 4-hr incubation period at 4°C, denatured DNA will precipitate and is removed by centrifugation for 10 min at 7000 rpm at 4°C. The supernatant is collected and polyethylene glycol (PEG 6000) as a 50% (w/v) solution is added to obtain a final concentration of 10% (w/v). Overnight incubation at 4°C precipitates the DNA, which is pelleted by centrifugation at 7000 rpm for 10 min at 4°C and dissolved in two portions of 25 ml T–E buffer. CsCl, 25 g, is dissolved in each DNA solution and 1.875 ml ethidium bromide (EtBr, 10 mg/ml) is added. Breakage of the DNA in the presence of EtBr under the influence of ultraviolet (UV) light is avoided by preparing the solution in a room equipped with yellow light. Buoyant density gradient centrifugation is performed in a 60-Ti rotor using polyallomer tubes (volume approximately 30 ml) at 32,000 rpm at 15°C for 60 hr. After centrifugation, two bands are visualized by exposure to UV light. The lowest band, consisting of supercoiled Ti-plasmid DNA, is collected and extracted five times with isoamyl alcohol saturated with 20 × SSC (1 × SSC is 0.15 *M* NaCl; 0.015 *M* Na-tri-citrate) to remove the EtBr. The DNA solution is dialyzed overnight at 4°C against 3% (w/v) NaCl in T–E buffer, extracted with phenol saturated with 3% (w/v) NaCl and thereafter with chloroform:isoamyl alcohol (50:1). This extraction cycle with phenol and chloroform is repeated once, and subsequently DNA is precipitated overnight from the water phase by adding 2 volumes of 96% cold ethanol and storing at −20°C. After centrifugation in 30 ml Corex tubes for 20 min at 10,000 rpm at 4°C, the DNA pellet is washed with 70% ethanol and 96% ethanol, dried *in vacuo,* and dissolved in 400 μl 1 m*M* Tris–0.4 m*M* EDTA, pH 7.0. Before use in DNA transformation experiments, the amount of DNA is determined by a colormetric assay, and the integrity of the plasmid is checked by agarose gel electrophoresis after restriction endonuclease digestion. The DNA solution is sterilized by the addition of a few drops of chloroform in a sterilized stoppered tube at least 24 hr prior to use. Sterility of the DNA solution is checked, after part of it is used in a DNA transformation experiment, by plating a few drops of the remaining solution onto a TY medium-containing Petri dish which is incubated at 29°C for several days.

III. DNA TRANSFORMATION PROCEDURE

A. Incubation of DNA and Protoplasts

In this procedure, a medium is used that is based on a medium used for somatic cell fusion experiments in our laboratory (Wullems *et al.*, 1980b). The medium has been adapted for application to DNA transformation experiments. The DNA transformation medium is called *F medium* and has the following composition: NaCl, 140 m*M*; KCl, 5 m*M*; Na_2HPO_4, 0.75

m*M*; glucose, 5 m*M*; and $CaCl_2 \cdot 2H_2O$, 125 m*M*. For the preparation of 100 ml F medium, all components except $CaCl_2$ are dissolved in 40 ml distilled water. The $CaCl_2$ is dissolved separately in another 40 ml water. Both clear solutions are mixed, and after adjustment of the pH to 7.0 the volume is brought to 100 ml. By this time, a fine precipitate of calcium phosphate is formed. This precipitate is not removed, since it is also thought to play a role in the DNA uptake, as do PEG and the high Ca^{2+} concentration. Solid PEG is dissolved in F medium up to a final concentration of 40% (w/v). After autoclaving the pH drops to about 5.5, the same pH as the K_3 medium in which the protoplasts are suspended.

In a 12-ml Kimax centrifuge tube, 1 ml protoplast suspension, containing 5×10^5 protoplasts, is mixed with 0.5 ml 40% (w/v) PEG 6000 in F medium and subsequently with 10 μg Ti-plasmid DNA (usually in a volume ranging from 25 to 50 μl), and 50 μg calf thymus DNA (50 μl) is added. Calf thymus DNA is required as a carrier DNA and is taken from a solution containing 1 mg/ml. The solution is sterilized in advance by a few drops of chloroform. After the addition of the various constituents, a massive clumping of the protoplasts occurs. The final concentration of the transformation-inducing agents in the incubation medium is 13.3% (w/v) PEG and approximately 40 m*M* Ca^{2+}. Incubation is performed for 30 min at 26°C in light (2000 lx), with occasional shaking to allow good contact between medium and protoplast clumps; the latter tend to float on the surface.

B. Postincubation and Culture

After a 30-min incubation period, the medium is diluted carefully with 2 ml F medium, lowering the PEG concentration to about 5.7% but raising the Ca^{2+} concentration to 89 m*M*. The suspension is left at room temperature (about 21°C) for 5 min, followed by another addition of 2 ml F medium and a 5-min rest. This treatment is repeated three more times until a total of 10 ml F medium has been added. It is clear that in this way the PEG concentration is lowered stepwise and the Ca^{2+} concentration is raised stepwise. The gradual addition of F medium is found to be extremely important. At the end of the postincubation period, the PEG concentration is about 1.6% (w/v) and the Ca^{2+} concentration is about 115 m*M*. The suspension is now centrifuged at 600 rpm. Because of the reduced sucrose concentration, the clumps and free protoplasts will sediment and the supernatant is easily removed by suction through a glass capillary using a peristaltic pump. Protoplasts are suspended in 10 ml K_3 medium (0.4 *M* sucrose) containing 250 μg/ml carbenicillin and supplemented with phytohormones, as described above. At this stage the large cell aggregates that were formed during the DNA transformation treatment have broken down into small clumps and many free protoplasts. Further washing to remove remaining PEG is not needed since the remaining PEG concentration is low enough not to affect protoplast viability and culture. Moreover, each centrifugation step gives rise to a loss of protoplasts of at least 10%. The protoplasts are transferred to Petri dishes immediately after they have been taken up in K_3 medium. Light microscopy examination reveals that cell survival after the transformation procedure is usually about 50%. Fusion of protoplasts is never observed under the applied conditions. If any fusion has taken place, it is estimated to be less than 5%. The protoplasts are cultured in the dark at 26°C for the first 24 hr.

IV. GROWTH AND SELECTION

The medium in which the protoplasts are cultured does contain phytohormones. However, the hormone level is rather reduced when compared to hormone levels used in culture media that are normally employed for the growth of normal callus tissue. Although nontransformed and therefore normal SR_1 tobacco protoplasts are able to grow in the medium we are using, it is also necessary, for the recovery of transformed protoplasts, to add phytohormones at the start (Wullems *et al.*, 1980a). The chosen level of phytohormones, however, allows a more rapid selection against nontransformed cells, when they are later cultured in the absence of phytohormones, than normal hormone levels. In the *in vitro* DNA transformation procedure, the tobacco cells are kept 1 month longer on medium containing the reduced phytohormone level than is usually done in the cocultivation procedure.

After the dark period, the Petri dishes with protoplasts are brought into the light (2000 lx for 12 hr/day) and further cultured at 26°C. Two weeks later, half of the content of a dish (about 5 ml) is transferred to a new dish, and to both dishes 5 ml of fresh K_3 medium is added. Two weeks later, the suspension of actively dividing cells is again diluted 1:10 with K_3 medium still containing phytohormones, but with a reduced sucrose concentration of 0.3 *M*. The medium is solidified with 0.3% agar. Small cell colonies are taken after 1 month of growth on these plates and placed on plastic Petri dishes containing a layer of phytohormone-free K_3 medium (K_3-H), 0.2 *M* sucrose, which is solidified with 0.5% agar. They are cultured for 1 month, and then the surviving calli are placed on fresh K_3-H, 0.2 *M* sucrose, 0.5% agar plates. After a culture period of 1 month, the calli which have still survived are called *putative transformants.* They are transferred and maintained on Linsmaier and Skoog (LS) medium lacking phytohormones up to the moment when they are examined biochemically for their transformed nature. LS medium has the following composition (in milligrams/liter): KNO_3, 1900; NH_4NO_3, 1650; $CaCl_2{\cdot}2H_2O$, 440; $MgSO_4{\cdot}7H_2O$, 370; KH_2PO_4, 170; KI, 0.83; H_3BO_3, 6.2; $MnSO_4{\cdot}4H_2O$, 22.3; $ZnSO_4{\cdot}4H_2O$, 8.6; $Na_2MoO_4{\cdot}2H_2O$, 0.25; $CuSO_4{\cdot}5H_2O$, 0.025; $CoCl_2{\cdot}6H_2O$, 0.025; Na_2EDTA, 37.3; $FeSO_4{\cdot}7H_2O$, 27.8; myo-inositol, 100; sucrose, 30,000; thiamine·HCl, 0.4; agar, 8000; adjusted to pH 5.6 with NaOH. Compared to K_3 medium, LS medium lacks nicotinic acid, pyridoxine·HCl, and xylose, whereas the thiamine·HCl and sucrose concentrations are much lower. In this respect, LS medium is more minimal than K_3 medium. The mineral salts are about the same, with slight variations.

V. CHARACTERIZATION OF PUTATIVE TRANSFORMANTS

Criteria for the transformed nature of calli obtained after the selection procedure are the continued growth on LS medium without phytohormones, the presence of LpDH activity, and the presence of T-DNA. The fact that calli are obtained with a capacity to grow continuously on medium lacking hormones is an indication of the transformed nature of these calli, but habituation may also have occurred spontaneously. Habituation is the phenomenon in which plant cells may acquire a phytohormone-independent growth trait during subculturing in the presence of phytohormones. So far, no habituated tissues have been isolated using SR_1 protoplasts and the described transformation procedure. An additional tumor-specific marker is the presence of LpDH. If calli have reached a size of approximately 0.5 cm^3, a piece can be used to assay the enzyme activity (Otten and Schilperoort, 1978). Although the presence of LpDH is a clear indication of the transformed state of a callus, its absence does not mean that a tissue is not transformed. Direct, conclusive evidence for transformation is obtained if the exogenously supplied DNA or fragments of it are detected as part of the plant DNA, that is, if T-DNA is present. This is done by Southern blot hybridization using restriction endonuclease-digested DNA of plant material and ^{32}P-labeled cloned T-region fragments (Thomashow *et al.*, 1980).

VI. RESULTS AND CONCLUSIONS

The method described above has been applied in our laboratory to transform tobacco protoplasts reproducibly with octopine-type Ti-plasmid

DNA. The transformation frequency appears to be rather low: On average, two to three transformants are obtained if one starts with 5×10^5 protoplasts. However, for a more accurate calculation of transformation frequency cell death, clumping and the number of colonies subjected to selection have to be taken into account. Usually the indicated number of transformants are found among about 3000 colonies that are placed on plates without hormones from K_3 medium containing 0.3 *M* sucrose, 0.3% agar, and phytohormones.

Among the transformants, variation in phenotype is observed. Three phenotypic classes are distinguished: class A, phytohormone independent, LpDH positive, and regeneration negative (regeneration in this case means development of shoots); class B, phytohormone independent, LpDH negative, and regeneration positive; class C, phytohormone dependent (!), LpDH positive, and regeneration negative. The last class was taken from K_3-H medium, in which it reached a sufficient size for the measurement of LpDH activity. Growth ceased on LS medium and could be stimulated again only by maintaining the callus on medium with phytohormones. T-DNA analysis showed the presence of sequences homologous to the T region of the Ti-plasmid in the nuclear DNA of all transformants. It revealed that T-DNA varied among the transformants, suggesting that the transformants have originated from independent transformation events. In transformed tobacco tissues obtained either by *in vivo* infection or by the cocultivation method, it is demonstrated that usually with only a few exceptions, a piece of T-DNA of a well-defined size is present (Thomashow *et al.*, 1980; Ooms *et al.*, 1982). This suggests that the T region may contain special sequences, called *border sequences,* which are used preferably for its integration in plant DNA when the bacterium is involved. In DNA transformation these sequences appear to be used less preferentially. The T-DNA detected in some DNA transformants extends beyond the known border sequences, whereas in others both smaller and more scrambled and scattered T-DNA is found. In DNA transformation, the integration of the exogenously supplied DNA seems to be less accurate than it is if agrobacteria transfer the T region into plant cells. It could well be that the border sequences act in the bacteria as recognition signals for transfer and integration of a defined piece of Ti-plasmid DNA. It has been demonstrated that genes of a *vir* region, present on the Ti-plasmid separated from the T region on its left side, are expressed in the bacteria for their virulence (Hille *et al.*, 1982; Klee *et al.*, 1982). This region has never been detected in crown gall cells, but may be involved in a T-region transfer mechanism whereby border sequences are important for specificity of the system.

Shoots that developed from transformants were subcultured separately as shoot cultures. They did not form roots and lacked apical dominance, which is found with transformed shoots obtained by the cocultivation method (Wullems *et al.*, 1981). However, after the shoots were grafted onto decapitated stems of healthy *N. tabacum* var. Samson plants in the greenhouse, mature flowering plants were obtained which were male sterile but set seed after cross-pollination with normal SR_1 pollen. Sexual transmission of T-DNA is a condition for the application of DNA transformation in genetic manipulation of plants.

If we compare *in vitro* DNA transformation with the cocultivation method, the latter has the advantage that with the use of the bacterium as a microinjector for DNA in plant cells, a distinct and fixed piece of DNA, T-DNA, is integrated. A new gene can be incorporated in the right position between the border sequences to allow integration followed by expression with rather high frequencies. The unpredictable varying sizes and scrambling of T-DNA that becomes integrated after DNA transformation may reduce the chance of integration of an intact new gene via Ti-plasmid DNA, especially when no strong selection for the gene is possible. The frequency of DNA transformation, moreover, is relatively low. However, if we aim to transform plant protoplasts directly with bacterial vectors other than Ti-plasmid–derived vectors, the described DNA transformation procedure might successfully be used, since no indications have been found so far that specific sequences are needed for DNA integration in this way. In addition, DNA transformation seems to be the only way for genetic manipulation of monocotyledons, since these are not susceptible to *A. tumefaciens*.

REFERENCES

Chilton, M.-D., Saiki, R. K., Yadav, N., Gordon, M. P., and Quétier, F. (1980). T-DNA from *Agrobacterium* Ti-plasmid is in the nuclear DNA fraction of crown gall tumor cells. *Proc. Natl. Acad. Sci. U.S.A.* **77,** 4060–4064.

Chilton, M.-D., Tepfer, D. A., Petit, A., David, C., Casse-Delbart, F., and Tempé, J. (1982). *Agrobacterium rhizogenes* inserts T-DNA into the genomes of the host plant root cells. *Nature (London)* **295,** 432–434.

Davey, M. R., Cocking, E. C., Freeman, J., Pearce, N., and Tudor, I. (1980). Transformation of *Petunia* protoplasts by isolated *Agrobacterium* plasmids. *Plant Sci. Lett.* **18,** 307–313.

De Greve, H., Leemans, J., Hernalsteens, J. P., Thia-Toong, L., De Beuckeleer, M., Willmitzer, L., Otten, L., Van Montagu, M., and Schell, J. (1982). Regeneration of normal and fertile plants that express octopine synthase, from tobacco crown galls after deletion of tumour-controlling functions. *Nature (London)* **300,** 752–755.

Dellaporta, S. L., and Fraley, R. F. (1981). Delivery of liposome-encapsulated nucleic acids into plant protoplasts. *PMB Newsl.* **2,** 59–66.

Draper, J., Davey, M. R., Freeman, J. P., Cocking, E. C., and Cox, B. J. (1982). Ti-plasmid homologous sequences present in tissues from *Agrobacterium* plasmid-transformed *Petunia* protoplasts. *Plant Cell Physiol.* **23,** 451–458.

Garfinkel, D. J., Simpson, R. B., Ream, L. W., White, F. F., Gordon, M. P., and Nester, E. W.

(1981). Genetic analysis of crown gall: Fine structure map of T-DNA by site-directed mutagenesis. *Cell* **27,** 143–155.

Gelvin, S. B., Thomashow, M. F., McPherson, J. C., Gordon, M. P., and Nester, E. W. (1982). Sizes and map positions of several plasmid-DNA-encoded transcripts in octopine-type crown gall tumors. *Proc. Natl. Acad. Sci. U.S.A.* **79,** 76–80.

Hille, J., Klasen, I., and Schilperoort, R. A. (1982). Construction and application of R prime plasmids, carrying different segments of an octopine Ti-plasmid from *Agrobacterium tumefaciens,* for complementation of vir genes. *Plasmid* **7,** 107–118.

Klee, H. J., Gordon, M. P., and Nester, E. W. (1982). Complementation analysis of *Agrobacterium tumefaciens* Ti-plasmid mutations affecting oncogenicity. *J. Bacteriol.* **150,** 327–331.

Krens, F. A., Molendijk, L., Wullems, G. J., and Schilperoort, R. A. (1982). *In vitro* transformation of plant protoplasts with Ti-plasmid DNA. *Nature (London)* **296,** 72–74.

Krens, F. A., Wullems, G. J., and Schilperoort, R. A. (1983). Transformation of plant protoplasts *in vitro*. *In* "Proceedings of the N.A.T.O. Advanced Studies Institute and F.E.B.S. Advanced Course on Structure and Function of Plant Genomes" (O. Ciferri and L. Dure, eds.), pp. 387–408. Plenum, New York.

Leemans, J., Deblaere, R., Willmitzer, L., De Greve, H., Hernalsteens, J. P., Van Montagu, M., and Schell, J. (1982). Genetic identification of functions of T_L-DNA transcripts in octopine crown galls. *EMBO J.* **1,** 147–152.

Márton, L., Wullems, G. J., Molendijk, L., and Schilperoort, R. A. (1979). *In vitro* transformation of cultured cells from *Nicotiana tabacum* by *Agrobacterium tumefaciens. Nature (London)* **277,** 129–131.

Memelink, J., Wullems, G. J., and Schilperoort, R. A. (1983). Nopaline T-DNA is maintained during regeneration and generative propagation of transformed tobacco plants. *Mol. Gen. Genet.* **190,** 516–522.

Murai, N., and Kemp, J. D. (1982). T-DNA of pTi-15955 from *Agrobacterium tumefaciens* is transcribed into a minimum of seven polyadenylated RNAs in a sunflower crown gall tumor. *Nucleic Acids Res.* **10,** 1679–1689.

Nitsch, J. P., and Nitsch, C. (1969). Haploid plants from pollen grains. *Science* **163,** 85–87.

Ooms, G., Hooijkaas, P. J. J., Moolenaar, G., and Schilperoort, R. A. (1981). Crown gall plant tumors of abnormal morphology, induced by *Agrobacterium tumefaciens* carrying mutated octopine Ti plasmids; analysis of T-DNA functions. *Gene* **14,** 33–50.

Ooms, G., Bakker, A., Molendijk, L., Wullems, G. J., Gordon, M. P., Nester, E. W., and Schilperoort, R. A. (1982). T-DNA organization in homogeneous and heterogeneous octopine-type crown gall tissues of *Nicotiana tabacum. Cell* **30,** 589–597.

Otten, L., and Schilperoort, R. A. (1978). A rapid microscale method for the detection of lysopine and nopaline dehydrogenase activities. *Biochim. Biophys. Acta* **527,** 497–500.

Otten, L., De Greve, H., Hernalsteens, J. P., Van Montagu, M., Schieder, O., Straub, J., and Schell, J. (1981). Mendelian transmissions of genes introduced in plants by the Ti-plasmids of *Agrobacterium tumefaciens. Mol. Gen. Genet.* **183,** 209–213.

Thomashow, M. F., Nutter, R., Montoya, A. L., Gordon, M. P., and Nester, E. W. (1980). Integration and organization of Ti-plasmid sequences in crown gall tumors. *Cell* **19,** 729–739.

Willmitzer, L., De Beuckeleer, M., Lemmers, M., Van Montagu, M., and Schell, J. (1980). DNA from Ti-plasmid present in nucleus and absent from plastids of crown gall plant cells. *Nature (London)* **287,** 359–361.

Willmitzer, L., Simons, G., and Schell, J. (1982a). The T_L-DNA in octopine crown gall tumours codes for seven well-defined polyadenylated transcripts. *EMBO J.* **1,** 139–146.

Willmitzer, L., Sanchez-Serrano, J., Buschfeld, E., and Schell, J. (1982b). DNA from *Agrobacterium rhizogenes* is transferred to and expressed in axenic hairy root plant tissues. *Mol. Gen. Genet.* **186,** 16–22.

Wullems, G. J., Márton, L., Molendijk, L., Krens, F., Ooms, G., Würzer-Figurelli, L., and

Schilperoort, R. A. (1980a). Genetic modification of plant cells by transformation and somatic hybridization. *In* "Advances in Protoplast Research" (L. Ferenczy, G. L. Farkas, and G. Lázár, eds.), pp. 407–424. Akadémiai Kiadó, Budapest.
Wullems, G. J., Molendijk, L., and Schilperoort, R. A. (1980b). The expression of tumour markers in intraspecific somatic hybrids of normal and crown gall cells from *Nicotiana tabacum*. *Theor. Appl. Genet.* **56,** 203–208.
Wullems, G. J., Molendijk, L., Ooms, G., and Schilperoort, R. A. (1981). Retention of tumor markers in F1 progeny plants from *in vitro* induced octopine and nopaline tumor tissues. *Cell* **24,** 719–727.

CHAPTER 60

Immobilization of Cultured Plant Cells and Protoplasts

P. Brodelius

Institute of Biotechnology
Swiss Federal Institute of Technology (ETH) Honggerberg
Zürich, Switzerland

I. INTRODUCTION

Immobilized biocatalysts have a great potential for biotechnological applications. Some of the major advantages of using such biocatalysts are as follows:

1. The biocatalyst is easily recovered and can be used over an extended period of time.
2. The desired product is easily separated from the catalyst.
3. The continuous operation of a process is readily achieved.
4. The immobilized biocatalyst often shows increased stability.

Immobilized enzymes and microbial cells have been used in various industrial processes for many years, for example, for the isomerization of

CELL CULTURE AND SOMATIC CELL
GENETICS OF PLANTS, VOL. 1

ISBN 0-12-715001-3

glucose to fructose and for the production of amino acids and semisynthetic penicillins (Brodelius, 1978).

An immobilized biocatalyst is obtained by binding the biocatalyst onto or within a solid support. The principal methods used to achieve immobilization are:

1. Covalent linkage to a polymer
2. Adsorption (ionic or hydrophobic bonding)
3. Entrapment within a polymer
4. Encapsulation (liposomes)
5. Cross-linking of the biocatalyst with a bi- or multifunctional reagent

During recent years there has been increasing interest in the immobilization of whole cells, since these can be utilized for the production and transformation of complex compounds. Viable cells with an intact metabolism are required for these complex reactions. The technique used for the immobilization of viable cells is almost exclusively entrapment in a polymeric network.

II. IMMOBILIZATION OF PLANT CELLS

The most widely used technique for the immobilization of cells with preserved viability has been their entrapment in alginate or carrageenan. One of the major advantages of using these polymers is the simplicity with which spherical particles can be obtained simply by dripping a polymer–cell suspension into a medium containing the appropriate cation. In the first reports on the successful immobilization of plant cells, that is, cells of *Catharanthus roseus, Morinda citrifolia,* and *Digitalis lanata,* alginate was used (Brodelius *et al.*, 1979, 1980).

There are a number of other polymers with desirable properties which are also useful for the immobilization of viable cells. However, the difficulties involved in the preparation of spherical particles when using these polymers have significantly hampered their wider use. Recently, a general method for the immobilization of cells in spherical particles of such polymers with preserved viability has been developed (Nilsson *et al.*, 1983). This method is based on a two-phase system consisting of an aqueous phase, that is, a cell–polymer suspension, and a hydrophobic phase, for example, vegetable oil or paraffin oil. The cell–polymer suspension is dispersed in the oil during continuous stirring, and after droplets of appropriate size have formed, the polymer is induced to form a gel.

For the immobilization of large, sensitive plant cells and protoplasts, only the most gentle procedures may be employed. Entrapment in various polysaccharides (e.g., alginate, carrageenan, agar, and agarose) has been successfully carried out (Brodelius *et al.*, 1979, 1980; Brodelius and Nilsson, 1980; Nilsson *et al.*, 1983). Some representative examples of the immobilization of plant cells in various matrices will be given below. A standard concentration of 20% cells (w/w) will be used in the examples, but the percentage of cells may be varied from 2–3% up to at least 50% (w/w). Only methods resulting in viable cell preparations will be discussed. Methods involving reactive chemicals, for example, entrapment in polyacrylamide or in glutaraldehyde cross-linked proteins, result in nonviable cell preparations (Brodelius and Nilsson, 1980).

A. Alginate

Alginate is a polysaccharide composed of glucuronic acid and mannuronic acid, and thus contains one carboxyl group on each sugar moiety. In the presence of calcium or other multivalent cations, alginate forms a gel due to the formation of ionic bridges. This gel can be solubilized by the addition of a calcium-complexing agent such as phosphate ion, citrate, or EDTA. Therefore, the immobilized cells can be released and studied in free suspension after a period of time in the immobilized state.

The concentration of sodium alginate used for immobilization may vary depending on the type of alginate used. For instance, a final concentration of 1–2% is appropriate when using alginate obtained from Sigma, whereas a somewhat higher concentration (2–4%) is required when using alginate (Manucol DH) supplied by Alginate Industries (Girven, England). The gel stability increases with increasing polymer concentrations; however, too high a concentration will result in a very viscous cell–alginate suspension, which may cause problems in bead formation. Thus, a compromise between gel stability and ease of preparation has to be found. It should also be pointed out that the gel obtained with the Sigma alginate is much more difficult to solubilize than the gel made from Manucol DH.

1. Small Scale

Small batches of alginate-trapped plant cells are readily prepared by the utilization of sterile plastic syringes, with or without a needle. If needles are used in order to obtain small beads, care must be taken to avoid their clogging by larger cell aggregates. Preferably, the needle should be cut to a length of a few millimeters to increase the speed of drop formation. If possible, the cell suspension can be filtered through a sterilized nylon net of appropriate mesh size to reduce clogging problems.

Procedure. A solution of sodium alginate (5% Manucol DH) is prepared in an appropriate medium (containing only small concentrations of calcium) and sterilized by autoclaving (20 min at 120°C). Excessive autoclaving should be avoided since the alginate is partly hydrolyzed during heating, which may result in a very weak gel. The plant cells are collected by centrifugation or filtration. Then the cells (2 g wet weight) are suspended in alginate (8 g) in a sterile 25-ml beaker using a sterile spatula. The suspension is subsequently poured into a plastic syringe (10 ml) without a piston, which is placed over an Erlenmeyer flask containing medium (50–100 ml) fortified with 50 m*M* $CaCl_2$. The viscous suspension drips slowly into the calcium-containing medium, which is gently stirred with a magnetic stirrer so that spherical beads can form. The suspension may also be gently pressed with the piston to increase the speed of drop formation. The formed beads are left in this solution for 30 min to allow the calcium to diffuse into their center. The beads are then collected by filtration, and washed with and transferred to an appropriate medium containing at least 5 m*M* $CaCl_2$, which is required to keep the beads intact. Normally, the beads (1–2 g wet weight) are incubated in 25-ml Erlenmeyer flasks containing medium (10 ml) on a rotary shaker (100 rpm). The bead size can, to some extent, be controlled using different-sized needles on the syringe. Bead size may range from 2 to 5 mm in diameter.

During "ripening," the beads will shrink slightly due to exclusion of water as the calcium ions diffuse into the beads and bind the polymer chains more tightly together. When quantitative experiments are carried out, the following procedure should be applied. Before the cell–alginate suspension is added to the medium, the flask containing the medium should be weighed. After addition of the suspension, it should be weighed again. The difference between the two weights will indicate how much cell–alginate suspension has been added. It is very difficult to add all of the prepared suspension (10.0 g in the example above). After washing, the collected beads can be weighed and a precise concentration of cells may be calculated.

2. Large Scale

Immobilization on a larger scale is carried out in principle, in the same way as described above. We have designed a device (schematically shown in Fig. 1), made of glass, for the preparation of up to 300 g of beads within a few minutes.

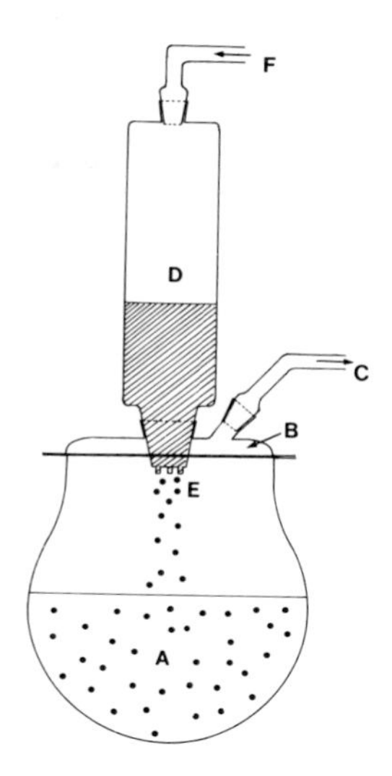

Fig. 1. Device for large-scale preparation of alginate-trapped plant cells. (A) Medium container, (B) lid, (C) air outlet, (D) reservoir for alginate–cell suspension, (E) six nipples (diameter 1 mm) for bead formation, (F) sterile air inlet.

Procedure. The entire device is weighed and autoclaved with the calcium-containing medium (500 ml) in the medium container (A). The cells (60 g wet weight) and alginate (240 g) are mixed and subsequently poured into the reservoir (D). A slight pressure of sterile air is applied so that beads of uniform size are formed. The medium is gently stirred during bead formation. Then the whole device is weighed and left for 30 min. Beads are subsequently collected by filtration, washed, weighed, and transferred to an appropriate medium containing at least 5 m*M* $CaCl_2$. From the weight data, the exact cell concentration can be calculated.

3. Two-Phase Method

Even though uniform spherical beads of alginate with trapped cells are obtained using the method described above, an alternative method based on a two-phase system may also be used. This method is recommended when smaller beads (0.1–1.0 mm) are desired.

Procedure. The cells (2 g wet weight) are suspended in alginate (8 g), and the suspension is subsequently poured under continuous magnetic stirring into soy oil (40 ml) that has been supplemented with an emulgator (0.1% GAFAC RS-410). The resulting emulsion is stirred until droplets of appropriate size have formed, and is then rapidly mixed with medium containing 50 m*M* $CaCl_2$ (250 ml). The alginate beads are allowed to stabilize for 15 min. Then they are collected on a sterile nylon net, and the oil is washed away with medium containing 5 m*M* $CaCl_2$. The beads containing the trapped cells are incubated on a rotary shaker or used in a bioreactor.

B. Carrageenan

Kappa-carrageenan is a polysulfonated polysaccharide that forms a relatively strong gel in the presence of potassium ions. Plant cells can be trapped in this polymer in a manner similar to that described for alginate, except that the polymer solution must be heated in order to be maintained in a liquid state. The beads formed by dripping the cell–carrageenan suspension into a medium containing potassium ions are, however, not as uniform in shape and size as the alginate beads. The temperature required for keeping carrageenan in a liquid state is also highly dependent upon the preparation of polymer used. The carrageenan supplied by Sigma (type III) requires 45–50°C for a 5% solution, whereas a low-gelling-temperature carrageenan from FMC (FMC Corp., Rockland, Maine), especially prepared for the immobilization of cells, requires only 30–35°C for a 5% solution. The three different procedures available for the immobilization of plant cells in carrageenan are described below.

1. Dripping Method

Procedure. Carrageenan (3% w/v) obtained from FMC is prepared in a heated 0.9% NaCl solution and autoclaved (20 min at 120°C). The NaCl solution is used since carrageenan is essentially insoluble in the presence of potassium ions, which are present in relatively high concentrations in most plant cell media. The plant cells (2 g wet weight) are suspended in carrageenan (8 g) at 35°C, and the mixture is subsequently dripped into a medium containing at least 0.3 *M* KCl. The carrageenan beads formed are left for 30 min in this solution and are then collected by filtration, washed, and transferred to the appropriate medium. Most plant cell media contain sufficient concentrations of potassium to keep the carrageenan beads intact.

2. Molding Method

Beads can be molded by pouring a warm cell–carrageenan suspension into a form. Two Teflon plates, one of which is tightly covered with holes (diameter 1–2 mm), can be used for the molding. The plates are held together by clamps.

Procedure. The cells (2 g wet weight) are suspended in 3% carrageenan (8 g) at 35°C, and the mixture is poured over the sterilized plates and carefully spread with a sterile spatula. After the gel solidifies, the plates are taken apart and the cylindrical beads are removed and placed in a medium containing 0.3 *M* KCl for 30 min. The beads are then washed and placed in an appropriate medium. These cylindrical beads can be used in batch experiments (e.g., 2 g beads in 10 ml medium), but they are not as convenient to use in a packed bed reactor. Furthermore, this method of immobilization can only be used to prepare relatively small quantities of immobilized cells.

3. Two-Phase Method

Carrageenan beads of uniform size and shape may be made by dispersing a cell–carrageenan suspension in soy oil by stirring.

Procedure. The plant cells (2 g wet weight) are suspended in 3% carrageenan (8 g) at 35°C, and the mixture is poured under constant magnetic stirring into soy oil (40 ml) maintained at the same temperature. When droplets of appropriate size have formed, the mixture is cooled on an ice bath under continuous stirring until the polymer has solidified. The mixture is then transferred to centrifuge tubes, after which medium containing at least 0.3 *M* KCl is added. The beads are spun down (for 2 min at 100 *g*), the upper oil phase and most of the aqueous phase are removed using an aspirator, and the washing procedure is repeated until no more oil is present. The beads are collected by filtration and fractionated, if necessary, by sieving on sterile metal screens. The size of the beads, which may range from 0.1 to 3.0 mm, can be adjusted by regulating the stirring speed. Alternatively, the beads can be collected and washed directly on a metal screen. This procedure can be easily scaled up keeping the ratio of cell–carrageenan suspension to oil constant (i.e., 1:4). These beads of uniform size and shape can be either incubated on a rotary shaker or used in a bioreactor.

C. Agarose

Agarose has the advantage that no counter-ion is required for gel stability. Therefore, immobilization can be carried out in any medium. There are many different agarose preparations on the market that can be used for the entrapment of plant cells but low-gelling-temperature types (e.g., type VII from Sigma or SeaPlaque from FMC, which both solidify at 28–30°C) are preferable. Three different methods available for the entrapment of cells in this polysaccharide are described below.

1. Block Method

Procedure. A solution of 3% agarose (Sigma type VII) is prepared in medium and autoclaved (for 20 min at 120°C). The plant cells (2 g wet weight) are suspended in agarose (8 g) at 35°C by mixing with a sterile spatula in a sterile 25-ml beaker. The mixture is allowed to solidify, and the block of agarose with trapped cells is disintegrated mechanically into small particles by pressing it through a sterile metal screen of an appropriate-sized mesh. The particles are washed, and the very small ones are removed. The washed agarose particles are transferred to the appropriate medium. This method results in a preparation containing a wide size distribution of irregularly shaped particles, which therefore can be used only in a batch procedure. It is, however, a very simple method of immobilization.

2. Molding Method

Cylindrical beads are molded with Teflon plates in the same manner as that described for carrageenan.

Procedure. The cells (2 g wet weight) are suspended in 3% agarose (8 g) at 35°C. The cell–agarose suspension is poured over the sterilized Teflon plates and carefully spread with a sterile spatula. After the gel has solidifed, the plates are taken apart and the cylindrical beads are taken out, washed, placed in medium, and incubated on a rotary shaker.

3. Two-Phase Method

Spherical beads of agarose with trapped cells can also be conveniently prepared in a two-phase system. Cells of *Daucus carota* trapped in agarose are shown in Fig. 2.

Procedure. The cells (2 g wet weight) are suspended in 3% agarose (8 g) at 35°C, and the suspension is dispersed in sterilized soy or paraffin oil (40 ml) by magnetic stirring. When droplets of appropriate size have formed, the mixture is cooled on an ice bath under continuous stirring until the polymer has solidified. The mixture is then transferred to centrifuge tubes, after which medium is added. The beads are spun down (for 2 min at 100 *g*), the upper oil phase and most of the aqueous phase are removed with an aspirator, and the washing procedure is repeated until no oil is present. The beads can, if necessary, be fractionated by sieving on sterile metal screens.

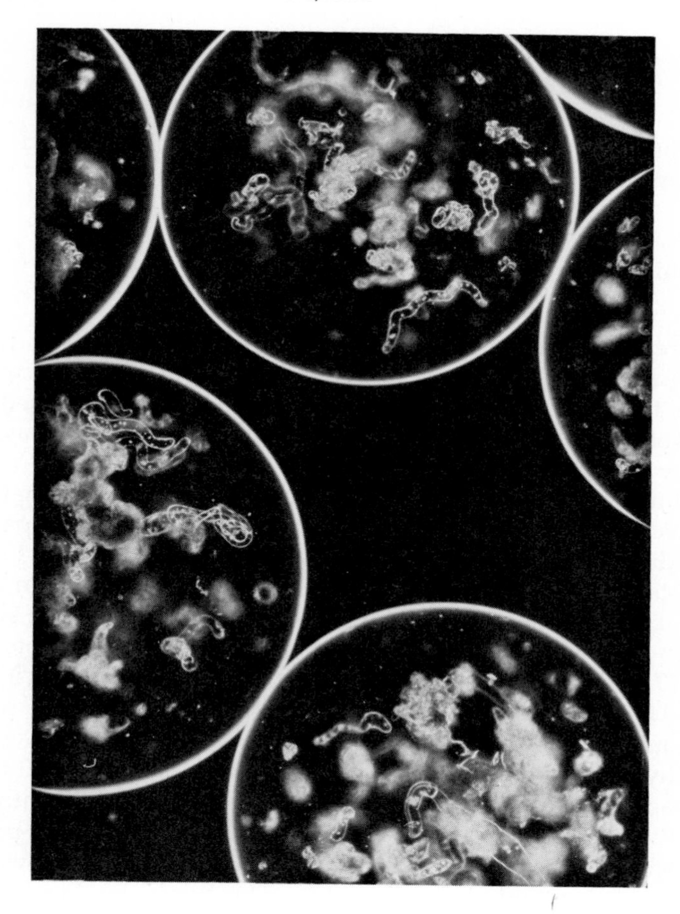

Fig. 2. Agarose beads (diameter 0.8 mm) containing cells of *Daucus carota* made according to the two-phase method.

D. Agar

Agar has the same advantage as agarose in that no counter-ion is required for gel stability. Standard Bactoagar from Difco can be used for immobilization, even though a relatively high temperature is required (around 45°C). The final concentration of agar within the beads may vary between 2 and 5%. The same methods of immobilization as those described for agarose, that is, the block method, the molding method, and the two-phase method, can be employed. The only modification required is an increase in temperature from 35° to 45°C.

III. IMMOBILIZATION OF PROTOPLASTS

Protoplasts are, of course, much more sensitive than intact cells, and only the most gentle entrapment methods may be employed.

A. Alginate

Alginate is probably one of the most gentle methods available for the immobilization of cells. This polymer is, however, somewhat complicated to use for the immobilization of protoplasts since calcium is used to stabilize the protoplasts during preparation. The calcium has to be washed away before the protoplasts are mixed with the polymer solution.

Procedure. Protoplasts are prepared in a medium containing 0.5–0.7 *M* sorbitol. A solution of 5% sodium alginate (Manucol DH) is prepared in the same medium without calcium and autoclaved (for 20 min at 120°C). The protoplasts are washed once with the calcium-free medium and centrifuged. The supernatant is removed, and the protoplasts are suspended in calcium-free medium (10^7 protoplasts per milliliter). The protoplast suspension (3 ml) is mixed with alginate (6 g) by gentle stirring with a sterile spatula. The mixture is poured into a 10-ml plastic syringe (without a piston) equipped with a cut needle. Then the suspension is allowed to drip into the hypertonic medium containing 50 m*M* $CaCl_2$ (50–100 ml) under gentle magnetic stirring. The viscous suspension may be gently pressed through the needle. The resulting alginate beads with trapped protoplasts are left for 30 min, and then collected by filtration and washed with hypertonic medium containing 5 m*M* $CaCl_2$. The beads are incubated in an appropriate medium.

B. Carrageenan

The same procedures as described for suspension cells can be employed for the immobilization of protoplasts in carrageenan. Only the low-gelling-temperature type of carrageenan may, however, be used with these sensitive cells.

Procedure. Protoplasts are prepared in a medium containing 0.5–0.7 *M* sorbitol. A solution of 3% carrageenan (FMC) is prepared in the same hypertonic medium (containing no potassium ions) and autoclaved (for 20 min at 120°C). The protoplast suspension (3 ml containing 10^7 protoplasts per milliliter) is mixed with carrageenan (6 g) at 30°C. Beads containing trapped protoplasts are made either by the molding or by the two-phase methods described above. Beads are collected by filtration and transferred to the hypertonic medium containing at least 0.3 *M* potassium ions. After 30 min, the beads are collected, washed, and incubated in an appropriate medium.

C. Agarose

The same procedures described for suspension cells can also be employed for the immobilization of protoplasts in agarose. The low-gelling-temperature agarose from Sigma (type VII) or FMC (SeaPlaque) may be used at 30°C, or an ultra-low-gelling-temperature agarose from FMC (SeaPrep) may be used at room temperature.

Procedure. Protoplasts are prepared in a medium containing 0.5–0.7 *M* sorbitol. A solution of 3% agarose (e.g., Sigma type VII) is prepared in the same hypertonic medium and autoclaved (for 20 min at 120°C). The protoplast suspension (3 ml containing 10^7 protoplasts per millimeter) is mixed with agarose (6 g) at 30°C. Beads containing trapped protoplasts are made either by the molding or by the two-phase methods described above. The beads are collected by filtration, washed, and transferred to an appropriate medium.

IV. CONCLUSIONS

Immobilized plant cells have shown several advantages over freely suspended cells in the production of biochemicals. These advantages include increased product yield, increased mechanical stability, and a prolonged stationary (production) phase. It has also been shown that immobilized plant cells can be intermittently permeabilized for the release of intracellularly stored products with retained viability and biosynthetic capacity (Brodelius and Nilsson, 1983). The potential biotechnological application of immobilized plant cells has been recently reviewed (Brodelius and Mosbach, 1982; Brodelius, 1983).

The potential use of immobilized plant cells and protoplasts in other areas of plant tissue culture has not yet been explored. The immobilization of these cells is, however, simple to carry out, and may prove to be a valuable tool in the further study of plant cells.

ACKNOWLEDGMENTS

The author wishes to express his gratitude to Ms. L. A. Clark for advice on usage.

REFERENCES

Brodelius, P. (1978). Industrial applications of immobilized biocatalysts. *Adv. Biochem. Eng.* **10,** 75–129.

Brodelius, P. (1983). Immobilized plant cells. *In* "Immobilized Cells and Organelles" (B. Mattiasson, ed.), pp. 27–55. CRC Press, Boca Raton, Florida.

Brodelius, P., and Mosbach, K. (1982). Immobilized plant cells. *Adv. Appl. Microbiol.* **28,** 1–26.

Brodelius, P., and Nilsson, K. (1980). Entrapment of plant cells in different matrices. A comparative study. *FEBS Lett.* **122,** 312–316.

Brodelius, P., and Nilsson, K. (1983). Permeabilization of immobilized plant cells, resulting in release of intracellularly stored products with preserved cell viability. *Eur. J. Appl. Microbiol. Biotechnol.* **17,** 275–280.

Brodelius, P., Deus, B., Mosbach, K., and Zenk, M. H. (1979). Immobilized plant cells for the production and transformation of natural products. *FEBS Lett.* **103,** 93–97.

Brodelius, P., Deus, B., Mosbach, K., and Zenk, M. H. (1980). The potential use of immobilized plant cells for the production and transformation of natural products. *Enzyme Eng.* **5,** 373–381.

Nilsson, K., Birnbaum, S., Flygare, S., Linse, L., Schroeder, U., Jeppsson, U., Larsson, P.-O., Mosbach, K., and Brodelius, P. (1983). A general method for the immobilization of cells with preserved viability. *Eur. J. Appl. Microbiol. Biotechnol.* **17,** 319–326.

CHAPTER 61

Mutagenesis of Cultured Cells

Patrick J. King

Friedrich Miescher Institute
Basel, Switzerland

I. INTRODUCTION

Various types of chemical and physical mutagens have been applied to callus, cell suspensions, or protoplasts in attempts to isolate mutants. Table I lists examples of traits isolated *in vitro* that were also expressed in regenerated plants, but not necessarily shown to be mutations by genetic analysis. The authors of less than half of these reports suggest that application of a mutagen enhanced "mutant frequency." However, the only data provided are of the incidence of variants in populations at the time that selection was applied. In only two cases (Schieder, 1976; Gebhardt *et al.*, 1981a) is it possible to calculate mutation rates per cell generation, and even here the data are very limited and not reliable. In two of the rare cases in which selected traits were shown to be due to mutation and inherited sexually (Bourgin, 1978; Hibberd and Green, 1982), the authors were unable to find any evidence for the effectiveness of the mutagens applied. Thus, there is little to be gained from the literature when faced with the questions of what is a suitable mutagen and how is it most effectively applied. Information is, however, available on the effects of irradiation (Ohyama *et al.*, 1974; Eapen, 1976; Werry and Stoffelsen, 1979; Werry, 1981) and chemical mutagens (Colijn *et al.*, 1979; Gebhardt *et al.*, 1981a,b; Negrutiu, 1981) on cell survival.

CELL CULTURE AND SOMATIC CELL
GENETICS OF PLANTS, VOL. 1

ISBN 0-12-715001-3

TABLE I

Mutagen Treatments of Cultured Cells Leading to Traits Expressed by Regenerated Plants[a]

Mutagen	Trait	Culture	Mutagen effective?	Reference
EMS	ETH^r	Suspension	No	Reisch and Bingham (1981)
	$NaCl^r$	Suspension	No	Nabors *et al.* (1980)
	MS^r	Protoplasts	?	Carlson (1973)
NaN_3	LT^r	Callus	?	Hibberd and Green (1982)
NEU	$KClO_3$	Suspension	Yes	Müller and Grafe (1978)
	$KClO_3$	Protoplasts	Yes	Márton *et al.* (1982)
	LM^r	Protoplasts	?	Cséplö and Maliga (1982)
MNNG	aux	Protoplasts	Yes	Gebhardt *et al.* (1981b)
	$5MT^r$	Suspension	Yes	Ranch *et al.* (1983)
UV	Val^r	Protoplasts	?	Bourgin (1978)
	INH^r	Suspension	?	Berlyn (1980)
X-ray	Pig.	Protoplasts	Yes	Schieder (1976)
	$PARQ^r$	Callus	?	Miller and Hughes (1980)
γ-ray	aux	Protoplasts	?	Sidorov *et al.* (1981)
	$KClO_3$	Protoplasts	Yes	Márton *et al.* (1982)

[a] Abbreviations: MS = methionine sulfoximine; LT = lysine/threonine; LM = lincomycin; INH = isonicotinic acid hydrazide; Pig. = leaf pigments; PARQ = paraquat; aux = auxotrophy; Val = valine; ETH = ethionine; 5MT = 5-methyltryptophan; EMS = ethyl methane-sulfonate; NEU = *N*-ethyl-*N*-nitrosourea; MNNG = *N*-methyl-*N'*-nitro-*N*-nitrosoguanidine; UV = short-wavelength ultraviolet irradiation; ? = data inconclusive or absent; r = resistant.

Some attention has been given to the dangers of handling chemical mutagens when treating plant cell cultures. For example, a special culture vessel was designed by Shillito *et al.* (1978) for aseptic treatment of suspension culture cells with ethyl methane-sulfonate (Fig. 1). Useful suggestions for the safe handling of mutagenic chemicals have been collated by Ehrenberg and Wachtmeister (1977).

II. APPLICATION OF *N*-METHYL-*N'*-NITRO-*N*-NITROSOGUANIDINE TO PROTOPLASTS

A. Freshly Isolated Protoplasts

N-Methyl-*N'*-nitro-*N*-nitrosoguanidine (MNNG), although a toxic alkylating mutagen, is unstable. It undergoes rapid denaturing reactions at both high and low pH and is readily decomposed by light. Solutions are most stable in the pH range 4.5–6.0. Their half-life at pH 7 is 7.5 hr.

Controlled, short-term MNNG treatment can be applied to small volumes of freshly isolated protoplasts as follows: Place 2–3 ml of protoplasts in wash medium, for example, 0.2 *M* $CaCl_2$, pH 5.6, in a sterile centrifuge tube. Add an equal volume of freshly prepared MNNG in wash medium (with MNNG at twice the required concentration). Seal the tube and lay it on its side for the required exposure time. Centrifuge to collect protoplasts, and wash three times by resuspension in wash medium and centrifugation.

B. Cultured Protoplasts

MNNG is supposedly more effective for yeast when applied to cycling rather than quiescent cells (Carter and Dawes, 1978). Although the data are limited, MNNG appears to be particularly effective when applied to *Hyoscyamus muticus* protoplasts 36 hr after isolation rather than immediately (Gebhardt *et al.*, 1981b). However, considerable loss of protoplasts can be expected if large volumes of protoplast cultures are transferred from

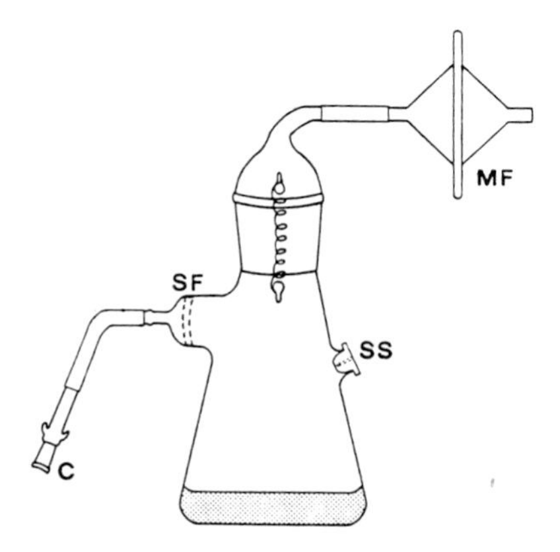

Fig. 1. A special culture vessel which permits the aseptic addition of mutagen via a self-sealing (SS) port using a hypodermic syringe and repeated aseptic washing of the cells with mutagen-free medium. For this purpose, an individual culture vessel is attached to one of the exit ports of a two-channel manifold system via the ground glass joint (C) of the side arm (this is protected from contamination prior to attachment by aluminum foil). Then with the vessel turned to the horizontal and with the side arm downward, the medium is drawn off via the sintered filter (SF) (this retains the cells). The vessel is then returned to the vertical, and the cells are washed off the sinter into the vessel by inflow of a predetermined volume of new mutagen-free medium. This process is repeated four times. Then the side arm is sealed, the culture vessel is returned to the reciprocal shaker, and the culture is incubated for 48 hr to allow recovery of the cells before submitting them to the chosen selection procedure. The microflow air filter (MF) allows entry or withdrawal of medium from the vessel without contamination. The manifold system allows nine such culture vessels to be washed for each autoclaving of the system. The whole system is placed in a fume cupboard and the air from the pump which draws medium from the vessel is led directly into the ventilation shaft. (From Shillito *et al.*, 1978.)

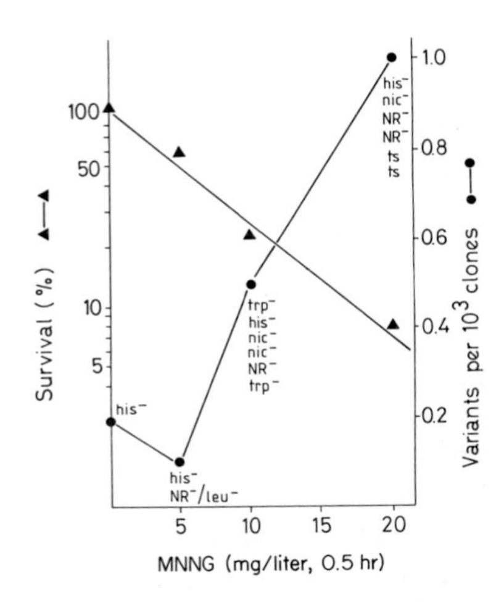

Fig. 2. The effect of MNNG on survival of *Hyoscyamus muticus* protoplasts after 44 hr of culture at 28°C (▲), and the relationship between MNNG dose and the number and type of conditional-lethal variants recovered (●). (Adapted from Gebhardt *et al.*, 1981b.)

Petri dishes to centrifuge tubes for MNNG treatment and washed as above, particularly as protoplasts with regenerating cell walls are prone to stick to the surface of Petri dishes. Because of its rapid loss of activity, protoplasts need not necessarily be washed free of MNNG, and the following procedure is recommended: Culture protoplasts at 10 times the final required density in 1-ml aliquots in Petri dishes. After a time interval sufficient to allow cells to begin cycling, add 1 ml MNNG in culture medium (at two times the required concentration) to each aliquot of cells. After the required exposure time, add 8 ml culture medium, bringing the final volume to 10 ml per dish, and culture further.

Particular attention must be given to the density of protoplasts during the MNNG exposure; the higher the cell density, the smaller the killing effect for any one MNNG concentration. Considerable variations in mutagen sensitivity can be expected between protoplast preparations. For short MNNG treatments, such as 30 min, the 95% kill concentration is about 20–30 mg/liter (Fig. 2). If the mutagen is to be left in the culture medium, the concentration must be reduced about 10-fold.

REFERENCES

Berlyn, M. B. (1980). Isolation and characterization of isonicotinic acid hydrazide-resistant mutants of *Nicotiana tabacum*. *Theor. Appl. Genet.* **58,** 19–26.

Bourgin, J.-P. (1978). Valine-resistant plants from *in vitro* selected tobacco cells. *Mol. Gen. Genet.* **161,** 225–230.

Carlson, P. S. (1973). Methionine sulfoximine-resistant mutants of tobacco. *Science* **180,** 1366–1368.

Carter, B. L. A., and Dawes, I. W. (1978). Nitrosoguanidine mutagenesis during the yeast cell cycle. *Mutat. Res.* **51,** 289–292.

Colijn, C. M., Kool, A. J., and Nijkamp, H. J. J. (1979). An effective chemical mutagenesis procedure for *Petunia hybrida* cell suspension cultures. *Theor. Appl. Genet.* **55,** 101–106.

Cséplö, A., and Maliga, P. (1982). Lincomycin resistance, a new type of maternally inherited mutation in *Nicotiana plumbaginifolia. Curr. Genet.* **6,** 105–109.

Eapen, S. (1976). Effect of gamma- and ultraviolet-irradiation on survival and totipotency of haploid tobacco cells in culture. *Protoplasma* **89,** 149–155.

Ehrenberg, L., and Wachtmeister, C. A. (1977). Handling of mutagenic chemicals: Experimental safety. *In* "Handbook of Mutagenicity Test Procedures" (B. J. Kilbey, ed.), pp. 411–418. Elsevier, Amsterdam.

Gebhardt, C., Strauss, A., and King, P. J. (1981a). Isolation of auxotrophic and temperature-sensitive variants using haploid plant protoplasts. *In* "Induced Mutations—A Tool in Plant Research," pp. 383–397. IAEA, Vienna.

Gebhardt, C., Schnebli, V., and King, P. J. (1981b). Isolation of biochemical mutants using haploid mesophyll protoplasts of *Hyoscyamus muticus.* II. Auxotrophic and temperature-sensitive clones. *Planta* **153,** 81–89.

Hibberd, K. A., and Green, C. E. (1982). Inheritance and expression of lysine plus threonine resistance selected in maize tissue culture. *Proc. Natl. Acad. Sci. U.S.A.* **79,** 559–563.

Márton, L., Dung, T. M., Mendel, R. R., and Maliga, P. (1982). Nitrate reductase deficient cell lines from haploid protoplast cultures of *Nicotiana plumbaginifolia. Mol. Gen. Genet.* **182,** 301–304.

Miller, O. K., and Hughes, K. W. (1980). Selection of paraquat-resistant variants of tobacco from cell cultures. *In Vitro* **16,** 1085–1091.

Müller, A. J., and Grafe, R. (1978). Isolation and characterisation of cell lines of *Nicotiana tabacum* lacking nitrate reductase. *Mol. Gen. Genet.* **161,** 67–76.

Nabors, M. W., Gibbs, S.E., Bernstein, C. S., and Meis, M. E. (1980). NaCl-tolerant tobacco plants from cultured cells. *Z. Pflanzenphysiol.* **97,** 13–17.

Negrutiu, I. (1981). Improved conditions for large-scale culture, mutagenesis and selection of haploid protoplasts of *Nicotiana plumbaginifolia* Viviani. *Z. Pflanzenphysiol.* **104,** 431–442.

Ohyama, K., Pelcher, L. E., and Gamborg, O. L. (1974). The effects of ultra-violet irradiation on survival and on nucleic acid and protein synthesis in plant protoplasts. *Radiat. Bot.* **14,** 343–346.

Ranch, J. P., Rick, S., Brotherton, J. E., and Widholm, J. M. (1983). The expression of 5-methyltryptophan-resistance in plants regenerated from resistant cell lines of *Datura innoxia. Plant Physiol.* **71,** 136–140.

Reisch, B., and Bingham, E. T. (1981). Plants from ethionine resistant alfalfa tissue cultures: Variation in growth and morphological characteristics. *Crop Sci.* **21,** 783–788.

Schieder, O. (1976). Isolation of mutants with altered pigments after irradiating haploid protoplasts from *Datura innoxia* Mill. with X-rays. *Mol. Gen. Genet.* **149,** 251–254.

Shillito, R. S., Robinson, N. E., and Street, H. E. (1978). Isolation and characterisation of mutant cell lines via plant cell cultures. *Proc. Int. Symp. Exp. Mutagen. Plants, 1976.*

Sidorov, V., Menczel, L., and Maliga, P. (1981). Isoleucine-requiring *Nicotiana* plant deficient in threonine deaminase. *Nature (London)* **294,** 87–88.

Werry, P. A. T. (1981). Induction by ionizing radiation of genetic markers for the development of *in vitro* genetic manipulation as a tool in crop plant improvement. *In* "Induced Mutations—A Tool in Plant Research," pp. 373–382. IAEA, Vienna.

Werry, P. A. T., and Stoffelsen, K. M. (1979). The effect of ionizing radiation on the survival of free plant cells cultivated in suspension cultures. *Int. J. Radiat. Biol. Relat. Stud. Phys., Chem. Med.* **35,** 293–298.

CHAPTER **62**

Cell Culture Procedures for Mutant Selection and Characterization in *Nicotiana plumbaginifolia*

Pal Maliga*

Institute of Plant Physiology
Biological Research Center
Hungarian Academy of Sciences
Szeged, Hungary

I. INTRODUCTION

Tissue culture has proven its value for the isolation of mutants in flowering plants. The use of protoplasts for mutant selection has many advantages over the use of other available tissue culture systems, such as callus cultures and suspension culture cells. Most experiments on mutant selection in protoplast cultures were carried out in a relatively small number of species in which protoplast culture is easy, and plant regeneration from cultured cells is routine. *Nicotiana plumbaginifolia* is one of these species.

*Present address: Advanced Genetic Sciences, Inc., P.O. Box 1373, Manhattan, Kansas 66502.

ISBN 0-12-715001-3

Other species used, and the problems of mutant selection in plant tissue culture, have been reviewed elsewhere (Maliga *et al.*, 1982a; Maliga, 1983; Bourgin, 1983).

In this chapter, plant-protoplast-plant tissue culture procedures which have been used to isolate and characterize a number of mutants, including auxotrophic, antibiotic-resistant, and pigment-deficient lines, will be summarized (Table I). The methods described have also been successfully applied in the related species *Nicotiana tabacum* (see references in Table I).

TABLE I

Genetic Lines Isolated in *Nicotiana plumbaginifolia* Cell Culture

Lines	References
Mutants	
Auxotrophs	Sidorov *et al.* (1981a); Sidorov and Maliga (1982)
Isoleucine (ILE401), leucine (LEU403)	
Uracil (URA401)	
Chlorate resistants	Márton *et al.* (1982a,b); Negrutiu *et al.* (1983)
Defective in the nitrate reductase apoenzyme (NA1, NA2, NA9, NA18, NA36)	
Defective in the molybdenum cofactor (NX1, NX9, NX21, NX24)	
Antibiotic resistants	Cséplö and Maliga (1982); P. Maliga (unpublished)
Lincomycin r. (LR400)	
Streptomycin r. (SR402)	
Chlorophyll deficients (e.g. A28)	Sidorov and Maliga (1982)
Lines with substituted cytoplasm	
Chloroplasts from the *N. tabacum* mutant SR1, e.g., Np3 (SR1); Np15 (SR1); inherit streptomycin resistance maternally	Maliga *et al.* (1982b); Menczel *et al.* (1982)
Chloroplasts from the *N. sylvestris* line CL105, e.g., Np47 (CL105); inherit lincomycin resistance maternally	A. Cséplö (unpublished)
Chloroplasts from a light-sensitive *N. tabacum* mutant	Sidorov *et al.* (1981b)
Somatic hybrids	
Intraspecific	
Hybrids of complementing nitrate reductase-deficient lines	Márton *et al.* (1982b)
Hybrids of auxotrophic- and pigment-deficient lines	Sidorov and Maliga (1982)
Interspecific	
N. tabacum and *N. plumbaginifolia*	Sidorov *et al.* (1981b); Maliga *et al.* (1982b); Menczel *et al.* (1982)
N. sylvestris and *N. plumbaginifolia*	A. Cséplö (unpublished)

II. PROCEDURES

A. Maintenance of Aseptically Grown Plants

Plants grown under aseptic conditions are a convenient source of protoplasts where they can be maintained for a long time (Section II,A,2). Aseptic plant cultures can be initiated from surface-sterilized seeds (Section II,A,1) or by propagating haploid plants derived from anther culture (Chapters 34–37, this volume).

1. Seed Sterilization and Germination

Place seeds in a sterile test tube. Wet the seeds with a drop of 70% ethanol, and sterilize them by adding 0.5% sodium hypochloride (3 min). Decant the sterilizing solution and wash the seeds five times in sterile distilled water. Let the seeds stand in the water (3 min) during the last two washes. Soak the seeds (1 hr) in gibberelic acid (GA_3; 0.5 mg/ml, filter sterilized), and transfer them into a Petri dish containing a wet (sterile distilled water) filter paper. Alternatively, seeds may be germinated on soft (0.6%) agar medium prepared with distilled water (Maliga *et al.*, 1982b). In order to obtain uniform germination, GA_3 treatment may also be necessary before sowing seeds in the greenhouse. In this case, there is no need for seed sterilization. GA_3 treatment is not needed in *N. tabacum*.

2. Maintenance of Plants in Culture

Plants can be maintained in sterile culture on RM salts (Linsmaier and Skoog, 1965) plus 3% sucrose (Table II). RM salts can be replaced by the commercial Murashige and Skoog (1962) salt mixture (MS salt). Incubate cultures at 28°C, 70% relative humidity, and illuminate them with 1000–3000 lx. Under short days (8 hr of illumination), cuttings form rosette plants with large leaves. Under long days (16 hr), cuttings form a flower stem on which fertile flowers and seeds are formed.

Plants can be multiplied by cuttings. To obtain a new plant, it is sufficient to insert a short stem with a single leaf into the medium. For rapid propagation, because of the long internodes, plants grown under long days are best. Supplementation of the medium with thiamine and inositol improves growth but induces polyploidy. Diploid plants occasionally appearing in haploid cultures can be identified by wider leaves (Sidorov and Maliga, 1982).

TABLE II

Media for Callus Culture and Plant Regeneration[a]

Use	Code	Thiamine (mg/liter)	Inositol (mg/liter)	BAP[b] (mg/liter)	NAA[c] (mg/liter)	IAA[d] (mg/liter)
Plant maintenance	RM	—	—	—	—	—
Callus culture—shoot induction	RMO	1	100	0.5	—	2
Shoot induction	RMOP	1	100	1	0.1	—
Shoot induction	RMB	1	100	1	—	—
Shoot elongation and root induction	P	—	—	—	—	—

[a] The medium in each case contained MS (RM) salts and 3% sucrose, and the pH was adjusted to 5.8. The only exception was the P medium, in which case the concentrations of KNO_3, NH_4NO_3, and $MgSO_4$ were reduced to one-fifth that of the MS (RM) salts.
[b] 6-Benzylaminopurine.
[c] 1-Naphthaleneacetic acid.
[d] Indoleacetic acid.

B. Protoplast Isolation

Genetic variability in leaf cells is low (compared to tissue culture cells). Therefore, leaf cells are an ideal source of protoplasts for mutant selection. Protoplast isolation from callus (or suspension culture cells) may be necessary for different purposes (e.g., organelle transfer); therefore, a short note on that topic will also be included (Section II,B,2).

1. Isolation of Leaf Mesophyll Protoplasts

Protoplasts may be prepared from the leaves of greenhouse plants or plants grown in sterile culture. Use of sterile plants is recommended since conditions for them can be better standardized. For highest plating efficiency, select the first two fully expanded leaves. If the plants are in good condition, older leaves may also give good results.

With greenhouse plants, surface sterilization of the leaves is necessary. In order to remove soil and other materials, wash the leaves under running tap water (30 sec). Soak the leaves in tap water containing 2 drops of a commercial detergent (3 min). Rinse the leaves under running tap water in order to remove the detergent (30 sec), dip them into 70% ethanol, and transfer them into 0.5% sodium hypochloride for sterilization (3 min). Rinse the leaves with sterile distilled water (five times) and proceed as with leaves from sterile plants.

Cut the leaves (greenhouse or from sterile culture) into narrow sections (1 mm) and place the sections (0.5 g/10 ml) into K_3 medium (Table III) containing 0.4 *M* sucrose as an osmotic stabilizer and the cell wall-degrading enzymes Cellulase Onozuka R-10 (1.5%) plus Macerozyme R-10 (0.2%), or Driselase (0.5%). Digestion should be carried out overnight (16–18 hr) at 28°C in darkness. Protoplasts are ready for purification if the leaf sections disintegrate after two to three gentle shakes.

In order to remove undigested leaf material, filter the protoplasts through a mesh (63 μm). Spin the suspension in a swing-out bucket at low speed (100–300 *g* for 2–3 min). In sucrose the protoplasts move to the surface. Remove the protoplasts from the top with a Pasteur pipette and transfer them into a clean tube. Add 10 volumes of W5 solution (154 m*M* NaCl; 125 m*M* $CaCl_2$; 5 m*M* KCl; 5 m*M* glucose; pH 5.6; Table IV) and spin at low speed (50–100 *g*). The protoplasts in W5 solution should sediment. Remove the supernatant, gently shake the tube to resuspend the protoplasts, and repeat the washing in W5 solution. (If the protoplasts are sensitive and burst, the second wash should be omitted.) Protoplast isolation from leaves is described elsewhere in more detail (Menczel *et al.*, 1981; Sidorov *et al.*, 1981b; Sidorov and Maliga, 1982; see Chapters 38–43, this volume).

2. Isolation of Callus Protoplasts

Protoplasts of nitrate reductase-deficient mutants were isolated from fast-growing callus on RMOP medium (for composition, refer to Table II) supplemented with 8.25 m*M* ammonium succinate. Protoplasts from non-mutant lines can be isolated from any fast-growing calli maintained, for example, on RMO medium. Callus should be incubated in K_3 medium (Table III) containing 0.4 *M* sucrose as an osmotic stabilizer and Onozuka Cellulase R-10 (1.5%) plus Macerozyme R-10 (0.5%). Protoplast isolation should be carried out as described for leaf mesophyll protoplasts in Section II,B,1.

C. Mutagenic Treatment

1. Treatment of Protoplasts by *N*-Ethyl-*N*-Nitrosourea

The mutagen *N*-ethyl-*N*-nitrosourea (NEU) was dissolved in the protoplast culture medium (K_3 medium with 0.4 *M* glucose as an osmotic stabilizer; Table III) and added to freshly isolated protoplasts which were in

TABLE III

The K_3 Protoplast Culture Medium

A. Stocks

I.	For 1 liter
$NaH_2PO_4 \cdot 2H_2O$	1.5 g
$CaCl_2 \cdot 2H_2O$	9.0 g
KNO_3	25.0 g
NH_4NO_3	2.5 g
$(NH_4)_2SO_4$	1.3 g

II. In 100 ml H_2O dissolve 557 mg $FeSO_4 \cdot 7H_2O$ and 745 mg Na_2EDTA

III. In 100 ml H_2O dissolve 1.25 mg $CuSO_4 \cdot 7H_2O$ and 1.25 mg $CoCl_2 \cdot 6H_2O$

IV.	For 100 ml
KI	7.5 mg
$MnSO_4 \cdot H_2O$	100.0 mg
$ZnSO_4 \cdot 7H_2O$	20.0 mg
H_3BO_3	30.0 mg
$Na_2MoO_4 \cdot 2H_2O$	2.5 mg
$MgSO_4 \cdot 7H_2O$	2500.0 mg

V.	For 100 ml
Thiamine·HCl	100 mg
Nicotinic acid	10 mg
Pyridoxine·HCl	10 mg
m-Inositol	1000 mg
Xylose	2500 mg

VI.	
2,4-Dichlorophenoxyacetic acid (2,4-D)	0.1 mg/liter
6-BAP	0.2 mg/liter
1-NAA	1.0 mg/liter

Dissolve separately 1 mg in 1 ml of 0.1 *M* KOH (2,4-D, NAA) or in 0.1 *M* HCl (BAP)

Storage of stocks: Stocks I, III, and IV can be autoclaved and stored at room temperature. Stock V should be filter sterilized and kept in the refrigerator (+4°C), or stored at −20°C. Store stocks II and VI in the refrigerator as is.

B. Preparation of the culture medium

	For 1 liter
Stock I	100 ml
Stock II	5 ml
Stock III	2 ml
Stock IV	10 ml
Stock V	10 ml
Stock VI	See above and Section II,D
Glucose or sucrose	See Section II,D

Before filter sterilizing, adjust the pH to 5.8 with KOH

C. Amount of osmotic stabilizer (g) for 100 ml

Moles/liter	0.1	0.2	0.3	0.4	0.5	0.6	0.7	0.8
Sucrose	3.42	6.84	10.26	13.68	17.30	20.72	24.14	27.56
Glucose	1.80	3.60	5.40	7.21	9.01	10.81	12.62	14.42
Mannitol	1.82	3.64	5.46	7.28	9.11	10.93	12.75	14.57

TABLE IV

Solutions for Protoplast Fusion and Purification

I. W5 solution for protoplast washing	For 500 ml
NaCl	4.5 g
Glucose	0.5 g
$CaCl_2 \cdot 2H_2O$	9.2 g
KCl	0.2 g
Sterilize by autoclaving; pH as is (5.6)	

II. W10 solution for postfusion washing	For 9 ml
A. Glucose	1.0 g
$CaCl_2 \cdot 2H_2O$	0.1 g
DMSO	1.0 ml
H_2O	7.0 ml
B. Glycine (0.3 *M*) dissolved in H_2O; pH adjusted to 10.5 with NaOH	
Filter-sterilize. Mix (9 parts stock A to 1 part stock B) only before use.	

III. PEG 40 for fusion induction	
Polyethylene glycol (MW: 6000)	8.0 g
Glucose	1.2 g
$CaCl_2 \cdot 2H_2O$	0.2 g
H_2O	11.0 ml
Autoclave for 15 min; pH as is (around 6.0)	

the same culture medium. The protoplasts were incubated until small calli grew. The mutagen was not washed out.

By the use of this mutagen, nitrate reductase-deficient (Márton *et al.*, 1982a), lincomycin-resistant (Cséplö and Maliga, 1982), and streptomycin-resistant (P. Maliga, unpublished) lines were obtained in protoplast culture. Use of a dose higher than 0.1 m*M* NEU is not advisable since it was found to induce sterility in the regenerated lincomycin-resistant plants (A. Cséplö, unpublished).

2. Treatment of Protoplasts by Gamma Irradiation (^{60}Co)

Freshly isolated protoplasts were suspended in the K_3 protoplast culture medium containing 0.4 *M* sucrose as an osmotic stabilizer (1–5 $\times$ 10^5 protoplasts per milliliter), irradiated in plastic screw-cap tubes in 10-ml batches using a ^{60}Co source (0.042 Gy/sec), floated by centrifugation, collected, and cultured in fresh protoplast culture medium (Márton *et al.*, 1982a). Survival data (Sidorov and Maliga, 1982) have been published after this treatment. Use of gamma-irradiated protoplasts in organelle transfer is described by Menczel (Chapter 49, this volume).

D. Protoplast Culture

High plating efficiencies (60–90%) can be obtained in both haploid and diploid protoplast cultures using K_3 medium with 0.4 *M* glucose (0.4 MG/K_3 medium) as an osmotic stabilizer. This medium was originally described by Kao *et al.* (1974), and in various modified forms (Nagy and Maliga, 1976) it is used as the general protoplast culture medium in our laboratory. The composition of the medium is given in Table III.

Sucrose (0.4 *M*) and mannitol (0.4 *M*) can also be used as osmotic stabilizers. If mannitol is used, 1% sucrose should be included in the medium, since mannitol is not utilized by *Nicotiana* cells as a carbon source.

Incubate the cultures at 28°C, 70% relative humidity. If the protoplast preparation is poor, the protoplasts burst upon exposure to light. For best results, first keep the cultures in the dark (3 days), and then transfer them to low light (100–500 lx).

Protoplasts can be cultured in liquid medium or directly plated in soft agar (0.6%). For plating in agar, mix 1 volume of liquid medium (containing the protoplasts) with 1 volume of cooled (46°C) agar medium (1.2%). Protoplasts in poor condition will burst upon plating in agar. It is best, therefore, to culture the protoplasts initially in liquid medium for 3–4 days at higher than optimum densities (10^5 protoplasts per milliliter) and then plate them in agar medium (same osmolarity, as explained below). The optimum protoplast plating density is around 4×10^4/ml (pale green color barely seen over a white background). Protoplasts plated at much higher or lower densities will not divide.

As colonies grow in size, they need more space and fresh medium. The need for dilution is indicated by the slowed growth and increase in cell size. Liquid cultures can be diluted by increasing the volume of culture. Fresh medium can be supplied by draining and replacing the liquid medium if one does not want to decrease the cellular density. Colonies in agar medium (e.g., 10 ml in a plate) can be diluted after preparing a fine suspension in liquid medium (50 ml) by repeated suction through a wide-mouth pipette. Mix the suspension with melted agar medium (50 ml, 1.2% agar) and pour 10 plates.

The concentration of glucose, the osmotic stabilizer, should be reduced during dilution in a stepwise manner. If the osmolarity is decreased too quickly, the cells die (cells plasmolyze by the next day and the cytoplasm turns brown). Protoplasts, however, can be grown into large calli without decreasing the concentration of the osmotic stabilizer (0.4 *M* glucose) in the culture medium, although the growth at this osmolarity is slower.

If the initial protoplast density is too high, the cultures should be diluted 4–10 times with a medium having the same osmolarity (0.4 MG/K_3 medium) during the first 10–15 days. Assuming that the protoplasts start dividing on the third or fourth day, the normal procedure would be dilution (5–10 times) of the initial culture (in 0.4 MG/K_3 medium) after 3–4 weeks directly onto 0.2 MG/K_3 medium. After an additional 3–5 weeks, the colonies are large enough to be transferred individually to a callus culture medium (Table II and Section II,E). A more careful approach would be to decrease the osmolarity in more steps, for example, to dilute the culture on day 10 fourfold with 0.4 mg/K_3 medium, on day 20 twofold with 0.3 MG/K_3, on day 30 twofold with 0.2 MG/K_3, and on day 40 tenfold with 0.2 MG/RMOP medium. Calli from this medium can be directly transferred to the regular callus culture medium.

E. Callus Culture and Plant Regeneration

Callus cultures are maintained on the RM basal medium, which contains RM (MS) salts, thiamine (1 mg/liter), inositol (100 mg/liter), sucrose (3%) and 0.7% agar, an auxin, and a cytokinin. Various modifications of this basal medium are summarized in Table II.

Callus cultures are induced and maintained on the RMO medium, on which the callus turns green and forms shoots. Shoot regeneration was obtained on RMOP medium (RM basal medium plus benzyladenine, 1 mg/liter; 1-NAA, 0.1 mg/liter) or on RMB medium (the same as RMOP, but with NAA omitted). If no shoots are formed on RMOP medium, try to subculture at least two more times on RMB medium.

Shoots formed on RMO, RMOP, or RMB media are short, and the leaves may be succulent. They will grow into normal plants (with roots) after transfer to the low-nitrogen P medium (RM salts but KNO_3, NH_4NO_3, and $MgSO_4$ reduced to one-fifth; sucrose 3%; agar 0.7%); (Table II). P medium can probably be replaced with any low nitrogen-low salt medium such as White's salts plus 3% sucrose. Since the shoots are at first small, one may need to transfer them to P medium with some callus at the base. As soon as the shoots are long enough (10 mm), cut them off and insert them into the same medium for rooting. The P medium is not suitable for long-term maintenance of plants.

Freshly rooted cuttings (four to five leaves, roots 3–4 cm long) are best for potting in soil. Before potting, remove agar from the roots. Initially (2 weeks), cover the plants with a transparent plastic foil and keep them in shade.

Incubate cultures at 28°C, 70% relative humidity, and illuminate them for 16 hr (500–2000 lx).

F. Protoplast Fusion Induction

Place 2 drops of a protoplast mixture into a plastic dish (35 mm in diameter) and allow the cells to settle for 20 min. Add a drop of PEG 40 solution (40% polyethylene glycol, MW = 6000; 0.3 *M* glucose; 66 m*M* $CaCl_2$; pH 6; Table IV). From 3 to 15 min later (treatment should be shorter if protoplasts burst), remove the liquid with a pipette and overlay the cells with 2 drops of W10 solution (Table IV). The W10 solution should be freshly prepared by mixing 9 parts of stock A (0.4 *M* glucose, 66 m*M* $CaCl_2$, 10% dimethylsulfoxide) with 1 part of stock B (0.3 *M* glycine-NaOH buffer, pH 10.5), both filter sterilized. After 20 min, dilute W10 solution by 1 ml K_3 medium containing 0.4 *M* glucose. Replace the medium 10 min later (or after 24 hr if the protoplasts are sensitive) by fresh K_3 medium. The method was described by Menczel *et al.* (1981).

III. RESULTS AND CONCLUSIONS

The tissue culture methods described for *N. plumbaginifolia* are based on a number of publications (Table I) and represent the version which we have tried and think is best. The list includes isolation of mutants and somatic hybrids, and transfer of chloroplasts and the cytoplasmic male sterility factor. Some of the methods described above are also mentioned in this volume in Chapters 49 and 58. Our experience, and that of others (see Bourgin, 1983), indicates that this species may be well suited as a model species for cellular genetic studies.

REFERENCES

Bourgin, J.-P. (1983). Selection of tobacco protoplast-derived cells for amino acids and regeneration of resistant plants. *In* "Genetic Engineering in Eucaryotes" (P. F. Lurguin and A. Kleinhofs, eds.), pp. 195–214. Plenum, New York.

Cséplö, A., and Maliga, P. (1982). Lincomycin resistance, a new type of maternally inherited mutation in *Nicotiana plumbaginifolia*. *Curr. Genet.* **6,** 105–110.

Kao, K. N., Constabel, N. F., Michayluk, M. R., and Gamborg, O. L. (1974). Plant protoplast fusion and growth of intergeneric hybrids. *Planta* **120,** 215–227.

Linsmaier, E. M., and Skoog, F. (1965). Organic growth factor requirements of tobacco tissue cultures. *Physiol. Plant.* **18,** 100–127.

Maliga, P. (1983). Protoplasts in mutant selection and characterization. *Int. Rev. Cytol., Suppl.* **16,** 161–167.

Maliga, P., Menczel, L., Sidorov, V. A., Márton, L., Dung, I. M., Lázár, G., Cséplö, A., Medgyesy, P., and Nagy, F. (1982a). Cell culture mutants and their uses. *In* "Plant Improvement and Somatic Cell Genetics" (I. K. Vasil, K. J. Frey, and W. R. Scowcroft, eds.), pp. 221–237. Academic Press, New York.

Maliga, P., Lörz, H., Lázár, G., and Nagy, F. (1982b). Cytoplast-protoplast fusion for interspecific chloroplast transfer in *Nicotiana. Mol. Gen. Genet.* **185,** 211–215.

Márton, L., Dung, T. M., Mendel, R. R., and Maliga, P. (1982a). Nitrate reductase deficient cell lines from haploid protoplast cultures of *Nicotiana plumbaginifolia. Mol. Gen. Genet.* **186,** 301–304.

Márton, L., Sidorov, V. A., Biasini, G., and Maliga, P. (1982b). Complementation in somatic hybrids indicates four types of nitrate reductase deficient lines in *Nicotiana plumbaginifolia. Mol. Gen. Genet.* **187,** 1–3.

Menczel, L., Nagy, F., Kiss, Z. S., and Maliga, P. (1981). Streptomycin resistant and sensitive somatic hybrids of *Nicotiana tabacum* and *Nicotiana knightiana:* Correlation of resistance to *N. tabacum* plastids. *Theor. Appl. Genet.* **59,** 191–195.

Menczel, L., Galiba, G., Nagy, F., and Maliga, P. (1982). Effect of radiation dosage on the efficiency of chloroplast transfer by protoplast fusion in *Nicotiana. Genetics* **100,** 487–495.

Murashige, T., and Skoog, F. (1962). A revised medium for rapid growth and bioassay with tobacco tissue cultures. *Physiol. Plant.* **15,** 437–497.

Nagy, J. I., and Maliga, P. (1976). Callus induction and plant regeneration from mesophyll protoplasts of *Nicotiana sylvestris. Z. Pflanzenphysiol.* **78,** 453–455.

Negrutiu, I., Dirks, R., and Jacobs, M. (1983). Regeneration of fully nitrate reductase-deficient mutants from protoplast culture of *Nicotiana plumbaginifolia. Theor. Appl. Genet.* **66,** 341–347.

Sidorov, V. A., and Maliga, P. (1982). Fusion-complementation of auxotrophic and chlorophyll-deficient lines isolated in haploid *Nicotiana plumbaginifolia* protoplast cultures. *Mol. Gen. Genet.* **186,** 328–332.

Sidorov, V. A., Menczel, L., and Maliga, P. (1981a). Isoleucine-requiring *Nicotiana* plant deficient in threonine deaminase. *Nature (London)* **294,** 87–88.

Sidorov, V. A., Menczel, L., Nagy, F., and Maliga, P. (1981b). Chloroplast transfer in *Nicotiana* based on metabolic complementation between irradiated and iodoacetate treated protoplasts. *Planta* **152,** 341–345.

CHAPTER **63**

Induction, Selection, and Characterization of Mutants in Carrot Cell Cultures

J. M. Widholm

Department of Agronomy
University of Illinois
Urbana, Illinois

I. INTRODUCTION

Carrot cells have been favorite tissue culture objects since the early demonstration of embryogenesis and subsequent plant formation from cultured cells by Steward *et al.* (1958). This ability, coupled with the capability of forming fine suspensions which can grow rapidly and be plated in solidified medium easily, has led to the use of carrot cells in many mutant selection experiments. Sung and Dudits (1981) have presented a comprehensive summary of somatic cell genetic studies with carrot. In this chapter, only a few selection experiments will be described in order to illustrate some of the many possible selection and characterization methods.

II. PROCEDURE

The selection for resistance is the simplest selection method. In the presence of an inhibitor, sensitive cells will not grow, but resistant ones in the

CELL CULTURE AND SOMATIC CELL
GENETICS OF PLANTS, VOL. 1

ISBN 0-12-715001-3

population do and thus can be selected from the general population. Growth-inhibitory amino acid analogs have been used in this laboratory to select many resistant carrot lines. The first selection utilized the tryptophan analog 5-methyltryptophan (5MT) (Widholm, 1972a). The first step in such selections is to determine the 5MT concentration which completely inhibits growth. In this case, 0.33 g fresh weight of cells was inoculated into 100 ml liquid medium [basal Murashige and Skoog (MS) medium (1962) with 0.4 mg/liter 2,4-dichlorophenoxyacetic acid (2,4-D) as auxin] containing a range of 5MT concentrations. The fresh weight of the cells added must be kept constant in such growth studies in order to have reproducible results. To determine the volume of cells to add, an aliquot is removed from the donor flask with a cut-off or wide-bore pipette and the cells are collected on a filter in a Buchner funnel under vacuum. The cells are scraped off the filter to be weighed. From the pipetted volume and the measured fresh weight, the volume to be pipetted into each flask to give 0.33 g fresh weight can be calculated.

The inoculated flasks were incubated on a reciprocating shaker for 10 days at 27–28°C and then were weighed by collecting the cells from each flask on a filter and determining the fresh weight. Such experiments showed that complete growth inhibition could be accomplished with 44 μM 5MT. Screening for resistant cells was carried out by inoculating 0.33 g fresh weight of cells into 100 ml 220-μM 5MT medium in 250-ml Erlenmeyer flasks and incubating for up to 60 days. Most of the flasks showed growth within this time. An estimate of the frequency of resistant cells in the original population was made by inoculating 10 mg fresh weight carrot cells (1.8×10^5 cells) in 10 ml of 220-μM 5MT medium in 25-ml Erlenmeyer flasks (Widholm, 1977a). Growth occurred in 2 of 63 flasks, which gives a resistance frequency of 1.8×10^{-7} assuming that there was one resistant cell present in each of the two flasks where growth occurred. Two mutagen treatments were also used in these experiments to increase the resistance frequency, as described below.

One gram fresh weight of cells was inoculated into 100 ml fresh liquid medium and was incubated for 2 days to allow the cells to reach early log phase. Then the mutagenic alkylating agent ethylmethane sulfonate was added to a concentration of 0.25% for 2–3 hr, after which the flasks were poured onto autoclaved Miracloth filters taped into plastic funnels. The collected cells were rinsed with fresh medium. Then a hole was cut in the bottom of the filter with a sterile scalpel, and the cells were rinsed with fresh medium into a clean flask. Rapidly growing cells (10 ml) were also placed in an open 10-cm Petri dish in a laminar flow hood 7 cm beneath an ultraviolet lamp (Mineralight model R-51, Ultra-Violet Products, Inc., San Gabriel, California) for periods of up to 10 min. Cells treated by both methods were examined microscopically 24 hr after the treatments. The excluded viability dye phenosafranine was used to measure viability (Widholm, 1972b). A drop of 0.1% phenosafranine in the culture medium was mixed with a drop of cells on a microscope slide. Cells which take up the dye and are stained red are dead, whereas cells whose membranes are intact exclude the dye and remain colorless. Fluorecein diacetate can also be used to detect live cells using a fluorescence microscope (Widholm, 1972b).

The desirable mutagen treatments usually kill ca. 50–70% of the cells. The treated cultures are allowed to grow for two or three cell divisions to fix the mutations, to recover from the treatments, and to dilute out dead cells. At this time, 1.8×10^5 cells were inoculated into 10 ml of medium containing 200 μM 5MT. Within the 60-day incubation period the untreated cells, as noted above, showed growth in 2 of 63 flasks, for a resistance frequency of 1.8×10^{-7}. Cells treated with ethylmethane sulfonate showed growth in 42 of 70 flasks (3.3×10^{-6}), whereas ultraviolet light-treated cells grew in 22 of 70 flasks (1.7×10^{-6}). Thus, in these experiments, known mutagenic treatments increased the 5MT resistance frequency by 10- to 20-fold, but it should also be noted that resistant cells can be selected from unmutagenized cell populations.

When growth occurred in the inhibitory selection medium, the cells were characterized in several ways. Usually the cells were reinoculated into the selection medium to determine if they were indeed resistant and were able to grow rapidly in this medium. Since selection in liquid medium may produce a mixture of resistant and sensitive cells, cloning or growing a single cell into a colony is desirable to produce a genetically identical population. One cell-cloning method involves picking out single cells with a micropipette under a dissecting microscope and placing them individually on a filter paper square placed on top of a normal carrot nurse culture callus, as was done with *P*-fluorophenylalanine- (PFP-) resistant carrot cells (Palmer and Widholm, 1975). After the cell had grown into a colony ca. 2 mm in diameter, it was placed into a small volume of liquid medium on the shaker. When this had grown sufficiently, the total volume was transferred into a larger volume of medium.

We have also used another cloning method with carrot cells which is based on the assumption that suspension-cultured cells divide at similar rates and that the clumps of cells break apart randomly with no reassociation (Hauptmann and Widholm, 1982). Thus, after so many cell generations (divisions), all cells in small cell clumps obtained by filtration through stainless steel filters should be of one type at a high probability. The model predicts that an individual clump will have originated from one cell (significant at the 0.05 level) if a 3-cell clump has grown for 19 generations, whereas a 13-cell clump must have grown for 58 generations. The filtered clumps were plated thinly on filter paper suspended above feeder cells in liquid medium by a stainless steel screen (Weber and Lark, 1979). This feeder plate system allows the cells to grow even at a low plating density. Other cell-cloning methods are also available; they will not be discussed here.

Cloned lines or cultures selected in liquid medium are tested for resistance to the inhibitor by inoculating them into a range of concentrations and determining the increase in fresh weight after a certain incubation period, as was done initially to find the completely inhibitory concentration. It is also important to determine the stability of the resistance by growing the cells for a significant period (50 generations) in a medium lacking the inhibitor. In general, we have found the carrot cell amino acid analog resistance to be stable and to be from 10- to 1000-fold above that of the wild-type cells.

Most of the selection we have done with carrot cells has used suspension cultures inoculated into completely inhibitory levels of inhibitor. In two studies, however, cells were inoculated into levels of the herbicides 2,4-D or glyphosate [*N*-(phosphonomethyl)glycine], which were only partially inhibitory (J. L. Killmer, E. D. Nafziger, and J. M. Widholm, unpublished). After the cells grew up to stationary phase, they were inoculated into a slightly higher concentration. This was continued for over a year until the line on 2,4-D showed increased tolerance of ca. 10-fold and the line on glyphosate 50-fold. The 2,4-D resistance was lost when the cells were grown away from the inhibitor for two transfers, but the glyphosate resistance was stable. This gradual selection might be expected to select for the gradual increase in, for instance, a detoxification enzyme which might not be stable.

Selection for resistance can also be carried out by plating cells in or on agar-solidified medium containing the inhibitor. We have been more successful using liquid medium for selecting resistant carrot cells, but carrot lines which are auxin autotrophic, that is, do not require an auxin for growth, have been selected when plated in solid medium (Widholm, 1977b). In these experiments, carrot cells were collected on sterile Miracloth filters and rinsed with 2,4-D–free medium. A hole was cut in the bottom of the filter with a sterile scalpel, and the cells were rinsed into a clean flask with 2,4-D–free medium. The cells were diluted to 0.5 g fresh weight in 100 ml medium, and 10 ml of this was pipetted into a 10-cm Petri dish. Then 10 ml of 2,4-D–free medium containing 1.2% molten agar at 45°C was added and mixed by rocking the plates with a circular motion. After cooling, the plates were sealed with Parafilm and incubated for ca. 45 days until some colonies formed.

Cell mutants can be characterized genetically if fertile plants can be regenerated from them. Although carrot cells are usually easily regenerated into complete plants, few of the selected lines have been studied in this way. In our laboratory, 32 plants were regenerated from one 5MT-resistant suspension culture (Widholm, 1974). Plant regeneration was induced by plating 0.5 ml of the suspension culture on top of 50 ml of MS medium containing 1 mg/liter indoleacetic acid (IAA) and kinetin and 1% agar in 125-ml Erlenmeyer flasks. After incubation for 1 month under continuous fluorescent light, the embryoids which began to form were placed onto fresh medium for complete plant formation. These plants were removed carefully from the agar medium and placed in autoclaved soil mix in pots. The plants were initially placed under low light and were covered with an inverted beaker or clear plastic bag to prevent desiccation. Once the plants are established, the light can be increased and the beaker or bag removed. Although the method used above did induce plant regeneration, the one used by Breton and Sung (1982) is more efficient. They removed the large clumps from a cell suspension culture with a 200-μm nylon filter and rinsed the smaller clumps by centrifugation into 2,4-D–free medium. Embryoids formed quickly in the 2,4-D–free medium, and complete plant formation occurred when these embryoids were placed on agar-solidified 2,4-D–free medium.

In our experiments (Widholm, 1974), in order to follow the 5MT resistance, cell cultures were initiated from the regenerated plants by surface sterilizing a 2-cm segment of a petiole in 1% sodium hypochlorite for 5 min, followed by rinsing at least two times in sterile distilled water. These pieces were then placed on agar-solidified MS medium containing 0.4 mg/liter 2,4-D. The callus which formed was placed in liquid medium to form suspension cultures, which were then tested for resistance to 5MT. In these studies, cultures initiated from 31 of the 32 regenerated plants retained resistance to 5MT. The 5MT resistance in this case was not due to the usual mechanism of tryptophan overproduction caused by a feedback-altered anthranilate synthase; rather, it was due to decreased uptake of 5MT as determined using [^{14}C]5MT (Widholm, 1974).

Although we wanted to determine if the 5MT resistance was inherited by progeny, a summary of our attempt to do this will show the difficulties of using carrots in genetic studies. First, carrot is a biennial, so the roots must be given a cold treatment to induce flowering. Roots which had grown to a diameter of ca. 3 cm were removed from the soil and washed; the tops were removed, and the roots were placed in a paper bag in a cold room (4°C) for 70 days. They were then repotted in soil in the greenhouse, where new leaves and flowers formed. The flowers are most easily pollinated using houseflies, which makes controlled pollination difficult unless male-sterile plants are used. We were not able to recover viable seeds from these regenerated plants even after placing houseflies in small cages with the flowering plants. This lack of success could be attributed to several factors, including sterility or improper growth conditions.

III. MECHANISMS OF RESISTANCE

It is usually desirable to determine the biochemical mechanism of the resistance. In general, resistance to compounds can be caused by lack of uptake, alteration of the site of action, or inactivation of the inhibitor. Lack of uptake can be tested for by comparing uptake by resistant and wild-type cells using radiolabeled inhibitor or by measuring the inhibitor chemically in the medium and in the cells. Inactivation of the inhibitor can best be done using radiolabeled inhibitor to determine its fate in resistant and wild-type cells. To determine if the active site within the cells is altered requires some knowledge of the mode of action of the inhibitor. In the case of amino acid analogs, the inhibition is usually caused by inhibiting an enzyme (5MT inhibits anthranilate synthase) or by incorporation into protein in place of the natural amino acid [PFP, azetidine-2-carboxylate (A2C), aminoethylcystine (AEC)]. In all cases, resistance can occur because of the overproduction of the corresponding free natural amino acid, so analysis of the free amino acids in resistant cells is normally carried out (Ranch *et al.*, 1983). If higher than normal levels of free amino acids are noted, then kinetic analysis of the feedback control enzyme in that biosynthetic pathway can be carried out, as was done for anthranilate synthase in the tryptophan biosynthetic pathway (Widholm, 1972a; Ranch *et al.*, 1983).

The resistance of cell lines can also be characterized by determining if it is expressed in fusion hybrids. Carrot cells carrying resistance to 5MT, A2C, and AEC will express these resistances in somatic hybrids formed by protoplast fusion (Harms *et al.*, 1981; Kameya *et al.*, 1981).

Although the evidence indicates that cell lines chosen by visual selection for differences in pigment level (carotenoids) were not mutants (Mok *et al.*, 1976), a description of the selection procedure seems appropriate. Callus was initiated from roots of carrot cultivars with red, dark orange, orange, light orange, yellow, and white coloration. Calli showing different pigmentation were obtained either by picking out visually regions of the growing callus with altered coloration or by plating suspension-cultured cells onto agar-solidified medium to obtain colonies which differed in coloration. In most cases, lines with different pigmentation could be derived from a single cultivar, except in the case of the white and yellow roots, where only yellow cultures were obtained. The coloration noted in the lines was correlated with the levels of lycopene, β-carotene, and xanthophyll. When plants were regenerated from lines with varying pigmentation, the resulting roots were always like the original source root and not like the callus. Several explanations for these results are possible. The pigment changes in the calli were not due to mutations, or if they were, the mutations were not expressed in the regenerated plants. It is also possible that plants were regenerated from cells in the calli which did not show the pigment alterations.

Breton and Sung (1982) used an interesting screening method to select temperature-sensitive variants which were blocked at different stages of cell growth and embryogenesis. Haploid suspension cultures were obtained from a haploid carrot plant from the laboratory of J. Straub. The use of these cultures, which remain largely haploid, should allow the selection of recessive mutations. This cell line grows as small cell clumps at 24°C in liquid medium with 2,4-D as the auxin. The suspension was filtered through 200-μm nylon filters, and the cell clumps in the filtrate were centrifuged down and rinsed once in 2,4-D–free medium before suspension in more 2,4-D–free medium. Incubation in the medium devoid of 2,4-D induces embryogenesis even at the higher temperature of 32°C used here. After 7 and 21 days, the suspensions were passed through 100-μm filters to remove embryos and growing cell clumps. The undifferentiated cells remaining in the filtrate were plated on agar and incubated at 24°C to induce growth and embryogenesis. Of the 177 colonies which formed, 104 were found to be escapes which grew normally when retested for growth and plant regeneration ability at both 24°C and 32°C. The rest, however, did show abnormalities of several types, including arrested embryogenesis at 32°C but not at 24°C, no embryogenesis at either temperature, embryogenesis in the presence of 2,4-D, slow growth at either temperature, or auxin autotrophy with no embryogenesis. Such lines should be useful in the study of embryogenesis and cell growth.

Carrot remains one of the most easily manipulated cell and protoplast systems. The relatively low chromosome number ($2n = 18$) and low amount of DNA per diploid nucleus (2.1 pg; Flavell *et al.*, 1974) are also advantageous. In comparison with many other species, however, carrot suffers since it is not a very valuable crop or one which can be easily studied genetically.

REFERENCES

Breton, A. M., and Sung, Z. R. (1982). Temperature-sensitive carrot variants impaired in somatic embryogenesis. *Dev. Biol.* **90,** 58–66.

Flavell, R. B., Bennett, M. D., Smith, J. B., and Smith, D. B. (1974). Genome size and the proportion of repeated nucleotide sequence DNA in plants. *Biochem. Genet.* **12,** 257–269.

Harms, C. T., Potrykus, I., and Widholm, J. M. (1981). Complementation and dominant expression of amino acid analogue resistance markers in somatic hybrid clones *Daucus carota* after protoplast fusion. *Z. Pflanzenphysiol.* **101,** 377–390.

Hauptmann, R. M., and Widholm, J. M. (1982). Cryostorage of cloned amino acid analog-resistant carrot and tobacco suspension cultures. *Plant Physiol.* **70,** 30–34.

Kameya, T., Horn, M. E., and Widholm, J. M. (1981). Hybrid shoot formation from fused *Daucus carota* and *D. capillifolius* protoplasts. *Z. Pflanzenphysiol.* **104,** 459–466.

Mok, M. C., Gabelman, W. H., and Skoog, F. (1976). Carotenoid synthesis in tissue cultures of *Daucus carota* L. *J. Am. Soc. Hortic. Sci.* **101,** 442–449.

Murashige, T., and Skoog, F. (1962). A revised medium for rapid growth and bio assays with tobacco tissue cultures. *Physiol. Plant.* **15,** 473–497.

Palmer, J. E., and Widholm, J. M. (1975). Characterization of carrot and tobacco cell cultures resistant to P-fluorophenylalanine. *Plant Physiol.* **56,** 233–238.

Ranch, J. P., Rick, S., Brotherton, J. E., and Widholm, J. M. (1983). Expression of 5-methyltryptophan resistance in plants regenerated from resistant cell lines of *Datura innoxia. Plant Physiol.* **71,** 136–140.

Steward, F. C., Mapes, M. D., and Smith, J. (1958). Growth and organized development of cultured cells. II. Organization in cultures grown from freely suspended cells. *Am. J. Bot.* **45,** 705–708.

Sung, Z. R., and Dudits, D. (1981). Carrot somatic cell genetics. *In* "Genetic Engineering in the Plant Sciences" (N. J. Panopoulos, ed.), pp. 11–37. Praeger, New York.

Weber, G., and Lark, K. G. (1979). An efficient plating system for rapid isolation of mutants from plant cell suspensions. *Theor. Appl. Genet.* **44,** 81–86.

Widholm, J. M. (1972a). Anthranilate synthetase from 5-methyltryptophan-susceptible and -resistant cultured *Daucus carota* cells. *Biochim. Biophys. Acta* **279,** 48–57.

Widholm, J. M. (1972b). The use of fluorescein diacetate and phenosafranine for determining viability of cultured plant cells. *Stain Technol.* **47,** 189–194.

Widholm, J. M. (1974). Cultured carrot cell mutants: 5-methyltryptophan-resistance trait carried from cell to plant and back. *Plant Sci. Let.* **3,** 323–330.

Widholm, J. M. (1977a). Isolation of biochemical mutants of cultured plant cells. *In* "Molecular Genetic Modification of Eucaryotes" (I. Rubenstein, R. L. Phillips, C. E. Green, and R. Desnick, eds.), pp. 57–65. Academic Press, New York.

Widholm, J. M. (1977b). Relation between auxin autotrophy and tryptophan accumulation in cultured plant cells. *Planta* **134,** 103–108.

CHAPTER **64**

Induction, Selection, and Characterization of Mutants in Maize Cell Cultures

Kenneth A. Hibberd

Molecular Genetics, Inc.
Minnetonka, Minnesota

I. INTRODUCTION

The development of plant tissue culture systems has permitted biologists to select for specific mutations at the cellular level and in a few cases to regenerate plants from the selected cell lines. However, there is considerable variation in the types of cell culture systems available for the hundreds of species grown *in vitro* and in the usefulness of each system for carrying out *in vitro* selections. The important criteria in cell culture systems for making cellular-level selections have included cultures which are undifferentiated, grow rapidly, have a friable nature, and can be induced to re-form into whole plants. Maize has proven to be one of the more challenging species in this regard, and only recently has a cell line been identified that will fulfill these requirements. Nevertheless, over the past 10 years a variety of cellular-level selections have been carried out. In several cases, plants have been regenerated and mutations shown to be expressed in whole plants.

Selections for resistance to the maize disease southern corn leaf blight were carried out using the host-specific phytotoxin from the causal orga-

CELL CULTURE AND SOMATIC CELL
GENETICS OF PLANTS, VOL. 1

ISBN 0-12-715001-3

nism *Helminthosporium maydis* (*Drechslera maydis*). The initial selections utilized a nonregenerable hard callus line in a t-cytoplasm (toxin sensitive) background (Gengenbach and Green, 1975). In later selections with a regenerable hard callus line, plantlets were obtained which were shown to be both phytotoxin and disease resistant (Gengenbach *et al.*, 1977; Brettell *et al.*, 1979).

Altering free amino acid levels in corn has been of considerable interest for both academic and practical reasons. Selections for overproduction of the essential amino acids of the aspartate family pathway have been attempted using the growth-inhibitory combination of lysine plus threonine in organized, regenerable maize culture. Increases in the level of free threonine pools were observed in both selected cell lines and kernels from progeny of regenerated plants (Hibberd *et al.*, 1980; Hibberd and Green, 1982). Variant maize lines have also been obtained using 1-azetidine-2-carboxylic acid, ethionine, and S-2-aminoethyl-L-cysteine analogs for the amino acids proline, methionine, and lysine, respectively (Strauss and King, 1979). Details, however, were limited. Recently, cell lines resistant to valine have been reported in maize (Hibberd, 1983). This selection utilized the recently developed friable embryogenic maize cultures (Green, 1982; Green *et al.*, 1983).

Antibiotic-resistant phenotype selections are potentially useful as cellular-level plant markers. Several aminopterin-resistant strains have been obtained from a friable but nonregenerable maize cell line (Shimamoto and Nelson, 1981). In addition, streptomycin resistance was observed in organized, regenerable maize cultures (Umbeck and Gengenbach, 1983). However, neither cell lines nor regenerated plantlets survived to allow characterization of the resistance.

II. PROCEDURES

The mutant selection process depends initially on devising a strategy for obtaining the desired mutant phenotypes. The mutants obtained thus far from plant cell culture, including disease, herbicide, stress and antibiotic resistance, and changes in basic metabolism, have generally been obtained through the use of positive selection systems. Positive selections are simply selections for the ability to grow or survive in the presence of the selection agent. Selections in maize described in the previous section are illustrative of this approach.

The development of selection strategies depends heavily on basic information at the cellular and biochemical levels concerning the traits of interest. For example, information on virulence factors produced and excreted by the pathogen would be very useful in developing selections for disease resistance. Also, basic knowledge on amino acid biosynthesis and regulation is necessary for devising selections for amino acid overproducers.

Developing appropriate cell culture systems is another consideration in carrying out *in vitro* selections. In maize there are several maintainable cell types, some of which appear to be appropriate for different selection needs. Both firm and friable callus have been developed which have or lack the capacity to regenerate plants. Selections in which the ultimate goal is an altered plant would require the use of callus types that re-form plants; however, this requirement would not necessarily hold with selections for cellular-level markers or metabolic variants. Organized tissue cultures are of value in selections when the desired characteristics are expressed primarily in the differentiated tissues. For example, selection for streptomycin resistance in maize was based on the appearance of callus sectors which were able to develop and maintain green leaflets in the light (Umbeck and Gengenbach, 1983).

The initiation and maintenance of maize tissue cultures, although not difficult, can be fairly tedious and labor-intensive compared to other plant tissue culture systems. Each maize culture type presently known differs in explant origins and maintenance requirements. Hard, nonregenerable maize calli can be induced from a variety of plant tissues including mature embryos, roots, and leaves from most varieties of maize. Once established, it will require the least amount of careful supervision. Friable, nonregenerable maize cultures have been induced from stem-leaf sections under high auxin conditions (4 mg/liter, 2,4-dichlorophenoxyacetic acid) for only a few varieties of maize, the most notable being Black Mexican sweet corn (see Green, 1977; Sheridan, 1982). Once stabilized, these cultures are very rapid growers on plates and in liquid. They can also tolerate some neglect. Regenerable cultures, both hard and friable types, in contrast, require frequent subculturing and a very uniform environment. The hard organized cultures have been initiated from the immature embryo (Green and Phillips, 1975). Although this issue type has been known for over 8 years, the number of varieties which will form it, can be maintained in culture for several months, and will re-form plants is still limited to perhaps a dozen. Recently, a stable maize tissue culture type has been developed from the variety A188 that is both friable and regenerable (Green, 1982). This tissue grows rapidly, is highly embryogenic, and can be induced to form regenerable suspensions (Green *et al.*, 1983).

Once appropriate cell cultures have been developed, it is important to examine the interaction between them and the potential selection agents. A critical concern is whether the selection agent is interacting specifically with the desired system. Will the selection agent select for the right type of mutant? The approach used to answer this question will depend on the particulars of each selection. It may involve comparing the effect of the selection agent on both cell culture and plant tissues. *Helminthosporium maydis* toxin, for example, was equally inhibitory to both plant- and callus-derived mitochondria (Gengenbach and Green, 1975). In the case of analog selections, the demonstration that the natural compound will reverse the analog effects is of paramount importance prior to initiating the selections.

A very helpful approach in characterizing a selection system is the use of naturally resistant varieties, if available, or even other species that have the desired resistance phenotype. Cultures from these types should permit one to determine not only if a particular selection agent is appropriate but whether a resistant phenotype is expressed *in vitro* at all. It can also be extremely useful in refining selection conditions.

Mutagens, including ethyl methane sulfonate (EMS), sodium azide, and *N*-methyl-*N*′-nitrosoguanidine (MNNG), have been used on maize cultures for the purpose of inducing mutations. However, their value in increasing the mutation frequency has been examined in only one system, that for aminopterin resistance. The number of recovered resistant lines in Black Mexican cultures was enhanced 100-fold with MNNG treatment, whereas EMS treatment gave no significant increase in the recovery frequency (Shimamoto and Nelson, 1981). In several selections, including those for phytotoxin and valine resistance, no mutagens were used (Gengenbach *et al.*, 1977; Hibberd, 1983). This suggests that in 10^6 to 10^8 cells, a reasonable number to have in one selection experiment, there is sufficient variation for at least some types of selection.

Selections can and have been carried out in liquid medium and on solid plates with maize cultures. The use of solid plates has predominated in maize, since the hard callus cultures grow poorly and the capacity to regenerate plants is often lost in a liquid environment. In solid plate selections, 20- to 50-mg pieces are placed on the selection medium and the surviving tissues transferred at 3- to 5-week intervals. Two simple selection processes have been used with maize. Straight selections are carried out at levels of the selection agent that are just above full inhibition. A second approach has been to select initially at sublethal levels and then to shift periodically to higher levels. Eventually, fully inhibitory levels are reached. Only one study tried to determine which approach was better. For aminopterin resistance in nonregenerable Black Mexican cultures, no significant difference was found for the recovery of variant lines between these two methods (Shimomoto and Nelson, 1981). The time needed for identifying variant types has ranged from 2 to 5 months. The longer times were required with slower-growing tissues.

Following selection, a major concern has been to demonstrate that the selected cell lines are in fact altered in their sensitivity to the selection agent. This has generally taken the form of a growth inhibition test. For example, following selection for valine resistance, cell lines were grown in the absence of valine for several doublings and then retested for growth at various valine levels (Hibberd, 1983).

Variant lines must also be phenotypically stable over time in order for them to be of great value. Tests at the cell culture level have generally been conducted by growing the variants away from the selection agent for an extended period of time and then retesting for altered sensitivity to that agent. In solid plate selections, resistance to *H. maydis* toxin and lysine plus threonine was shown to be stable in cultures for 4 and 12 months, respectively, without selection pressure (Gengenbach and Green, 1975; Hibberd *et al.*, 1980). With amimopterin selections in liquid, both stable and unstable lines were found after 90 days in the absence of selection pressure (Shimamoto and Nelson, 1981).

Expression of traits selected in culture in whole plants is of considerable interest. *In vitro* selections are again of little value if they do not function at the plant level. Plants have been regenerated from maize cell lines selected for *H. maydis* toxin, lysine plus threonine, and valine resistance, and the progeny shown to express the resistance phenotypes (Gengenbach *et al.*, 1977; Hibberd and Green, 1982; Hibberd, 1983). All three traits appear to be dominant in both cell cultures and whole plants, but differ in their form of inheritance. Phytotoxin resistance is maternally inherited, whereas valine and lysine plus threonine resistances are nuclear traits.

Depending on the purpose for a selection, it may be important to determine the biochemical or physiological basis for the selected lesion. In maize selections, alterations in mitochondrial sensitivity to toxin, an altered aspartokinase, and variations in dihydrofolate reductase were observed in variants selected for phytotoxin, lysine plus threonine, and aminopterin resistances, respectively (Gengenbach *et al.*, 1977; Hibberd *et al.*, 1980; Shimamoto and Nelson, 1981).

III. CONCLUSIONS

Much of the research in plant cell selections has been aimed at answering basic questions concerning the nature of selected mutations. These include the types of mutants that can be selected, their expression in culture and in whole plants, stability, and type of inheritance. To some degree, variants selected in maize cultures have been useful in answering these questions, especially in the area of expression of mutations in whole plants.

Selection systems in maize have been fairly limited due to the nature of the regenerable maize cultures available until recently. The hard, highly organized types of callus had permitted the use of only simple positive selection approaches. This will probably change with the development of the regenerable friable maize cell cultures that form fine suspensions and can be induced to re-form whole plants. These properties should make maize cell cultures a useful model system for cellular selections and DNA transfer technologies.

REFERENCES

Brettell, R. I. S., Goddard, B. V. D., and Ingram, D. S. (1979). Selections of Tms-cytoplasm maize tissue cultures resistant to *Drechslera maydis* T-toxin. *Maydica* **24,** 203–213.

Gengenbach, B. G., and Green, C. E. (1975). Selection of T-cytoplasm maize callus cultures resistant to *Helminthosporium maydis* race T pathotoxoin. *Crop Sci.* **15,** 645–649.

Gengenbach, B. G., Green, C. E., and Donovan, C. M. (1977). Inheritance of selected pathotoxin resistance in maize plants regenerated from cell cultures. *Proc. Natl. Acad. Sci. U.S.A.* **74,** 5113–5117.

Green, C. E. (1977). Prospects for crop improvement in the field of cell culture. *HortScience* **12,** 131–134.

Green, C. E. (1982). Somatic embryogenesis and plant regeneration from friable callus of *Zea mays*. *In* "Plant Tissue Culture 1982" (A. Fujiwara, ed.), p. 107. Maruzen, Tokyo.

Green, C. E., and Phillips, R. L. (1975). Plant regeneration from tissue cultures of maize. *Crop Sci.* **15,** 417–421.

Green, C. E., Armstrong, C. L., and Anderson, P. C. (1983). Somatic cell genetic systems in corn. *Miami Winter Symp.* **20** (in press).

Hibberd, K. A. (1983). Value resistance selected in maize tissue culture. *Plant Mol. Newsl.* (abstr.) (in press).

Hibberd, K. A., and Green, C. E. (1982). Inheritance and expression of lysine plus threonine resistance selected in maize tissue culture. *Proc. Natl. Acad. Sci. U.S.A.* **79,** 559–563.

Hibberd, K. A., Walter, T., Green, C. E., and Gengenbach, B. G. (1980). Selection and characterization of a feedback-insensitive tissue culture of maize. *Planta* **148,** 183–187.

Sheridan, W. F. (1982). Black Mexican sweet corn: Its uses for tissue cultures. *In* "Maize for Biological Research" (W. F. Sheridan, ed.), pp. 385–388. Univ. of North Dakota Press, Grand Forks.

Shimamoto, K., and Nelson, O. E. (1981). Isolation and characterization of aminopterin-resistant cell lines in maize. *Planta* **153,** 436–442.

Strauss, A., and King, P. J. (1979). Selection of amino acid analog resistant corn cell lines. *Maize Gen. Coop. Newsl.* **53,** 14.

Umbeck, P. F., and Gengenbach, B. G. (1983). Streptomycin and other inhibitors as selection agents in corn tissue cultures. *Crop Sci.* (in press).

CHAPTER **65**

Elimination of Viruses*

K. K. Kartha

Plant Biotechnology Institute
National Research Council
Saskatoon, Saskatchewan, Canada

I. INTRODUCTION

Plant diseases are caused by fungi, bacteria, viruses, mycoplasma-like agents, and nematodes. Effective control measures are available to combat most of the diseases, except for those caused by viruses and mycoplasma-like agents. Viral diseases, causing serious yield losses, are present in virtually all seed-propagated as well as vegetatively propagated crop species. In the absence of effective therapeutic chemicals capable of eradicating viruses from infected plants, tissue culture techniques, especially meristem culture, have been employed in the elimination of viral infection and the production of disease-free plants (for reviews, see Hollings, 1965; Quak, 1977; Walkey, 1978; Kartha, 1981). The application of plant tissue culture methods to the elimination of viral infection in plants began after the observation by Limmaset and Cornuet (1949) that the virus titer in a plant systemically invaded by viruses decreases as the shoot meristem is reached. Morel and Martin (1952) confirmed this observation by regenerating virus-

*NRCC No. 22970.

CELL CULTURE AND SOMATIC CELL
GENETICS OF PLANTS, VOL. 1

ISBN 0-12-715001-3

free dahlia plants from *in vitro* cultured meristems isolated from virus-infected plants. These encouraging findings formed the basis for the now well-established domain of meristem culture-mediated virus elimination.

Various *in vitro* techniques, incorporation of chemicals to suppress or inactivate viruses, gamma radiation, and somatic cell hybridization by fusion of plant protoplasts have been attempted to eliminate viral infections from tissues or their regenerants.

II. METHODS OF VIRUS ELIMINATION

Of the several methods employed in the elimination of viruses in plants, meristem culture alone or in combination with heat treatment still ranks as the most promising method applicable to a large array of plant species.

A. Meristem Culture

A clear understanding of the viral disease in question, with particular reference to the methods of transmission of the virus, its host range, and physical and chemical properties, is needed to determine with certainty the success of virus elimination through any *in vitro* techniques. In most of the systemic viral diseases, the virus often establishes a gradient in the plant system, with the rapidly growing apical regions containing the least quantity of the virus. The success of elimination of virus through *in vitro* culture of meristems, therefore, depends upon the types of plant viruses, particular host–virus combinations and, most importantly, the size of the meristem cultured *in vitro*. The larger the size of the meristem cultured, the greater will be the number of plants regenerated, but the number of virus-free plants obtained will be inversely proportional to the size of the meristem cultured. For example, in attempts to eliminate cassava (*Manihot esculenta*) mosaic disease, it was found that up to 60% of the plantlets regenerated from meristems 0.4 mm in length isolated from sprouted cuttings of infected plants were mosaic disease free. The plants were regenerated from meristems cultured on Murashige and Skoog's (MS) medium (1962) supplemented with 0.5 μM benzylaminopurine (BAP), 1.0 μM naphthaleneacetic acid (NAA), and 0.1 μM gibberellic acid (GA_3) at 26°C and 4000 lx intensity (Kartha *et al.*, 1974). The regenerated plants were grown in a greenhouse or a growth chamber, and freedom from mosaic disease agents was determined by grafting the scions from regenerated plants onto healthy but susceptible cassava stocks at monthly intervals for a period of up to 6 months. In contrast, all the plants regenerated from meristems exceeding 0.4 mm in length were indexed to be diseased (Kartha and Gamborg, 1975). Pea seed-borne mosaic virus was eliminated in high frequency (90–100%) from over 100 breeding lines of *Pisum sativum* by culturing shoot apical meristems 0.4–0.5 mm in length (see Chapter 13, this volume, for the methodology). Since pea seed-borne mosaic virus is mechanically transmitted, indexing for the presence of absence of the virus was carried out by inoculating the leaf extract (=standard extract: 1.0 g of leaf tissue macerated in 1.0 ml of 0.1 *M* phosphate buffer) obtained from meristem-derived plants onto the leaves of the local lesion host, *Chenopodium amaranticolor,* and the systemic host, *Pisum sativum.* A negative reaction to inoculation indicated freedom from virus. The indexing was repeated several times during the various growth stages of the plants.

Work done with potato (*Solanum tuberosum*) and several other types of plant species also emphasizes the importance of the size of meristems in relation to the frequency with which virus-free plants are regenerated. Kassanis and Varma (1967) demonstrated that of 20 plantlets regenerated from a total of 196 meristems cultured from potatoes infected with potato virus X (PVX) and potato virus S (PVS), only 19 were virus free, and only when meristems 0.1 mm in length with or without a leaf primordium were cultured. On the other hand, Accatino (1966) included two leaf primordia on the meristem explants cultured *in vitro.* Of the 18 plants regenerated, all were free of potato leaf roll virus (PLRV), potato virus Y (PVY), and potato virus M (PVM), but 6 contained PVX and 15 PVS. These studies indicate not only the importance of the size of the meristem in the successful elimination of viruses but also the role certain host–virus combinations play in determining the success of virus elimination. According to Mellor and Stace-Smith (1977), some of the potato viruses—in order of increasing difficulty of elimination—are PLRV, potato virus A (PVA), PVY, aucuba mosaic virus, PVM, PVS, and spindle tuber (viroid) (PSTV).

The *in vitro* techniques used for culturing meristems for the purpose of eliminating viruses are essentially the same as those used for propagation purposes, except that explants of smaller size (0.3–0.5 mm) are used. The instruments need sterilization in 70% ethanol after each step in the dissection process as a precaution against accidentally contaminating the meristem explant with the virus. Meristems can be isolated from apices of stems, tuber sprouts, leaf axils, sprouted buds of cuttings, or germinated seeds. Thorough disinfection is recommended if the meristems originate from underground plant organs such as tubers, rhizomes, bulbs, or corms.

B. Heat Treatment and Meristem Culture

Heat treatment of the infected plants prior to the excision of meristems is especially advantageous when viruses are difficult to eradicate by meristem culture alone or when large numbers of virus-free plants are needed. Growing the infected plants at elevated temperatures (35–37°C) for 3–4 weeks permits faster vegetative growth and in some cases complete suppression of symptoms on the young foliage. The enhanced vegetative growth of the plant and the rapid cell division in the shoot apices restrict virus multiplication and movement of the viral particles into the meristem region, resulting in a larger virus-free zone in the apices. As had been explained earlier, since larger explants regenerate easily *in vitro* into plants, this technique facilitates the production of large numbers of virus-free plants as well. For example, cassava meristems isolated from mosaic-diseased cuttings grown at 35°C for 30 days at 16-hr photoperiods, 4000 lx intensity, and 70% relative humidity regenerated into plants which were all indexed to be mosaic disease free, as opposed to only 60% mosaic disease-free plants regenerated from non-heat-treated plants. Moreover, meristems up to 0.8 mm in length could be cultured to produce healthy progenies from the heat-treated plants (Kartha and Gamborg, 1975). Similarly, the "frog skin" disease of cassava, of suspected viral etiology, has been eliminated from five cassava cultivars by a combination of heat treatment and meristem culture. The diseased cuttings were grown for 3–4 weeks at an alternating temperature of 40°C (day) and 35°C (night). Shoot apical meristems 0.4–0.5 mm in length cultured on cassava meristem culture medium (Kartha *et al.*, 1974) produced plantlets which were indexed to be disease free (W. M. Roca, personal communication). The application of heat treatment and meristem culture in the elimination of viruses in several plants has been reviewed in detail (Quak, 1977; Mellor and Stace-Smith, 1977). As an alternative to heat treatment, meristems isolated from cucumber mosaic and alfalfa mosaic virus-infected plants were grown at 32° or 34°C, and the viruses were eradicated or their titer considerably reduced (Walkey and Cooper, 1975). However, it should be pointed out that although many viruses could be eliminated by heat treatment and meristem culture, there are a number of viruses which are very difficult to eradicate by existing methods.

C. Suppression of Viruses by Chemicals

Since plant hormones play an integral part in any *in vitro* propagation system, it was suggested that the addition of cytokinins and other growth-promoting substances to the culture medium may suppress virus in infected tissue cultures (Quak, 1961). However, conflicting results were subsequently observed; in some instances virus multiplication was suppressed, whereas in others it was stimulated (Milo and Srivastava, 1969; Omura and Wakimoto, 1978). Similarly, the effect of certain antimetabolites incorporated into the culture media was studied in relation to their role in the inactivation of viruses with limited or no success. These compounds included malachite green (Norris, 1954), thiouracil (Vasti, 1973), and acetylsalicylic acid (White, 1979).

Recent studies have suggested that suppression of virus in cultured plant tissues may be effected by the incorporation of the nucleoside analog ribavirin (syn. virazole) into the culture medium (Shepard, 1977; Cassells and Long, 1980). This broad-spectrum antiviral agent has been shown to act against animal (Harris and Robins, 1980) and plant viruses (Schuster, 1976; Kluge and Marcinka, 1979; Hansen, 1979). It has been found to be effective in eradicating PVX in tobacco plantlets regenerated from mesophyll protoplasts (Shepard, 1977). Recently, Simpkins *et al.* (1981) reexamined the role of ribavirin in the suppression of virus in cultured plant tissues. They found that concentrations of cucumber mosaic (CMV) and alfalfa mosaic virus present in various types of plant tissues were considerably less on medium supplemented with 50–100 mg/liter ribavirin. The CMV concentration decreased within 24 hr in infected cultures treated with the drug. Surprisingly, virus-free plantlets could be regenerated from CMV-infected meristem tips irrespective of the presence or absence of ribavirin. Contrary to Quak's suggestions (Quak, 1961), Simpkins *et al.* (1981) noted that kinetin at concentrations up to 25.6 mg/liter had no persistent antiviral effect.

D. Other *in Vitro* Methods

Based on the uneven distribution of viruses in plants, the occurrence of populations of uninfected cells even in severely infected plants, and the totipotency of plant cells in culture, methods have been devised to produce virus-free plants or plants with transient viral resistance. Virus-free tobacco plantlets have been regenerated *in vitro* from tobacco mosaic virus (TMV)-infected callus cultures (Hansen and Hildebrandt, 1966). Shepard (1975) succeeded in regenerating virus-free tobacco plants after mass isolation of mesophyll protoplasts from plants systemically infected with PVX. These studies are based on the "chance selection" of uninfected cells from a population of infected cells.

Employing a tissue culture system, Murakishi and Carlson (1976) regenerated virus-free plants from dark green islands of TMV-infected tobacco leaves. Explants isolated from the dark green islands (areas) of TMV-infected tobacco leaves were aseptically cultured on Linsmaier and Skoog (LS) medium (1965) containing 3.0 mg/liter 3-indoleacetic acid (IAA) and 0.3 mg/liter kinetin. After several weeks, the tissue pieces were transferred to LS medium supplemented with 0.3 mg/liter IAA and 10.0 mg/liter N^6-(2-isopentenyl)adenine (2iP) to induce shoot formation, followed by transfer to the medium with 1.0 mg/liter IAA to induce root differentiation. About 50% of the plants so regenerated were indexed to be virus free. Leaf sap from such virus-free plantlets apparently contained a factor which was inhibitory to TMV infection. In subsequent studies they employed gamma radiation for *in vitro* selection of *Nicotiana sylvestris* variants with limited resistance to TMV. *Nicotiana sylvestris* plants inoculated with a yellow strain of TMV (TMV-flavum) were exposed to 500 rads of acute gamma radiation, and the leaf strips cultured *in vitro* produced two types of colonies, yellow containing virus, and green apparently healthy. Of the 3210 calli assayed, approximately 5% were virus free, and after regeneration, 0.2% were resistant at the plant stage. The observed resistance, measured on the basis of restricted virus multiplication and movement, resulted in a 3- to 8-week delay in symptom expression in both the seedling progeny and their rooted cuttings (Murakishi and Carlson, 1982).

White (1982) cultured TMV-induced lesions of *N. tomentosa* on LS medium supplemented with IAA (0.3–3.0 mg/liter) and kinetin (0.2 mg/liter) or 2iP (10.0 mg/liter). After 1 month or more, calli were transferred to LS medium containing IAA (0.3 mg/liter) and 2iP (10.0 mg/liter) to induce shoot formation, and subsequently to medium lacking hormones to induce root formation. About 60% of the regenerated plants were virus free.

In another study, of the 57 plants regenerated from the callus derived from immature spindle tissue above the apical meristem of sugarcane plants, systemically infected with sugarcane mosaic virus, 55 were virus free even after 6 months of growth in a greenhouse. Yet, sugarcane mosaic virus was present in many, if not all, of the explants from which virus-free plants were regenerated (Dean, 1982). With sugarcane, *in vitro* manipulation has resulted in the production of clones with desirable agronomic traits (Heinz and Mee, 1971). Some of the clones derived through tissue culture also possessed higher levels of resistance to fungal and viral diseases (Heinz *et al.*, 1977). One clone, Pindar-70-31, was especially resistant to both fiji and downy mildew diseases while maintaining the equivalent yield potential of the donor parent (Nickell, 1977). Studies carried out for over 4 years on the stability of the subclones of the Pindar variety revealed that the subclones were noncarriers for both of the disease (Krishnamurthy, 1982).

Other types of tissues used to produce virus-free plants include nucellus (citrus), anthers (Abo-El-Nil and Hildebrandt, 1971), and floral meristems (Walkey *et al.*, 1974).

Somatic cell hybridization by fusion of plant protoplasts shows great promise, at least in transferring genes governing resistance to viral diseases from one plant to other susceptible plant. Although examples of such gene transfer are limited, the technique has great potential. For example, somatic hybrid plants produced by the fusion of mesophyll protoplasts of *N. tabacum* and *N. rustica* were resistant to artificially inoculated TMV (Nago, 1978). Recently, Butenko *et al.* (1982) obtained somatic hybrids between cultivated potato (*Solanum tuberosum*) and a wild species (*S. chacoense*). The somatic hybrids exhibited absolute field resistance to infection by PVY.

III. CONCLUSION

Various examples discussed in this chapter clearly attest to the fact that tissue culture techniques can be successfully employed in the elimination of viral pathogens and the production of virus-free plants. In the absence of effective chemotherapeutic agents, *in vitro* techniques will continue to be a versatile tool in the elimination of viral infection in plants. Establishment of freedom from virus depends greatly on the reliability of the methods employed in the detection of viruses. Such methods are based on the nature and properties of the virus and its host range, since some viruses are mechanically transmitted whereas others are transmitted only through insect vectors and other biological agents. For a detailed account of the assay, detection, and diagnosis of viruses, the reader is referred to Matthews (1981).

REFERENCES

Abo El-Nil, M. M., and Hildebrandt, A. C. (1971). Differentiation of virus-symptomless geranium plants from anthers callus. *Plant Dis. Rep.* **55,** 1017–1020.

Accatino, P. (1966). Papa corahila libre de virus mediante cultivo de meristemas. *Agric. Tec. (Santiago)* **26,** 34–39.

Butenko, R., Kuchko, A., and Komarnitsky, I. (1982). Some features of somatic hybrids between *Solanum tuberosum* and *S. chacoense* and its F1 sexual progeny. *In* "Plant Tissue Culture 1982" (A. Fujiwara, ed.), pp. 643–644. Maruzen, Tokyo.

Cassels, A. C., and Long, R. D. (1980). The regeneration of virus-free plants from cucumber mosaic virus and potato virus Y infected tobacco plants cultured in the presence of virazole. *Z. Naturforsch., C: Biosci.* **35C,** 350–351.

Dean, J. L. (1982). Failure of sugarcane mosaic virus to survive in cultured sugarcane tissue. *Plant Dis.* **66,** 1060–1061.

Hansen, J. A. (1979). Inhibition of apple chlorotic leaf spot in *Chenopodium quinoa* by ribavirin. *Plant Dis. Rep.* **63,** 17–20.

Hansen, J. A., and Hildebrandt, A. C. (1966). The distribution of tobacco mosaic virus in plant callus cultures. *Virology* **28,** 15–21.

Harris, S., and Robins, R. K. (1980). Ribavirin: Structure and antiviral relationships. *In* "Ribavirin" (R. A. Smith and W. Kirkpatrick, eds.), pp. 1–22. Academic Press, New York.

Heinz, D. J., and Mee, G. W. P. (1971). Morphologic, cytogenetic and enzymatic variation in *Saccharum* species hybrid clones derived from callus tissue. *Am. J. Bot.* **58,** 257–262.

Heinz, D. J., Krishnamurthy, M., Nickell, L. G., and Maretzki, A. (1977). Cell, tissue and organ culture in sugarcane improvement. *In* "Applied and Fundamental Aspects of Plant Cell, Tissue, and Organ Culture" (J. Reinert and Y. P. S. Bajaj, eds.), pp. 3–17. Springer-Verlag, Berlin and New York.

Hollings, M. (1965). Disease control through virus-free stock. *Annu. Rev. Phytopathol.* **3,** 367–396.

Kartha, K. K. (1981). Tissue culture techniques for virus elimination and germplasm preservation. *Genet. Eng. Crop Improve., Rockefeller Found. Conf. 1980* pp. 123–141.

Kartha, K. K., and Gamborg, O. L. (1975). Elimination of cassava mosaic disease by meristem culture. *Phytopathology* **65,** 826–828.

Kartha, K. K., Gamborg, O. L., Constabel, F., and Shyluk, J. P. (1974). Regeneration of cassava plants from shoot apical meristems. *Plant Sci. Lett.* **2,** 107–113.

Kassanis, B., and Varma, A. (1967). The production of virus-free clones of some British potato varieties. *Ann. Appl. Biol.* **59,** 447–450.

Kluge, S., and Marcinka, K. (1979). The effect of polyacrylic acid and virazole on the replication and component infection of red clover mottle virus. *Acta Virol.* **23,** 148–152.

Krishnamurthy, M. (1982). Disease resistance in sugarcane developed through tissue culture. *In* "Plant Tissue Culture 1982" (A. Fujwara, ed.), pp. 769–770. Maruzen, Tokyo.

Limmaset, P., and Cornuet, P. (1949). Recherche de virus de la mosaïque du tabac (*Marmor tabaci* Holmes) dans les méristèmes des plantes infectées. *C. R. Hebd. Seances Acad. Sci.* **228,** 1971–1972.

Linsmaier, E. M., and Skoog, F. (1965). Organic growth factor requirements of tobacco tissue cultures. *Physiol. Plant.* **18,**100–127.

Matthews, R. E. F. (1981). "Plant Virology," 2nd ed. Academic Press, New York.

Mellor, F. C., and Stace-Smith, R. (1977). Virus-free potatoes by tissue culture. *In* "Applied and Fundamental Aspects of Plant Cell, Tissue and Organ Culture" (J. Reinert and Y. P. S. Bajaj, eds.), pp. 616–635. Springer-Verlag, Berlin and New York.

Milo, G. E., and Srivastava, B. I. S. (1969). Effect of cytokinins on tobacco mosaic virus production in tobacco pith cultures. *Virology* **39,** 621–623.

Morel, G., and Martin, C. (1952). Guérison de dahlias atteints d'une maladie à virus. *C. R. Hebd. Seances Acad. Sci.* **235,** 1324–1325.

Murakishi, H. H., and Carlson, P. S. (1976). Regeneration of virus-free plants from dark-green islands of tobacco mosaic virus-infected tobacco leaves. *Phytopathology* **66,** 931–932.

Murakishi, H. H., and Carlson, P. S. (1982). *In vitro* selection of *Nicotiana sylvestris* variants with limited resistance to TMV. *Plant Cell Rep.* **1,** 94–97.

Murashige, T., and Skoog, F. (1962). A revised medium for rapid growth and bioassays with tobacco tissue cultures. *Physiol. Plant.* **15,** 473–497.

Nago, T. (1978). Somatic hybridization by fusion of protoplasts. I. The combination of *Nicotiana tabacum* and *Nicotiana rustica. Jpn. J. Crop Sci.* **47,** 491–498.

Nickell, L. G. (1977). Crop improvement in sugarcane: Studies using *in vitro* methods. *Crop Sci.* **17,** 717–719.

Norris, D. O. (1954). Development of virus-free stock of Green Mountain by treatment with malachite green. *Aust. J. Agric. Res.* **5,** 658–663.

Omura, T., and Wakimoto, S. (1978). Effect of plant hormones on tobacco mosaic virus concentrations on tobacco tissue cultures. *J. Fac. Agric., Kyushu Univ.* **22,** 211–219.

Quak, F. (1961). Heat treatment and substances inhibiting virus multiplication in meristem culture to obtain virus-free plants. *Adv. Hortic. Sci. Their Appl., Proc. Int. Hortic. Congr., 15th, 1958* Vol. 1, pp. 144–148.

Quak, F. (1977). Meristem culture and virus-free plants. *In* "Applied and Fundamental Aspects of Plant Cell, Tissue and Organ Culture" (J. Reinert and Y. P. S. Bajaj, eds.), pp. 598–615. Springer-Verlag, Berlin and New York.

Schuster, G. (1976). Wirkung on 1-B-D-ribofuranosyl-1,2,4-triazole-3-carboxamide (Virazole) auf die Vermehrung systemischer Viren in *Nicotiana tabacum* 'Samsun.' *Ber. Inst. Tabakforsch. Dresden* **23,** 21–36.

Shepard, J. F. (1975). Regeneration of plants from protoplasts of potato virus X-infected tobacco leaves. *Virology* **66,** 492–501.

Shepard, J. F. (1977). Regeneration of plants from protoplasts of potato virus X-infected tobacco leaves. II. Influence of Virazole on the frequency of infection. *Virology* **78,** 261–266.

Simpkins, I., Walkey, D. G. A., and Neely, H. A. (1981). Chemical suppression of virus in cultured plant tissues. *Ann. Appl. Biol.* **99,** 161–169.

Vasti, S. M. (1973). Effect of antiviral chemicals on production of virus X-free potato tubers. *Pak. J. Bot.* **5,** 139–142.

Walkey, D. G. A. (1978). *In vitro* methods for virus elimination. *In* "Frontiers of Plant Tissue Culture 1978" (T. A. Thorpe, ed.), pp. 245–254. Univ. of Calgary Press, Calgary, Alberta,Canada.

Walkey, D. G. A., and Cooper, V. C. (1975). Effect of temperature on virus eradication and growth of infected tissue cultures. *Ann. Appl. Biol.* **80,** 185–190.

Walkey, D. G. A., Cooper, V. C., and Crisp, P. (1974). The production of virus-free cauliflowers by tissue culture. *J. Hortic. Sci.* **49,** 273–275.

White, J. L. (1982). Regeneration of virus-free plants from yellow-green areas and TMV-induced enations of *Nicotiana tomentosa. Phytopathology* **72,** 866–867.

White, R. F. (1979). Acetylsalicylic acid (aspirin) induces resistance to tobacco mosaic virus in tobacco. *Virology* **99,** 410–412.

CHAPTER **66**

Cocultures of Plant and Bacterial Cells

Minocher Reporter

Charles F. Kettering Research Laboratory
Yellow Springs, Ohio

I. INTRODUCTION

A majority of publications on the use of plant cells cultured in association with microbes during the past decade have dealt with some aspect of symbiosis or the production of phytotoxins. These topics have been reviewed by Earle (1978) and Torrey (1978). The approaches developed by the phytopathologists have emphasized the use of bacterial culture filtrates from host-specific pathogenic microbes on cultured plant cells or protoplasts derived from these cells (see Chapter 67, this volume). The particular lesion caused by the bacterial filtrate on the plant cells in each case has been used as a bioassay to aid in the isolation of the toxin. This approach has been successful in the majority of cases in which the pathology of the lesions is distinct. The problems of nonspecific interactions between cultured plant cells and bacteria are avoided. For example, Samaddar and Scheffer (1968) classified the effects of *Helminthosporium victoriae* on cell membranes of oat protoplasts by using preparations from susceptible and

CELL CULTURE AND SOMATIC CELL
GENETICS OF PLANTS, VOL. 1

ISBN 0-12-715001-3

resistant plants. Further explanations of the effects of toxin on electropotentials and electrogenic pumps of the plasma membranes were later obtained with single cells from coleoptiles of oats, sorghum, and maize (Gardner *et al.*, 1974). Another recent example is the isolation of tumor-inducing (Ti) and root-inducing (Ri) plasmids from *Agrobacterium*, which has now made it possible to change the expression of plant cells in culture (Chapters 58 and 59, this volume). Introduction of a single T-DNA mutation of the nopaline Ti-plasmid pTiT37 has enabled it to be introduced into tobacco cells in culture and to regenerate plants (Chilton *et al.*, 1983).

Studies with directly mixed nitrogen-fixing bacteria and plant organ, callus, or cell cultures have indicated that such associations are difficult to simplify. Reviews on this subject have been provided by Davey and Cocking (1979), Giles and Vasil (1980), and Gresshoff and Mohpatra (1981). This difficulty stems from the many types of changes which cells of host plants as well as bacteria naturally undergo during symbiosis and even during associative-type nitrogen fixation in which the bacterium resides within the rhizosphere of the host or within the intercellular spaces between host cells (Giles and Vasil, 1980; Berg *et al.*, 1979; Pence *et al.*, 1982). The majority of the *in vivo* changes have not been clearly exhibited in mixed cultures. (It is now claimed that this is possible if L-form microorganisms are used; see Paton, 1982). In *Rhizobium* alone, a large number of extrachromosomal and chromosomal genes are involved in initiation of symbiosis and in the functions of nitrogen-fixing bacteroids (Forrai *et al.*, 1983). A small number of these changes have been demonstrated in the rhizobia used with the mixed cultures, and fewer papers indicate changes in plant cell cytoplasm of cells infected in mixed cultures (Hardy and Holsten, 1972; Hermina and Reporter, 1977).

Many of the *in vitro* studies used agar as substratum on which to place bacteria and plant cells under study. Alternatively, products were included in the agar medium and bacteria or plant cells placed on top for further *in vitro* studies of the influence of these products on cells. Under these conditions, materials being exchanged were difficult to isolate (LaRue *et al.*, 1975; DeMoranville *et al.*, 1981). However, many interesting experiments were presented by this approach (e.g., Child and Kurz, 1978).

When mixed cultures of plant cells and bacteria are maintained *in vitro*, destruction or inhibition of either the plant cells or bacteria is usually noticed (e.g., Ozawa and Yamaguchi, 1979, or Verma *et al.*, 1978) and the exceptions are difficult to work with (Hermina and Reporter, 1977). Transfilter cultures have often been preferred in order to avoid cell growth inhibition in directly mixed cultures.

Transfilter cultures of bacteria and plant cells have been used to study exchange of materials. These methods have limitations in common with other *in vitro* systems since the associations are artificial. Three methods of growing bacteria (usually rhizobia) in suspension coculture with callus-derived plant cells have been used in our laboratory, and two of these have been published (Hermina and Reporter, 1975; Reporter, 1976; Bednarski and Reporter, 1978). A small transfilter apparatus, easy to manipulate and assemble, has been used by Lustig *et al.* (1980a). The use of these methods on coculture of bacteria and plant cells using interposing filters is described in this chapter.

II. APPARATUS FOR TRANSFILTER CULTURES

A. Available Designs

Transfilter cultures have also been useful for ecological-type *in vitro* studies between different types of cells and organisms. An apparatus labeled "Ecologen" (New Brunswick, New Jersey) was introduced some years ago for transfilter studies. This apparatus was placed on a reciprocating shaker. The central square stainless steel chamber of the apparatus had four round 6-cm-diameter filters leading into four screw-on glass vessels. The apparatus was sterilized in a regular sterilizer, usually after being filled with the desired media. Cells were then introduced into the different chambers as desired. The apparatus had ports for sampling each chamber separately and quickly. The screw-on vessels were, however, difficult to seal, and the thin glass often cracked during tightening of the chambers. Atmospheric control above different vessels could not be easily achieved.

A number of parabiotic, all-glass chambers are also commercially available from Bellco Glass, Inc., Vineland, New Jersey. These are based on Katsuta or Karush designs (see Bellco Catalog BGE-6, pp. 10, 11, 33). The capacity of each chamber is limited and, depending on the model, varies from 1 to 15 ml. These chambers are satisfactory for many types of transfilter studies. The membranes commonly used are 25 mm in diameter and are clamped between glass chambers using high-vacuum silicone stopcock grease on the sides. This latter feature could be disadvantageous if sampling of low concentrations of materials of interest dissolved in the fluid media is important since constituents of the stopcock grease may cause interference.

B. Transfilter Culture of Callus and Bacteria without Agar

A square chamber design made from polyvinyl chloride was used for studies of callus cultures of plant cells separated from bacteria by a dialysis membrane by Lustig *et al.* (1980a,b). The dialysis membrane was tightly fixed between flat chamber sides by screws in the flanges to the chambers. The whole transfilter apparatus (each half 20 × 30 × 30 mm outside and 6 ml in volume inside) was placed inside a larger vessel whose atmosphere could be monitored.

A similar design was also adapted in our laboratory. Our design used the central pair of connecting chambers (10 ml each) from the remains of old double cell (DCIII) Plexiglas models (see Section II,C). The chambers were tightened by screwing one end of a round box tightly onto the assembled chambers. The sides of the box were open, and access to the chambers in which cells were growing was made from the sides after removal of two Teflon plugs. Access needles for sampling gas were also provided in this adaptation by R. Velliquette of our laboratory. The self-contained apparatus could be placed in larger vessels, or the atmosphere of the chambers could be controlled independently at various time intervals for later sampling (Fig. 1). The use of thick-walled Plexiglas made it possible to sterilize the chambers by autoclaving as well as by chemical means. Lustig *et al.* (1980a) used 1% benzalkonium chloride, followed by copious washing with sterile water. However, autoclaving with membranes in place was easily accomplished when the chambers were filled with distilled water. The water was discarded after sterilization and replaced by bacterial cells or callus tissue. Since no shaking was involved and cells were discretely massed, cross-membrane infection was minimal. From 2 to 3 ml of fluid was placed on each side of the membrane for better material exchange, as shown in Fig. 1.

C. Transfilter Cultures Using Liquid Media

The original double chambers used in our laboratory for liquid cocultures of different cell types were designed with the following points in mind.

1. They would be subjected to repeated autoclaving. A polycarbonate (General Electric) version with thick walls which were originally polished to translucence with coconut oil has been used for 7 years. One version of this model (DCIII) is shown in Fig 2A and 2B. The blocks of polycarbonate were heated in a sterilizer before machining in order to reduce later stresses. The clear plastics used in earlier models (DCI and DCII) were found to show stress cracks at pressure points.

2. To avoid stress and loosening of the filter assembly after repeated use, threading of the chambers was avoided. A screw pressure-type assembly with a pressure plate on the outside was favored. This also meant that filters would not tear from twisting deformation. The filter size was chosen

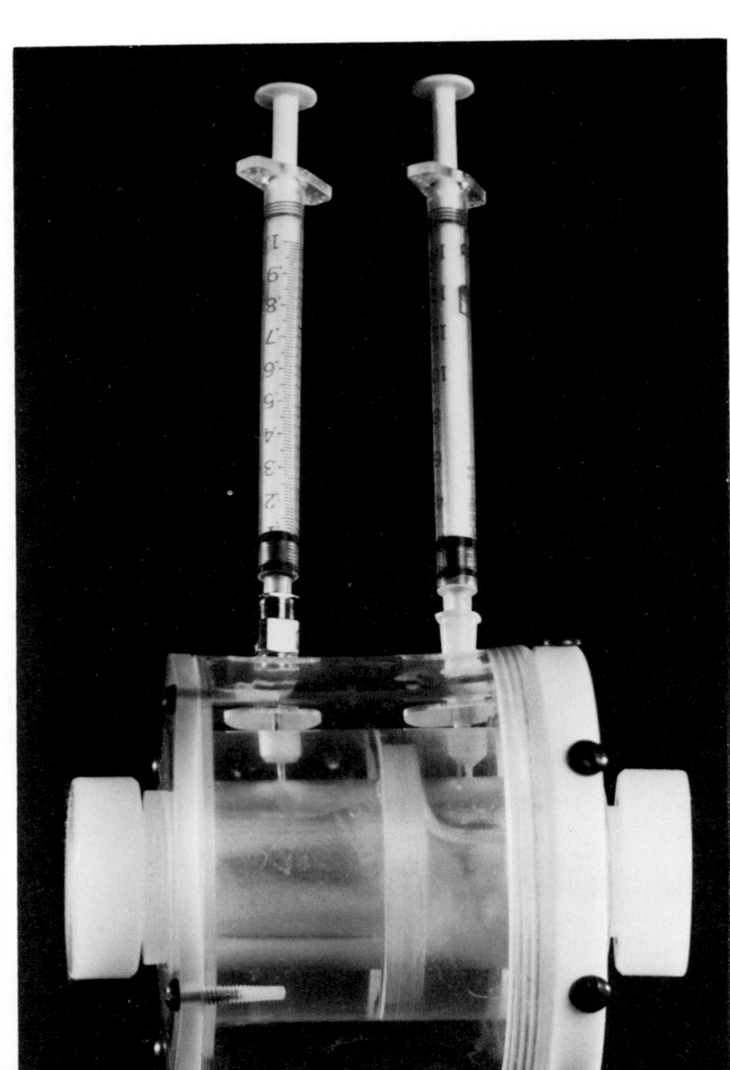

Fig. 1. Adaptation of small chambers for transfilter culture of callus and bacteria. The chambers were made by pressing dialysis tubing between two small interlocking open-end cylinders (see Fig. 2). The cylinders were tightened together after being placed inside a larger box by means of a Teflon screw cap. Openings at the ends of the apparatus allowed access to the two chambers. These openings were closed with Teflon plugs after the small chambers were loaded with plant cells or bacteria on either side of the dialysis tubing. Syringe needles on the side allowed injection or withdrawal of gas samples.

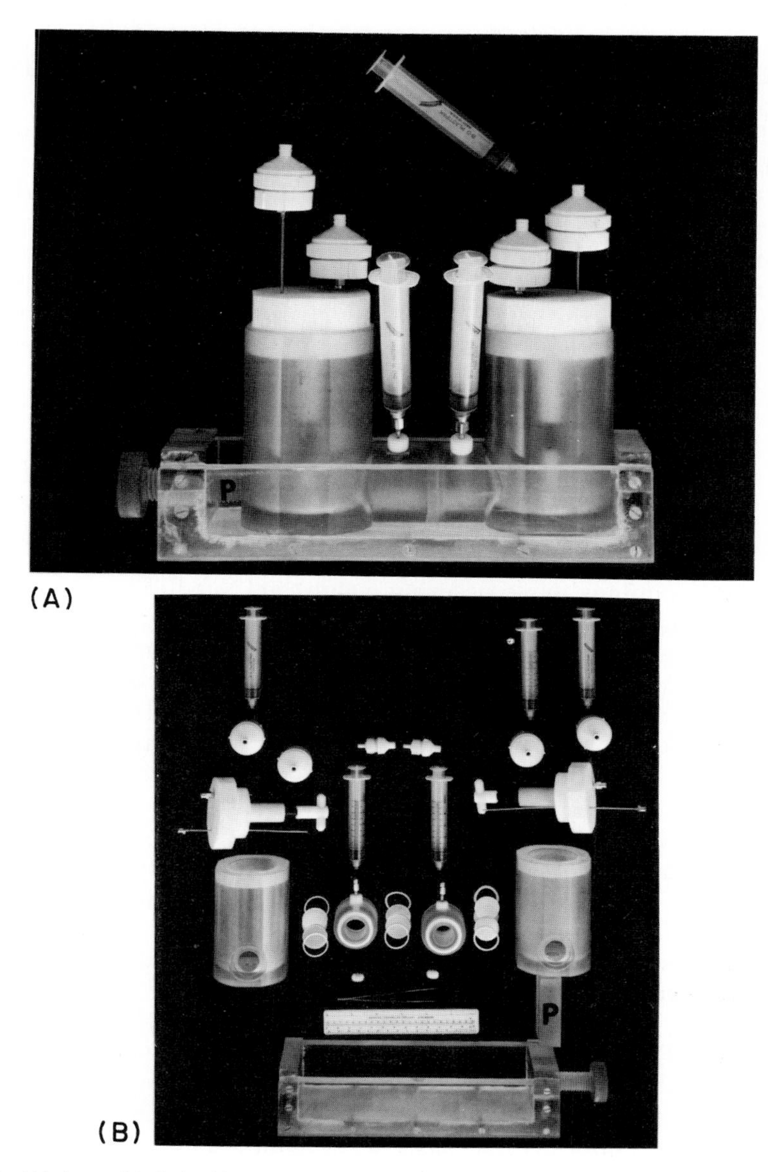

Fig. 2. (A) Assembled double-cell apparatus. The pressure plate (P) distributes screw pressure evenly along the small interconnecting chambers. (B) The double-cell apparatus before assembly. The small interconnecting chambers with dialysis tubing substituted for filters is used for the assembly shown in Fig. 1.

so that 25-mm commercial filters and filter screens could be used together with Teflon (du Pont) or Viton o-rings, which are commercially available. A variety of filter and membrane combinations have been used.

3. The filters were placed at the bottom to assure exchange of only dissolved gas and solutes. The train of three filters or membranes via two small (10-ml) chambers enabled sampling of the media near each side. The chamber design was flexible, and only one small chamber with a single filter in place could also be used when an appropriate blank was placed on the outside of the assembly to enable the pressure screw to be tightened. The syringes above the small connecting chamber were placed after venting of the chambers was completed, and air was removed by syringes set in these needles atop the chambers.

4. To obtain better material exchange, Teflon-coated magnets were used inside the small chambers. The magnetic stirrers for the larger chambers were suspended 1.5 cm from the bottom to avoid damage to suspended cells. The use of a dual-magnet stir table assured the smooth operation of all four magnets.

5. The chamber tops of Teflon made machining easy and allowed for a number of openings while assuring a tight fit. Needles with membrane filters were used for gas inlet and exit. These needles could also be modified for taking out cell samples. Sampling was carried out after removal of the apparatus from the incubator to a sterile airflow hood. When necessary, oxygen was monitored via a sterilized oxygen electrode. A set point on the electrode output allowed metering of air or oxygen as needed by the cells. Alternatively, the electrode opening was closed with a rubber stopper, as in Fig. 2. The bacteria were gased with nitrogen, and their oxygen was also monitored. These large chambers were gas tight, and introduction of gas into both chambers had to be balanced in order to avoid surges of suspension media between the chambers or up the gas inlet needles after accidental buildup of gas pressure.

Adequate aeration and renewal of 30% of the plant medium on the first 2 days permitted use of heavier suspensions of rapidly dividing cells with six to eight cells per clump. The dry weight of soybean cells in suspension varied from 1.3 to 1.7 mg/ml, optical density (OD_{540}) near 2; the mitotic index varied from 15 to 16% at the start to less than 4% by day 7 of a typical experiment, with the oxygen level in the plant cell chamber near 18%.

The bacteria (*Rhizobium japonicum*) were rediluted from late log phase to an OD_{660} of 0.4–0.5, $2–5 \times 10^8$ cells per millimeter at the start, and reached an OD of 0.8–1.2 by day 7, $2–3 \times 10^9$ cells per milliliter, with oxygen levels of 2–5%. Our usual media are now based on B5 medium without nitrate, but with ammonia (2 m*M*) and 0.08% casein hydrolysate. The pH of the plant side is 5.7 and that of the bacterial side is 6.2.

III. CELLS BAGGED IN DIALYSIS TUBING

A. Double-Bag Method

The dialysis cultures of microorganisms reviewed by Schultz and Gerhardt (1969) have proved useful in working out methods for coculture of plant cells with bacteria. The version of the method used for our requirements needed some practice. A laminar flow hood was essential for carrying out a variety of manipulations. The method was originally used to indicate that rhizobia of different species elaborate compound(s) capable of passing through Spectrapore tubing (Spectrum Industries, Los Angeles, California) with an MW cutoff of 3500. This material presumably stimulated the plant cells outside the bags in suspension culture (shaking rate 110 rpm at 26°C) to make plant cell-conditioned medium. Fractions separated from this medium were found to aid expression of nitrogenase activity *in vitro* in cultures of rhizobia which could not be derepressed for nitrogenase activity using chemically defined media (Bednarski and Reporter, 1978; Storey and Reporter, 1980).

The bacteria were placed at high cell density ($3–4 \times 10^9$/ml and 4 ml per bag) in a Spectrapore tubing no. 3. Bacteria from regular shake cultures in late log were centrifuged, redispersed, and then used in the bags. The use of a single layer of tubing was not considered sufficient to avoid containing the rhizobia, and a second larger tubing with an MW cutoff of 6500 was used. The double bag proved fortuitously to set conditions for nitrogenase derepression of rhizobia in the smaller inner bag. These bacteria had high respiratory activity. They also elaborated exopolysaccharides which coated the inside of the small bag and restricted material exchange. Presumably, these bacteria existed under microaerophilic conditions after 3–4 days of coculture.

The dialysis membrane bags used for these experiments were soaked in 0.1 m*M* EDTA for 4–6 hr, and the EDTA was removed by copious washing. The bags were knotted at one end and sterilized in water. During sterilization, small funnels and test tubes were used to keep the open ends separate and convenient to handle aseptically. The tubings were clipped shut after filling by the use of Spectrapore tubing clips and presterilized hemostats. Extensive washing of the completed dialysis bags was carried out by sequential dipping in sterile distilled water in a series of beakers. The double bags (two per flask) were then placed in a series of Erlenmeyer flasks containing 100 ml of plant cells in suspension culture. The assembled Erlenmeyer flasks were then shaken at 110 rpm in a New Brunswick shaker at 26°C. The bags were sampled at regular intervals to test derepression of nitrogenase in the treated rhizobia. Fluid contents of the outer and inner bags were assayed separately since bacteria were sometimes inevitably released from the inner bag. In experiments lasting for up to 10 days, bacteria were seldom found outside of the second, larger dialysis bag. The conditioned media obtained in this manner from the suspension cultures of plant cells were found difficult to sterilize by passage through 0.2-μm filters even after 1 hr of centrifugation at 100,000 *g*. Conditioned media obtained by this method contained carbohydrate materials that comigrated with active copper-containing peptides isolated from plant cells (M. Reporter, unpublished). These copper-containing peptides, presently under study, aid in the derepression of nitrogenase in many types of rhizobia (Storey and Reporter, 1980; Reporter *et al.*, 1981).

B. Single-Bag Method

Another version in which bacteria have been placed in dialysis tubing with plant cells in suspension culture outside was used at the Australian National University by Gresshoff. The plant cells were grown in suspension culture in Erlenmeyer flasks. After a desired level of plant cell mass had accumulated, the flask was capped with a rubber stopper in which a glass chimney was inserted. The top of the chimney was plugged with cotton. The bottom of the chimney had a dialysis bag attached. The bag was long enough to reach the bottom of the Erlenmeyer flask. Bacteria were suspended in the bottom of this dialysis bag. The flasks were then placed on gyrotary shakers. Bacteria could be conveniently sampled via the glass chimney and the plant cells after removal of the rubber stopper. Lowered oxygen tension, if desired, was obtained after gasing of the bag with argon or nitrogen and closing of the chimney with a rubber stopper. Manipulations were carried out in a sterile laminar flow hood (Gresshoff *et al.*, 1981).

Conditioned medium obtained by this method was difficult to obtain reproducibly (Gresshoff *et al.*, 1981). Active fractions for use in nitrogenase derepression assays were also obtained after gel filtration to remove inhibitors in the medium conditioned in this manner.

IV. COCULTURES OF IMMOBILIZED PLANT AND MICROBIAL CELLS

Cocultures of separately immobilized plant and microbial cells in bead form were used in our laboratory following the publication by Brodelius *et al.* (1979). Again, the purpose of our experiments (M. Reporter, unpublished) was to use the diffusion products of plant cells to derepress nitro-

genase from *Rhizobium*. Suspension cultures of soybean cells were screened to remove the large clusters of cells and retain single cells or clusters of up to 12–15 cells. The cells were collected after centrifugation and resuspended in B5 medium containing 2% or 1% alginate (Kelco Co., San Diego, California) and 2 m*M* $CaCl_2$. The lower alginate gave more open and softer beads. Initial attempts to form beads by dripping suspended cells into the same medium, but containing 50 m*M* $CaCl_2$, gave beads of irregular size. The drip rate of cells from a separatory funnel used for this purpose into the stirred high-calcium medium was affected by the size of the plant cell clusters. This problem was avoided in later runs by pumping the alginate medium containing plant cells via a peristaltic pump (Polystaltic by Buchler Instruments) and using presterilized Technicon Autoanalyzer tubing (internal diameter, 2 mm). Bead size could be controlled between 2.5 and 4.0 mm by changing pump speed and drop height. Bacterial cells were also made into beads in a similar fashion but using 2.5% alginate. The beads were washed extensively in appropriate normal medium before being mixed in minimal (1 ml) volumes of succinate (20 m*M*) and sealed in serum bottles for further incubation. Acetylene reduction was noted after coculture for 5 days. The initial oxygen concentration of 2.0% decreased to 0.3% in bottles that showed acetylene reduction activity. The oxygen was estimated by gas chromatography and the change in color of myoglobin (2 m*M*) that was preincorporated into the beads during fabrication. Using a constant number (20) of bacterial beads, each with 10^9 cells per milliliter of alginate medium, and increasing the number of beads of soybean cells (0–30), increasing amounts of acetylene reduction were shown by day 5. The increasing number of plant cells also consumed oxygen at a faster rate, and the included myoglobin presumably compensated for the rate of oxygen flow to the immobilized cells. When it was deemed useful to liquefy the alginate inside the beads, the procedure of Lim and Sun (1980) was used. The beads were treated with polylysine (0.02–0.03%, MW 35,000) for 10 min. The charge from polylysine aided in keeping beads intact for further manipulations. The beads were then treated with 1% $CaCl_2$ washed in water and then in 0.2% polyethyleneimine, and finally with sodium citrate (isotonic). Alternatively, cells could be freed by treatment with 1 *M* potassium phosphate (pH 6.5) any time during the experiment.
methods (Venkatsubramanian, 1979) (see Chapter 60, this volume).

V. CONCLUSIONS

Further work on the identification of both bacterial and plant substances which eventually aid in derepression of nitrogenase will be continued in our laboratory. Concentrated bacteria and plant cell products elaborated *in vitro* in minimal fluid volumes would be useful at this time. The use of small chambers (Lustig *et al.*, 1980a; Fig. 1) and immobilized cells will be helpful for this purpose.

It is hoped that these experiments will aid in the identification of substances from a variety of plant–microorganism associations, including phytotoxins and secondary metabolites.

REFERENCES

Bednarski, M. A., and Reporter, M. (1978). Expression of rhizobial nitrogenase: Influence of plant cell-conditioned medium. *Appl. Environ. Microbiol.* **36,** 115–120.

Berg, R. H., Vasil, V., and Vasil, I. K. (1979). The biology of *Azospirillum*-sugarcane association. II. Ultrastructure. *Protoplasma* **101,** 143–163.

Brodelius, P., Deus, B., Mosbach, K., and Zenk, M. H. (1979). Immobilized plant cells for the production and transformation of natural products. *FEBS Lett.* **103,** 97–97.

Child, J. J., and Kurz, W. G. W. (1978). Inducing effect of plant cells on nitrogenase activity by *Spirillum* and *Rhizobium* in vitro. *Can. J. Microbiol.* **24,** 143–148.

Chilton, M. D., DeFremond, A., Fraley, R., Barton, K., Matzke, A. J. M., Binns, A. N., Byrne, M., Koplow, J., David, C., and Tempé, J. (1983). Ti and Ri plasmids as vectors for genetic engineering of higher plants. *Miami Winter Symp.* **15,** 14 (abstr.).

Davey, M. R., and Cocking, E. C. (1979). Tissue and cell cultures and bacterial nitrogen fixation. *In* "Recent Advances in Biological Nitrogen Fixation" (N. S. Subba Rao, ed.), pp. 281–324. Oxford Publ., New Delhi.

DeMoranville, C. J., Kaminski, A. R., Barnett, N. M., Bottino, P. J., and Blevins, D. G. (1981). Substances from cultured soybean cells which stimulate or inhibit acetylene reduction by free-living *Rhizobium japonicum*. *Physiol. Plant.* **52,** 53–58.

Earle, E. D. (1978). Phytotoxin studies with plant cells and protoplasts. *In* "Frontiers of Plant Tissue Culture 1978" (T. A. Thorpe, ed.), pp. 363–372. Univ. of Calgary Press, Calgary, Alberta, Canada.

Forrai, T., Vincze, E., Banfalvi, Z., Kiss, G., Randhawa, G., and Kondorosi, A. (1983). Localization of symbiotic mutations in *Rhizobium meliloti*. *J. Bacteriol.* **153,** 635–643.

Gardner, J. M., Scheffer, R. P., and Higinbotham, N. (1974). Effects of host specific toxins on electropotentials of plant cells. *Plant Physiol.* **54,** 246–249.

Giles, K. L., and Vasil, I. K. (1980). Nitrogen fixation and plant tissue culture. *Int. Rev. Cytol., Suppl.* **11B,** 81–97.

Gresshoff, P. M., and Mohpatra, S. S. (1981). Legume cell and tissue culture. *In* "Tissue Culture of Economically Important Plants" (A. N. Rao, ed.), pp. 11–24.

Gresshoff, P. M., Carroll, B., Mohpatra, S. S., Reporter, M., Shine, J., and Rolfe, B. G. (1981). Host factors control of nitrogenase function. *In* "Current Perspectives in Nitrogen Fixation" (A. H. Gibson and W. E. Newton, eds.), pp. 209–212. Aust. Acad. Sci., Canberra.

Hardy, R. W. F., and Holsten, R. D. (1972). Symbiotic fixation of atmospheric nitrogen. U.S. Patent 3,704,546.

Hermina, N., and Reporter, M. (1975). Acetylene reduction by transfilter suspension cultures of *Rhizobium japonicum*. *Biochem. Biophys. Res. Commun.* **64,** 1126–1133.

Hermina, N., and Reporter, M. (1977). Root hair cell enhancement in tissue cultures from soybean roots: A useful model system. *Plant Physiol.* **59,** 97–102.

LaRue, T. A., Kurz, W. G. W., and Child, J. J. (1975). Methods for growing nitrogen-fixing bacteria separated from plant cells. *Can. J. Microbiol.* **21,** 1884–1886.

Lim, F., and Sun, A. M. (1980). Microencapsulated islets as bioartificial endocrine pancreas. *Science* **210,** 908–910.

Lustig, B., Plischke, W., and Hess, D. (1980a). Induction of nitrogenase activity associations between *Portulaca grandiflora* and *Rhizobium:* A simple transfilter apparatus for studying nitrogen fixation. *Z. Pflanzenphysiol.* **98,** 277–281.

Lustig, B., Plischke, W., and Hess, D. (1980b). Nitrogenase activity in a transfilter culture of rhizobia with a non-leguminous plant callus culture: Transfer of fixed $^{15}N_2$ from bacteria to *Portulaca* callus. *Experientia* **36,** 1386–1387.

Ozawa, T., and Yamaguichi, M. (1979). Inhibition of soybean cell growth by adsorption of *Rhizobium japonicum. Plant Physiol.* **64,** 65–68.

Paton, A. L. (1982). The induction of nonpathogenic infections in plants. *Proc. Int. Congr. Microbiol., 13th, 1982.*

Pence, V. C., Novick, N. J., Ozias-Akins, P., and Vasil, I. K. (1982). Induction of nitrogenase activity in *Azosprillum brasilense* by conditioned medium from cell suspension cultures of *Pennisetum americanum* (Pearl millet) and *Panicum* maximum. *Z. Pflanzenphysiol.* **106,** 139–147.

Reporter, M. (1976). Synergetic cultures of *Glycine max* root cells and rhizobia separated by membrane filters. *Plant Physiol.* **57,** 651–655.

Reporter, M., Mort, A., and Calvert, H. (1981). Exopolysaccharides are lost from *Rhizobium* before detection of nitrogenase *in vitro. Abstr., North Am. Rhizobium Conf., 8th, 1981* p. 58.

Samaddar, K. R., and Scheffer, R. P. (1968). Effect of specific toxin in *Helminthosporium victoriae* on host cell membranes. *Plant Physiol.* **43,** 21–28.

Schultz, J. S., and Gerhardt, P. (1969). Dialysis culture of microorganisms: Theory and results. *Bacteriol. Rev.* **33,** 1–47.

Storey, R., and Reporter, M. (1980). Plant peptidoglucans affecting phenotypic expression of rhizobial nitrogenase. *Aust. J. Plant Physiol.* **7,** 251–260.

Torrey, J. G. (1978). *In vitro* methods in the study of symbiosis. *In* "Frontiers of Plant Tissue Culture 1978" (T. A. Thorpe, ed.), pp. 373–380. Univ. of Calgary Press, Calgary, Alberta, Canada.

Venkatsubramanian, K. (1979). Immobilized microbial cells. *ACS Symp. Ser.* **106,** 1–12.

Verma, D., Hunter, N., and Bal, A. K. (1978). Asymbiotic association of *Rhizobium* with pea epicotyls treated with plant hormone. *Planta* **138,** 107–110.

CHAPTER **67**

Isolation and Bioassay of Fungal Phytotoxins

Jonathan D. Walton
Elizabeth D. Earle

Department of Plant Breeding and Biometry
Cornell University
Ithaca, New York

I. INTRODUCTION

Microorganisms produce an immense variety of secondary metabolites that are useful for commercial and scientific purposes. We will use the term *phytotoxin* to describe those secondary metabolites with molecular weights less than about 2000 that can perturb the normal growth of higher plants. Most phytotoxins inhibit plant growth and development, but some stimulate growth at certain concentrations. Although we will restrict our discussion to fungal phytotoxins, many of our points are also applicable to extraction and bioassay of bacterial phytotoxins. An excellent volume of reviews on many aspects of phytotoxins has recently appeared (Durbin, 1981a).

Fungal phytotoxins have been studied by plant biologists for several reasons (Durbin, 1981b). In a few cases, it has been shown that they are

CELL CULTURE AND SOMATIC CELL
GENETICS OF PLANTS, VOL. 1

ISBN 0-12-715001-3

essential for the phytopathogenicity of the fungi that produce them. They can be useful as inhibitors or metabolite analogs for studying processes such as photosynthesis or the mode of action of hormones. They can also be used as selection agents for tissue culture work.

A nonspecific phytotoxin will affect most or all plant species in a similar way at a similar concentration, although the producing fungus itself may be quite restricted in habitat or plant host. Presumably, nonspecific toxins affect metabolic processes common to all or most plants. Nonspecific toxins such as fusicoccin (Marré, 1979) are useful for physiological experiments since no particular species or genotype is required. Nonspecific toxins are generally active at micrograms per milliliter concentrations.

Host-specific toxins are conceptually important to the field of plant pathology because they have the same species, variety, or gene specificity as the producing fungi and can thus account for the restricted host range of these fungi. Host-specific toxins are known with activity against particular genotypes of oat, maize (cytoplasmic and nuclear), sorghum, sugarcane, pear, apple, mandarin orange, tomato, rough lemon, and strawberry (Kono *et al.*, 1981). Some host-specific toxins (e.g., victorin and T toxin) are active at nanogram per milliliter concentrations, and others at microgram per milliliter concentrations.

II. PREPARATION OF FUNGAL PHYTOTOXINS

Fungal phytotoxins for research purposes can either be purchased (Tentoxin: Sigma Chemical Co., St. Louis, Missouri), obtained from other investigators, or produced in the laboratory. Fungal cultures can be obtained from the American Type Culture Collection (Mycology Department, Rockville, Maryland), from other investigators (permits to ship fungi, especially plant pathogens, should be obtained from the local state department of agriculture), or from soil or infected plant material (Tuite, 1969; Stevens, 1974). Identification or confirmation of the fungal species is usually based on spore morphology (Barron, 1968; Ainsworth *et al.*, 1973).

Growing fungi in the laboratory requires equipment similar to that used for growing plant tissue cultures. Although "dusty spore" formers such as *Penicillium* and *Aspergillus* have given fungi a bad reputation as culture contaminants, many fungi sporulate poorly in culture, and their spores are specialized for purposes other than massive production and rapid air dispersal. We routinely grow species of the plant pathogen *Helminthosporium*, which produces few and large spores, in a tissue culture laboratory.

Fungi can be maintained on agar using any of several media, such as commercial potato dextrose agar (PDA) or oatmeal agar (Tuite, 1969). Some fungi stop producing secondary metabolites such as toxins after repeated subculture on artificial medium. To avoid this possibility, plant pathogenic fungi can be stored desiccated as lesions on plant material and reisolated as needed.

Fungal phytotoxins have been isolated from infected plants, from fungal mycelium, or, most commonly, from liquid medium inoculated with fungus and allowed to sit for 14–28 days ("culture filtrates"). The reader is referred to Shaw (1981) for a discussion of the environmental and nutritional factors that can be important in the production of fungal toxins in culture. An important consideration is whether toxin production is facultative or constitutive; if the former, then culture conditions that induce toxin production must be found, for example, by addition of host plant material or a special nutrient to the culture medium (Wheeler and Gantz, 1981; Kohmoto *et al.*, 1979; Larkin and Scowcroft, 1981).

A. Isolation

For critical experiments it is best to use purified preparations of a toxin, but substantial excellent work has been done with partially purified preparations. For procedures in which the active agent is unknown, crude preparations may be preferred, for example, when selecting tissue cultures resistant to cell-free preparations of pathogens. In any case, the compound(s) of interest should be removed from the majority of contaminating chemicals such as the residual culture medium. The following preliminary steps are effective and easily done in any biological laboratory.

1. Filtration. Filtration of inoculated culture media through cheesecloth, nylon mesh, and crude cellulose paper such as Whatman no. 1 will remove fungal mycelium, spores, and precipitates of medium salts, macromolecules, and so on. Membrane filters will remove very fine particulate matter including contaminating microorganisms, but are slower and more expensive, and will sometimes retain low molecular weight polysaccharides and peptides.

2. Concentration. Most toxins are isolated from substrates in which they are present in very dilute concentrations, so large volumes of starting material are required for reasonable final yields. Concentration of the toxin-containing solutions reduces the filtrate to a more convenient volume. The simplest technique for concentration is rotary evaporation under reduced pressure. Evaporation is faster at elevated temperatures, but ther-

mal instability of the toxin may be a problem. Lyophilization has also been used to concentrate culture filtrates.

3. Alcohol precipitation. Most (all?) low molecular weight phytotoxins are alcohol soluble. Cold 50–80% methanol or ethanol precipitates high molecular weight macromolecules, which can then be removed by filtration. Removal of contaminating high molecular weight compounds before solvent extraction minimizes formation of emulsions. Methanol or ethanol should be removed by evaporation before solvent extraction.

4. Solvent extraction. If the compound of interest is soluble in a water-immiscible organic solvent such as ether or chloroform, extraction of culture filtrates with that solvent can remove a large proportion of the dry weight, especially culture medium salts. The pH may influence partitioning into organic solvents of compounds with ionizable functional groups such as carboxylic acids and amines, and this can be used to advantage.

B. Purification

Further purification of phytotoxins requires more sophisticated techniques, the most popular being thin-layer chromatography (TLC), adsorption chromatography, gel filtration, ion-exchange chromatography, and high-pressure liquid chromatography (HPLC). TLC and adsorption chromatography using, for example, silica, charcoal, or alumina, are inexpensive, relatively easy and fast, and have moderate resolution (Pringle and Braun, 1957; Kohmoto *et al.*, 1979). They are excellent initial purification procedures. TLC is more reproducible than adsorption chromatography but has a lower capacity. Samples in organic solvents such as chloroform can be applied directly to both. Conventional gel filtration separates molecules according to size and is not very suitable for separating low molecular weight compounds. Newer types of gel filtration media such as Sephadex LH (Pharmacia Co., Piscataway, New Jersey) combine gel filtration and adsorption and partition chromatography.

Ion-exchange chromatography separates molecules on the basis of any ionizable functional groups. The column is loaded under conditions of ionic strength and pH that cause the compound of interest to be retained; the conditions are then changed to release it as specifically as possible. Extremes of pH at which the toxin might be unstable and high salt concentrations in the eluted sample are potential problems.

The technique of choice for toxin purification is HPLC. It has extremely good resolution and reproducibility, and a large literature of protocols is accumulating for separation of many types of toxins, antibiotics, and other secondary metabolites. Because of its high resolution, retention time on HPLC can provide *prima facie* evidence that two compounds are identical. Modern HPLC systems are equipped with detectors for ultraviolet (UV) and visible light absorption, refractive index, and/or fluorescence. Columns of many different types are available, although the conventional reverse-phase columns are the most versatile and convenient. HPLC columns are classified as analytical or preparative, depending on their size and concomitant capacity, but even analytical columns can be loaded with 0.5–1.0 mg material in up to a 2.0-ml volume. By using polar organic solvents and volatile buffers such as ammonia and trifluoroacetic acid, HPLC can be used as a final purification step. The major disadvantages of HPLC are the initial expense and the time needed to develop new separation protocols. Techniques for separation of many low molecular weight organic molecules have been developed (Johnson, 1977), but as yet HPLC has been used only occasionally for purification of phytotoxins (Wolpert and Dunkle, 1980; Walton *et al.*, 1982; Walton and Earle, 1984).

III. BIOASSAY OF FUNGAL PHYTOTOXINS

Bioassays are sometimes more sensitive, reliable, and convenient than chemical assays (Yoder, 1981). Choice of a suitable bioassay depends on the goals of the investigator. Preliminary experiments are usually done by monitoring some gross parameter such as cell death or inhibition of growth. This may be sufficient for following the purification of a compound or for using toxins as selection agents in tissue culture. As the action of a toxin at the cellular and subcellular levels becomes known, new bioassays with advantages over cruder ones may be devised. Bioassays such as root growth inhibition are more prone to interference by contaminating molecules in partially purified toxin preparations than an assay based on, for example, inhibition of a particular enzymatic reaction.

A good bioassay should be reproducible, quantitative, and reliable; require small amounts of the available phytotoxin preparation; require small amounts of test material; and be as specific as possible for the compound of interest. We will discuss in detail two widely applicable bioassays.

A. Root Growth Bioassay

The root growth bioassay has been used to screen for inhibitory phytotoxins, especially from plant pathogens (Scheffer and Ullstrup, 1965; Walton *et al.*, 1982). It is fast, easy, and semiquantitative. Since roots grow relatively quickly, this assay is quite sensitive if the assay is run for 2 or more days. It will respond to inhibitors that cause general cell death or inhibit cell elongation, differentiation, or division. It is not suitable for toxins that have the wrong tissue specificity, for example, photosynthesis inhibitors such as tentoxin.

Protocol

Use 6–10 seeds per treatment with two to three replicates. Germinate the seeds in darkness for 36–64 hr, depending on the species, wrapped loosely in damp paper towels. Pregermination of the seeds is not essential but allows selection for healthy, uniform seedlings. For quantitative work, each root should be measured at the time it is put into the toxin solution, so that its growth can be measured as the final length minus the initial length. Alternatively, only roots with lengths between two values can be used, or no measurement made at all.

Pipette the toxin solution to be tested onto a piece of 9-cm-diameter Whatman no. 1 filter paper (the paper prevents sloshing of the solution and ensures uniform seedling–solution contact), allow the solvent to air dry (important only if the solvent is noxious to roots), and place the filter paper in a standard 100-mm glass Petri plate (solutions tend not to disperse evenly in plastic plates). Add 5 ml distilled water or 10 m*M* KH_2PO_4, pH 6.0, to each plate. If the phytotoxin is not water soluble, 1% dimethylsulfoxide (DMSO) or 1% ethanol can be added.

Add the roots last. They should grow along the paper and thus stay aerated but in constant contact with the toxin. For maize, three or four seeds per plate are optimal. Roots can grow around the inner wall of the plate without being inhibited.

Allow growth to occur for 24–72 hr in darkness. Longer times are less convenient but give larger absolute differences in elongation between control and toxin treatments. If the initial length is between about 20 and 40 mm, primary root growth in maize is linear for about 72 hr, but then begins to slow down. The root growth bioassay can be qualitative (growth or no growth) or semiquantitative (Walton *et al.*, 1982).

B. Protoplast Bioassay

Assays involving protoplasts require very small amounts of phytotoxin preparations and plant material. They avoid problems associated with poor penetration of tissues, permit fairly rapid testing of many different samples, and require no prior information about the nature of toxin action on sensitive cells. However, assays using protoplasts are technically more sophisticated than root bioassays. Moreover, they may not be sensitive to toxins that act only on tissues or that affect cell types or biological activities not seen in the protoplast population used (Earle, 1978).

Protocol

Isolate the protoplasts by standard methods from leaves of the species or genotypes to be used (Chapters 38–43, this volume). Suspend the protoplasts either in a nutrient medium or in a simple osmoticum such as 0.5 *M* sorbitol plus 10 m*M* $CaCl_2$ plus 5 m*M* 2-(*N*-morpholino)-ethanesulfonic acid (MES) buffer, pH 6.0 (Earle *et al.*, 1978). Mix the protoplasts with aliquots of toxin in containers such as 35-mm disposable plastic Petri dishes ($1–2 \times 10^5$ protoplasts in 1–2 ml of medium per plate) or 24-well multiwell plates ($1–5 \times 10^4$ protoplasts in 0.5 ml of medium). Incubate protoplasts both in the light and in the dark to determine whether toxin action is influenced by culture conditions.

Protoplast preparations can be examined *in situ* with an inverted microscope (150–300X) or sampled for examination in a hemocytometer. Turgor, shape, size, integrity, and distribution of chloroplasts and other cytological features of toxin-treated protoplasts can be compared to those of control populations either qualitatively or quantitatively. Sometimes toxin-treated protoplasts are completely collapsed within 1–3 days, whereas control protoplasts remain spherical and turgid (Earle *et al.*, 1978). In such cases, rapid *in situ* screening of large numbers of toxin fractions in multiwell plates is feasible.

Possible quantitative measures of toxin effects include protoplast diameter and the number or percentage of protoplasts that are healthy as judged by morphology, staining with fluorescein diacetate, exclusion of Evans Blue stain, cytoplasmic streaming, and so on. With species whose protoplasts divide readily, toxin effects on division and colony formation can be noted. However, in many cases, effects even on undivided protoplasts are obvious within a few days.

With T toxin and maize, bioassays based on microscopic observation of protoplasts proved as sensitive as those following toxin effects on isolated mitochondria (Gregory *et al.*, 1980). However, assays of toxin effects on biochemical activities of protoplasts (e.g., inhibition of uptake or incorporation of labeled precursors) may be faster, more sensitive, or more convenient (Breiman and Galun, 1981).

IV. CONCLUSIONS

Exposure of plant cell and tissue cultures to phytotoxins can serve several purposes (Earle, 1978). The features that make protoplasts convenient for toxin bioassays (e.g., rapid penetration of toxin, sensitivity, ease of sampling and microscopic observation) have also made them valuable material for physiological, biochemical, and ultrastructural studies of toxin action (Rancillac *et al.*, 1976; Walton *et al.*, 1979; York *et al.*, 1980). Partially purified preparations of host-specific toxins have generally been used in such experiments.

Interest in phytotoxins as screening agents for *in vitro* selection of disease-resistant material dates back to the work of Carlson (1973) with tobacco. The most significant accomplishment in this area is selection of maize tissue resistant to culture filtrates from *Helminthosporium maydis* race T and subsequent regeneration of plants with maternally inherited resistance to T toxin and to *H. maydis* race T (Gengenbach *et al.*, 1977; Brettell *et al.*, 1980). The plants recovered were not agronomically useful since they no longer showed the desirable Texas type of male sterility, but they have provided material of considerable interest for studies of the molecular basis of disease resistance (Dixon *et al.*, 1982). More recent attempts to use fungal toxins for *in vitro* selection have involved potato cultures and *Phytophthora infestans* and *Fusarium oxysporum* toxins (Behnke, 1980a,b), and *Brassica napus* cultures and *Phoma lingam* toxins (Sacristan, 1982). Plants regenerated from material resistant to the toxic culture filtrates showed some degree of enhanced resistance to the relevant pathogens, but the agronomic significance and genetic basis of such resistance are not yet clear. It is notable that in several cases (Brettell *et al.*, 1980; Sacristan, 1982) some plants regenerated from unselected control cultures also showed enhanced resistance.

Phytotoxins also have potential applications in genetic manipulations such as protoplast fusion. Fusion of maize protoplasts sensitive to T toxin with resistant protoplasts gives resistant fusion products (Earle, 1982). Unfortunately, use of the many host-specific *Helminthosporium* toxins as selective agents after fusion is limited by current difficulties in obtaining division of cereal protoplasts. Aviv and Galun (1980) have used resistance to tentoxin in the analysis of cybrid plants recovered after fusion of resistant and sensitive tobacco protoplasts.

REFERENCES

Ainsworth, G. C., Sparrow, F. K., and Sussman, A. S., eds. (1973). "The Fungi," Vol. 4A. Academic Press, New York.

Aviv, D., and Galun, E. (1980) Restoration of fertility in cytoplasmic male sterile (CMS) *Nicotiana sylvestris* by fusion with X-irradiated *N. tabacum* protoplasts. *Theor. Appl. Genet.* **58,** 121–127.

Barron, G. L. (1968). "The Genera of Hyphomycetes from Soil." Williams & Wilkins, Baltimore, Maryland.

Behnke, M. (1980a). General resistance to late blight of *Solanum tuberosum* plants regenerated from callus resistant to culture filtrates of *Phytophthora infestans*. *Theor. Appl. Genet.* **56,** 151–152.

Behnke, M. (1980b). Selection of dihaploid potato callus for resistance to the culture filtrate of *Fusarium oxysporum*. *Z. Pflanzenzuecht.* **85,** 254–258.

Breiman, A., and Galun, E. (1981). Plant protoplasts as tools in quantitative assays of phytotoxic compounds from culture filtrates of *Phytophthora citrophthora*. *Physiol. Plant Pathol.* **19,** 181–191.

Brettell, R. I. S., Thomas, E., and Ingram, D. S. (1980). Reversion of Texas male-sterile cytoplasm maize in culture to give fertile T-toxin resistant plants. *Theor. Appl. Genet.* **58,** 55–58.

Carlson, P. S. (1973). Methionine sulfoximine-resistant mutants of tobacco. *Science* **180,** 1366–1368.

Dixon, L. K., Leaver, C. J., Brettell, R. I. S., and Gengenbach, B. G. (1982). Mitochondrial sensitivity to *Drechslera maydis* T-toxin and the synthesis of a variant mitochondrial polypeptide in plants derived from maize tissue cultures with Texas male-sterile cytoplasm. *Theor. Appl. Genet.* **63,** 75–80.

Durbin, R. D., ed. (1981a). "Toxins in Plant Disease." Academic Press, New York.

Durbin, R. D. (1981b). Applications. *In* "Toxins in Plant Disease" (R. D. Durbin, ed.), pp. 495–505. Academic Press, New York.

Earle, E. D. (1978). Phytotoxin studies with plant cells and protoplasts. *In* "Frontiers of Plant Tissue Culture 1978" (T. A. Thorpe, ed.), pp. 363–372. Univ. of Calgary Press, Calgary, Alberta, Canada.

Earle, E. D. (1982). Cytoplasm-specific effects of *Helminthosporium maydis* race T toxin on corn protoplasts and mitochondria. *In* "Variability in Plants Regenerated from Tissue Culture" (E. D. Earle and Y. Demarly, eds.), pp. 351–367. Praeger, New York.

Earle, E. D., Gracen, V. E., Yoder, O. C., and Gemmill, K. P. (1978). Cytoplasm-specific effects of *Helminthosporium maydis* race T toxin on survival of corn mesophyll protoplasts. *Plant Physiol.* **61,** 420–424.

Gengenbach, B. G., Green, C. E., and Donovan, C. M. (1977). Inheritance of selected pathotoxin resistance in maize plants regenerated from cell cultures. *Proc. Natl. Acad. Sci. U.S.A.* **74,** 5113–5117.

Gregory, P., Earle, E. D., and Gracen, V. E. (1980). Effects of purified *Helminthosporium maydis* Race T toxin on the structure and function of corn mitochondria and protoplasts. *Plant Physiol.* **66,** 477–481.

Johnson, E. L. (1977). "Liquid Chromatography Bibliography." Varian, Palo Alto, California.

Kohmoto, K., Scheffer, R. P., and Whiteside, J. O. (1979). Host-selective toxins from *Alternaria citri*. *Phytopathology* **69,** 667–671.

Kono, Y., Knoche, H. W., and Daly, J. M. (1981). Structure: Fungal host-specific. *In* "Toxins in Plant Disease" (R. D. Durbin, ed.), pp. 221–257. Academic Press, New York.

Larkin, P. J., and Scowcroft, W. R. (1981). Eyespot disease of sugarcane. Induction of host-specific toxin and its interaction with leaf cells. *Plant Physiol.* **67,** 408–414.

Marré, E. (1979). Fusicoccin: A tool in plant physiology. *Annu. Rev. Plant Physiol.* **30,** 273–288.

Pringle, R. B., and Braun, A. C. (1957). The isolation of the toxin of *Helminthosporium victoriae*. *Phytopathology* **47,** 369–371.

Rancillac, M., Kaur-Sawhney, R., Staskawicz, B., and Galston, A. W. (1976). Effects of cycloheximide and kinetin pretreatments on responses of susceptible and resistant *Avena* leaf protoplasts to the phytotoxin victorin. *Plant Cell Physiol.* **17,** 987–995.

Sacristan, M. D. (1982). Resistance response to *Phoma lingam* of plants regenerated from selected cell and embryogenic cultures of haploid *Brassica napus*. *Theor. Appl. Genet.* **61,** 193–200.

Scheffer, R. P., and Ullstrup, A. J. (1965). A host-specific toxic metabolite from *Helminthosporium carbonum*. *Phytopathology* **55,** 1037–1038.

Shaw, P. D. (1981). Production and isolation. *In* "Toxins in Plant Disease" (R. D. Durbin, ed.), pp. 21–44. Academic Press, New York.

Stevens, R. B., ed. (1974). "Mycology Handbook." Univ. of Washington Press, Seattle.

Tuite, J. (1969). "Plant Pathological Methods: Fungi and Bacteria." Burgess, Minneapolis, Minnesota.

Walton, J. D., and Earle, E. D. (1984). Characterization of the host-specific phytotoxin victorin by high pressure liquid chromatography. *Plant Sci. Lett.* **34** (in press).

Walton, J. D., Earle, E. D., Yoder, O. C., and Spanswick, R. M. (1979). Reduction of ATP levels in susceptible maize mesophyll protoplasts by *Helminthosporium maydis* race T toxin. *Plant Physiol.* **63,** 806–810.

Walton, J. D., Earle, E. D., and Gibson, B. W. (1982). Purification and structure of the host-specific toxin from *Helminthosporium carbonum* race 1. *Biochem. Biophys. Res. Commun.* **107,** 785–794.

Wheeler, H., and Gantz, D. (1981). Victorin production stimulated by oat flakes. *Phytopathology* **71,** 853–854.

Wolpert, T. J., and Dunkle, L. D. (1980). Purification and partial characterization of host-specific toxins produced by *Periconia circinata*. *Phytopathology* **70,** 872–876.

Yoder, O. C. (1981). Assay. *In* "Toxins in Plant Disease" (R. D. Durbin, ed.), pp. 45–78. Academic Press, New York.

York, D. W., Earle, E. D., and Gracen, V. E. (1980). Ultrastructural effects of *Helminthosporium maydis* race T toxin on isolated corn mitochondria and mitochondria within corn protoplasts. *Can. J. Bot.* **58,** 1562–1570.

CHAPTER **68**

Freeze Preservation of Cells

Lyndsey A. Withers

Department of Agriculture and Horticulture
School of Agriculture
University of Nottingham
Sutton Bonington, Loughborough, England

I. INTRODUCTION

There is a general requirement for good, reproducible germplasm storage methods throughout plant tissue culture, but the need is particularly pressing in biotechnology and genetic conservation (Withers, 1980a, 1983). Genetic instability, risks of loss through microbial contamination or equipment failure, and the sheer expense of maintaining stock cultures are all contributory factors. Although a range of storage techniques involving growth limitation and freeze preservation exists for organized cultures, only the latter is available at present for cell cultures.

CELL CULTURE AND SOMATIC CELL
GENETICS OF PLANTS, VOL. 1

ISBN 0-12-715001-3

The first success was reported in 1968 by Quatrano, who froze cells of flax (*Linum usitatissimum*) to a temperature of −50°C while maintaining viability at a level of 14%. This achievement must be noted with reservations since, as subsequent work has shown, a positive viability test response does not necessarily guarantee recovery of a culture capable of growth by an increase in cell number. Further, as is apparent from both theoretical and practical treatments of freeze preservation, −50°C is not an adequately low temperature for stable, long-term storage (Meryman and Williams, 1982). Nonetheless, during the decade following Quatrano's report, his general approach (slow freezing after treatment with a cryoprotectant) was vindicated by the publication of studies in which 10 more species were frozen to the temperature of liquid nitrogen, surviving in terms of viability test response and/or recovery growth.

From 1978 to the present, the number of species succumbing to freeze preservation as cells has expanded to over 30, and routine techniques can now be recommended (see Withers, 1982, 1983, 1984 for details and references). Where once deficiencies lay in the methodology, it is now probably fair to say that takeup of techniques lags behind their development.

II. PROCEDURE

A. Background

In the development of freeze preservation procedures, much attention has been given to the importance of exogenous factors. Thus, it is necessary to optimize freezing and thawing rates and cryoprotectant treatments. From early critical examinations by Nag and Street (1973, 1975a,b) based upon the classical roots of cryobiology (Meryman, 1965; Meryman and Williams, 1982) and more recent ones by Withers and King (1980; Withers, 1980b) are derived the common approaches of many current studies. However, probably more important than the examination of exogenous factors has been the emergence of the role played by the physiological condition of the cells before freezing and after thawing.

We now know that even given ideal conditions during freeze preservation, success will not be assured unless cultures are pregrown and recovered under appropriate conditions. Again, these factors are given early mention by Nag and Street (1973, 1975a,b). Empirical examination continued in several studies (Kartha *et al.*, 1982; Maddox *et al.*, 1983; Withers and King, 1979, 1980; Withers and Street, 1977a,b), physiological analysis being tackled by the work of Sala, Cella, and colleagues (Cella *et al.*, 1978, 1982; Sala *et al.*, 1979) and Pritchard and colleagues (Pritchard *et al.*, 1982). Electron microscopic studies (e.g., Pritchard *et al.*, 1982; Withers, 1978; Withers and Davey, 1978) have played some part in the development of techniques, although in corroboration rather than innovation.

B. Basic Procedure

The freeze preservation procedure to be recommended here is based upon that of Withers and King (1980; summarized in Fig. 1), which has proved its reproducibility in a wide range of test species (Withers, 1983, 1984). Accepting the fact that there is still room for improving the technique in general and making fine adjustments for particularly difficult species, the author will give alternative treatments here wherever they are considered helpful. The procedure is broken down into the stages described in the following sections. (Further details of these stages, with relevant references to supporting work, can be found in Withers, 1980a,c,d, 1982, 1983, 1984).

C. Choice of Cultures and Pregrowth Conditions

A close correlation is found between growth cycle stage and survival potential, the relatively small and/or highly cytoplasmic cells of the late lag phase or early exponential phase having the highest freeze tolerance. However, in some species, further intervention is necessary, involving frequent subculturing to minimize lag and stationary phases, and pregrowth in a supplemented medium. Accordingly, exponentially growing cells should be transferred to a passage of 4–7 days in medium supplemented with 6% (w/v) mannitol. Alternative additives reported in the literature for individual species include 6–15% (w/v) sorbitol (Maddox *et al.*, 1983; Weber *et al.*, 1983), 5% (v/v) dimethylsulfoxide (DMSO; Kartha *et al.*, 1982), and 10% (w/v) proline (Withers and King, 1980).

Culture in supplemented medium may depress growth and induce some darkening of the cells. These effects are not intrinsically serious, but viability should be monitored, for example, by fluorescein diacetate staining (see Section II,I,1) if there is any apparent toxicity. An occasional effect of pregrowth in supplemented medium is an increase in aggregate size. Filtering may be employed to overcome both this and a naturally high degree of aggregation.

To avoid falling below the minimum cell density permissive of growth, cells should be concentrated by sedimentation immediately before preser-

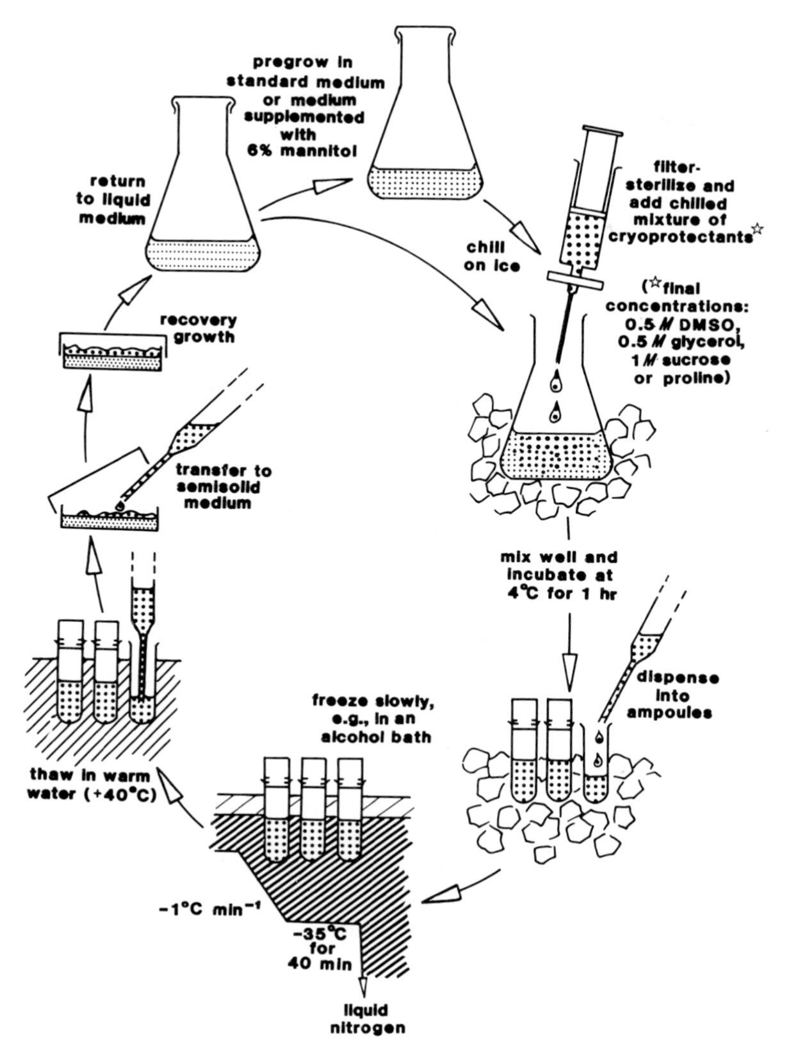

Fig. 1. Routine freeze preservation procedure for cell cultures. A freezing unit appropriate for carrying out this procedure is shown in detail in Fig. 2A. (Based on Withers and King, 1980.)

vation, taking into account dilution by cryoprotectants and a possible viability loss of up to ca. 50% during freezing.

D. Cryoprotection

The application of cryoprotectant compounds is essential to the survival of frozen cells. Cryoprotectants in combination are more generally successful than single compounds (Finkle and Ulrich, 1979), and a mixture of 0.5 *M* DMSO, 0.5 *M* glycerol, and 1 *M* sucrose or proline is recommended. A mixture of DMSO and glycerol alone, each at ca. 10% (v/v), is suggested as an alternative in several studies, but only in one case (Maddox *et al.*, 1983) is there evidence to suggest superiority over the three-component mixture. Occasionally, DMSO at 5–10% (v/v) may be effective. In one exceptional case where extreme sensitivity to DMSO was found (Weber *et al.*, 1983), sorbitol at ca. 18% (w/v) was adopted as cryoprotectant.

Cryoprotectants are prepared at double strength in standard culture medium, not water (see Withers, 1980b), their pH adjusted back to that of the unadulterated culture medium, and then sterilized by microfiltration. (autoclaving will cause caramelization). A universally suitable method of cryoprotectant application is not apparent, but good results have been found with the following convenient procedure: Equal volumes of cell suspension and cryoprotectant solution are chilled to ca. 4°C. The dense, viscous cryoprotectant solution is then poured into the cell suspension (not vice versa), which is gently swirled to ensure mixing and then placed on a rotary shaker at 4°C for 1 hr to permit cryoprotectant uptake.

The cryoprotected cells are transferred to containers for freezing. Presterilized polypropylene ampoules of 2-ml capacity (normally half filled in use) are particularly suitable. Although glass containers may be used with great care (Kartha *et al.*, 1982), they are not recommended since they may explode if liquid nitrogen penetration occurs. Heat-sealed plastic-aluminium foil envelopes may offer some practical advantages over ampoules (Takeuchi *et al.*, 1980; L. A. Withers, unpublished observations).

Careful identification of samples is essential. Once frozen, they look very similar, and close examination is neither easy nor wise (they may thaw out!). Indelible marking of polypropylene ampoules is best achieved by etching with a hot needle. This, followed by organized location in a strictly cataloged refrigerator (see below), should permit easy retrieval.

E. Freezing

Cells require protective dehydration by slow or stepwise freezing to minimize ice damage (Meryman and Williams, 1982). A freezing program involving cooling at ca. 1°C/min to −35°C, followed by holding at that temperature for ca. 40 min before plunging into liquid nitrogen, is recommended as a starting point. Minor adjustments of holding temperature ranging from −30 to −40°C and a holding time ranging from 0 to 60 min may be beneficial, but there is little evidence to support the necessity for freezing slowly below ca. −40°C. Purpose-built freezing units are available.

However, considerable success has been achieved using the improvized units illustrated in Fig. 2, which can offer adequately reproducible cooling rates.

F. Storage

Stored specimens must be maintained at a sufficiently low temperature, that is, below −150°C. This can be achieved satisfactorily only by the use of a liquid nitrogen-cooled refrigerator. Of the various models available, one in which the specimens are held in the cold vapor phase and located in stacks of drawers is preferred.

G. Thawing

Specimens should normally be thawed rapidly to avoid damage by ice recrystallization. This is best achieved by dropping three or four ampoules into jars containing ca. 200 ml sterile distilled water held at +40°C in a water bath. The sterile water minimizes the risk of introducing microbial contaminants, and the use of a covered jar facilitates agitation of the thawing specimens while containing any violence resulting from the thawing process. (Liquid nitrogen may have penetrated the ampoules, causing sudden expansion upon warming.) Thawed ampoules should be transferred to an ice bath in readiness for returning to culture and viability testing.

Exceptionally, some specimens may be unharmed by and even benefit from slow thawing (Withers, 1979, 1980b). If damage due to factors such as overdehydration during cryoprotection and freezing is suspected, then slow thawing by placing the specimen ampoule in the sterile airflow of the work station may be worth investigating.

H. Postthaw Treatments and Recovery

In wide-ranging studies, Withers and King (1979, 1980; Withers, 1979, 1980b, 1983) have found consistently that washing of freshly thawed cells is never beneficial and is often highly deleterious. This is supported by evidence of sensitivity during the early postthaw period to deplasmolysis injury and loss of solutes through leaky membranes (Cella *et al.*, 1978, 1982). However, accepting the fact that future studies may reveal some benefits of washing (e.g., removal of cryoprotectants to prevent toxicity in

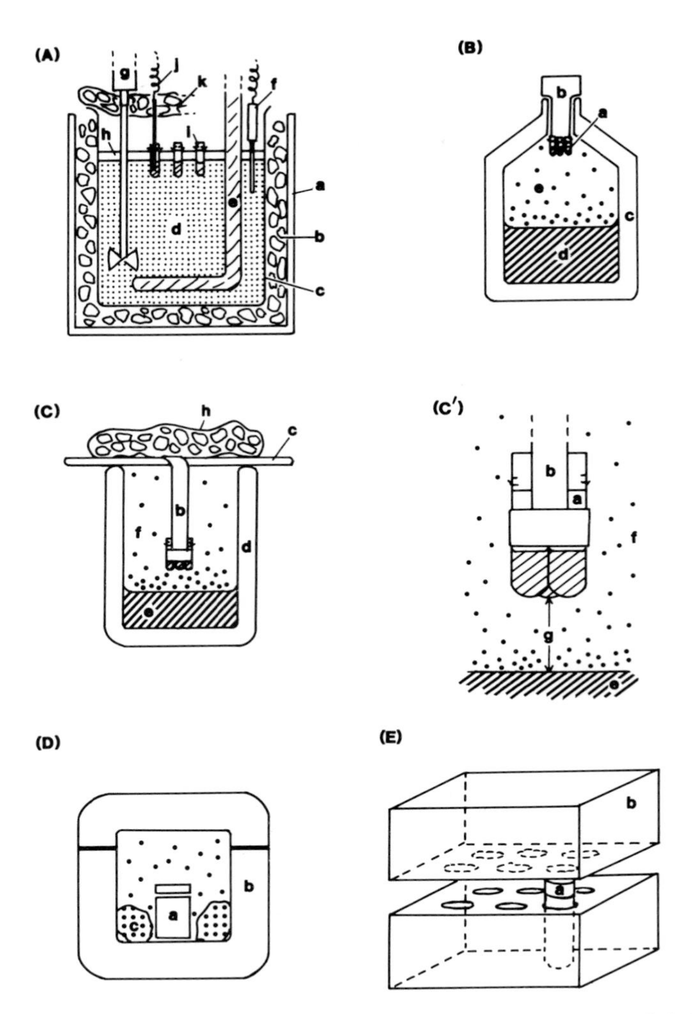

Fig. 2. Improvised freezing units. (A) The unit consists of an electrically cooled alcohol bath on the surface of which are floated the specimens (a: plastic bin; b: polystyrene packing material; c: glass beaker or metal cannister; d: alcohol, e.g., industrial methylated spirit; e: dip cooler; f: thermostat to cooler; g: stirrer; h: polystyrene raft; i: specimen ampoule; j: thermometer for recording specimen temperature; k: bag of polystyrene packing material). (B) This unit and the one shown in (C) achieve slow cooling by exposing the specimens to cold nitrogen vapor above a body of liquid nitrogen. The specimen ampoules (a) are attached to the plug (b) of a Dewar vessel (c). The vessel contains liquid nitrogen (d), which generates a cold atmosphere (e). The commercial apparatus used, for example, by Nag and Street (1973) (Union Carbide BF-6 Biological Freezer) works on this simple principle. (C) The specimen ampoules (a, detailed in C′) are held together by adhesive tape (b) and suspended from a glass rod (c) placed across the neck of a vacuum flask (d). The flask contains liquid nitrogen (e). The specimens are located in the cold nitrogen gas (f) at a known distance (g) from the surface of

long-term exposure), it is noted that damaging effects may be reduced by washing with warm medium (Finkle and Ulrich, 1982), washing with medium supplemented with the same additives as the pregrowth medium (Maddox *et al.*, 1983), or slow dilution of the medium supplements (Weber *et al.*, 1983).

Thawed cells should be simply pipetted or poured, with the cryoprotectant-containing medium, onto a plate of standard semisolid medium. Recovery may be accelerated by pushing cells together into a mound using a spatula and, after a few days, draining away any unincorporated liquid medium (P. J. King, personal communication). Reports of recovery on semisolid medium are balanced by a similar number involving liquid medium. However, in terms of rate of recovery and maintenance of viability, it would appear that in critical comparisons, the former is more successful (Withers and King, 1979).

In one reported instance, liquid culture was favored since it was found to avoid a callus stage in the recovery process (Maddox *et al.*, 1983). This point requires further examination since, in the experience of the author, cells recovering in a liquid medium layer over solidified medium tend to form a slurry which quickly disperses when returned to liquid medium. Cells can usually be resuspended in liquid medium within 2 weeks of thawing, cell division having resumed within the first 2–3 days (see Withers, 1980c,d).

When the percentage of viability is low, or when it is suspected that the thawed cells are seriously impaired physiologically, standard medium may be inadequate for recovery. The use of feeder layers from which the recovering cells are separated by a foam pad barrier or wire screen has been reported by Hauptmann and Widholm (1982), and Maddox and colleagues (1983) have employed a medium appropriate for recovery from very low cell densities.

I. Viability Testing

Recovery from freeze preservation is usually self-evident, a plate of semisolid medium overlain with cells in liquid medium developing a visible

---

the liquid nitrogen. The rate of cooling increases with the reduction of distance g. The cold environment is maintained by covering the flask with a bag of polystyrene insulating material (h). (D) The specimens are placed within a vacuum flask (a) inside an insulated cabinet (b) containing solid CO_2 (c). An electrically cooled refrigerator running at ca. −70°C would also be suitable. (E) The specimens (a) are placed within a polystyrene block (b) constructed in two halves and drilled to the dimensions of the ampoules. The block is transferred to a refrigerator (as in D) for cooling. In all cases, the specimens are cooled slowly to an intermediate temperature (usually −35 to −40°C) and then plunged into liquid nitrogen [(A) Adapted from Withers and King, 1980; (D) adapted from Sala *et al.*, 1979; (E) adapted from Maddox *et al.*, 1983.]

"lawn" of recovering cells within a few days. Nonetheless, rapid viability tests such as the following can be useful to aid evaluation of a storage procedure.

1. Fluorescein diacetate staining. One drop of cell suspension is placed on a slide and mixed with 1 drop of fluorescein diacetate solution [stock solution of 0.1% (w/v) in acetone diluted 1:50 in culture medium immediately before use]. The slide is viewed under a light microscope. All cells are visible in white light, whereas only viable cells are visible by fluorescence in UV light. The staining reaction occurs within ca. 2 min (Widholm, 1972).
2. Evans' blue staining and phenosafranine staining. In these tests, lethally damaged cells are revealed by their failure to exclude the dyes (Widholm, 1972; Gaff and Okong'o-Ogola, 1971). In each case, 1 drop of the stain [0.025% (w/v) of Evans' Blue; 0.1% (w/v) of phenosafranine, both in liquid medium] is added to 1 drop of cell suspension. The staining reaction develops in 5–10 min.
3. Tetrazolium chloride reduction. Respiration in recovering cells can be estimated by their capacity to reduce 2,3,5-triphenyl tetrazolium chloride (TTC; Towill and Mazur, 1974). Cells suspended in liquid medium are mixed in a narrow test tube with an equal volume of 1% (w/v) TTC in 0.05 *M* phosphate buffer, pH 7.5. The mixture is incubated for ca. 20 hr at 20°C in the dark. The red color (formazan) which develops, marking viable cells, can be extracted in 95% ethanol and estimated spectrophotometrically (by absorbance at a 485-nm wavelength of light).

Viability tests should be applied with caution. The TTC test may seriously under- or overestimate the potential recovery. Staining reactions may be difficult to discern in large cell masses, and there may not be a clear distinction between live and dead cells during the early postthaw period. Consequently, the tests need corroboration by measurement of parameters of growth whenever possible.

J. A Note on Callus and Protoplast Cultures

A detailed treatment of the freeze preservation of callus and protoplast cultures is outside the scope of this chapter. However, a brief mention is justified in view of the structural and methodological affinities of these cultures to cells. Callus cultures, although sharing similar freezing requirements, are less amenable to freeze preservation than are cell suspensions. Yet, unlike cells and protoplasts, they can be stored in the short to medium term by other techniques. In limited investigations, callus cultures have been successfully maintained at a reduced temperature, under low oxygen tension, by desiccation and even by immersion in mineral oil (see Withers, 1980b,c,d, 1982, 1983).

However, the little information which we do have on the freeze preservation of protoplasts (as opposed to the more frequent biophysical studies) presents a more optimistic picture. A procedure involving cryoprotection with a mixture of 5% (v/v) DMSO and 10% (w/v) glucose, and freezing at 1 to 2°C/min before plunging into liquid nitrogen and then thawing rapidly, appears to be sufficiently widely successful to form the basis of a routine method, given further development (Takeuchi *et al.*, 1982; see also Hauptmann and Widholm, 1982; Weber *et al.*, 1983).

III. RESULTS AND CONCLUSIONS

A. Success in the Application of Freeze Preservation Techniques

As indicated in Section I, cells of over 30 species can now be freeze preserved and successfully returned to culture. This may seem a modest number when compared with the vast range of species currently under general investigation and reflected in the other chapters in this volume. However, when viewed against the early difficulties and the relatively little effort involved, the gathering momentum is manifest. In addition to the increasing number of species preserved, the percentage of viabilities recorded after thawing and the rates of recovery are much higher than in earlier reports. It is now reasonable to expect up to, say, 75% postthaw viability and a resumption of growth within ca. 2 days. This being so, it is perhaps surprising that few laboratories are actively using freeze preservation for their stock cultures.

Of relevance in the promotion of interest in freeze preservation is the relative security it offers to users. The record for stability in the ultra-low-temperature preservation of a range of biological materials is very good (see Withers, 1980a, 1982). Further, several studies have demonstrated retention by freeze-preserved cell cultures of morphogenic potential (e.g., Nag and Street, 1973; Withers, 1978) and specific biochemical characters (P. J. King, personal communication; Hauptmann and Widholm, 1982; Weber *et al.*, 1983).

B. The Future

In spite of relative optimism concerning the future of freeze preservation as a means of germplasm storage, it is emphasized here that research must continue toward the refinement of freeze preservation methodology and extension of the range of species which can be preserved in this way. Further, scientists need to be made aware of the potential range of applications of storage procedures.

In the biotechnology laboratory, freeze preservation can be of enormous value in maintaining experimental material awaiting screening and samples from time course treatments, as well as stock cultures. The stage has now been reached when material which would otherwise be discarded due to pressures on resources could certainly be stored without reservations. Some would advocate freeze preservation of other, more precious material. This is recommended here, with the caveat that stocks be duplicated, preferably on a second site, and that backup, growing cultures be maintained for the immediate future, until the working of storage techniques can be evaluated in each laboratory.

In the near future, the value of unique cell cultures will surely become recognized and higher plant genotypes will need to be distributed in the routine manner that currently applies to microbial cultures. Accordingly, it will be necessary for distribution techniques and networks to be developed, and for scientists to have access to detailed documentation on the characteristics, location, and availability of cultures (see Withers, 1984).

ACKNOWLEDGMENTS

The author acknowledges with gratitude the support of the Science and Engineering Research Council.

REFERENCES

Cella, R., Sala, F., Nielsen, E., Rollo, F., and Parisi, B. (1978). Cellular events during the regrowth phase after thawing of freeze-preserved rice cells. *Fed. Eur. Soc. Plant Physiol. 1978* pp. 127–128 (abstr.).

Cella, R., Colombo, R., Galli, M. G., Nielsen, E., Rollo, F., and Sala, F. (1982). Freeze-preservation of *Oryza sativa* L. cells: A physiological study of freeze-thawed cells. *Physiol. Plant.* **55**, 279–284.

Finkle, B. J., and Ulrich, J. M. (1979). Effect of cryoprotectants in combination on the survival of frozen sugarcane cells. *Plant Physiol.* **63,** 598–604.

Finkle, B. J., and Ulrich, J. M. (1982). Cryoprotectant removal temperature as a factor in the survival of frozen rice and sugarcane cells. *Cryobiology* **19,** 329–335.

Gaff, D. F., and Okong'o-Ogola, O. (1971). The use of non-penetrating pigments for testing survival of cells. *J. Exp. Bot.* **22,** 756–758.

Hauptmann, R. M., and Widholm, J. M. (1982). Cryostorage of cloned amino-acid analog-resistant carrot and tobacco suspension cultures. *Plant Physiol.* **70,** 30–34.

Kartha, K. K., Leung, N. L., Gaudet-LaPrairie, P., and Constabel, F. (1982). Cryopreservation of periwinkle, *Catharanthus roseus* cells cultured *in vitro. Plant Cell Rep.* **1,** 135–138.

Maddox, A., Gonsalves, F., and Shields, R. (1983). Successful preservation of plant cell cultures at liquid nitrogen temperatures. *Plant Sci. Lett.* **28,** 157–162.

Meryman, H. T. (1965). "Cryobiology." Academic Press, New York.

Meryman, H. T., and Williams, R. J. (1982). The mechanisms of freezing injury and natural tolerance, and the principles of artificial cryoprotection. *In* "Crop Genetic Resources—The Conservation of Difficult Material" (L. A. Withers and J. T. Williams, eds.), pp. 5–37. IUBS/IBPGR, Paris.

Nag, K. K., and Street, H. E. (1973). Carrot embryogenesis from frozen cultured cells. *Nature (London)* **245,** 270–272.

Nag, K. K., and Street, H. E. (1975a). Freeze-preservation of cultured plant cells. I. The pretreatment phase. *Physiol. Plant.* **34,** 254–260.

Nag, K. K., and Street, H. E. (1975b). Freeze-preservation of cultured plant cells. II. The freezing and thawing phases. *Physiol. Plant.* **34,** 261–265.

Pritchard, H. W., Grout, B. W. W., Reid, D. S., and Short, K. C. (1982). The effects of growth under water stress on the structure, metabolism and cryopreservation of cultured sycamore cells. *In* "The Biophysics of Water" (F. Franks and S. Mathias, eds.), pp. 315–318. Wiley, New York.

Quatrano, R. S. (1968). Freeze-preservation of cultured flax cells utilizing DMSO. *Plant Physiol.* **43,** 2057–2061.

Sala, F., Cella, R., and Rollo, F. (1979). Freeze-preservation of rice cells. *Physiol. Plant.* **45,** 170–176.

Takeuchi, M., Matsushima, H., and Sugawara, Y. (1980). Long-term freeze-preservation of protoplasts of carrot and *Marchantia. Cryo-Lett.* **1,** 519–524.

Takeuchi, M., Matsushima, H., and Sugawara, Y. (1982). Totipotency and viability of protoplasts after long-term freeze-preservation. *In* "Plant Tissue Culture 1982" (A. Fujiwara, ed.), pp. 797–798. Maruzen, Tokyo.

Towill, L. E., and Mazur, P. (1974). Studies on the reduction of 2,3,5-triphenyl tetrazolium chloride as a viability assay for plant tissue cultures. *Can. J. Bot.* **53,** 1097–1102.

Weber, G., Roth, E. J., and Schweiger, H.-G. (1983). Storage of cell suspensions and protoplasts of *Glycine max.* (L.) Merr., *Brassica napus* (L.), *Datura innoxia* (Mill.), and *Daucus carota* (L.) by freezing. *Z. Pflanzenphysiol.* **109,** 23–29.

Widholm, J. M. (1972). The use of fluorescein diacetate and phenosafranine for determining viability of cultured plant cells. *Stain Technol.* **47,** 189–194.

Withers, L. A. (1978). A fine-structural study of the freeze-preservation of plant tissue cultures. II. The thawed state. *Protoplasma* **94,** 235–247.

Withers, L. A. (1979). Freeze-preservation of somatic embryos and clonal plantlets of carrot (*Daucus carota* L.). *Plant Physiol.* **63,** 460–467.

Withers, L. A. (1980a). "Tissue Culture Storage for Genetic Conservation," IBPGR Publ. No. AGP:IBPGR/80/8. Int. Board Plant Genet. Resour., Rome.

Withers, L. A. (1980b). The cryopreservation of higher plant tissue and cell cultures—an overview with some current observations and future thoughts. *Cryo-Lett.* **1,** 239–250.

Withers, L. A. (1980c). Preservation of germplasm. *Int. Rev. Cytol., Suppl.* **11B,** 101–136.
Withers, L. A. (1980d). Low temperature storage of plant tissue. *Adv. Biochem. Eng.* **18,** 102–150.
Withers, L. A. (1982). Storage of plant tissue cultures. *In* "Crop Genetic Resources—The Conservation of Difficult Material" (L. A. Withers and J. T. Williams, eds.), pp. 49–82. IUBS/IBPGR, Paris.
Withers, L. A. (1983). Germplasm storage in plant biotechnology. *Semin. Ser.—Soc. Exp. Biol.* **18,** 187–218.
Withers, L. A. (1984). *In vitro* techniques for germplasm storage. *Proc. Eucarpia Congr., 10th, 1983* (in press).
Withers, L. A., and Davey, M. R. (1978). A fine-structural study of the freeze-preservation of plant tissue cultures. I. The frozen state. *Protoplasma* **94,** 207–219.
Withers, L. A., and King, P. J. (1979). Proline—a novel cryoprotectant for the freeze-preservation of cultured cells of *Zea mays* L. *Plant Physiol.* **64,** 675–678.
Withers, L. A., and King, P. J. (1980). A simple freezing unit and cryopreservation method for plant cell suspensions. *Cryo-Lett.* **1,** 213–220.
Withers, L. A., and Street, H. E. (1977a). Freeze-preservation of cultured plant cells. III. The pregrowth phase. *Physiol. Plant.* **39,** 171–178.
Withers, L. A., and Street, H. E. (1977b). Freeze-preservation of plant cell cultures. *In* "Plant Tissue Culture and Its Bio-technological Application" (W. Barz, E. Reinhard, and M.-H. Zenk, eds.), pp. 226–244. Springer-Verlag, Berlin and New York.

CHAPTER **69**

Freeze Preservation of Meristems*

K. K. Kartha

Plant Biotechnology Institute
National Research Council
Saskatoon, Saskatchewan, Canada

I. INTRODUCTION

Prolonged preservation of plant material should guarantee genetic stability, that is, it should prevent progressive changes in the genome. Storage of plant material at an ultralow temperature, such as that of liquid nitrogen (−196°C), is one of the most promising approaches being pursued to achieve this goal. The underlying principle of this approach is that at the temperature of liquid nitrogen all metabolic activities of living cells are approaching a standstill, and the material could thus be preserved for indefinite periods of time.

The realization of the importance of plant meristems in producing disease-free plants in a genetically stable condition, coupled with the possibilities of mass propagation, marked the beginning of freeze preservation

*NRCC No. 23137.

CELL CULTURE AND SOMATIC CELL
GENETICS OF PLANTS, VOL. 1

ISBN 0-12-715001-3

(cryopreservation) studies with meristems or shoot tips as a means of germplasm preservation. The first report on the successful freeze preservation of shoot tips was made with carnation (Seibert, 1976). During the last few years, meristems or shoot tips of a number of plant species have been successfully cryopreserved (Kartha, 1981, 1982a). Several procedures by which plant meristems are cryopreserved are discussed in this chapter.

II. REQUIREMENTS

A. Plant Material

One of the basic requirements prior to cryopreservation of plant meristems is the availability or development of an efficient and reliable technique for faithful regeneration of plants from *in vitro* cultured meristems or shoot tips. For vegetatively propagated crops, it is ideal to maintain the collections *in vitro* either as proliferating shoots or as plantlets, thus ensuring continued availability of explants (meristems) in a physiologically uniform condition.

B. Equipment

1. Instruments for dissection of shoot meristems (Chapter 13, this volume).
2. Cryopreservation equipment: Requirements for cryopreservation equipment depend upon the method by which the meristems are frozen. Freezing by regulated slow cooling requires that the cooling rates be precisely monitored. A number of programmable freezing machines facilitating linear cooling rates are available. The cryopreservation equipment itself consists of a programmable temperature controller, a freezing chamber, and a temperature chart recorder (Kartha, 1982b). The main advantage of using commercially available freezing equipment rather than that designed in some laboratories is that the former permits universality of application, whereas the latter is subjected to the ingenuity of the experimenter.
3. Low-pressure liquid nitrogen cylinder.
4. Cryogenic storage facilities.
5. A water bath for thawing the frozen samples.

C. Chemicals

1. Cryoprotectants such as dimethylsulfoxide (DMSO), glycerol, sucrose, and so on.
2. Chemicals and hormones required for the preparation of culture media.

III. PROCEDURE

A. Isolation of Meristems and Preculture

The shoot apical or lateral meristems are aseptically isolated from germinated seeds, or from plants maintained *in vitro* or in a greenhouse. Details concerning the isolation and culture of meristems are discussed in Chapters 13 and 14, this volume. Immediately after isolation, the meristems, 0.4–0.5 mm in length, are precultured on shoot regeneration medium supplemented with DMSO and incubated under appropriate physical environmental conditions. For example, pea meristems from germinated seeds and strawberry meristems isolated from *in vitro*-propagated shoots are precultured for 2 days on shoot regeneration media [pea, B5 medium plus 0.5 μ*M* benzylaminopurine (BAP); strawberry, Murashige and Skoog (MS) medium plus 1 μ*M* each of BAP, indolebutyric acid (IBA), and 0.1 μ*M* gibberellic acid (GA_3)] supplemented with 5% (v/v) DMSO (Kartha *et al.*, 1979, 1980). The strawberry meristems are incubated at 26°C, 16-hr photoperiods at 4000 lx intensity, whereas the pea meristems are incubated under diurnal temperature regimes. Similarly, preculture of shoot tips for 1–2 days on nutrient medium alone has also been shown to be beneficial in some potato species (Grout and Henshaw, 1978; Towill, 1981a,b).

B. Application of Cryoprotectants

Cryoprotectants, that is, chemicals which protect the cells from freezing and thawing injury, are necessary for the successful cryopreservation of meristems of non-cold-hardy species. Although a number of chemicals such as DMSO, glycerol, sugars, and other hydroxylic compounds are available, the one most beneficial for the cryopreservation of meristems is

DMSO. Because of its toxicity at high concentrations, a lower level of DMSO (5–10%) is commonly used. In a few instances, a combination of cryoprotectants has proven to increase the rate of survival.

The 110-ml jars, each containing 20 meristems, are transferred to an ice bath in a laminar airflow cabinet. The cryoprotectant solution, at twice the final concentration is made up in culture medium devoid of growth hormones and is kept chilled. The meristems are aseptically transferred from the jars to known volumes of chilled, cryoprotectant-free culture medium, to which equal volumes of the cryoprotectant solution are added stepwise over a period of 30 min until the required final concentration is reached. The final concentration of DMSO used for pea and strawberry meristems is 5% (v/v), 10% for potatoes. The meristems are equilibrated for an additional period of 10 min. At this stage the meristems are ready for freezing, and the manner in which this is accomplished depends entirely on the freezing methods employed.

C. Freezing Methods

1. Slow Freezing

Slow freezing permits the efflux of cellular water and facilitates extracellular freezing as a consequence of the imposed reduction in temperature. The resulting protective dehydration is attained at temperatures of about −30 to −40°C and is governed by the cooling rate, the type of cryoprotectant used, and the permeability of the membrane to water. Therefore, freezing by the slow method requires precisely controlled cooling rates. For this purpose, programmable freezing machines are extremely desirable. About 20 DMSO-treated meristems are aseptically transferred to 1.2 ml prescored cryogenic glass ampoules, along with 1 ml of the DMSO solution. All of the ampoules except one are flame sealed and transferred to the freezing chamber. A temperature probe for monitoring the cooling rates is introduced into the unsealed ampoule. The meristems are frozen at slow cooling rates monitored by the temperature controller down to temperatures of −30 to −40°C, and the cooling program is simultaneously recorded in the chart recorder. At the termination of freezing, for example, at −40°C, the ampoules are removed from the freezing chamber and stored in liquid nitrogen.

2. Rapid Freezing

Rapid freezing is carried out by directly immersing the cryoprotectant-treated meristems in liquid nitrogen, thus avoiding the necessity for programmable freezing equipment. Cooling rates on the order of several hundred degrees per minute are obtained by freezing the sample rapidly. It is anticipated that during rapid freezing, viability of meristem cells is maintained by preventing the growth of intracellular ice crystals by rapidly passing the tissue through the temperature zone in which lethal ice crystal growth occurs (Luyet's mechanism, quoted by Seibert and Wetherbee, 1977). Meristems of a few plant species have successfully been frozen by this method (Kartha, 1981, 1982a). A simple rapid freezing method that consists of mounting cryoprotectant-treated meristems on hypodermic needles followed by immersion in liquid nitrogen has been developed for potato meristems (Grout and Henshaw, 1978). Similarly, ampoules containing meristems and cryoprotectant solution are immersed in liquid nitrogen in order to attain elevated cooling rates. Although rapid freezing methods may be applicable to a very limited number of plant species, a generalization concerning the universality of their application to all species cannot be made. On the other hand, slow freezing has resulted in successful cryopreservation of a number of plant species, including those which have been frozen by rapid freezing methods.

3. Droplet Freezing

This technique has been developed for the cryopreservation of *Manihot esculenta* meristems (Kartha *et al.*, 1982). The principle of the technique is essentially the same as that of slow freezing. However, preculturing of the meristems on medium supplemented with cryoprotectants is not required. The treatment of meristems with cryoprotectants is carried out as explained earlier, except that here 15% DMSO plus 3% sucrose is used. The cryoprotectant solution is dispensed in 2- to 3-μl droplets over an 18-μm strip of aluminum foil contained in a sterile plastic Petri dish, and a cryoprotectant-treated meristem is placed on each droplet, totaling about 20–30 meristems per dish. The Petri dish is then transferred to the freezing chamber, cooled at a rate of 0.5°C/min to −25 or −30°C, and subsequently stored in liquid nitrogen.

IV. STORAGE, THAWING, AND RECULTURE

Viability can be preserved indefinitely if meristems are stored in liquid nitrogen. A number of storage containers are commercially available for this purpose. Care should be taken to ensure that the meristems are stored in the immersion phase of liquid nitrogen. Accordingly, periodic replenishment of liquid nitrogen should be carried out to compensate for the loss due to static evaporation.

Thawing is a very important component in the cryopreservation steps. Rapid thawing in a water bath held at 37 or 39°C until the phase change from solid to liquid is completed is often the practice of choice. After thawing is complete, the meristems are gradually washed free of the cryoprotectants, avoiding sudden deplasmolysis, and recultured on appropriate nutrient medium. The viability is determined on the basis of the number of meristems exhibiting regrowth and plant regeneration.

V. RESULTS AND CONCLUSION

Employing various methods of freezing, successful cryopreservation has been accomplished with a number of plant species (Table I). Comparison of slow and rapid freezing with meristems of pea, chickpea, strawberry, and cassava revealed that the viability is retained better under slow-freezing regimes. The success of rapid freezing appears to be species dependent. The material which responds to rapid freezing responds equally well or better to slow freezing, as has been shown in the case of potato (Towill, 1981a,b, 1983). Under a slow-freezing regime, better survival has been noted with those meristems which prior to freezing had been precultured

TABLE I

Proven Freezing Methods for the Cryopreservation of Meristems or Shoot Tips

Species	Freezing method	Reference
Apple (*Malus* spp.)	Slow	Sakai and Nishiyama (1978)
Carnation (*Dianthus caryophyllus*)	Rapid	Seibert (1976)
Cassava (*Manihot esculenta*)	Slow	Kartha *et al.* (1982)
	Rapid	Bajaj (1979)
Pea (*Pisum sativum*)	Slow	Kartha *et al.* (1979)
Peanut (*Arachis hypogaea*)	Rapid	Bajaj (1979)
Potato		
Solaunum goniocalyx	Rapid	Grout and Henshaw (1978)
Solanum etuberosum	Slow	Towill (1981a)
Solanum tuberosum	Slow	Towill (1981b, 1983)
Strawberry (*Fragaria* × *ananassa*)	Slow	Kartha *et al.* (1980)
		Sakai *et al.* (1978)
Tomato (*Lycopersicon esculentum*)	Liquid nitrogen vapor freezing	Grout *et al.* (1978)

on DMSO-supplemented nutrient medium. Although survival has been obtained at cooling rates in the range of 0.5–1.0°C/min, there appears to be an optimal cooling rate at which maximal viability is retained. This rate is species specific and has been determined to be 0.6°C/min for pea (60%) and chickpea (40%), 0.84°C/min for strawberry (95%), and 0.5°C/min for cassava (16–86%). Similarly, the terminal freezing temperature (prefreezing) to which meristems are frozen prior to liquid nitrogen storage also influences the rate of viability. Although −30 to 40°C may be considered as a safe range for meristems of most of the species examined, cassava meristems require prefreezing only to −25 or −30°C. Even after 2 years of storage in liquid nitrogen, more than 60% of pea and 75% of strawberry meristems can be regenerated into plants and grown to maturity.

The cryopreservation methods described here should be considered only as basic guidelines. Various freezing methods and cryobiological parameters should be critically examined while a cryopreservation protocol is devised for unexplored material.

REFERENCES

Bajaj, Y. P. S. (1979). Freeze-preservation of *Arachis hypogaea* and *Cicer arietinum*. *Indian J. Exp. Biol.* **17,** 1405–1407.

Grout, B. W. W., and Henshaw, G. G. (1978). Freeze-preservation of potato shoot-tip cultures. *Ann. Bot. (London)* [N.S.] **42,** 1227–1229.

Grout, B. W. W., Westcott, R. J., and Henshaw, G. G. (1978). Survival of shoot meristems of tomato seedlings frozen in liquid nitrogen. *Cryobiology* **15,** 478–483.

Kartha, K. K. (1981). Meristem culture and cryopreservation-methods and applications. *In* "Plant Tissue Culture" (T. A. Thorpe, ed.), pp. 181–211. Academic Press, New York.

Kartha, K. K. (1982a). Cryopreservation of germplasm using meristem and tissue culture. *In* "Application of Plant Cell and Tissue Culture in Agriculture and Industry" (D. T. Tomes, B. E. Ellis, P. M. Harney, K. J. Kasha, and R. L. Peterson, eds.), pp. 139–161. Univ. of Guelph, Guelph, Ontario, Canada.

Kartha, K. K. (1982b). Cryopreservation of plant meristems and cells. *In* "Plant Tissue Culture Methods" (L. R. Wetter and F. Constabel, eds.), 2nd rev. ed., NRCC No. 19876, pp. 25–33. National Research Council of Canada, Ottawa.

Kartha, K. K., Leung, N. L., and Gamborg, O. L. (1979). Freeze-preservation of pea meristems in liquid nitrogen and subsequent plant regeneration. *Plant Sci. Lett.* **15,** 7–15.

Kartha, K. K., Leung, N. L., and Pahl, K. (1980). Cryopreservation of strawberry meristems and mass propagation of plantlets. *J. Am. Soc. Hortic. Sci.* **105,** 481–484.

Kartha, K. K., Leung, N. L., and Mroginski, L. A. (1982). *In vitro* growth responses and plant regeneration from cryopreserved meristems of cassava (*Manihot esculenta* Crantz). *Z. Pflanzenphysiol.* **107,** 133–140.

Sakai, A., and Nishiyama, Y. (1978). Cryopreservation of winter-vegetative buds of hardy fruit trees in liquid nitrogen. *HortScience* **13,** 225–227.

Sakai, A., Yamakawa, M., Sakata, D., Harada, T., and Yakuwa, T. (1978). Development of a whole plant from an excised strawberry runner apex frozen to −196°C. *Low Temp. Sci., Ser. B* **36,** 31–38.

Seibert, M. (1976). Shoot initiation from carnation shoot apices frozen to −196°C. *Science* **191,** 1178–1179.

Seibert, M., and Wetherbee, P. J. (1977). Increased survival and differentiation of frozen herbaceous plant organs through cold treatment. *Plant Physiol.* **59,** 1043–1046.

Towill, L. E. (1981a). *Solanum etuberosum:* A model system for studying the cryobiology of shoot-tips in the tuber-bearing *Solanum* species. *Plant Sci. Lett.* **20,** 315–324.

Towill, L. E. (1981b). Survival at low temperatures of shoot-tips from cultivars of *Solanum tuberosum* group *Tuberosum. Cryo-Lett.* **2,** 373–380.

Towill, L. E. (1983). Improved survival after cryogenic exposure of shoot-tips derived from *in vitro* plantlet cultures of potato. *Cryobiology* **20,** 567–573.

CHAPTER **70**

Selection of Cell Lines for High Yields of Secondary Metabolites

Yasuyuki Yamada

Research Center for Cell and Tissue Culture
Kyoto University
Kyoto, Japan

I. BACKGROUND FOR THE SELECTION OF HIGH-YIELD CELLS

Plant cells divide and grow continuously year round in cell cultures, even though in nature plant growth is regulated by the seasons. This continuous growth of plant cells is one of the most beneficial results of plant cell culture. Using cell culture techniques, scientists are trying to obtain commercial and medicinal compounds which some plants produce in an organ or tissue.

These compounds of economic value are mainly secondary metabolites: pigments, alkaloids, and steroids. Primary metabolites (amino acids, proteins, nucleic acids, and some vitamins) can usually be obtained much faster by the culture of microorganisms, particularly bacteria. Today, the production of primary and secondary metabolites of economic importance is one of the major objectives of plant cell culture scientists.

There are, however, some drawbacks in the plant cell culture method.

CELL CULTURE AND SOMATIC CELL
GENETICS OF PLANTS, VOL. 1

ISBN 0-12-715001-3

Cell cultures produce only small amounts of secondary metabolites, and one or more of the metabolizing enzymes normally found in intact plants may be present in only very small amounts, or absent, in the corresponding cultured cells.

Recently, a number of cell lines that produce larger amounts of secondary metabolites than their intact parent plants have been isolated by the selection of high-yield cells. These include pigments (Nishi *et al.*, 1974; Mizukami *et al.*, 1978; Yamamoto *et al.*, 1982), vitamins (Ikeda *et al.*, 1976; Hagimori *et al.*, 1978; Yamada and Watanabe, 1980; Watanabe *et al.*, 1982; Watanabe and Yamada, 1982), alkaloids (Zenk *et al.*, 1977; Ogino *et al.*, 1978; Sato *et al.*, 1982), and steroids (Alfermann *et al.*, 1977).

The results of these studies are evidence that cultured cells are heterogeneous in their ability to produce useful compounds. They are mixtures of producing and nonproducing cells for any one compound. The reason for the presence of different productive potentials is an interesting topic in itself, but our purpose here is to outline the methods by which specific cells with high productivity can be selected.

Culture in a defined medium under constant conditions represents a crude selection of cells, as only the most suitable will survive in a medium under given conditions. Because knowledge of what constitutes the best media for the growth of various plant cells is imperfect, the researcher must survey known defined media and previously successful culture conditions, and must then choose that medium and those conditions which give the best selection of high-yield cells from the experimental material.

II. SELECTION METHODOLOGY

A. Cell-Aggregate Cloning

Bacteria are prokaryotes and usually single-cell organisms, whereas higher plants are eukaryotes and are multicellular. In higher plants, specific secondary metabolites are found in a particular organ that varies with the compound.

In the selection of bacterial cells that produce useful compounds, the cells are spread over an agar selection medium. The clones produced by a single bacterium are selected for high production of the desired compound. In the selection of high-yield plant cells, single-cell cloning can be used to select the first, most promising specific cells. But with an established cell line that produces relatively large amounts of a secondary metabolite, it is better to clone small cell aggregates of 10–100 cells when specific cells are desired (Yamamoto *et al.*, 1982). This technique of aggregate cloning produces very stable cell lines that give compounds of high quality in quantity.

An outline of the cell-aggregate selection method for the production of a red pigment (Yamamoto *et al.*, 1982) is shown in Fig. 1. The original *Euphorbia millii* calluses that produced the red pigment, cyanidin monoglucoside, were induced from leaf cells. The calluses formed were mottled red, white, and faint green.

Each callus was split into many segments, one segment consisting of 10–100 cells. These segments were cultured on Murashige-Skoog's (MS) basal medium (Murashige and Skoog, 1962) with 10^{-6} *M* 2,4-dichlorophenoxyacetic acid (2,4-D), 0.2% (w/v) malt extract, and 2% (w/v) sucrose at 28°C under fluorescent light of 6000 lx for 10 days. Each segment that grew was divided into two aggregates, one to be used for subculture and the other for the quantitative analysis of the pigment.

The reddest of the aggregates for subculture were placed in an empty Petri dish. These aggregates were divided into several segments, of which the reddest pieces (10–100 cells) were again transplanted in the basal medium under the original conditions. This selection procedure was repeated 28 times.

The two characteristic terms ($\bar{c}$ and c^{max}) in Fig. 2 show how much the

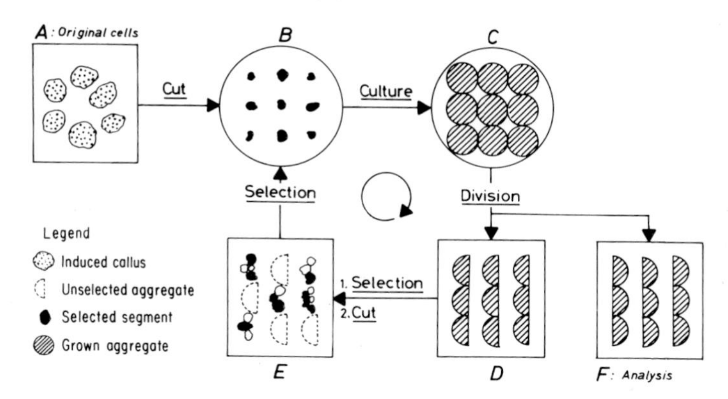

Fig. 1. Selection method for high-yield cells. (A) The original *Euphorbia millii* calluses were divided into 128 segments. Each segment was placed on agar medium in separate sections of a nine-section Petri dish and recorded. (B) Segments were cultured at 28°C under light (6000 lx) for 10 days. (C) Each segment that grew was divided into two cell aggregates: one (D) for subculture, the other (F) for the quantitative analysis of the pigment. (D) The reddest of the nine aggregates were placed in an empty Petri dish (E). (E) Each of the red aggregates was divided into several segments, of which the reddest pieces were recorded and then placed on agar medium (B). (From Yamamoto *et al.*, 1982.)

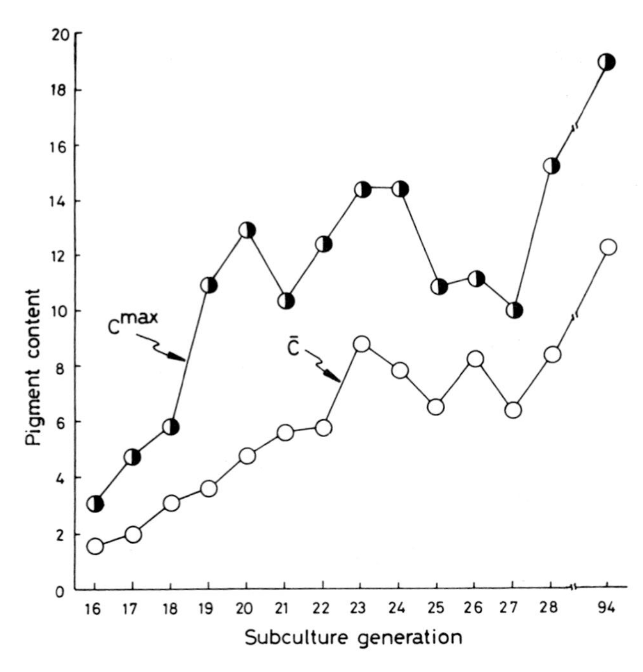

Fig. 2. Trends for the mean value ($\bar{c}$) and the maximum value (c^{max}) in the frequency distribution of cell aggregates that had varied pigment contents during the 16th–94th passages. Cell aggregates were subcultured at 10-day intervals on MS agar medium with 2,4-D (10^{-6} *M*), malt extract (0.2% w/v), and sucrose (2% w/v) at 28°C under light (6000 lx). Cell aggregates were cloned and selected through the 56th subculture. After the 56th subculture, only cultures of the cell lines were done. (Based on Yamamoto *et al.*, 1982.)

pigment content increased during subsequent subcultures. After 23 selections the mean value ($\bar{c}$) for pigment content increased sevenfold, and the maximum value (c^{max}) was 12 times the pigment content of the original callus. After the 56th selection, the choosing of cell aggregates was stopped, but the culture of the selected cell lines was continued through the 94th subculture.

The production of pigment in these subcultured cells has continued to increase slightly over a period of 2 years, even though no further selection of cells has been made.

Recently, there have been several reports on the establishment of lines that give high yields of ubiquinone 10 and biotin (Matsumoto *et al.*, 1980, 1981; Watanabe *et al.*, 1982; Watanabe and Yamada, 1982). These compounds are both vitamins and are primary metabolites. Both of the cell lines that produce high yields of these primary metabolites were selected by the same cell-aggregate method of cloning that was used to produce the

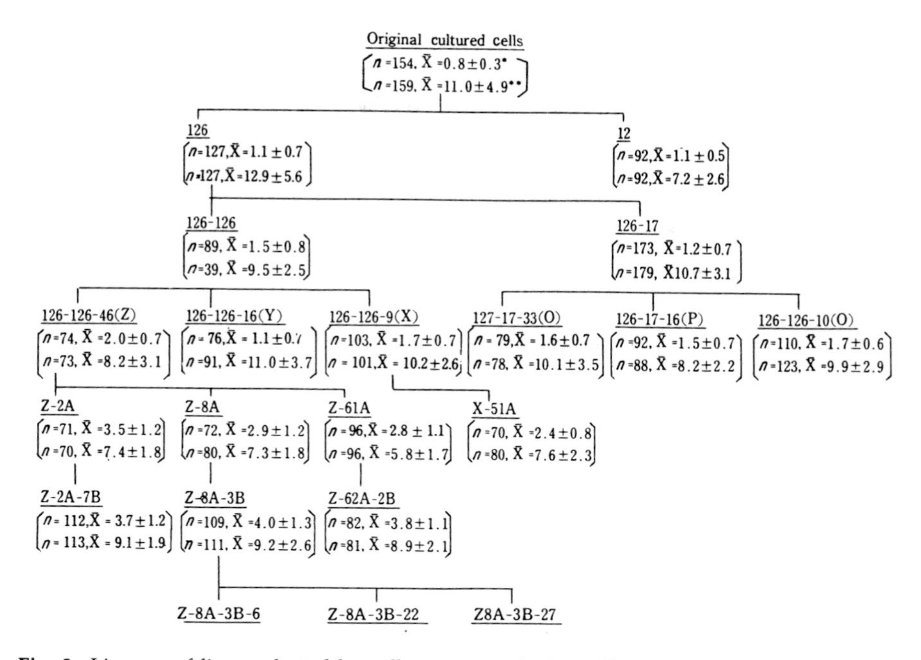

Fig. 3. Lineage of lines selected by cell-aggregate cloning. The number of clones examined (n), and the mean values ($\bar{X}$) plus or minus the standard deviations (SD) of the UQ content (*) and the growth rate (**) at each cell cloning are given in parentheses. The unit of growth is expressed as the ratio of harvested callus tissue (grams of fresh weight) and the UQ content in micrograms per 100 mg fresh weight of callus tissue. (From Matsumoto *et al.*, 1981.)

secondary metabolite red pigment. This production of primary metabolites is a breakthrough in cell culture techniques.

In the case of the tobacco cell line that gives ubiquinone 10, the cells were filtered aseptically through double stainless steel filters (125 and 250 μm). The filtrate contained many cell aggregates composed of about 10 cells, which were then plated in Petri dishes containing the colony formation medium. This selection was repeated 20 times to establish the high-yield cell line (Fig. 3).

B. Single-Cell Cloning

Theoretically, single-cell cloning should be the best method for isolating cells that yield large amounts of useful compounds. Practically, however, many clones started from a single cell are heterogeneous in their ability to produce useful compounds. Therefore, further selection is necessary.

Many protoplasts were isolated from cells that produced large amounts of berberine, and each protoplast produced a single clone. After several subcultures, the productivity of cells in the same clone became heterogeneous; some cells were yellow (berberine), others white (no berberine). The chromosome numbers of cells from the same clones also varied widely (Y. Yamada, M. Mino, and Y. Tanaka, unpublished).

As a result, most of the established clones that originated from a single cell did not give cell lines that yielded larger amounts of berberine than the original source. At present, it is not clear whether heterogeneous productivity among cells in a single-cell clone is a phenomenon only of berberine-producing cells. High berberine-producing cell lines could be obtained by cell-aggregate cloning.

III. CHOOSING USEFUL COMPOUNDS AS STANDARDS FOR SELECTION

The basic selection method has been described above. But there is another important step, which is to determine the relative amount of useful compound present in a small cell-aggregate clone.

A. Visible Compounds Selectable by Sight and Microscopy

When pigments, or compounds that fluoresce, are desired, cells with high yields can be selected by sight alone or with the aid of a fluorescence microscope. The selection of cells with pigments that are readily visible is very easy.

B. Invisible Compounds Selectable by Various Techniques

Should a compound be colorless, it may be possible to make it visible by changing the pH of the solution or by adding a compound that will form a color complex. Most alkaloids, steroids, and vitamins are colorless in cells, and since the addition of a compound or a change in pH may harm the cells, a simple method of detection that can be used without harm with many types of clones is desirable.

A Selection Method Combining Different Techniques

High-yield cells can be obtained by a method that combines different simple techniques. An example is the selection of cells that produce large amounts of tropane alkaloid.

a. Dragendorff Reagent Test. When a callus has grown large enough for use in a quick alkaloid assay, a small piece of it is squashed tightly between the folds of a piece of filter paper. By this means, various intracellular substances are squeezed out, some of which are absorbed by the paper. The remaining cell material is scraped away, and then Dragendorff's reagent (specific reaction with alkaloids) is sprayed where the cells have been (Ogino *et al.*, 1978; Yamada and Hashimoto, 1982). Calluses that give a strongly positive test can be identified by this technique.

b. Thin-Layer Chromatography. Selected cell lines that react with Dragendorff's reagent can be used for another simple test. Cell lines that contain an alkaloid that shows the same Rf value as that of an authentic tropane alkaloid can be identified by thin-layer chromatography (TLC) (Yamada and Hashimoto, 1982).

c. Gas–Liquid Chromatography and Gas Chromatography–Mass Spectrometry. Cultured cells that produce tropane alkaloids can be identified by comparing the Rf values of selected cells with those of authentic tropane alkaloids on two-dimensional TLC and by comparing the retention times on gas–liquid chromatography (GLC) and the gas chromatography–mass spectroscopy (GC–MS) spectra (Hashimoto and Yamada, 1983). Identification of alkaloids by TLC alone is not sufficient because compounds with very similar Rf values cannot be distinguished, and measurement of alkaloid contents by colorimetric analysis on TLC is not accurate enough.

TLC, however, can be used for the preliminary selection of cell lines that produce alkaloids. But for the selection of the few specific cell lines that give high yields of alkaloids, additional and more accurate identifications by GLC and GC–MS are necessary.

d. The Radioimmunoassay. TLC and GLC are practical but relatively insensitive methods of identification. A radioimmunoassay (RIA) that combines very high sensitivity in the p-mol range with high specificity has been reported for use with indole alkaloids (Zenk *et al.*, 1977). As an assay method, RIA is very accurate.

In this method, the immunogen of the target compound must be synthesized and antibodies formed in a rabbit. In addition, the target compound must be labeled with a radioisotope (usually tritium). The amount of compound present in the sample is then calculated from the dilution effect of the labeled compounds. Where this special RIA technique is applicable, it is a very good selection method for detecting a small amount of compound.

REFERENCES

Alfermann, A. W., Boy, H. M., Doller, P. C., Hagedorn, W., Heins, M., Wahl, J., and Reinhard, E. (1977). Biotransformation of cardiac glycosides by plant cell cultures. *In* "Plant Tissue Culture and Its Biotechnological Application" (W. Barz, E. Reinhard, and M. H. Zenk, eds.), pp. 125–141. Springer-Verlag, Berlin and New York.

Hagimori, M., Matsumoto, T., and Noguchi, M. (1978). Isolation and identification of ubiquinone 9 from cultured cells of Safflower (*Carthamus tinctorius* L.). *Agric. Biol. Chem.* **42,** 499–500.

Hashimoto, T., and Yamada, Y. (1983). Scopolamine production in suspension cultures and redifferentiated roots of *Hyoscyamus niger. Planta Med.* **47,** 195–199.

Ikeda, T., Matsumoto, T., and Noguchi, M. (1976). Formation of ubiquinone by tobacco plant cells in suspension culture. *Phytochemistry* **15,** 568–569.

Matsumoto, T., Ikeda, T., Kanno, N., Kisaki, T., and Noguchi, M. (1980). Selection of high ubiquinone 10-producing strains of tobacco cultured cells by cell cloning technique. *Agric. Biol. Chem.* **44,** 967–969.

Matsumoto, T., Kanno, N., Ikeda, T., Obi, Y., Kisaki, T., and Noguchi, M. (1981). Selection of cultured tobacco cell strains producing high levels of ubiquinone 10 by a cell cloning technique. *Agric. Biol. Chem.* **45,** 1627–1633.

Mizukami, H., Konoshima, M., and Tabata, M. (1978). Variation in pigment production in *Lithospermum erythrorhizon* callus cultures. *Phytochemistry* **17,** 95–97.

Murashige, T., and Skoog, F. (1962). A revised medium for rapid growth and bioassay with tobacco tissue cultures. *Physiol. Plant.* **15,** 473–497.

Nishi, A., Yoshida, A., Mori, M., and Sugano, N. (1974). Isolation of variant carrot cell lines with altered pigmentation. *Phytochemistry* **13,** 1653–1656.

Ogino, T., Hiraoka, N., and Tabata, M. (1978). Selection of high nicotine-producing cell lines of tobacco callus by single-cell cloning. *Phytochemistry* **17,** 1907–1910.

Sato, F., Endo, T., Hashimoto, T., and Yamada, Y. (1982). Production of berberine in cultured *Coptis japonica* cells. *In* "Plant Tissue Culture 1982" (A. Fujiwara, ed.), pp. 319–32O. Maruzen, Tokyo.

Watanabe, K., and Yamada, Y. (1982). Selection of variants with high levels of biotin from cultured green *Lavandula vera* cells irradiated with gamma rays. *Plant Cell Physiol.* **23,** 1453–1456.

Watanabe, K., Yano, S., and Yamada, Y. (1982). The selection of cultured plant cell lines producing high levels of biotin. *Phytochemistry* **21,** 513–516.

Yamada, Y., and Hashimoto, T. (1982). Production of tropane alkaloids in cultured cells of *Hyoscyamus niger. Plant Cell Rep.* **1,** 101–103.

Yamada, Y., and Watanabe, K. (1980). Selection of high vitamin B_6 producing strains in cultured green cells. *Agric. Biol. Chem.* **44,** 2683–2687.

Yamamoto, Y., Mizuguchi, R., and Yamada, Y. (1982). Selection of a high and stable pigment-producing strain in cultured *Euphorbia millii* cells. *Theor. Appl. Genet.* **61,** 113–116.

Zenk, M. H., El-Shagi, H., Arens, H., Stöckigt, J., Weiler, E. W., and Dues, B. (1977). Formation of the indole alkaloids serpentine and ajmalicine in cell suspension cultures of *Catharanthus roseus. In* "Plant Tissue Culture and Its Biotechnological Application" (W. Barz, E. Reinhard, and M. H. Zenk, eds.), pp. 37–43. Springer-Verlag, Berlin and New York.

CHAPTER **71**

Isolation and Analysis of Terpenoids

Joseph H. Lui

Biotechnology Department
The Goodyear Tire and Rubber Company
Akron, Ohio

I. INTRODUCTION

Terpenoids are a diverse class of secondary metabolites that are widely distributed throughout the plant kingdom (Banthorpe and Charlwood, 1980). Terpenoid compounds are defined as a group of natural products whose structure may be derived from the isoprene unit (C_5). The principal classes of naturally occurring terpenoids are: hemiterpenes (C_5), monoterpenes (C_{10}), sesquiterpenes (C_{15}), diterpenes (C_{20}), triterpenes (C_{30}), tetraterpenes (C_{40}), and polyterpenes (C_{5xn}).

The chemistry (Newman, 1972), biosynthesis (Mann, 1978), and distribution (Banthorpe and Charlwood, 1980) of terpenoids in the plant kingdom have been reviewed. This chapter will discuss the techniques (steam distillation and solvent extraction) used for extraction and isolation of terpenoids. Techniques for identification and analysis by chromatographic (thin-layer chromatography, gas–liquid chromatography, high-performance liquid chromatography, gel permeation chromatography) and spectroscopic (mass spectroscopy, nuclear magnetic resonance, infrared spectroscopy) methods will also be discussed. A special class of compounds, steroids, which are derived from the triterpenes by degradation of the

CELL CULTURE AND SOMATIC CELL
GENETICS OF PLANTS, VOL. 1

ISBN 0-12-715001-3

original C_{30} molecule, as well as terpene glycosides, will be excluded from this chapter.

II. PROCEDURE

A. Extraction

There are two basic techniques which may be used separately or in combination for obtaining terpenoids from plant material: steam distillation and solvent extraction. The choice of techniques is governed by the nature of the compound of interest.

1. Steam Distillation

Steam distillation of plant material can be employed frequently as a means of separation for the highly volatile low molecular weight terpenoids such as hemiterpenes and monoterpenes, as well as sesquiterpenes. The same procedure and equipment used for steam distillation may be used for fractional distillation, except that a source of steam in the apparatus is required. After distillation, the steam distillate is separated. The aqueous layer is saturated with salt and extracted with nonpolar solvents such as light petroleum, cyclohexane, or hexane. The combined oil and solvent extracts are dried, and the solvent is evaporated by fractional distillation.

2. Solvent Extraction

Since most of the terpenoid compounds are nonpolar in nature, direct extraction of plant material with nonpolar solvents is often used. The extraction is usually carried out at room temperature. The filtered extract is evaporated under reduced pressure for analysis. The solvent extraction method is not only used for nonvolatile terpenoid compounds but is also used extensively in the perfume industry for the extraction of essential oils from flowers. It is especially valuable when the oil is heat sensitive because the steam distillation method gives unsatisfactory results.

Extracts of many plant samples contain large amounts of fats, oils, and wax. It is necessary to clean up or separate the residues (terpenoids) from the sample extract for further identification and quantitation. A number of techniques can be employed. The cleanup method may include column chromatography, thin-layer chromatography, or other chromatographic methods. The method can be simplified depending upon the chemical properties of the terpenoid. For example, essential oil can easily be crystallized out from the mother liquid by forming a thiourea derivative (Stahl, 1969). Use of a commercial SEP-PAK cartridge (Water Associates, Milford, Massachusetts) can significantly reduce the number of steps and the total time required for sample cleanup.

B. Analysis

1. Thin-Layer Chromatography

Depending on their chemical nature, terpenoids can be classified as terpene hydrocarbons, terpene alcohols, terpene ketones, terpene aldehydes, or terpene esters.

Terpene hydrocarbon compounds are nonpolar in nature. Because of the low polarity of these substances, solvents with low polarity should be used as the mobile phase for separation in thin-layer chromatography (TLC) (Stahl, 1969; Kirchner, 1978). Some commonly used solvents are hexane, cyclohexane, and isopentane. They may be used alone or in combination with a small proportion of a polar solvent such as acetone. Silica gel and alumina, as well as silicic acid, have been found to be suitable adsorbents.

Terpene alcohols, terpene ketones, terpene aldehydes, and terpene esters are polar in nature. Combinations of nonpolar solvents (hexane, cyclohexane, benzene, chloroform) and polar solvents (ethyl acetate, methanol, acetone, etc.) are generally employed as the mobile phase. Silica gel, alumina, and silica gel impregnated with silver nitrate have been found to be appropriate adsorbents for analysis.

Colored and fluorescent spots on plates can be easily located by white and ultraviolet (UV) light. Other spots can be identified by spraying or vaporizing with detecting reagents. The terpenoid compounds can be detected with different reagents depending on the nature of the substituents (which are usually hydroxyl and carbonyl groups). A number of detecting reagents have been used. Among these are anisaldehyde-sulfuric acid, fluorescein-bromine, chlorosulfonic acid-acetic acid, and diphenyl picrylhydrazyl, as well as various salts such as antimony chloride and stannic chloride.

2. Gas–Liquid Chromatography

The application of gas–liquid chromatography (GLC) (Risby *et al.*, 1982) using a liquid stationary phase with various polarities and solid carriers having low catalytic activity proved satisfactory for separation and analysis of most of the terpenoid compounds. However, many of the terpenoids have low stability against thermal and catalytic effects. Formation of artifacts may sometimes occur especially during GLC analysis. Therefore, special attention must be paid to thermal and/or catalytic decomposition, stability of the column, and reproducibility of the retention time in order to obtain good qualitative and quantitative results.

A capillary column is recommended for separation and analysis. Glass is generally chosen as the material for the construction of columns due to its inertness compared to other common column materials such as copper and stainless steel. Column packings for GLC analysis of most terpenoids have routinely used diatomaceous earth supports and a variety of both polar and nonpolar liquid phases. Supports such as a Chromosorb W, Chromosorb G, and Gas Chrom Q are preferred. Liquid phases such as SE-30, OV101, FFAP, and Carbowax are recommended. FFAP was found to be the most suitable liquid phase for the separation of terpenoid hydrocarbons.

In most cases, isothermal conditions were employed. However, programmed-temperature GLC is sometimes useful in separating a variety of terpenoids from the same extraction sample. A program proceeding from ca. 200–300°C with a 3–5°C/min rate of increase after an initial isothermal period is generally employed. A great variety of devices have been used to detect the presence of the separated terpenoid components in the column effluent. Of these, the hydrogen flame ionization detector (FID) is the most suitable.

3. High-Performance Liquid Chromatography

The application of high-performance liquid chromatography (HPLC) (Kingston, 1979; Scott, 1976) to the analysis of terpenoid compounds has increased significantly in the last few years. Separation of terpenoids using HPLC is possible on any of the commercially available instruments. However, it is desirable that the system have a gradient elution capacity.

In general, the most useful packing materials appear to be microparticles of porous silica such as μ Porasil, Zorbox SIL, or Lichromsorb, and reverse-phase supports such as μ Bondapak C_{18}, Vydac ODS, or Zorbox ODS C_{18}. Normal phase solvents routinely contain *n*-hexane, cyclohexane, or petroleum ether with low or increasing concentrations of polar solvents such as ethyl acetate, acetone, and so on. Reverse-phase systems generally utilize mixtures of water or buffers in alcohols or acetonitrile. Most of the terpenoids may be detected by either UV absorption or fluorescence.

4. Nuclear Magnetic Resonance

The application of nuclear magnetic resonance (NMR) (Leyden and Cox, 1977) to the structural analysis of terpenoids is now well established. Most of the terpenoid compounds are significantly soluble in the commonly used solvent deuteriochloroform (CDC13) for direct NMR analysis. A substantial amount of ^{13}C NMR data (The Sadtler Standard Spectra, Sadtler Research Laboratories, Philadelphia, Pennsylvania) has now been established for the various major groups of terpenoids, and this method has taken its place along with ^{1}H NMR as a structural tool. A study using ^{13}C NMR in natural abundance suggests that this technique is a very powerful method for the determination of the natural rubber (*cis*-1,4-polyisoprene) content in guayule bushes (Hayman *et al.*, 1982; Visintainer *et al.*, 1981). In comparison to other analytical methods, the results are less variable and can be obtained more rapidly.

5. Mass Spectroscopy

The instability, solvation, and small quantities of substance available from the biological materials, as well as the limitations of conventional microanalysis, make it very difficult to deduce the correct molecular formula for an unknown terpenoid. This problem has been circumvented by the application of (MS) (Budde and Elchelberger, 1979) particularly gas chromatography–mass spectroscopy, to the structural elucidation of terpenoids. The molecular composition is given by accurate mass determination. Further evidence of a structure may be obtained by consideration of the fragmentation process. Today MS is the routine technique most widely applied to terpenoid compounds for structure determination. The mass spectra of many terpenoids have been published (Heller and Milne, 1978).

6. Miscellaneous

There are a number of useful methods or techniques other than those mentioned above that may be applied for analysis of terpenoid compounds. For example, spectrophotometric x-ray analysis is coming into routine use for structure determination (Ladd and Palmer, 1977). It played an important part in the solution of several stereochemical problems in 1982. Absorption in the UV and visible regions (185–800 mμ) is particularly useful in the detection of conjugation and the determination of its extent and nature (Rao, 1975). As would be expected of molecules containing conjugated double bonds, the UV or visible spectra give the best indication of the nature of the polyene system and are important for the analysis of carotenoid compounds. Infrared spectroscopy (IR) is extremely important in determining the nature of the ancillary functional group present in terpenoids (Nakanishi and Solomon, 1977). Another chromatographic method, gel permeation chromatography (GPC), is a useful characterization technique for large-molecule polyterpenoids (Altgelt and Segal, 1971). The molecular weight and molecular weight distribution of these components can be easily determined.

III. CONCLUSIONS

The study of terpenoid chemistry and the analysis of terpenoid compounds have now been revolutionized by the widespread and highly successful application of physicochemical techniques which were made possible by rapid developments in commercial instrumentation. For example, high-performance TLC has now taken its place along with gas and liquid chromatography as a rapid, efficient, and quantitative method. The wide variety of developmental techniques permits the separation of complex mixtures of terpenoid compounds, and numerous detection methods extend the usefulness of TLC. The use of modern instruments such as an automatic sample applicator, densitometer, and so on makes TLC an even more useful analytical tool (Issaq, 1981). Other instruments such as high-field ^{1}H NMR and ^{13}C NMR have substantially increased the total number of known terpenoids and have also provided accurate screening methods for natural rubber.

Many papers have been published on various terpenoid compounds when chromatographic and spectroscopic methods have assisted in the separation, identification, and quantitation of its constituents ("A Special Periodical Report—Terpenoids and Steroids," **1–11,** 1971–1982, The Chemical Society, Burlington House, London). Given space constraints, it is not possible to discuss here the detailed theory and procedures underlying these methods. But the preceding review of their applications to terpenoids should help the reader to appreciate their significance and provide a general guideline for separation and analysis.

ACKNOWLEDGMENTS

The author thanks Dr. L. K. Hunt for critical reading and discussion of this manuscript.

REFERENCES

Altgelt, K. H., and Segal, L. (1971). "Gel Permeation Chromatography." Dekker, New York.

Banthorpe, D. V., and Charlwood, B. V. (1980). The Isoprenoids. *Encycl. Plant Physiol., New Ser.* **8,** 185–215.

Budde, W. L., and Elchelberger, J. W. (1979). "Organic Analysis Using Gas Chromatography/Mass Spectrometry—A Techniques and Procedures Manual." Ann Arbor Sci. Publ., Ann Arbor, Michigan.

Hayman, E., Yokoyama, H., and Schuster, R. (1982). Carbon-13 nuclear magnetic resonance determination of rubber in guayule (*Parthenium argentatum*). *J. Agric. Food Chem.* **30,** 399–401.

Heller, S. R., and Milne, G. W. A. (1978). "EPA/NIH Mass Spectral Data Base," Vols. 1–4 and Supplement. U.S. Govt. Printing Office, Washington, D.C.

Issaq, H. J. (1981). Modern advances in thin-layer chromatography. *In* "Separation and Purification Methods" (C. J. Van Oss, E. Grushka, and N. H. Sweed, eds.), pp. 73–116. Dekker, New York.

Kingston, D. G. (1979). High performance liquid chromatography of natural products. *J. Nat. Prod.* **42**(3), 273–260.

Kirchner, J. G. (1978). Thin layer chromatography. *Tech. Chem. (N.Y.)* **14,** 897–937.

Ladd, M. F. C., and Palmer, R. A. (1977). "Structure Determination by X-Ray Crystallography." Plenum, New York.

Leyden, D. E., and Cox, R. H. (1977). "Analytical Applications of NMR." Wiley (Interscience), New York.

Mann, J. (1978). "Secondary Metabolism." Oxford Univ. Press, London and New York.

Nakanishi, K., and Solomon, P. H. (1977). "Infrared Absorption Spectroscopy." Holden-Day, San Francisco, California.

Newman, A. A. (1972). "Chemistry of Terpenes and Terpenoids." Academic Press, New York.

Rao, C. N. R. (1975). "Ultra-Violet and Visible Spectroscopy—Chemical Applications." Butterworth, London.

Risby, T. H., Field, L. R., Yang, F. J., and Cram, S. P. (1982). Gas chromatography. *Anal. Chem.* **40**(5), 410R–428R.

Scott, R. P. W. (1976). Contemporary liquid chromatography. *Tech. Chem. (N.Y.)* **11.**

Stahl, E. (1969). "Thin Layer Chromatography—A Laboratory Handbook," pp. 201–250. Springer-Verlag, Berlin and New York.

Visintainer, J., Beebe, D. H., Myers, J. W., and Hirst, R. C. (1981). Determination of rubber content in guayule bushes by carbon-13-nuclear magnetic resonance spectrometry. *Anal. Chem.* **53**(11), 1570–1572.

CHAPTER **72**

Isolation and Analysis of Alkaloids*

W. G. W. Kurz

Plant Biotechnology Institute
National Research Council
Saskatoon, Saskatchewan, Canada

I. INTRODUCTION

The pharmacological activities and relative scarcity of a number of alkaloids have created considerable interest over the past decade, interest focusing on the biosynthesis and biotransformation of these compounds by plant cell cultures. Alkaloids are a diverse group of compounds which include derivatives of some amines and nitrogen heterocycles. In general, they show a weak alkaline reaction and are capable of forming salts.

Thus far, tropane, nicotine, purine, indole, morphinane, quinoline, and cephalotaxus alkaloids have been isolated from plant cell cultures, and results of these investigations have been summarized in reviews (Kurz and

*NRCC No. 20573.

CELL CULTURE AND SOMATIC CELL
GENETICS OF PLANTS, VOL. 1

ISBN 0-12-715001-3

Constabel, 1979a,b; Reinhard and Alfermann, 1980; Barz and Ellis, 1981).

The majority of extraction and analytical techniques for the compounds in question have been developed for whole plants, and in some cases only slight changes of these procedures were necessary for use with plant cell cultures.

Particular attention has been given to the indole and morphinane alkaloids because of their medical importance.

II. INDOLE ALKALOIDS

Extraction Procedure

Cells (8–10 g) are separated from the medium by filtration. After being washed with water, the cells are suspended in methanol (100 ml), stirred for 5 min, and filtered under slight suction. The cells are further washed with boiling methanol (3 × 100 ml) and discarded. The combined methanol extract is concentrated *in vacuo* and the residue partitioned between ethyl acetate (30 ml) and 1 *N* HCl (30 ml). The ethyl acetate portion is removed, washed with 1 *N* HCl (100 ml), and discarded. The combined aqueous portion is neutralized with sodium bicarbonate, adjusted to pH 10 with 10 *N* NaOH, and then extracted with ethyl acetate (2 × 60 ml). The combined ethyl acetate fraction is evaporated to dryness, and the residue is triturated with ethyl acetate or dichloromethane and filtered or decanted away from the insoluble material. Evaporation of the filtrate affords the indole alkaloids. For analysis the alkaloids are generally dissolved in 5 ml ethyl acetate (Constabel *et al.*, 1981).

If freeze-dried cells are used, the extraction procedure is the same, except that the cells, after being suspended in methanol, are placed in an ultrasonic bath for 4 hr instead of being stirred for 5 min.

III. SEPARATION AND IDENTIFICATION OF THE DIFFERENT ALKALOIDS

Primarily four methods for the separation and/or identification of indole alkaloids are available. These are thin-layer chromatography (TLC), high-performance liquid chromatography (HPLC), mass spectrometry (MS), and radioimmunoassay (RIA). Of these assays, TLC should be used only for separation and tentative identification, as the complexity of the alkaloid systems and their respective retention times and/or color reactions on chromatoplates can lead to misleading interpretations. Whenever possible, actual isolation and characterization of the formed alkaloids by appropriate comparison with authentic samples should be attempted.

A. Thin-Layer Chromatography

Forty microliters of sample is spotted with the help of a microsyringe on Polygram Silicagel plates without gypsum (Macherey-Nagel Co., Düren, Federal Republic of Germany) (layer 0.25 mm), and the chromatogram is eluted in a solvent system of methanol:ethylacetate (1:9). The alkaloids are detected by means of ceric ammonium sulfate spray reagent (Farnsworth *et al.*, 1964). The reagent is made by dissolving 1 g ceric ammonium sulfate in 100 ml syrupy *o*-phosphoric acid under gradual heating. Before use, the reagent is cooled to room temperature. For identification the chromatogram is cospotted with pure samples of the expected alkaloids (Constabel *et al.*, 1981).

B. High-Performance Liquid Chromatography

A typical procedure for analytical detection is as follows: 10 μl of sample is chromatographed over Lichrosorb RP-8 (particle-size 10 μm) (E. Merck, Darmstadt, Federal Republic of Germany) in a stainless steel column (4.6 × 250 mm) with methanol/H_2O [0.0025 *M* tetrabutylammonium phosphate (TBAP)] as the mobile phase. The composition of the mobile phase varies as indicated below:

Time	MeOH (%)	H_2O/TBAP (%)
0 min	55	45
6 min	58	42
20 min	68	32
25 min	68	32
30 min	68	32

A flow rate of 2 ml/min at 30°C with a resulting pressure of approximately 2500 psi is used. For ultraviolet (UV) detection a wavelength of 254 nm and a sensitivity of 0.04 absorbence units full scale is used.

A typical procedure for preparative separation is as follows (Kurz *et al.*, 1980): 200 mg of alkaloid mixture is chromatographed over Porasil B (Waters Associates, Milford, Massachusetts), 150 g, in a stainless steel column (25 × 300 mm) with H_2O:MeCN (68:32) containing 0.1% Et_3N modifier, at a flow rate of 10 ml/min. A total of 40 × 25 ml fractions are collected, and the column is eluted with MeCN (400 ml, fraction 41). The 41 fractions are analyzed by HPLC on reverse-phase packing, using H_2O:MeCN (62:38) containing 0.1% Et_3N at 4 ml/min and detection wavelengths of 254 and 280 nm. Identification of alkaloids is achieved by comparison with reference samples.

C. Mass Spectrometric Analysis

Alkaloid samples are analyzed on a solid probe. The probe is heated ballistically from room temperature to 350°C. Mass spectra are scanned from 40 to 650 in 2 sec. Single ion traces should be done for all ions of interest. For electron impact ionization the ionization voltage should be 70 eV.

D. Radioimmunoassay

RIA methods are very specific and sensitive, and detection limits of a few nanograms have been achieved in indole alkaloid analysis (Kutney *et al.*, 1980).

Female white rabbits are immunized subcutaneously with either 1, 5, or 20 mg of [^{3}H]alkaloid–bovine serum albumin (BSA) conjugate (Behringwerke Co., Federal Republic of Germany) emulsified in Freund's adjuvant (Miles Laboratories). Booster injections of the serum dosage are administered at 2-week intervals. Blood is collected via the ear veins 10 days after inoculation, and the antigen-binding capacity of the serum is determined by serial dilution. After the fourth booster injection, significant antibody against the alkaloid is detected in the rabbits. The antibody affinity is found to increase following each of the three subsequent booster injections, with the highest antigen binding provided by the antisera from the rabbit immunized with 1-ng doses of conjugate. The antigen-binding capacity of the sera is determined by a modification of the usual $(NH_4)_2SO_4$ precipitation method. Thus, the following reagents are pipetted into a test tube in an ice bath: 0.3 ml phosphate-buffered saline (PBS) (0.9% NaCL, 0.01 *M* PH_4^{3-}, pH 6.8) with 0.04% BSA added, 0.1 ml [^{3}H]alkaloid in PBS (~10^{-7} mmol, ca. 8500 cpm), and 0.1 ml serially diluted antibody solution. Blanks for determination of nonspecific binding contain an additional 0.1 ml PBS instead of the antibody solution. The total assay volume is 500 μl, and each assay should be performed in duplicate. The samples are incubated at 4°C for 20 hr. Normal rabbit serum (50 μl) is then added to each sample, followed by saturated $(NH_4)_2SO_4$ (0.5 ml) to precipitate the antibody–tracer complex. The precipitate is separated by centrifugation and washed with 50% $(NH_4)_2SO_4$ (1 ml). The residual antibody–antigen complex is dissolved in H_2O (1 ml) and counted for radioactivity by liquid scintillation. All suitable antisera are pooled and diluted to give a final working titer of 1:80 under constant assay conditions so as to bind 45% of a fixed mass of radioactive hapten. The total assay volume is 500 μl. The assay is carried out at 0°C. To increase the sensitivity of the assay, the standard alkaloid solution is incubated with the antiserum solution for 1 hr before addition of [^{3}H]alkaloid solution. Each sample is then incubated at 4°C for 20 hr, and the antibody-bound [^{3}H]alkaloid is separated from the free alkaloid by the $(NH_4)_2SO_4$ sulfate method described above. With this procedure, nonspecific binding of less than 4% is attained. The frozen antiserum reveals no deterioration in activity even upon long standing.

IV. MORPHINANE ALKALOIDS

A. Extraction Procedure

The cells (~10 g) are separated from the medium by filtration and washed with water. The filtrate is extracted with ethyl acetate (2 × 50 ml) and discarded. The cells are extracted with boiling methanol (3 × 100 ml), filtered, and discarded. The methanol and ethyl acetate portions are combined and evaporated *in vacuo*. The residue is partitioned between ethyl acetate (50 ml) and 1 *N* HCl (50 ml). After separation, the ethyl acetate portion is washed with 1 *N* HCl (10 ml) and discarded. The combined aqueous layer is adjusted to pH 7.6 with sodium bicarbonate and extracted with ethyl acetate (2 × 25 ml). The combined organic portion is evaporated *in vacuo* and the residue dissolved in 5 ml ethyl acetate for further analysis (Tam *et al.*, 1980).

B. Gas Chromatographic Analysis

Gas chromatography (GC) should be performed on a gas chromatograph equipped with a flame ionization detector (FID) using a capillary column

(WCOT, 8.5 m) packed with OV-101. A splitless injection system is employed, with 2 μl as the injection volume for the sample examined. Solutions of the samples to be examined are subjected to GC using a column temperature programmed at 150–240°C at 4°C/min, with helium at 100 ml/min as the carrier gas. The injector and detector should be held at 200°C and 250°C, respectively (Tam *et al.*, 1982).

C. Gas Chromatographic–Mass Spectrometric Analysis

For confirmation of compounds tentatively identified with GC, a gas chromatographic–mass spectrometric (GC–MS) analysis may be necessary. In this case, the mass spectra should be recorded under conditions in which the transfer line and the jet separator are maintained at ca. 280°C and the GC injector temperature at 275°C. For electron impact ionization, the ionization voltage should be 70 eV and scanning should be done repetitively at 2.5 sec per scan (Tam *et al.*, 1982).

D. High-Performance Liquid Chromatographic Analysis

HPLC allows a simple and rapid quantitative analysis of the major alkaloids by direct isocrative HPLC on a reverse-phase partition mode column without using ion-pair reagents. A 10-μl sample is chromatographed over either medium polar Nucleosil 10 CN or nonpolar Nucleosil 10 C_{18} (Macherey, Nagel and Co., Düren, Federal Republic of Germany) in a stainless steel column (4 × 300 mm) with a mobile phase of 1% ammonium acetate (pH 5.8):acetonitrile:dioxane (80:10:10) for the former and 1% ammonium acetate (pH 5.8):acetonitrite (70:30) for the latter packing. The flow rate for both systems is 1.5 ml/min. The wavelength is set at 254 nm (Nobuhara *et al.*, 1980).

E. Radioimmunoassay

An RIA kit for morphine is commercially available through Roche Diagnostics. The protocol specified in the test kit is applied to analysis of plant cell material, except that the volume of saturated $(NH_4)_2SO_4$ for antibody precipitation is increased to 140% of the kit value (Hodges and Rapoport, 1982).

REFERENCES

Barz, W., and Ellis, B. (1981). Plant cell cultures and their biotechnological potential. *Ber. Dtsch. Bot. Ges.* **94,** 1–26.

Constabel, F., Rambold, S., Chatson, K. B., Kurz, W. G. W., and Kutney, J. P. (1981). Alkaloid production in *Catharanthus roseus* (L.) G. Don. VI. Variation in alkaloid spectra of cell lines derived from one single leaf. *Plant Cell Rep.* **1,** 3–5.

Farnsworth, N. R., Blomster, R. N., Damratoski, D., Meer, W. A., and Camarato, L. V. (1964). Studies on *Catharanthus* alkaloids. VI. Evaluation by means of thin-layer chromatography and ceric ammonium sulfate spray reagent. *J. Nat. Prod.* **27,** 302–314.

Hodges, C. C., and Rapoport, H. (1982). Morphinan alkaloids in callus cultures of *Papaver somniferum*. *J. Nat. Prod.* **45,** 481–485.

Kurz, W. G. W., and Constabel, F. (1979a). Plant cell cultures, a potential source of pharmaceuticals. *Adv. Appl. Microbiol.* **25,** 209–240.

Kurz, W. G. W., and Constabel, F. (1979b). Plant cell suspension cultures and their biosynthetic potential. *In* "Microbial Technology" (H. J. Peppler and D. Perlman, eds.), 2nd ed., Vol. 1, pp. 389–416. Academic Press, New York.

Kurz, W. G. W., Chatson, K. B., Constabel, F., Kutney, J. P., Choi, L. S. L., Kolodziejczyk, P., Sleigh, S. K., Stuart, K. L., and Worth, B. R. (1980). Alkaloid production in *Catharanthus roseus* cell cultures. IV. Characterization of the 953 cell line. *Helv. Chim. Acta* **63,** 1891–1896.

Kutney, J. P., Choi, L. S. L., and Worth, B. (1980). Radioimmunoassay determination of vindoline. *Phytochemistry* **19,** 2083–2087.

Nobuhara, Y., Hirano, S., Namba, K., and Hashimoto, M. (1980). Separation and determination of opium alkaloids by high-performance liquid chromatography. *J. Chromatogr.* **190,** 251–255.

Reinhard, E., and Alfermann, A. W. (1980). Biotransformation by plant cell cultures. *Adv. Biochem. Eng.* **16,** 50–83.

Tam, W. H. J., Constabel, F., and Kurz, W. G. W. (1980). Codeine from cell suspension cultures of *Papaver somniferum* L. cv. Marianne. *Phytochemistry* **19,** 486–487.

Tam, W. H. J., Kurz, W. G. W., Constabel, F., and Chatson, K. B. (1982). Biotransformation of thebaine by cell suspension cultures of *Papaver somniferum* L. cv. Marianne. *Phytochemistry* **21,** 253–255.

CHAPTER **73**

Protein Extraction and Analysis

L. R. Wetter

Plant Biotechnology Institute
National Research Council
Saskatoon, Saskatchewan, Canada

I. INTRODUCTION

Studies related to plant proteins require special extraction procedures and frequently some method for determining their concentration. Many investigations of higher plants and their culture products, for example, cell cultures, include studies of their enzyme systems. Whether these studies are involved with the metabolism or the genetics of the cell, invariably enzyme systems will be investigated, therefore requiring extraction of the enzyme from a complex living system.

In many protein studies, especially those embracing enzymes, it is important that the extraction conditions be such that maximum yields are obtained and that a minimum of denaturation occurs. Since plant material, particularly green tissue, contains many phenolic compounds as well as tannins, which affect the isolation and stability of enzymes, special attention must be given when dealing with such material. These special problems can be alleviated by employing various techniques and additives to inhibit the reaction of phenolic compounds and tannins with protein, as well as the deleterious effect of plant phenolases (Loomis, 1974; Rhodes,

CELL CULTURE AND SOMATIC CELL
GENETICS OF PLANTS, VOL. 1

ISBN 0-12-715001-3

1977). In addition to dealing with special problems, one must ensure that the extraction medium meets all the conditions required for maintaining enzyme stability outside the cell.

Often it is necessary to ascertain the protein concentration of an extract, for example, when one wants a comparison of the enzyme activity of various tissues of a plant, a comparison of the protein content in organized plant tissue, callus, and cell suspension cultures, or a comparison of isozyme electrophoretic patterns derived from various samples. There are several different methods for determining the protein content of an extract. The Dumus and Kjeldahl assays are based on the quantitative estimation of nitrogen. The Folin phenol method is based on the color development of aromatic amino acids found in protein. Still other methods are based on the dye-binding capacity of proteins (Flores, 1978; Sedmak and Grossberg, 1977).

The assay eventually chosen will often depend on the experimenter's needs. One must consider the accuracy desired, the number of samples to be assayed, and the speed desired. Ideally, a simple, rapid, accurate assay is the goal. One developed by Lowry and co-workers (1951) is frequently employed to satisfy the above criteria; however, it must be modified to fill the special needs of plant extracts.

The procedures described in this chapter are those developed in the author's laboratory primarily for the study of isozyme systems in plant cell suspensions, in calli, and for various organs and tissues of higher plants, such as leaves (Wetter, 1982; Wetter and Dyck, 1983).

II. EXTRACTION OF PLANT MATERIAL

The extraction procedure described here was designed primarily to attain maximum enzyme activity. This was achieved in part by controlling the pH and by using additives such as reducing agents as necessary to maintain a reducing environment. All operations were conducted at 5°C and carried out as quickly as practical. The size of the sample often dictated the method employed for extraction. When samples were small, semimicro techniques were utilized.

A. Procedure

The extraction buffer is prepared by dissolving 24.23 g tris(hydroxymethyl)aminomethane (Tris, 0.2 *M*) in approximately 500 ml distilled water. The pH is adjusted to 8.5 with HCl. Then 342.3 g sucrose is dissolved in the buffer, after which distilled water is added to a total volume of 1 liter of solution (Gamborg *et al.*, 1979). Enough buffer is prepared to meet a week's requirements. The buffer is stored at 5°C, and 2-mercaptoethanol (4.38 g or 2.9 ml per liter of buffer, equivalent to 0.056 *M*) is added just prior to the extraction. When working with green plant material, it was found that 0.077 *M* 2-mercaptoethanol was required to prevent extensive browning of the extract. In many cases, dithiothreitol (DTT) at the 0.026 *M* level can be substituted for 2-mercaptoethanol.

The cells from a suspension culture or callus are collected on Miracloth on a Millipore filter and washed exhaustively with 3% (w/v) mannitol. Leaves are usually washed thoroughly in distilled water.

Cells or calli can be extracted in several different ways. The French pressure cell and various tissue grinders are employed. For cell or callus samples of 1 g or less, the French pressure cell can be used; samples as small as 10 mg can be ruptured. The washed cells are added to the cold pressure cell, together with cold extraction medium (1 ml per gram of wet weight), and ruptured at 140 MPa (20,000 psi) in an Aminco laboratory press. The ruptured cells are collected in centrifuge tubes and centrifuged at 30,000 *g* for 1 hr at 5°C. The protein extract is stored under nitrogen at 5°C (Wetter, 1977). If no French pressure cell is available, the samples can be extracted in a Polytron PT 10-35 homogenizer. The cell samples are placed in 15 × 100-mm Corex centrifuge tubes together with approximately 1 ml extraction medium and a drop of *n*-octanol to inhibit excessive foaming. The mixture is ground for 2 min at top speed, keeping the tube in an ice water bath. The sample is centrifuged as indicated above.

Glass tissue homogenizers (available in a variety of sizes) can be employed when working with "soft" material such as cells from suspension cultures, callus, and soft, spongy leaves. They are, however, not very effective on "hard" tissue such as leaves obtained from cereals or grasses.

B. Results and Conclusions

This method of extraction has worked very well for the preparation of enzymes to be studied by gel electrophoresis. In studies related to the somatic hybridization of soybean and *Nicotiana*, the isozyme patterns proved that fusion had been achieved (Wetter, 1977; Chien *et al.*, 1982). Its use has not been restricted to cell and callus systems; it can be utilized equally effectively in whole plant systems, such as plants derived from the somatic hybridization of *Nicotiana rustica* and *N. tabacum* (Douglas *et al.*, 1981).

There are many techniques and modifications, far too numerous to mention in this chapter, that can be employed in the preparation of protein extracts from plants. One must keep in mind that most plants contain many kinds of secondary products that may seriously affect the extraction of proteins, particularly active enzyme systems. Perhaps the most troublesome are those that produce phenolic compounds, such as phenylpropanoids, flavonoids, tannins, and the enzyme phenolase, which reacts with phenolic compounds, resulting in browning. Various additives are employed in an attempt to control the deleterious consequences of these compounds. Thiols such as 2-mercaptoethanol and dithiothreitol, and reducing agents such as metabisulfite and cysteine, are employed to inhibit phenolase and its reaction with phenolics. Use is also made of polymeric agents which absorb the tannins and thus remove them from the extract. Polyvinylpyrrolidone, particularly the insoluble form known as Polyclar AT, is perhaps one of the more effective. A recent report suggests that absorbent polystyrenes may be used in the preparation of active plant enzyme extracts (Loomis *et al.*, 1979). The reader is referred to two excellent reviews that deal with these particularly thorny problems (Loomis, 1974; Rhodes, 1977).

Kelley and Adams (1977) had to employ a complex extraction mixture in order to obtain active enzyme preparations from juniper. Their standard extraction mixture consisted of 0.1 *M* Tris-malate buffer (pH 7) containing 0.2 *M* sodium tetraborate, 0.25 *M* sodium ascorbate, 0.02 *M* sodium metabisulfite, 0.02 *M* sodium diethyldithiocarbamate, 0.01 *M* germanium oxide, and 10% dimethylsulfoxide. In addition, they found that the utilization of polyvinylpyrrolidone powder and additional extractions with *n*-butanol and/or diethyl ether improved the yield of active enzyme. Kuo *et al.* (1982) reported that the stability of nitrate reductase isolated from barley leaves was dependent on the various components in the extraction medium. These included not only the correct pH and composition of the buffer but also other additives such as sulfhydryl reagents, molybdate, flavin adenine dinucleotide, and ethylenediaminetetraacetic acid.

The preparation of acetone powders is often employed to achieve high yields of undenatured protein which is free of pigments and low molecular weight contaminants. Ibrahim and Cavia (1975) used this procedure to obtain active protein from intact and cultured plant tissues of a number of plants. These powders frequently produce excellent yields of protein which might not be achieved when working with plant cultures of high water and low protein content.

A variety of compounds are added to extraction media to meet special problems encountered by the experimenter. Frequently a chelating agent, as ethylenediaminetetraacetic acid, is added. It is primarily intended to immobilize metal ions which can inactivate the enzyme in question. Other sensitive enzyme systems can be protected by the addition of high concentrations of foreign proteins; the one most commonly used is bovine serum albumin (Loomis, 1974; Torres and Tisserat, 1980). Crude protein mixtures often contain potent proteases which will destroy the protein of interest. These can be inhibited by organic compounds such as phenylmethane–sulfonylfluoride (Lönnendonker and Schieder, 1980).

Finally, one should keep in mind that speed during extraction is essential; once the cells have been ruptured, the longer it takes to extract a sample, the more likely it is that protein damage will occur. In most cases, it is important to keep the temperature low; this will retard any undesirable reactions which could cause inactivation.

Kelley and Adams (1977) summarize the importance of extraction media nicely: "A general guideline which might be used is that the best results will be obtained from the 'cleanest' extracts prepared with the fewest inhibitory chemicals and simplest extraction procedure."

III. ESTIMATION OF PROTEIN

The primary objective in doing a protein assay is to keep it relatively simple and still give a good measure of protein content. The procedure described here does this by combining a protein precipitation step with a colorimetric assay that is rapid and simple. The precipitation step is employed to separate the protein quantitatively from the interfering components often found in plant extracts.

A. Procedure

1. The Precipitation of Protein

The method presented here was developed by Bensadoun and Weinstein (1976). The protein extract (in our case 10–50 μl, depending on the protein content) is placed in a 15-ml conical centrifuge tube (either glass or disposable tubes are utilized). It is important to employ conical tubes, as these facilitate the packing of the minute quantity of protein. Three milliliters of distilled water is added, followed by 25 μl of 2% (w/w) sodium deoxycholate solution. The contents of the tube are mixed thoroughly using a Vortex mixer and left to stand at room temperature for 15 min. Now 1 ml of 24% (w/v) trichloroacetic acid solution is added, again mixing well. The tubes are centrifuged at 1300 *g* for 30 min at room temperature. A bench-top centrifuge with a horizontal head is recommended, as this will ensure that the protein precipitate is firmly packed into the tip of the tube. After centrifugation, the tubes are turned upside down and allowed to drain. Usually one precipitation step is sufficient, but if the blanks are high this step should be repeated.

2. The Assay of Protein

The protein assay is based on the Folin phenol method originally developed by Lowry *et al.* (1951); however, the one described here is based on a modification described by Miller (1959). Two solutions are required and are made up as follows: *Reagent A* consists of 10 parts of 10% (w/v) sodium carbonate in 0.5 *N* sodium hydroxide, 0.5 parts of 2% (w/v) potassium tartrate solution, and 0.5 parts of 1% (w/v) copper sulfate pentahydrate ($CuSO_4 \cdot 5H_2O$) solution which contains 1 drop of concentrated sulfuric acid for every 100 ml of solution. It is important to mix reagent A as follows: The tartrate and copper solutions are combined, and then the carbonate solution is added to the tartrate–copper mixture. This reagent is prepared fresh every day. *Reagent B* is made up of 1 part of Folin reagent (purchased from any chemical company) in 10 parts of distilled water.

The assay is performed by adding 1 ml of reagent A to the drained tube containing the precipitated protein. The precipitate is then dissolved by thorough mixing; again, it is recommended that a Vortex mixer be employed. Make sure that the precipitate is completely dissolved. One milliliter of water is now added; again, the contents are well mixed and allowed to stand at room temperature for 15 min. Three milliliters of reagent B is added to the tube as forcibly as practical. In the author's laboratory, a preset syringe (with the needle removed) is employed. The Folin phenol reagent is ejected into the protein solution with one stroke of the plunger. Again, the sample is thoroughly mixed and the tubes are transferred to a water bath set at 50°C, where the contents are heated for 15 min. Then 1.5 ml of distilled water is added to bring the total volume to 6.5 ml, mixed, and cooled to room temperature. The color is now fully developed and is read in the spectrophotometer at 625 nm.

The spectrophotometer readings are compared to a calibration curve obtained by using a standard protein solution prepared from bovine serum albumin.

B. Results and Conclusions

It is important to establish that the components employed in the extraction medium do not interfere with the protein assay. This can be done by performing the assay on the extraction medium. An example of what can happen is taken from the author's experience. Before the above method was adopted, a method based on the biuret reaction described by Itzhaki and Gill (1964) was employed. On testing, it was noted that buffers containing tris(hydroxymethyl)aminomethane gave a positive color reaction, resulting in abnormally high readings. For this reason, an effective precipitation step is very important. With this step, one can remove the water-soluble interfering compounds. Testing of the extraction medium is particularly important when complex mixtures are employed.

Loomis (1974) indicates that there are a variety of pitfalls one must be aware of when selecting a protein assay for plant material.

The method described above has proven to be extremely useful. It is possible to determine accurately the protein content of an extract obtained from as little as 100 mg of callus or cell suspension culture. Reproducible results can be obtained on extracts equivalent to as little as 3 mg of cellular material.

ACKNOWLEDGMENTS

The author thanks Mr. John Dyck for his valuable technical assistance in developing the techniques described in this chapter.

REFERENCES

Bensadoun, A., and Weinstein, D. (1976). Assay of proteins in the presence of interfering materials. *Anal. Biochem.* **70,** 241–250.

Chien, Y.-C., Kao, K. N., and Wetter, L. R. (1982). Chromosomal and isozyme studies of *Nicotiana tabacum- Glycine max* hybrid cell lines. *Theor. Appl. Genet.* **62,** 301–304.

Douglas, G. C., Wetter, L. R., Nakamura, C., Keller, W. A., and Setterfield, G. (1981). Somatic hybridization between *Nicotiana rustica* and *N. tabacum.* III. Biochemical, morphological, and cytological analysis of somatic hybrids. *Can. J. Bot.* **59,** 228–237.

Flores, R. (1978). A rapid and reproducible assay for quantitative estimation of proteins using bromophenol blue. *Anal. Biochem.* **88,** 605–611.

Gamborg, O. L., Shyluk, J. P., Fowke, L. C., Wetter, L. R., and Evans, D. (1979). Plant regeneration from protoplasts and cell cultures of *Nicotiana tabacum* sulfur mutants (Su/Su). *Z. Pflanzenphysiol.* **95,** 255–264.

Ibrahim, R. K., and Cavia, E. (1975). Acrylamide gel electrophoresis of proteins from intact and cultured plant tissues. *Can. J. Bot.* **53,** 517–519.

Itzhaki, R. F., and Gill, D. M. (1964). A micro-biuret method for estimating proteins. *Anal. Biochem.* **9,** 401–410.

Kelley, W. A., and Adams, R. P. (1977). Preparation of extracts from juniper leaves for electrophoresis. *Phytochemistry* **16,** 513–516.

Kuo, T.-M., Warner, R. L., and Kleinhofs, A. (1982). *In vitro* stability of nitrate reductase from barley leaves. *Phytochemistry* **21,** 531–533.

Lönnendonker, N., and Schieder, O. (1980). Amylase isoenzymes of the genus *Datura* as a simple method for an early identification of somatic hybrids. *Plant Sci. Lett.* **17,** 135–139.

Loomis, W. D. (1974). Overcoming problems of phenolics and quinones in the isolation of plant enzymes and organelles. *In* "Methods in Enzymology" (S. Fleischer and L. Packer, eds.), Vol. 31, Part A, pp. 528–545. Academic Press, New York.

Loomis, W. D., Lile, J. D., Sandstrom, R. P., and Burbott, A. J. (1979). Absorbent polystyrene as an aid in plant enzyme isolation. *Phytochemistry* **18,** 1049–1054.

Lowry, O. H., Rosebrough, N. J., Farr, A. L., and Randall, R. L. (1951). Protein measurement with the Folin phenol reagent. *J. Biol. Chem.* **193,** 265–275.

Miller, G. L. (1959). Protein determination for large numbers of samples. *Anal. Chem.* **31,** 964.

Rhodes, M. J. C. (1977). The extraction and purification of enzymes from plant tissues. *In* "Regulation of Enzyme Synthesis and Activity in Higher Plants" (H. Smith, ed.), pp. 245–269. Academic Press, New York.

Sedmak, J. J., and Grossberg, S. E. (1977). A rapid, sensitive and versatile assay for protein using coomassie brilliant blue G-250. *Anal. Biochem.* **79,** 544–552.

Torres, A. M., and Tisserat, B. (1980). Leaf isozymes as genetic markers in date palms. *Am. J. Bot.* **67,** 162–167.

Wetter, L. R. (1977). Isoenzyme patterns in soybean-*Nicotiana* somatic hybrid cell lines. *Mol. Gen. Genet.* **150,** 231–235.

Wetter, L. R. (1982). Isozyme analyses of cultured plants cells. *In* "Plant Tissue Culture Methods" (L. R. Wetter and F. Constabel, eds.), 2nd rev. ed., NRCC No. 19876, pp. 105–111. National Research Council of Canada, Ottawa.

Wetter, L. R., and Dyck, J. (1983). Isoenzyme analysis of cultured cells and somatic hybrids. *In* "Handbook of Plant Cell Culture" (D. A. Evans, W. R. Sharp, P. V. Ammirato, and Y. Yamada, eds.), Vol. 1, pp. 607–628. Macmillan, New York.

CHAPTER **74**

Isolation and Analysis of Plant Growth Regulators

Kerry T. Hubick
David M. Reid

Plant Physiology Research Group
Department of Biology
University of Calgary
Calgary, Alberta, Canada

CELL CULTURE AND SOMATIC CELL
GENETICS OF PLANTS, VOL. 1

ISBN 0-12-715001-3

I. INTRODUCTION

There are only five major groups of plant growth regulators (PGR) [auxin, gibberellins (GA), cytokinins, (CK), abscisic acid (ABA) and related compounds, and ethylene]. ABA and ethylene are single compounds, but there is a complex array of structurally similar compounds within the GA, auxin, and CK groups. A compound is defined as a member of these three groups on the basis of both biological activity and molecular structure, making the number of compounds in these groups very large. For example, GAs number over 60 different molecular species, not including conjugated forms.

It is no surprise, then, that there are many techniques for the estimation of endogenous levels of these compounds. Because of the bewildering array of PGR and methods of extraction/quantification, it will not be possible either to describe all of the methods or to go into great detail for any one method. Our aim, rather, is to direct the reader to those methods we consider to be reliable and practical. In addition, we will try to guide readers away from some of the numerous pitfalls that exist in this area of research.

The PGR were originally discovered because of their growth-promoting or inhibitory activities, so early quantification techniques depended on using the response of biological systems, that is, the bioassay, to a PGR-containing fraction isolated from a plant. These techniques have been extensively reviewed and have many limitations; therefore, they will be included in this chapter only when necessary. Limitations of the bioassay include the time required to perform the assay, the varying sensitivity to compounds of the same PGR class, the possibility of conversion of inactive precursors into active PGR during bioassay, the lack of specificity, especially in bioassays of inhibitors or mixtures of inhibitory and promotive PGR, and lack of precise quantification.

Because of the inadequacies of bioassay, new physicochemical techniques for identification and quantification of PGR have been designed. A combination of bioassay with physicochemical techniques still shows much promise in the accurate analysis of the wide spectrum of GAs and CKs. Measurements of indoleacetic acid (IAA) by bioassay have generally been superseded by physicochemical techniques. In determinations of ethylene and ABA, the bioassay is rarely needed.

The use of high-performance liquid chromatography (HPLC) has become widespread in the purification of PGR. HPLC provides a technique that is relatively fast and more reproducible than older techniques such as thin-layer chromatography (TLC). Also, recoveries of 100% can be achieved with HPLC, but this is not likely with TLC because of irreversible adsorption unless the strictest care is taken when eluting R_f bands (Saunders, 1978). A selection of a wide range of columns, both normal and reverse phase, and any number of solvents and solvent mixtures, provide great resolving power. Reeve and Crozier (1980) have lauded the merits of the resolving power of HPLC in providing PGR pure enough to enable unequivocal identification, and this is in accord with the findings of recent reviews on PGR analysis (Brenner, 1979, 1981; Walton, 1980). An added benefit of HPLC is detection without derivatization.

The investigator may be interested in more than one PGR or group of PGR at one time in one plant tissue. Ideally, purification schemes should be devised to allow such analyses. This is indeed possible using an HPLC system (Brenner, 1979). In addition to modern physicochemical techniques, there is the immunoassay (Weiler, 1982). Immunoassay increasingly shows greater promise but still has shortcomings, which will be discussed later.

Quantification and identification of PGR (except with immunoassay) first requires fairly extensive extraction and purification. The amount of PGR is then estimated, ideally with a system that is both specific for that PGR and sensitive enough to measure the very low amounts of PGR found in plant tissue. Sensitivity ranges for modern physicochemical techniques are in the nanogram to picogram range.

We agree with the above authors on the necessity, whenever possible, of analyzing PGR using physicochemical techniques. Therefore, this chapter will be concerned with the more recent literature, selecting only those methods that we consider to be most efficient in the analysis of PGR.

II. GENERAL POINTS TO BE CONSIDERED IN DESIGNING A SELECTIVE SCHEME FOR EXTRACTION AND QUANTIFICATION OF PLANT GROWTH REGULATORS

A. Extraction

1. Use fresh or lyophilized (freeze-dried) tissue. Rapid freezing in liquid nitrogen is essential prior to lyophilization and can also be used before extraction of fresh tissue. Fresh tissue should not be stored at −20°C for subsequent extraction, but ideally should be extracted immediately after harvest.
2. Avoid contamination from previously used synthetic or natural PGR. All glass apparatus should be acid washed, and silylation of glass apparatus is a useful precaution (Cargile *et al.*, 1979; Brenner, 1979).
3. Avoid high temperature and bright lights during the extraction process, particularly for IAA.
4. It is unwise to leave PGR in aqueous solution at extremes of pH

(high or low) for extended periods of time (several hours). This can lead to hydrolysis of conjugates and isomerization (ABA, GAs) or epimerization (GAs).

5. With HPLC, always use guard columns to avoid unnecessarily shortening the column life. Precolumns are also recommended, especially if the elution solvent is at a pH greater than 7. Neither type of precolumn significantly reduces the efficiency of the HPLC column.

6. In a purification scheme, attempt to use as many different types of procedures as possible (e.g., rather than do two sequential adsorptive chromatographic steps, carry out only one and follow it with partition chromatography.)

7. Do not unnecessarily extract parts of plants or tissues in which you have no interest. This will result in extra purification steps. For instance, if only the differentiating zone of a callus tissue is of interest, extract only that part.

8. Carry out extractions and purifications as rapidly as is practical.

9. Store extracts for as short a time as possible in dry (lyophilized) form or dissolved in dry organic solvent.

10. Always remember that the final isolated compound may be an artifact produced during the procedure. Points 1–4, 7, and 8 and the use of antioxidants (see Sections III, B, D) will reduce this possibility (see also Yokota *et al.*, 1980).

B. Quantification

1. When practical, avoid the use of bioassays as precise quantitative tools. On the other hand, they are useful as inexpensive diagnostic methods, especially in the early stages of purification.

2. So-called identification of a PGR based on inadequate purification and utilization of one nonspecific detection method such as gas chromatography (GC) with a flame ionization detector (FID) is meaningless. In such a situation, a bioassay result is much more useful.

3. When bioassay is used for GA, CK, and auxins, it is vital to assay the extracts in a wide range of dilutions. Often the highest concentrations of the extracts have a toxic effect in the assay system due to inhibitory material which masks the response of the stimulatory PGR. As more dilute extracts are bioassayed, they may be found to elicit a stimulatory response. At this stage, one has diluted the inhibitory material to an ineffective concentration, leaving only the action of promotive PGR. The highest value for the PGR content from this series of dilutions is taken as the best estimate for the PGR content. For details of the technique, see Reid *et al.* (1974, p. 54).

4. Ideally, combinations of HPLC, gas chromatography–mass spectrometry (GC–MS), and bioassay give the most reliable results.

5. If the purification/quantification procedure used is complex, with many steps, the chances of losing the PGR are greatly increased. We strongly recommend use of an internal marker (^{14}C-, ^{3}H-, or ^{2}H-labeled isotopes). This allows one to estimate losses at any stage of the procedure. A tracer must have as high specific radioactivity as possible. This ensures that the tracer does not appreciably add to the mass of the endogenous PGR. A scintillation counter is necessary to measure the level of ^{14}C or ^{3}H, and a mass spectrometer is required for ^{2}H-labeled compounds (Fig. 1).

III. EXTRACTION FROM TISSUE

One aspect of PGR analysis that is often overlooked is removal of the substance from the tissue and transfer into a suitable solvent. Incomplete

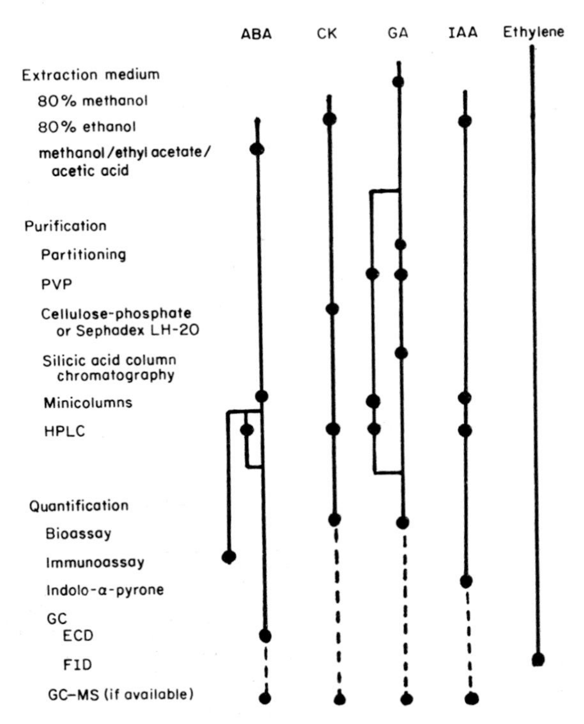

Fig. 1. A scheme showing a range of methods available for extraction, purification, and quantification of PGR. Vertical lines interspersed with ——●—— indicate the suggested sequence of procedures to be followed. The cleaner the extract, the fewer procedures will be needed.

or nonreproducible extraction leads to incorrect estimation of the endogenous PGR. Organic solvents remove PGR from tissue and, quite frequently, methanol is the extraction solvent of choice. Other solvents and water-organic solvent mixtures are also used (Brenner, 1981).

With GAs, CKs, auxins, and ABA, the plant tissue is usually homogenized in the appropriate solvent in a cold (2–4°C) Waring blendor for 2–3 min. Sonication can also be useful. The ideal solvent extracts the PGR with as few other compounds as possible. In addition, it does not cause chemical transformation of the PGR and does not later require evaporation at high temperatures. After extraction, the homogenate is filtered with a Buchner funnel to separate the dissolved PGR from the cellular debris. Occasionally, centrifugation can be useful to get rid of fine particulates.

A. Gibberellins

GAs and GA conjugates can be efficiently extracted with 80% (v/v) aqueous methanol, usually 10 ml per gram fresh weight, although a range of different solvent mixtures has been used (Russell, 1975). That the extraction solvent for use with GA extractions can affect the accuracy of the estimate of endogenous GAs was illustrated by Railton and Rechav (1979), who found that 80% methanol released more "GA-like" activity measured by the lettuce hypocotyl bioassay than a mixture of ethyl acetate:methanol:formic acid (50:50:10 v/v) and much more than Triton X-100. Avoid prelonged exposure of GA to acidic conditions, which are known to cause ring rearrangement and other effects (MacMillan *et al.*, 1960).

B. Abscisic Acid

Aqueous methanol is a good extractant for ABA and its conjugates, but only if the tissue is ground thoroughly and allowed to stand for some time. Otherwise the methanol:ethyl acetate:formic acid mixture is more effective than aqueous methanol (Loveys, 1977). We find methanol:ethyl acetate:acetic acid (50:50:1 v/v) plus 20 mg/l butylated hydroxytoluene (BHT) mixture with the acetic acid substitution for formic acid (to avoid ABA degradation; Loveys, 1977) very effective (Hubick and Reid, 1980). When aqueous methanol is used to extract ABA, it should be acidified to pH 5 (Little *et al.*, 1978) to avoid the formation of ABA methylester and release of the sugar moiety from conjugated ABA (Milborrow and Mallaby, 1975). For this reason, if ABA only is to be considered, acetone or 80% acetone is often the extraction solvent chosen (Walton, 1980). However, ABA isomerizes most easily to *trans, trans*-ABA in acidic solution (Saunders, 1978), so the extraction mixture should not be exposed to strong light during the extraction procedure.

C. Cytokinins

Cytokinins are most frequently extracted with 80% aqueous methanol or ethanol. Horgan (1978) suggests that degradation of CK nucleotides may occur with this solvent because of incomplete inactivation of some nonspecific phosphatases. A better solvent for CK extraction is methanol:chloroform:formic acid (12:5:3 v/v) at −20°C for 24 hr, although adding ethanol at 0°C for 12 hr to soft tissues such as callus tissues until the ethanol is 80% in relation to tissue water does not result in CK nucleotide breakdown. Even then the relative amounts of AMP and adenosine extracted from tissue callus vary with the extraction solvent (see Horgan, 1978).

D. Auxins

The situation is complex in regard to the choice of extraction solvent for IAA. Both methanol and ethanol can convert IAA to its methyl ester. Methanol converted 10% of the endogenous IAA to its methyl ester; ethanol converted 2% to the ethyl ester (Allen *et al.*, 1982). Alcohol extraction of IAA also promotes the conversion of indolepyruvic acid (IPyA) to IAA (Atsumi *et al.*, 1976), and the conversion occurs overnight in methanol (Hemberg and Tillberg, 1980). Hemberg and Tillberg (1980) carefully considered the effect of the extractant on the yield of IAA from several different plant tissues. Methanol, diethyl ether, acetone, and methylene chloride were tested for different time periods. The amount of IAA extracted varied with the temperature and time. Low-temperature (below 0°C) extraction is a good idea. Aqueous methanol generally gave the highest amount of IAA, but this may have been due to release of IAA from conjugates as much as to the excellence of the extractant (Hangarter and Good, 1981). Too little attention is often paid to the extraction solvent (Horgan, 1978). As Hemberg and Tillberg (1980) proposed for IAA, a comparison of any PGR levels in experiments is valid only if the same extraction solvent, extraction period, and extraction temperature have been used. It follows, then, that the efficiency of solvents should be tested at the outset of experiments with new tissues.

Recovery of IAA can sometimes be extremely poor and variable. Little *et al.* (1978) reported up to fivefold differences in recovery of IAA from *Picea*, with as little as 0.5% recovery in some cases. Thus, precautions must be taken with the extraction of auxins.

1. To reduce losses, antioxidants should be added to the initial extraction medium. Antioxidants such as BHT or sodium diethyl dithiocarbamate added to the extraction medium (Durley *et al.*, 1982; Martin *et al.*, 1980) and subsequent stages (Moritoshi *et al.*, 1980) greatly increase auxin recovery.
2. To allow for estimation of variable losses, labeled tracers can be added at the same stage. Compounds with the highest possible specific radioactivity and with chemical structures as similar as possible to those being extracted should be employed. More recently, since the availability of [^{2}H]ABA (Rivier and Pilet, 1981) or [^{2}H]IAA (Allen *et al.*, 1982), stable isotope dilution techniques have been devised. Although allowing accurate estimation of these compounds without the need for a radioisotope license, they do require access to GC–MS facilities, which is not likely to be routine in many laboratories.
3. IAA can also be lost from ether extracts through sublimation during low-pressure evaporation of the solvent (Mann and Jaworski, 1970). If ether is used, it should be peroxide free and evaporated at atmospheric pressure.
4. High irradiation conditions should be avoided during auxin extractions.

E. Ethylene

Since ethylene is a gas, the methods used with it are very different from those used with the other PGR. Some simple methods used for collecting ethylene produced and evolved by plants are described below.

1. Extraction from Tissue

We do not recommend the method of Beyer and Morgan (1970), as the solution of ammonium sulfate into which the tissue is placed can sometimes absorb and later release ethylene into subsequent samples. A simpler method is to place the tissue into a small glass vial containing about twice the volume of the tissue and seal it with a serum cap. The needle of a gas-tight syringe is then inserted through the serum cap, and the syringe plunger is withdrawn so that the combined volume of the space within the vial and syringe is five times the volume of the vial alone. Hold the syringe out for 4 min and then allow it to return to the zero position. Immediately remove 1–5 ml of the vial head space for ethylene analysis by GC (Fabijan *et al.*, 1981). One precaution is necessary: The procedure must be carried out within a few minutes, as wound ethylene will start to be produced by the tissue within 20–30 min (Abeles, 1973).

2. Accumulation of Ethylene in the Head Space of an Enclosed Volume

Tissues are placed in a sterile (microorganisms can absorb and/or produce ethylene) sealed container. Serum caps or glass in combination with a small quantity of high-vacuum silicone grease make good seals. Do not heat or sterilize rubber or plastics, as they can evolve ethylene. The head space is then regularly sampled for ethylene. Large concentrations of ethylene or carbon dioxide should not be allowed to build up, as they may feed back and influence growth and development (Abeles, 1973). Short periods (a few hours) of sealing interspersed with longer periods of free gas movement are ideal. Experimenters should be aware that recently autoclaved agar may produce some ethylene for a few days.

3. Ethylene Collections in a Gas Flow System

Tissues are kept in containers through which a controlled flow of gas is passed. After moving over the tissue, the gas can be passed through an ethylene-absorbing agent such as 0.25 *M* mercuric perchlorate (Abeles, 1973, p. 21 for method of preparation). The mercuric perchlorate can be stored at 2–4°C in airtight glass containers. Subsequently, it is placed in a small serum-capped vial into which an equal volume of 4 *M* NaCl is injected (Abeles, 1973). This liberates the ethylene. However, it is best to keep stirring the vials at room temperature for at least 1 hr to ensure that all of the ethylene has evolved. Great care should be taken with mercuric perchlorate, as it is toxic and, when dry, is highly explosive. The following precautions are imperative: Make up only small volumes at any one time, use no ground-glass joints, and wash spills with excess water.

An alternative way of trapping small amounts of ethylene in large volumes of gas is to take 50-ml gas samples and slowly pass them through a small U-shaped metal column packed with Poropak S precooled to −80°C. This traps only hydrocarbons. Later, ethylene can be released by heating the column to 100°C for 3 min and then allowing the gas to flow directly into a GC column. This is achieved by attaching the U-trapping column to the GC between the carrier gas line and the injector so that the carrier gas sweeps the liberated ethylene directly onto the GC analytical column (Eastwell *et al.*, 1978; DeGreef *et al.*, 1976). This allows one to measure very low concentrations of ethylene and increases the sensitivity of the GC. It is essential that the U-shaped trapping column have welded fittings and be connected to a high-pressure, gas-tight switching valve, as the heating and cooling produce large changes in gas volume that can cause normal Swagelock fittings to leak.

IV. PURIFICATION

A. Initial Purification

The next step is often the partitioning of the extract. However, partitioning of crude aqueous plant extracts can lead to troublesome emulsions (Singh *et al.*, 1979; Sandberg *et al.*, 1981a). In an effort to avoid partitioning of crude extracts, we devised a purification scheme using prepacked silica columns to remove most of the nonpolar pigmentation from samples for ABA estimation. Others have used C_{18}- (octadecylsilazane-) bonded silica columns for purifying ABA and other PGR (Pierce and Raschke, 1980, 1981; Dumbroff and Walker, 1981; Bennett *et al.*, 1981; Knegt *et al.*, 1981; Loveys and Milborrow, 1981; Suzuki *et al.*, 1981; Koshioka *et al.*, 1983; Lewis and Visscher, 1983).

B. Subsequent Purification

At this stage, one must ask whether purification has been adequate for subsequent analysis. Most quantification methods, even immunoassay (Weiler, 1982), require at least one or two purification steps before reliable final analysis. It is difficult to judge if there has been adequate purification. A common mistake is the injection of an extract with excess dry weight onto an analytical HPLC column. This greatly reduces column life and chromatographic resolution. If nonspecific analytical detectors are used, such as ultraviolet (UV) detector with HPLC, or FID with GC, a dirty extract will simply produce a chromatogram with so many peaks that "one cannot see the wood for the trees."

Often the crude extract is subjected to a series of partitioning steps with organic solvents and the aqueous extract. Protonation of the carboxyl groups of acid PGR (ABA, IAA, GAs) by alteration of the pH of the aqueous phase allows one to affect the solubility of the PGR in various solvents. The final step is to partition the PGR into an organic solvent that can be easily reduced to a small volume with a rotary evaporator.

The solvent selected for a group of PGR depends on the partitioning coefficient between the aqueous extract and a particular organic solvent. The judicious choice of a solvent series is important in deciding the eventual recoveries. Also, the choice of solvents can eliminate some compounds that have similar molecular structures and that interfere with subsequent quantification techniques. A good example of this occurs in the purification of IAA. Atsumi *et al.* (1976) found that by substituting methylene chloride ($MeCl_2$) for ethyl ether when partitioning acidic aqueous extracts for purification of IAA, the contaminating IPyA was not carried along with the IAA. Also, IPyA can be converted to IAA in alcoholic solution. The extraction efficiency for IAA by $MeCl_2$ was 91% but only 18% for IPyA, whereas it was 99% and 92%, respectively, by ethyl ether. Thus, the possibility of conversion of IPyA to IAA and overestimation of IAA was avoided. Partition coefficients have been published for the various PGR and should be consulted for assistance in planning a purification scheme if it is to include partitioning. Partition steps always consist of shaking aqueous extracts with organic solvents one to five times and combining the organic fractions from each partitioning. A beginner's list for distribution coefficients is as follows: ABA (Ciha *et al.*, 1977), CKs (Letham, 1974), GA's (Durley and Pharis, 1972), and IAA (Atsumi *et al.*, 1976).

There are alternatives to partitioning. The effectiveness of partitioning steps is limited by several factors. With the increasing cost of organic solvents, the expense of using large volumes of solvent for each extract has required modifications in the partitioning steps of previous extraction techniques. Solvent partitioning is also slow due to the time taken in shaking each extract and the formation of inevitable emulsions. Singh *et al.* (1979) reported large losses of an ABA radioisotope tracer due to partitioning. One aid in removal of emulsions is to centrifuge the extracts.

The use of prepacked cartridges packed with silica gel instead of partitioning avoids some of the problems (Hubick and Reid, 1980). By careful selection of solvent mixtures, a plant extract loaded onto a silica gel Sep-pak (Waters Associates) could have most of the nonpolar impurities eluted while allowing the ABA to remain on the cartridge for later elution by solvents of increasing polarities. Since the introduction of Sep-paks in purification of PGR (Hubick and Reid, 1980), similar procedures, generally using reverse-phase material have arisen for ABA and other PGR (GAs: Suzuki *et al.*, 1981; Koshioka *et al.*, 1983; IAA: Knegt *et al.*, 1981; CKs: MacDonald *et al.*, 1981). Purchase of the silica gel or C_{18}-bonded silica gel in bulk and preparation of minicolumns in pipettes can eliminate much of the cost associated with purchase of the cartridges.

C. Preparative Chromatography

Preparative chromatography is usually necessary. It is possible to estimate the amount of ABA (Hubick and Reid, 1980; Knegt *et al.*, 1981; Lewis and Visscher, 1983) or IAA (Knegt *et al.*, 1981) after purification by techniques using silica gel or reverse-phase cartridges. However, the possibility of interfering substances contributing to the measured peak in gas liquid chromatography–electron capture detector (GLC–ECD) cannot be overlooked (Brenner, 1981), so further purification is advised. Ideally, estimates of the PGR of interest should be undertaken at each purification step to allow the best estimation to be achieved by successive approximation (Reeve and Crozier, 1980). However, the time involved in routinely undertaking such a task is formidable. A more reasonable approach would be to attempt such a procedure at the outset, when the purification scheme for each particular tissue is being devised.

Selection of a chromatography column depends upon the PGR being studied, although some materials are suitable for more than one PGR. The acquisition of a high-performance liquid chromatograph (HPLC) is well worth the initial investment. However, the use of less expensive gravity-type columns has not been completely replaced by HPLC, so some of the more recent reports of techniques involving older procedures are worth mentioning. This is particularly true when the technique allows simultaneous purification of several groups of PGR.

Older types of column chromatography are still useful for purification of GAs and CKs. An ion-exchange step in the purification of CKs is obligatory (Horgan, 1978). Cellulose phosphate, a cation exchanger, is very suitable for this procedure. It gives recoveries of greater than 90% of zeatin and related compounds if washed with water at pH 3 and followed with 1 *M* NH_4OH (Horgan, 1978). We have used cellulose phosphate as a purification step, allowing the separation and purification of CKs, GAs, and ABA from one plant extract at one time. A Sephadex LH-20 column purification step after an ion-exchange column is an inexpensive alternative to HPLC (Hutton and van Staden, 1981).

The use of polyvinylpyrrolidone (PVP) for purification is recommended for GAs, CKs, ABA, and IAA, but the benefits must outweigh the drawbacks of elution with aqueous buffer with nonvolatile salts (Horgan, 1978). PVP prepurification before HPLC is recommended for CK (Stahly and Buchanan, 1982), as well as for ABA and IAA (Durley *et al.*, 1982). Combination of PVP with Sephadex LH-20 in one column is useful in increasing the purification of a PGR in just one step. Recently, Sandberg *et al.* (1981a,b) found that combination of 25 cm PVP with 20 cm Sephadex LH-20 in one column enabled adequate purification for the estimation of GAs, IAA, and ABA.

The purification of GAs often involves the use of either charcoal:celite and/or silicic acid column chromatography in addition to PVP purification (Russell, 1975). The gradient elution technique of Durley *et al.* (1972) is regularly used by GA investigators. However, the requirement for PVP or charcoal:celite columns can be eliminated by the use of reverse-phase C_{18} minicolumns (Suzuki *et al.*, 1981; Koshioka *et al.*, 1983). Therefore, several options are available to the investigator depending on the PGR being studied. If more than one group of PGR is under investigation, a combination of PVP and Sephadex LH-20 or cellulose phosphate column chromatography may be required to separate the various PGR groups. This could then be followed by HPLC using appropriate elution conditions for the individual groups of PGR and subsequent quantification.

It must be emphasized that for each new tissue, the investigator will have to determine empirically the most appropriate purification procedure. We strongly recommend the use of trace amounts of radioactively labeled PGR during this preliminary stage of testing a new technique with a novel tissue.

It is important to prepurify adequately a plant extract before performing HPLC. This is important in maintaining adequate column life. Purification schemes such as those prescribed by Arteca *et al.* (1980), in which HPLC on reverse-phase C_{18}-bonded columns precedes PVP column chromatography and subsequent HPLC, should shorten the life of the C_{18} column dramatically. Cargile *et al.* (1979) and Zeevart (1980) noticed a gradual deterioration of reverse-phase columns over time. Inadequate sample prepurification will further promote deterioration.

V. QUANTIFICATION

Once the PGR has been adequately purified, one can proceed with its quantification. The technique of choice will depend on several factors, an important one being the cost. Bioassay is possible for all PGR, but this should be done only if another superior physicochemical technique is unavailable. Although inexpensive in terms of equipment, bioassay tends to be very labor intensive, and a simple analysis may take days to weeks. Only a few bioassay techniques are included in this chapter.

A. Auxins

1. Biological Techniques

Early IAA quantification used the *Avena* coleoptile curvature test (Audus, 1972). However, this technique is very tedious. Bioassay of IAA is not sufficient as a quantification technique (Yamaki *et al.*, 1974). Immunoassays have been developed for IAA, including a nonradioactive solid-phase enzyme immunoassay (Weiler *et al.*, 1981) and an RIA (Pengelly *et al.*, 1981; Weiler, 1981). Weiler (1982) pointed out that an early problem with RIA of IAA was the instability of IAA and the radiolabeled IAA. Pengelly *et al.* (1981) found that it was necessary first to validate an RIA for a particular tissue by comparing the ability of the RIA to estimate IAA with purification involving HPLC followed by GC–MS before the RIA could be relied upon to give accurate results. Column chromatography with DEAE- and LH-20 Sephadex was required to prepurify the extracts. Weiler *et al.* (1981) have designed a nonradioactive enzyme immunoassay in which the extracts are methylated to IAAMe and avoid the instability problem. However, the problems with cross-reactivity have not yet been completely solved, and some IAA analogs will cross-react with the IAA antibodies (Weiler, 1981), particularly when the samples are not sufficiently purified (Weiler, 1982). The necessity for prepurification does not yet give immunoassay much of an advantage over other physicochemical techniques.

2. High-Performance Liquid Chromatography Techniques

Several reports suggest that the combination of HPLC with sensitive detectors enables accurate quantification of IAA. IAA may be separated from 4-chloro-IAA (4-C1-IAA), 5-hydroxy-IAA (5-OH-IAA), and other impurities which can interfere with either bioassay or fluorescence determinations (Sandberg *et al.*, 1981a,b). However, it should be remembered that selection of partitioning solvents can also eliminate many of these compounds (Atsumi *et al.*, 1976). Some authors (Mousdale, 1981; Arteca *et al.*, 1980) suggest that IAA can be quantified by HPLC–UV spectrometry alone. However, it is probably better to exploit the fluorescence characteristics of IAA for quantification instead of UV (Crozier *et al.*, 1980; Sandberg *et al.*, 1981a,b; Durley *et al.*, 1982). Crozier *et al.* (1980), and Durley *et al.* (1982) needed two HPLC steps to obtain adequate IAA resolution for quantification by fluorescence detectors. Also, the tissue sample must be small enough to avoid masking the IAA peak (Crozier *et al.*, 1980; Sandberg *et al.*, 1981a,b). Analysis of IAA by an electrochemical detector has also been reported.

3. Indolo-α-Pyrone Technique

Probably, the most popular technique for IAA estimation is the indolo-α-pyrone derivatization of IAA (Stoessl and Venis, 1970; Eliasson *et al.*, 1976; Mousdale *et al.*, 1978), which exploits the fluorescence of the derivative. However, this technique is not without problems. The pyrone formed from

IAA and acetic anhydride is unstable, so the fluorescence must be read immediately after the reaction has been completed (Knegt *et al.*, 1981; Mousdale *et al.*, 1978). Also, 4-Cl-IAA, 5-OH-IAA, and impurities interfere with the reaction (Sandberg *et al.*, 1981a,b). The impurities in pine extracts eliminate this technique for IAA determination (Sandberg *et al.*, 1981a,b). The impurities should be removed by adequate prepurification of extracts, and the variability in fluorescence intensity can be minimized by adding BHT at all steps of the purification procedure and even at the IAA derivatization step (Moritoshi *et al.*, 1980).

4. Gas Chromatography

Identification of IAA in tissue extracts is achieved with GC–MS. Recently, Allen *et al.* (1982) have reported a technique using deuterated internal standards of IAA derivatives. The technique gives IAA identification but has some problems, not the least of which is access to GC–MS equipment. The pentafluorobenzyl esters of IAA can be used with the ECD for IAA quantification (Epstein and Cohen, 1981). The advantage of the technique is that it combines HPLC with the ECD detection of both IAA and ABA.

Martin *et al.* (1980) have proposed using the thermionic-specific detector (TSD) for IAA quantification. The detection method is sensitive to 5 pg IAA, similar to ECD detection of the pentafluorobenzyl esters (Epstein and Cohen, 1981), and compares with 100–500 pg for HPLC with on-line fluorescence or electrochemical detectors (Durley *et al.*, 1982), although Crozier *et al.* (1980) achieved sensitivity to 1 pg using HPLC with a fluorescence online detector. The advantage of the TSD is that the IAA is simply derivatized with diazomethane and does not require the use of unfamiliar reagents.

As mentioned previously, with each new tissue, the various techniques should be tested beforehand to decide which one is most appropriate to that tissue.

B. Cytokinins

Quantification of CK is generally still dependent on bioassay. The soybean hypocotyl bioassay (Manos and Goldthwaite, 1976) is very reliable and inexpensive to set up, but requires a 2-week incubation period and good aseptic techniques. More recently, immunoassays have been developed for CKs (MacDonald *et al.*, 1981; Weiler, 1982). Combination of HPLC with RIA shortened the analysis time from weeks to hours and the RIA has a detection limit of 100–200 pg (MacDonald *et al.*, 1981). A problem with immunoassay for CKs is not lack of specificity but too much specificity. The specificity of the assay depends on whether the antisera are produced against ribosylzeatin or isopentenyladenosine because the two types of antisera give different estimates for the CK content of an HPLC fraction (MacDonald *et al.*, 1981). Thus, if only one or two CKs are to be studied, the RIA may be useful enough. Since prepurification by TLC or HPLC is still required (Weiler, 1982), this method of quantification probably has little advantage over physicochemical techniques.

1. High-Performance Liquid Chromatography

Quantification of CKs using the UV detector is possible (Stahly and Buchanan, 1982). However, the UV detector, as with other PGR, has inadequate specificity for positive identification (Reeve and Crozier, 1980), and attempts by Stahly and Buchanan (1982) to quantify zeatin-riboside from small apple fruits were unsuccessful because of interfering UV peaks.

2. Gas Chromatography

Derivatization of CK for GLC is possible. The trimethylsilyl ethers can be prepared, but care must be taken to keep them dry (Horgan, 1978; Martin *et al.*, 1981) in order to assure reproducible derivatization. Permethylation of CK may be more suitable (Summons *et al.*, 1980; Rivier *et al.*, 1981; Martin *et al.*, 1981). Martin *et al.* (1981) found that permethylating the CK molecule to a stable derivative before HPLC allowed superior HPLC purification, resulting in extracts pure enough for GC–MS. The most accurate quantification of CK is then achieved by GC–MS (Entsch *et al.*, 1980). Zelleke *et al.* (1980) have proposed a method for CK quantification using TSD and permethylation. The technique required adequate prepurification to avoid interfering impurities by increased the sensitivity to the CK derivatives.

The major problem with permethylation of plant extracts is the danger involved in using sodium hydride, which is explosive if allowed to come into contact with moisture. Probably the best method of CK quantification on a routine basis remains bioassay, until the chemical derivatizations and immunoassay become more simple and reliable.

C. Gibberellins

As with CKs, GA quantification relies heavily on bioassay techniques. Physicochemical techniques are generally used only for unequivocal identification, and RIA has not been perfected.

1. Radioimmunoassay

There is as yet only one published immunoassay for GAs (Weiler and Wieczorek, 1981). Several problems must be overcome before the assay will be routinely used in laboratories other than those dedicated to GA analysis. First, the assay depends on the immunological reaction between antibodies and a radioactive tracer labeled with ^{125}I. Not many laboratories have a license for a gamma emitter, nor are they interested in working with hazardous isotopes. Second, the radioactive tracer must be prepared from a very precise protocol. Even then, the tracer is stable for only a few months. Finally, the antibodies produced by Weiler and Wieczorek (1981) were reactive only with GA_3, with some cross-reactivity with GA_1, GA_4, GA_7, and GA_9. This severely limits the range of GAs that can be quantified properly. Apparently, assays are being developed for other GAs (Weiler, 1982). Although the RIA of Weiler and Wieczorek (1981) is very sensitive (4–80 pg), the disadvantages will preclude its general use until more improvements are made.

2. Gas Chromatography

The trimethylsilyl derivatives of GAs, which are first methylated with diazomethane, can be identified and quantified by GC–MS (Metzger and Zeevart, 1980a,b,c). GAs can also be permethylated (Rivier *et al.*, 1981). This technique is more suitable than silylation because trimethylsilyl ethers are less stable than permethylated derivatives under the higher column temperature required for GLC.

3. Bioassay

Although subject to inadequacies, bioassay is still the preferred method for GA quantification. Reeve and Crozier (1978) suggest that a combination of the barley aleurone, Tangin-bozu dwarf rice, and lettuce hypocotyl assays (Reeve and Crozier, 1975) gives good results. The Tangin-bozu dwarf rice microdrop bioassay (Kaufmann *et al.*, 1976) is routinely used in our laboratory.

D. Abscisic Acid

Because ABA is a single molecule rather than a group of compounds like the GAs or CKs, the quantification techniques for producing adequately purified extracts are relatively straightforward. Bioassay is not recommended because of inherent problems when measuring inhibitory compounds (Zeevart, 1979). Zeevart (1979) has summarized the various techniques for ABA quantification, and we are of the same opinion concerning the relative importance of ABA purification and quantification techniques. The most sensitive quantification technique is GC with the ECD, with a sensitivity limit of 5–50 pg (Zeevart, 1979). We find that the ECD can be made more sensitive by using it together with fused silica capillary columns.

1. Immunoassay

Both enzyme immunoassay and RIAs have been developed for ABA (Weiler, 1982). Quantification with these techniques shows promise because of the specificity of immunoassay and the fact that the investigator is interested in only one compound (ABA). The earlier problems with cross-reactivity with ABA analogs and derivatives with anti-ABA sera (Walton *et al.*, 1979; Weiler, 1979) apparently have been overcome (Weiler, 1982). However, it is advisable to check the accuracy of the immunoassay technique by physicochemical means, as has been advised for IAA (Pengelly *et al.*, 1981). The major advantage of ABA immunoassay is the lack of need for extensive purification before the assay (Weiler, 1982).

2. High-Performance Liquid Chromatography

As with IAA and CK, techniques have been proposed that estimate ABA levels using HPLC with a UV monitor. ABA absorbs strongly at 254 nm, but is only one of numerous compounds that do so (Reeve and Crozier, 1980; Brenner, 1981). We find that, when using gradient elution of reverse-phase columns, the ABA peak is masked by numerous impurities even after partitioning and silicic acid chromatography. Ciha *et al.* (1977) found that HPLC with UV detection was not specific enough to quantify ABA adequately from soybean. However, Cargile *et al.* (1979) and Durley *et al.* (1982) used reverse-phase HPLC followed by normal-phase HPLC to measure ABA. Durley *et al.* (1982) checked the purity of the putative ABA peaks with GLC. During and Bachmann (1975) used an ion-exchange HPLC column for ABA quantification. The lack of specificity of the UV detector warrants very careful checking of any new type of plant extract before ABA quantification may be trusted. Any variation in composition of the plant extract could introduce interfering UV peaks.

One possible modification to improve reliability of the measurement using HPLC alone is first to purify the free acid with a simple cleanup such as with a silica (Hubick and Reid, 1980) or C_{18} (Lewis and Visscher, 1983) minicolumn, methylate the appropriate fraction, and then run the fraction on HPLC to quantify the ABAMe. In this type of procedure, all nonacidic impurities associated originally with the ABA will separate from the ABA methyl ester, which is retained longer on reverse-phase columns eluted with methanol and water mixtures. We have found that if this procedure is used, the ABAMe peak is removed from the majority of interfering UV peaks. A subsequent quantification step is still advised. Rerunning on HPLC may suffice, but it is just as easy to prepare the sample for GC after this step as it is to attempt to quantify the ABAMe using HPLC.

3. Gas Chromatography

Derivatization of ABA for GC is a fast and simple technique. It gives the increased resolution afforded by GC after HPLC for final ABA quantification an edge over other techniques. Taylor *et al.* (1981) describe an efficient technique involving HPLC combined with GC–ECD. For ABA, ECD is at least two orders of magnitude more sensitive than FID and is a more selective detector. The carboxyl group of ABA is easily derivatized with ethereal diazomethane at room temperature. Diazomethane must be stored cold (−20°C) in a cork-stoppered (not ground-glass) container due to its explosive nature. A point that is seldom mentioned in the literature is the variability in methylation of impure extracts. Depending on the "dirtiness" or moisture content of the extract, only a portion of the ABA may be methylated. One way to ensure complete methylation is to retain the yellow color for at least 10 min, and then remove the ether under nitrogen gas (1 min) and remethylate until the color persists. If the extract is yellow to begin with, the reaction is complete if no more nitrogen gas bubbles are evolved.

Subsequent separation of the (possibly) underivatized ABA from the ABAMe by a chromatography step (as suggested in the last paragraph) ensures the most accurate quantification of ABA if an internal standard was added at the extraction step. ABA can also be silylated with appropriate reagents (Davis *et al.*, 1968). However, the procedure is not as reliable as methylation with diazomethane because of inconsistent silylation and possible breakdown of the derivatives (Most *et al.*, 1970; Saunders, 1978).

GC–MS as a quantification technique for ABA has been reported several times. Monitoring of the base ion at 190 m/e was used to quantify ABA in at least two situations (Railton *et al.*, 1974; Little *et al.*, 1978). Recently, the use of [^{2}H]ABA in a stable isotope dilution technique has also been used to measure ABA in roots (Rivier *et al.*, 1977; Rivier and Pilet, 1981; Pilet and Rivier, 1981).

E. Ethylene

Since the advent of GC, ethylene has become by far the easiest of the PGR to quantify. GC with FID is the method of choice. Most modern chromatographs can easily and accurately measure a few parts per billion (ppb) (as low as 1×10^{-12} m^3/dm^3 in 1 cm^2 of air). Although the gas can easily be separated from the common gases such as other hydrocarbons—acetylene, methane, ethane (the last is often present with ethylene)—care should be taken to use more than one column packing to make sure that the peak thought to be ethylene is not confused with acetylene or ethane. The best source of information on ethylene is Abeles (1973). We recommend using stainless steel GC columns packed with any of the following: Poropak N, Q, T, or alumina silica gel. With the Poropaks, oven temperatures of 60–80°C should be used; with alumina and silica gel, 110°C. Nitrogen is a good carrier gas. Only 1–3 min are needed to measure ethylene levels using GC. After a large number of injections over a span of 1–2 hr, it might be necessary to increase column temperature briefly (10–15 min) by 20% in order to drive off water vapor and other volatiles from the column.

Problems may arise in areas with heavy air pollution since ethylene is a significant component of the air near many industrial or combustion processes. The investigator should separately monitor the ambient air.

Airtight glass syringes with Teflon plungers are recommended. Less expensive medical plastic syringes can be used but should not be allowed to come into contact with high levels of ethylene or high temperatures. From 1 to 5 ml of air can be injected into a 2-m-long × 3-mm O.D. column (Ward *et al.*, 1978; Huxter *et al.*, 1979).

VI. CONCLUSIONS

The purification and quantification of PGR can now be attempted with less effort than in the past. Numerous chromatography steps are no longer necessary to achieve adequate resolution from impurities. Newer types of detectors with increased selectivity for PGR or their derivatives can be added to a high-resolution chromatographic step for quantification.

The advent of HPLC in PGR analysis allows purification with relative speed and ease. Brenner (1979) suggested that several groups of PGR can be prepurified using a reverse-phase HPLC column. At that time, few attempts had been made to purify more than one group of PGR from one extract. Brenner (1979) suggested using the following method as a starting point for such an analysis: elution of an extract from a reverse-phase HPLC column with a linear gradient of 0.1 *N* aqueous acetic acid to 0.1 *N* acetic acid in 50% ethanol (v/v), with water delivered over a 25-min period. A large-bore column should be used to avoid overloading. This type of purification should follow some sort of quick prepurification of the original extract to maintain the efficiency of the HPLC column as long as possible. We recommend either a silica or reverse-phase minicolumn eluted with appropriate solvents.

When possible, GC–MS is an ideal method of quantification, but GC with the appropriate detector, chosen to be as selective and sensitive as possible (e.g., ECD for ABA or TSD for IAA or CKs), can, with care, be a powerful tool.

In the future, a more improved immunoassay may replace many of the physicochemical analysis techniques for large-scale routine analysis of PGR. The combination of HPLC with immunoassay probably will be the most popular choice for PGR analysis.

REFERENCES

Abeles, F. B. (1973). "Ethylene in Plant Biology." Academic Press, New York.

Allen, R. F., Rivier, L., and Pilet, P. E. (1982). Quantification of indol-3-yl acetic acid in pea and maize seedlings by gas chromatography-mass spectrometry. *Phytochemistry* **21,** 525–530.

Arteca, R. N., Poovaiah, B. W., and Smith, O. E. (1980). Use of high performance liquid chromatography for the determination of endogenous hormone levels in *Solanum tuberosum* L. subjected to carbon dioxide enrichment of the root zone. *Plant Physiol.* **65,** 1216–1219.

Atsumi, S., Kuraishi, S., and Hayashi, T. (1976). An improvement of auxin extraction procedure and its application to cultured plant cells. *Planta* **129,** 245–247.

Audus, L. J. (1972). "Plant Growth Substances," Vol. 1, pp. 24–35. International Textbook Co., London.

Belke, C. J., Sjut, V., and Brenner, M. L. (1980). Metabolism and phloem transport of [2-^{14}C]ABA in tomato leaves. *Plant Physiol.* **65,** Suppl., 94.

Bennett, R. D., Norman, S. M., and Maier, V. P. (1981). Biosynthesis of abscisic acid from [1,2-$^{13}C_2$]acetate in *Cercospora rosicola*. *Phytochemistry* **10,** 2343–2344.

Beyer, E. M., Jr., and Morgan, P. W. (1970). A method for determining the concentration of ethylene in the gas phase of vegetative plant tissues. *Plant Physiol.* **46,** 353–354.

Brenner, M. L. (1979). Advances in analytical methods for plant growth substance analysis. *ACS Symp. Ser.* **3,** 215–244.

Brenner, M. L. (1981). Modern methods for plant growth substance analysis. *Annu. Rev. Plant Physiol.* **32,** 511–538.

Cargile, N. L., Borchert, R., and McChesney, J. D. (1979). Analysis of abscisic acid by high-performance liquid chromatography. *Anal. Biochem.* **97,** 331–339.

Ciha, A. J., Brenner, M. L., and Brun, W. (1977). Rapid separation and quantification of abscisic acid from plant tissues using high performance liquid chromatography. *Plant Physiol.* **59,** 821–826.

Crozier, A., Loferski, K., Zaerr, J. B., and Morris, R. O. (1980). Analysis of picogram quantities of indole-3-acetic acid by high performance liquid chromatography-fluorescence procedures. *Planta* **150,** 366–370.

Davis, L. A., Heinz, D. E., and Addicott, F. T. (1968). Gas-liquid chromatography of trimethylsilyl derivatives of abscisic acid and other plant hormones. *Plant Physiol.* **43,** 1389–1394.

DeGreef, J., DeProft, M., and DeWinter, F. (1976). Gas chromatographic determination of ethylene in large air volumes at the fractional parts-per-billion level. *Anal. Chem.* **48,** 38–41.

Dumbroff, E. B., and Walker, M. A. (1981). The analysis of abscisic acid—an update. *Plant Physiol.* **67,** Suppl., 109.

During, H., and Bachmann, O. (1975). Abscisic acid analysis in *Vitis vinifera* in the period of endogenous bud dormancy by high pressure liquid chromatography. *Physiol. Plant.* **34,** 201–203.

Durley, R. C., and Pharis, R. P. (1972). Partition coefficients of 27 gibberellins. *Phytochemistry* **11,** 317–326.

Durley, R. C., Kannangara, T., and Simpson, G. M. (1982). Leaf analysis for abscisic, phaseic and 3-indolylacetic acids by high-performance liquid chromatography. *J. Chromatogr.* **236,** 181–188.

Eastwell, K. C., Bassi, P. W., and Spencer, M. E. (1978). Comparison and evaluation of methods for removal of ethylene and other hydrocarbons from air for biological studies. *Plant Physiol.* **62,** 723–726.

Eliasson, L., Stromquist, L.-H., and Tillberg, E. (1976). Reliability of the indolo- -pyrone fluorescence method for indole-3-acetic acid determination in crude plant extracts. *Physiol. Plant.* **36,** 16–19.

Entsch, B., Letham, D. S., Parker, C. W., Summons, R. E., and Gollnow, B. I. (1980). Metabolites of cytokinins. *In* "Plant Growth Substances 1979" (F. Skoog, ed.), pp. 109–118. Springer-Verlag, Berlin and New York.

Epstein, E., and Cohen, J. D. (1981). Microscale preparation of pentafluorobenzyl esters. Electron-capture gas chromatographic detection of indole-3-acetic acid from plants. *J. Chromatogr.* **209,** 413–420.

Fabijan, D., Taylor, J. S., and Reid, D. M. (1981). Adventitious rooting in hypocotyls of sunflower (*Helianthus annuus*) seedling. II. Action of gibberellins, cytokinins, auxins and ethylene. *Physiol. Plant.* **53,** 589–597.

Hangarter, R. P., and Good, N. E. (1981). Evidence that IAA conjugates are slow-release sources of free IAA in plant tissues. *Plant Physiol.* **68,** 1424–1427.

Hardin, J. M., and Stutte, C. A. (1981). Analysis of plant hormones using high-performance liquid chromatography. *J. Chromatogr.* **208,** 124–128.

Hemberg, T., and Tillberg, E. (1980). The influence of the extraction procedure on yield of indole-3-acetic acid in plant extracts. *Physiol. Plant.* **50,** 176–182.

Horgan, R. (1978). Analytical procedures for cytokinins. *In* "Isolation of Plant Growth Substances" (J. R. Hillman, ed.), pp. 97–114. Cambridge Univ. Press, London and New York.

Hubick, K. T., and Reid, D. M. (1980). A rapid method for the extraction and analysis of abscisic acid from plant tissue. *Plant Physiol.* **65,** 523–525.

Hutton, M. J., and Van Staden, J. (1981). An efficient column chromatographic method for separating cytokinins. *Ann. Bot. (London)* [N.S.] **46,** 527–529.

Huxter, T. J., Reid, D. M., and Thorpe, T. A. (1979). Ethylene production by tobacco (*Nicotiana tabacum*) callus. *Physiol. Plant.* **46,** 374–380.

Kaufmann, P. B., Ghosheh, N. S., Nakosteen, L., Pharis, R. P., Durley, R. C., and Morf, W. (1976). Analysis of native gibberellins in the internode, nodes, leaves, and inflorescences of developing *Avena* plants. *Plant Physiol.* **58,** 131–134.

Knegt, E., Vermeer, E., and Bruinsma, J. (1981). The combined determination of indolyl-3-acetic and abscisic acids in plant materials. *Anal. Biochem.* **114,** 362–366.

Koshioka, M., Takeno, K., Beall, F. D., and Pharis, R. P. (1983). Purification and separation of plant gibberellins from their precursors and glucosyl conjugates. *Plant Physiol.* **73,** 398–406.

Letham, D. S. (1974). Regulators of cell division in plant tissues. XXI. Distribution coefficients for cytokinins. *Planta* **118,** 361–364.

Lewis, R. W., and Visscher, S. N. (1983). A simplified purification method for the analysis of abscisic acid. *Plant Growth Regul.* **1,** 25–30.

Little, C. H. A., Heald, J. K., and Browning, G. (1978). Identification and measurement of indoleacetic and abscisic acids in the cambial region of *Picea sitchensis* (Bong.) Carr. by combined gas chromatography-mass spectrometry. *Planta* **139,** 133–138.

Loveys, B. R. (1977). The intracellular location of abscisic acid in stressed and non-stressed leaf tissue. *Physiol. Plant.* **40,** 6–10.

Loveys, B. R., and Milborrow, B. V. (1981). Isolation and characterization of 1'-O-abscisic acid-D-glucopyranoside from vegetative tomato tissue. *Aust. J. Plant Physiol.* **8,** 571–589.

MacDonald, E. M. S., Akiyoshi, D. E., and Morris, R. O. (1981). Combined high-performance liquid chromatography-radioimmunoassay for cytokinins. *J. Chromatogr.* **214,** 101–109.

McDougall, J., and Hillman, J. R. (1978). Analysis of indole-3-acetic acid using GC–MS techniques. *In* "Isolation of Plant Growth Substances" (J. R. Hillman, ed.), pp. 1–25. Cambridge Univ. Press, London and New York.

MacMillan, J., Seaton, J. C., and Suter, P. J. (1960). Plant hormone. I. Isolation of gibberellin A_1 and gibberellin A_5 from *Phaseolus multiflora. Tetrahedron* **11,** 60–66.

Mann, J. D., and Jaworski, E. G. (1970). Minimizing loss of indole acetic acid during purification of plant extracts. *Planta* **92,** 285–291.

Manos, P. J., and Goldthwaite, J. (1976). An improved cytokinin bioassay using cultured soybean hypocotyl sections. *Plant Physiol.* **57,** 894–897.

Martin, G., Horgan, R., and Scott, I. M. (1981). High-performance liquid chromatographic analysis of permethylated cytokinins. *J. Chromatogr.* **219,** 167–170.

Martin, G. C., Nishijima, C., and Labavitch, J. M. (1980). Analysis of indoleacetic acid by the nitrogen-phosphorus detector gas chromatograph. *J. Am. Soc. Hortic. Sci.* **105,** 46–50.

Metzger, J. D., and Zeevart, J. A. D. (1980a). Identification of six endogenous gibberellins in spinach shoots. *Plant Physiol.* **65,** 623–626.

Metzger, J. D., and Zeevart, J. A. D. (1980b). Comparison of the levels of six endogenous gibberellins in roots and shoots of spinach in relation to photoperiod. *Plant Physiol.* **66,** 679–683.

Metzger, J. D., and Zeevart, J. A. D. (1980c). Effect of photoperiod on the levels of endogenous gibberellins in spinach as measured by combined gas chromatography-selected ion current monitoring. *Plant Physiol.* **66,** 844–846.

Milborrow, B. V., and Mallaby, R. (1975). Occurrence of methy (+)- abscisate as an artefact of extraction. *J. Exp. Bot.* **26,** 741–748.

Moritoshi, I., Yu, R. S.-T., and Carr, D. J. (1980). Improved procedure for the estimation of nanogram quantities of indole-3-acetic acid in plant extracts using the indolo-α-pyrone fluorescence method. *Plant Physiol.* **66,** 1099–1105.

Most, B. H., Gaskin, P., and MacMillan, J. (1970). The occurrence of abscisic acid in inhibitors B_1, and C from immature fruit of *Ceratonia siliqua* L. (carob) and in commercial carob syrup. *Planta* **92,** 41–49.

Mousdale, D. M. A. (1980). Criteria for assessing the accuracy of the pyrone fluorimetric assay for indole-3-acetic acid. *J. Exp. Bot.* **121,** 515–523.

Mousdale, D. M. A. (1981). Reversed-phase ion-pair high-performance liquid chromatography of the plant hormones indolyl-3-acetic and abscisic acid. *J. Chromatogr.* **209,** 489–493.

Mousdale, D. M. A., Butcher, D. M., and Powell, R. G. (1978). Spectrophotofluorimetric methods of determining indole-3-acetic acid. *In* "Isolation of Plant Growth Substances" (J. R. Hillman, ed.), pp. 27–40. Cambridge Univ. Press, London and New York.

Pengelly, W. L., Bandurski, R. S., and Schulze, A. (1981). Validation of a radioimmunoassay for indole-3-acetic acid using gas chromatograph-selected ion monitoring-mass spectrometry. *Plant Physiol.* **68,** 96–98.

Pierce, M., and Raschke, K. (1980). Correlation between loss of turgor and accumulation of abscisic acid in detached leaves. *Planta* **148,** 174–182.

Pierce, M., and Raschke, K. (1981). Synthesis and metabolism of abscisic acid in detached leaves of *Phaseolus vulgaris* L. after loss and recovery of turgor. *Planta* **153,** 156–165.

Pilet, P. E., and Rivier, L. (1981). Abscisic acid distribution in horizontal maize root segments. *Planta* **153,** 453–458.

Railton, I. D., and Rechav, M. (1979). Efficiency of extraction of gibberellin-like substances from chloroplasts of *Pisum sativum* L. *Plant Sci. Lett.* **14,** 75–78.

Railton, I. D., Reid, D. M., Gaskin, P., and MacMillan, J. (1974). Characterization abscisic acid in chloroplasts of *Pisum sativum* L. c.v. Alaska by combined gas-chromatography-mass spectrometry. *Planta* **117,** 170–182.

Reeve, D. R., and Crozier, A. (1975). Gibberellin bioassays. *In* "Gibberellins and Plant Growth" (H. N. Krishnamoorthy, ed.), pp. 35–64. Wiley Eastern, New Delhi.

Reeve, D. R., and Crozier, A. (1978). The analysis of gibberellins by high performance liquid chromatography. *In* "Isolation of Plant Growth Substances" (J. R. Hillman, ed.), pp. 41–77. Cambridge Univ. Press, London and New York.

Reeve, D. R., and Crozier, A. (1980). Quantitative analysis of plant hormones. *Encycl. Plant Physiol., New Ser.* **9,** 203–309.

Reid, D. M., Pharis, R. P., and Roberts, D. W. A. (1974). Effect of four temperature regimes on the gibberellin content of winter wheat cv. Kharkov. *Physiol. Plant.* **30,** 53–57.

Rivier, L., and Pilet, P. E. (1981). Abscisic acid levels in the root tips of seven *Zea mays* varieties. *Phytochemistry* **20,** 17–19.

Rivier, L., Milon, H., and Pilet, P. E. (1977). Gas-chromatography-mas spectrometric determinations of abscisic acid levels in the cap and the apex of maize roots. *Planta* **134,** 23–27.

Rivier, L., Gaskin, P., Alsone, K. S., and MacMillan, J. (1981). GC–MS identification of endogenous gibberellins and gibberellin conjugates as their permethylated derivatives. *Phytochemistry* **4,** 687–692.

Russell, S. (1975). Extraction, purification and chemistry of gibberellins. *In* "Gibberellins and Plant Growth" (H. N. Krishnamoorthy, ed.), pp. 1–34. Wiley Eastern, New Delhi.

Sandberg, G., Andersson, B., and Dunberg, A. (1981a). Identification of 3-indoleacetic acid in *Pinus sylvestris* L. by gas chromatography-mass spectrometry, and quantitative analysis by ion-pair reversed-phase liquid chromatography with spectrofluorimetric detection. *J. Chromatogr.* **205,** 125–137.

Sandberg, G., Dunberg, A., and Oden, P.-C. (1981b). Chromatography of acid phytohormones on columns of Sephadex LH-20 and insoluble poly-*N*-vinylpyrrolidone, and application to the analysis of conifer extracts. *Physiol. Plant.* **53,** 219–224.

Saunders, P. F. (1978). The identification and quantitative analysis of abscisic acid in plant extracts. *In* "Isolation of Plant Growth Substances" (J. R. Hillman, ed.), pp. 115–134. Cambridge Univ. Press, London and New York.

Singh, B. N., Galson, E., Dashek, W., and Walton, D. C. (1979). Abscisic acid levels and metabolism in the leaf epidermal tissue of *Tulipa gesneriana* L. and *Commelina communis* L. *Planta* **146,** 135–138.

Stahly, E. A., and Buchanan, D. A. (1982). High-performance liquid chromatographic procedure for separation and quantification of zeatin and zeatin riboside from pears, peaches and apples. *J. Chromatogr.* **235,** 453–459.

Stoessl, A., and Venis, M. A. (1970). Determination of submicrogram levels of indole-3-acetic acetic acid: A new, highly specific method. *Anal. Biochem.* **34,** 344–351.

Summons, R. E., Entsch, B., Letham, D. S., Gollnow, B. I., and MacLeod, J. K. (1980). Regulators of cell division in plant tissues. XXVIII. Metabolites of zeatin in sweet-corn kernels: Purifications and identifications using high-performance liquid chromatography and chemical-ionization mass spectrometry. *Planta* **147,** 422–434.

Suzuki, Y., Kurogochi, S., Murofushi, Y. O., and Takahashi, N. (1981). Seasonal changes of GA_1,GA_{19} and abscisic acid in three rice cultivars. *Plant Cell Physiol.* **22,** 1085–1093.

Sweetser, P. B., and Vatvars, A. (1976). High-performance liquid chromatographic analysis of abscisic acid in plant extracts. *Anal. Biochem.* **71,** 68–78.

Taylor, J. S., Reid, D. M., and Pharis, R. P. (1981). Mutual antagonism of sulfur dioxide and abscisic acid in their effect on stomatal aperture in broad bean (*Vicia faba*) epidermal strips. *Plant Physiol.* **68,** 1504–1507.

Walton, D. C. (1980). Biochemistry and physiology of abscisic acid. *Annu. Rev. Plant Physiol.* **31,** 453–489.

Walton, D. C., Dashek, W., and Galson, E. (1979). A radioimmunoassay for abscisic acid. *Planta* **146,** 139–145.

Ward, T. M., Wright, M., Roberts, J. A., Self, R., and Osborne, D. J. (1978). Analytical procedures for the assay and identification of ethylene. *In* "Isolation of Plant Growth Substances" (J. R. Hillman, ed.), pp. 134–151. Cambridge Univ. Press, London and New York.

Weiler, E. W. (1979). Radioimmunoassay for the determination of free and conjugated abscisic acid. *Planta* **144,** 255–263.

Weiler, E. W. (1981). Radioimmunoassay for pico-mol quantities of indole-3-acetic acid for use with highly stable [^{125}I]- and [^{3}H]IAA derivatives as radiotracers. *Planta* **153,** 319–325.

Weiler, E. W. (1982). Plant hormone immunoassay. *Physiol. Plant.* **54,** 230–234.

Weiler, E. W., and Wieczorek, U. (1981). Determination of femto-mol quantities of gibberellic acid by radioimmunoassay. *Planta* **152,** 159–167.

Weiler, E. W., Jourdan, P. S., and Conrad, W. (1981). Levels of indole-3-acetic acid in intact and decapitated coleoptiles as determined by a specific and highly sensitive solid-phase enzyme immunoassay. *Planta* **153,** 561–571.

Yamaki, T., Shimojo, E., Yoshizawa, K., and Namekawa, K. (1974). Auxin in cancer tissue. *Plant Growth Subst. Proc. Int. Conf., 8th, 1973* pp. 44–51.

Yokota, T., Murofushi, N., and Takahashi, N. (1980). Extraction, purification, and identification. *Encycl. Plant Physiol., New Ser.* **9,** 113–202.

Zeevart, J. A. D. (1979). Chemical and biological aspects of abscisic acid. *ACS Symp. Ser.* **3,** 99–114.

Zeevart, J. A. D. (1980). Changes in the levels of abscisic acid and its metabolites in excised leaf blades of *Xanthium strumarium* during and after water stress. *Plant Physiol.* **66,** 672–678.

Zelleke, A., Martin, G. C., and Labavitch, J. M. (1980). Detection of cytokinins using a gas chromatograph equipped with a sensitive nitrogen-phosphorus detector. *J. Am. Soc. Hortic. Sci.* **105,** 50–53.

CHAPTER **75**

Plastic Embedding for Light Microscopy

Claudia Botti
Indra K. Vasil

Department of Botany
University of Florida
Gainesville, Florida

I. INTRODUCTION

The traditional procedure for preparing material to be studied under the light microscope has been, for many years, the paraffin embedding technique. However, with the introduction of glycol methacrylate (a water-soluble acrylic ester) for electron microscopy and enzyme histochemistry (Rosenberg *et al.*, 1960), it was recognized, by comparison, that the paraffin method could produce a high degree of distortion and shrinkage artifact (Butler, 1979). Nuclei and cell walls are generally well preserved, but the cytoplasm may be badly disrupted during fixation and embedding (Feder and O'Brien, 1968). Paraffin-embedded tissues have also been found to be unsatisfactory in the study of enzyme activity due to the denaturing effect produced during fixation, dehydration, and use of paraffin at 60°C (Higuchi *et al.*, 1979). Furthermore, it is difficult to obtain sections which are less than 5 μm thick.

CELL CULTURE AND SOMATIC CELL
GENETICS OF PLANTS, VOL. 1

ISBN 0-12-715001-3

Feder and O'Brien (1968) introduced the use of glycol methacrylate for light microscopy; since then, it has been generally recognized as a superior embedding material for cell structure preservation. Sections 0.5–8.0 μm thick may be obtained and easily stained with conventional histological stains. Nevertheless, the use of commercial glycol methacrylate presents certain difficulties due to the variation in quality from different sources, and to the presence of hydroquinone to inhibit premature polymerization (Bennett *et al.*, 1976) and methacrylic acid (Frater, 1979). Both impurities should be removed prior to polymerization.

The JB-4 embedding kit (Polysciences, Inc., 1976) has been developed to replace the glycol methacrylate kit and the paraffin technique. The composition of JB-4 is unrevealed, but it probably contains hydroxyethyl methacrylate (Higuchi *et al.*, 1979).

II. PROCEDURE

The JB-4 embedding kit consists of three components to provide stability of the kit during shipment (Polysciences, Inc., 1976):

Solution A, a formulated water-soluble monomer
Solution B, the activator or polymerizer
Component C, the catalyzer

A. Fixation

The tissues embedded in JB-4 utilize all fixatives, including OsO_4 and aqueous buffered fixatives. Glutaraldehyde buffered close to neutrality is recommended by the manufacturer for tissue structure preservation. In our experiments, formalin–acetic acid–alcohol (FAA) was used as a fixative with good results.

B. Dehydration

Since JB-4 is water soluble, there is no need to dehydrate the tissues completely before infiltration. Samples were dehydrated in 70, 80, and 95% ethanol for 30–60 min each. The specimens were taken directly from 95% ethanol to catalyzed solution A.

C. Infiltration

A 25-ml portion of solution A was catalyzed by adding 225 mg of the catalyst (component C) while stirring, until the solid was completely dissolved (approximately 15 min at room temperature with a magnetic stirrer). The catalyzed solution A was then used for infiltration of dehydrated tissues. For each series of embeddings, fresh solution A was used. The tissues were left for at least 3 hr in solution A plus catalyzer, but leaving them overnight or longer caused no apparent damage.

D. Embedding

One part by volume of solution B was added to 25 parts by volume of new catalyzed solution A while stirring. The infiltrated tissue was embedded in this mixture of solution A plus catalyzer plus solution B in a closed Beem capsule. This was done in a fume hood, as the solutions are toxic. Oxygen inhibits the reaction and must be excluded to obtain satisfactory block formation. Therefore, the capsules were filled to the top with the mixture, the tissue was dropped in, and the capsule was closed (one or two small bubbles on the top did not prevent polymerization). Orientation of the tissue was very difficult at this stage. Orientation was easier once the block had polymerized. The tip of the capsule containing the tissue was cut and fixed to the same block at the correct angle of orientation. This could be done quite easily under a stereo or dissecting microscope, using a strong glue.

E. Sectioning

Sections 2–4 μm were easily cut with a Sorvall MT2-B "Porter Blum" ultramicrotome, using *dry* glass knives. *No water* was used to receive the sections. Handling and mounting of the dry sections were slow and presented certain difficulties at the beginning. Each section was picked up with a thin needle and placed over a small drop of water on a clean glass slide. In order to produce serial sections, it was easier to place several small water drops on the glass slide, with a syringe, before sectioning of the material was begun. Each section was then dropped onto a single drop of water. Plastic sections on the edge of a dry knife may curl or fold because they are affected by the humidity in the laboratory, and it is sometimes difficult to place them on the glass slide. However, upon contact with water, they expand and flatten. Once the sections are on water, they should not be touched or they will instantly form an unusable string of plastic. The slides were dried at room temperature or at 35–40°C over a hot plate for at least 3 hr before staining.

F. Staining

Successful staining was accomplished with basic and acid water-soluble dyes. Some basic dyes slightly stained the matrix, but such staining could be removed by rinsing in absolute alcohol. Care was taken, though, because the alcohol also removed stain from the tissue. The Periodic acid-Schiff stain, counterstained with aniline blue, was found to give the best results.

1. Dry the slides with sections on a hot plate and allow them to cool.
2. Oxidize in 0.5% Periodic acid for 10 min.
3. Rinse in running tap water for 10 min.
4. Stain in Schiff's reagent for 15 min.
5. Wash well in running tap water and allow the stain to develop a dark pink color.
6. Wash with 2% sodium bisulfite for 2 min.
7. Rinse with running tap water for 10 min.
8. Counterstain with aniline blue (1% aniline blue in 7% v/v acetic acid) for 4–6 min, checking the staining under the microscope.
9. Wash in running tap water until the water is clear (the plastic will remain stained blue).
10. Let the slides dry.
11. Mount cover glasses with Permount.

This staining procedure gave very good contrast between polysaccharides (cell walls and starch grains), which stained bright pink, and proteins (nucleus and cytoplasm), which stained blue. Successful results were also obtained using toluidine blue (0.5% aqueous solution) alone for 3–4 min, but the contrast was not as sharp as with Schiff's reagent and aniline blue.

III. RESULTS AND CONCLUSIONS

Except for the initial difficulty in handling sections, which made it a tedious and slow process, and for the toxicity of the solutions, the JB-4 embedding procedure has a number of advantages over the usual paraffin method. Tissue embedded in this plastic allows sectioning down to 1 μm with negligible tissue distortion, allows the use of many fixatives, and shortens the dehydration, infiltration, and embedding processes to less than 8 hr. Sections adhere directly to the slides, using water only as a spreading medium, and staining is quick and simple, allowing the use of most aqueous histological dyes. The stained matrix may, in certain cases, make it difficult to obtain good photographs, but this problem can be partially solved by using a special filter (Zeiss Interferenz-Breitbandfilter grün) during photomicrography and a dark filter during the printing process. Tissues prepared by this method show fine cytoplasmic detail and good structure preservation (Botti and Vasil, 1983, 1984; Ho and Vasil, 1983).

REFERENCES

Bennett, H. S., Wyrick, A. D., Lee, S. W., and McDaniel, J. H. (1976). Science and art in preparing tissues embedded in plastic for light microscopy, with special reference to glycol methacrylate, glass knives and simple stains. *Stain Technol.* **51,** 71–97.

Botti, C. B., and Vasil, I. K. (1983). Plant regeneration by somatic embryogenesis from mature embryos of *Pennisetum americanum* (L.) K. Schum. *Z. Pflanzenphysiol.* **111,** 319–325.

Botti, C. B., and Vasil, I. K. (1984). The ontogeny of somatic embryos of *Pennisetum americanum* (L.) K. Schum. II. In cultured immature inflorescences. *Can. J. Bot.* (in press).

Butler, J. K. (1979). Methods of improved light microscope microtomy. *Stain Technol.* **54,** 53–69.

Feder, N., and O'Brien, T. P. (1968). Plant microtechnique: Some principles and new methods. *Am. J. Bot.* **55,** 123–142.

Frater, R. (1979). Rapid removal of acid from glycol methacrylate for improved histological embedding. *Stain Technol.* **54,** 241–243.

Higuchi, S., Suga, M., Dannenberg, A. M., Jr., and Schofield, B. H. (1979). Histochemical demonstration of enzyme activities in plastic and paraffin embedded tissue sections. *Stain Technol.* **54,** 5–12.

Ho, W. J., and Vasil, I. K. (1983). Somatic embryogenesis in sugarcane (*Saccharum officinarum* L.). I. The morphology and physiology of callus information and ontogeny of somatic embryos. *Protoplasma* (in press).

Polysciences, Inc. (1976). "JB-4 Embedding Kit," Data Sheet 123. Polysciences, Inc., Warrington, Pennsylvania.

Rosenberg, M., Bartl, P., and Lesko, J. (1960). Water-soluble methacrylate as an embedding medium for the preparation of ultrathin sections. *J. Ultrastruct. Res.* **4,** 298–303.

CHAPTER **76**

Histological and Histochemical Staining Procedures

Edward C. Yeung

Department of Biology
University of Calgary
Calgary, Alberta, Canada

I. INTRODUCTION

Structural investigations have contributed significantly to plant tissue culture research in areas such as organogenesis (Thorpe, 1980; Yeung *et al.*, 1981). Some of these advances are made possible through a better understanding of stain action and new and improved microscopic methods. Successful application and interpretation of the staining reaction depend on a thorough understanding of the procedure involved and the "willing-

CELL CULTURE AND SOMATIC CELL
GENETICS OF PLANTS, VOL. 1

ISBN 0-12-715001-3

ness to think in the combined terms of morphology, physiology, and biochemistry" (Jensen, 1962). It is important to note that different experimental materials may require modifications of existing procedures. Fixation, processing, and staining conditions should be optimized for one's own needs. Failure to obtain the desired staining reaction can be due to a number of factors, such as (1) improper fixation and processing conditions and (2) extraction of cellular components during tissue preparation. Thus, before any inferences or conclusions are drawn, one should conduct parallel investigations using fresh plant tissues whenever possible. Also, known or familiar tissue should be used to test the suitability of a new procedure, and control treatments should be run if a claim for specificity of a reaction is to be made.

In recent years, plastic embedding methods have become the methods of choice for structural investigations. Due to the hardness of the plastic blocks, thin sections, that is, 0.5–2 μm, can be obtained, thus improving the resolution of the specimen. These new embedding procedures enable one to obtain high-quality specimens giving good clarity, contrast, and resolution for light microscopic examination. The most common plastic embedding media for light microscopy are based on 2-hydroxyethylmethacrylate (glycol methacrylate, GMA) and epoxy resins. For methacrylate, the formulations according to Feder and O'Brien (1968) and PolySciences (Inc., Warrington, Pennsylvania) (the JB-4 plastic) are the most popular for botanical specimens (Chapter 75, this volume). The advantage of GMA is that, due to the hydrophilic nature of the polymer, most of the staining methods developed for paraffin-embedded materials can be used with only minor modifications. Furthermore, the plastic does not interfere with the staining procedure; therefore, its removal prior to staining is not necessary. Epoxy resins are primarily used as embedding media for electron microscopy. However, in recent years, procedures have become available for the use of these resins for light microscopic examinations, such as the method of Kosakai (1973). However, epoxy sections are much more difficult to stain than GMA sections. It is speculated that the staining sites within the tissues may have reacted with the epoxy resins (O'Brien and McCully, 1981), thus reducing their staining capacity. Claims of specific staining reactions involving epoxy-type sections should be interpreted with care.

In this chapter, selected histochemical staining methods for the major macromolecules will be outlined and procedures for enzyme histochemistry will be introduced. Emphasis is placed on GMA sections. For additional staining procedures and theoretical considerations, consult the following monographs: Chayen *et al.* (1973), Lillie (1977), O'Brien and McCully (1981), and Pearse (1980).

II. GENERAL AND HISTOCHEMICAL STAINING PROCEDURES

A. General and Histochemical Staining Using Toluidine Blue O for Glycol Methacrylate Sections

Stain GMA sections for 1–5 min in 0.05% toluidine blue O (C.I.52040) in benzoate buffer, pH 4.4. Destain in slow running water, briefly rinse slides in distilled water, shake off excess water, and dry on a hot plate (35–40°C) (Feder and O'Brien, 1968). Metachromatic colors of the stain can be restored by breathing on the sections just prior to the application of mountants such as DePeX, Euparal, and Permount (O'Brien and McCully, 1981). Background staining of the plastic varies from batch to batch. If a high background stain is present, it is advantageous to lower the pH of the toluidine blue O solution to 2.5 (no lower), using an acetate buffer or water acidified with dilute hydrochloric acid (Tippet and O'Brien, 1975). It may then be necessary to increase the concentration of toluidine blue 0 to 0.1% and increase the staining time. Results: polyphosphates, polysulfates, and polycarboxylic acids are stained red or reddish purple; lignin and some polyphenols, green or blue-green; RNA, purple; DNA, blue or blue-green (O'Brien *et al.*, 1964; Feder and Wolf, 1965).

B. General Histological Staining of Epoxy Sections

Stain sections in solution A for up to 30 min at 60°C in Coplin jars. Rinse thoroughly in distilled water and counterstain with solution B for up to 2 min at 60°C. Rinse in distilled water and dry on a slide warmer. Pass quickly through three changes of xylene and mount. The staining time varies according to the thickness of the sections and the type of epoxy resin used. The optimal staining time should be determined by trial and error. The composition of the staining solutions is as follows: solution A—0.13 g methylene blue (C.I.52015), 0.02 g Azure A (C.I.52005), 10 ml glycerol, 10 ml methanol, 30 ml 0.1 *M* phosphate buffer, pH 6.9, and 50 ml distilled water; solution B—0.1 g safranin O (C.I.50240) in 100 ml 0.2 *M* Tris buffer, pH 9.0. The staining solutions should be filtered before use. Results: A variety of colors result from this stain (for details, see Warmke and Lee, 1976).

C. The Periodic Acid-Schiff (PAS) Reaction for Polysaccharides

This staining reaction (Feder and O'Brien, 1968; Chayen *et al.*, 1973) can be performed successfully on both methacrylate and epoxy sections. Slides are placed in a saturated solution of dimedone (5,5-dimethylcyclohexane-1,3-dione), an aldehyde blocking agent, overnight at room temperature. Rinse in slow running water for 10 min and place the slide in a 1% Periodic acid solution for 10 min for GMA sections and 20 min for epoxy sections. Wash the sections in slow running water for 5 min and place them in Schiff's reagent for 30 min. Schiff's reagent can be prepared by dissolving 1 g basic fuchsin (C.I. 42510) in 200 ml boiling distilled water. Cool to 50°C and add 30 ml 1 *N* HCl and then 3 g potassium metabisulfite. Shake for 2 min. Be sure that the flask or container is tightly capped. Leave in the dark for 24 hr and then add 1 g decolorizing activated charcoal (Norit). Shake or stir for 5 min and filter quickly through filter paper. The solution should be clear and colorless. Store it at 4°C when not in use. This solution can be stored for months in a well-capped dark bottle. One can also obtain Schiff's reagent directly from commercial supply companies. After staining, the slides are transferred into three successive baths of a 0.5% solution of sodium or potassium metabisulfite in a 1% dilution of concentrated HCl for 2 min each. Gently wash the slides in slow running water for 5–10 min. Counterstain them if desired, dry, and mount them. Results: Starch and some polysaccharide wall components are stained red. Callose and cellulose are not stained (O'Brien and Thimann, 1967; Smith and McCully, 1978a).

D. Calcofluor White Staining of Cell Wall Components

In this procedure (Hughes and McCully, 1975), stain sections in 0.1% calcofluor white M2R in distilled water for 1–2 min. Wash them briefly in water, dry, and mount them. This procedure works well with both GMA and epoxy sections, although the staining intensity of the epoxy sections tends to be lower. Examine the sections using a fluorescence microscope with a near-ultraviolet light source such as the HBO 50W superpressure mercury lamp. Results: Most cell walls and extracellular mucilages give a strong pale blue fluorescence when viewed with the following Zeiss filter combination: G365 exciter filter, FT 420 chromatic splitter, and LP 418 barrier filter or its equivalent.

E. Aniline Blue-Induced Fluorescence of Callose

This method (Smith and McCully, 1978a,b) works best with GMA sections. Pretreat the slide by staining with either the PAS reaction, as described in Section II,C, or a 0.05% toluidine blue O solution, as described in Section II,A, to cut down on the autofluorescence of the specimen. Mount the sections in 0.05% (w/v) water-soluble aniline blue (C.I. 42755) in 0.067 *M* phosphate buffer, pH 8.5. Stain for 10 min and examine with a fluorescence microscope, as described in Section II,D. The staining solution is preferably freshly prepared or should be stored in a dark bottle when not in use. Results: Callose fluoresces yellowish white. For a critical discussion concerning the specificity of the reaction, see Smith and McCully (1978a).

F. The Feulgen Reaction for DNA

This is one of the best-documented histochemical reactions and is specific to DNA. Carry out the aldehyde blockage reaction as described in Section II,C. Rinse the slides in slow running water for 5 min. Hydrolyze the sections in 1.0 *N* HCl at 37°C for 3 hr and place them directly into Schiff's reagent for 30 min. Transfer them into the metabisulfite solution, as in the PAS reaction. Rinse in slow running water for about 5 min, counterstain if necessary, dry, and mount. Incubation in HCl at 37°C instead of 60°C allows for better time control when evolving conditions for new fixatives or different tissues (Cole and Ellinger, 1981). However, with familiar tissues, one can hydrolyze the tissues using 1 *N* HCl at 60°C for approximately 10 min. The staining intensity in epoxy sections tends to be weaker than in GMA sections. Results: Chromatin (DNA) is stained red.

G. Methyl Green Pyronin Y Stain for DNA and RNA

This procedure (Cole and Ellinger, 1981) applies only to GMA sections. Stain sections in methyl green pyronin Y at room temperature for 2 hr. Blot off excess stain, rinse in two changes of *n*-butanol for 5 min each, dry, and mount. The staining solution is made by mixing 1 part methyl green (C.I.42590) solution and 1 part pyronin Y (C.I.45005) solution with 2 parts of distilled water just before use. Both solutions need to be purified by chloroform extraction, according to Kurnick (1955). Results: Polymerized DNA stains green. Depolymerized DNA and RNA stain red.

H. Amido Black 10B and Coomassie Brilliant Blue as Protein Stains

Although amido black and Coomassie brilliant blue have been widely used as protein stains in paper chromatography and gel electrophoresis, their specificity toward proteins in histological specimens is still not clear. However, we find that protein bodies in seeds show strong affinity for these stains. In order to use them as protein stains, one should carry out control experiments using protease digestion. In general, these stains are excellent cytoplasmic stains and are especially useful as counterstains after the PAS treatment. Stain slides in 1% amido black (C.I. 20470) or 0.25% Coomassie brilliant blue G250 (C.I. 42655) or Coomassie blue R250 (C.I. 42660) in 7% acetic acid for 10 min at 40°C. Rinse and destain in 7% acetic acid, dry, and mount (Fisher, 1968). For GMA sections, only amido black can be used because there is no appreciable background staining after rinsing in acetic acid. However, with Coomassie brilliant blue, the background stain in the plastic is impossible to remove. In the case of epoxy sections, there is no appreciable background with either stain. Coomassie brilliant blue is preferred over amido black due to its greater staining intensity. Results: Cytoplasm stains blue.

I. Benzpyrene Method for Lipids

Using this method (Berg, 1951; Chayen *et al.*, 1973; Jensen, 1962), stain sections for 20 min in a benzpyrene solution. The staining solution is made up as follows: Prepare a saturated solution of caffeine in water at 20°C (about 1.5% caffeine). To 100 ml of filtered caffeine solution add 2 mg 3,4-benzpyrene, and keep at 37°C for 2 days. Filter the solution, add 100 ml distilled water, let stand for 2 hr, and refilter. The solution will remain stable when stored in a dark bottle for several months. After staining the sections, rinse the slides briefly in distilled water, mount them in water, and examine them under the fluorescence microscope with the near-ultraviolet light, as described in Section II,D. Results: Lipids fluoresce silver to bluish white or even yellow, depending on the concentration of the benzpyrene in the lipid. The fluorescence fades very rapidly. Note: Great care should be taken in the handling of benzpyrene because it is carcinogenic.

J. Nile Blue A Method for Lipids

This procedure (Jensen, 1962; Chayen *et al.*, 1973) applies to GMA sections only. Stain the sections in a 1% aqueous solution of Nile blue A (C.I.

51180) at 37°C for 5 min. Wash them quickly in warm water and differentiate them for 30 sec in 1% warm acetic acid. Wash the sections quickly again in warm water and mount them in either glycerol or water. Complete drying of sections prior to mounting will destroy the color differentiation of the stain. Results: Neutral lipids stain red. Phospholipids and fatty acids stain blue, but the blue color is not specific and will stain other cellular components.

III. AN INTRODUCTION TO ENZYME HISTOCHEMISTRY OF GLYCOL METHACRYLATE (JB-4) SECTIONS

With the appropriate catalyst, GMA can be polymerized at low temperatures under ultraviolet irradiation. Since tissue blocks can be processed at low temperature, some enzyme activity can be retained and detected subsequently using appropriate staining procedures. The introduction of the JB-4 resin simplifies the procedures because it can be polymerized at low temperature without the use of an ultraviolet source. Enzyme activity has been demonstrated using JB-4 as the embedding medium (Higuchi *et al.*, 1979; Nusbickel and Swartz, 1979; see also Data sheet No. 123, Poly-Sciences, Inc.).

The following procedures have been successfully carried out in the author's laboratory (see also Chapter 75, this volume). Small tissue blocks such as 1-mm-thick pieces are fixed in a solution of 2% formaldehyde and 1% glutaraldehyde in phosphate buffer, pH 7.2, for 2–3 hr at 4°C. Wash the blocks gently in cold buffer, with three changes over a 2-hr period. Fixation conditions are critical for the preservation of enzyme activity (for additional information and discussion, see Sexton and Hall, 1978; Hayat, 1981). After washing, if necessary, tissues should be gently evacuated to remove air using a vacuum pump. Then place them directly into solution A with the catalyst of the JB-4 plastic for 4 hr. Change the solution once and allow it to infiltrate the tissues for approximately 12 hr at 4°C. For highly vacuolated tissues, cells may be plasmolyzed when placed directly into solution A of the JB-4 plastic. In this case, tissues can first be dehydrated with a graded GMA series, that is, 30%, 50%, and 70%, for 1 hr each (GMA formulation according to Feder and O'Brien, 1968). Then place the tissues into solution A with the catalyst, as described above. The following day, the tissues can be embedded, using either Beem or gelatin capsules, and allowed to polymerize in a freezer at about −18°C; alternatively, they can simply be placed in the freezer compartment of the refrigerator for 24 hr. Air should be excluded from the capsules in order to ensure proper polymerization of the blocks. Once polymerized, the blocks can be sectioned with a dry glass knife, and the sections placed on a drop of water. Excess water is then removed with filter paper in order to facilitate the drying of sections on slides. The sections are allowed to air dry at room temperature and should be stained at once. The above processing procedures are intended only to serve as a guide. One should optimize the conditions according to one's own needs.

For enzyme staining, the following incubation solutions can be used:

1. Acid phosphatase. Dissolve 20 mg sodium naphthol AS–MX phosphate in 1 ml *N,N*-dimethylformamide and mix it with 50 ml 0.1 *M* acetate buffer, pH 5.5. Add 50 mg of the coupling agent (i.e., Fast garnet GBC, Fast blue B, Fast blue BB, or Fast blue RR) to the above buffer, mix thoroughly, filter, and use immediately. Incubate the sections for 30 min at 37°C (Burstone, 1962).
2. Esterase. Dissolve 20 mg α-naphthyl acetate in 0.5 ml acetone and mix it with 40 ml 0.1 *M* phosphate buffer at pH 7.4. Then dissolve 40 mg Fast blue B salt in this solution. Filter the solution and use it immediately (Chayen *et al.*, 1973).
3. Peroxidase. Dissolve 10 mg 3,3′-diaminobenzidine tetrahydrochloride in 10 ml 0.1 *M* Tris-HCl buffer, pH 7.2. Add 10 ml 0.2% freshly prepared hydrogen peroxide (prepared from the 30% stock solution). Mix it well, filter the solution, and use it immediately (Lojda *et al.*, 1979).

Parallel control experiments, such as using the incubation solution without the substrate, should always be conducted. For additional theoretical and practical considerations, consult the laboratory manual by Lojda *et al.* (1979).

ACKNOWLEDGMENTS

The technical assistance of Ms. Susan Blackman is gratefully acknowledged. The author would also like to thank Dr. S. Y. Zee, Department of Botany, University of Hong Kong, for introducing him to the enzyme histochemical techniques. This work was supported by Natural Sciences and Engineering Research Council of Canada Grant A6704.

REFERENCES

Berg, N. O. (1951). A histological study of masked lipids. *Acta Pathol. Microbiol. Scand., Suppl.* **40**, 1–192.

Burstone, M. S. (1962). "Enzyme Histochemistry and Its Application in the Study of Neoplasms." Academic Press, New York.

Chayen, J., Bitensky, L., and Butcher, R. G. (1973). "Practical Histochemistry." Wiley, New York.

Cole, M. B., Jr., and Ellinger, J. (1981). Glycol methacrylate in light microscopy: nucleic acid cytochemistry. *J. Microsc. (Oxford)* **123,** 75–88.

Feder, N., and O'Brien, T. P. (1968). Plant microtechnique: Some principles and new methods. *Am. J. Bot.* **55,** 123–142.

Feder, N., and Wolf, M. K. (1965). Studies on nucleic acid metachromasy. II. Metachromatic and orthochromatic staining by toluidine blue of nucleic acids in tissue sections. *J. Cell Biol.* **27,** 327–336.

Fisher, D. B. (1968). Protein staining for ribboned epon sections for light microscopy. *Histochemie* **16,** 92–96.

Hayat, M. A. (1981). "Fixation for Electron Microscopy." Academic Press, New York.

Higuchi, S., Suga, M., Dannenberg, A. M., Jr., and Schofield, B. H. (1979). Histochemical demonstration of enzyme activities in plastic and paraffin embedded tissue sections. *Stain Technol.* **54,** 5–12.

Hughes, J., and McCully, M. E. (1975). The use of an optical brightener in the study of plant structure. *Stain Technol.* **50,** 319–329.

Jensen, W. A. (1962). "Botanical Histochemistry." Freeman, San Francisco, California.

Kosakai, H. (1973). Epoxy embedding, sectioning and staining of plant material for light microscopy. *Stain Technol.* **48,** 111–115.

Kurnick, N. B. (1955). Pyronin Y in the methyl green-pyronin histological stain. *Stain Technol.* **30,** 213–230.

Lillie, R. D., ed. (1977). "H. J. Conn's Biological Stains." Williams & Wilkins, Baltimore, Maryland.

Lojda, Z., Gossrau, R., and Schiebler, T. H. (1979). "Enzyme Histochemistry: A Laboratory Manual." Springer-Verlag, Berlin and New York.

Nusbickel, F. R., and Swartz, W. J. (1979). Enzyme histochemical investigation of glycol methacrylate embedded chick embryonic tissue. *Histochem. J.* **11,** 197–203.

O'Brien, T. P., and McCully, M. E. (1981). "The Study of Plant Structure: Principles and Selected Methods." Termarcarphi Pty. Ltd., Melbourne, Australia.

O'Brien, T. P., and Thimann, K. V. (1967). Observations on the fine structure of the oat coleoptile. III. Correlated light and electron microscopy of the vascular tissues. *Protoplasma* **63,** 443–478.

O'Brien, T. P., Feder, N., and McCully, M. E. (1964). Polychromatic staining of plant cell walls by Toluidine blue O. *Protoplasma* **59,** 368–373.

Pearse, A. G. E. (1980). "Histochemistry: Theoretical and Applied," 4th ed., Vol. 1. Churchill-Livingstone, Edinburgh and London.

Sexton, R., and Hall, J. L. (1978). Enzyme cytochemistry. *In* "Electron Microscopy and Cytochemistry of Plant Cells" (J. L. Hall, ed.), pp. 63–147. Elsevier/North-Holland Biomedical Press, Amsterdam.

Smith, M. M., and McCully, M. E. (1978a). A critical evaluation of the specificity of aniline blue induced fluorescence. *Protoplasma* **95,** 229–254.

Smith, M. M., and McCully, M. E. (1978b). Enhancing aniline blue fluorescent staining of cell wall structures. *Stain Technol.* **53,** 79–85.

Thorpe, T. A. (1980). Organogenesis *in vitro:* Structural, physiological and biochemical aspects. *Int. Rev. Cytol., Suppl.* **11A,** 71–111.

Tippet, J. T., and O'Brien, T. P. (1975). Procedure for purifying 2-hydroxyethyl methacrylate and some methods for using it impure in plant histology. *Lab. Pract.* **24,** 239–240.

Warmke, H. E., and Lee, S.-L. J. (1976). Improved staining procedures for semithin epoxy sections of plant tissues. *Stain Technol.* **51,** 179–185.

Yeung, E. C., Aitken, J., Biondi, S., and Thorpe, T. A. (1981). Shoot histogenesis in cotyledon explants of radiata pine. *Bot. Gaz. (Chicago)* **142,** 494–501.

CHAPTER **77**

Staining and Nuclear Cytology of Cultured Cells

Alan R. Gould

Plant Genetics Department
Central Research
Pfizer, Inc.
Groton, Connecticut

I. INTRODUCTION

Cytological studies on plant cells in culture have never achieved the level of popularity enjoyed by biochemical or molecular biological approaches. This is unfortunate because cytology can provide information at the single-cell or cell-population level which is unavailable with other invasive or destructive methods. The tedium of making many painstaking observations at the microscope, and the fact that many preparative procedures in cytology contain more art than science, has generally discouraged the rapid development of new technical advances. However, new computer-based methods (image analysis, scanning microspectrophotometry, flow cytometry and sorting) and the gradual spread of techniques from animal cell biology (fluorescence microscopy, chromosome banding, *in situ* hybridization of nucleic acids) seem to have sparked new interest in cytological staining of plant tissue cultures.

CELL CULTURE AND SOMATIC CELL
GENETICS OF PLANTS, VOL. 1

ISBN 0-12-715001-3

There are marked differences between the range of cell types found in whole plants and the phenotypes expressed by plant cells in culture. Thus, cytological methods used for cells in whole plants must often be modified if they are to be successfully applied to tissue culture material. This is not always the case. Some cultured tissues and organs so closely resemble their whole-plant analogs that excellent results can be obtained with well-known cytological protocols. Root tip meristems are a good example of this situation. More often, however, a tried and trusted general technique will give disappointing results with plant tissue cultures. Therefore, this chapter will focus on cytological methods which work well with cultured plant cells, and an attempt will be made to outline problem areas and critical steps in staining protocols. The methods to be reviewed are exclusively for light microscopy and will mainly concern nuclear cytology. Three previous reviews which describe other cytological and cytogenetic techniques for plant tissue cultures or whole plants are recommended: Evans and Reed (1981), Sharma and Sharma (1980), and Darlington and LaCour (1976).

In cytological studies, no factor is more important than the state of the cells or tissues before fixation. Plant tissue cultures used for nuclear cytology should be in the exponential part of their growth cycle to ensure that all stages of cell division are represented at reasonably high frequencies. The classic cell line type of culture (as discussed by King, 1980) may be homogeneous in terms of cell type, but can also be kinetically heterogeneous, that is, there may be nondividing, Go, or Q-cell subpopulations which are biochemically and cytologically distinct (Chu and Lark, 1976; Ashmore and Gould, 1979; Giroud, 1982). Thus, plant cell lines in culture should be subjected to a rigorously defined and controlled passage length between subcultures so that cells do not arrest in stationary phase for long periods of time. It has been this author's experience that passage lengths of 4–7 days with dilution ratios of 1:20 (for carrot) to 1:5 (for corn or tobacco) keep cell line-type populations in virtually continuous division. So-called callus cultures of many species are often complex mixtures of differentiating and differentiated cells. For this type of material, frequent transfers and sometimes visual selection of particular cell and tissue types are necessary. Certainly, only plant tissue cultures with high viability should be used for cytological work, to avoid artifacts due to dead cells.

II. PRETREATMENTS

Prefixation treatments of cells and tissues are intended to improve the quality of, or bring out special features in, the final cytological preparation.

For cytogenetic work, chromosome contractants and/or mitotic arrestants are commonly employed. These compounds are used to produce high frequencies of mitotic figures, chromosome contraction, good separation of chromosomes, and resolution of centromeres and secondary constrictions. For these purposes, methods using colchicine, 8-hydroxyquinoline (8-HQ), mono-bromonaphthalene, *p*-dichlorobenzene, and cold treatments appear to be most successful with cultured plant cells (Table I). Neither the commercial product Colcemid (deacetylated methyl colchicine) nor vinblastine sulfate seems to give good results in plant systems. Before setting out protocols for various pretreatments, it is worthwhile to examine the differential effects of certain methods.

A certain degree of chromosome contraction aids in the preparation of nonoverlapping spreads, especially in species with long chromosomes. Constrictions in chromosomes are also emphasized by moderate contraction, but if chromosomes are treated for too long a period or at too high a concentration of the contracting agent, considerable loss of resolution can occur. This is especially true of spreads being prepared for chromosome banding, because too much contraction reduces or even cancels interband distances. Colchicine should not be overzealously applied to plant material for this reason.

The mechanism of mitotic arrest involves spindle disruption and changes

TABLE I

Pretreatment Protocols[a]

Agent	Protocol
Colchicine	Make a fresh stock solution of 10 mg/ml colchicine. Add stock to growing culture in standard medium to give a final concentration of 0.05–0.1%. Incubate for 3 hr in a shaking incubator at 26°C in the dark. Wash cells thoroughly in medium before fixation. The most common problem is overcontraction.
p-Dichlorobenzene	Place a few crystals in growth medium and heat gently until a molten globule forms. Cool the solution. Incubate cells in this saturated solution for 4 hr at room temperature with shaking. This method is best for species with many small chromosomes. A similar technique can be used with mono-bromonaphthalene.
Cold	Incubate culture at 1–2°C for 18–24 hr, then fix. For some species 5–10°C is more suitable and shorter treatment times can be used. This is a good method if only slight contraction is required.
8-HQ	Take up 29 mg 8-HQ crystals in a few drops of 95% ethanol and then dilute to 100 ml with water or medium (effective concentration 2 m*M*). Incubate cells in this solution for 1–4 hr at room temperature.

[a] In all cases, cultures should be in the exponential phase of the growth cycle.

in cytoplasmic viscosity caused by depolymerization of microtubules. Different treatments produce different effects based on the level of cytoplasmic disorganization induced. For instance, colchicine treatments usually cause a characteristic scattering of chromosomes in squash preparations, whereas 8-HQ seems to be less disruptive and maintains the arrangement of chromosomes on the metaphase plate. 8-HQ works particularly well with long chromosomes and is very effective for the definition of centromeres and secondary constrictions. The main negative consideration with 8-HQ is that it can produce large separations in satellited chromosomes, so that satellites can be lost or unassignable to particular chromosomes. The contractant *p*-dichlorobenzene is another chemical which resolves chromosome constrictions, and it has been of special value with cultures of species with many small chromosomes (Bayliss and Gould, 1974). When unsatisfactory results are obtained with common contractants or arrestants, cold treatments are very often helpful, especially if very slight and controlled chromosome shortening is desired.

For plant cells not grown in suspension culture (i.e., callus), it will be necessary to infiltrate the tissue under vacuum to ensure penetration of the pretreatment chemicals. In all cases, species-specific effects are likely, and so trial and error with several concentrations, incubation times, and pretreatments are recommended to establish the best method.

Other special pretreatments are required for some techniques. For instance, the ability to stain sister chromatids differentially requires preincubation of cultures with bromodeoxyuridine for two cell cycles. This ensures that for each chromosome one chromatid has both strands of its DNA containing the substituted brominated base analog, whereas the other chromatid is substituted in only one of its DNA strands. Such sophisticated techniques have generally not yet been applied to plant cells in culture. Also, this particular example requires a knowledge of cell cycle time, a parameter which has been measured for only about a dozen plant species in tissue culture (Gould, 1984). Information on such special techniques is available elsewhere (Sharma and Sharma, 1980; Darlington and LaCour, 1976).

III. FIXATION

The most widely preferred fixative for nuclear cytology under the light microscope has been acetic alcohol (variously referred to as *Farmer's fixative, Carnoy's fluid,* or *three in one*). This fixative should always be freshly prepared just prior to use by carefully adding 25 ml concentrated (glacial) acetic acid to 75 ml 95% ethanol. There should be at least 20 volumes of fixative per volume of tissue, and for storage longer than 24 hr, the tissue should be washed and resuspended in 70% ethanol and stored below 10°C. Extended fixation in acetic–alcohol should be avoided since under certain conditions this can cause extraction of DNA from the sample. Some simple modifications to this fixative are sometimes useful. Substitution of methanol for ethanol, or propionic acid for acetic acid, can yield improved results with some species and cell types. The fixative known as Carnoy's II contains absolute ethanol, chloroform, and glacial acetic acid in the ratio 6:3:1, and is suitable for cultured cells with waxy or oily constituents.

A fixation problem which is common and peculiar to plant cells in culture concerns starch grains. Even with excellently prepared material, the presence of highly refractive starch grains can obscure the detail over interphase and mitotic nuclei. Starch grains are often present in suspension cultured plant cells, and need to be dispersed by fixation in 50% aqueous formic acid. This treatment seems to break down or solubilize the starch grains overnight, and after washing out the formic acid fixative, the common staining methods can be applied.

In a minority of cases, fixation in nonacid conditions may be required; thus, the traditional alcoholic–acid mixtures are unsuitable. For example, in a method for quantitative staining of nuclear histone proteins with the dye Fast Green F.C.F., an overnight fixation in 10% neutral buffered formalin was used (Gould, 1975). This fixative is made up with 1.182 g Na_2HPO_4, 1.122 g KH_2PO_4, 62 ml fresh stock formaldehyde solution, and distilled water to 250 ml.

Finally, it should be emphasized that some of the more complex fixatives which have been used for whole plant material (e.g., Newcomer's Fluid: isopropyl alcohol, propionic acid, petroleum ether, acetone, and dioxane in the ratio 6:3:1:1:1) have no particular advantage over acetic–alcohol for most fixations involving tissue culture cells. Also, it should be recognized that all fixatives are by their very design extremely toxic, and may give off very dangerous vapors. All fixations should therefore take place in a fume hood. The fixation of higher plant protoplasts presents special problems. Protoplasts are fragile and must be artificially buffered with an osmoticum to avoid swelling and bursting. Once fixation is well advanced, the semipermeability of protoplast membranes is abolished and osmotic buffering is no longer needed. The initial shock as the fixative is applied to osmotically balanced protoplasts is the critical step which needs modification. Two methods have been used successfully in this author's experience.

Fixation at similar osmotic potential can be easily done by bringing a standard fixative up to a suitable potential by the addition of sucrose, sorbitol, or mannitol. For acetic–ethanol fixation, freshly made fixative (3 parts ethanol:1 part glacial acetic) is diluted 3 parts fixative to 1 part water, and sorbitol is dissolved in this at 13% (w/v). Protoplasts can be gently suspended in this fixative with little breakage. After 24 hr it is best to replace the sorbitol-buffered fixative with freshly made acetic–ethanol without sorbitol.

A second method involves stepwise addition of a fixative composed of ethanol diluted to 70% with a protoplast buffer solution of 600 m*M* KCl, 2 m*M* NH_4NO_3, 3 m*M* $CaCl_2$, and 1 m*M* KH_2PO_4. Protoplasts are resuspended in 0.2 ml protoplast buffer, and the fixative is added one drop at a time, with very gentle mixing between each addition. This method has been found to be very useful in fluorescent staining of protoplasts.

IV. STAINING

Staining protocols for cultured plant cells may be conveniently divided into four types: staining in a tube before squashing on slides; staining on the slide before squashing; squashing fixed but unstained material onto slides and then staining; air drying without squashing. The first three methods require squashing at some stage, and as the technique used to squash samples onto microscope slides is of paramount importance for the quality of the completed preparation, a squash protocol follows in detail.

The material to be squashed is suspended in 45% acetic acid for 30 min to soften the cell walls. Some workers prefer to use 25% acetic acid for longer periods. A small amount of material in acetic acid is placed on a clean slide, and a thick glass coverslip is placed over the sample. If too much material is put on the slide, it will be difficult to flatten it effectively and cells will not be spread into a monolayer. The next stage is absolutely critical. The coverslip is tapped firmly with a brass rod to spread the cells out (a pencil can be substituted, but firmer pressure can be achieved with a metal rod). The pressure should occur at 90° to the coverslip, which should not be allowed to shift around on the slide, as this can roll cells over and cause clumping of the chromosomes. After 5–10 firm taps, a piece of filter paper is placed over the coverslip and the squash is made with heavy pressure with the thumb or with a rubber stopper.

The slide (coverslip up) is then placed on a block of dry ice (solid CO_2) for 5 min or until it is completely frozen, and then the coverslip is carefully pried off with a scalpel blade. Depending on which protocol is being used, the slide is then either dipped in absolute ethanol for 5 sec and dried or dehydrated in a suitable organic solvent series. The dry ice freezing method derives from the protocol of Conger and Fairchild (1953), but a dip in liquid nitrogen will also rapidly freeze squash preparations so that the coverslip can be removed.

Slides produced in this manner can be made permanent by dehydration in a suitable solvent series. For the popular mounting medium "Euparal" a four-step ethanol series of 70%, 95%, absolute, absolute (5 min for each step) is suitable. For Gurr's XAM mountant, a four-step series of 95% ethanol, 3:1 ethanol xylene, 1:3 ethanol xylene, absolute xylene is necessary.

A. In-Tube Methods

1. Feulgen Staining for Material in Tubes

This widely used staining method can produce excellent results in both qualitative and quantitative work. Fixation in acetic–ethanol is suitable, and 50% formic acid is also a fixative compatible with the Feulgen staining protocol. Fixation in formaldehyde-based reagents is recommended if very long (over 7 days) fixing times cannot be avoided. The stain solution itself can be obtained commercially, but better results are usually achieved with freshly made laboratory preparations.

Traditionally, the raw material for Feulgen stain has been basic fuchsin, a mixture of three dyes: *p*-rosaniline, magenta, and new magenta. However, some suppliers stock only *p*-rosaniline, which is inferior in terms of staining intensity, but which is often refered to as "basic fuchsin." Good results have been obtained in this laboratory with Fuchsin "Fuchsine for microscopy" from the Fluka Co. (Switzerland).

The Feulgen stain preparation is prepared as follows:

1. Dissolve 1 g basic fuchsin in 200 ml boiling glass distilled water.
2. Stir well and allow to cool to 50°C.
3. Filter and add 30 ml 1 *N* HCl to the filtrate.
4. Add 3 g potassium metabisulfite.
5. Stir well in an airtight container and store in the dark at about 10°C overnight.
6. Stir well and add 0.5 g activated charcoal.
7. Stir rapidly and filter after 1 min.
8. Repeat the charcoal filtering step until the filtrate is pale straw to colorless.
9. Do not continue the charcoal treatment far beyond the point of full decoloration.

The stain should be stored in the dark, below 10°C, and has a shelf life of no more than 6 weeks.

The Feulgen staining method is as follows:

1. Fixed cells are rehydrated by two sequential centrifugations and resuspensions in distilled water. Cells are spun down once more, the supernatant is removed, and the cell sample is hydrolysed in one of two different ways: hot hydrolysis in 1 *N* HCl at 60°C for 8–12 min or cold hydrolysis in 5 *N* HCl for several hours (the time depends on the species and tissue type). After hydrolysis, the cells are immediately spun down and resuspended in Feulgen stain. Staining continues for 2 hr. Then the sample is centrifuged and resuspended in 45% acetic acid for 20 min. Finally, squash preparations are made.
2. If the cytoplasm stains lightly, a resuspension in SO_2 water for a few minutes will remove nonspecifically bound dye. SO_2 water is a fresh 1:1 mixture of 1 *N* H_2SO_4 and 10% potassium metabisulfite solution, and should be used prior to the 45% acetic acid step.

2. Combined Feulgen Methods

For samples which stain poorly with the standard Feulgen method, two useful modifications are available. The first approach is to squash the Feulgen-stained preparations in 1% aceto–carmine, which greatly increases the contrast of nuclei and chromosomes against the cytoplasmic background. Aceto–carmine is prepared by dissolving 1 g solid carmine (natural red 4) in 100 ml 45% acetic acid. The solution is brought to boiling, cooled, and filtered. The stain is quite stable and can be used at lower concentrations by diluting with 45% acetic acid.

A second method combines Feulgen staining and Giemsa staining, is particularly suited to small chromosomes, and has been claimed to produce C-bands (Gostev and Asker, 1979). Samples can be stained by the standard in-tube Feulgen method. Squashes are made in 45% acetic acid, the slides are frozen, coverslips are removed, and after a 5-sec dip in absolute ethanol the slides are air-dried. The slides are then incubated in 2X SSC (double-strength standard saline citrate; see Section IV,C,1) for 5–6 hr at room temperature. After repeated washing in distilled water, the slides are stained in 2% Giemsa in phosphate buffer (1/15 *M*, pH 6.8) until satisfactory contrast is achieved. The banding produced by this method is not usually as pronounced as the effect produced by the standard Giemsa C-band technique.

3. Fluorescent Stains for Material in Tubes

DNA-specific fluorochromes such as mithramycin, Hoechst 33258, Hoechst 33342, or propidium iodide have a distinct advantage over the

Feulgen method in that no hydrolysis step is needed to produce good nuclear and chromosomal staining. Laloue *et al.* (1980) describe a rapid method using the *bis*-benzimidazole derivative Hoechst 33258. Suspension cultured cells are fixed for 30 min in fresh acetic–ethanol, and are then rehydrated by repeated distilled water rinses. Cells are then resuspended in a 1 μg/ml solution of Hoechst 33258 in citric acid–phosphate buffer (0.1 *M*, pH 5). After 30–60 min, cells can be squashed [addition of glycerol to 50% (v/v) is recommended] on slides and immediately observed. For improved spreading, the cell preparation can be softened by incubation in 0.5% pectinase in the staining solution. In case samples need to be stored after staining, 70% ethanol has been found to be suitable for resuspension of well-stained cells.

The related dye Hoechst 33342 has been used in this laboratory to stain a range of cultured plant cells. A 100X stock (1 m*M*) is prepared by dissolving 5.6 mg of the dye in 10 ml of a compatible isotonic diluent (the usual growth medium is generally suitable). Then 0.1 ml of stock is added to 10 ml of cell suspension, and staining is allowed to proceed at room temperature in the dark for 1–2 hr. Cells are then washed repeatedly in medium, squashed on slides, and immediately observed. Improved flatness can be attained by pectinase treatments as set out for Hoechst 33258 above.

Propidium iodide and ethidium bromide are also useful stains for fluorescent visualization of nuclei. However, there have been few applications in the cytology of cultured plant cells.

Excellent nuclear staining can be achieved with the dye mithramycin (aureolic acid). Cells or protoplasts are fixed in 70% ethanol. A suitable fixation mixture for protoplasts was described in Section II. After fixation on ice for 1 hr, the sample is gently spun down and resuspended in an equal volume of mithramycin solution. The presence of magnesium ion is essential for staining. Then 3.05 g $MgCl_2 \cdot 6H_2O$ is dissolved in 1 liter of 1 *N* saline (irrigation saline). Finally, 25 ml of this solution is used to dissolve 2.5 mg mithramycin. The stain solution should be kept below 10°C in the dark and has a shelf life of about 2 weeks.

B. On-Slide Stain–Squash Methods

1. Lacto–Propionic Orcein

The lacto–propionic orcein method is applied to unfixed material from culture, and is particularly useful for the rapid determination of the mitotic index or preliminary estimations of chromosome number. The stain is prepared by stirring 1 g synthetic orcein in 100 ml of a 1:1 mixture of lactic acid and propionic acid, with gentle heating. After 2 hr the stain is filtered and is ready for use. The stain has a very long shelf life but sometimes requires filtering if precipitates form.

Small samples of callus or suspension culture cells are placed on a clean microscope slide, and the tissue is dried by touching it with a clean piece of filter paper. This removes water from the tissue, which appears characteristically white and glassy when enough moisture has been absorbed. A drop of stain is then added to the tissue, and after 2 min the slide is warmed in the flame of a spirit burner until the preparation is just too hot to be placed on the back of the hand. Then a coverslip is placed over the sample, and the preparation is tapped to spread the cells out and then squashed under a filter paper. The coverslip is sealed around the edges with clear nail varnish, and the slide is then ready for examination. Staining intensity can be varied by altering the length of the heating step. Slides made in this way will last for only 2–3 hr before drying out.

2. Acetic Orcein, Formic Orcein

For samples fixed in acetic–ethanol or formic acid, a filtered 1% solution of synthetic orcein in 45% acetic acid or 50% formic acid will efficiently stain nuclei and chromosomes of cultured plant cells. The fixed material is placed in a drop of stain on a clean slide, and after 10 min the slide is heated, tapped, and squashed, as in the previous protocol. However, preparations can be made permanent by the CO_2 freezing method outlined for Feulgen-stained material.

3. Carbol–Fuchsin

Carbol–fuchsin reagent is a useful alternative for tissues and especially protoplasts, which are difficult to stain with the standard Feulgen, orcein, or carmine protocols. The stain is prepared according to Kao (1975):

Stock A: 3% basic fuchsin in 70% ethanol
Stock B: 1 to 9 dilution of stock A in 5% aqueous phenol
Working stock: 45 ml stock B plus 6 ml glacial acetic acid plus 6 ml 37% aqueous formaldehyde (formalin)

A modified version of the carbol–fuchsin stain is a 2–10% solution of the working stock in 45% acetic acid and 1.8% sorbitol. Kao (1975) recommends different stain concentrations for different species, and a trial-and-error approach is recommended with unfamiliar material.

Cells are stained on the slide after softening in 45% acetic acid, the carbol–fuchsin is added dropwise, and staining is complete within 30 min. After tapping and squashing, the preparations can be made semipermanent by sealing with wax or varnish, but CO_2 freezing and permanent mounting methods are also compatible with carbol–fuchsin staining. The carbol–fuchsin method has been most useful in the present author's laboratory for species with small, very heterochromatic chromosomes. Whereas such chromosomes appear overstained and fuzzy with orcein-based techniques, a combination of light carbol–fuchsin staining and phase-contrast microscopy can reveal centromeres and even band patterns under ideal conditions (i.e., very flat preparations with well-separated chromosomes).

C. Staining Presquashed Material on Slides

1. Giemsa C-Banding

Chromosome banding techniques allow rather precise analysis of chromosome instability in plant tissue cultures (Ashmore and Gould, 1981; Gould, 1982), but the Giemsa C-banding method is technically somewhat demanding, and has not yet been widely applied to plant cells in culture. For this reason, the following C-banding protocol is described in great detail to enable inexperienced cytologists to obtain banded chromosomes by avoiding mistakes of omission.

Squash preparations of freshly fixed cells should be made as described earlier, taking care not to overcontract chromosomes. Isolated chromosomes or metaphases in broken cells appear to band with increased contrast, and so the tapping stage of slide preparation should be especially vigorous. Fixed samples should not be more than 7 days old, but once the unstained squash preparation is made, good results can be obtained even with month-old slides, if the slides are stored in a desiccator.

The Giemsa C-banding method is as follows:

1. Hydrolyze unstained squash preparations in saturated barium hydroxide [$Ba(OH)_2$] solution at 45°C for 1–30 min. Wash the slides thoroughly in distilled water. Incubate them in double strength saline-citrate (2X SSC) at 60°C for 2 hr. Wash in distilled water. Stain the slides with Giemsa stain 2–10% in 1/15 *M* phosphate buffer until good contrast is obtained for C-positive regions in interphase nuclei.
2. Wash in distilled water and air-dry in dust free area.
3. Observe the slides without a coverslip by direct application of immersion oil to the slide.

The most critical step in the C-band protocol is the alkaline hydrolysis in $Ba(OH)_2$. Solutions of $Ba(OH)_2$ are very unstable due to their reaction with atmospheric CO_2 to form insoluble barium carbonate. Therefore, the following precautions are essential for the preparation of saturated $Ba(OH)_2$

solutions. Always use freshly distilled water or boil it just prior to use. Add excess solid $Ba(OH)_2$ to distilled water so that there is no air space in the container. Add a magnetic stirring bar and seal the container; again make sure that there is no air in the container. Stir the solution overnight at room temperature. Allow the saturated $Ba(OH)_2$ solution to settle for at least 1 hr before use. Bring the saturated solution to 45°C in a water bath, and immediately before hydrolyzing the samples, remove the white scum which forms on the surface (the scum is probably mainly insoluble barium carbonate).

The other steps are less critical, and some variations are necessary with different species and tissue types. Suspension cultured cells usually require shorter alkaline hydrolysis in $Ba(OH)_2$ (1–5 min) than root tip meristem cells (a full 30 min). The 2X SSC solution (0.3 *M* NaCl, 0.03 *M* Na citrate, pH 7.4) can be stored as a 10X SSC concentrate and diluted just before use. Variable results have been obtained with Giemsa stain from different sources. Giemsa is a mixture of at least three dyes: methylene blue, its oxidation products (azurs), and eosin. Different brands of Giemsa stain can contain widely varying proportions of these three components, and so it is wise to identify a brand which gives satisfactory results and to use it exclusively. Good results have been obtained in this laboratory with Gibco diagnostic Giemsa stain and Gurr's Giemsa R66.

2. Other Banding Techniques

Although other banding techniques have been successfully applied to whole plant material (e.g., N-banding and Q-banding), plant tissue culture research has yet to benefit from these methods (Gerlach, 1977; Vosa, 1970). N-banding, supposedly specific for nucleolar organizers, involves incubation in 1 *M* NaH_2PO_4 (pH 4.15) at 95°C for 3–5 min and then staining in Giemsa. Q-banding is a fluorescent technique for picking out heterochromatin using quinacrine mustard as the staining agent.

The puzzling lack of G-bands on plant chromosomes has been explained in terms of severe compaction of plant as compared to animal chromosomes (Greilhuber, 1977), but this explanation has been refuted recently by Anderson *et al.* (1982). Although this volume is dedicated to laboratory techniques, speculations on such problems as the absence of G-bands on plant chromosomes are not out of place. It is, for example, notable that the hypotonic swelling technique which produces such excellent results in G-banding with cultured animal cells has yet to be adapted for use with plant material. Presumably it would be relatively easy to produce hypotonic swelling in live protoplasts before fixation.

With G-bands presently unavailable and C-bands few in number and limited in some plant species to telomeres and centromeres, it has become necessary to look for alternatives which will provide higher-resolution mapping of plant chromosomes at the light microscopic level. Two approaches worthy of further study with plant tissue culture systems are orcein- (O-) banding (Sharma and Sharma, 1980) and the treatment of plant chromosomes with endonucleases to produce specific patterns of nicks or cuts (Subrahmanyam *et al.*, 1976). Neither approach has yielded published information in the field of plant tissue culture, and this area represents a worthwhile opportunity for development of important new techniques.

D. Air-Drying Techniques

In the parlance of animal cell cytology, *air drying* refers to the technique of placing drops of cell suspension in organic solvents onto microscope slides and then blowing air gently over the slides to dry the fluid. The method produces very flat chromosome spreads in which all chromosomes in a metaphase are at or very near the same plane of focus. With plant cells, the method is less successful due to their rigid cell walls. However, with fixed protoplasts, a similar air-drying approach can yield flat preparations which have been very useful for autoradiographic analysis of nuclei in protoplasts (Gould, 1979). Protoplasts fixed in 75% acetic–ethanol in water with sorbitol at a final concentration of 11% are washed twice in fresh acetic–ethanol to remove the sorbitol. The sample is then rehydrated by resuspension in water, and the standard Feulgen staining protocol is followed. Finally, the Feulgen-stained protoplasts are resuspended to give a dense suspension in 25% acetic acid. Drops of this suspension are placed on clean slides, and the preparation is allowed to evaporate to dryness. Subsequently, autoradiographic film can be applied. If air drying is applied to fixed but unstained protoplasts, staining can be done after autoradiographs have been developed and fixed by dipping the slide in 2% Giemsa stain until the nuclei show good contrast. This technique is often referred to as *staining through the emulsion.*

In a recent paper, Murata and Orton (1983) refer to air-drying techniques applied to plant chromosome analysis (Mouras *et al.*, 1978; Murata, 1983). The further development of such techniques will certainly play a crucial role in cytological analyses of plant tissue cultures.

REFERENCES

Anderson, L. K., Stack, S. M., and Mitchell, J. B. (1982). An investigation of the basis of a current hypothesis for the lack of G-banding in plant chromosomes. *Exp. Cell Res.* **138,** 433–436.

Ashmore, S. E., and Gould, A. R. (1979). Cell cycle analysis of tumor-derived cultures of *Crepis capillaris:* A kinetic analogy with proliferation of animal tumor cells. *Protoplasma* **101,** 217–230.

Ashmore, S. E., and Gould, A. R. (1981). Karyotype evolution in a tumor-derived plant tissue culture analysed by Giemsa C-banding. *Protoplasma* **106,** 197–308.

Bayliss, M. W., and Gould, A. R. (1974). Studies on the growth in culture of plant cells. XVIII. Nuclear cytology of *Acer pseudoplatanus* suspension cultures. *J. Exp. Bot.* **25,** 772–783.

Chu, Y., and Lark, K.G. (1976). Cell cycle parameters of soybean (*Glycine max* L.) cells growing in suspension culture: Suitability of the system for genetic studies. *Planta* **132,** 259–268.

Conger, A. D., and Fairchild, L. M. (1953). A quick freeze method for making smear slides permanent. *Stain Technol.* **28,** 281–283.

Darlington, D. C., and LaCour, L. F. (1976). "The Handling of Chromosomes," 6th ed. Wiley, New York.

Evans, D. A., and Reed, S. M. (1981). Cytogenetic techniques. *In* "Plant Tissue Culture: Methods and Applications in Agriculture" (T. A. Thorpe, ed.), pp. 213–240. Academic Press, New York.

Gerlach, W. L. (1977). N-banded karyotypes of wheat species. *Chromosoma* **62,** 49–56.

Giroud, F. (1982). Cell nucleus pattern-analysis-geometric and densitometric features, automatic cell phase identification. *Biol. Cell.* **44,** 177–188.

Gostev, A., and Asker, S. (1979). A C-banding method for small plant chromosomes. *Hereditas* **91,** 140–143.

Gould, A. R. (1975). Experimental studies on the cell cycle of cultured sycamore cells. Ph.D. Thesis, Univ. of Leicester, England.

Gould, A. R. (1979). Combined microspectrophotometry and automated quantitative autoradiography applied to the analysis of the plant cell cycle. *J. Cell Sci.* **39,** 235–245.

Gould, A. R. (1982). Chromosome instability in plant tissue cultures studied with banding techniques. *In* "Plant Tissue Culture 1982" (A. Fujiwara, ed.), Maruzen, Tokyo.

Gould, A. R. (1984). Control of the cell cycle in cultured plant cells. *CRC Crit. Rev. Plant Sci.* (in press).

Greilhuber, J. (1977). Why plant chromosomes do not show G-bands. *Theor. Appl. Genet.* **50,** 121–124.

Kao, K. N. (1975). A nuclear staining method for plant protoplasts. *In* "Plant Tissue Culture Methods" (O. L. Gamborg and L. R. Wetter, eds.), pp. 60–62. National Research Council of Canada, Saskatoon.

King, P. J. (1980). Cell proliferation and growth in suspension cultures. *Int. Rev. Cytol., Suppl.* **11A,** 25–53.

Laloue, M., Courtois, D., and Manigault, P. (1980). Convenient and rapid fluorescent staining of plant cell nuclei with "33258" Hoechst. *Plant Sci. Lett.* **17,** 175–179.

Mouras, A., Salesses, G., and Lutz, A. (1978). Sur l'utilisation des protoplastes in cytologie amélioration d'une méthode récente en vie de l'identification des chromosomes mitotiques des genres *Nicotiana* et *Prunus. Caryologia* **31,** 117–127.

Murata, M. (1983). Staining air-dried protoplasts for observation of plant chromosomes. *Stain Technol.* (in press).

Murata, M., and Orton, T. J. (1983). Chromosome structural changes in cultured celery cells. *In Vitro* **19,** 83–89.

Sharma, A. K., and Sharma, A. (1980). "Chromosome Techniques. Theory and Practice," 3rd ed. Butterworth, London.

Subrahmanyam, N. C., Gould, A. R., and Doy, C. H. (1976). Cleavage of plant chromosomes by restriction endonucleases. *Plant Sci. Lett.* **6,** 203–208.

Vosa, C. G. (1970). Heterochromatin recognition with fluorochromes. *Chromosoma* **30,** 366–372.

CHAPTER **78**

Chromosome Analysis

R. L. Phillips
A. S. Wang

Department of Agronomy and Plant Genetics
University of Minnesota
St. Paul, Minnesota

I. INTRODUCTION

Karyotypic analyses commonly are performed as an adjunct to plant cell and tissue culture studies. Such studies are seldom done principally for the purpose of determining the cytogenetic behavior of cultured cells or regenerated plants. Chromosome analyses, therefore, usually have not been carried out in a manner designed to generate maximum information. Most chromosomal variation studies have been based on mitotic analyses and usually involve small numbers of plants. Sometimes chromosomal variation is suggested solely on the basis of sterility. Relatively few species are represented among reports with cytological results on regenerated plants, and most of these employ mitotic analyses. This chapter is intended to provide ideas for gaining more genetic–cytogenetic information in the course of cell and tissue culture studies. Most of the points discussed come to mind because of our experiences with plant cell and tissue cultures, and are not intended to represent a comprehensive plan but rather helpful

ISBN 0-12-715001-3

hints for cytogenetic analysis. Our experience has been mostly with corn and oats; the ideas and cautionary notes in this chapter are presented with one or both of these species in mind. Most of the points, however, can be readily extended to other species of interest. Staining procedures are not discussed in this chapter (see Chapters 76 and 77, this volume).

II. WHY CHROMOSOME ANALYSIS?

The need for understanding the cytological–cytogenetic behavior of cultured cells, regenerated plants, and derived progenies cannot be overemphasized. Even when the primary goal of the study is not cytological, there are many reasons why the cytogenetic behavior of the material should be assessed. We first need to recognize that essentially all types of detectable cytogenetic changes have been reported from a variety of tissue culture systems. These encompass ploidy changes (including mixoploidy), aneuploidy of various types, a wide range of nuclear constitutions from multipolar mitoses, structural modifications (including duplications, deficiencies, dicentrics, isochromosomes, interchanges, and inversions), and chromosomal chimeras (D'Amato, 1977, 1978; Sunderland, 1973, 1977; Skirvin, 1978; Bayliss, 1980).

Variants selected in culture may not be the result of point mutations but have a chromosomal anomaly as the basis of the phenotype. The variant phenotype may be the result of a dosage effect due to a partial or whole chromosome duplication or deletion. Combining chromosome analysis with selection experiments may provide valuable information not only on the genetic basis of variants but also on the expected transmission and segregation patterns. Variation may not be stably heritable and thus may give the appearance of an epigenetic response. Given the frequent chromosomal alterations that occur in certain culture systems, a single-gene variant associated with a change in chromosome number or structure may segregate in a complex genetic ratio. A trait first thought to be controlled by two or more genes may actually be a single-gene trait that segregates in a more complicated pattern due to linkage with a chromosome alteration. Sectoring of such chromosome changes within the reproductive tissues can lead to complicated genetic ratios and, as discussed later, even affect the generation in which segregation will be detected. Many phenotypic variants described to date from a wide variety of tissue culture systems could be caused by chromosomal alterations. On the other hand, the phenotype of the regenerated plant may not always reveal chromosomal abnormalities. We believe that it is wise to plan in advance to couple cytology with selection experiments.

Much of the evidence on nuclear stability in cell and tissue cultures and in regenerated plants deals with polyploid species. Relatively less is known about chromosomal stability in cultured cells or regenerated plants of diploids. For example, Torrey (1967) studied the karyotype of pea (*Pisum sativum*) cultures and regenerated roots over a period of 7 years. Higher chromosome numbers and greater aneuploid frequencies occurred with older cultures. *Petunia parviflora* plants regenerated from protoplasts included six diploids, two monosomics, and one tetraploid (Sink and Power, 1977). Novak (1980) analyzed 264 clones derived from onion (*Allium sativum*) tissue cultures and found that 55.4% were diploid, 31.8% chimeric, 2.2% tetraploid, and 10.6% aneuploid. Increasing numbers of tetraploid and various aneuploid cells occurred in long-term callus cultures of *Allium* (Novak, 1981). Similar instability has been reported recently for celery callus cultures (Browers and Orton, 1982). In a study on *Hordeum vulgare, H. jubatum,* and the interspecific hybrid, Orton (1980a) concluded that chromosomal variability also was generated at the point of regeneration and later in development. One of the most stable systems appears to be in the diploid *Lilium longiflorum*. Sheridan (1974, 1975) reported the karyotype to be stable over 7 years. Other authors (Sunderland, 1973; Butcher *et al.*, 1975) pointed out several instances in which the variability present in cultures did not appreciably change with time. Pearl millet (*Pennisetum americanum*) plants regenerated via somatic embryogenesis were shown to possess the normal diploid chromosome number (Vasil and Vasil, 1981b).

Studies on chromosomal variability in maize include those of LaRue (1949) and Straus (1954, 1958) with cultures initiated from endosperm tissue. Cells with various chromosomal anomalies were almost as common as those with normal chromosome numbers. Cell suspension cultures of diploid Black Mexican sweet corn have been shown to contain many tetraploid and some aneuploid cells (Brar *et al.*, 1979; Mi *et al.*, 1982). Balzan (1978) reported a high frequency of cells with abnormal chromosome numbers in tissue cultures derived from mesocotyl tissue of maize-inbred SI 104. Edallo *et al.* (1981) found 92% and 84.5% diploid cells ($2n = 20$) in tissue cultures initiated from immature embryos of inbreds W64A and S65, respectively. About half of the nondiploid cells were tetraploid, and most of the remaining cells had 31–39 chromosomes, suggesting chromosome elimination after tetraploidization. A few cells with 21–30 chromosomes were also observed. They found only one tetraploid and one trisomic, however, among 110 regenerated plants (108 were diploids). Our work on maize plants regenerated from cultures initiated from immature embryos suggests that over 90% of the plants are of normal karyotype and have normal meiotic behavior (Green, 1977; Green *et al.*, 1977; McCoy and Phillips, 1982). Such maize tissue cultures may have considerable genetic instability while maintaining a high degree of cytogenetic stability.

In contrast, oats ($2n = 6X = 42$) have been shown in our laboratories to have considerable chromosomal variation among regenerated plants (Cummings *et al.*, 1976; McCoy *et al.*, 1978, 1982). The 1982 study involved the meiotic analysis of 655 regenerated plants. An unexpected result was the high frequency of partial chromosome loss, principally the generation of apparent telocentric chromosomes. The higher frequency of aneuploidy recovered in the regenerated oat plants compared to the maize system may be related to the high degree of genetic buffering in at least certain polyploids that permits more normal cell function in an aneuploid state.

III. MITOTIC VERSUS MEIOTIC ANALYSIS

Most chromosome analyses performed in relation to plant cell and tissue culture studies have involved mitotic cells. Mitotic analysis of cultured cells can readily reveal information on polyploidy, aneuploidy, or gross chromosomal deletions or duplications. Polyploid and aneuploid regenerated plants or shoots have occurred in cell or tissue cultures of many species (D'Amato, 1977). Plants with a doubled chromosome number were produced from over 100 sugarcane clones (Heinz *et al.*, 1977). Nine octoploids were found among 200 regenerated alfalfa plants from tissue cultures initiated from tetraploid alfalfa (*Medicago sativa*) by Saunders and Bingham (1972). Mattingly and Collins (1974) initiated cultures from the leaf midvein of a nullihaploid tobacco plant and derived a nullisomic regenerant apparently through spontaneous chromosome doubling in culture. Nullihaploid tobacco plants can also be produced through the *in vitro* culture of anthers from monosomic plants (Moore and Collins, 1982). Sacristan (1971) showed that haploid cultures of *Crepis capillaris* produced a higher frequency of chromosome rearrangements than a diploid culture. Meiotic analysis of 16 regenerated sorghum (*Sorghum bicolor*) plants indicated that 11 were diploid and 5 were tetraploid (Brar *et al.*, 1979). Dulieu (1972) found a genotype effect in which regenerants of one variety of tobacco were diploid, whereas those of another were tetraploid.

One advantage of mitotic compared with meiotic analysis is the wide array of available tissues. Investigators should be aware of meristematic areas other than root tips that are useful for mitotic chromosome studies. Ear shoot and anther tapetal cells have been useful in maize. In plants, Feulgen staining assists in recognizing areas of high meristematic activity because of the usual darker staining resulting from a denser cell population. Areas of higher mitotic activity in callus tissue can also be identified by their more intense Feulgen staining. Chromosomes may be recognized on the basis of length, position of nucleolus organizer region(s), distribution of heterochromatin, and centromere position. The centromere divides the chromosome into two arms; the long/short arm ratio is more distinctive than the length for a particular chromosome due to differences in stretching or condensation. Asymmetrical pericentric inversions or unequal chromosomal translocations may also be detectable using standard cytological techniques. New chromosome-banding procedures theoretically allow the detection of less obvious translocations, inversions, duplications, and deletions. Unfortunately, banding of plant chromosomes generally is not as revealing as it is with animal chromosomes; therefore, it is of more limited value except for detecting certain heterochromatic blocks by C-banding. The mitotic index of callus or suspension cultures usually is rather low, making mitotic analysis difficult especially for the less experienced cytologist.

A problem more prevalent in mitotic than meiotic analysis is that of broken cells. Techniques for making good preparations with somatic tissue usually involve considerable pressure applied to the coverslip in order to achieve maximum chromosome spreading. Pressure tends to break cells, with a subsequent loss of chromosomes. Investigators need to be aware of this problem and interpret chromosome counts carefully. Cells with fewer than the expected chromosome number are suspect; even apparently intact cells should be interpreted with caution. Of course, the more cells observed with a specific chromosome constitution, the stronger the evidence for the existence of such cells in the culture.

Callus tissue or the somatic or meiotic tissue of regenerated plants may be sectored for a chromosome anomaly. Sectoring has been reported in durum wheat (Bennici and D'Amato, 1978; Lupi *et al.*, 1981), tobacco (Sacristan and Melchers, 1969; Carlson and Chaleff, 1975; Ogura, 1976), *Brassica* (Horak, 1972), onion (Novak, 1980), barley (Mix *et al.*, 1978), sugarcane (Heinz and Mee, 1971; Liu and Chen, 1976), ryegrass hybrid (Ahloowalia, 1976), *Lycopersicum peruvianum* (Ramulu *et al.*, 1976a,b), and maize (McCoy and Phillips, 1982). The fascinating study by Lupi *et al.* (1981) with durum wheat (*Triticum durum*) showed that the majority of plants regenerated from mesocotyl-derived tissue cultures contained both normal and aneuploid cells even though all pollen mother cells of regenerated plants had normal chromosome behavior and karyotype.

D'Amato *et al.* (1980) stated that "the sometimes extensive chromosomal mosaicism of the regenerated plants is a clear-cut proof of the multicellular origin of adventitious buds in culture." Vasil (1981) reported that most species of cereals and grasses studied thus far undergo plant regeneration via a multicellular origin of shoot meristem formation. He stated that one should not expect such regenerants to be genetically uniform. We have relatively little information on how many cells participate in differentiating the apex of maize regenerants. Histological studies (Springer *et al.*, 1979) suggest a multicellular origin for maize regenerants from immature embryo-initiated cultures (Green and Phillips, 1975). Perhaps the sectoring observed in maize regenerants (McCoy and Phillips, 1982) is a function of variability among cultured cells coupled with a multicellular origin of regenerated plantlets. Sectoring was not observed in the oat studies (McCoy *et al.*, 1982). Two or three panicles from tillers of 61 regenerated plants were analyzed; only one regenerant did not have concordance among tillers. Thus, analyzing more cells and more portions of the tissue will enhance the chances of making a proper interpretation. Sectoring appears to be more common in tissue culture-derived plants than in seed-grown plants. Mixoploid root tips, for example, may be common in plantlets derived from haploid tissue cultures (Green and Donovan, 1978; Dhaliwal and King, 1979).

Limited information is available on the meiotic behavior of regenerated plants. Meiotic studies on regenerated tobacco (*Nicotiana*) plants show the existence of abnormal pairing and disjunction, as well as chromosome numbers ranging from 14 to 60 (Zagorska *et al.*, 1974), univalent metaphase I and lagging anaphase I chromosomes (Ogura, 1976), and polyploidy (Tabata *et al.*, 1968; Murashige and Nakano, 1966). Regenerated plants from sugarcane (*Saccharum*) species hybrids possessed large variation in chromosome number, as did the parent donor clone (Heinz and Mee, 1971). Meiotic analysis enhances the probability of detecting certain changes and allows a more complete cytogenetic description. Whenever possible and appropriate, meiotic analysis should be performed. Meiotic analysis should include the scoring of various stages of meiosis I and II, as well as postmeiotic divisions, because each provides the opportunity to notice additional changes. Meiotic analysis allows evaluation of pairing so that asynaptic or desynaptic mutants can be recognized. Heteromorphic chromosome pairs may occur, suggesting the loss or addition of chromosome arm segments. Translocations can be evaluated at diakinesis for ring-, chain-, or pair-forming capacity that provides insights on break positions. Chromosome types such as isochromosomes are more readily recognized in meiosis. Paracentric inversions can also be more readily recognized by bridge and fragment formation in anaphase I or II. The presence of micronuclei at the quartet stage is a good indicator of univalent or fragment formation. Although meiotic analysis is not appropriate for certain purposes, the information obtained is generally much more meaningful than that obtained from mitotic analysis. Meiotic analyses should be encouraged. In preparation for meiotic (or mitotic) analysis, one should remember that what is abnormal is always relative to what is normal for that species, genotype, environmental conditions, and so on. Time must first be expended in determining what is the normal cytological behavior for that strain grown under appropriate conditions. This review of normal behavior should be done in advance of studying cultured cells or regenerated plants. The investigator needs to know what is the normal variability, for example, in such meiotic features as chromosome appearance, pairing and distribution, micronuclei frequency at the quartet stage, and nucleolar constitution.

IV. GENOTYPIC DIFFERENCES AND GENETIC STOCKS

Genotypic differences exist for nearly every plant trait. This is no less true for traits of interest to plant cell and tissue culturists. Genotypic differences in the capacity to regenerate plants are expected on the basis of past experience (Green and Phillips, 1975). Chromosome variation is also a function of genotype (McCoy *et al.*, 1982). One should be careful not to extrapolate to the entire species based on experience with one genotype (cultivar). Cytogenetic differences among genotypes can be dramatic. McCoy *et al.* (1982) reported 31% cytogenetically abnormal plants among regenerants of the oat cultivar Tippecanoe versus 61% for the cultivar Lodi. Telocentric chromosomes were observed in tissue cultures of wheat (*Triticum aestivum*) aneuploids, particularly from nulli-5B tetra-5D cultures (Shimada *et al.*, 1974). Subsequently, tissue cultures were initiated from wheat lines carrying telocentrics with known degrees of instability (Inomata *et al.*, 1976). Although not discussed in detail, culture age, media, tissue source, degree of meristematic organization, karyotype variation of donor tissue, and ploidy level can also affect the degree of cytogenetic variation. Comparative cytogenetic studies must include controls for as many of these variables as possible.

A variety of genetic–cytogenetic stocks are readily available for several plant species and can be advantageously employed for many purposes. Heterozygous dominant or recessive gene markers may facilitate the detection of chromosome variation throughout much of plant development or in specific tissues depending on the marker. These markers can save considerable cytology and allow concentrated study on the proper tissue, stage, or strain as indicated by marker behavior. Again, appropriate control plants should be grown. Certain chromosome aberrations, such as deletions for increasing the chances of recovering a recessive mutation in a specific chromosome segment, or translocations for testing the occurrence of somatic reduction followed by doubling, may also be useful. Aneuploid stocks can be employed in many ways, such as the use of monosomics for recovering recessive mutations in specific chromosomes. Other genetic markers are valuable for producing cytogenetic stocks of interest, such as the maize rx-1 deletion that produces monosomic and trisomic progeny (Weber, 1983) and the maize indeterminate gametophyte (*ig*) gene that increases the frequency of haploidy from 0.1% to 2–3%, with paternal haploid occurrence rising from 0.001% to about 2% (Rhodes and Green, 1981).

Genetic stocks specially developed for plant cell and tissue culture studies are needed for all species. Multiple marker stocks, with marker genes on several if not all chromosomes, are of great value, allowing many chromosomes to be genetically monitored simultaneously. The production of near-isogenic regenerable lines that carry the genetic–cytogenetic marker of interest would not only allow the use of that marker but would also reduce the confounding of genotype-specific effects. Such markers in regenerable genetic backgrounds may allow the detection of a relatively rare cytogenetic event and probably should be employed whenever possible. For example, the rare exchange of genomic segments in an interspecific hybrid or an alien addition line might first be detected by the use of a genetic marker and subsequently shown by cytological analysis. Some unexpected segregations might occur in interspecific hybrid cultures and their regenerants, such as the intact haploid genomes of one or both donor species segregating out, as observed by Orton (1980b) in *Hordeum vulgare*/*H. jubatum* F_1 hybrid cultures. Alternatively, a large range of aneuploid types may be recovered with chromosomes coming from both genomes, as was common in the regenerants of the *Hordeum* interspecific cultures (Orton and Steidl, 1980; Orton, 1980a).

V. UNEXPECTED DEVELOPMENTAL BEHAVIORS

Knowledge of the developmental ontogeny of plants derived from seed or vegetative propagation is invaluable in understanding the consequences of anomalies arising in cultured cells and regenerated plants. An investigator interested in chromosome analysis should be as familiar as possible with the cellular development of the species of interest. In most cases, only the general concepts of plant development will be known for the species. In other cases, more complete definition of cell number and cell lineages for specific tissues will be known (Johri and Coe, 1982). Such knowledge is useful in understanding chimeras for phenotypic traits or chromosome variations. For example, the location and sector size of a chromosome alteration in the maize tassel have important implications in terms of how early the alteration occurred in development; sector information may also bear on whether or not the plant had a multicellular origin in the culture. Also in maize, for example, developmental studies suggest that the cells for the ear versus tassel in the main culm are already set apart in the embryo of the mature kernel. However, cells for the ear versus tassel in the tiller are not yet separate in the mature embryo. Thus, sectors occurring in the main culm during development may not include both the ear and the tassel, whereas these organs are expected to be concordant in tillers.

Additional cytological parameters may deviate from normal in cultured cells or regenerated plants. Coenocytic cells containing several genomes, a high degree of asynchrony within anthers, spindle abnormalities including tripolar spindles and various uneven divisions, and $2n$ gamete formation have been noted. These cytological deviations have implications in terms of sterility and transmission frequencies.

The tissue explant source is also important to consider in anticipating chromosomal variations. The polysomatic condition of sugarcane is well known and is believed to account for some of the observed variation (Heinz and Mee, 1971). All species probably have variations in cell ploidy levels in various tissues. Cytological examination of donor tissue provides useful information for discovering the source of variation. D'Amato *et al.* (1980) stated that nuclear fragmentation followed by mitosis in the initial stages of initiating tissue cultures is much more frequent and important than previously suspected.

The mode of regeneration should also be considered. For certain species, regeneration may be accomplished via organogenesis or embryogenesis. Although not much information is available, the regeneration process probably determines certain developmental outcomes. For example, plantlets derived by embryogenesis may not be of multicellular callus origin. If sectoring in regenerated plants sometimes is a manifestation of a multicellular origin, plants derived via embryogenesis may be less likely to possess chimeras if bipolar embryonic development is of single cell origin (Vasil and Vasil, 1981a,b; Vasil, 1981).

VI. CRYPTIC GENOMIC DEVIATIONS

Molecular biological techniques provide opportunities to examine somatic cells in other than the mitotic phase of the cell cycle. *In situ* hybridization of molecular probes specific to certain chromosomes would allow the detection of dosage changes among interphase as well as mitotic cells. Dosage changes may reflect whole or partial chromosome changes or perhaps a form of gene amplification (Skokut and Filner, 1980; Yamaya and Filner, 1981), compensation, magnification, reduction, or transposition. Presently, *in situ* hybridization is limited to gene sequences present in multiple copies; techniques developed for other systems no doubt will soon be adapted to plants and allow the recognition of low copy number genes and possibly even unique sequences. In maize, for example, chromosome 6 can be monitored by a 18 S ribosomal RNA (rRNA) probe (Phillips *et al.*, 1979) and chromosome 2 can be monitored by a 5 S rRNA probe (Mascia *et al.*, 1981). Cell suspensions or callus tissue can also be monitored for specific DNA sequence modifications or changes in sequence multiplicity. Sequence modifications among regenerated maize plants have also been detected via mitochondrial DNA endonuclease restriction patterns (Gengenbach *et al.*, 1981; Pring *et al.*, 1981; Umbeck and Gengenbach, 1983; Kemble *et al.*, 1982). Chromosome variations among interphase cells may also be reflected by the number of nucleoli or the number of particular heterochromatic blocks (Dhaliwal and King, 1979). Large numbers of interphase cells are usually available because of the low mitotic index and represent an underutilized source of valuable information.

Species that are auto- or allopolyploid in nature may produce increased variation among regenerants due to a chromosome substitution mechanism (Larkin and Scowcroft, 1983). These substitutions may be difficult to detect cytologically where the chromosomes are similar in morphology.

VII. HANDLING DONOR PLANT SOURCES AND SUBSEQUENT REGENERANTS

One of the keys of plant genetic–cytogenetic analysis is the use of a precise pedigree handling system. Every plant should be individually identified prior to serving directly as an explant source or as a parent for later tissue culture use (e.g., as a source of immature embryos or inflorescences). Bulking seed, pollen, and so on are never a good idea from the point of view of genetic or chromosome analysis. Unusual behavior of a specific plant or the progeny of a specific plant is often of immense interest; it is generally important to trace the pedigree to specific individuals in order to determine the genetic basis of the unusual behavior. Although considerable effort is entailed, we believe cell lines should also be pedigreed to elucidate the origin of interesting genetic changes. Complete cell line pedigrees are seldom maintained because of the extra effort involved in labeling, but certain genetic hypotheses could be ruled out if that information were available. Genetic–cytogenetic analysis would be facilitated if regenerated plants were also identified with their respective cell line. For example, several variants of a specific phenotype segregating in progenies of regenerants are more likely to be controlled by different loci if they trace back to independent cell lines.

Control seed-grown plants are another important consideration in order to test for the occurrence of certain alterations in the source material. In using immature maize embryos, a good procedure is to remove only one-half of the ear for extracting embryos and to leave the other half to mature. Thus, seed-grown plants can be compared with regenerants originating from embryos from exactly the same plant and ear. Each plant species and culture protocol will require a variation on the scheme, but control plants not derived from tissue culture should be part of the experiment.

Microsporocytes can usually be readily sampled from the regenerated plants, although the age of the plant may not be the same as that of seed-propagated plants. If possible, the desired method is one which does not sacrifice the plant. In maize, for example, one can fix part of the tassel and leave another portion intact. Crosses can be made later using the same plant. In addition, pollen can be analyzed from the remaining branches of the tassel. Similar objectives can be achieved by sampling tillers, as on a cereal plant such as oats or wheat. Pollen analysis, especially in diploids, is a sensitive indicator of chromosome alterations. In order to detect sectoring, pollen sterility determinations should be made using various portions of the inflorescence. Pollen of varying sizes as well as aborted pollen should be scored, because that information can be related to the type of aberration (e.g., 50% smaller, starch-filled pollen in maize suggests the presence of a duplication).

The objectives of the study may determine whether the primary regenerant (R0) or first (R1) or second (R2) selfed generations should be examined cytologically. Analysis of R1 or R2 progeny without studying the R0 will reflect only those chromosomal alterations that were transmissible and that occurred within the sample analyzed. Chromosome analysis of the R0 plants gives a better indication of the array of cytological and cytogenetic changes occurring among the regenerated plants. How accurately the R0 array reflects the frequency of changes among the cultured cells is not certain. Gross chromosomal changes may be of similar frequency among R0 plants compared with cultured cells (McCoy and Phillips, 1982).

Plant phenotypes among R1 and R2 progenies may be especially indicative of certain karyotypic changes. For example, progeny of monosomic oat or wheat plants may segregate for characteristic nullisomic phenotypes. Segregation of unusual phenotypes in progeny should be carefully noted.

If sectoring occurs, such as in maize regenerants, the R2 generation may be the first to show segregation for certain recessive mutations due to discordance between the male and female inflorescences (McCoy and Phillips, 1982). If the recovery of variability is an important objective of the tissue culture program, evidence on sectoring may indicate the generations that need to be evaluated for maximum mutant recovery frequencies.

VIII. FUTURE OF CHROMOSOME ANALYSIS

Additional techniques to improve the precision of chromosome analysis will become available as the fields of cytogenetics and molecular biology advance. Molecular probes will be available for every chromosome in certain species. *In situ* hybridization techniques will eventually be developed to allow the use of unique sequences as probes. Banding and other cytological analysis procedures will no doubt continue to improve the longitudinal differentiation of metaphase and prophase chromosomes. Premature chromosome condensation protocols for plants (Dudits *et al.*, 1982) will allow the visualization of interphase chromosomes.

The goal is to understand the mechanisms underlying chromosomal/genetic instability that appears to be a common feature of many cell systems. Perhaps we will learn to control the instability in order to achieve more effective somatic cell genetic systems and selection protocols. On the other hand, the spontaneous variability that is obtained with today's procedures may be of extreme value in providing (1) novel materials for breeding (especially if the variability involves quantitative as well as qualitative traits), (2) an array of cytogenetic stocks in common genetic backgrounds, (3) new methods for mapping genes or other genetic analyses, or (4) genetic exchanges that might generate enhanced recombination or genomic transfers in interspecific hybrids.

REFERENCES

Ahloowahlia, B. S. (1976). Chromosomal changes in parasexually produced ryegrass. *In* "Current Chromosomal Research" (K. Jones and P. E. Brandham, eds.), pp. 115–122. Elsevier/North-Holland Biomedical Press, Amsterdam.

Balzan, R. (1978). Karyotype instability in tissue cultures derived from the mesocotyl of *Zea mays* seedlings. *Caryologia* **31,** 75–87.

Bayliss, M. W. (1980). Chromosomal variation in plant tissues in culture. *Int. Rev. Cytol., Suppl.* **11A,** 113–144.

Bennici, A., and D'Amato, F. (1978). *In vitro* regeneration of durum wheat plants. *Z. Pflanzenzuecht.* **81,** 305–311.

Brar, D.S., Rambold, S., Gamborg, O., and Constabel, F. (1979). Tissue culture of corn and *Sorghum. Z. Pflanzenphysiol.* **95,** 377–388.

Browers, M. A., and Orton, T. J. (1982). A factorial study of chromosomal variability in callus cultures of celery (*Apium graveolens*). *Plant Sci. Lett.* **26,** 65–73.

Butcher, D. N., Sogene, A. K., and Tommerup, I. C. (1975). Factors influencing changes in ploidy and nuclear DNA levels in cells from normal crown gall and habituated cultures of *Helianthus annuus L. Protoplasma* **86,** 295–308.

Carlson, P. S., and Chaleff, R. S. (1975). Heterogeneous associations of cells formed *in vitro. In* "Genetic Manipulations with Plant Material" (L. Ledoux, ed.), pp. 245–261. Plenum, New York.

Cummings, D. P., Green, C. E., and Stuthman, D. D. (1976). Callus induction and plant regeneration in oats. *Crop Sci.* **16,** 465–470.

D'Amato, F. (1977). Cytogenetics of differentiation in tissue and cell cultures. *In* "Applied and Fundamental Aspects of Plant Cell, Tissue, and Organ Culture" (J. Reinert and Y. P. S. Bajaj, eds.), pp. 343–357. Springer-Verlag, Berlin and New York.

D'Amato, F. (1978). Chromosome number variation in cultured cells and regenerated plants. *In* "Frontiers of Plant Tissue Culture 1978" (T. A. Thorpe, ed.), pp. 287–295. Univ. of Calgary Press, Calgary, Alberta, Canada.

D'Amato, F., Bennici, A., Cionini, P. G., Baroncelli, S., and Lupi, M. C. (1980). Nuclear fragmentation followed by mitosis as mechanism for wide chromosome number variation in tissue cultures: Its implications for plant regeneration. *In* "Plant Cell Cultures: Results and Perspectives" (F. Sala, B. Parisi, R. Cella, and O. Ciferri, eds.), pp. 67–72. Elsevier/ North-Holland Biomedical Press, Amsterdam.

Dhaliwal, H. S., and King, P. J. (1979). Ploidy analysis of haploid-derived tissue cultures of *Zea mays* by chromocentre counting. *Maydica* **24,** 103–112.

Dudits, D., Szabados, L., and Hadloczky, G. (1982). Premature chromosome condensation in plant cells and its potential use in genetic manipulation. *In* "Premature Chromosome Condensation" (P. N. Rao, R. T. Johnson, and K. Sperling, eds.), pp. 359–369. Academic Press, New York.

Dulieu, H. (1972). The combination of cell and tissue culture with mutagenesis for the induction and isolation of morphological or developmental mutants. *Phytomorphology* **22,** 283–296.

Edallo, S., Zucchinali, C., Perenzin, M., and Salamini, F. (1981). Chromosomal variation and frequency of spontaneous mutation associated with *in vitro* culture and plant regeneration in maize. *Maydica* **26,** 39–56.

Gengenbach, B. G., Connelly, J. A., Pring, D. R., and Conde, M. F. (1981). Mitochondrial DNA variation in maize plants regenerated during tissue culture selection. *Theor. Appl. Genet.* **59,** 161–167.

Green, C. E. (1977). Prospects for crop improvement in the field of cell culture. *HortScience* **12,** 131–134.

Green, C.E., and Donovan, C. M. (1978). Regeneration of monoploid plants from tissue cultures of maize. *In* "Frontiers of Plant Tissue Culture 1978" (T. A. Thorpe, ed.), Abstracts, p. 157. Univ. of Calgary Press, Calgary, Alberta, Canada.

Green, C. E., and Phillips, R. L. (1975). Plant regeneration from tissue cultures of maize. *Crop Sci.* **15,** 417–421.

Green, C. E., Phillips, R. L., and Wang, A. S. (1977). Cytological analysis of plants regenerated from maize tissue cultures. *Maize Genet. Coop. News Lett.* **51,** 53–54.

Heinz, D. J., and Mee, G. W. P. (1971). Morphologic, cytogenetic, and enzymatic variation in *Saccharum* species hybrid clones derived from callus tissue. *Am. J. Bot.* **58,** 257–262.

Heinz, D. J., Krishnamurthi, M., Nickell, L. G., and Maretzki, A. (1977). Cell, tissue, and organ culture in sugarcane improvement. *In* "Plant Cell, Tissue, and Organ Culture" (J. Reinert and Y. P. S. Bajaj, eds.), pp. 3–17. Springer-Verlag, Berlin and New York.

Horak, J. (1972). Ploidy chimeras in plants regenerated from the tissue cultures of *Brassica oleracea* L. *Biol. Plant.* **14,** 423–426.

Inomata, N., Okamoto, M., and Asami, H. (1976). Behavior of unstable telocentric chromosomes in cultured callus cells of Chinese Spring wheat (*Triticum aestivum* L.). *Jpn. J. Genet.* **51,** 223–228.

Johri, M. M., and Coe, E. H., Jr. (1982). Genetic approaches to meristem organization. *In* "Maize for Biological Research" (W. F. Sheridan, ed.), pp. 301–310. Univ. of North Dakota Press, Grand Forks.

Kemble, R. J., Flavell, R. B., and Brettell, R. I. S. (1982). Mitochondrial DNA analysis of fertile and sterile maize plants derived from tissue cultures with the Texas male sterile cytoplasm. *Theor. Appl. Genet.* **62,** 213–217.

Larkin, P. J., and Scowcroft, W. R. (1983). Somaclonal variation and crop improvement. *In* "Genetic Engineering of Plants" (A. Hollaender, C. Meredith, and T. Kosuge, eds.), pp. 289–314. Plenum, New York.

LaRue, C. D. (1949). Cultures of the endosperm of maize. *Am. J. Bot.* **36,** 798.

Liu, M.-C., and Chen, W.-H. (1976). Tissue and cell culture as aids to sugarcane breeding. I. Creation of genetic variation through callus culture. *Euphytica* **25,** 393–403.

Lupi, M. C., Bennici, A., Baroncelli, S., Gennai, D., and D'Amato, F. (1981). In vitro regeneration of *Durum* wheat plants. II. Diplontic selection in aneusomatic plants. *Z. Pflanzenzuecht.* **87,** 167–171.

McCoy, T. J., and Phillips, R. L. (1982). Chromosome stability in maize (*Zea mays*) tissue cultures and sectoring in some regenerated plants. *Can. J. Genet. Cytol.* **24,** 559–565.

McCoy, T. J., Phillips, R. L., and Cummings, D. P. (1978). Cytogenetic variability in plants regenerated from tissue cultures of oats (*Avena sativa*). *In* "Frontiers of Plant Tissue Culture 1978" (T. A. Thorpe, ed.), Abstracts, p. 101. Univ. of Calgary Press, Calgary, Alberta, Canada.

McCoy, T. J., Phillips, R. L., and Rines, H. W. (1982). Cytogenetic variation in tissue culture regenerated plants of *Avena sativa:* High frequency of partial chromosome loss. *Can. J. Genet. Cytol.* **24,** 37–50.

Mascia, P. N., Rubenstein, I., Phillips, R. L., Wang, A. S., and Xiang, L. Z. (1981). Localization of the 5S rRNA genes and evidence for diversity in the 5S rDNA region of maize. *Gene* **15,** 7–20.

Mattingly, C. F., and Collins, G. B. (1974). The use of anther-derived haploids in *Nicotiana*. III. Isolation of nullisomics from monosomic lines. *Chromosoma* **46,** 29–36.

Mi, C. C., Wang, A. S., and Phillips, R. L. (1982). Partial synchronization of maize cells in liquid suspension culture. *Maize Genet. Coop. News Lett.* **56,** 142–144.

Mix, G., Wilson, H. M., and Foroughi-Wehr, B. (1978). The cytological status of plants of *Hordeum vulgare* L. regenerated from microspore callus. *Z. Pflanzenzuecht.* **80,** 89–99.

Moore, G. A., and Collins, G. B. (1982). Isolation of nullihaploids from diverse genotypes of *Nicotiana tabacum*. *J. Hered.* **73,** 192–196.

Murashige, T., and Nakano, R. (1966). Tissue culture as a potential tool in obtaining polyploid plants. *J. Hered.* **57,** 115–118.

Novák, F. J. (1980). Phenotype and cytological status of plants regenerated from callus cultures of *Allium sativum* L. *Z. Pflanzenzuecht.* **84,** 250–260.

Novák, F. J. (1981). Chromosomal characteristics of long-term callus cultures of *Allium sativum* L. *Cytologia* **46,** 371–379.

Ogura, H. (1976). The cytological chimeras in original regenerates from tobacco tissue cultures and in their offsprings. *Jpn. J. Genet.* **51,** 161–174.

Orton, T. J. (1980a). Chromosomal variability in tissue cultures and regenerated plants of *Hordeum*. *Theor. Appl. Genet.* **56,** 101–112.
Orton, T. J. (1980b). Haploid barley regenerated from callus cultures of *Hordeum vulgare* × *H. jubatum*. *J. Hered.* **71,** 280–282.
Orton, T. J., and Steidl, R. P. (1980). Cytogenetic analysis of plants regenerated from colchicine-treated callus cultures of an interspecific *Hordeum* hybrid. *Theor. Appl. Genet.* **57,** 89–95.
Phillips, R. L., Wang, A. S., Rubenstein, I., and Park, W. D. (1979). Hybridization of ribosomal RNA to maize chromosomes. *Maydica* **24,** 7–21.
Pring, D. R., Conde, M. F., and Gengenbach, B. G. (1981). Cytoplasmic genome variability in tissue culture-derived plants. *Environ. Exp. Bot.* **21,** 369–377.
Ramulu, K. S., Devreux, M., Ancora, G., and Laneri, U. (1976a). Chimerism in *Lycopersicum peruvianum* plants regenerated from *in vitro* cultures of anthers and stem internodes. *Z. Pflanzenzuecht.* **76,** 299–319.
Ramulu, K. S., Devreux, M., and deMartinis, P. (1976b). Origin and genetic analysis of plants regenerated *in vitro* from periclinal chimeras of *Lycopersicum peruvianum*. *Z. Pflanzenzuecht.* **77,** 112–120.
Rhodes, C., and Green, C. E. (1981). An *in vitro* detection system for monoploid maize tissue cultures. *Maize Genet. Coop. News Lett.* **55,** 93–94.
Sacristan, M. D. (1971). Karyotypic changes in callus cultures from haploid and diploid plants of *Crepis capillaris* (L.) Wallr. *Chromosoma* **33,** 273–283.
Sacristan, M. D., and Melchers, G. (1969). The caryological analysis of plants regenerated from tumorous and other callus cultures of tobacco. *Mol. Gen. Genet.* **105,** 317–333.
Saunders, J. W., and Bingham, E. T. (1972). Production of alfalfa plants from callus tissue. *Crop Sci.* **12,** 804–808.
Sheridan, W. F. (1974). Long term callus cultures of *Lilium:* Relative stability of the karyotype. *J. Cell Biol.* **63,** 313a.
Sheridan, W. F. (1975). Plant regeneration and chromosome stability in tissue cultures. *In* "Genetic Manipulations with Plant Materials" (L. Ledoux, ed.), pp. 263–295. Plenum, New York.
Shimada, T., Inomata, N., Okamoto, M., and Asami, H. (1974). Telocentric chromosomes obtained from calluses of nulli-5B tetra-5D of Chinese Spring wheat. *Wheat Inf. Ser.* **39,** 11–12.
Sink, K. C., and Power, J. B. (1977). The isolation, culture, and regeneration of leaf protoplasts of *Petunia parviflora* Juss. *Plant Sci. Lett.* **10,** 335–340.
Skirvin, R. M. (1978). Natural and induced variation in tissue culture. *Euphytica* **27,** 241–266.
Skokut, T. A., and Filner, P. (1980). Slow adaptive changes in urease levels of tobacco cells cultured on urea and other nitrogen sources. *Plant Physiol.* **65,** 995–1003.
Springer, W. D., Green, C. E., and Kohn, K. A. (1979). A histological examination of tissue culture initiation from immature embryos of maize. *Protoplasma* **101,** 269–281.
Straus, J. (1954). Maize endosperm tissue grown *in vitro*. II. Morphology and cytology. *Am. J. Bot.* **41,** 833–839.
Straus, J. (1958). Spontaneous changes in corn endosperm tissue cultures. *Science* **128,** 537–538.
Sunderland, N. (1973). Nuclear cytology. *In* "Plant Tissue and Cell Culture" (H. E. Street, ed.), pp. 161–190. Blackwell, Oxford.
Sunderland, N. (1977). Nuclear cytology. *In* "Plant Cell and Tissue Culture" (H. E. Street, ed.), Vol. 2, pp. 177–205. Univ. of California Press, Berkeley.
Tabata, M., Yamamoto, H., and Hiraoka, N. (1968). Chromosome constitution and nicotine formation of mature plants derived from cultured pith of tobacco. *Jpn. J. Genet.* **43,** 319–322.

Torrey, J. G. (1967). Morphogenesis in relation to chromosomal constitution in long-term plant tissue cultures. *Physiol. Plant.* **20,** 265–275.

Umbeck, P. F., and Gengenbach, B. G. (1983). Reversion of male-sterile T-cytoplasm maize to male fertility in tissue culture. *Crop Sci.* **23,** 584–588.

Vasil, I. K. (1981). Plant cell culture and somatic cell genetics of cereals and grasses. *Plant Mol. Biol. Newsl.* **2**(1), 9–23.

Vasil, V., and Vasil, I. K. (1981a). Somatic embryogenesis and plant regeneration from tissue culture of *Pennisetum americanum,* and *P. americanum* × *P. purpureum* hybrid. *Am. J. Bot.* **68,** 864–872.

Vasil, V., and Vasil, I. K. (1981b). Somatic embryogenesis and plant regeneration from suspension culture of pearl millet (*Pennisetum americanum*). *Ann. Bot. (London)* [N.S.] **47,** 669–678.

Weber, D. F. (1983). Monosomic analysis in diploid crop plants. *In* "Cytogenetics of Crop Plants" (M. S. Swaminathan, P. K. Gupta, and U. Sinha, eds.), pp. 351–378. MacMillan of India Ltd., New Delhi.

Yamaya, T., and Filner, P. (1981). Resistance to acetohydroxamatic acquired by slow adaptive increases in urease in cultured tobacco cells. *Plant Physiol.* **67,** 1133–1140.

Zagorska, N. A., Shamina, Z. B., and Butenko, R. G. (1974). The relationship of morphogenetic potency of tobacco tissue culture and its cytogenetic features. *Biol. Plant.* **16,** 262–274.

CHAPTER **79**

Preparation of Cultured Cells and Tissues for Transmission Electron Microscopy

Larry C. Fowke

Department of Biology
University of Saskatchewan
Saskatoon, Saskatchewan, Canada

I. INTRODUCTION

The transmission electron microscope (TEM) is a basic tool for studies of cell structure. Coupled with thin sectioning techniques, it has been used routinely to examine the fine structure of cultured plant material. In particular, it has been used extensively to study the structure of cultured plant cells and protoplasts (reviewed by Fowke, 1978; Fowke and Gamborg, 1980). This chapter provides detailed methods for the preparation of cultured plant material for examination in the TEM. The techniques described can be applied to samples from whole plants or organs, as well as suspension-cultured plant cells. In addition to presenting the basic electron microscopic thin sectioning techniques, this chapter describes methods for the routine preparation of high-quality plastic sections for examination of cell structure with the light microscope.

The development of adequate techniques for preparing plant material for

CELL CULTURE AND SOMATIC CELL
GENETICS OF PLANTS, VOL. 1

ISBN 0-12-715001-3

the TEM occurred primarily in the 1960s. Vastly improved methods of fixation resulted from the introduction of the fixative glutaraldehyde in 1963 (Sabatini *et al.*, 1963). The standard fixation methods used today involve double fixation with glutaraldehyde and osmium tetroxide. Highly vacuolated plant cells are particularly fragile and difficult to preserve. The most successful methods for handling such cells were introduced by Feder and O'Brien (1968). They achieved excellent results by controlling the speed and temperature of the dehydration and infiltration steps. Their methods, with modifications, have been successfully applied to a wide variety of plant material, including cultured cells and protoplasts, and form the basis for the methodology outlined in this chapter (see O'Brien and McCully, 1981, for further details). Epoxy resins such as Epon or Araldite have been used extensively for embedding plant specimens. However, the introduction of a low-viscosity resin by Spurr (1969) has facilitated the infiltration and embedding of many specimens.

The TEM can also be used to examine negatively stained preparations. Negative staining is a rapid, simple, and very effective method for examining small specimens. It has been used successfully for investigations of cellulose microfibril deposition on protoplast surfaces (Herth and Meyer, 1977), organelles on the inner surface of the plant plasma membrane (Doohan and Palevitz, 1980; Van der Valk and Fowke, 1981), and isolated plant cell organelles (Mersey *et al.*, 1982). This chapter includes a basic method for negative staining which can be applied to a variety of plant specimens.

II. PROCEDURES

A. Thin Sectioning Methods for Plant Cells and Tissues

The following methods have proven satisfactory for routine preparation of plant specimens for examination in the TEM. Unless otherwise stated, all materials and reagents can be obtained from electron microscope suppliers (see Section IV).

1. Fixation

1. Place tissue in vials containing 1.5% glutaraldehyde in 0.025 *M* sodium phosphate buffer (pH 6.8) for 30 min at room temperature.

Buffer	
0.1 *M* Na_2HPO_4	1 part
0.1 *M* NaH_2PO_4	1 part
Distilled H_2O	6 parts
Fixative	
50% glutaraldehyde	1.5 ml
0.025 *M* buffer	48.5 ml

Note: The tissue may have to be aspirated to remove trapped air. The fixation and dehydration of suspension-cultured plant cells can be carried out in centrifuge tubes using gentle centrifugation to facilitate solution changes.

2. Transfer to 3% glutaraldehyde in the same buffer for 2 hr at room temperature.
3. Transfer tissue to 0.025 *M* buffer and place on ice. Give at least three changes of buffer over 1–12 hr at 0°C.
4. Postfix in 1% osmium tetroxide (OsO_4) in the same buffer for 1 hr at room temperature or overnight at 0°C.

1% OsO_4	
Stock 2% OsO_4	1 part
0.05 *M* buffer	1 part

5. Give two rinses in distilled water over 60 min at 0°C.

2. Dehydration

The dehydration techniques used with Araldite and Spurr's resin differ and are therefore dealt with separately.

Dehydration for Araldite

1. Transfer tissue to pure methyl cellosolve by 25% increments at 0°C (ca. 15 min per change) and leave for 3–24 hr with two changes at 0°C.
2. Transfer to absolute ethanol and leave for 3–24 hr at 0°C with two changes. Note: Specimens can be stored in absolute alcohol in a freezer (−25°C) for prolonged periods.
3. Pour off most of the ethanol and add propylene oxide dropwise over ca. 6 hr at 0°C until the propylene oxide:ethanol ratio is 2:1.
4. Transfer to pure propylene oxide and leave overnight at 0°C.
5. Bring to room temperature and give two changes of propylene oxide.

Dehydration for Spurr's Resin

1. Transfer tissue to pure acetone by 10% increments at 0°C (ca. 15 min per change) and leave for 3–24 hr with two changes at 0°C.
2. Bring to room temperature and give two changes of acetone.

3. Infiltration and Embedding

Araldite resin	
Araldite 502	25.0 g
Dodecenyl succinic anhydride (DDSA)	19.0 g
DMP-30 accelerator	0.7 g
Spurr's resin	
ERL-4206	10.0 g
Diglycidyl ether of polypropylene glycol (DER) 736	6.0 g
Nonenyl succinic anhydride (NSA)	26.0 g
S-1 accelerator	0.4 g

Note: i. The components can be weighed directly into a disposable plastic cup on a top-loading pan balance. The mixture should be stirred with a glass rod before use (Araldite requires thorough mixing).
ii. All epoxy resins and catalysts are skin irritants and potential carcinogens; they should be handled with care (see O'Brien and McCully, 1981, p. 6.67, for specific precautions).

1. Pour off most of the solvent (propylene oxide or acetone depending upon the embedding medium to be used). Add a 50:50 mixture of solvent and embedding resin dropwise over ca. 6 hr until enough resin is present to ensure that the tissue will be covered by resin when the solvent is removed.
2. Cover vials with aluminum foil containing pinholes and evaporate the solvent in a fume hood overnight.
3. Once all of the solvent has evaporated, transfer specimens to fresh resin and allow them to soak for 1–2 days.
4. Transfer specimens to fresh resin in embedding containers (e.g., plastic molds, aluminum weighing dishes, gelatin capsules, Beem capsules, silicone rubber molds).
5. Polymerize at appropriate temperature (Araldite, 60°C for 24–48 hr; Spurr's resin, 70°C for 8–20 hr).

Note: The following procedure for flat embedding facilitates the selection of cells for subsequent ultrastructural study. Spray an even coat of release agent MS 122 (Miller-Stephenson Chemical Co., Ltd., Toronto, Ontario, Canada) on the inner surface of a clean glass Petri dish, dry, and wipe thoroughly with soft tissue. Spread cells in a thin layer of fresh resin (ca. 2 mm deep) and polymerize in the normal manner. The glass Petri dish can then be broken away, leaving a thin, flat sheet of plastic with excellent optical qualities. Cells can be selected using a light microscope, removed from the block with a fine coping saw, and mounted for ultramicrotomy.

4. Sectioning and Staining

Sections cut for electron microscopy should be mounted on grids and stained with uranyl acetate and lead citrate by conventional techniques (see Hayat, 1970; O'Brien and McCully, 1981, p. 6.102).

For light microscopy, thicker sections (0.5–1.0 μm) can be cut from the same material and stained as follows.

1. Transfer sections in a drop of water with a small platinum loop to distilled water on a clean glass slide.
2. Heat gently over an alcohol lamp to spread and dry the sections.
3. Place a drop of toluidine blue (1% w/v in 1% borax solution) on the sections, heat gently, rinse, and dry.
4. Mount in immersion oil under a coverslip. The coverslip can be ringed with nail polish to make a permanent slide; however, the stain may fade with time.

B. Negative Staining

Negative staining involves the staining of the background support film rather than the specimen itself. It is a particularly useful technique for studying tiny specimens with the TEM. The method described below has been used successfully in our laboratory to examine cell organelles. A number of other negative staining techniques are available (Haschemeyer and Meyers, 1972).

1. Film Preparation

Coated grids should be prepared in the normal manner. Plastic films such as Formvar or collodion are more hydrophilic but less stable than carbon films. The spreading of stain on carbon films can be facilitated by adding a wetting agent to the stain (e.g., 0.01–1.0% serum albumin). Alternatively, hydrophilic carbon films can be prepared by removing surface hydrocarbon molecules by ion bombardment. The Edwards vacuum coating unit, for example, is equipped with a "Plasmaglo" attachment for such purposes.

2. Fixation

1. Fix specimens in buffered 1.5–3.0% glutaraldehyde as described in Section II,A,1 above.
2. Give three changes of distilled water.

3. Staining

1. Apply a small drop of suspension to a coated grid and leave for 1–2 min. It may be necessary to prepare 10-fold dilutions of the suspension to determine the best specimen density.
2. Drain the drop using a torn edge of Whatman no. 1 filter paper to remove excess liquid, and immediately apply a drop of 1% aqueous phosphotungstic acid (pH 7.0) to the grid. Do not allow the grid surface to dry before application of the stain.
3. After 30 sec, remove the stain using torn filter paper, allow the grid to dry, and examine in the TEM.

III. RESULTS AND CONCLUSIONS

Excellent preservation of plant cells has been achieved using the preparative methods described in this chapter (Figs. 1, 2). These methods are designed to be used with highly vacuolated, fragile plant cells. They involve the very gradual dehydration and infiltration of the tissue prior to embedding. It is important to note that the procedures described can be considerably shortened and streamlined when working with tissues composed primarily of nonvacuolated cells (e.g., meristematic cells). In fact, when initiating a structural study, it is advisable to determine which steps can be shortened without affecting the quality of specimen preservation. When using Araldite, it may be possible to eliminate the use of methyl cellosolve and dehydrate directly with ethanol. The transition from ethanol to propylene oxide may also be shortened by stepwise rather than dropwise addition of propylene oxide. Similarly, the gradual dropwise infiltration with resin may not be necessary with all tissues.

The preparation of adjacent thick and thin sections permits the examination of specimens with both light and electron microscopes. Considerable structural detail can be derived from light microscopic examination of thick plastic sections stained with toluidine blue. In addition, the thick sections can be used to monitor the progress of thin sectioning for electron microscopy.

The basic methods presented in this chapter have also been applied to studies of plant protoplasts (Fowke, 1982). The critical step for successful preservation of protoplasts seems to be the initial fixation in glutaraldehyde. The best fixation was achieved using low concentrations of glutaraldehyde (1.0–1.5%) prepared either in the culture medium or in a mannitol or sorbitol solution of the same osmolality. Very gradual dehydration

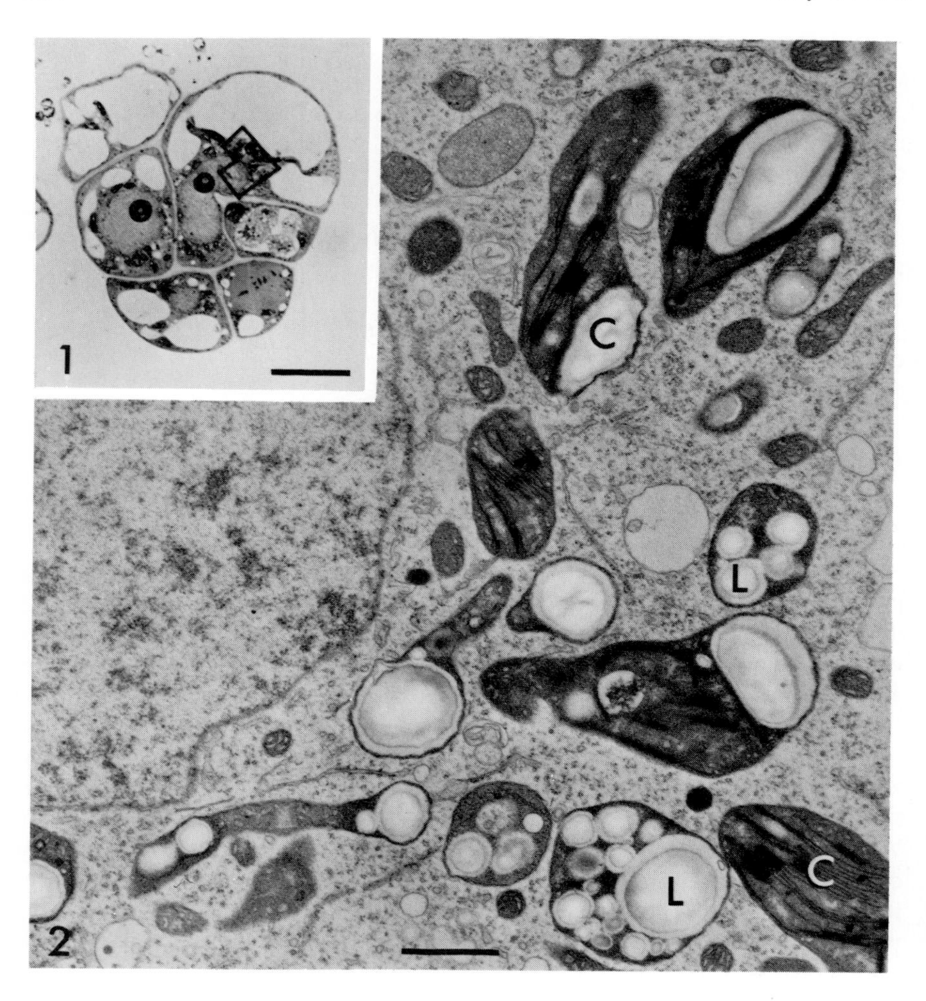

Fig. 1. Light micrograph showing a 4.5-day-old clump of hybrid cells derived from the fusion of protoplasts from suspension-cultured soybean cells and tobacco leaf cells. An electron micrograph of an area adjacent to that outlined in black is shown in Fig. 2. Bar = 20 μm.

Fig. 2. Electron micrograph of section adjacent to the area outlined in black in Fig. 1. The hybrid cytoplasm contains both soybean leukoplasts (L) and tobacco chloroplasts (C). Bar = 1 μm.

through ethanol to propylene oxide and slow infiltration with Araldite seemed essential for minimizing the collapse of protoplasts.

The negative staining method described above has been used routinely to stain coated vesicles isolated from cultured plant cells (Mersey *et al.*, 1982; Fig. 3). Because it is a relatively simple technique, it can be used to monitor the presence of coated vesicles in fractions during the actual isola-

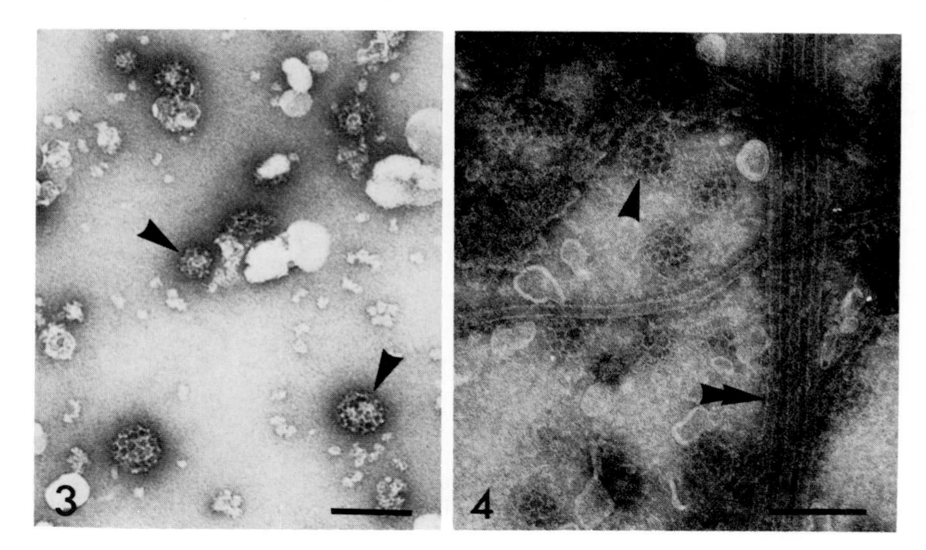

Fig. 3. Electron micrograph of negatively stained coated vesicles (arrows) isolated from suspension-cultured tobacco cells (see Mersey *et al.*, 1982, for details). Bar = 200 nm.
Fig. 4. Electron micrograph of a negatively stained protoplast plasma membrane fragment. Microtubules (double arrow) and coated vesicles (single arrow) are evident on the inner surface of the plasma membrane (see Van der Valk and Fowke, 1981, for details). Bar = 200 nm.

tion procedure. The resulting stained preparations provide valuable information concerning the structure of plant coated vesicles. The negative staining technique has also been used to study the structure and distribution of cell organelles associated with the inner surface of the plasma membrane of plant protoplasts (Doohan and Palevitz, 1980; Van der Valk and Fowke, 1981; Fig. 4).

APPENDIX

Electron Microscope Supply Companies

Ernest F. Fulham, Inc.
P. O. Box 444
Schenectady, New York 12301

Ladd Research Industries, Inc.
P. O. Box 1005
Burlington, Vermont 05402

J. B. EM Services, Inc.
P. O. Box 693
Pointe-Claire
Dorval, Quebec, Canada H9R 4S8

(continued)

APPENDIX (*Continued*)

Pelco International
P. O. Box 510
Tustin, California 92680

Polysciences, Inc.
Paul Valley Industrial Park
Warrington, Pennsylvania 18976

ACKNOWLEDGMENTS

The micrographs in this chapter have resulted from collaborative research involving Drs. Y. C. Chien, F. Constabel, B. G. Mersey, P. J. Rennie, and P. Van der Valk. Continued support from the Natural Sciences and Engineering Research Council of Canada is gratefully acknowledged.

REFERENCES

Doohan, M. E., and Palevitz, B. A. (1980). Microtubules and coated vesicles in guard-cell protoplasts of *Allium cepa* L. *Planta* **149,** 389–401.

Feder, N., and O'Brien, T. P. (1968). Plant microtechnique: Some principles and new methods. *Am. J. Bot.* **55,** 123–142.

Fowke, L. C. (1978). Ultrastructure of isolated and cultured protoplasts. *In* "Frontiers of Plant Tissue Culture 1978" (T. A. Thorpe, ed.), pp. 223–233. Univ. of Calgary Press, Calgary, Alberta, Canada.

Fowke, L. C. (1982). Electron microscopy of protoplasts. *In* "Plant Tissue Culture Methods" (L. R. Wetter and F. Constabel, eds.), 2nd rev. ed., NRCC No. 19876 pp. 72–82. National Research Council of Canada, Ottawa.

Fowke, L. C., and Gamborg, O. L. (1980). Applications of protoplasts to the study of plant cells. *Int. Rev. Cytol.* **68,** 9–51.

Haschemeyer, R. H., and Meyers, R. J. (1972). Negative staining. *In* "Principles and Techniques of Electron Microscopy" (M. A. Hayat, ed.), Vol. 2, pp. 101–147. Van Nostrand-Reinhold, Princeton, New Jersey.

Hayat, M. A., ed. (1970). "Principles and Techniques of Electron Microscopy," Vol. 1. Van Nostrand-Reinhold, Princeton, New Jersey.

Herth, W., and Meyer, Y. (1977). Ultrastructural and chemical analysis of the wall fibrils synthesized by tobacco mesophyll protoplasts. *Biol. Cell.* **30,** 33–40.

Mersey, B. G., Fowke, L. C., Constabel, F., and Newcomb, E. H. (1982). Preparation of a coated vesicle-enriched fraction from plant cells. *Exp. Cell Res.* **141,** 459–463.

O'Brien, T. P., and McCully, M. E. (1981). "The Study of Plant Structure, Principles and Selected Methods." Termarcarphi Pty. Ltd., Melbourne, Australia.

Sabatini, D. D., Bensch, K., and Barrnett, R. J. (1963). Cytochemistry and electron microscopy. The preservation of cellular ultrastructure and enzymatic activity by aldehyde fixation. *J. Cell Biol.* **17,** 19–58.

Spurr, A. R. (1969). A low-viscosity epoxy resin embedding medium for electron microscopy. *J. Ultrastruct. Res.* **26,** 31–43.

Van der Valk, P., and Fowke, L. C. (1981). Ultrastructural aspects of coated vesicles in tobacco protoplasts. *Can. J. Bot.* **59,** 1307–1313.

CHAPTER **80**

Preparation of Cultured Tissues for Scanning Electron Microscopy

Vimla Vasil
Indra K. Vasil

Department of Botany
University of Florida
Gainesville, Florida

I. INTRODUCTION

The use of the scanning electron microscope (SEM) for examining the external manifestations of various developmental processes in plants, both *in vivo* and *in vitro*, is becoming increasingly common and useful. The SEM complements both the optical microscope and the transmission electron microscope (TEM), and can be compared to the stereoscopic binocular microscope. However, it is vastly superior to the latter in resolution, in the range of magnifications at which it can be operated, and, most importantly, in its great depth of field. Its ability to "look" directly at the external features of a specimen with great depth of field at high magnification and high resolution has made the SEM a valuable tool in our studies of morphogenesis in tissue cultures of cereals and grasses.

The basic principle behind the SEM, as with the TEM, is an electron-optical column in which an electron beam, generated in the electron gun under high vacuum, is focused at the surface of the specimen. The beam scans the surface, and when it detects electrons reflected off the specimen

CELL CULTURE AND SOMATIC CELL
GENETICS OF PLANTS, VOL. 1

ISBN 0-12-715001-3

surface, an image is created on the screen of a cathode ray tube. The operation of the specimen chamber under high vacuum necessitates preparation of the specimen in such a manner that it is completely dry and can withstand vacuum without structural damage or alteration. Improvements in the operation and resolution of the SEM, which became commercially available only in the early 1960s, now make it possible to obtain high-quality micrographs from a variety of materials, provided the specimen has been prepared properly and adequately.

The preparation of the specimen depends on its nature. Although a range of techniques are currently available, no single one can be applied equally to all specimens. The embryogenic tissue cultures of the Gramineae described in our laboratory are rather compact and opaque. When samples are adequately fixed and dehydrated, critical-point drying (CPD) followed by coating with gold has consistently given good results with all of the species investigated.

II. SELECTION, FIXATION, AND DEHYDRATION

Selection of a suitable specimen is the first important requirement. Cultures must first be examined carefully under a dissecting microscope, and the area of interest should be excised and removed. Unnecessary portions of the specimen should be trimmed away since only small areas of the specimen can be examined at high magnification.

The specimen is then fixed in order to preserve the structure and arrangement of cells without shrinkage, collapse, or swelling during subsequent treatments. The overall quality of fixation is controlled not only by the type of fixative used but also by the temperature, pH, and osmolarity of fixative solutions. Rapid penetration and distribution of the fixative to all parts of the specimen is desirable. The fixative should be isotonic in order to avoid osmotic shock to the surface layers of the specimen, since these are the object of study. Accordingly, the osmolarity of the fixative as well as the buffer should be controlled.

The following schedule of fixation and dehydration was found to be suitable for embryogenic tissue cultures of the Gramineae:

1. Fix the specimens in glass vials in 2.0–2.5% (v/v) glutaraldehyde in 0.1 *M* sodium–cacodylate buffer at pH 7.2–7.5 for 1–2 hr at room temperature.
2. Wash with three changes of buffer for 30 min.
3. Fix in 1% (w/v) osmium tetroxide in buffer for 1–3 hr at room temperature or overnight in a refrigerator.
4. Wash in buffer as in (2).

The specimen is then dehydrated in an organic solvent, such as ethanol or acetone, to remove all traces of water. Removal of water is critical since its presence can not only contaminate the SEM column but also cause serious damage to the specimen due to surface tension forces of water during evaporation. Dehydration must be carried out gradually by immersion for 1 hr each in 25, 50, 75, and 100% ethanol. Two additional changes of 100% ethanol are made at the end. Extremely delicate tissues may require finer increments in the dehydration series to prevent tissue collapse.

III. CRITICAL-POINT DRYING

Following fixation and dehydration, the samples are dried. The most common procedure in practice is CPD, and it gave excellent results. This method makes use of the property of liquids to change to a gaseous phase with no abrupt change in properties. At a certain temperature and pressure, together known as the *critical point*, the phase boundary between the liquid and gaseous states disappears. Thus, there are no surface tension forces when the liquid volatilizes.

Balzers' CPD apparatus, model CPD 010, was used. In it liquid CO_2 was used as a transitional fluid, since it has a more acceptable critical temperature (T_c 31°C) and pressure (P_c 1072 psc).

The dehydrated specimens are carefully transferred to small containers, such as wire baskets or other porous containers, and placed in the pressure-sealed specimen chamber or "bomb" of the CPD unit. Fresh ethanol (100%) is added to cover the surface of the specimen. With all the valves shut off, the chamber is securely closed and precooled to well below room temperature (5–10°C). Liquid CO_2 is flushed into the specimen chamber slowly from a cylinder under pressure until the chamber is filled. The magnetic stirrer switch is activated to provide good mixing of the ethanol and liquid CO_2. After 5 min of mixing, a valve is opened to drain the chamber slowly, until the specimen is barely covered by the liquid, and closed once again. This sequence of flushing liquid CO_2 into the specimen chamber, followed by soaking of specimens in liquid CO_2 for 5 min, and finally draining of the liquid CO_2 is repeated three to seven times depending upon the volume and number of specimens in the drying basket. After all of the ethanol has been completely exchanged with liquid CO_2, the chamber is filled to barely cover the specimens, making sure that a liquid–gas interphase is present. Then the samples are brought above the CO_2 critical point. With all the valves closed on the CPD as well as on the

CO_2 tank, the chamber is heated by adjusting the thermostat to 40°C and activating the heating switch. The actual drying of the specimen begins when the temperature and pressure in the chamber rise. Near the critical point, turbulence inside the chamber becomes optically apparent. At the critical point the gas–liquid interphase disappears and only interference streaks are visible. After a temperature of 40°C and a pressure of 1100–1200 psi have been reached, the chamber can be vented slowly through a valve (to avoid adiabatic cooling, the rate should not exceed 100 psi/min). The temperature should be maintained above the critical point to avoid cooling and condensation of the CO_2 gas until the pressure gauge reaches 1 atm. The specimen chamber is then opened and the basket removed.

If coating does not follow immediately, the dried specimens should be stored in a desiccator to avoid reabsorption of moisture from the air.

IV. MOUNTING AND SPUTTER COATING

Aluminum as well as brass stubs can be used for mounting of the specimens. It is possible to mount several small specimens on a single stub. The basket containing the dried specimens is opened over a piece of filter paper, and the specimens are carefully positioned on the stub to expose areas of interest, with the aid of a dissecting microscope. The specimen must be in close, tight contact with the stub to avoid electron charge buildup, distortion of the image, or detachment and loss. The specimen is attached to the stub by an adhesive material which is also electrically conductive and thus is able to convey electrons from the specimen to the stub. We have used liquid carbon black for larger specimens and double-stick tape for smaller ones. When liquid adhesive is used, it should be allowed to dry out a little before the specimen is placed on it in order to avoid its absorption or spreading over the specimen. The double-stick tape should be attached to the stub firmly and evenly so that no air bubbles are trapped.

The final step before the stub can be loaded in the SEM is the coating of the specimen with a thin layer of a conductive material such as gold or gold–palladium. The coating material conducts charge and heat from the specimen to the stub to avoid charging and beam damage artifacts. Direct current coating was done with a thin plate of gold in a Eiko Engineering IB_2 Sputter Coater. The principle behind the various sputter coaters is more or less the same. The apparatus consists of a glass vacuum sputtering chamber supplied with an argon inlet port and a mechanical pump. The optimum vacuum for coating is stabilized against the mechanical pump by the continued, slow inlet of gas, usually argon, into the chamber.

The specimen stub is placed on the stage in the sputtering chamber. A relatively low vacuum is established in the specimen chamber by the mechanical pump. With the low vacuum, the only gas molecules remaining in the chamber are air or large, inert molecules such as argon or nitrogen. When such molecules are ionized and excited to a rapidly moving state by high-voltage current for short durations (3–6 min), they bombard the gold (the target material) mounted above the stage in the sputtering chamber, dislodging particles of it, which are then deposited on the specimen surface, forming a thin coating. To minimize specimen heat damage, we used high-voltage current twice for 3 min each, rather than only once for 6 min. The stubs with the specimens are removed after the vacuum pump has been shut off, and are ready to be examined and photographed in the SEM.

V. CONCLUSIONS

This method for the preparation of specimens for scanning electron microscopy was used to study somatic embryogenesis in a variety of materials and species of the Gramineae (Haydu and Vasil, 1981; Lu and Vasil, 1981; Vasil and Vasil, 1982a,b; Wang and Vasil, 1982; Botti and Vasil, 1984; Ho and Vasil, 1983; Lu *et al.*, 1982, 1983). The examination and photography of specimens with the high resolution, greater depth of field, and high magnification of this technique provided a new dimension and insight in the understanding and interpretation of the development of somatic embryos and plants *in vitro.*

REFERENCES

Botti, C., and Vasil, I. K. (1984). The ontogeny of somatic embryos of *Pennisetum americanum* (L.) K. Schum. II. In cultured immature inflorescences. *Can. J. Bot.* (in press).

Haydu, Z., and Vasil, I. K. (1981). Somatic embryogenesis and plant regeneration from leaf tissues and anthers of *Pennisetum purpureum* Schum. *Theor. Appl. Genet.* **59,** 269–273.

Ho, W., and Vasil, I. K. (1983). Somatic embryogenesis in sugarcane (*Saccharum officinarum* L.). I. The morphology and physiology of callus formation and ontogeny of somatic embryos. *Protoplasma* **118,** 169–180.

Lu, C., and Vasil, I. K. (1981). Somatic embryogenesis and plant regeneration from leaf tissues of *Panicum maximum* Jacq. *Theor. Appl. Genet.* **59,** 275–280.

Lu, C., Vasil, I. K., and Ozias-Akins, P. (1982). Somatic embryogenesis in *Zea mays* L. *Theor. Appl. Genet.* **62,** 109–112.

Lu, C., Vasil, V., and Vasil, I. K. (1983). Improved efficiency of somatic embryogenesis and plant regeneration in tissue cultures of maize (*Zea mays* L.). *Theor. Appl. Genet.* **62,** 285–290.

Vasil, V., and Vasil, I. K. (1982a). Characterization of an embryogenic cell suspension culture derived from inflorescences of *Pennisetum americanum* (pearl millet, Gramineae). *Am. J. Bot.* **69,** 1441–1449.

Vasil, V., and Vasil, I. K. (1982b). The ontogeny of somatic embryos of *Pennisetum americanum* (L.) K. Schum.: In cultured immature embryos. *Bot. Gaz. (Chicago)* **143,** 454–465.

Wang, D., and Vasil, I. K. (1982). Somatic embryogenesis and plant regeneration from inflorescence segments of *Pennisetum purpureum* Schum. (Napier or Elephant grass). *Plant Sci. Lett.* **25,** 147–154.

CHAPTER **81**

Microspectrophotometric Analysis*

Jerome P. Miksche
Sukhraj S. Dhillon

Department of Botany
North Carolina State University
Raleigh, North Carolina

I. INTRODUCTION

Microspectrophotometry is an optical method for measuring chemicals or macromolecules in tissues, single cells, or parts of cells in a way that is irreplaceable by any other approach. The original interest in microspectrophotometry was chiefly oriented toward DNA determinations. Initial techniques utilized ultraviolet light with prohibitively expensive reflecting or quartz microscopic optics. Ultimately, visible wavelength methods with specific stains and glass optics were developed which popularized microspectrophotometric instrumentation.

The Feulgen method was the first quantitative staining procedure and evolved from Feulgen and Rossenbeck's (1924) discovery of a positive

*Publication No. 9040 of the Journal Series of the North Carolina Agricultural Research Service, Raleigh, North Carolina. Research reported in this paper was supported in part by National Science Foundation BMS 74-21120 and by the USDA/SEA research contract number 12-14-7001-1154.

ISBN 0-12-715001-3

Schiff test for aldehydes after acid hydrolysis of thymus nucleic acid. The technique has been under almost continuous refinement since its discovery and now ranks as the most quantitative cytochemical method. Feulgen microspectrophotometry assumes that the Feulgen stain reaction is proportional to the amount of DNA. The correlation between Feulgen stain absorbance and amount of nuclear DNA has been justified in numerous tests (Miksche, 1967; Banerjee and Sharma, 1979; Dhillon *et al.*, 1980). However, alterations in the stoichiometry of the Feulgen reaction have been reported in response to different states of chromatin compaction or condensation (Mayall and Mendelsohn, 1967; Garcia, 1970; Noeske, 1971).

The Feulgen–DNA complex can be measured by a single-beam microspectrophotometer or by a modern scanning unit. The two-wavelength method (Ornstein, 1952; Patau, 1952) and the flying spot scanning system (Berlyn and Miksche, 1976; Dhillon *et al.*, 1983) are the most commonly used techniques, which reduce the distributional error (see Dhillon *et al.*, 1983, for the principles and technical aspects of microspectrophotometry and errors involved in performing the techniques).

This chapter provides a workable laboratory technique for the use of microspectrophotometry.

II. PROCEDURES

The methods of quantitatively determining a cellular substance or cellular stain under the microscope are now well established. The Feulgen method represents a good example to explain the technique.

A. Cell and Tissue Preparation

The procedure begins with fixation of tissue. The commonly used fixative is Carnoy's no. 2 (6:3:1 ethanol–chloroform–acetic acid), and a 2-hr fixing time is used for most of the plant material. The fixative should be prepared just before use to avoid formation of ketones which may interfere with fixation. Another fixative used quite often for cytological studies is Farmer's fluid (3:1 ethanol–acetic acid). For the study of a particular cell component, the type of fixative as well as the fixation time, however, may vary (Berlyn and Miksche, 1976). For example, fixation in 10% formalin for 3–6 hr is appropriate for the study of basic histone proteins. The fixed material is generally transferred to 70% ethanol and stored in a refrigerator until further processed. The material is subsequently squashed after hydration to water through 70, 50, and 30% ethanol. The breaking down of material into isolated cells is accomplished by tapping with a glass rod or applying pressure on a coverslip (Conger and Fairchild, 1953). If the cells are difficult to separate, use of pectinase is suggested, according to Berlyn *et al.* (1979). For pectinase treatment, the material stored in 70% ethanol is hydrated to water and then incubated in 1% pectinase in 0.01 *M* sodium citrate (pH 5.0) overnight at 30°C. The material is refrigerated at 5°C to inhibit enzyme action and processed immediately. This method reduces the interference of variable DNase activity associated with the purity of the pectinase enzyme (Berlyn *et al.*, 1979). Normally, the macerated cells readily adhere to the slide after drying. However, if cell loss occurs during staining, a nonformalin adhesive of the gelatin type is recommended. The adhesive consists of 2.5 g gelatin dissolved in 500 ml distilled water at 30–35°C, to which is added 0.25 g chromium potassium sulfate (chrome alum). Ethanol-precleaned glass slides are immersed in the adhesive and allowed to dry in a vertical position, leaving a thin, uniform film of adhesive on the slides. The tissue can be squashed in a drop of water on the slide, or, if microtome ribbon sections are used, these can be flattened by floating on a drop of water and dried before further processing. Unlike formalin-containing adhesives such as Haupt's, this adhesive does not interfere with Feulgen staining (Greenwood and Berlyn, 1968). For microtomy, different samples of tissues within the experiment must be processed through embedment and staining in the same manner so that meaningful comparisons can be made.

B. Stain Preparation

The DNA stain called Schiff's reagent is prepared according to the procedure of Berlyn and Miksche (1976). Two grams basic fuchsin and 3.8 g sodium or potassium metabisulfite are added to 200 ml 0.15 *N* HCl and shaken in the dark for 2 hr on a mechanical shaker or stirred with a magnetic stirrer. One gram of fresh activated charcoal is added to the stain and then shaken continuously for 2 min. The reagent is suction filtered through Whatman no. 1 filter paper, and if the filtrate is not completely clear, the charcoal treatment is repeated. The reagent can be stored in an airtight container with minimum air space for several months in the refrigerator. This is in contrast to the 1 or 2 weeks' storage recommended in the literature. The storing conditions and preparation procedures perhaps are the reason for longer storage. When a white precipitate forms or if the reagent turns pink, it is not adequate for quantitative staining purposes and should be discarded.

C. Staining Procedure

Prior to staining, the tissue is first hydrolyzed in 5 *N* HCl at 25°C for a period of time that will produce a stain with the deepest intensity. Depending upon the species and type of fixation, the hydrolysis time is usually 15–45 min. The HCl treatment selectively removes the purines from deoxyribose, freeing functional aldehyde groups. In acid solution (0.15 *N* HCl) with excess SO_2 (from sodium metabisulfite), basic fuchsin is converted to the Schiff reagent. The hydrolyzed tissue (apurinic DNA) is then incubated with the Schiff reagent for 2 hr, whereupon two aldehyde groups condense with each molecule of Schiff's reagent, and the dye complex is converted to a stable, magenta-colored form (Wieland and Scheuing, 1921).

The detailed procedures of Feulgen staining are essentially modified from Leuchtenberger (1958). These should be followed closely to avoid the staining variation problems discussed later. The procedure is as follows:

1. Dried slides with squashed tissue are carried to distilled water through 95, 70, 50, and 30% ethanol series with retention of about 2 min in each series. If paraffin sections are used, two changes of xylene for about 5 min each are necessary to dissolve the paraffin before passing through an ethanol series to water.
2. Hydrolyze tissue in 5 *N* HCl at 25°C for the required time. (If not critical, 30 min on the average works for most tissues.)
3. Wash HCl with one quick change of distilled water and stop the HCl reaction by keeping slides in ice cold (5°C) distilled water for 2 min.
4. Place slides in Schiff's reagent for 2 hr in the dark at room temperature.
5. Rinse in distilled water three times for 2 min each.
6. Change in fresh bleaching solution (1:1:18 1 *N* HCl–10% sodium or potassium metabisulfite-distilled water) three times for 10 min each.
7. Wash in running tap water for 5 min.
8. Rinse in distilled water and dehydrate through 30, 50, 70, and 95% ethanol for 2 min each and three changes of absolute ethanol for 5 min each.
9. Air-dry and store slides in a slide box. Mount them in refractive index oil (1.556) using a coverslip of no. 1 thickness before reading on a microspectrophotometer. Alternatively, the tissue can be mounted in synthetic mounting media. However, unclean mounting media can interfere with light absorption, and some mounting media that contain water may cause stain fading, as discussed below.

Detectable fading of stains within 2 days has been observed in studies such as those by Dewse and Potter (1975) and Teoh and Rees (1976). In contrast to these studies, we did not detect fading of stains in nuclei followed for up to 110 days. The slides were prepared according to the procedures outlined above, except that at the end of the dehydration series, three changes for 5 min each of xylene were necessary to mount in Preservaslide. The studies on different mounting media using chicken erythrocytes indicated that improper dehydration of tissue may be a possible cause of stain fading (S. S. Dhillon and G. P. Berlyn, unpublished). Out of six mounting media (Preservaslide, Permount, Canada balsam, Euparal, Diaphane, and Uvak), only cells mounted in Uvak displayed marked fading in the presence of light. In Uvak which contains water, the mounted cells were not dehydrated, and therefore, the Feulgen-stained erythrocyte nuclei lost more than 70% of the stain in 6 days. In other mounting media with complete dehydration, stain fading was not observed when followed for up to 14 days.

D. Internal Standard

An internal standard is used to reduce slide-to-slide variation and to calculate the amount of DNA in picograms. Nuclei from mature chicken erythrocytes serve as an accurate reference standard because their DNA is nonreplicated (2C) (Dhillon *et al.*, 1977). Further data on the advantages of using chicken erythrocytes as an internal standard, and the risks involved in using plant nuclei, are presented and discussed by Dhillon *et al.* (1983) and Miksche *et al.* (1979). For the preparation of chicken erythrocyte slides, heparin is mixed with the freshly drawn blood (0.6 mg/dl) to prevent coagulation. The blood is smeared only on the right one-third of the slide and air dried, leaving the rest of the slide clean for experimental material. Each slide now carries the chicken blood internal standard and the experimental material side by side. Thus, the experimental material and the erythrocytes are stained together, and slide-to-slide variation is reduced. All the slides with blood smears are fixed for 1 hr in Carnoy's no. 2 fixative and are stored in the refrigerator until used. The absolute amount of DNA in each chicken erythrocyte nucleus is determined by interference microscopy (Berlyn and Cecich, 1976; Berlyn and Miksche, 1976; Dhillon *et al.*, 1980). The absolute amount of DNA per experimental nucleus in picograms is calculated as follows:

$$\text{Experimental DNA (pg)} = \frac{\text{erythrocyte DNA (pg)}}{\text{erythrocyte DNA (Feulgen units)}} \times \text{experimental DNA (Feulgen units)}$$

E. Absorption Measurements

For microspectrophotometric measurements, two commonly used methods are the two-wavelength method for microspectrophotometers of the nonscanning type and the flying spot scanning method for modern automatic scanning microspectrophotometers.

For two-wavelength microspectrophotometry, the wavelength settings for the chicken erythrocyte internal standard and the experimental material are determined on homogeneous nuclei using the plug method (Berlyn and Miksche, 1976). First, the wavelength with the maximum absorption (e.g., λb = 560 nm) should be noted and then the wavelength that corresponds to half of the maximal value, either from a spectral curve or by calculation (e.g., λa = 503 nm), should be determined. These two wavelengths can be checked by the zero-slope test (Berlyn and Miksche, 1976). Nuclear transmissions are determined at each of the two selected wavelengths, and tables can be used to avoid lengthy calculations to compute relative Feulgen absorption (Mendelsohn, 1958).

For a scanning unit, the wavelength of maximum absorption is determined. The maximal absorption peak for Feulgen-stained DNA, for example, is between 550 and 570 nm. With a Vickers M86 scanning microdensitometer, the specimen is scanned in a raster fashion by a flying spot consisting of a small beam of light for which the material exhibits maximal absorption. During a scan, over 120,000 spot measurements of light intensity are taken and stored in a digital computer. At the end of a scan, which requires approximately 10 sec, these values are electronically integrated to give a figure proportional to the amount of stained material in the specimen.

III. APPLICATIONS AND CONCLUSIONS

Microspectrophotometry has many applications in cell culture and somatic cell genetics, such as checking the occurrence of polyploidy, aneuploidy, and DNA amplification in cultured plant cells. Polyploidy and aneuploidy that do not occur *in vivo* have been found to occur in cultures. Various combinations of phytohormones and other constituents have been shown to affect the degree of polyploidy (Bayliss, 1977; Ghosh and Gadgil, 1979; Banks-Izen and Polito, 1980). Even in genetically conservative plant groups such as the gymnosperms, there are reports on the occurrence of polyploidy *in vitro* (Huhtinen, 1976; Patel and Berlyn, 1982). Several studies have reported aneuploid cells that are successfully grown in culture and lose morphogenetic potential (Murashige and Nakano, 1967; Torrey, 1977; Skirvin, 1978).

Microspectrophotometry is an effective technique used to measure DNA changes in individual nuclei of undifferentiated and differentiated tissues in culture and to check the genetic stability of cultured plant cells. The method can also be used for identification of fused cells during somatic hybridization. The fused cells should have twice as much DNA per nucleus as that of parental nuclei used for somatic hybridization. Dhillon *et al.* (1983) suggested specific patterns of DNA variation in relation to growth and development. For instance, amplification of DNA in expanding leaves and DNA declination in senescing leaves had been noted in several plant species.

In addition to determining DNA levels in individual nuclei, modern microspectrophotometers such as the Vickers M86 can be used to study changes in the density of chromatin in relation to undifferentiated and differentiated states of cultured cells. Dhillon and Miksche (1981, 1982, 1983) observed an increase in the proportion of denser chromatin (heterochromatin) in dormant and senescing tissue over that observed in active tissue. The heterochromatin, which is often associated with an absence of gene expression (Walbot and Goldberg, 1979), would be expected to be higher in nonfunctioning dormant and senescing nuclei. The technique of studying chromatin density change can be applied to determine if the increased proportion of heterochromatin is associated with tissue that fails to differentiate in the presence of hormonal combinations that are suggested to induce differentiation.

Among other applications, microspectrophotometry can be very useful in determining ploidy levels of specific cells of an organism (Nagl, 1978; Dhillon and Miksche, 1982, 1983); the degree of aneuploidy in a cell population; the genome size of a species (Bennett and Smith, 1976; Dhillon, 1980; Dhillon *et al.*, 1980); studies of the cell cycle by determining cells in G_1, G_2, or the S phase (Vendrely, 1971); the effect of aging on DNA levels (Dhillon and Miksche, 1981); the variation in the diploid genome within (Miksche, 1971) and between (Miksche, 1967) species; and the DNA determination of individual chromosomes.

In addition to DNA determinations, microspectrophotometry has been used for a wide variety of analytical techniques. Some of the commonly used microspectrophotometric methods are determinations of RNA, protein, lignin, cellulose, phytochrome, and other macromolecules on a per-cell basis (Cecich *et al.*, 1972; Mackenzie *et al.*, 1978; Dhillon and Miksche, 1983). The technique has also been extended to the study of enzymes, hormones, and immunochemistry (Bahr, 1979). Berlyn (1969) described the use of microspectrophotometry for determination of relative wood fiber density. Many of the techniques mentioned above can be exploited for use in tissue culture as well as *in vivo* studies to understand the processes of growth and development.

REFERENCES

Bahr, G. F. (1979). Frontiers of quantitative cytochemistry; a review of recent developments and potentials. *Anal. Quant. Cytol.* **1,** 1–19.

Banerjee, M., and Sharma, A. K. (1979). Variations in DNA content. *Experientia* **35,** 42–43.

Banks-Izen, M. S., and Polito, V. S. (1980). Changes in ploidy level in calluses derived from two growth phases of *Hedera helix* L., the English ivy. *Plant Sci. Lett.* **18,** 161–167.

Bayliss, M. W. (1977). The effect of 2,4 D on growth and mitosis in suspension cultures of *Daucus carota*. *Plant Sci. Lett.* **8,** 99–103.

Bennett, M. D., and Smith, J. B. (1976). Nuclear DNA amounts in angiosperms. *Philos. Trans. R. Soc. London, Ser. B* **274,** 227–274.

Berlyn, G. P. (1969). Microspectrophotometric investigations of free space in plant cell walls. *Am. J. Bot.* **56,** 498–506.

Berlyn, G. P., and Cecich, R. A. (1976). Optical techniques for measuring DNA quantity. *In* "Modern Methods in Forest Genetics" (J. P. Miksche, ed.), pp. 1–18. Springer-Verlag, Berlin and New York.

Berlyn, G. P., and Miksche, J. P. (1976). "Botanical Microtechnique and Cytochemistry." Iowa State Univ. Press, Ames.

Berlyn, G. P., Dhillon, S. S., and Miksche, J. P. (1979). Feulgen cytophotometry of pine nuclei. II. Effect of pectinase used in cell separation. *Stain Technol.* **54,** 201–204.

Cecich, R. A., Lersten, N. R., and Miksche, J. P. (1972). A cytophotometric study of nucleic acids and proteins in the shoot apex of white spruce. *Am. J. Bot.* **59,** 442–449.

Conger, A. D., and Fairchild, L. M. (1953). A quick-freeze method for making smear slides permanent. *Stain Technol.* **28,** 281–283.

Dewse, C. D., and Potter, C. G. (1975). Influence of light and mounting medium on the fading of stain. *Stain Technol.* **50,** 301–306.

Dhillon, S. S. (1980). Nuclear volume, chromosome size and DNA content relationships in three species of *Pinus*. *Cytologia* **45,** 555–560.

Dhillon, S. S., and Miksche, J. P. (1981). DNA changes during sequential leaf senescence of tobacco (*Nicotiana tabacum* L.) *Physiol. Plant.* **51,** 291–298.

Dhillon, S. S., and Miksche, J. P. (1982). DNA content and heterochromatin variations in various tissues of peanut (*Arachis hypogaea*). *Am. J. Bot.* **69,** 219–226.

Dhillon, S. S., and Miksche, J. P. (1983). DNA, RNA, protein, and heterochromatin changes during embryo development and germination of soybean (*Glycine max* L.). *Histochem. J.* **15,** 21–37.

Dhillon, S. S., Berlyn, G. P., and Miksche, J. P. (1977). Requirement of an internal standard for microspectrophotometric measurements of DNA. *Am. J. Bot.* **64,** 117–121.

Dhillon, S. S., Rake, A. V., and Miksche, J. P. (1980). Reassociation kinetics and cytophotometric characterization of peanut (*Arachis hypogaea* L.) DNA. *Plant Physiol.* **65,** 1121–1127.

Dhillon, S. S., Miksche, J. P., and Cecich, R. A. (1983). Microspectrophotometric applications in plant science research. *In* "New Frontiers in Food Microstructure" (D. B. Bechtel, ed.), pp. 27–69. Am. Assoc. Cereal Chem., St. Paul, Minnesota.

Feulgen, R., and Rossenbeck, H. (1924). Mikroskopische-chemischer nachweis einer nucleinsäure von typus der thymonucleinsä und die darauf beruhende elektiue färbung von zellkerzen in mikroskopischen präparaten. *Hoppe-Seyler's Z. Physiol. Chem.* **135,** 203–248.

Garcia, A. M. (1970). Stoichiometry of dye binding versus degree of chromatin coiling. *In* "Introduction to Quantitative Cytochemistry-II" (G. L. Wied and G. F. Bahr, eds.), pp. 153–170. Academic Press, New York.

Ghosh, A., and Gadgil, V. N. (1979). Shift in ploidy level of callus tissue. A function of growth substances. *Indian J. Exp. Biol.* **17,** 562–564.

Greenwood, M. S., and Berlyn, G. P. (1968). Feulgen cytophotometry of pine nuclei; effect of fixation, role of formalin. *Stain Technol.* **43,** 111–117.

Huhtinen, O. (1976). *In vitro* culture of haploid tissue of trees. *Proc. IUFRO World Congr., 1976* pp. 28–30.

Leuchtenberger, C. (1958). Quantitative determination of DNA in cells by Feulgen microspectrophotometry. *Gen. Cytochem. Methods* **1,** 219–278.

Mackenzie, J. M., Briggs, W. R., and Pratt, L. H. (1978). Intracellular phytochrome distribution as a function of its molecular form and of its destruction. *Am. J. Bot.* **65,** 671–676.

Mayall, B. H., and Mendelsohn, M. L. (1967). Chromatin and chromosome compaction, and the stoichiometry of DNA staining. *J. Cell Biol.* **35,** 88A.

Mendelsohn, M. L. (1958). The two-wavelength method of microspectrophotometry. II. A set of tables to facilitate the calculations. *J. Biophys. Biochem. Cytol.* **4,** 415–424.

Miksche, J. P. (1967). Variation in DNA content of several gymnosperms. *Can. J. Genet. Cytol.* **9,** 717–722.

Miksche, J. P. (1971). Intraspecific variation of DNA per cell between *Picea sitchensis* (Bong.) Carr. provenances. *Chromosoma* **32,** 343–352.

Miksche, J. P., Dhillon, S. S., Berlyn, G. P., and Landauer, K. J. (1979). Nonspecific light loss and intrinsic DNA variation problems associated with Feulgen DNA cytophotometry. *J. Histochem. Cytochem.* **27,** 1377–1379.

Murashige, T., and Nakano, R. (1967). Chromosomal complement as a determinant of the morphogenetic potential of tobacco cells. *Am. J. Bot.* **54,** 963–970.

Nagl, W. (1978). "Endopolyploidy and Polyteny in Differentiation and Evolution." North-Holland Publ., Amsterdam.

Noeske, K. (1971). Discrepancies between cytophotometric Feulgen values and deoxyribonucleic acid content. *J. Histochem. Cytochem.* **19,** 169.

Ornstein, L. (1952). The distributional error in microspectrophotometry. *Lab. Invest.* **2,** 250–265.

Patau, K. (1952). Absorption microphotometry of irregular-shaped objects. *Chromosoma* **5,** 341–362.

Patel, K. R., and Berlyn, G. P. (1982). Genetic instability of multiple buds of *Pinus coulteri* regenerated from tissue culture. *Can. J. For. Res.* **12,** 93–101.

Skirvin, R. M. (1978). Natural and induced variation in tissue culture. *Euphytica* **27,** 241–266.

Teoh, S. B., and Rees, H. (1976). Nuclear DNA amounts in populations of *Picea* and *Pinus* species. *Heredity* **36,** 123–137.

Torrey, J. G. (1977). Cytodifferentiation in cultured cells and tissues. *HortScience* **12,** 138–139.

Vendrely, C. (1971). Cytophotometry and histochemistry of the cell cycle. *In* "Cell Cycle and Cancer" (R. Baserga, ed.), pp. 227–268. Decker, New York.

Walbot, V., and Goldberg, R. B. (1979). Plant genome organization and its relationship to classical plant genetics. *In* "Nucleic Acids in Plants" (T. C. Hall and J. W. Davies, eds.), pp. 3–40. CRC Press, Boca Raton, Florida.

Wieland, H., and Scheuing, G. (1921). Die fuchsin-schweflige säure und ihre färbreaktion mit aldehyden. *Ber. Dtsch. Chem. Ges.* **54,** 2527–2555.

CHAPTER **82**

Cell Cycle Analysis by Conventional Methods

Alan R. Gould

Plant Genetics Department
Central Research
Pfizer, Inc.
Groton, Connecticut

I. INTRODUCTION

Comparative growth curves based on fresh or dry weight, or on cell number doubling times, are often used in plant tissue culture research in attempts to understand the factors controlling cell division. However, a more sophisticated approach involves the analysis of the cell life cycle and its functional partition into G_1, S, G_2, and mitotic phases. Cell cycle analysis can determine at which points control is exerted, either via exogenous manipulations of the culture environment or by the complex endogenous systems of cellular regulation. The cell cycle approach was first applied to plant cells in culture by Eriksson (1967), who analyzed the mitotic cycle of suspensions of *Haplopappus gracilis.* Since then the accumulation of cell cycle data from plant cell cultures has been very slow, with only about a

CELL CULTURE AND SOMATIC CELL
GENETICS OF PLANTS, VOL. 1

ISBN 0-12-715001-3

dozen different species represented in the literature: *Acer pseudoplatanus* (Gould *et al.*, 1974; Rembur, 1974), *Daucus carota* (Bayliss, 1975), *Glycine max* (Chu and Lark, 1976), *Datura innoxia* (Blaschke *et al.*, 1978), *Crepis capillaris* tumor (Ashmore and Gould, 1979), *Catharanthus roseus* (Courtois and Guern, 1980), *Haplopappus gracilis* (Gould, 1977), *Zinnia elegans* (Fukuda and Komamine, 1981), *Nicotiana tabacum* (Bezdek and Vyskot, 1981), *Nicotiana sylvestris* (Gould, 1984), *Brachycome dichromosomatica* (Gould, 1984), and *Zea mays* (Gould, 1984).

Cultures of these species have been analyzed by methods which give estimates of the duration of the phases of the cell cycle (i.e., G_1, S, G_2, and mitosis). Another type of analysis describes the distribution of cell populations through the cell cycle but does not estimate phase durations. Both duration and distribution analyses will be described in this chapter, and special attention will be paid to problems inherent in interpretation of kinetic data obtained from plant tissue culture systems.

II. AGE DISTRIBUTIONS AND GROWTH FRACTIONS

Before attempting to analyze the cell cycle in a plant tissue culture, it is essential to define the division kinetics of the cell population under study. Generally, the three kinds of proliferating populations that can be dealt with by the less complex methods of cell cycle analysis are steady state, asynchronous, and synchronous. Steady-state populations are most often encountered in chemostats, but only a limited amount of cell cycle work has been published on plant cell populations with steady-state kinetics (Gould *et al.*, 1981a). Steady-state populations are relatively easy to analyze because they have linear age distributions. This means that in a steady-state population, the *average rate of entry of cells* (the cell flux) into each cell cycle compartment (G_1, S, G_2, and mitosis) is the same for each compartment. Thus, if 20% of a steady-state population is in S phase and the total duration of the cell cycle is 25 hr, then the S-phase duration is 5 hr (i.e., 20% of 25 hr). This simple relationship does not hold for exponentially increasing asynchronous cultures. At any point in time in an exponential asynchronous culture, twice as many cells are entering G_1 as are entering mitosis (i.e., for every cell entering mitosis, two cells exit mitosis). It therefore follows that the cell flux into S phase is greater than the cell flux into mitosis, and generally there must be more young cells than old cells in the population. Thus, to calculate the duration of (for example) S phase from the relative frequency of S-phase cells in an exponential asynchronous population, correction factors for the skewed age distribution must be applied. The correction factors for such calculations are displayed in Table

I. Thus, if all cells in a population are cycling, it is possible to calculate the duration of each phase from the proportions of the cell population in each phase (i.e., with the correction factors in Table I, a distribution analysis can generate phase duration data).

Cell populations displaying unusual growth kinetics, such as "partial" synchrony, mitotic oscillation or periodic growth (Everett *et al.*, 1981; Wang *et al.*, 1981; Gould, 1984), or heterogeneous mixtures of slowly and rapidly cycling subpopulations, can rarely be satisfactorily analyzed for cell cycle durations. Truly synchronous cell populations, on the other hand, should by definition have very narrow age distributions because all cells traverse the cycle in concert. Thus, with a culture which is cell cycle synchronized, cell cycle analysis is reduced to a simple time course using tritiated thymidine and autoradiography to chart S phase, and cytological analysis to time and locate mitosis. Unfortunately, many so-called synchronized plant cell cultures are more likely to be suffering from disturbed proliferation kinetics (Gould, 1984), and so this simple approach to cycle analysis is often invalid.

Plant cell populations in which all cells are growing and dividing are said to have a growth fraction (GF) of 1.0, but such populations are not common

TABLE I

Correction Factors for Frequency–Duration Calculations for Exponentially Increasing Asynchronous Cell Populations

Cell cycle phase	Frequency–duration relation
Mitosis (M)	$\frac{N_M}{N_0} = \ln 2 \frac{t_M}{T_C}$
G_1	$\frac{N(G_1 + ½M)}{N_0} = 2(1 - \exp \frac{t_1 \ln 2}{T_C})$
S	$\frac{N_S}{N_0} = [\exp t_2 \frac{\ln 2}{T_C}][\exp(t_S \frac{\ln 2}{T_C}) - 1]$
G_2	$\frac{N(G_2 + ½M)}{N_0} = (\exp t_2 \frac{\ln 2}{T_C}) - 1$

where T_C = duration of the cell cycle
t_1 = duration of G_1 + ½ M
t_2 = duration of G_2 + ½ M
t_S = duration of S
t_M = duration of mitosis
N_0 = number of cells in the population
N_M = number of cells in mitosis
N_S = number of cells in S phase

even in cultures which seem to be increasing their cell number exponentially. In practice, populations with a GF > 0.95 can be treated as though all cells are cycling. Two kinds of analysis yield estimates of GF; one uses tritiated thymidine ([^{3}H]Tdr) in a pulse-chase labeling experiment, and the other entails a continuous labeling approach. Only the derivation of GF estimates will be described here; the complete cell cycle analyses possible with pulse-chase and continuous labeling are discussed in the following sections.

Continuous application of [^{3}H]Tdr to an exponentially dividing culture with a GF of 1.0 will cause all cells in the culture to show nuclear label in autoradiographs after a time equal to the cell population doubling time (T_D) minus the duration of S phase. If a cell population contains a significant proportion of quiescent or noncycling cells, then the plot of the percentage of labeled nuclei against the time of continuous labeling will not reach 100%. This plot will be linear until all cycling cells show some nuclear label ($T_C - t_S$, where T_C is the total cycle time and t_S is the duration of S phase). Thereafter the gradient of the plot should decrease, and the percentage of labeled nuclei will make an asymptotic approach to 100% as the unlabeled quiescent cells are progressively diluted out by the labeled dividing cells. Such dilution demonstrates that GF, in this instance, is increasing. The GF can be estimated by identifying the end of the linear accumulation of labeled nuclei. For example, if 90% of nuclei are labeled at this point, then the GF *at this point* is roughly 0.9. However, at the start of continuous labeling, the GF would have been lower, probably on the order of 0.82 (45 cells doubled to 90 cells over the period in question, while 10 cells were noncycling). Again, this makes the point that the GF is not a constant for a given cell population, and can be quoted only at a specific point in time.

A more satisfactory derivation of GF estimations can be made from comparisons between actual and theoretical values for the proportion of cells in S phase during exponential growth. The actual proportion can be obtained from the labeling index (LI), which is the fraction of the population showing grain in autoradiograph preparations after a culture has been briefly exposed to [^{3}H]Tdr. The theoretical proportion *assumes* that all cells are cycling and derives the duration of S phase from direct measurement (i.e., by FLM, the fraction of labeled mitoses, described later). With the corrections for exponential kinetics (Table I), the theoretical proportion can be calculated from the measured duration. The GF can then be calculated from the following relationship:

$$\text{GF} = \text{actual S proportion (LI)} \div \text{theoretical S proportion}$$

Thus, for a population with an LI of 0.15 and a calculated theoretical S proportion of 0.2, the GF is 0.75. Similar approaches are possible using the mitotic index (MI) (the fraction of the population in mitosis) or the proportion of cells in G_2, rather than the LI. Then

$$GF = MI \div \text{theoretical mitotic fraction}$$

or

$$GF = \text{actual } G_2 \text{ proportion} \div \text{theoretical } G_2 \text{ proportion}$$

In both cases, the theoretical proportions would be calculated from measured durations converted to proportions, assuming that all cells are cycling.

The essential point with populations with low GFs is that the population doubling time T_D will be longer than the total cell cycle time T_C. $T_D = T_C$ only if the GF = 1.0.

III. ANALYSIS METHODS

A. Fraction of Labeled Mitoses

The fraction of labeled mitoses (FLM) analysis was first presented in a fully developed form by Quastler and Sherman (1959). Figure 1 displays idealized and more realistic results of an FLM experiment. A pulse of [^{3}H]Tdr (1×10^{-8} *M*, 1–85 kBq/ml) supplied to a rapidly dividing asynchronous cell population will be incorporated into the nuclear DNA of replicating S-phase cells. Thus, directly after the [^{3}H]Tdr pulse, only interphase cells in S phase will show silver grains over their nuclei in autoradiograph preparations. A chase of unlabeled thymidine (5×10^{-5} *M*, i.e., 5000-fold higher than the pulse) applied directly after 30 min of pulse labeling ensures that insignificant amounts of label will be incorporated after the pulse (Gould *et al.*, 1974). When the S-phase cells labeled during the pulse enter mitosis after traversing G_2, mitotic figures showing grain in autoradiographs will appear. The fraction of mitotic figures which show such labeling rises to a peak value as the cohort of labeled cells passes through S. Subsequently the FLM falls as the labeled cohort passes into G_1; but if the cohort is not severely attenuated as it passes through the next cell cycle, a second wave of labeled mitoses will appear. The double peaked pattern of the FLM curve allows the derivation of durations for all of the cell cycle phases, as illustrated in Fig. 1.

The G_2-phase duration is often the most accurately determined value in

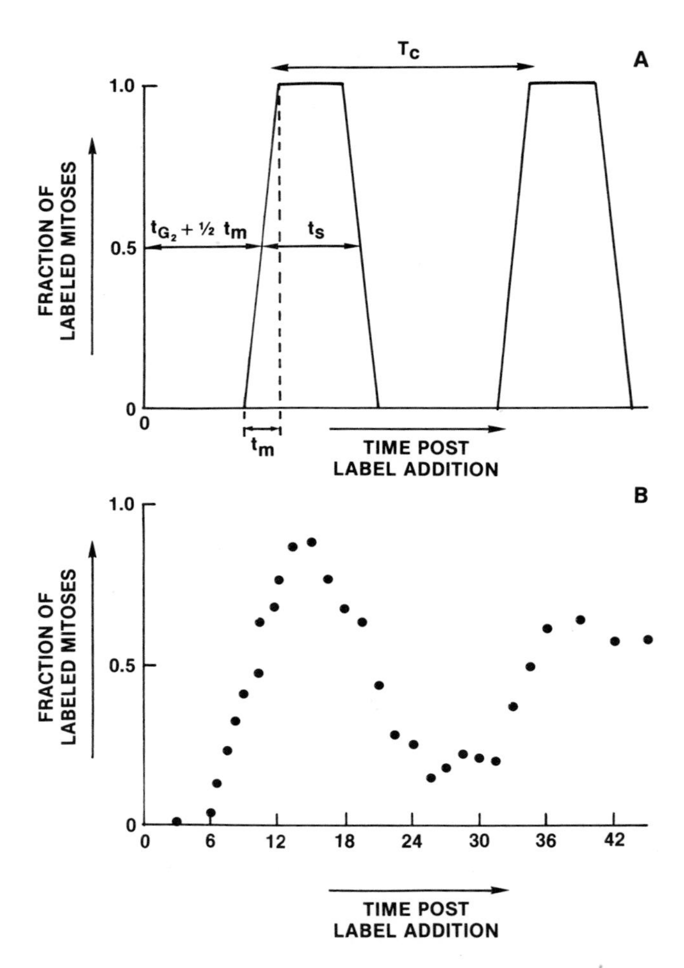

Fig. 1. (A) Idealized FLM curve showing the derivation of durations for all cell cycle phases except G_1. The G_1 duration t_{G_1} is obtained by the difference $T_C - t_{G_2} - t_M - t_S$, where T_C = total cell cycle time, t_{G_1} = G_1 duration, t_S = S-phase duration, t_{G_2} = G_2 duration, and t_M = duration of mitosis. (B) More realistic FLM plot with cell cycle parameters similar to the idealized diagram in (A). Note that t_m is difficult to estimate. Also, the second peak is low and broad. Even these difficulties are acceptable; often the second peak does not appear due to individual cell differences in cycle time. The least reliable estimate is T_C (and thus also t_{G_1}).

an FLM analysis, and the S-phase duration estimated by this method can also usually be accepted with confidence. The values derived from the position of the second peak (T_C, the total cycle time; t_{G_1}, the duration of G_1) are less reliable, and in some cases are unavailable because no second peak appears. Other methods must then be applied.

B. Continuous Labeling

Autoradiographic analysis of cell populations during continuous application of [^{3}H]Tdr also generates cell cycle duration data (Fig. 2). Repeated doses of [^{3}H]Tdr are used to ensure a steady rate of label incorporation

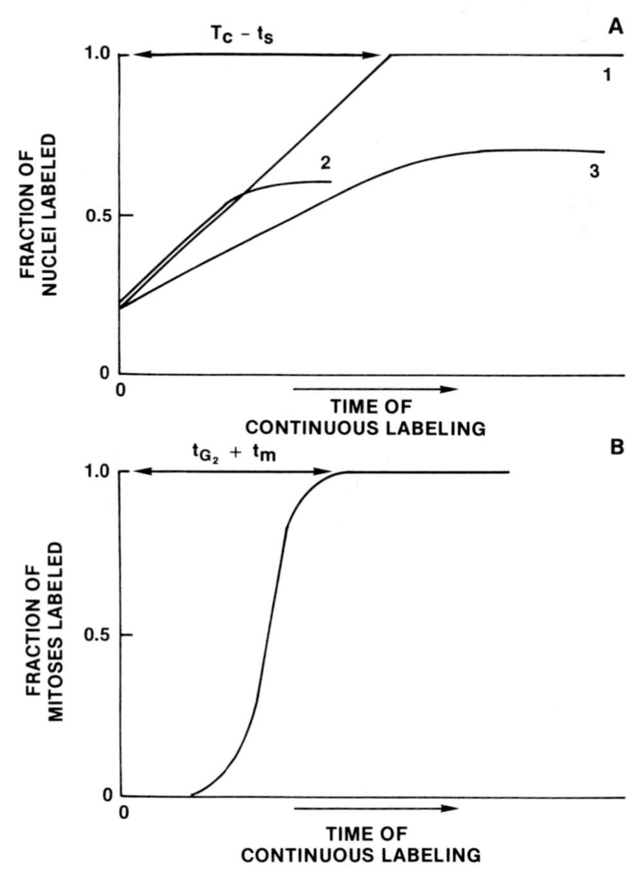

Fig. 2. (A) Accumulation of labeled nuclei during continuous labeling with [^{3}H]Tdr. Line 1 represents the ideal case in which all cells are cycling and the label supply is not exhausted. In this case, $T_C - t_S$ can be derived easily. Line 2 represents the case in which all cells are cycling but label is not continuously available. $T_C - t_S$ can be derived by extrapolating the linear part of the plot. Line 3 is characteristic of growth fractions less than 1.0. The fraction of labeled nuclei will never reach 1.0, but it is sometimes possible to derive $T_C - t_S$ if the point at which the plot declines from linearity can be found. (B) Accumulation of labeled mitoses during continuous labeling with [^{3}H]Tdr. The plot is analogous to the first part of the FLM curve (Fig. 1). GFs less than 1.0 will cause problems similar to those for the plot of labeled nuclei.

throughout the analysis. Doses of 1×10^{-8} mol/liter (1.85 kBq/ml) [^{3}H]Tdr every 6 hr are usually effective in continuous labeling work (Ashmore and Gould, 1979). G_2 durations are calculated, as for the FLM analysis, by plotting the accumulation of labeled mitoses against time (Fig. 2). S-phase duration can be derived from the plot of the fraction of labeled cells against time. For a population in which all cells are cycling (GF = 1.0), continuous labeling should give a totally labeled cell population after a time equal to the total cell cycle duration minus the S-phase duration ($T_C - t_S$). T_C can be obtained from the population doubling time T_D because GF = 1.0, and so t_S can be derived. The duration of mitosis can be obtained from the MI and then G_1 duration can be calculated from $T_C = t_{G_1} + t_S + t_{G_2} + t_M$.

With populations with low GFs, suitable corrections must be made because the linear portion of the fraction of labeled cells plot will not reach unity. $T_C - t_S$ for the cycling population will be given by the point at which linearity is lost (i.e., when steady accumulation of labeled cells slows down).

C. Double Labeling

A rapid method for determining the duration of S phase circumvents the rather tedious FLM experiment and also avoids problems with low GFs. A dividing culture is first labeled with [^{3}H]Tdr for a standard time "Q" (1 or 2 hr is suitable for plant cell suspensions). Then the culture is quickly exposed to [^{14}C]Tdr and fixed for preparation of autoradiographs. After suitable exposure, the autoradiographs are scored for the number of cells showing only ^{3}H labeling and the number of cells showing ^{14}C labeling. The duration of S phase is then the product of the standard [^{3}H]Tdr labeling time and the number of ^{14}C nuclei divided by the ^{3}H nuclei. For example, with a 2-hr standard [^{3}H]Tdr labeling time, a ^{14}C LI of 0.4 (400 ^{14}C-labeled cells per 1000 cells counted), and a ^{3}H LI of 0.1, S phase would be estimated at an 8-hr duration.

The difference between ^{14}C and ^{3}H labeling is easily scored in autoradiographs because of the stronger β emission of ^{14}C, which leads to greater path lengths and more scattered silver grains. The reasoning behind this method is illustrated in Fig. 3.

An expansion of this method, which allows estimation of all cycle phase durations, has been applied to root tips of *Tradescantia paludosa* by Wimber and Quastler (1963).

D. Mak's Method

This combination of autoradiography and microspectrophotometry of Feulgen-stained nuclei (Mak, 1965) allows a distribution analysis. If the GF, age distribution, and total cycle time are known, estimates of phase duration can also be derived from the frequency of cells in each cycle compartment.

A dividing culture is labeled for 30 min with [^{3}H]Tdr. A sample is then fixed, and from it a Feulgen-stained autoradiograph preparation is made. Microscopic examination allows nuclei to be sorted into three cytological classes: Labeled (S phase), mitotic, and unlabeled interphase (G_1 or G_2). Unlabeled interphase cells are then divided into G_1 and G_2 classes by their DNA content (i.e., Feulgen staining density) by measurements on single nuclei made with a microspectrophotometer. Thus, unlabeled interphases with low DNA content are in G_1; unlabeled interphases with high DNA content are in G_2. The method is not suitable for populations containing diploid–polyploid cell mixtures, because it is not possible to distinguish between G_2 diploids and G_1 tetraploids. For exponentially dividing cultures with a single ploidy mode and GF = 1.0, the corrections in Table I and the T_D allow calculation of all phase durations.

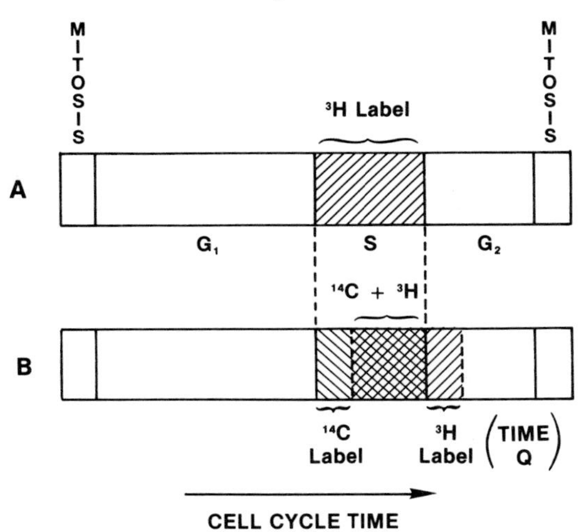

Fig. 3. Double-labeling approach for the rapid analysis of S-phase duration (t_S). (A) The initial condition immediately after application of [^{3}H]Tdr: only cells in S phase show label over their nuclei. (B) The condition after a period of time Q: some cells have left S and show ^{3}H label. Thymidine-^{14}C is added after time Q, and all S-phase nuclei incorporate ^{14}C label. Cells are fixed, and autoradiographs are prepared. Differentiation of ^{14}C and ^{3}H label can be done by grain density and/or distribution in the autoradiograph. The t_S is then simply calculated from the duration of ^{3}H labeling (Q) and the relative fractions of cells showing ^{14}C or only ^{3}H label; that is,

$$t_S = Q \times \frac{^{14}\text{C-labeled fraction}}{^{3}\text{H-only fraction}}$$

(B) is idealized for a *pulse* of ^{3}H label. If ^{3}H is continuously available for all of time Q, there will be no cohort of exclusively ^{14}C-labeled cells as shown in early S. In either case, t_S can be calculated.

E. Probes Other Than Tritiated Thymidine

If Mak's method is revised by the use of radiolabeled probes other than [^{3}H]Tdr, and the grain density in autoradiographs is quantified (either automatically with a densitometer or manually by grain counting), cell cycle analysis can be expanded beyond simple phase duration estimations (Gould, 1979). For example, with tritiated tobacco mosiac virus (Gould *et al.*, 1981b) and tritiated purified DNA (Gould and Ashmore, 1982) as probes, a picture of variation in membrane charge through the cell cycle has been suggested. The experimental approach is simple: The cell population is exposed to some tritiated probe, which will be specifically bound to or incorporated into cells or protoplasts. A sample of the population is then fixed, and the sample is Feulgen stained and used to make autoradiograph preparations on microscope slides. In properly exposed and developed microautoradiographs, the amount of probe bound will then be expressed as the number of silver grains associated with each cell or protoplast. Additionally, the cell cycle state of each cell or protoplast can be assessed in terms of nuclear DNA content (relative Feulgen staining density) measured by microspectrophotometry (Gould, 1979). Practically, two different measurement systems have been used to derive the silver grain density and Feulgen staining density of the same protoplast. The first system relies on the different absorption spectra of Feulgen-stained nuclei and silver grains: The "specular" absorption of silver grains is relatively insensitive to measurement at different wavelengths, whereas a Feulgen-stained nucleus has a well-defined absorption maximum at about 570 nm. Thus, a measurement over a cell taken at 570 nm will give an absorption value that includes both nuclear and silver grain absorption. The same measured area read at 670 nm will give a value that is almost entirely due to silver grains because the Feulgen-stained nucleus absorbs almost no light at that wavelength. This two-wavelength method is most suitable for species with low nuclear DNA content.

The second method makes both measurements at the absorption peak (570 nm) for Feulgen but compares the absorption values obtained before and after removal of silver grains from the autoradiograph. Thus, cell cycle position is obtained from the second measurement (no grains present), and an estimate of grain number is derived from the first minus the second measurement [i.e., (nucleus + grain) − nucleus = grain absorbance]. A good correlation has been demonstrated between absorbance and grain number in microautoradiographs (Gould, 1979).

F. Collection (Stathmokinetic) Methods

Colchicine is a stathmokinetic agent; that is, it will arrest dividing cells in metaphase. Under ideal conditions, the rate of entry of cells into metaphase can be found by measuring the progressive increase in the proportion of metaphases in a population after application of a stathmokinetic agent. A modification of this method by Puck and Steffen (1963) also applies a continuous labeling regime with [^{3}H]Tdr, which allows estimation of all cell cycle phase durations. These sophisticated techniques have not been generally applied to plant cell populations, but the interested reader is directed to the excellent review on cell population kinetics by Aherne *et al.* (1977) for an overview of stathmokinesis.

IV. SUMMARY

In general, no single method of analysis applied in isolation yields completely satisfactory cell cycle phase duration data. However, by applying several methods to the same population, reliable estimates can be generated, because the strengths of one analysis often complement the weaknesses of another, and vice versa. A combined method has been proposed (Gould, 1984) which takes advantage of the strong points of several analytical approaches. In this method, T_D is estimated from cell number data during the exponential phase of culture growth; GF is derived by continuous labeling; duration of mitosis (t_M) is calculated from the mitotic index; t_{G_2}, the G_2 duration, is measured in a short FLM experiment which plots only the ascending limb of the first peak; S-phase duration (t_S) is derived from the LI; G_1 is calculated by difference (i.e., $t_{G_1} = T_D - t_M - t_{G_2} - t_S$). As emphasized repeatedly above, T_D can be substituted for T_C only if GF = 1.0. Otherwise, all of these derivations have been explained in Section III.

REFERENCES

Aherne, W. A., Camplejohn, R. S., and Wright, N. A. (1977). "An Introduction to Cell Population Kinetics." Arnold, London.

Ashmore, S. E., and Gould, A. R. (1979). Cell cycle anlaysis of tumour-derived cultures of *Crepis capillaris:* A kinetic analogy with proliferation of animal tumour cells. *Protoplasma* **101,** 217–230.

Bayliss, M. W. (1975). The duration of the cell cycle of *Daucus carota* in *in vivo* and *in vitro. Exp. Cell Res.* **92,** 31–38.

Bezdek, M., and Vyskot, B. (1981). DNA synthesis in cytokinin autotrophic tobacco cells. Effect of bromodeoxyuridine, fluorodeoxyuridine and kinetin. *Planta* **152,** 215–224.

Blaschke, J. R., Forche, E., and Neumann, K. H. (1978). Investigations on the cell cycle of haploid and diploid tissue cultures of *Datura innoxia* Mill. and its synchronization. *Planta* **144,** 7–12.

Chu, Y., and Lark, K. G. (1976). Cell cycle parameters of soybean (*Glycine max*) cells growing in suspension culture: Suitability of the system for genetic studies. *Planta* **132,** 259–268.

Courtois, D., and Guern, J. (1980). Temperature response of *Catharanthus roseus* cells cultivated in liquid medium. *Plant Sci. Lett.* **17,** 473–482.

Eriksson, T. (1967). Duration of the mitotic cycle in cell cultures of *Haplopappus gracilis. Physiol. Plant.* **20,** 348–354.

Everett, N. P., Wang, T. L., Gould, A. R., and Street, H. E. (1981). Studies on the control of the cell cycle in cultured plant cells. II. Effects of 2,4-dichlorophenoxyacetic acid (2,4-D). *Protoplasma* **106,** 15–22.

Fukuda, H., and Komamine, A. (1981). Relationship between tracheary element differentiation and DNA synthesis in single cells isolated from the mesophyll of *Zinnia elegans*—analysis by inhibitors of DNA synthesis. *Plant Cell Physiol.* **22,** 41–49.

Gould, A. R. (1977). Temperature response of the cell cycle of *Haplopappus gracilis* in suspension culture and its significance to the G_1 transition probability model. *Planta* **137,** 29–36.

Gould, A. R. (1979). Combined microspectrophotometry and automated quantitative autoradiography applied to the analysis of the plant cell cycle. *J. Cell Sci.* **39,** 235–245.

Gould, A. R. (1984). Control of the cell cycle in cultured plant cells. *CRC Crit. Rev. Plant Sci.* (in press).

Gould, A. R., and Ashmore, S. E. (1982). Interaction of purified DNA with plant protoplasts of different cell cycle stage: The concept of a competent phase for plant cell transformation. *Theor. Appl. Genet.* **64,** 7–12.

Gould, A. R., Bayliss, M. W., and Street, H. E. (1974). Studies on the growth in culture of plant cells. XVII. Analysis of the cell cycle of asynchronously dividing *Acer pseudoplatanus* L. cells in suspension culture. *J. Exp. Bot.* **25,** 468–478.

Gould, A. R., Everett, N. P., Wang, T. L., and Street, H. E. (1981a). Studies on the control of the cell cycle in cultured plant cells. I. Effects of nutrient limitation and nutrient starvation. *Protoplasma* **106,** 1–13.

Gould, A. R., Ashmore, S. E., and Gibbs, A. J. (1981b). Cell cycle related changes in the quantity of TMV virions bound to protoplasts of *Nicotiana sylvestris. Protoplasma* **108,** 211–223.

Mak, S. (1965). Mammalian cell cycle analysis using microspectrophotometry combined with autoradiography. *Exp. Cell Res.* **39,** 286–308.

Puck, T. T., and Steffen, J. (1963). Life cycle analysis of mammalian cells. I. A method for localising metabolic events within the life cycle and its application to the action of Colcemid and sublethal doses of x-irradiation. *Biophys. J.* **3,** 379–397.

Quastler, H., and Sherman, F. G. (1959). Cell population kinetics in the intestinal epithelium of the mouse. *Exp. Cell Res.* **17,** 420–438.

Rembur, J. (1974). Cycle cellulaire et teneurs en ADN nucleaire de cellules en suspension de l'*Acer pseudoplatanus* en phase exponentielle de croissance. Hétérogenéité de la culture. *Can. J. Bot.* **52,** 1535–1543.

Wang, T. L., Everett, N. P., Gould, A. R., and Street, H. E. (1981). Studies on the control of the cell cycle in cultured plant cells. III. The effects of cytokinin. *Protoplasma* **106,** 23–25.

Wimber, D. E., and Quastler, H. (1963). A ^{14}C- and ^{3}H-thymidine double labelling technique in the study of cell proliferation in *Tradescantia* root tips. *Exp. Cell Res.* **30,** 8–22.

CHAPTER **83**

Flow Cytometric Analysis of the Cell Cycle

David W. Galbraith

School of Biological Sciences
University of Nebraska at Lincoln
Lincoln, Nebraska

I. INTRODUCTION

Central to an understanding of plant growth and development is an understanding of the operation of the cell cycle. In eukaryotic cell systems that are active in cell growth and division, the cell cycle is defined as a series of linked temporal phases. These comprise mitosis (M), followed by a period preceding DNA synthesis (termed G_1), DNA synthesis (S), and a second period (G_2) prior to the next mitosis.

The use of flow cytometry for the analysis of the cell cycle in animal cell populations has been extensively reviewed (Bohmer, 1982; Gray and Coffino, 1979; Gray *et al.*, 1979). The technique relies on the observation that the nuclear DNA content reflects the position of the cell within the cell cycle. Thus, a diploid cell in G_1 will contain a 2C nuclear DNA level. In G_2 the nuclear DNA level will be 4C, and in S the level will be intermediate between that of 2C and 4C. The operation of the cell cycle within cellular

ISBN 0-12-715001-3

populations can therefore be followed by the use of methods that accurately determine the relative proportions of cells containing these different nuclear DNA amounts. For animal and yeast cell systems, such methods have included flow cytometry. This involves the staining of individual cells or isolated nuclei with DNA-specific fluorochromes. Under suitable conditions, the emission of fluorescence can be shown to be proportional to the amount of DNA present. The cells or nuclei are then constrained to pass in a precisely defined stream through the focus of a continuous-wave tunable laser. The amount of emitted fluorescence is quantitated for each cell. The fluorescence values are digitized and collected in the form of histograms in the memory of a microprocessor. The major advantages of flow cytometry are its high accuracy, convenience, and rapidity.

An obvious prerequisite for flow cytometry is that the cells of interest be in the form of single-cell suspensions. Whereas most animal and yeast cell types either exist in the form of single cells or can be converted into cell suspensions by the use of mild enzymatic treatments, plant tissues normally comprise complex three-dimensional structures. At first sight, this would appear to preclude the use of flow cytometry for the estimation of plant nuclear DNA levels.

In our studies, we have developed a variety of experimental techniques for the production of populations of protoplasts or nuclei from plant tissues, suitable for the measurement of nuclear DNA levels by flow cytometry (Galbraith and Shields, 1982; Galbraith *et al.*, 1983). This chapter presents the methods involved in the use of flow cytometry for measurements of the total cellular and nuclear DNA contents, provides typical data obtained with these methods, and suggests examples of systems for which such measurements should be extremely useful. Finally, this chapter focuses upon the analysis of the cell cycle, and proposes methods for the measurement of phase transit times and for the identification of controlling factors and the occurrence of subpopulations of noncycling cells.

II. PROCEDURES

A. Flow Cytometric Analysis of Protoplasts

1. Preparation of Protoplasts

In our studies, we have commonly employed leaf protoplasts of *Nicotiana tabacum* cv. *Xanthi*. These are prepared according to a modification of the method of Chupeau *et al.* (1978). Fully expanded leaves from young plants grown under greenhouse conditions are washed in tap water for 15–30 min. They are sterilized by immersion in 70% ethanol for 30 sec and 33% commercial bleach for 15 min. All further operations are carried out in a laminar sterile airflow hood. The residual bleach is removed by two sterile water rinses. The leaves are trimmed to remove the veins, midrib, and all bleached tissue areas, and are transferred into osmoticum (solution 1 in the appendix). The leaves are sliced into 30 × 2-mm pieces and are transferred into a 10-ml polysaccharidase solution contained in 250-ml Erlenmeyer flasks. The flasks are evacuated for 90 sec using a vacuum aspirator. Incubation is carried out for a period of 15–18 hr at 25°C in darkness, with reciprocal shaking at 20 excursions per minute. For more rapid production of protoplasts, we have employed the technique described by Nagata and Ishii (1979). We find that a preplasmolysis of the leaf fragments in osmoticum for 8–10 hr prior to addition of the polysaccharidase solutions can improve yields. Following incubation, the protoplasts are separated from large pieces of debris by filtration through six layers of cheesecloth. The protoplast suspension is then centrifuged at 100 *g* for 5 min. The protoplast pellet is resuspended in 20 ml of a 25% (w/v) sucrose solution (solution 3 in the Appendix). The suspension is transferred into a 50-ml polycarbonate centrifuge tube, overlaid with an equal volume of osmoticum, and centrifuged at 100 *g* for 5 min. The intact protoplasts are recovered from the interface, following dilution with osmoticum, by centrifugation at 100 *g* for 5 min. The protoplasts are finally subjected to two cycles of resuspension and centrifugation in osmoticum. The protoplasts can now be transferred into culture in 5-cm diameter sterile plastic Petri dishes after resuspension in 5 ml of growth medium (solution 4 in the Appendix).

2. Protoplast Fixation

A sample containing 1.5×10^5 protoplasts is centrifuged at 50 *g* for 4 min. The protoplast pellet is resuspended in 2 ml ice cold, freshly prepared fixative. This comprises 1% sorbitol dissolved in ethanol–acetic acid–water (18:3:2 v/v). After 5 min the protoplasts are pelleted at 50 *g* for 4 min using a 10-ml conical Pyrex centrifuge tube. The protoplasts are resuspended in 2 ml ice cold 70% ethanol and are subsequently recovered by centrifugation at 50 *g* for 4 min. The protoplast pellet is resuspended in 2 ml staining solution (solution 6 in the Appendix). If required, approximately 0.7×10^5 chicken red blood cells (CRBC), separately fixed according to the protocol listed above, can be added prior to staining. The CRBC are initially collected from mature chickens by wing vein puncture and can be stored in Alsever's solution at 4°C for up to 4 weeks without noticeable deterioration.

3. Protoplast Lysis

A sample containing 5×10^4 protoplasts is centrifuged at 50 *g* for 5 min. The supernatant is discarded. All further operations are carried out on ice. If required, approximately 4×10^4 unfixed CRBC can be added at this point. The protoplasts are then resuspended in a hypotonic staining solution (solution 6 in the Appendix) containing mithramycin. After 15 min, the suspension of released nuclei is filtered through a 15-μm nylon mesh before analysis.

In some studies, we have employed osmotic lysis and fixation sequentially in order to produce suspensions of fixed nuclei. For this procedure, the nuclei released by osmotic lysis are concentrated by centrifugation at 1000 *g* for 5 min. The nuclear pellet is then resuspended in 2 ml fixative. All further operations are as described in Section II,A,2 except that the force of centrifugation is increased to 1000 *g*.

4. Operation of the Flow Cytometer

Flow cytometry involves the rapid sequential analysis of the degree of individual fluorescence possessed by the suspensions of protoplasts or nuclei. Since the degree of fluorescence is linearly related to the DNA content of the individuals within the populations, flow cytometry can be used to provide accurate measures of the cell cycle activity and of the total nuclear and protoplast DNA contents.

The population of protoplasts or nuclei is constrained by the sheath fluid (solution 5 in the Appendix) to pass through a flow cell containing a jeweled orifice 72 μm in diameter (Fig. 1). The fluid stream intersects the focus of the laser beam. As the protoplasts or nuclei pass through the focus, they absorb light, which is subsequently reemitted in the form of fluorescence. The fluorescence signals comprise a rapid series of pulses, which are converted into corresponding direct current (DC) voltages by means of a photomultiplier located orthogonally to the laser light path and the fluid flow direction. The voltage pulses are amplified and converted into binary equivalents by means of an analog-to-digital converter (ADC). The binary information is stored in the form of 256-channel, one-dimensional histograms in the memory of a microprocessor.

For the analysis of protoplasts and nuclei stained with mithramycin, the flow cytometer is operated at a wavelength of 457 nm using a laser output of 200 mW. Scattered laser light is eliminated from the fluorescence detector by use of barrier filters LP510 and LP515. We accumulate histograms to a total count of $5\text{–}10 \times 10^3$ at a flow rate of 50/sec. This provides optimal accuracy and convenience of operation: The total numbers accumulated are sufficient for statistical accuracy, and the period of accumulation is low

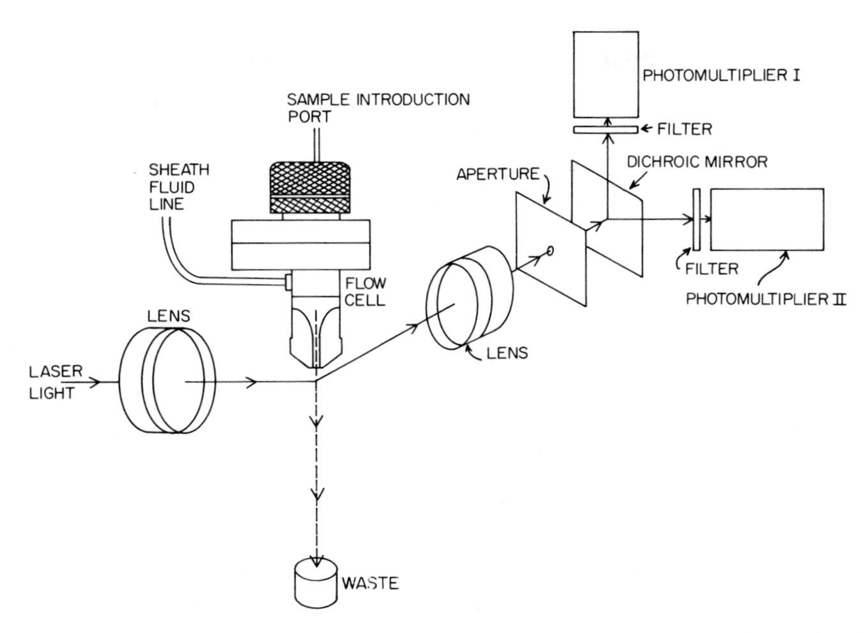

Fig. 1. Operation of the flow cytometer in schematic form. The fluorescence of the individual particles passing through the focus of the laser beam is measured by means of one of two photomultipliers. These values are automatically digitized and stored in the memory of a microprocessor.

enough to eliminate error due to machine drift. For cell cycle analyses, the high voltage of the photomultiplier and the gain of the fluorescence amplifier are adjusted so that the modes of the peaks of fluorescence corresponding to the G_1 and G_2 DNA contents fall approximately in channels 50 and 100. This allows the identification of the emergence of higher ploidy levels within the population, such as that peak, at channel 200, which would represent tetraploid G_2 nuclei or protoplasts.

For histograms of this type, the coefficients of variation of the major peaks of DNA content and the corresponding cell cycle parameters are automatically computed using the extended analysis system (E.A.Sy.1). The operating characteristics of these programs are available from the manufacturer (Coulter Electronics, Hialeah, Florida). Parameters computed include the proportions of G_1, S, and G_2 individuals within the populations.

For the accurate measurement of total nuclear or protoplast DNA contents, the high voltage and gain settings are adjusted so that the mode of the fluorescent peak corresponding to the CRBC is at the highest possible channel number, consistent with the retention of the protoplast or nuclear fluorescence peak on the same 256-channel histogram. The DNA value of the unknown can then be calculated by simple division of the modal channel number of the fluorescent peak by the corresponding value of that of CRBC.

B. Flow Cytometric Analysis of Intact Tissues

The observation that plant nuclei remained intact under conditions of osmotic lysis of the protoplasts led us to consider the possibility of mechanical release of nuclei from intact plant tissues using a modification of homogenization techniques designed for the isolation of fragile plant organelles such as dictyosomes (Ray, 1977). The following method yields good results with a wide variety of plant species and tissues (Galbraith *et al.*, 1983).

Tissue samples are excised from intact, healthy plants. Samples can include leaf, root, stem, petal, and anther tissues. The tissue samples are cleaned by thorough rinsing in water, and the fresh weight is determined. As little as 10 mg tissue can be employed for a single flow cytometric measurement (5×10^3 nuclei). The tissue samples can be stored under cool, humid conditions for several days prior to analysis. All further operations are carried out in a walk-in cold room maintained at 4°C. The selected tissues are placed on ice in 5-cm-diameter plastic Petri dishes. A sample size of 0.1 g is convenient; to it is added 1.7 ml osmotic lysis solution (solution 5 in the Appendix). The tissue is vigorously chopped by hand using a standard single-edge razor blade for 2–3 min. This results in a fine tissue homogenate from which the tissue debris is removed by filtration through a 60-μm nylon mesh. A small (0.2-ml) aliquot of the filtrate is mixed with 1.8 ml staining solution (solution 6 in the Appendix). After 15 min the nuclei are filtered through a 15-μm nylon mesh prior to analysis. Flow cytometry is carried out as described in Section II,A,4.

III. RESULTS AND CONCLUSIONS

A. Analysis of Protoplasts

1. Noncycling Systems

Figure 2 illustrates typical flow cytometric histograms obtained with *Nicotiana tabacum* leaf protoplasts subjected to osmotic lysis in the presence of mithramycin. In Fig. 2A the gain of the photomultiplier has been adjusted

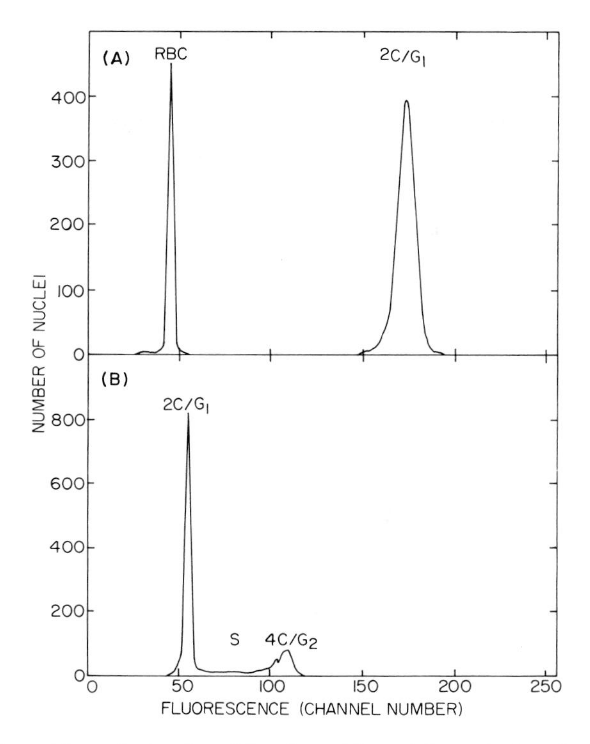

Fig. 2. Flow cytometric analysis of nuclei, released from tobacco leaf protoplasts by osmotic lysis, after staining with mithramycin. In (A), the gain of the photomultiplier is increased, and chicken red blood cells (RBC) included as an internal standard, to allow accurate measurement of the $2C/G_1$ DNA content of the tobacco nuclei. In (B), the gain of the photomultiplier is reduced and chicken RBC are omitted to allow analysis of the cell cycle within the leaf protoplast population.

to permit maximal accuracy in the measurement of the 2C nuclear DNA content of the protoplasts. The internal CRBC standard gives a peak of fluorescence at around channel 45. The peak corresponding to the G_1 (2C) nuclei of the protoplasts appears at around channel 180. In Fig. 2B the gain of the photomultiplier has been reduced so that the peak of fluorescence of the 2C (G_1) appears at channel 55 and the corresponding 4C (G_2) peak at channel 110. In this latter histogram, which is used for calculation of the proportions of protoplasts within the various phases of the cell cycle, a CRBC internal standard is unnecessary and has therefore been omitted. The appearances of histograms of protoplasts subjected to fixation or osmotic lysis followed by fixation are essentially identical to those of Fig. 2. The coefficients of variation (CV) of the G_1 peak and the DNA contents of the nuclei and protoplasts for a representative experiment are listed in

TABLE I

Flow Cytometric Parameters for Leaf Protoplasts of *Nicotiana tabacum*

Sample	G_1 peak		CRBC peak		Calculated DNA content (pg)
	Location (channel)	CV (%)	Location (channel)	CV (%)	
Fixed protoplasts	151	6.0	32	6.5	10.99
Nuclei from osmotic lysis	187	4.4	45	3.4	9.68
Nuclei fixed after osmotic lysis	161	4.7	41	6.0	9.25

Table I. The CVs are within the ranges obtained with animal cell systems. Our data indicate the high degree of reproducibility of the method. The standard deviations of independent measurements of protoplast or nuclear DNA contents were typically within 1–2% of the mean. From the data of Table I, we can calculate that nuclear DNA comprises 84% of the total cellular DNA. The identification of the main peak of fluorescence as corresponding to the 2C (diploid) DNA content is confirmed by analysis of the nuclear DNA content of root tip cells (Galbraith *et al.*, 1983; Table II). Further, the identification of the 2C DNA peak as representing cells arrested in G_1 is confirmed by analysis of protoplasts entering into the cell cycle (Galbraith and Shields, 1982; Fig. 3). Leaf protoplasts incubated in growth media for up to 24 hr provide essentially unchanged histograms. Beyond this point the initiation of DNA synthesis can be observed. We

TABLE II

Flow Cytometric Parameters for Nuclei Prepared from Intact Organs of *Nicotiana tabacum*

Sample	Structure	G_1 peak		CRBC peak		Calculated DNA content (pg)[a]
		Location (channel)	CV (%)	Location (channel)	CV (%)	
N. tabacum (Petit Havana SR1)	Leaf	171	4.5	43	5.9	9.38
	Root	175	4.2	44	4.2	9.39
N. tabacum (Petit Havana SR1) haploid[b]	Leaf	149	4.3	77	4.6	4.57

[a]Standard deviations for independent measurements were within 4% of the mean DNA content.
[b]Haploid plantlets were obtained by conventional anther culture.

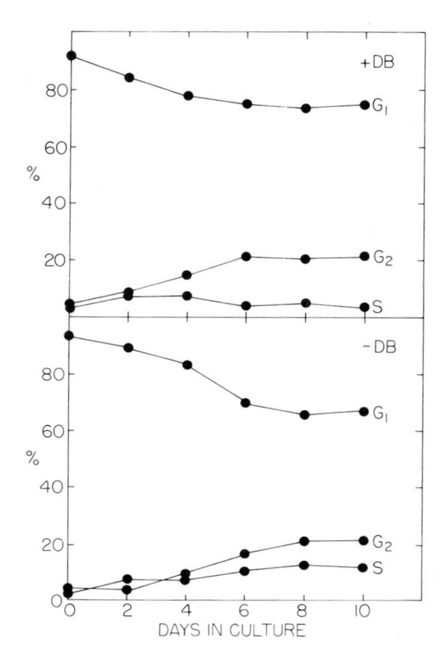

Fig. 3. Changes in the proportion of nuclei representative of the various phases of the cell cycle during culture of protoplast populations in the presence and absence of DB, as measured by flow cytometry.

thus conclude that the protoplasts prepared from tobacco leaf tissues are largely arrested in G_1.

2. Cycling Systems

Plant cell systems active in the cell cycle can be conveniently studied by flow cytometry in two ways. The first is the use of leaf protoplasts transferred into culture, and the second is the use of tissue (suspension) cultures. In both cases, strategies have to be employed to avoid problems unique to plant cells. These include the fact that the protoplasts initiate cell wall synthesis during culture, which prevents the release of nuclei by procedures of osmotic lysis (Section II,A,3). A second problem relates to the process of cell division. In higher plant systems, the daughter cells do not separate following cytokinesis and deposition of the cell plate. This makes it impossible to assign individual DNA contents to each daughter cell from the aggregate fluorescence signal. Further, the onset of cell expansion coupled with cell division means that the sizes of cells and cell clusters rapidly approach the diameter of the flow cell tip.

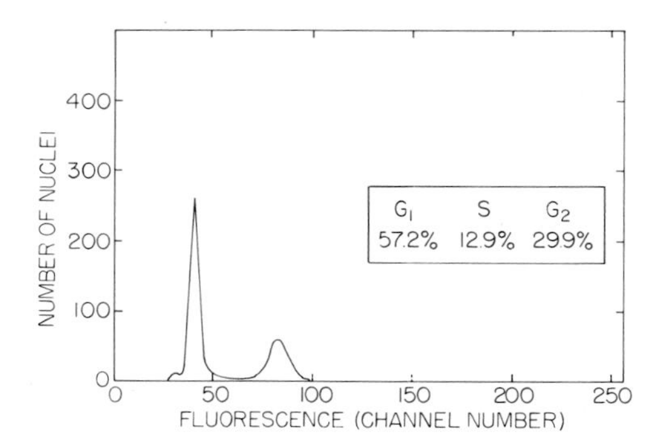

Fig. 4. Distribution of the proportions of nuclei representative of the various phases of the cell cycle contained in populations of protoplasts prepared from rapidly dividing *Nicotiana tabacum* tissue (suspension cultures), as measured by flow cytometry.

The strategies that we have developed to avoid these problems involve either the use of 2,6-dichlorobenzonitrile (DB) or the conversion of the growing cells and cell clusters back into protoplasts using polysaccharidases. The former treatment almost completely inhibits cell wall deposition during culture (Galbraith and Shields, 1982). Figure 3 compares these two methods in the analysis of the cell cycle in cultured leaf protoplasts of *N. tabacum.* Flow cytometric analysis of the nuclear DNA content was carried out following osmotic lysis in the presence of mithramycin, as described in Section II,A,2. The onset of DNA synthesis in the isolated protoplasts during culture results in the appearance of S- and G_2-phase nuclei. A subsequent mitosis results in the production of more G_1 nuclei. This onset of a true cycle explains the observed trend of the G_1 and G_2 phase proportions toward limiting values. Such limiting values are displayed by protoplasts prepared from suspension cultured cells (Fig. 4). If it is assumed that all of the cells in the population are homogeneous, asynchronous, and cycling at equal rates, the data indicate that G_1, S, and G_2 phases occupy 57.2, 12.9, and 29.9% of the cell cycle, respectively.

B. Analysis of Intact Tissues

We have obtained considerable data using the method of gentle homogenization to release nuclei from intact tissue prior to staining and analysis of DNA content by flow cytometry (Galbraith *et al.*, 1983). From these data we have been able to draw the following conclusions. The flow cytometric method is highly accurate. reproducible, and convenient. It is applicable to a wide range of mono- and dicotyledenous plants. We have not yet found any species to be intractable to this method. It is, moreover, applicable to a wide range of tissues within individual plants. The flow cytometric parameters obtained for *N. tabacum* are summarized in Table II.

C. Establishment of Cell Cycle Parameters

As is pointed out clearly by Bohmer (1982), flow cytometric analyses of the type described above provide only a static measurement of the cell cycle in the protoplast or tissue samples. For a complete description of the cell cycle, including the respective G_1-, S-, and G_2-phase durations, further information is required, particularly with respect to the degree of variation of cycle time and the proportions of noncycling (arrested) cells within the populations. Such information has been obtained in animal cell systems by the use of bromodeoxyuridine quenching of 33258 Hoechst fluorescence as a measurement of the onset of S phase during culture (Bohmer, 1982). I anticipate that this approach will be of considerable future value for the determination of the proportion of noncycling cells within plant protoplast populations. However, we have developed one alternative approach unique to plant cell systems that can answer the same question. This approach relies on the observation that leaf protoplasts, initially arrested in G_1, that are incubated in the presence of DB undertake S phase and mitosis normally but cannot synthesize a cell wall (Galbraith and Shields, 1982). The consequence is that the protoplasts do not undergo cytokinesis; this results, after the first round of the cell cycle, in the appearance of binucleate protoplasts. The two nuclei progress through the cell cycle coordinately during subsequent rounds of the cell cycle, to give rise to tetranucleate and octonucleate protoplasts. Corresponding analysis of flow cytometric histograms of the DNA content of fixed protoplast samples reveals the progression of the protoplasts during culture from a DNA level characteristic of 2C to levels of 4C, 8C, and subsequently 16C. The fact that the protoplasts do not divide means that the operation of the cell cycle can be dissected according to the total protoplast DNA content given by flow cytometric analysis after fixation. The first round of the cycle corresponds to the progression of the protoplast DNA content from 2C to 4C, the second round from 4C to 8C, and the third from 8C to 18C. For convenience these can be simultaneously displayed on single histograms by use of a logarithmic abscissa. Our experiments (Galbraith and Shields, 1982) indicate that none of the protoplasts cultured in the presence of DB remain arrested in G_0–G_1, and therefore that all of the leaf protoplasts, in terms of the operation of the cell cycle, represent a single population. With this approach it should be possible to determine, using protoplasts either initially arrested (those from leaf tissues) or nonsynchronously cycling (those from suspension cultures), the factors controlling progression through the cell cycle. Such factors may include phytohormone and nutrient deprivation, as well as chemical treatments such as the use of hydroxyurea, colchicine, and aphidicolin. Finally, we believe that the above approach will pemit an evaluation of the plant cell cycle with respect to the applicability of the probabalistic model of cell cycle progression (Smith and Martin, 1973).

APPENDIX

Solutions Employed in Flow Cytometric Analyses

1. Osmoticum for preparation of tobacco leaf protoplasts
   130 g mannitol
   TO salts and hormones (Chupeau *et al.*, 1978) to 1 liter
   Adjust pH to 5.5 with 0.1 *M* NaOH.
   The final solution is filtered through a 0.22-μm filter prior to use to remove any insoluble material.
2. Enzyme solution used in preparation of tobacco leaf protoplasts
   1.0 g Cellulysin
   0.5 g Driselase
   0.2 g Macerase
   Dissolve to a final volume of 1 liter with solution 1.
   This enzyme solution is clarified by centrifugation at 10,000 *g* for 10 min and is sterilized by passage through a Millipore Millex-GS-type SLGS0250S disposable ultrafiltration unit.
3. Sucrose solution used for protoplast purification
   250 g sucrose
   Dissolve in an aqueous solution of TO salts and hormones (Chupeau *et al.*, 1978) to a final volume of 1 liter. This solution should be sterilized only by ultrafiltration.
4. Protoplast growth medium
   KOM growth medium (Galbraith and Mauch, 1980) is based on that of Kao (1977), modified to contain 68.4 g/liter glucose and 50.5 g/liter mannitol.
   This solution should be sterilized only by ultrafiltration. It is photosensitive but can conveniently be stored as 20-ml aliquots at −20°C.
5. Protoplast lysis solution
   45 m*M* magnesium chloride
   30 m*M* sodium citrate
   20 m*M* morpholino propane sulfonic acid
   0.1% Triton X-100
   Adjust to pH 7.0 using 0.1 *M* sodium hydroxide.
   This solution is always subjected to ultrafiltration using a Millipore-type GSTF filter to remove small particles.
6. Mithramycin staining solution
   This solution is nominally 0.1 g/liter mithramycin dissolved in solution 5. For standardization we measure the optical density of the solution at 600 nm, adjusting the dilution to produce an absorbance of 0.6. It is stored as small aliquots at −20°C.

REFERENCES

Bohmer, R.-M. (1982). Flow cytometry and cell proliferation kinetics. *Prog. Histochem. Cytochem.* **14,** 1–65.

Chupeau, Y., Missonier, C., Hommel, M.-C., and Goujard, J. (1978). Somatic hybrids of plants by fusion of protoplasts. *Mol. Gen. Genet.* **165,** 239–245.

Galbraith, D. W., and Mauch, T. J. (1980). Identification of fusion of plant protoplasts. II. Conditions for the reproducible fluorescence labelling of protoplasts derived from mesophyll tissue. *Z. Pflanzenphysiol.* **98,** 129–140.

Galbraith, D. W., and Shields, B. A. (1982). The effects of inhibitors of cell wall synthesis on tobacco protoplast development. *Physiol. Plant.* **55,** 25–30.

Galbraith, D. W., Harkins, K. R., Maddox, J. M., Ayres, N. M., Sharma, D. P., and Firoozabady, E. (1983). Rapid flow cytometric analysis of the cell cycle in intact plant tissues. *Science* **220,** 1049–1051.

Gray, J. W., and Coffino, P. (1979). Cell cycle analysis by flow cytometry. *In* "Methods in Enzymology" (W. B. Jakoby and I. H. Pastan, eds.), Vol. 58, pp. 233–248. Academic Press, New York.

Gray, J. W., Dean, P. N., and Mendelsohn, M. L. (1979). Quantitative cell cycle analysis. *In* "Flow Cytometry and Sorting" (M. R. Melamed, P. F. Mullaney, and M. L. Mendelsohn, eds.), pp. 383–408. Wiley, New York.

Kao, K. N. (1977). Chromosomal behavior in somatic hybrids of soybean-Nicotiana glauca. *Mol. Gen. Genet.* **150,** 225–230.

Nagata, T., and Ishii, S. (1979). A rapid method for isolation of mesophyll protoplasts. *Can. J. Bot.* **57,** 1820–1823.

Ray, P. M. (1977). Auxin-binding sites of maize coleoptiles are localized on membranes of the endoplasmic reticulum. *Plant Physiol.* **59,** 594–599.

Smith, J. A., and Martin, L. A. (1973). Do cells cycle? *Proc. Natl. Acad. Sci. U.S.A.* **70,** 1263–1267.

CHAPTER **84**

Autoradiography

Edward C. Yeung

Department of Biology
University of Calgary
Calgary, Alberta, Canada

I. INTRODUCTION

Autoradiography is a technique used for the detection and localization of radioactive compounds within biological specimens. After the specimen has incorporated a supplied radioactive isotope into macromolecules, it is placed in contact with a photographic emulsion. Ionizing radiation emitted during radioactive decay of the label causes a chemical change in the silver halide crystals of the photographic emulsion. Upon treatment with a developing agent, those crystals affected by the radiation are converted into metallic silver grains which are visible in the light microscope. The silver grains serve as the means of detecting radioactivity. Also, because of their number and distribution, they provide information on the relative activity of the cells and tissues under study. The process of producing this "picture" is called *autoradiography,* and the picture itself is called an *autoradiogram* (Rogers, 1979).

Autoradiography has been used to study various aspects of plant cell

ISBN 0-12-715001-3

biology. In the area of plant tissue culture, this method has been used to study cell cycle events (Eriksson, 1967), organogenesis (Tran Thanh Van *et al.*, 1974), and aspects of pollen and anther culture (Raghavan, 1979).

In this chapter, the most common procedure is described—the liquid emulsion dipping technique for insoluble macromolecules. This technique applies only to those macromolecules that are insoluble in the wide range of organic and aqueous solvents to which they are exposed during the processing of tissues. For detailed theoretical and practical considerations, as well as the interpretation of autoradiograms, consult the monograph "Techniques of Autoradiography" by Rogers (1979).

II. CHOICE OF EMULSION

The most common sources of liquid emulsion are Eastman Kodak (Rochester, New York) and Ilford, Ltd. (Ilford, Essex, United Kingdom). Kodak and Ilford emulsions differ in several properties, such as viscosity, safelight requirements, sensitivity to repeated melting and drying, and the required conditions for exposure. The techniques that have been described in the literature for Kodak emulsions do not give optimal results with Ilford products, and vice versa. It follows that one must select a technique appropriate to the source from which emulsions will normally be obtained (Rogers, 1979).

Kodak manufactures four nuclear emulsions: NTB, NTB-2, NTB-3, and special product type 129-01. The first three are designed for light microscope autoradiography and the last one for electron microscope autoradiography. These products differ in grain size and sensitivity. NTB-2 is commonly used in light microscope autoradiography. This emulsion is especially suited for tritium-labeled compounds. It may also be used for ^{14}C-labeled compounds. Ilford manufactures three basic types of nuclear emulsions: G, K, and L. Within each major type, a range of grain sizes and sensitivities are also available. Type K is primarily designed for the light microscope, and type L is intended for electron microscopic examinations. Type K2 is the most common one employed for light microscopic examinations of tritiated compounds, and types K5 and G5 are suitable for ^{14}C-labeled materials. For detailed information, consult data sheets provided by these manufacturers. In general, the choice of emulsion depends on the labeled precursor being used and the degree of resolution required.

III. LABELING PROCEDURES AND PREPARATION OF TISSUES

Cells and tissues are labeled under appropriate conditions according to the experimental design. The usual dose of radioactive precursor ranges from 0.5 to 10 μC/ml. Higher concentrations have also been used (for details, see Herrmann and Abel, 1972). After labeling, the plant material is washed, dehydrated, and embedded in the usual manner: For the paraffin method, see Jensen (1962); for plastic embedding methods, see O'Brien and McCully (1981). Appropriate precautions should be taken when using the radioactive materials. Rubber gloves should be worn while handling the radioactive solutions, the fixatives, and the first washing of dehydrating fluid. These solutions should be discarded in the liquid radioactive waste container after use.

Following embedding, serial sections of the desired thickness are obtained. The sections are then mounted on cleaned glass slides pretreated with adhesive. Slides used in the mounting of sections should be precleaned by soaking overnight in a chromate–sulfuric acid solution, and then washed thoroughly with distilled water and subbed without drying. The subbing solution is prepared by dissolving completely in 1 liter of warm distilled water 5.0 g gelatin; then add 0.5 g chrome alum (chromium potassium sulfate). After the solution has cooled, it is filtered through Whatman no. 1 filter paper. The cleaned slides are then dipped in it and allowed to dry in a dust-free area. The subbing solution may be stored at 5°C for 48 hr but should then be discarded (Pappas, 1971). In addition, for paraffin sections, Haupt's adhesive can also be used (see Jensen, 1962). After sections have been securely attached to slides, for the paraffin method, remove the paraffin with xylene and gradually rehydrate the sections through an alcohol dilution series to water just prior to dipping. For the plastic-embedded materials, slides can be dipped without further treatment.

A. Darkroom Preparations and Dipping Techniques

The following equipment and materials are needed in the darkroom prior to the dipping of slides:

1. Photographic emulsion.
2. Porcelain spoon, plastic forceps, or glass rod for transferring emulsion.
3. Container for melting emulsion such as a beaker.
4. Water bath stabilized at 43°C.

5. Slide-drying racks. Test tube stands are good, with the slides allowed to drain on moist blotting paper such as paper towels or filter paper in the rack.

6. Safelights equipped with Wratten filter no. 2, with a 15-W bulb (maximum) for Kodak emulsion or another safelight according to the specifications of the manufacturer.

7. Small black slide boxes with a drying agent for storing dipped slides. Place a clean slide near one end of the slide box so as to create a small compartment. Fill this small compartment with Drierite. Be careful not to tip the slide box.

8. Scissors and black electrician's tape.

9. Blank dust-free slides for removing air bubbles from the surface of the liquid emulsion. These can also be used to check the background silver grain counts of the emulsion.

10. Experimental slides to be dipped.

11. Beaker and double-distilled water.

12. A large light-tight box.

A darkroom used in autoradiography should be lightproof, with a double door to serve as a light trap. The design of such a darkroom is discussed by Bogoroch (1972) and Rogers (1979). Any equipment used in the darkroom should also be lightproofed. For example, the indicator light of the water bath should be covered with black electrician's tape to prevent any accidental exposure of the emulsion. The safelight should be kept as far from the working area as possible. Care should be taken in arranging the working area. The location of all materials and equipment should be memorized since subsequent procedures are carried out in almost total darkness.

The emulsion used is solid at room temperature and will melt when warmed to 43°C. For tritium autoradiography, the emulsion is usually diluted 1:1 with distilled water. This diluted emulsion will produce a thinner coat on the slides. The energy of the β particles of tritiated compounds is limited. Therefore, in a thick coat of emulsion, the upper portion of the emulsion is beyond the range of the β particles emitted by the preparation. A thick layer also decreases the resolution of the specimen and creates the potential problem of high background silver grain counts. However, in the case of ^{14}C compounds, it is advisable to use the emulsion without dilution due to the higher energy of the β particles.

B. Melting of Emulsion

Make two marks on the beaker so that they divide it into two portions of equal volume. Fill to the lower mark with distilled water and set into the water bath for about 10 min. With only safelights on, gently spoon out the emulsion from its container with the porcelain spoon. Fill the beaker up to the second mark. One can check this by holding the beaker up to the safelight, with the hand shielding most of it. Place the beaker back in the water bath and wait for approximately 20 min. Gently stir the diluted emulsion with the porcelain spoon or glass rod to ensure complete mixing. The size of the beaker used depends on the amount of emulsion required. Electron Microscope Sciences, Fort Washington, Pennsylvania, has a small slide-coating cup—Dip Miser, specially designed for the dipping of slides. This device can conserve quite a bit of emulsion. If such a device is used, the liquid emulsion is poured gently into this cup. Before dipping the experimental slides, dip a few blank slides slowly up and down to remove bubbles from the surface of the emulsion. When most of the bubbles are gone, the emulsion is ready for use.

In almost total darkness, dip the slides according to the order in which you have placed them. Hold one or two slides (back to back) vertically at one end and gently immerse them in the dipping chamber. Wait for a few seconds and slowly withdraw them from the container. Touch the slides momentarily to the edge of the container and then to a piece of moist blotting paper to drain off excess emulsion at the bottom of the slide. If a single slide is dipped, the excess emulsion on the back should be wiped off with a moist paper towel. If two slides are dipped back to back, they should be separated from one another at this time. Place the slide vertically into the slide-drying rack with moist absorbing paper underneath and allow the slides to dry slowly. Place the slides into the rack systematically so that their order is maintained. The safelight should be turned off during the drying period. After all the preparations have been dried (allowing at least 1 hr, very rapid drying can produce high background), they can be placed into small black slide boxes. The slide boxes are further secured with electrician's tape. All of the slide boxes are then stored in the large lighttight box (e.g., a large cardboard box). The box is then placed in a 4°C cold room or refrigerator for exposure. The total exposure time depends on the amount of labeled precursor incorporated into the tissue; it can range from a few days to several weeks. However, one can develop test slides, and judging from the amount of silver grains present, one can calculate the proper time of exposure for the rest of the slides.

IV. DEVELOPMENT AND STAINING OF SLIDES

Process the slides according to the manufacturer's specifications. The following is a procedure suitable for Kodak NTB2 Nuclear Track emulsions. The slides are removed from the cold room or the refrigerator, are allowed to reach room temperature and, working under a safelight, are processed as follows:

Kodak D-19 developer (1:1 dilution)	4 min
Distilled water stop bath	10 sec
Kodak fixer (undiluted)	5 min
Wash very gently in running water	15 min

Rinse in distilled water and dry slowly in a dust-free atmosphere if necessary or proceed directly to staining.

All solutions should be kept at 15°C. Slides are gently agitated a few times during development. Acid stop baths should not be used since they could result in the formation of microscopic bubbles in the emulsion layer. Also, do not use rapid (ammonium) fixers since they could result in loss of developed silver grains. The above processing procedures serve only as a guide. The developer dilutions, development times, and processing temperatures can be modified to suit individual practice and experimental requirements (Kodak information sheet P64—Kodak materials for light microscope autoradiography). When processing large numbers of slides, we have found that the Tissue-tek II slide-staining set is very desirable because 25 slides can be processed as a single unit, thus increasing the efficiency of the process.

After the slides have been properly washed, they are ready to be stained. For paraffin material, slides can be stained without drying for about 1 min in a 0.1% freshly prepared aqueous solution of toluidine blue O. Destain the slides in 70% alcohol since the emulsion will be deeply stained. Then dehydrate them rapidly in two changes of *n*-butanol, clear in xylene, and mount. For glycol methacrylate sections, the slides can be stained according to the method of Feder and O'Brien (1968). Stain sections for 1–5 min in a 0.05% toluidine blue O solution in benzoate buffer, pH 4.4, destain in water, dry, and mount. For epoxy sections, slides can be stained with 0.5% toluidine blue O in 0.1% Sodium carbonate (pH 11.1) for 1–10 min at 37°C, destain in water, dry, and mount (O'Brien and McCully, 1981). For certain staining procedures such as the Fuelgen technique, the slides must be stained before dipping because the acid hydrolysis step will remove deposited silver grains from the emulsion. Not all stains are compatible with the liquid emulsion due to chemographic effects (Rogers, 1979).

V. GENERAL COMMENTS

The above procedure is limited to those macromolecules that are insoluble in a wide range of organic and aqueous solvents. For those that are soluble during processing, special techniques are required to avoid extraction or displacement of the labeled substances during the preparation and processing of the specimens and the dipping of emulsion. In general, in these cases, either freeze drying or freeze substitution techniques must be employed, and special fixation precautions may also be necessary during the processing of tissues. For both theoretical and practical considerations of these procedures, consult the following references: Boyenval and Droz (1976), Fisher (1972), Fisher and Housley (1972), Harvey (1982), Lüttge (1972), O'Brien and McCully (1981), Pearse (1980), and Roth and Stumpf (1969).

REFERENCES

Bogoroch, R. (1972). Liquid emulsion autoradiography. *In* "Autoradiography for Biologists" (P. B. Gahan, ed.), pp. 65–94. Academic Press, New York.

Boyenval, J., and Droz, B. (1976). Cryo-ultramicrotomy combined with radioautography for the detection of lipids. *J. Microsc. Biol. Cell.* **27,** 129–132.

Eriksson, T. (1967). Duration of the mitotic cycle in cell cultures of *Haplopappus gracilis*. *Physiol. Plant.* **20,** 348–354.

Feder, N., and O'Brien, T. P. (1968). Plant microtechnique: Some principles and new methods. *Am. J. Bot.* **55,** 123–142.

Fisher, D. B. (1972). Artifacts in the embedment of water-soluble compounds for light microscopy. *Plant Physiol.* **49,** 161–165.

Fisher, D. B., and Housley, T. L. (1972). The retention of water-soluble compounds during freeze-substitution and microautoradiography. *Plant Physiol.* **49,** 166–171.

Harvey, D. M. R. (1982). Freeze-substitution. *J. Microsc. (Oxford)* **127,** 209–221.

Herrmann, R. G., and Abel, W. O. (1972). Microautoradiography of organic compounds insoluble in a wide range of polar and non-polar solvents. *In* "Microautoradiography and Electron Probe Analysis: Their Application to Plant Physiology" (U. Lüttge, ed.), pp. 123–165. Springer-Verlag, Berlin and New York.

Jensen, W. A. (1962). "Botanical Histochemistry." Freeman, San Francisco, California.

Lüttge, U., ed. (1972). "Microautoradiography and Electron Probe Analysis: Their Application to Plant Physiology." Springer-Verlag, Berlin and New York.

O'Brien, T. P., and McCully, M. E. (1981). "The Study of Plant Structure: Principles and Selected Methods." Termarcarphi Pty. Ltd., Melbourne, Australia.

Pappas, P. W. (1971). The use of a chrome alum-gelatin (subbing) solution as a general adhesive for paraffin sections. *Stain Technol.* **46,** 121–124.

Pearse, A. G. E. (1980). "Histochemistry: Theoretical and Applied," 4th ed., Vol. 1. Churchill-Livingstone, Edinburgh and London.

Raghavan, V. (1979). Embryogenic determination and ribonucleic acid synthesis in pollen grains of *Hyoscyamus niger* (henbane). *Am. J. Bot.* **66,** 36–39.

Rogers, A. W. (1979). "Techniques of Autoradiography," 3rd ed. Elsevier/North-Holland Biomedical Press, Amsterdam.

Roth, L. J., and Stumpf, W. E., ed. (1969). "Autoradiography of Diffusible Substances." Academic Press, New York.

Tran Thanh Van, M., Chlyah, H., and Chlyah, A. (1974). Regulation of organogenesis in thin layers of epidermal and sub-epidermal cells. *In* "Tissue Culture and Plant Science 1974" (H. E. Street, ed.), pp. 101–139. Academic Press, London.

CHAPTER **85**

Immunofluorescence Techniques for Studies of Plant Microtubules

Larry C. Fowke*
Daina Simmonds†
Pieter van der Valk**
George Setterfield†

**Department of Biology*
University of Saskatchewan
Saskatoon, Saskatchewan, Canada

†Department of Biology
Carleton University
Ottawa, Ontario, Canada

***Institute of Human Genetics*
Free University of Amsterdam
Amsterdam, The Netherlands

I. INTRODUCTION

The immunofluorescence technique is based on the high specificity of immune reactions. Antibodies labeled with fluorescent dye, usually fluorescein isothiocyanate (FITC) or rhodamine, are used to localize macromolecules in cells with the fluorescence microscope. This technique has been used extensively to investigate the distribution of microtubules in animal cells (reviewed by Brinkley *et al.*, 1980), and recently has been applied to studies of plant microtubules as well. Microtubules are involved

CELL CULTURE AND SOMATIC CELL
GENETICS OF PLANTS, VOL. 1

ISBN 0-12-715001-3

in a number of important functions in plants including cell division, cell wall formation, and motility (Gunning and Hardham, 1982). Immunofluorescence techniques are particularly useful because they provide a rapid means for visualizing the overall three-dimensional distribution of microtubules in a cell. Relatively large samples of cells are easily examined. Use of electron microscopy as an alternative method of determining three-dimensional relationships requires time-consuming serial sectioning and is restricted to small samples.

Plant microtubules have been studied using indirect immunofluorescence, which involves an initial treatment with antibodies and subseqeunt application of a fluorescent-labeled secondary antibody. This technique was first used to study plant microtubules in the spindle of liquid endosperm cells (Franke *et al.*, 1977) and naked algal gametes (Weber *et al.*, 1977). Marchant (1978) used immunofluorescence to identify microtubules on the inner surface of plasma membrane fragments derived from protoplasts of *Mougeotia;* similar results were obtained in studies of plasma membrane fragments from higher plant protoplasts (Lloyd *et al.*, 1980; Van der Valk *et al.*, 1980).

Recently, methods have been devised for permeabilizing cell walls to allow uptake of antibodies into plant cells. Staining of elongated suspension-cultured plant cells with fluorescent antitubulin following a brief treatment with cell wall-degrading enzymes has clearly demonstrated the hooplike arrangement of cortical microtubules (Lloyd *et al.*, 1980; Simmonds *et al.*, 1982a). The various microtubule distributions found in cells in the mitotic cycle have also been visualized in cultured cells (Simmonds *et al.*, 1982a,b). Wick *et al.* (1981) have devised methods which permit the study of microtubules in permeabilized, isolated root cells which have been stabilized so as to retain their original shape. It is therefore possible to relate microtubule distribution directly to cell division stage and cell shape during root development. The microtubules of protoplasts regenerating walls in culture have also been demonstrated by immunofluorescence (Simmonds *et al.*, 1982b).

This chapter provides immunofluorescence methods for the study of plant microtubules. The first method described can be used to stain microtubules attached to the inner surface of plasma membrane fragments derived from plant protoplasts. The other methods can be applied to suspension-cultured plant cells and protoplasts regenerating walls.

II. PROCEDURES

Primary antitubulin antisera used in the following procedures can be prepared according to well-established methods (e.g., Connolly *et al.*, 1978;

Weber and Osborn, 1979; Brinkley *et al.*, 1980). Secondary FITC-conjugated goat antirabbit antisera can be purchased from a number of sources (e.g., Dako, Cedarlane Laboratories Ltd., Hornby, Ontario, Canada; Hyland, Baxter Laboratories of Canada, Malton, Ontario, Canada; Miles–Yeda Ltd., Rexdale, Ontario, Canada). The composition of all solutions is listed after the procedures described.

A. Protoplast Plasma Membrane Fragments

Protoplasts were prepared from either suspension-cultured cells or leaf mesophyll cells (Chapters 38–45, this volume; Constabel, 1982).

1. Place large drops of protoplasts suspended in wash buffer on individual polylysine-coated coverslips (11 × 22 mm) and allow protoplasts to settle on the glass surface (ca. 20 min).
2. Gently dip coverslips in a large volume (ca. 100 ml) of wash buffer to dislodge protoplasts which have not stuck to the glass and then dip them in a similar volume of microtubule-stabilizing buffer (MtSB). Note: Coverslips should be held horizontally to retain a layer of liquid over the protoplasts and prevent their collapse.
3. Place the coverslips face down over depression slides containing MtSB for 20 min to allow the protoplasts to burst.
4. Dip the coverslips in three changes of MtSB and then wash them in a stream of MtSB to remove protoplast contents. Large fragments of protoplast plasma membrane will remain attached to the coverslip surface.
5. Fix the coverslips in 3.7% formaldehyde in MtSB for 10 min.
6. Dip the coverslips in three changes of phosphate-buffered saline (PBS) and then wash them in a stream of PBS.
7. Drain the coverslips slightly and place them in a humid chamber. Spread 50–100 μl primary antitubulin antiserum over each coverslip and incubate for 45–60 min. The appropriate concentration of antiserum must be determined by serial dilution.
8. Drain the serum. Dip the coverslips in three changes of PBS and then wash them in a stream of PBS.
9. Drain the coverslips slightly and place them in a humid chamber. Spread 50–100 μl secondary antibody over each coverslip and incubate for 30–45 min. The appropriate dilution of secondary antibody varies according to the source (e.g., the following dilutions have proved satisfactory: Hyland, 1:5 with PBS; Dako, 1:50 with PBS).
10. Dip the coverslips in three changes of PBS and then wash them in a stream of PBS.
11. Mount the coverslips on a slide in PBS containing 50% glycerol, pH 7.0. The coverslips can be ringed with clear nail polish and stored in a refrigerator.

12. Examine the coverslips in a fluorescence microscope using filters appropriate for FITC. Preparations were photographed using a 63× planapo oil immersion lens [numerical aperture (NA) = 1.4].

Solutions were as follows:

1. Wash buffer
   0.7 *M* sorbitol
   3.0 m*M* 2-[*N*-morpholino]ethanesulfonic acid (MES) buffer
   6.0 m*M* $CaCl_2 \cdot 2H_2O$
   0.7 m*M* NaH_2PO_4
   Dilute 1:1 with cell culture medium.
2. Polylysine
   0.1 g Polylysine (MW ≃ 80,000)
   100 ml H_2O
   Dip half coverslips in polylysine, rinse in wash buffer, drain, and apply protoplasts while the coverslips are still wet.
3. MtSB
   100 m*M* PIPES buffer
   1.0 m*M* $MgSO_4$
   2.0 m*M* EGTA
   Adjust to pH 6.9.
4. PBS
   0.14 *M* NaCl
   2.7 m*M* KCl
   8.0 m*M* Na_2HPO_4
   1.5 m*M* KH_2PO_4
   Adjust to pH 7.0.

B. Cultured Cells

Procedures for cultured cells were adapted from animal workers (Osborn *et al.*, 1978; Bershadsky *et al.*, 1978; Rogers *et al.*, 1981) and developed for *Vicia hajastana* Grossh. cells (Simmonds *et al.*, 1982a). Microtubules in *Solanum melongena* L. cell cultures stained well with this method. However, some modifications were required to stain microtubules in cultured tuber slices of *Helianthus tuberosus* L. Other species may require modification of details for optimal staining.

1. Use 1 ml of 2- to 4-day cell suspension culture. Allow the cells to settle, pipette off most of the cell-free medium, and resuspend the cells in an equal volume of MtSB-1 solution. Allow the cells to settle.

2. Pipette off most of the MtSB-1 solution and add 5 ml enzyme solution. Resuspend the cells and allow them to stand for 10–15 min. Resuspend the cells gently once during this period.

3. Pipette off the enzyme solution and rinse twice with 2 ml MtSB-1 solution by allowing the cells to settle and pipetting off the supernatant fluid.

4. Using a Pasteur pipette, gently layer the cells on coverslips (11 × 22 mm) coated with polylysine. Allow 5–8 min for cells to settle and adhere. Note: For the remaining part of the procedure, the coverslips (with attached cells) should be held horizontally when transferred from one solution to the next. Disposable plastic weighing boats are convenient receptacles for solutions.

5. Extract in MtSB-2 plus 1% Triton X-100 for 15 min.

6. Extract in MtSB-3 plus 1% Triton X-100 for 60 min.

7. Rinse in MtSB-3 four times for 1 min each.

8. Fix in 1% glutaraldehyde in MtSB-3 for 10 min.

9. Treat with 1 mg/ml $NaBH_4$ in PBS three times for 3 min each (to reduce free aldehyde groups).

10. Rinse in PBS three times for 3 min each.

11. Spread 75 μl primary antitubulin antiserum over each coverslip and incubate for 45 min in a humid chamber. The appropriate concentration of the antiserum must be determined by serial dilution.

12. Rinse in PBS three times for 4 min each.

13. Spread 75 μl secondary antibody over each coverslip and incubate for 45 min in a humid chamber. Miles secondary antibody has been used satisfactorily when diluted 1:60 with PBS.

14. Rinse in PBS three times for 4 min each.

15. Mount coverslips in PBS containing 50% glycerol, pH 9.0. Samples can be stored in the refrigerator for several weeks. Fading of fluorescence during observation can be reduced by adding 0.1% diphenylamine to the mounting medium.

16. Examine the coverslips in fluorescence microscope using filters appropriate for FITC.

C. Regenerating Protoplasts

The method used for cultured cells has been modified and applied successfully to *V. hajastana* protoplasts cultured for up to 24 hr in 8p medium (Kao and Michayluk, 1975).

1. Wash protoplasts with protoplast microtubule-stabilizing buffer (PMtSB) (MtSB-1 containing 0.15 *M* sorbitol and 0.15 *M* mannitol) by centrifuging for 5 min at 100 *g*.

2. Prefix for 60 min by resuspending protoplasts in 3% paraformaldehyde in PMtSB.

3. Wash with PMtSB by centrifuging two times.

4. Layer protoplasts on coverslips coated with polylysine and allow them to settle for 5–8 min.
5. Extract in MtSB-3 plus 1% Triton X-100 for 60 min.
6. Follow steps 7–16 for cultured cells to complete the procedure.

Solutions were as follows:

1. MtSB-1
   0.1 *M* PIPES buffer
   3.0 m*M* EGTA
   Adjust to pH 6.9.
2. Enzyme solution
   0.2 g Cellulase (Onozuka R-10, Kanematsu-Gosha Canada, Ltd., Vancouver, British Columbia, Canada).
   0.4 g Hemicellulase (Rhozyme, Rohm and Haas Co. Canada, Ltd., West Hill, Ontario, Canada).
   0.2 g Pectinase (Sigma Chemical Co., St. Louis, Missouri).
   10.0 ml H_2O
   Dissolve for 5 min and centrifuge at 10,000 *g* for 10 min.
   Mix:
     5.0 ml enzyme supernatant
     5.0 ml 0.1% gelatin solution (predissolved by heating)
     10.0 ml 2× concentrated MtSB-1, adjusted to pH 6.0.
     Adjust final pH to 6.1.
3. Polylysine
   0.1 g Polylysine (MW ≃ 330,000)
   100 ml H_2O
   Spread 1 drop on a coverslip and air-dry. Repeat. Dip in water and air-dry.
4. MtSB-2 (adapted from Bershadsky *et al.*, 1978)
   50.0 m*M* imidazole
   50.0 m*M* KCl
   0.5 m*M* $MgCl_2$
   1.0 m*M* EGTA
   0.1 m*M* EDTA
   1.0 m*M* 2-mercaptoethanol
   2.0 *M* glycerol
   Dissolve and adjust to pH 6.7.
5. MtSB-3
   As for MtSB-2, except that the amount of glycerol is raised to 4.0 *M*.
6. PBS
   0.13 *M* NaCl
   5.1 m*M* Na_2HPO_4
   1.56 m*M* KH_2PO_4
   Adjust to pH 7.0.

III. RESULTS AND CONCLUSIONS

The immunofluorescence technique has been used successfully to examine microtubules on protoplast plasma membrane fragments from both tobacco and soybean suspension cultures. This technique provides a relatively rapid means of observing the general frequency and overall distribution of microtubules associated with the plasma membrane (Fig. 1). It may be possible to utilize other fluorescent antibodies to study a variety of

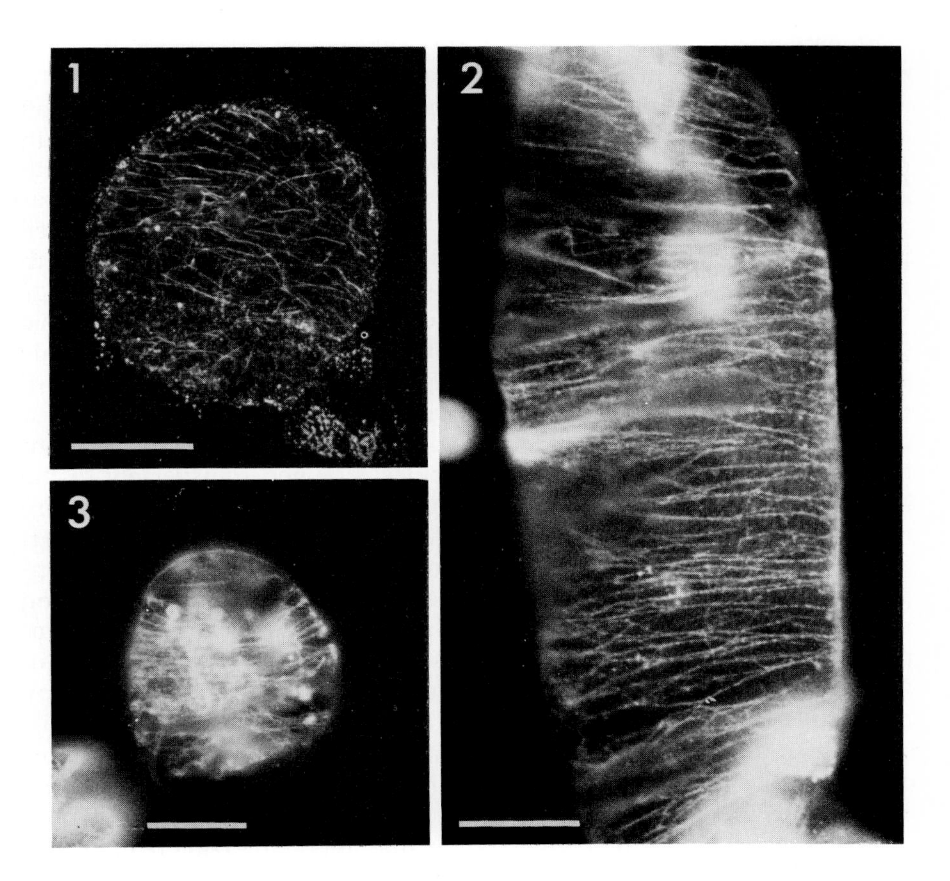

Fig. 1. Immunofluorescence of microtubules attached to the inner surface of a plasma membrane fragment from a tobacco cell culture protoplast. Bar = 20 μm.

Fig. 2. Immunofluorescence of microtubules in an elongating suspension-cultured cell of *Vicia hajastana*. Bar = 20 μm.

Fig. 3. Immunofluorescence of microtubules in a *Vicia hajastana* protoplast cultured for 24 hr. Note that the protoplast (cell) has organized microtubules and has begun to establish an asymmetric shape. Bar = 20 μm.

macromolecules on the inner surface of the plasma membrane (e.g., microtubule-associated proteins, microfilament actin, coated vesicle clathrin). Fragments of protoplast plasma membrane can also be prepared directly on electron microscope grids, stained, and examined in the transmission electron microscope (see Chapter 79, this volume). Such preparations provide considerable information regarding the frequency, distribution, and length of microtubules, as well as details concerning the structure and distribution of coated vesicles on the plasma membrane.

The microtubule-staining technique developed for *V. hajastana* cells was modified for the more fragile protoplasts. Prefixation with paraformaldehyde prevents protoplast rupture during subsequent processing. It is also possible to prefix cultured cells providing the formaldehyde treatment is applied after step 4 (method B). In prefixed preparations, permeabilization and extraction are carried out directly in MtSB-3 as in step 5, method C. In method B, in which cells are not prefixed, the short period in MtSB-2 with 2.0 *M* glycerol (step 5) is important. Direct immersion in MtSB-3 causes plasmolysis. Microtubule organization in prefixed and nonprefixed cell preparations is similar, but more cells adhere to the coverslip with prefixation.

The immunofluorescence technique of staining microtubules in whole plant cells and groups of cells makes it possible to examine microtubule arrays within individual cells and in adjoining cells during early stages of differentiation. It is also possible to relate microtubule organization to changes in cell morphology. Elongating cells of *V. hajastana* cultures contain arrays of parallel cortical microtubules arranged transversely to the long axis (Fig. 2). In dividing cells, the interphase cortical microtubules are not present during mitosis when preprophase band, spindle, and phragmoplast arrays of microtubules are found (Simmonds *et al.*, 1982a,b). Application of this staining method increases the probability of detecting rarely or briefly occurring microtubule arrangements because large numbers of cells can be scanned. Preprophase bands are rare in tissue culture in which they have not been detected with conventional electron microscopy methods (Fowke *et al.*, 1974).

In preparations in which cell walls remain intact, microtubule inhibitors can be used to study the role of microtubules in microfibril orientation and control of cell shape (Simmonds *et al.*, 1982a,b). This approach can be extended to examine depolymerization and repolymerization of microtubules in order to gain insight into microtubule initiation sites and other control mechanisms. Behavior of microtubules may also be followed as organized wall forms on protoplasts and cells take on asymmetric form (Fig. 3) and should provide information about wall–cytoskeleton relations.

ACKNOWLEDGMENTS

We wish to acknowledge the excellent technical assistance of Ms. P. J. Rennie. Antitubulin for the studies of whole cells has been generously supplied by K. Rodgers and D. L. Brown. Financial support from the Natural Sciences and Engineering Research Council of Canada is gratefully acknowledged.

REFERENCES

Bershadsky, A. D., Gelfand, V. I., Svitkina, T. M., and Tint, I. S. (1978). Microtubules in mouse embryo fibroblasts extracted with Triton X-100. *Cell Biol. Int. Rep.* **2,** 425–432.

Brinkley, B. R., Fistel, S. H., Marcum, J. M., and Pardue, R. L. (1980). Microtubules in cultured cells; indirect immunofluorescent staining with tubulin antibody. *Int. Rev. Cytol.* **63,** 59–95.

Connolly, J. A., Kalnins, V. I., Cleveland, D. W., and Kirschner, M. W. (1978). Intracellular localization of the high molecular weight microtubule accessory protein by indirect immunofluorescence. *J. Cell Biol.* **76,** 781–786.

Constabel, F. (1982). Isolation and culture of plant protoplasts. *In* "Plant Tissue Culture Methods" (L. R. Wetter and F. Constabel, eds.), 2nd rev. ed., NRCC No. 19876, pp. 38–48. National Research Council of Canada, Ottawa.

Fowke, L. C., Bech-Hansen, C. W., Constabel, F., and Gamborg, O. L. (1974). A comparative study on the ultrastructure of cultured cells and protoplasts of soybean during cell division. *Protoplasma* **81,** 189–203.

Franke, W. W., Seib, E., Osborn, M., Weber, K., Herth, W., and Falk, H. (1977). Tubulin-containing structures in the anastral mitotic apparatus of endosperm cells of the plant *Leucojum aestivum* as revealed by immunofluorescence microscopy. *Cytobiologie* **15,** 24–48.

Gunning, B. E. S., and Hardham, A. R. (1982). Microtubules. *Ann. Rev. Plant Physiol.* **33,** 651–698.

Kao, K. N., and Michayluk, M. R. (1975). Nutritional requirements for growth of *V. hajastana* cells and protoplasts at a very low population density in liquid media. *Planta* **126,** 105–110.

Lloyd, C. W., Slabas, A. R., Powell, A. J., and Lowe, S. B. (1980). Microtubules, protoplasts and plant cell shape. An immunofluorescent study. *Planta* **147,** 500–506.

Marchant, H. (1978). Microtubules associated with the plasma membrane isolated from protoplasts of the green alga *Mougeotia*. *Exp. Cell Res.* **115,** 25–30.

Osborn, M., Webster, R. E., and Weber, K. (1978). Individual microtubules viewed by immunofluorescence and electron microscopy in the same PtK2 cell. *J. Cell Biol.* **77,** R27–R34.

Rogers, K. A., Khoshbaf, M. A., and Brown, D. L. (1981). Relationships of microtubule organization in lymphocytes to the capping of immunoglobulin. *Eur. J. Cell Biol.* **24,** 1–8.

Simmonds, D., Setterfield, G., Tanchak, M., Brown, D. L., and Rogers, K. A. (1982a). Microtubule organization in cultured plant cells. *In* "Plant Tissue Culture 1982" (A. Fujiwara ed.), pp. 31–34. Maruzen, Tokyo.

Simmonds, D., Setterfield, G., Tanchak, M., Brown, D. L., and Rogers, K. A. (1982b). Microtubule organization and form of dividing and expanding plant cells. *J. Cell Biol.* **95**(2), 336a.

Van der Valk, P., Rennie, P. J., Connolly, J. A., and Fowke, L. C. (1980). Distribution of cortical microtubules in tobacco protoplasts. An immunofluorescence microscopic and ultrastructural study. *Protoplasma* **105,** 27–43.

Weber, K., and Osborn, M. (1979). The intracellular display of microtubular structures revealed by indirect immunofluorescence microscopy. *In* "Microtubules" (K. Roberts and J. S. Hyams, eds.), pp. 279–313. Academic Press, New York.

Weber, K., Osborn, M., Franke, W. W., Seib, E., Scheer, U., and Herth, W. (1977). Identification of microtubular structures in diverse plant and animal cells by immunological cross-reaction revealed in immunofluorescence microscopy using antibody against tubulin from porcine brain. *Cytobiologie* **15,** 285–302.

Wick, S. M., Seagull, R. W., Osborn, M., Weber, K., and Gunning, B. E. S. (1981). Immunofluorescence microscopy of organized microtubule arrays in structurally stabilized meristematic plant cells. *J. Cell Biol.* **89,** 685–690.

Index

D

E

F

G

J

K

N

O

P

S

U

V

W

X

Y

Z